linear and combinatorial programming

board of advisors, engineering

linear and combinatorial programming

KATTA G. MURTY
Department of Industrial and Operations Engineering
University of Michigan

JOHN WILEY & SONS, New York • Chichester • Brisbane • Toronto

Library of Congress Cataloging in Publication Data:

Murty, Katta G 1936–
Linear programming.

Includes bibliographies and index.
1. Linear programming. I. Title.
T57.74.M87 519.7′2 76-7047
ISBN 0-471-57370-1

Printed in the United States of America

10 9 8 7 6 5

To Vijaya, Vani, and my parents,
Adilakshmi and Narayanaswamy

preface

This book evolved from lecture notes for undergraduate and graduate courses on linear and integer programming that I taught at the University of Michigan for the last six years. The courses were attended by students in industrial engineering, operations research, business school, accounting, mathematics, statistics, economics, computer science, natural resources, electrical engineering, civil engineering, and other branches of engineering. On this mixed group of students, the lecture notes proved quite popular. This favorable reaction and encouragement from Wiley prompted me to write this book.

The objectives of the book are as follows.

1. To discuss the various methods used to model problems as linear programs and integer programs and to help to develop skill in modeling.
2. To discuss the theory of linear programming systematically in an elementary but rigorous manner so that a student without much mathematical background but with good motivation can understand it.
3. To discuss clearly the algorithms for solving linear and integer programs, including the branch and bound approaches, which have proved to be the most successful methods thus far developed for solving combinatorial optimization problems.
4. To help to develop skill in using algorithms intelligently to solve problems.

The only background required to study this book is familiarity with matrix algebra (which is gained in an undergraduate course), especially the concept of linear independence.

Various methods for formulating problems as linear programs are discussed in Chapter 1. The discussion through section 1.1.4 is very elementary. Sections 1.1.5 to 1.1.8 discuss more complicated methods and can be omitted in a first reading of the book. Chapter 2 discusses the simplex method for solving linear programs using canonical tableaus.

Chapter 3 presents the necessary geometrical and linear algebra concepts for studying linear programs, and explains how the simplex method operates on them. Related algorithms in linear algebra (e.g., the algorithm for testing linear independence) are all discussed here. Results on convex polyhedra that are used in studying linear programming are all proved in this chapter. Sections 3.1 to 3.10

consider these mathematical results and can be omitted by the reader who is interested only in the computational aspects. Sections 3.11 and 3.12 discuss the simplex method using matrix operations. Section 3.14 contains a method for formulating Phase I problems using only one artificial variable. Chapter 4 deals with the duality theory of linear programming. The economic arguments behind duality are clearly discussed. I clearly establish that the relative cost coefficients in the primal simplex algorithm are the values of the dual slack variables, and that the optimality criterion is the dual feasibility criterion. Chapter 5 discusses the revised simplex method with both the explicit form of the inverse and the product form of the inverse. Chapter 6 considers the dual simplex method using both the canonical tableaus and the inverse tableaus.

Chapters 7 and 8 discuss postoptimality analysis in linear programming. Chapter 9 considers degeneracy. The geometry of degenerate linear programs is discussed in section 9.1, which can be omitted by readers who are not mathematically inclined. Chapter 10 deals with bounded variable linear programs.

In Chapter 11 we deal with the primal algorithm for solving transportation problems. Sections 11.1 and 11.3 contain the mathematical results for this problem. Section 11.4 discusses the recent developments in the use of tree labeling methods. Chapter 12 is a summary of network algorithms in common use.

Chapter 13 is devoted to modeling using integer variables. In Chapter 14 we discuss the geometry of integer programming and cutting plane methods. Chapter 15 deals with the branch and bound approach. After discussing the general nature of the branch and bound approach, various examples are worked out in detail to illustrate the development of branch and bound algorithms.

Chapter 16 outlines recent developments in linear complementarity theory. Chapter 17 shows that *LU* decomposition and Cholesky factorization can be used to improve the numerical accuracy of the results obtained by the simplex method. Chapter 18 briefly describes the efficiency of the simplex method. The decomposition principle of linear programming is discussed in Appendix 1. Bender's decomposition for mixed integer programs is discussed in Appendix 2.

The many problems in the book can be assigned as exercises. They are of three types.

1. Formulation problems: The reader is likely to find some of the formulation problems in Chapters 1 and 13 difficult upon first reading.
2. Numerical problems: These require the application of the algorithms discussed in the book. Solving these problems will give insight into the ways in which algorithms work.
3. Proofs of statements and problems that require the development of an algorithm: These problems are likely to be hard; they normally require a deeper understanding of the material discussed in the text.

Problems that are listed in the middle of a chapter can normally be solved using the methods discussed in the section in which they appear. A new sequence of problem numbers begins with each chapter.

An undergraduate course on "Introduction to Optimization" can be taught by using the material in Chapter 1 (excluding sections 1.1.5 to 1.1.8), Chapter 2, sections 4.1 to 4.4 and section 4.5.10, section 11.2, section 12.4.1, sections 8.2, 8.5, and sections 15.1 to 15.7. A graduate course in linear programming can be taught using Chapters 1 to 11. To teach the course in a mathematical style, I recommend emphasis on the mathematical theory discussed in Chapter 3. The same course can be taught to emphasize applications of the material, especially the formulation, algorithmic, and the computational aspects of programming. The material from Chapter 12 to the end of the book can be the basis for a second course on optimization. It can also be used as a basis to teach a course on flows in networks, integer programming, or combinatorial optimization. A highly motivated student can use this book for self-study to learn optimization methods. Operations researchers and people who need optimization in their professional work can use this textbook as a reference for details of the methods or for developing new algorithms.

References are listed at the end of each chapter in some of the later chapters. References in linear programming are provided at the end of the book. Because of lack of space, the list of references was kept very brief. Most references used in the preparation of the material have been cited.

Katta Gopalakrishna Murty
University of Michigan
Ann Arbor
1976

acknowledgments

I thank Professors G. B. Dantzig, D. Gale, R. M. Oliver, O. L. Mangasarian, and R. M. Vanslyke from whom I learned mathematical programming at the University of California, Berkeley.

Students in my courses in the last six years at the University of Michigan have given many suggestions for improving earlier versions of this material. The manuscript was reviewed by several reviewers who offered many useful suggestions to streamline the presentation. The present version has benefited from all of their suggestions. E. Gainer's "Notes on LP Using Cholesky Factorization" and the suggestions of Dr. M. Saunders proved to be very useful in preparing Chapter 17.

I spent the summer of 1974 at Bell Labs in Holmdel where I had several useful discussions on these topics with the technical staff there, in particular, R. Saigal, M. Segal, J. Suurballe, and P. Unger.

The final version of the manuscript was completed during my sabbatical year at the Indian Statistical Institute, Madras. I thank the University of Michigan for this sabbatical, the American Institute of Indian Studies for providing travel funds, and the Indian Statistical Institute (particularly S. M. Sundara Raju and Professor C. R. Rao) for inviting me to spend the year with them. The various versions of the manuscript were typed by Geraldine Cox and Susan Parry at the University of Michigan, R. Santhanam at the Indian Statistical Institute, Madras and by many other typists too numerous to mention by name at these and other places. I thank them all for their untiring efforts.

My colleagues C. R. Prasad and C. S. Ramakrishnan, at the Indian Statistical Institute, Madras and my brother K. Sai have read the final draft carefully and made very valuable comments. I am very grateful to the US Educational Foundation in India for a Fulbright travel grant and to the late S. C. Sen of the Indian Statistical Institute who initially made it possible for me to go to the United States for higher studies. I am equally grateful to my uncle, P. Krishnamurty, for providing timely help during a critical period in my education.

Finally, I thank T. R. Poston, editor at Wiley, and his consultant Professor Don Phillips for their encouragement.

K. G. M.

contents

NOTATION

1 FORMULATION OF LINEAR PROGRAMS 1

1.1 Formulation of linear programs 1
1.2 Solving linear programs in two variables 19

2 THE SIMPLEX METHOD 35

2.1 Introduction 35
2.2 Transforming the problem into the standard form 36
2.3 Canonical tableau 40
2.4 Phase I with a full artificial basis 44
2.5 The simplex algorithm 46
2.6 Outline of the simplex method for a general linear program 52
2.7 The Big-M Method 65

3 THE GEOMETRY OF THE SIMPLEX METHOD 75

3.1 Euclidean vector spaces: Definitions and geometrical concepts 75
3.2 Matrices 84
3.3 Linear independence of a set of vectors and simultaneous linear equations 86
3.4 Basic feasible solutions and extreme points of convex polyhedra 97
3.5 The usefulness of basic feasible solutions in linear programming 102
3.6 The edge path traced by the simplex algorithm 109
3.7 Homogeneous solutions, unboundedness, resolution theorems 114
3.8 The main geometrical argument in the simplex algorithm 124
3.9 The dimension of a convex polyhedron. 124
3.10 Alternate optimum feasible solutions and faces of convex polyhedra 125
3.11 Pivot matrices 128
3.12 Canonical tableaus in matrix notation 130
3.13 Why is an algorithm required to solve linear programs? 131
3.14 Phase I of the simplex method using one artificial variable 136

4 DUALITY IN LINEAR PROGRAMMING 147

4.1 Introduction 147
4.2 An example of a dual problem 147
4.3 Dual variables are the prices of the items 149
4.4 How to write the dual of a general linear program 150
4.5 Duality theory for linear programming 155
4.6 Other interpretations and applications of duality 169

5 REVISED SIMPLEX METHOD 183

5.1 Introduction 183
5.2 Revised simplex algorithm with the explicit form of inverse when Phase II can begin directly 183
5.3 Revised simplex method using Phase I and Phase II 187
5.4 Revised simplex method using the product form of the inverse 192
5.5 Advantages of the product form implementation over the explicit form implementation 197

6 THE DUAL SIMPLEX METHOD 201

6.1 Introduction 201
6.2 Dual simplex algorithm when a dual feasible basis is known initially 201
6.3 Disadvantages and advantages of the dual simplex algorithm 208
6.4 Dual simplex method when a dual feasible basis is not known at the start 208
6.5 Comparison of the primal and the dual simplex methods 215

7 PARAMETRIC LINEAR PROGRAMS 219

7.1 Introduction 219
7.2 Analysis of the parametric cost problem given an optimum basis for some λ 221
7.3 To find an optimum basis for the parametric cost problem 224
7.4 Summary of the results on the parametric cost problem 224
7.5 Analysis of the parametric right-hand side problem given an optimum basis for some λ 225
7.6 To find an initial optimum basis for the parametric right-hand side problem, given a feasible basis for some λ 226
7.7 To find a feasible basis for the parametric right-hand side problem 226
7.8 Summary of the results on the parametric right-hand side problem 227
7.9 Convex and concave function on the real line 227

8 SENSITIVITY ANALYSIS 243

8.1 Introduction 243
8.2 Introducing a new activity 244
8.3 Introducing an additional inequality constraint 245
8.4 Introducing an additional equality constraint 248
8.5 Cost ranging of a nonbasic cost coefficient 249
8.6 Cost ranging of a basic cost coefficient 250

8.7 Right hand side ranging 251
8.8 Changes in the input-output coefficients in a nonbasic column vector 252
8.9 Change in a basic input-output coefficient 253
8.10 Practical applications of sensitivity analysis 253

9 DEGENERACY IN LINEAR PROGRAMMING 259

9.1 Geometry of degeneracy 259
9.2 Resolution of cycling under degeneracy 263
9.3 Summary of the simplex algorithm using the lexico minimum ratio rule 267
9.4 Do computer codes use the lexico minimum ratio rule? 268

10 BOUNDED VARIABLE LINEAR PROGRAMS 271

10.1 Introduction 271
10.2 Bounded variable problems 271
10.3 Simplex method using working bases 274

11 PRIMAL ALGORITHM FOR THE TRANSPORTATION PROBLEM 289

11.1 The balanced transportation problem: Theory 289
11.2 Revised primal simplex algorithm for the balanced transportation problem 299
11.3 The unimodualarity property 307
11.4 Labeling methods in the primal transportation algorithm 309
11.5 Sensitivity analysis in the transportation problem 321
11.6 Transportation problems with inequality constraints 325
11.7 Bounded variable transportation problems 329

12 NETWORK ALGORITHMS 341

12.1 Introduction: Networks 341
12.2 Notation 342
12.3 Single commodity maximum flow problems 348
12.4 The primal-dual approaches for the assignment and transportation problems 360
12.5 Single commodity minimum cost flow problems 373
12.6 Shortest route problems 382
12.7 Minimum spanning tree problems 390
12.8 Multicommodity flow problems 391
12.9 Other network algorithms 392

13 FORMULATION OF INTEGER AND COMBINATORIAL PROGRAMMING PROBLEMS 397

13.1 Introduction 397
13.2 Formulation examples 398

14 CUTTING PLANE METHODS FOR INTEGER PROGRAMMING 419

14.1 Introduction 419
14.2 Fractional cutting plane method for pure integer programs 426
14.3 Cutting plane methods for mixed integer programs 432
14.4 Other cutting plane methods 433
14.5 How efficient are cutting plane methods? 434

15 THE BRANCH AND BOUND APPROACH 437

15.1 Introduction 437
15.2 The lower bounding strategy 438
15.3 The branching strategy 440
15.4 The search strategy 441
15.5 Requirements for efficiency 445
15.6 The 0-1 knapsack problem 446
15.7 The traveling salesman problem 449
15.8 The general mixed integer linear program 456
15.9 The set representation problem 458
15.10 0-1 problems 467
15.11 Advantages and limitations 469
15.12 Ranking methods 471
15.13 Notes on the branch and bound approach 478

16 COMPLEMENTARITY PROBLEMS 481

16.1 Introduction 481
16.2 Geometric interpretations 482
16.3 Applications in linear programming 485
16.4 Quadratic programming 486
16.5 Two person games 493
16.6 Complementary pivot algorithm 495
16.7 Conditions under which the algorithm works 503
16.8 Remarks 508

17 NUMERICALLY STABLE FORMS OF THE SIMPLEX METHOD 521

17.1 Review 521
17.2 The *LU* decomposition 522
17.3 The Cholesky factorization 528
17.4 Reinversions 533
17.5 Use in computer codes 534

18 COMPUTATIONAL EFFICIENCY 535

18.1 How efficient is the simplex algorithm? 535

APPENDIX 1
THE DECOMPOSITION PRINCIPLE OF LINEAR PROGRAMMING 541

APPENDIX 2
BENDER'S DECOMPOSITION FOR MIXED INTEGER PROGRAMS 553

APPENDIX 3
CRAMER'S RULE 557

SELECTED REFERENCES IN LINEAR PROGRAMMING 559

INDEX 561

notation

M	A large positive number.
M	A square matrix used in defining a linear complementarity problem.
$\mathcal{N}$	Set of points in a graph or a network.
$\mathcal{A}$	Set of lines (arcs or edges) in a network.
G	A graph or a network.
E	Node-arc incidence matrix of a directed network.
$\bar{E}$	The node arc incidence matrix of a connected directed network, with the row corresponding to the root node deleted.
$\mathbf{A}_i$	Set of point j such that (i, j) is an arc in a network.
$\mathbf{B}_i$	Set of points j such that (j, i) is an arc in a network.
$\mathscr{s}$	Source in a network.
$\mathscr{t}$	Sink in a network.
$\mathscr{a}$	an assignment.
τ	An ECR (Edge Covering Route) in an undirected network.
τ	A tour in a traveling salesman problem.
X	A set of points in a network. Frequently it is either a set of points used in defining a cutset, or a set of labeled nodes.
$\bar{\mathbf{X}}$	All the points in the network not included in the set **X**.
T	A spanning tree in a network.
$\mathscr{P}$	A path in a network.
e	A vector in which all coordinates are equal to 1. e_r is a vector like this in $\mathbf{R}^r$.
e	Edge in a graph or network.
(i, j)	An arc in a network.
$(i; j)$	An edge in a graph or network.
x, y	Points in a network.
α-, β-, γ-, a-, and b- arcs	In section 12.5, these designations denote the kitter status of an arc.
w, z	Unless otherwise defined, these are usually the Phase I and Phase II objective functions, respectively, in the simplex method.

w, z — Variables in a linear complementarity problem.

(w_i, z_i) — A complementary pair of variables in a linear complementarity problem.

(q, M) — The linear complementarity problem in which the data are the square matrix M and the column vector q. See section 16.2.

$\mathscr{C}(M)$ — Set of complementary cones corresponding to the matrix M in a linear complementarity problem. See section 16.2.

x, y — Vectors or variables

E, A, B, X, P, F, D, U, L — Unless otherwise specified, these symbols usually denote matrices or vectors.

B — Unless otherwise specified, this symbol usually denotes a basis for a linear program in which the constraints are a system of linear equations in nonnegative variables.

B^{-1} — Inverse of the matrix B.

I — Unit matrix or identity matrix. It is a square matrix in which all diagonal entries are 1 and all off-diagonal entries are 0. See section 3.2.

M, N, E, P, Q, F, **Γ, D, I, U, H, J** — These symbols usually denote sets that are defined in that section or chapter.

$\mathbf{A}(\tilde{x})$ — A set of column vectors associated with a feasible solution, $\tilde{x}$, in a linear program in which the constraints are a system of equations in nonnegative variables. See section 9.1.

$\boldsymbol{B}$ — Augmented basis in a linear program when an additional constraint is introduced. See sections 8.3, 8.4.

$\mathbf{K}^{\Delta}$ — The convex hull of the extreme points of a convex polyhedron **K**. See section 3.7.

$\mathbf{K}^{\angle}$ — The cone of homogeneous solutions corresponding to a convex polyhedron **K**. See section 3.7.

c — This symbol usually denotes the vector of original cost coefficients.

x_B — The vector of basic variables associated with a basis B for a linear program.

d_B, c_B — The row vectors of the Phase I and Phase II cost coefficients respectively associated with a basis B for a linear program.

b — Unless otherwise specified, this symbol usually denotes the right-hand side constant vector in a linear program.

$\bar{a}_{ij}, \bar{b}_i$ — Updated entries in a canonical tableau for a linear program.

$\bar{d}_j, \bar{c}_j$ — Unless otherwise specified, these are usually the Phase I and Phase II relative cost coefficients, respectively, in the simplex method.

π — Unless otherwise specified, this denotes the vector of dual variables.

u_i, v_j — In a transportation model, these denote the dual variables associated with the ith source and the jth demand center, respectively. In other models these denote slack variables or other variables, as defined there.

P(j), S(j), EB(j), YB(j) — Indices used in the transportation algorithm. See section 11.4.

m, n In chapter 11, these denote the number of sources and demand centers respectively in a transportation model.

$\mathscr{B}$ A basic set of cells in a transportation array for a balanced transportation problem.

$G_{\mathscr{B}}$ The rooted tree corresponding to a basic set of cells $\mathscr{B}$ in a transportation problem.

G_{Δ} The graph corresponding to a set of cells, Δ, in a transportation array.

$\mathscr{A}_{\Delta}$ The set of edges in the graph G_{Δ}.

$\mathscr{A}_{\mathscr{B}}$ The set of edges in the graph $G_{\mathscr{B}}$.

$[\alpha]$ The largest integer less than or equal to the real number α.

$\langle\alpha\rangle$ The smallest integer greater than or equal to the real number α.

& In section 15.9, this symbol is used to denote the UCP operator.

$\sum$ Summation sign.

∞ Infinity.

$\in$ Set inclusion symbol. If **F** is a set, "$F_1 \in \mathbf{F}$" means that "F_1 is an element of **F**." Also "$F_2 \notin \mathbf{F}$" means that "F_2 is not an element of **F**."

$\subset$ Subset symbol. If **E**, **Γ** are two sets, "$\mathbf{E} \subset \mathbf{\Gamma}$" means that "**E** is a subset of **Γ** or that every element in **E** is also an element of **Γ**."

$\cup$ Set union symbol. If **D**, **H** are two sets, $\mathbf{D} \cup \mathbf{H}$ is the set of all elements that are either in **D** or in **H** or in both **D** and **H**.

$\cap$ Set intersection symbol. If **D** and **H** are two sets, $\mathbf{D} \cap \mathbf{H}$ is the set of all elements that are in both **D** and **H**.

$\emptyset$ The empty set. The set containing no elements.

$\setminus$ Set difference symbol. If **D** and **H** are two sets, $\mathbf{D}\setminus\mathbf{H}$ is the set of all elements of **D** that are not in **H**.

$\{\}$ Set brackets. The notation $\{x\text{: some property}\}$ represents the set of all elements, x, satisfying the property mentioned after the ":"

Minimum $\{\}$ The minimum number among the set of numbers appearing inside the set brackets. Maximum $\{\}$ has a similar meaning. If the set is empty we will adopt the convention that the minimum in it is $+\infty$ and the maximum in it is $-\infty$.

$\mathbf{R}^n$ Real Euclidean n-dimensional vector space. It is the set of all *ordered* vectors $(x_1, \ldots, x_n)$ where each x_j is a real number, with the usual operations of addition and scalar multiplication defined on it.

$|\alpha|$ Absolute value of the real number α

$\|x\|$ Euclidean norm of a vector $x \in \mathbf{R}^n$
If $x = (x_1, \ldots, x_n)$, $\|x\| = +\sqrt{x_1^2 + \cdots + x_n^2}$.

$\geqq, \geq, >$ The symbols representing order relationships among vectors in $\mathbf{R}^n$. See section 3.1.

$\succ$ Lexicographically greater than. See section 9.2.3.

$A_{i.}$ The ith row vector of the matrix A.

$A_{.j}$ The jth column vector of the matrix A.

x_1^+, x_1^-	See section 1.1.7.
iff	If and only if
LP	Linear program
BFS	Basic feasible solution
FAP	Flow augmenting path. See section 12.2.
UCP	Unordered Cartesian product. See section 15.9.3.
PSD	Positive semidefinite
PD	Positive definite
LCP	Linear complimentarity problem
LIFO	"Last in first out" selection strategy.
L.B.	Lower bound. See section 15.6.
B.V.	Branching variable. See section 15.6.
C.P.	Candidate problem. See section 15.6.
B.C.	Branching constraint. See section 15.9.10.
θ-loop	A minimal linearly dependent set of cells in a transportation array. See section 11.1.6.
P-matrix	See section 16.7.
Q-matrix	See section 16.7.
$(i.j)$	This refers to the jth equation in the ith chapter. Equations are numbered serially in each chapter.
$i.j$; $i.j.k$	The sections are numbered serially in each chapter. "$i.j$" refers to section j in Chapter i. "$i.j.k$" refers to subsection k in section $i.j$.
Figure $i.j$	The jth figure in Chapter i. The figures are numbered serially in this manner in each chapter.
Reference [i]	The ith reference in the list of references given at the end of the chapter.
Problem $i.j$	The jth problem in Chapter i. Problems are numbered serially in each chapter.
$n!$	n factorial. Defined only for nonnegative integers, $0! = 1$. And $n!$ is the product of all the positive integers from 1 to n, whenever n is a positive integer.
$\binom{n}{r}$	Defined only for nonnegative integers n, r. It is the number of distinct subsets of r objects from a set of n distinct objects. It is equal to $(n!)/((r!)(n-r)!)$.
Incidence vector	When considering the set $\mathbf{\Gamma} = \{1, 2, \ldots, n\}$ if $\mathbf{E} \subset \mathbf{\Gamma}$, the incidence vector of $\mathbf{E}$ is $x = (x_1, \ldots, x_n)$ where $x_j = 1$ if $j \in \mathbf{E}$ and $x_j = 0$ otherwise.
Bounded set	A subset $\mathbf{S} \subset \mathbf{R}^n$ is *bounded* if there exists a finite real number α such that $\|x\| \leqq \alpha$, for all $x \in \mathbf{S}$.
Superscript ^{T}	Denotes transposition. See section 3.2. A^{T} is the transpose of the matrix A. If x is a column vector, x^{T} is the same vector written as a row vector and vice versa. Column vectors are printed as transposes of row vectors to conserve space in the text.
Superscripts	We use superscripts to enumerate vectors or matrices or ele-

ments in any set. When considering a set of vectors, in $\mathbf{R}^n$, x^r may be used to denote the rth vector in the set, and it will be the vector $(x_1^r, \ldots, x_n^r)$. In a similar manner, while considering a sequence of matrices, the symbol P^r may be used to denote the rth matrix in the sequence. Superscripts should not be confused with exponents and these are distinguished by different type styles.

Exponents In the symbol ε^{r}, $\mathbf{r}$ is the exponent. $\varepsilon^{\mathrm{r}} = \varepsilon \times \varepsilon \times \cdots \times \varepsilon$, where there are $\mathbf{r}$ ε's in this product. Notice the difference in type style between superscripts and exponents.

Affine function An affine function of the variables (or parameters) $\lambda_1, \ldots, \lambda_r$, is a function of the form $c_0 + c_1\lambda_1 + \cdots + c_r\lambda_r$, where $c_0, c_1, \ldots, c_r$ are given real numbers.

Consistent system of equations A system of simultaneous linear equations, say $Ax = b$, for which at least one solution, x, exists. The system is *inconsistent*, if it has no solutions. See section 3.3.3.

Redundant equations In a system of simultaneous linear equations, a redundant equation is a linear combination of the others. See sections 2.6.2 and 3.3.3.

Basis for $\mathbf{R}^n$ A set of n vectors in $\mathbf{R}^n$ that is a linearly independent set.

Basis, basic vector, for an LP These terms are *only* defined for LPs in which the constraints are a system of equations in nonnegative variables. Before talking about a basis or a basic vector for an LP, first transform it so that the constraints in it are a system of equations in nonnegative variables (see section 2.2). Eliminate all the redundant equations from the system. Let the resulting system of constraints be $Ax = b$, $x \geqq 0$. In this system the variable x_j is associated with the column $A_{.j}$. Let A be of order $m \times n$. Any non-singular square submatrix of A of order m is called a *basis* for this LP. A *basic vector* for this LP is the vector of variables associated with the columns in a basis. See sections 2.3.1 and 3.5.1.

Pos $\{A_1, \ldots, A_k\}$ If $A_1, \ldots, A_k$ are vector in $\mathbf{R}^n$ then $\text{pos}\{A_1, \ldots, A_k\} = \{y: y = \alpha_1 A_1 + \cdots + \alpha_k A_k, \alpha_1 \geqq 0, \alpha_2 \geqq 0, \ldots, \alpha_k \geqq 0\}$.

Cardinality Defined only for sets. The cardinality of a set is the number of elements in it.

Tight or slack constraints Let $\bar{x}$ be a point in $\mathbf{R}^n$ satisfying an inequality constraint $f(x) \geqq 0$. If $f(\bar{x}) > 0$, the constraint "$f(x) \geqq 0$" is said to be *slack* at $\bar{x}$. If $f(\bar{x}) = 0$, the constraint "$f(x) \geqq 0$" is said to be *tight* or *active* at $\bar{x}$.

Maximization In chapter 2 and the following chapters, whenever a function $f(x)$ has to be maximized subject to some conditions, we look at the equivalent problem of minimizing $-f(x)$ subject to the same conditions. Both problems have the same set of optimum solutions, and the maximum value of $f(x) = -$minimum value of $(-f(x))$. Hence the algorithms discussed there are stated in terms of solving minimization problems.

1 formulation of linear programs

1.1 FORMULATION OF LINEAR PROGRAMS

.1.1 Introduction

Linear programming deals with problems in which a linear objective function is to be optimized (i.e., either maximized or minimized) subject to linear equality and inequality constraints and sign restrictions on the variables. To formulate a real life problem as a linear program is an art in itself. Even though there are excellent methods for solving a problem once it is formulated as a linear program, there is little theory to help in formulating the problems in this way. In real life problems several approximations may have to be made before they can be modeled as linear programs. The basic principles will be illustrated by considering a simple example. We will use the abbreviation "LP" for "Linear program."

1.1.2 Product Mix Problem

A company makes two kinds of fertilizers, called Hi-phosphate and Lo-phosphate. Three basic raw materials are used in manufacturing these fertilizers in this manner:

Raw material	Tons of raw material required to manufacture one ton: Hi-phosphate	Lo-phosphate	Maximum amount of raw materials available per month (tons)
1	2	1	1500
2	1	1	1200
3	1	0	500
Selling price per ton of fertilizer	\$15	\$10	

How much of each fertilizer should the company manufacture to maximize its gross monthly sales revenue?

This problem will be formulated using two different approaches, the *direct approach* and the *input-output approach.*

1.1.3 The Direct Approach

Step 1 Prepare a list of all the decision variables in the problem. This list must be complete in the sense that if an optimum solution providing the values of each of the variables is obtained, the decision maker should be able to translate it into an optimum policy that he can implement.

In the problem of section 1.1.2 the variables are:

x_1 = the tons of Hi-phosphate to be manufactured

x_2 = the tons of Lo-phosphate to be manufactured

Associated with each variable in the problem is an *activity* that the decision maker can perform. In this problem there are two such activities:

Activity 1: to manufacture one ton of Hi-phosphate

Activity 2: to manufacture one ton of Lo-phosphate

The variables in the problem just define the *levels* at which these activities are carried out.

When the constraints binding the variables are written, it will be clear that there are additional variables (called *slack variables* in what follows) imposed on the problem by the inequality constraints. Slack variables will be discussed in greater detail later. Formulation of the problem as an LP requires the following assumptions:

(i) PROPORTIONALITY ASSUMPTION

It requires two tons of raw material 1 to manufacture one ton of Hi-phosphate. The *proportionality assumption* implies that $2x_1$ tons of raw material 1 are required to manufacture x_1 tons of Hi-phosphate for any $x_1 \geqq 0$.

In general the proportionality assumption guarantees that if a_{ij} units of the ith item are consumed (or produced) in carrying out activity j at unit level, then $a_{ij}x_j$ units of this item are consumed (or produced) in carrying out activity j at level x_j for any $x_j \geqq 0$.

Also, by the proportionality assumption, since one ton of Hi-phosphate fertilizer sells for \$15, x_1 tons of the same sells for \$$15x_1$ for any $x_1 \geqq 0$.

(ii) ADDITIVITY ASSUMPTION

It requires two tons of raw material 1 to manufacture one ton of Hi-phosphate and one ton of the same raw material to manufacture one ton of Lo-phosphate. The *additivity assumption* implies that $2x_1 + x_2$ tons of raw material 1 is required to manufacture x_1 tons of Hi-phosphate and x_2 tons of Lo-phosphate for any $x_1 \geqq 0, x_2 \geqq 0$.

The additivity assumption generally implies that the total consumption (or production) of an item is equal to the sum of the various quantities of the item consumed (or produced) in carrying out each individual activity at its specified level.

The additivity assumption implies that that the objective function is *separable in the variables*, that is, if the variables in the model are $x_1, \ldots, x_n$, and the objective function is $z(x_1, \ldots, x_n) = z(x)$, then $z(x)$ can be written as a sum of n functions each of which involves only one variable in the model [i.e., as $z_1(x_1) + \cdots + z_n(x_n)$, where $z_j(x_j)$ is the contribution of the variable x_j to the objective function].

(iii) CONTINUITY OF VARIATION ASSUMPTION

It is assumed that each variable in the model can take all the real values in its range of variation.

In real life problems, some variables may be restricted to take only integer values (e.g., if the variable represents the number of empty buses transported from one location to another). Such restrictions make the problem an *integer programming problem.* Integer programs are much harder to solve than continuous variable LPs. They will be considered in Chapters 13 to 15.

The Implications of the Proportionality and Additivity Assumptions In most real life problems, the proportionality and additivity assumptions may not hold exactly. For example, the selling price per unit normally decreases in a nonlinear fashion as the number of units purchased increases. Thus, in formulating real life problems as LPs, several approximations may have to be made. It has been found that most real life problems can be approached very closely by suitable linear approximations. This, and the relative ease with which LPs can be solved, have made it possible for linear programming to find a vast number of applications.

The proportionality and the additivity assumptions automatically imply that all the constraints in the problem are either linear equations or inequalities. They also imply that the objective function is linear.

Step 2 Write down all the constraints and the objective function in the problem.

Nonnegativity Restrictions on the Variables The variables x_1, x_2 have to be nonnegative to make any practical sense. In linear programming models in general the nonnegativity restriction on the variables is a natural restriction that occurs because certain activities (manufacturing a product, etc.) can only be carried out at nonnegative levels.

Make a list of all the *items* that lead to constraints in the problem. In the fertilizer problem of section 1.1.2 each raw material leads to a constraint; for example, the amount of raw material 1 used is $2x_1 + x_2$ tons and this cannot exceed 1500 tons. This imposes the constraint:

$$2x_1 + x_2 \leqq 1500$$

Since this inequality just compares the amount of raw material 1 used to the amount available, it is called a *material balance inequality*. The material balance equations or inequalities corresponding to the various items in the problem are the constraints of the problem.

Each material balance inequality in the model contains in itself the definition of another nonnegative variable known as a *slack variable*. For example, the constraint imposed by the restriction on the amount of raw material 1 available, namely $2x_1 + x_2 \leqq 1500$, can be written as $1500 - 2x_1 - x_2 \geqq 0$. If we define $x_3 = 1500 - 2x_1 - x_2$, then x_3 is a slack variable representing the amount of raw material 1 remaining unutilized, and the constraint can be written in the equivalent form:

$$2x_1 + x_2 + x_3 = 1500, \qquad x_3 \geqq 0$$

Now write down the objective function. By the proportionality and additivity assumptions, this is guaranteed to be a linear function of the variables. This completes the formulation of the problem as an LP. It is as follows:

$$\begin{array}{llrl} \text{Maximize} & z(x) = 15x_1 + 10x_2 & & \\ \text{Subject to} & 2x_1 + & x_2 \leqq 1500 & \\ & x_1 + & x_2 \leqq 1200 & \\ & x_1 & \leqq 500 & \\ & x_1 \geqq 0, & x_2 \geqq 0 & \end{array} \tag{1.1}$$

Another Example: A Blending Problem

A refinery takes four raw gasolines, blends them, and produces three types of fuel. Here is the data.

Raw gas type	Octane rating	Available barrels per day	Price per barrel
1	68	4000	$1.02
2	86	5050	1.15
3	91	7100	1.35
4	99	4300	2.75

Fuel blend type	Minimum octane rating	Selling price ($/barrel)	Demand pattern
1	95	5.15	At most 10,000 barrels/day
2	90	3.95	Any amount can be sold
3	85	2.99	At least 15,000 barrels/day

The company sells raw gasolines not used in making fuels at $2.95/barrel if its octane rating is over 90 and at $1.85/barrel if its octane rating is less than 90. How can the company maximize its total daily profit?

In formulating this problem, we find that the variables are

x_{ij} = barrels of raw gasoline of type i used in making fuel type j per day for $i = 1$ to 4 and $j = 1, 2, 3$

and

y_i = barrels of raw gasoline type i sold as it is in the market

The octane rating of fuel type 1 should be greater than or equal to 95. Assuming that the proportionality and additivity assumptions hold for the octane rating of the blend, the octane rating of this fuel, if a positive quantity of this fuel is made, is $(68x_{11} + 86x_{21} + 91x_{31} + 99x_{41})/(x_{11} + x_{21} + x_{31} + x_{41})$. Thus, if a positive quantity of this fuel is made, the constraint on its octane rating requires

$$(68x_{11} + 86x_{21} + 91x_{31} + 99x_{41})/(x_{11} + x_{21} + x_{31} + x_{41}) \geqq 95$$

or, equivalently,

$$68x_{11} + 86x_{21} + 91x_{31} + 99x_{41} - 95(x_{11} + x_{21} + x_{31} + x_{41}) \geqq 0 \tag{1.2}$$

The inequality (1.2) will also hold when the amount of this fuel manufactured is zero. Hence, the octane rating constraint on this fuel may be represented by (1.2). If we proceed in a similar manner, the problem is

Maximize

$$\begin{aligned}&5.15(x_{11} + x_{21} + x_{31} + x_{41}) + 3.95(x_{12} + x_{22} + x_{32} + x_{42})\\ &+2.99(x_{13} + x_{23} + x_{33} + x_{43}) + y_1(1.85 - 1.02)\\ &+y_2(1.85 - 1.15) + y_3(2.95 - 1.35)\\ &+y_4(2.95 - 2.75) - 1.02(x_{11} + x_{12} + x_{13})\\ &-1.15(x_{21} + x_{22} + x_{23}) - 1.35(x_{31} + x_{32} + x_{33})\\ &-2.75(x_{41} + x_{42} + x_{43})\end{aligned}$$

Subject to

$$\begin{aligned}68x_{11} + 86x_{21} + 91x_{31} + 99x_{41} - 95(x_{11} + x_{21} + x_{31} + x_{41}) &\geqq 0\\ 68x_{12} + 86x_{22} + 91x_{32} + 99x_{42} - 90(x_{12} + x_{22} + x_{32} + x_{42}) &\geqq 0\\ 68x_{13} + 86x_{23} + 91x_{33} + 99x_{43} - 85(x_{13} + x_{23} + x_{33} + x_{43}) &\geqq 0\end{aligned}$$

$$\begin{aligned}x_{11} + x_{12} + x_{13} + y_1 &= 4000\\ x_{21} + x_{22} + x_{23} + y_2 &= 5050\\ x_{31} + x_{32} + x_{33} + y_3 &= 7100\\ x_{41} + x_{42} + x_{43} + y_4 &= 4300\\ x_{11} + x_{21} + x_{31} + x_{41} &\leqq 10{,}000\\ x_{13} + x_{23} + x_{33} + x_{43} &\geqq 15{,}000\end{aligned}$$

and

$$x_{ij}, y_i \geqq 0 \text{ for all } i \text{ and } j$$

By defining the variables as the amounts of various ingredients in the blend, blending problems can be formulated in this manner. However, if the variables are defined to be the proportions of the various ingredients in the blend, remember to include in the model the constraint that the sum of the proportions of the various ingredients in a blend should be equal to one.

1.1.4 The Input-Output Approach

In this procedure the lists of all the possible *activities* and the *items* defining the constraints are first prepared. The list of *activities* should include all the possible actions that the decision maker can perform in the problem. Units should be fixed for measuring the level at which each activity is carried out. The levels at which the activities are performed define the variables in the problem.

An *item* in the problem is any material or resource on which there is either a requirement or a limit on its availability or use. Each item leads to a constraint in the problem. The list of items should include an item called *dollar* that defines the objective function. If money is required to carry out an activity at unit level, dollar is an *input* for carrying out this activity. If money is produced as a result of carrying out an activity, dollar is an *output from* it. Even if the optimized item is different from money, we will use the term *dollar* to describe it.

To write the constraints in tabular form, define x_j as the level at which the jth activity is carried out, and assign the jth column in the tableau to correspond to this activity. Likewise associate a row of the tableau with each item. If a_{ij} units of item i are required as an input for carrying out the jth activity at unit level, enter $+a_{ij}$ in the (i, j) position (i.e., in the ith row and jth column) of the tableau. On the other hand, if a_{ij} units of item i are produced as an output by carrying out the jth activity at unit level enter $-a_{ij}$ in the (i, j) position of the tableau. Thus, inputs to activities are entered with a "+" sign, and outputs from activities are entered with a "−" sign ($\xrightarrow{+}\square\xrightarrow{-}$).

In the objective row corresponding to the item "dollar" all costs are inputs and, hence, are entered with a "+" sign, and all profits are outputs and are considered as *negative costs* and are entered with a "−" sign.

The constraints are completed by entering the limits on the availability or requirements of the items on a right-hand side column of the tableau. The objective row gives the coefficients of the objective function that is to be minimized.

When formulated in this manner, the fertilizer product mix problem leads to the following tableau. The blank entries in the tableau are zeros.

	Activities and Levels						
	To make one ton		To leave one ton unutilized of raw material				
	Hi-phosphate	Lo-phosphate	1	2	3		Right-hand side constants
Items	x_1	x_2	x_3	x_4	x_5		
Raw material 1	2	1	1			=	1500
Raw material 2	1	1		1		=	1200
Raw material 3	1				1	=	500
Dollars cost	−15	−10				=	$z(x)$ minimize

This is equivalent to the model formulated earlier. x_3, x_4, x_5 are the slack variables corresponding to the inequality constraints obtained in the earlier model.

Example: A Multiperiod Problem

A nationalized company can either manufacture capital goods or consumer goods. The company's production is being planned over a five period horizon. At the beginning of period 1 the company has 100 units of capital goods and 50 units of consumer goods in stock. Each unit of capital goods can be used to produce *either* three new units of capital goods at the expense of three units of consumer goods to be provided from stock *or* 10 units of consumer goods, over an uninterrupted two period interval, at the end of which the unit of capital goods again becomes available for use. Once capital goods are produced they can be used for production in subsequent periods. If necessary, capital goods can also be left idle. The company will sell everything in its stock at the beginning of period 6 and close up. The company can also sell either capital goods or consumer goods in any period according to the following price schedule.

	Price in monetary units per unit goods during the period	
Period	Capital goods	Consumer goods
1	4	0.1
2	6	0.15
3	8	0.15
4	6	0.15
5	3	0.1
6	3	0.1

Which production plan will best maximize the total returns?

We will formulate this problem by using the input-output approach. During each period the company can possibly engage in the following activities.

1. Assign one unit of capital goods to manufacture three units of capital goods over the next two periods.
2. Assign one unit of capital goods to manufacture 10 units of consumer goods over the next two periods.
3. Leave one unit of capital goods idle until next period.
4. Sell one unit of capital goods.
5. Sell one unit of consumer goods from stock.
6. Carry over one unit of consumer goods in stock until next period.

Of course the company cannot engage in activities 1 or 2 during period 5 because the company has to close down at the beginning of period 6. Let x_{ij} be the level at which activity j is carried out during the ith period.

The *items* are the capital goods and the consumer goods in stock in the various periods. Here is the list of items:

Capital goods available at the beginning of period i,
Consumer goods available at the beginning of period i for various values of i.

Clearly the item *capital goods at the beginning of period* 3 is an input into activities 1, 2, 3, 4 during period 3 and it is an output from activity 1 during period 1.

Also one unit of capital goods assigned to produce new capital goods at the beginning of period 1 finishes this job by the end of period 2 and becomes available for use again at the beginning of period 3. Thus the output from the activity of assigning one unit of capital goods at the beginning of period 1 to produce capital goods is four units of capital goods (itself and the three new units of capital goods it has produced) at the beginning of period 3. Figure 1.1

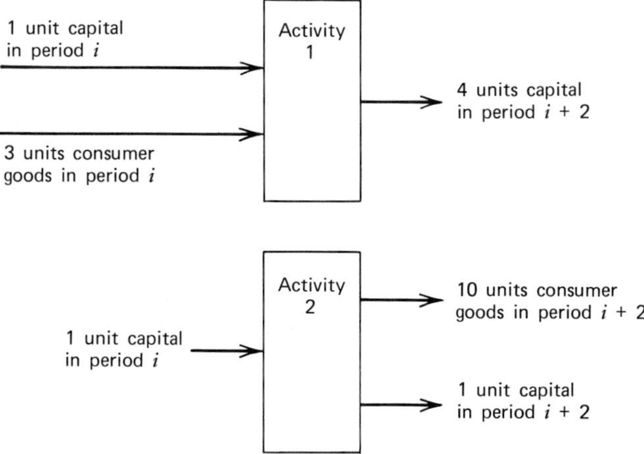

Figure 1.1

represents some of the input-output features in this problem. From these considerations we arrive at the formulation of the problem in Tableau 1.1. Each of the constraints in the problem is a material balance equation; this formulation could have been obtained also by direct arguments.

Tableau 1.1.

Item	Activities and their levels																														
	Period 1						Period 2						Period 3						Period 4						Period 5				Beginning of period 6		
	1	2	3	4	5	6	1	2	3	4	5	6	1	2	3	4	5	6	1	2	3	4	5	6	3	4	5	6	4	5	
	x_{11}	x_{12}	x_{13}	x_{14}	x_{15}	x_{16}	x_{21}	x_{22}	x_{23}	x_{24}	x_{25}	x_{26}	x_{31}	x_{32}	x_{33}	x_{34}	x_{35}	x_{36}	x_{41}	x_{42}	x_{43}	x_{44}	x_{45}	x_{46}	x_{53}	x_{54}	x_{55}	x_{56}	x_{64}	x_{65}	
Capital in 1	1	1	1	1																											= 100
Consumer goods in 1	3				1	1																									= 50
Capital in 2			−1				1	1	1	1																					= 0
Consumer goods in 2						−1	3				1	1																			= 0
Capital in 3	−4	−1							−1				1	1	1	1															= 0
Consumer goods in 3		−10										−1	3				1	1													= 0
Capital in 4							−4	−1							−1				1	1	1	1									= 0
Consumer goods in 4								−10										−1	3				1	1							= 0
Capital in 5													−4	−1							−1				1	1					= 0
Consumer goods in 5														−10										−1			1	1			= 0
Capital at beginning of 6																			−4	−1					−1				1		= 0
Consumer goods at beginning of 6																				−10								−1		1	= 0
$ = objective to be minimized				−4	−0.1					−6	−0.15					−8	−0.15					−6	−0.15			−3	−0.1		−3	−0.1	

Problems

Formulate the following problems as LPs.

1.1 A nonferrous metals corporation manufactures four different alloys from two basic metals. The requirements are:

Metal	Proportion of the metal in alloy				Total supply of metal per day
	1	2	3	4	
1	0.5	0.6	0.3	0.1	6 tons
2	0.5	0.4	0.7	0.9	4 tons
Selling price of alloy per ton	$10	15	18	40	

Determine the optimal product mix to maximize gross revenue.

1.2 DIET PROBLEM. A farmer can use two different grains in his chicken feed. The relevant data is given below.

Grain type	Nutrient units/kilogram grain			Cost per kilogram of grain
	Starch	Protein	Vitamins	
1	5	4	2	$0.60
2	7	2	1	$0.35
Minimum number of units of nutrient required per day	8	15	3	

Determine the minimal cost of the chicken feed.

1.3 TRANSPORTATION PROBLEM. A steel company has two mines and three steel plants. Here is the relevant data.

From mine	Unit cost of shipping ore from mine to steel plant (cents per ton)			Amount of ore available at mine (tons)
	1	2	3	
1	9	16	28	103
2	14	29	19	197
Tons of ore required at steel plant	71	133	96	

Set up the model describing the shipping of ore from mines to the steel plants that minimizes the total shipping bill.

1.4 ASSIGNMENT PROBLEM. A club of sociologists consists of five men and five women. It is assumed that the amount of happiness that the ith man and the jth women derive by spending a period of time x_{ij} together is $c_{ij}x_{ij}$, where c_{ij} are tabulated below.

$i \backslash j$	1	2	3	4	5
1	78	−16	19	25	83
2	99	98	87	16	92
3	86	19	39	88	17
4	−20	99	88	79	65
5	67	98	90	48	60

Determine what fraction of his lifetime each man should spend with each woman in order to maximize the sum of everyone's happiness in the club.

The objective in this model is to maximize the club's happiness. In writing down the objective function as a function of the decision variables, we notice that the proportionality and the additivity assumptions are perhaps very inappropriate. Discuss why.

1.5 BLENDING PROBLEM. A huge skyscraper is to be painted. The paint to be used can be obtained by blending four raw paints and two thinners. Data on them is:

Data	Paint type				Thinner type	
	1	2	3	4	1	2
Cost (dollars/gallon)	9	7	5.75	4	3	1.85
Viscosity (CP)	900	780	620	375	2	25
Vapor pressure (PSI)	0.2	0.4	0.6	0.8	12.0	8.0
Brilliance content (grams/gallon)	30	20	50	10	0	0
Durability content (grams/gallon)	2000	1500	1000	500	0	0

Requirements on the blend are:

Property	Requirement	Reason
Viscosity	$\geqq 400$	So that a single coat would do.
Brilliance content	Between 15 and 30	So that surface is properly reflective
Vapor pressure	Between 2 and 4 PSI	For proper drying in time.
Durability content	$\geqq 575$	

Determine an optimal blend.

1.1.5 Minimizing a Separable Piecewise Linear Convex Objective Function

We assumed that the contribution of a variable, say x_1, to the objective function varies linearly with x_1, as the term c_1x_1 (Figure 1.2). However, in most real life problems this assumption is too rigid. It is probably more realistic to assume that the contribution of the variable x_1 to the cost function varies in a piecewise linear manner. If this is a piecewise linear convex function in x_1, it can be depicted as in Figure 1.3. Here the slope is $c_1{}^1$ in the range $0 \leqq x_1 \leqq x_1{}^1$, $c_1{}^2$ in the range $x_1{}^1 \leqq x_1 \leqq x_1{}^2$, and $c_1{}^3$ in the range $x_1{}^2 \leqq x_1$ etc., where $x_1{}^1$, $x_1{}^2$, etc. are the values of the variable x_1 at which the contribution of x_1 changes slope. This is a convex function if $c_1{}^1 < c_1{}^2 < c_1{}^3$, etc., that is, iff the slope is monotone increasing in x_1 (see Figure 1.3).

It is a concave function iff the slope is monotone decreasing in x_1 (see Figure 1.4).

Suppose the additivity and the continuity of the variables assumptions hold. Suppose the constraints are all linear, and the cost function (which is to be minimized) is separable in the variables and piecewise linear. Then the proportionality assumption is violated for the cost function. However, if the piecewise linear function is convex, the problem can still be modeled as an LP.

Let $z_1(x_1)$ denote the contribution of x_1 to the objective function. Let $0 < x_1{}^1 < x_1{}^2 < \cdots < x_1{}^r < \infty$ be the points at which $z_1(x_1)$ changes slope, and let the slope in the interval $x_1{}^{t-1} \leqq x_1 \leqq x_1{}^t$ be $c_1{}^t$ for $t = 1$ to $r + 1$, where $x_1{}^0 = 0$, $x_1^{r+1} = \infty$.

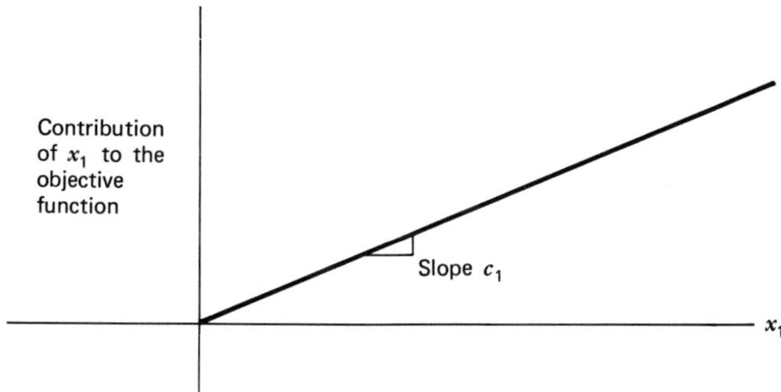

Figure 1.2

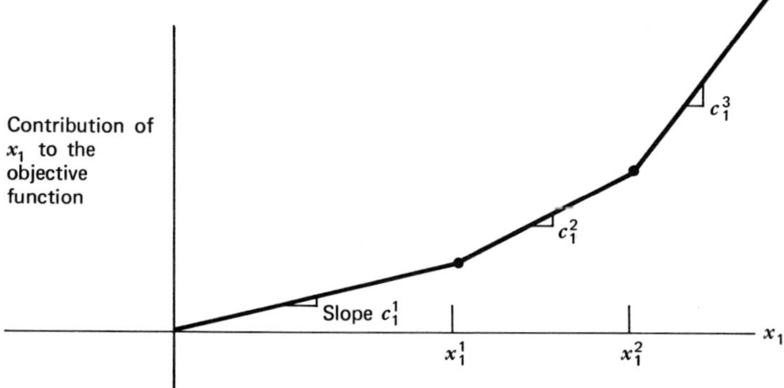

Figure 1.3 Piecewise Linear Convex Function.

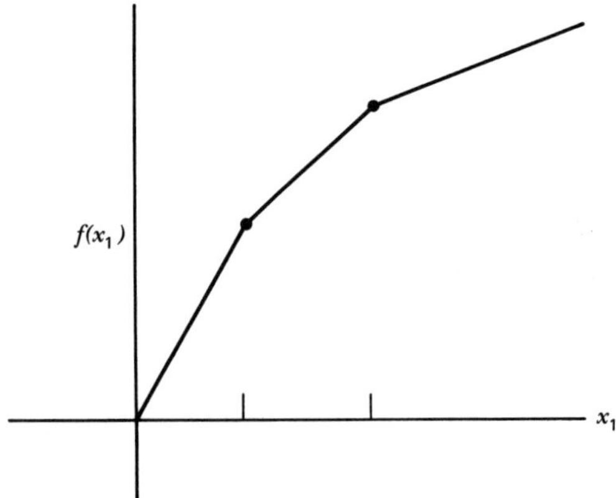

Figure 1.4 Piecewise Linear Concave Function.

The intervals within which $z_1(x_1)$ is linear are 0 to $x_1{}^1$, $x_1{}^1$ to $x_1{}^2, \ldots, x_1{}^r$ to ∞. Let y_t be the portion of x_1 lying in the tth interval, x_1^{t-1} to $x_1{}^t$, (i.e., y_t is the length of the overlap of the interval 0 to x_1 with the interval x_1^{t-1} to $x_1{}^t$), $t = 1$ to $r + 1$. When defined in this manner, the new variables $y_1, \ldots, y_{r+1}$ partition x_1 as $y_1 + \cdots + y_{r+1}$. They are subject to the constraints:

$$\begin{aligned} 0 &\leqq y_1 \leqq x_1{}^1 \\ 0 &\leqq y_2 \leqq x_1{}^2 - x_1{}^1 \\ &\vdots \\ 0 &\leqq y_r \leqq x^r - x_1^{r-1} \\ 0 &\leqq y_{r+1} \end{aligned} \tag{1.3}$$

and

for every t, if $y_t > 0$, then each of y_j is equal to its upper bound $x_1{}^j - x_1^{j-1}$, for all $j < t$. (1.4)

When the variables $y_1, \ldots, y_{r+1}$ are defined in this manner, $z_1(x_1)$ is clearly equal to $c_1{}^1 y_1 + \cdots + c_1^{r+1} y_{r+1}$.

Example

Consider the case where $z_1(x_1)$ has the following slopes:

Interval of x_1	Slope
0–10	3
10–25	5
25–∞	7

In this case x_1 will be partitioned as $y_1 + y_2 + y_3$, where

$$\begin{aligned} 0 &\leqq y_1 \leqq 10 \\ 0 &\leqq y_2 \leqq 15 \\ 0 &\leqq y_3 \end{aligned}$$

and

$$y_2 > 0 \quad \text{implies} \quad y_1 = 10$$

$$y_3 > 0 \quad \text{implies} \quad y_1 = 10 \text{ and } y_2 = 15$$

Then $z_1(x_1)$ can be expressed as $3y_1 + 5y_2 + 7y_3$.

The variable x_1 can now be eliminated from the model by substituting $x_1 = y_1 + \cdots + y_{r+1}$ wherever it appears in the model. In the objective function $z_1(x_1)$ is replaced by $c_1{}^1 y_1 + \cdots + c_1^{r+1} y_{r+1}$. Notice that the new objective function is linear in the new variables. The constraints (1.3) on the new variables are included with the other constraints in the model. The following facts guarantee that in any optimum solution of the transformed model, the constraints (1.4) are automatically satisfied.

1. We are trying to minimize the overall objective function.
2. The slopes discussed satisfy the conditions $c_1{}^1 < c_1{}^2 < \cdots < c_1^{r+1}$.

If the slopes do not satisfy condition 2, and the objective function has to be minimized, then the constraints (1.4) on the new variables have to be specifically included in the model; since these constraints are not linear constraints, the transformed model is not a linear programming model.

If the contribution of some other variable to the objective function is also piecewise linear, partition that variable too in a similar manner. If the original separable piecewise linear objective function to be minimized is convex, constraints on the new variables of the type (1.4) can be ignored in the transformed model, and the transformed model is a linear programming model in terms of the new variables.

SUMMARY

A problem in which a separable piecewise linear objective function has to be minimized subject to linear constraints on the variables can be transformed into an LP as above, if the objective function is convex (i.e., its slopes are monotone increasing in each variable).

The same method can be used to transform a problem in which a separable piecewise linear objective function has to be maximized subject to linear constraints on the variables, into an LP, if the objective function is concave (i.e., its slopes are monotone decreasing in each variable).

For an example, see section 1.1.8.

Problems

1.6 Explain why the above statements are true.

1.7 A firm manufactures four products called P_1, P_2, P_3, P_4. Product P_1 can be sold at a profit of \$10 per ton up to a quantity of 10 tons. Quantities of P_1 over 10 tons but not more than 25 tons can be sold at a profit of \$7 per ton. Quantities beyond 25 tons earn a profit of only \$5 per ton.

Product P_2 yields a profit of \$8 per ton up to 7 tons. Quantities of P_2 above 7 tons yield a profit of only \$4 per ton. Everyone who buys P_2 also buys P_4 to go along with it. This is described later. Both products P_1, P_2 can be sold in unlimited amounts.

P_3 is a by-product obtained while producing P_1. Up to 10 tons of P_3 can be sold at \$2 profit per ton. However, beyond 10 tons there is no market for P_3, and since it can't be stored, it has to be disposed of at a cost to the firm of \$3 per ton.

P_4 is a by-product obtained while producing P_2. Also, P_4 can be produced independently. Every customer who buys θ tons of P_2 has to buy $\theta/2$ tons of P_4 to go along with it for every $\theta \geqq 0$. Also, P_4 has an independent market in unlimited quantities. One ton of P_4 yields a profit of \$3 per ton, if it is sold along with P_2. One ton of P_4 sold independently yields a profit of \$2.50 per ton.

Production of one ton of P_1 requires one hour of machine 1 time plus two hours of machine 2 time. One ton of P_2 requires two hours of machine 1 time plus three hours of machine 2 time. Each ton of P_1 produced automatically delivers 3/2 tons of P_3 as a by-product without any additional work. Each ton of P_2 produced yields 1/4 tons of P_4 as a by-product without any additional work. To produce one ton of P_4 independently requires three hours of machine 3 time.

The company has 96 hours of machine 1 time, 120 hours of machine 2 time, and 240 hours of machine 3 time available. The company wishes to maximize its total net profit. Formulate this problem as an LP and justify your formulation.

1.1.6 Minimizing the Maximum of Several Linear Functions

Consider an optimization problem that is like an LP with the exception that there are k different linear cost functions instead of one. Suppose the objective is to minimize the maximum of these k linear cost functions. Let $x = (x_1, \ldots, x_n)^{\mathrm{T}}$ represent the column vector of variables in the problem and let $c^1x, \ldots, c^kx$ be the k linear cost functions. We have to find an x that minimizes

$$f(x) = \text{maximum } (c^1x, c^2x, \ldots, c^kx)$$

subject to the constraints in the problem. We do not have $f(x)$ given by an explicit formula, but for any given vector x, $f(x)$ can be computed easily from the above equation. That's why the function $f(x)$ defined in this manner is known as the *pointwise maximum (or supremium) of the functions* $c^1x, \ldots, c^kx$.

Problems like this often appear in practice. For example, suppose we want to finish a project in minimal time. There may be k different agencies working independently on segments of this project. Let x be the vector of variables in the problem. c^rx may represent the time taken by the rth agency by adopting the solution vector x. Then the overall time taken for the project is $f(x)$, defined above.

Such problems can be transformed into LPs. Define a new variable x_{n+1} and introduce these constraints.

$$\begin{aligned} x_{n+1} - c^1x &\geqq 0 \\ &\vdots \\ x_{n+1} - c^kx &\geqq 0 \end{aligned}$$

Let $X = (x_1, \ldots, x_n, x_{n+1})^{\mathrm{T}}$. Augment the constraints in the original problem with these constraints. Then minimize $z(X) = x_{n+1}$, subject to all the constraints. The new objective function is linear in the variables and, hence, the new problem is an LP. If $\hat{X} = (\hat{x}_1, \ldots, \hat{x}_n, \hat{x}_{n+1})^{\mathrm{T}}$ is an optimum solution of this new LP, $\hat{x} = (\hat{x}_1, \ldots, \hat{x}_n)^{\mathrm{T}}$ is an optimum solution, of the original problem and $\hat{x}_{n+1} = f(\hat{x})$ is the minimum value of $f(x)$ in the original problem.

Problem

1.8 Justify the preceding statement.

Example

Consider the problem:

$$\begin{aligned} \text{Minimize } f(x) &= \text{maximum } (2x_1 - 3x_2, -7x_1 + 8x_2) \\ \text{Subject to } \quad & x_1 - 3 \cdot 5x_2 \geqq 8 \\ & 9x_1 + 15x_2 \geqq 100 \\ & x_1 \text{ and } x_2 \geqq 0 \end{aligned}$$

When transformed as above, this problem becomes the LP

$$\begin{array}{lrl}\text{Minimize} & x_3 & \\ \text{Subject to} & x_3 - 2x_1 + 3x_2 & \geqq 0 \\ & x_3 + 7x_1 - 8x_2 & \geqq 0 \\ & x_1 - 3 \cdot 5x_2 & \geqq 8 \\ & 9x_1 + 15x_2 & \geqq 100 \\ & x_1 \text{ and } x_2 & \geqq 0 \end{array}$$

Problem

1.9 Consider the problem in which ore has to be transported from m mines to n plants. The amount of ore available at the ith mine is a_i tons and the amount of ore required at the jth plant is b_j tons. $\sum_i a_i = \sum_j b_j$. Exactly one truck can be used to make these shipments, and the truck is restricted to making no more than one trip from a mine to a plant. What is the minimum capacity truck that can do this job? What is the optimal shipping schedule? Formulate this as an LP.

1.1.7 Minimizing a Nonnegative Weighted Sum of Absolute Values

In some cases we may have to solve an optimization problem of the form: Find $x = (x_1, \ldots, x_n)$ to

$$\text{Minimize } V(x) = \sum_{j=1}^{n} c_j|x_j| \tag{1.5}$$

$$\text{Subject to } \sum_{j=1}^{n} a_{ij}x_j = b_i \qquad \text{for } i = 1 \text{ to } m$$

where $|x_j|$ is the absolute value of x_j and all the c_j are given nonnegative numbers.

Problems like this can be transformed into LPs. The variable x_j is unrestricted in sign in the problem. Every real number can be expressed as the difference of two nonnegative numbers. Hence, we can express the variable x_j as the difference of two nonegative variables, as in

$$\begin{array}{l} x_j = x_j^+ - x_j^- \\ \qquad x_j^+ \geqq 0 \qquad x_j^- \geqq 0 \end{array} \tag{1.6}$$

By defining

$$\begin{array}{rll} x_j^+ & = 0 & \text{if } x_j \leqq 0 \\ & = x_j & \text{if } x_j > 0 \\ x_j^- & = 0 & \text{if } x_j \geqq 0 \\ & = -x_j & \text{if } x_j < 0 \end{array} \tag{1.7}$$

we can even guarantee that these variables x_j^+ and x_j^- can be chosen to satisfy (1.6) and

$$x_j^+ x_j^- = 0 \tag{1.8}$$

that is, at least one of x_j^+ and x_j^- is always zero. When defined in this manner, x_j^+ is known as the *positive part* and x_j^- as the *negative part* of the unrestricted variable x_j.

If x_j^+ and x_j^- satisfy both (1.6) and (1.8), clearly

$$|x_j| = x_j^+ + x_j^-$$

By using this equation, (1.5) is transformed into this problem: Find

$$x^+ = (x_1^+, \ldots, x_n^+) \quad \text{and} \quad x_n^- = (x_1^-, \ldots, x_n^-)$$

to

$$\text{Minimize} \quad U(x^+, x^-) = \sum_{j=1}^{n} c_j(x_j^+ + x_j^-)$$

$$\text{Subject to} \quad \sum_{j=1}^{n} a_{ij}(x_j^+ - x_j^-) = b_i, \quad \text{for } i = 1 \text{ to } m \tag{1.9}$$

$$x_j^+ \geqq 0, x_j^- \geqq 0 \quad \text{for all } j$$

$$x_j^+ x_j^- = 0 \quad \text{for all } j \tag{1.10}$$

Let $\tilde{x}^+, \tilde{x}^-$ be any feasible solution to (1.9) that may or may not satisfy (1.10). Let:

$$\tilde{v}_j = \min(\tilde{x}_j^+, \tilde{x}_j^-) \quad \text{for all } j$$

$$\hat{x}_j^+ = \tilde{x}_j^+ - \tilde{v}_j \quad \text{for all } j$$

$$\hat{x}_j^- = \tilde{x}_j^- - \tilde{v}_j \quad \text{for all } j$$

Clearly $(\hat{x}^+, \hat{x}^-)$ is also a feasible solution to (1.9), and since all $c_j \geqq 0$,

$$U(\hat{x}^+, \hat{x}^-) \leqq U(\tilde{x}^+, \tilde{x}^-) \tag{1.11}$$

and $(\hat{x}^+, \hat{x}^-)$ obviously satisfies (1.10). Thus the property that $c_j \geqq 0$ for all j implies that there is an optimum solution to (1.9) that automatically satisfies (1.10). When the LP (1.9) is solved by the simplex method, an optimum solution for it that satisfies (1.10) automatically will be obtained provided (1.9) has a feasible solution. From the optimum solution of (1.9) an optimum solution of (1.5) is obtained by using (1.6). Hence (1.5) is equivalent to the LP (1.9).

Problem

1.10 If some $c_j < 0$, show that (1.11) may not hold and hence the transformation of (1.5) into the LP (1.9) will not work.

1.1.8 Modeling with Excesses and Shortages

In many practical problems (e.g., inventory modeling), we are required to compare a linear function of the problem variables, say, $\sum a_j x_j$, (which in inventory models may be *the amount of production*) with a known constant, say, b (which will correspond to the *demand* in inventory models).

If $\sum a_j x_j > b$, there is an *excess situation* and a cost is incurred, which is equal to $c_1(\sum a_j x_j - b)$. In inventory models this will be the *holding cost.*

If $\sum a_j x_j < b$, there is a *shortage situation* and a penalty or cost is incurred, which is equal to $c_2(b - \sum a_j x_j)$. In inventory models this will be the *penalty for shortage.*

There is no cost due to excess in a shortage situation; and likewise, there is no penalty for shortage in an excess situation. Assume that the problem is to minimize the total cost. The actual excess and the shortage can be represented by expressing the real number $\sum a_j x_j - b$ as a difference of two nonnegative numbers as in section 1.1.7. Clearly the excess is the positive part of $(\sum a_j x_j - b)$ and the shortage is its negative part. Hence, if y_1 and y_2 are such that

$$\begin{aligned} \textstyle\sum a_j x_j - y_1 + y_2 &= b \\ y_1 \geqq 0, \quad y_2 &\geqq 0 \end{aligned} \tag{1.12}$$

$$y_1 y_2 = 0 \tag{1.13}$$

y_1 is the excess and y_2 is the shortage. Hence the cost incurred is $c_1 y_1 + c_2 y_2$. Since the aim is to minimize the total cost, using arguments similar to those in section 1.1.7, it can be shown that the constraint (1.13) will be satisfied automatically by an optimum solution to the problem obtained by the simplex method, if $c_1 + c_2 \geqq 0$.

Assuming that $c_1 + c_2 \geqq 0$, we can therefore represent the total contribution of the excess and shortage costs as $c_1 y_1 + c_2 y_2$ where y_1 and y_2 are subject to the constraints (1.12). This models the problem correctly as long as $c_1 + c_2 \geqq 0$. If $c_1 + c_2 < 0$, it is not possible to model this problem as an LP.

Problem

1.11 Justify the above model.

Example

Consider the following problem. A firm has m cement manufacturing plants and n cement selling markets. During a season the ith plant can manufacture at most a_i tons of cement during *regular production time* at a cost of r_i dollars per ton for $i = 1$ to m.

In addition to *regular production*, the ith plant can manufacture an amount of at most b_i tons of cement by running *overtime* at a cost of s_i dollars per ton (where $s_i > r_i$) for $i = 1$ to m.

Note A plant cannot operate on *overtime* unless the *regular time* is fully used up.

The transportation cost from the ith plant to the jth market is c_{ij} dollars per ton of cement. All the cement manufactured during a season can be transported to the market in the same season.

At the jth market there is a *demand* of d_j tons of cement for $j = 1$ to n. There is no restriction that the *demand* should be met; however, each ton of *demand met* in market j fetches a price of α_j dollars and each ton of demand left unsatisfied at market j results in a penalty of δ_j dollars to the firm.

If the total amount of cement shipped to market j exceeds the demand at that market, the excess can be sold to a warehouse at a price of γ_j dollars per ton (note $\gamma_j < \alpha_j$), for $j = 1$ to n, in unlimited quantities.

The problem is to determine the optimal production-distribution-marketing pattern that minimizes the firm's net cost during a season.

We will now formulate this problem as an LP. Consider a season. The variables in the problem are

x_{ij} = tons of cement shipped from the ith plant to the jth market

x_i = total tons of cement manufactured at the ith plant

y_{1i} = total tons of cement manufactured at the ith plant in regular time

y_{2i} = total tons of cement manufactured at the ith plant in overtime

u_j = excess amount in tons (amount supplied above the demand) at the jth market

v_j = shortage of cement in tons (amount of unfulfilled demand) at the jth market

The production cost for producing x_i tons of cement at the ith plant is a piecewise linear convex function as in Figure 1.5.

As in section 1.1.5 we formulate this by expressing x_i as $y_{1i} + y_{2i}$, with the restrictions, $0 \leqq y_{1i} \leqq a_i$, $0 \leqq y_{2i} \leqq b_i$, and the production cost will be $r_i y_{1i} + s_i y_{2i}$. Similarly the shortage and excess amounts at the jth market are obtained by expressing the difference between the total supply and demand for cement at this market as the difference of excess

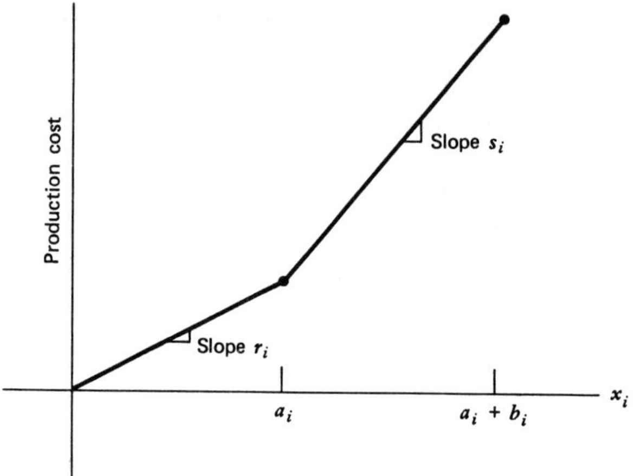

Figure 1.5

and shortage, both of which are nonnegative numbers, that is,

$$\sum_{i=1}^{m} x_{ij} - d_j = u_j - v_j$$

and then the net cost at this market can be expressed as

$$\delta_j v_j - \alpha_j(d_j - v_j) - \gamma_j u_j$$

Hence eliminating the constant terms from the objective function, the overall problem is

$$\text{Minimize} \quad \sum_{i=1}^{m} (r_i y_{1i} + s_i y_{2i}) + \sum_{j=1}^{n} ((\delta_j + \alpha_j)v_j - \gamma_j u_j) + \sum_{i=1}^{m} \sum_{j=1}^{n} c_{ij} x_{ij}$$

$$\begin{aligned} \text{Subject to} \quad & \sum_{j=1}^{n} x_{ij} - y_{1i} - y_{2i} = 0 && \text{for } i = 1 \text{ to } m \\ & \sum_{i=1}^{m} x_{ij} - u_j + v_j = d_j && \text{for } j = 1 \text{ to } n \\ & y_{1i} \leqq a_i && \text{for } i = 1 \text{ to } m \\ & y_{2i} \leqq b_i && \text{for } i = 1 \text{ to } m \\ & x_{ij}, y_{1i}, y_{2i}, u_j, v_j \geqq 0 && \text{for all } i, j \end{aligned}$$

Problems

1.12 In a laboratory the yield in a chemical reaction was measured at various temperatures. The data is given below.

Temperature t	Yield y
−5	80
−3	92
−1	96
0	98
1	100
3	105
5	115
7	130
10	150

It is expected that the yield can be approximated by a cubic expression in terms of the temperature, t, namely, $a_0 + a_1 t + a_2 t^2 + a_3 t^3$. Find the optimum values of the coefficients a_0, a_1, a_2, a_3 that minimize the sum of the absolute deviations of the actual yield from the cubic fit, at the values of t used in the experiment. Formulate this as an LP.

1.13 There are three types of machines that can be used to make a chemical. The chemical can be manufactured at four different purity levels. Here is the data.

Machine type	Capacity: production per day	Production costs per day if used to manufacture purity			
		1	2	3	4
1	50 tons	$900	1000	1250	1500
2	40 tons	600	750	1000	1050
3	25 tons	600	700	800	900
Monthly demand for chemical at purity level (tons)		150	300	90	175
Penalty per ton short		40	60	75	90

The chemical is used by the company internally; if the company does not manufacture enough of it, it can be bought at a price, which is called the penalty for shortage. Assume that there are 30 production days per month at the company. Formulate the problem of minimizing the overall cost as an LP.

1.2 SOLVING LINEAR PROGRAMS IN TWO VARIABLES

LPs involving only two variables can be solved by drawing a diagram on the Cartesian plane. Given an LP, a *feasible solution* is a vector that specifies a value for each variable in the problem, such that when these values are substituted for the variables all the constraints and sign restrictions are satisfied. An *optimum solution* is a feasible solution that either maximizes the objective function or minimizes it (depending on whether it is required to maximize it or minimize it in the problem) on the set of feasible solutions.

To solve an LP involving only two variables, the set of feasible solutions is first identified on the two dimensional Cartesian plane. The optimum solution is then identified by tracing the line corresponding to the set of feasible solutions that give a specific value to the objective function and then moving this line parallel to itself.

As an example consider the fertilizer product mix problem (1.1). The constraint $2x_1 + x_2 \leqq 1500$ requires that any feasible solution (x_1, x_2) to the problem should be on one side of the line $2x_1 + x_2 = 1500$, the side that contains the origin (because the origin makes $2x_1 + x_2 = 0 < 1500$). This side is indicated by an arrow on the line in Figure 1.6. Likewise, all the constraints can be represented on the diagram. The set of feasible solutions is the set of points on the plane satisfying all the constraints; this is the shaded region in Figure 1.6.

For any real value of z_0, $z(x) = z_0$ represents all the points on a straight line in the two dimensional plane. Changing the value of z_0 translates the entire line to another parallel line. Pick an arbitrary value of z_0 and draw the line $z(x) = z_0$ on the diagram. If this line intersects the set of feasible solutions, the points of intersection are the feasible solutions that give the value z_0 to the objective function.

If the line does not intersect the set of feasible solutions, check whether increasing z_0, or decreasing it, translates it to a parallel line that is closer to the set of feasible

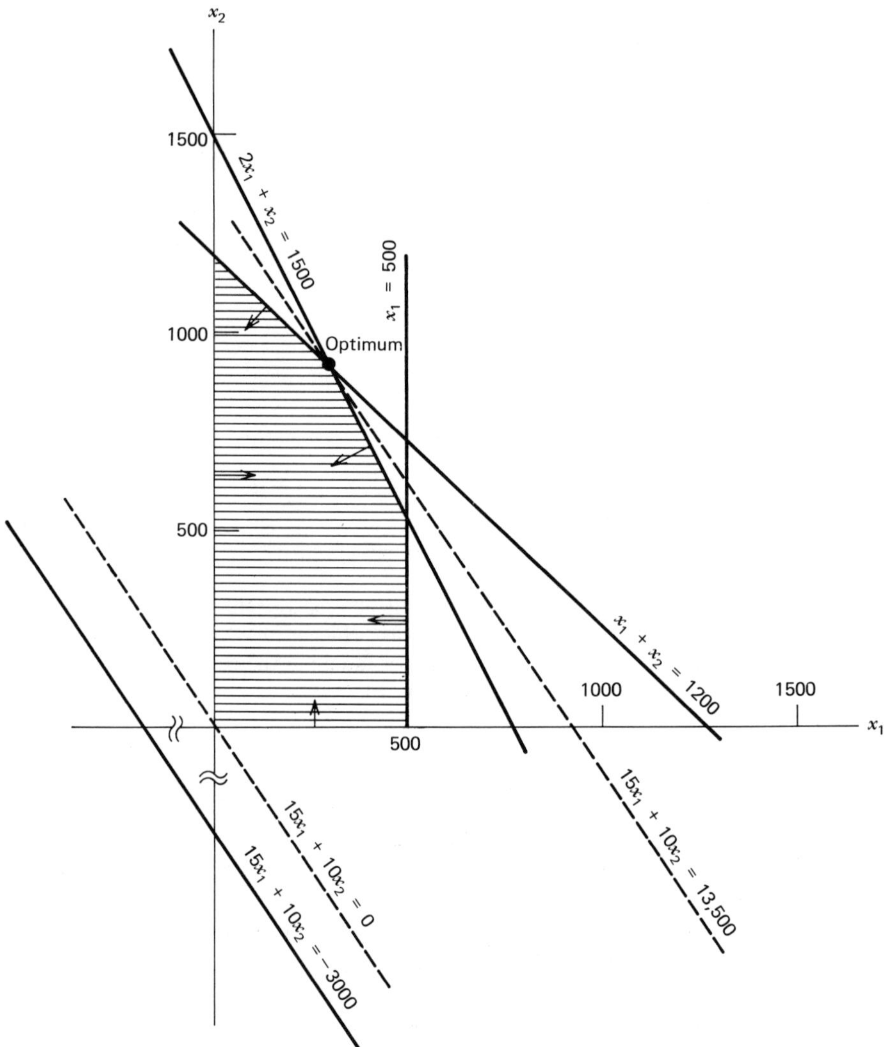

Figure 1.6 Fertilizer Product Mix Problem.

solutions. Change the value of z_0 appropriately until a value for z_0, say z_0', is found such that the line $z(x) = z_0'$ has a nonempty intersection with the set of feasible solutions. Then go to the next step.

In the fertilizer problem, if $z_0 = -3000$, the line $z(x) = -3000$ does not intersect the set of feasible solutions at all. Also, the line $z(x) = z_0$ moves closer to the set of feasible solutions as z_0 increases from -3000, and when $z_0 = 0$, the line

$$z(x) = 15x_1 + 10x_2 = 0$$

has a nonempty intersection with the set of feasible solutions.

If $z(x)$ has to be maximized, try to move the line $z(x) = z_0$ in a parallel fashion by increasing the value of z_0 steadily from z_0' as far as possible while still intersecting the set of feasible solutions. If $\hat{z}_0$ is the maximum value of z_0 obtained in this process, it is the maximum value of $z(x)$ in the problem and the set of optimum feasible solutions is the set of feasible solutions that lie on the line $z(x) = \hat{z}_0$.

On the other hand, if the line $z(x) = z_0$ has a nonempty intersection with the set of feasible solutions for every $z_0 \geqq z'_0$, then $z(x)$ is *unbounded above* on that set. In this case $z(x)$ can be made to diverge to $+\infty$ and the problem has no finite optimum solution.

If the aim is to minimize $z(x)$, then steadily decrease the value of z_0 from z'_0 and apply the same kind of arguments.

For the fertilizer problem, as z_0 is increased from 0, the line $15x_1 + 10x_2 = z_0$ moves up until $\hat{z}_0 = 13{,}500$. For any value of $z_0 > 13{,}500$ the line $z(x) = z_0$ does not intersect the set of feasible solutions. Thus, the optimum objective value is \$13,500 and the optimum solution of the problem is $(x_1, x_2) = (300, 900)$.

Problem

1.14 Solve the following LPs by drawing diagrams.

(a)

$$\begin{array}{ll} \text{Maximize} & 30x_1 + 20x_2 \\ \text{Subject to} & x_1 + x_2 \geqq 1 \\ & x_1 - x_2 \geqq -1 \\ & 3x_1 + 2x_2 \leqq 6 \\ & x_1 - 2x_2 \leqq 1 \\ & x_1 \geqq 0, x_2 \geqq 0 \end{array}$$

(b)

$$\begin{array}{ll} \text{Minimize} & 2x_1 - x_2 \\ \text{Subject to} & x_1 + x_2 \geqq 10 \\ & -10x_1 + x_2 \leqq 10 \\ & -4x_1 + x_2 \leqq 20 \\ & x_1 + 4x_2 \geqq 20 \\ & x_1 \geqq 0, x_2 \geqq 0 \end{array}$$

(c)

$$\begin{array}{ll} \text{Minimize} & x_1 + 3x_2 \\ \text{Subject to} & 2x_1 + x_2 \geqq 10 \\ & -x_1 + x_2 \leqq 20 \\ & x_1 - 2x_2 \leqq 10 \\ & x_1 + x_2 \leqq 30 \\ & x_1 \geqq 0, x_2 \geqq 0 \end{array}$$

(d)

$$\begin{array}{ll} \text{Minimize} & 10x_1 + 3x_2 \\ \text{Subject to} & x_1 + x_2 \geqq 20 \\ & x_1 \leqq 6 \\ & x_1 \geqq 2 \\ & x_2 \leqq 12 \\ & x_2 \geqq 1 \end{array}$$

(e)

$$\begin{array}{ll} \text{Maximize} & 2x_1 - 2x_2 \\ \text{Subject to} & -2x_1 + x_2 \leqq 2 \\ & x_1 - x_2 \leqq 1 \\ & x_1 \geqq 0, x_2 \geqq 0 \end{array}$$

(f) In the following problem show that the variables x_3, x_4, x_5 play the roles of slack variables. Show that the constraints in the problem can be transformed into inequality constraints in the variables x_1, x_2, by eliminating these slack variables. Use this and solve this problem by drawing a diagram.

$$\begin{array}{ll} \text{Minimize} & x_1 - 3x_2 \\ \text{Subject to} & 2x_1 + x_2 + x_3 = 20 \\ & x_1 - x_4 = 7 \\ & x_2 - x_5 = 17 \\ & x_j \geqq 0 \quad \text{for all } j = 1 \text{ to } 5 \end{array}$$

Note LPs in higher dimensional spaces ($\geqq 3$) cannot be solved by drawing diagrams, but the simplex algorithm that will be discussed in Chapter 2 provides a systematic procedure for solving them. However, we will gain a lot of intuition on how the simplex algorithm works by referring to and visualizing the concepts of geometry.

PROBLEMS

Formulate Problems 1.15 to 1.34 as LPs.

1.15 A forestry company has four sites on which they grow trees. They are considering four species of trees, the pines, spruces, walnuts and other hardwoods. Data on the problem is given below.

Site number	Area available at site (kiloacres)	Expected annual yield from species (cubic meters per kiloacre)				Expected annual revenue from species (money units per kiloacre)			
		Pine	Spruce	Walnut	Hardwood	Pine	Spruce	Walnut	Hardwood
1	1500	17	14	10	9	16	12	20	18
2	1700	15	16	12	11	14	13	24	20
3	900	13	12	14	8	17	10	28	20
4	600	10	11	8	6	12	11	18	17
Minimal required expected annual yield (kilo cubic meters)		22.5	9	4.8	3.5				

How much area should they devote to the growing of various species in the various sites?

1.16 A farmer has to purchase the following quantities of fertilizer:

Fertilizer type	Minimal quantity required (tons)
1	185
2	50
3	50
4	200
5	185

He can buy these quantities from four different shops, subject to the following capacities.

Shop number	Maximum amount (all types combined) that shop can supply
1	350 tons
2	225
3	195
4	275

These are prices of the various types of fertilizer at the various shops.

At shop	Price in money units/ton of fertilizer type				
	1	2	3	4	5
1	4.50	1.39	2.99	3.19	0.99
2	4.25	1.78	3.10	3.50	1.23
3	4.75	1.99	2.40	3.25	1.24
4	4.13	1.25	3.12	2.98	1.10

How can the farmer fulfill his fertilizer requirements at minimal cost?

1.17 A product can be made in three sizes, large, medium, and small, which yield a net unit profit of $12, 10, and 9, respectively. The company has three centers where this product can be manufactured and these centers have a capacity of turning out 550, 750, and 275 units of the product per day, respectively, regardless of the size or combination of sizes involved. Manufacturing this product requires cooling water and each unit of large, medium, and small sizes produced require 21, 17, and 9 gallons of water, respectively. The centers 1, 2, and 3 have 10,000, 7000, and 4200 gallons of cooling water available per day, respectively. Market studies indicate that there is a market for 700, 900, and 450 units of the large, medium, and small sizes, respectively per day. By company policy, the fraction, (scheduled production)/(center's capacity), must be the same at all the centers. How many units of each of the sizes should be produced at the various centers in order to maximize the profit?

1.18 A housewife wants to fix a naturally vitamin-rich cake made out of blended fruit for supper. Data on the available types of fruit is given below.

Fruit type	Number of units of nutrient/kilogram of fruit		Cost ($/kg) of fruit	Restrictions on fruit use (kg)	
	Vitamin A	Vitamin C		Minimum	Maximum
1	2	0	0.25	2	10
2	0	3	0.31	3	6
3	2	4	0.48	0	7
4	1	2	0.21	5	20
5	3	2	0.19	0	5
Minimum nutrient requirement (units)	41	80			

To keep the cake palatable she has to make sure that the use of the various fruits is within the bounds specified. Formulate the problem of finding the minimum cost blend that satisfies all of these constraints.

By simple linear transformations on the variables show that this LP can be transformed into an equivalent LP in which the lower bounds on all the variables are zero.

1.19 THREE DIMENSIONAL TRANSPORTATION PROBLEM. A firm transports wood logged from its forest stands through one of its own depots to a customer's depot. They have m forest stands, p depots and n customers. c_{ijk} is the cost of transporting logged wood from the ith forest stand to the kth customer through the jth depot in dollars/cubic meter, for $i = 1$ to m, $j = 1$ to p, $k = 1$ to n. a_i is the annual production of wood in the ith forest stand in cubic meters. b_j is the amount of wood that the jth depot can handle in cubic meters per year. d_k is the demand for wood by the kth customer in cubic meters per year. It is required to minimize the overall annual transportation bill subject to all the restrictions.

1.20 A company makes a blend consisting of two chemicals, 1 and 2, in the ratio 5:2 by weight These chemicals, can be manufactured by three different processes using two different raw materials and a fuel. Here is the production data.

Process	Requirements per unit time			Output per unit time	
	Raw material 1 (units)	Raw material 2 (units)	Fuel (units)	Chemical 1 (units)	Chemical 2 (units)
1	9	5	50	9	6
2	6	8	75	7	10
3	4	11	100	10	6
Amount available	200	400	1850		

For how much time should each process be run in order to maximize the total amount of *blend* manufactured?

1.21 In constructing a hydrological model, it is required to obtain the expected runoff, denoted by R_i during the ith period, as a linear function of the observed precipitation. From hydrological considerations the expected runoff depends on the precipitation during that period and the previous two periods. So the model for expected runoff is

$$R_i = b_0 p_i + b_1 p_{i-1} + b_2 p_{i-2}$$

where p_i = precipitation during the ith period; b_0, b_1, b_2 are the coefficients that are required to be estimated. These coefficients have to satisfy the following constraints from hydrological considerations

$$b_0 + b_1 + b_2 = 1$$
$$b_0 \geqq b_1 \geqq b_2 \geqq 0$$

The following data is available.

Period	1	2	3	4	5	6	7	8	9	10	11	12
Precipitation (inch hours)	3.8	4.4	5.7	5.2	7.7	6.0	5.4	5.7	5.5	2.5	0.8	0.4
Runoff (acre feet)	0.05	0.35	1.0	2.1	3.7	4.2	4.3	4.4	4.3	4.2	3.6	2.7

Obtain the best estimates for b_0, b_1, b_2, if the objectives are:

(i) To minimize the sum of absolute deviations

$$\sum_i |R_i - b_0 p_i - b_1 p_{i-1} - b_2 p_{i-2}|$$

(ii) To minimize the maximum absolute deviation

$$\operatorname*{Max}_{i} |R_i - b_0 p_i - b_1 p_{i-1} - b_2 p_{i-2}|$$

—(*R. Deininger*)

1.22

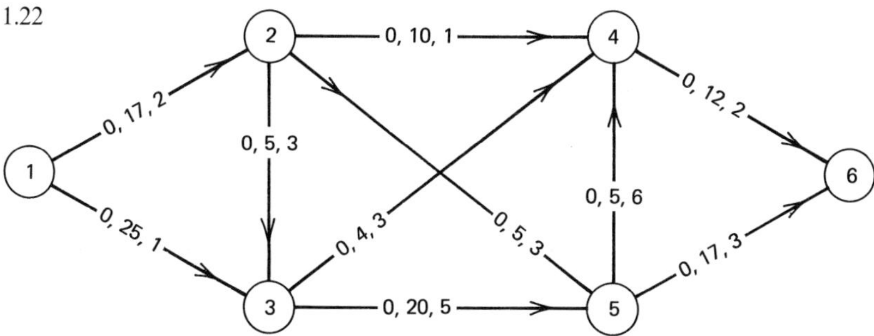

Figure 1.7

Each circle in Figure 1.7 is a city. Material can be shipped from city i to city j only if there is a directed arc from city i to city j as ⓘ—ℓ_{ij}, k_{ij}, c_{ij}→ⓙ. The three numbers on the arc are the least amount you can ship along this arc, the maximum tonnage you can ship along this arc, and the cost in dollars/ton shipped along this arc, respectively. There is 27 tons of material at city 1, and all of it should be shipped to city 6 at minimal total cost. All material should originate from city 1 and end up in city 6. At any of the intermediate cities the amount of material reaching should be equal to the amount of material leaving.

1.23 A European company has three coke oven plants, code named 1, 2, 3. The coal comes from four different sources; USA, Ruhr, Lorraine, and Saar. The plants produce coke, which may be classified into two categories, metallurgical coke and coke screenings; they also produce coke oven gas. The coke oven plants are operated by heating them with either blast furnace gas or coke oven gas. The process flow chart is given in Figure 1.8. The production of coke oven gas and coke depends on the coal used.

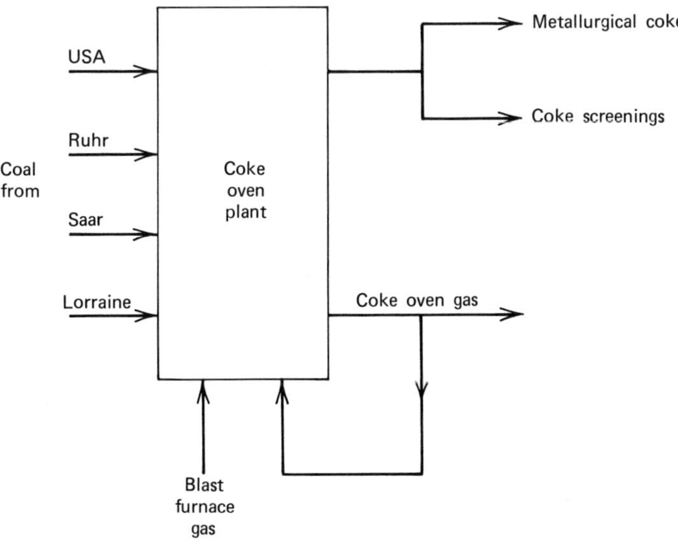

Figure 1.8

Coal source	Coke oven gas* produced Kth/ton of coal	Coke tons/ton of coal
USA	1.08	0.88
Ruhr	1.10	0.80
Saar	1.25	0.74
Lorraine	1.45	0.72

* The amount of coke oven gas produced is measured by its heat content. Kth is a kilothermie, where thermie is the amount of heat needed to raise the temperature of 1 ton of water up by 1 degree centigrade.

The proportion of coke screenings in the coke produced depends on the coal used and the plant where it is used.

Proportion of coke screenings in the coke produced

At plant	Using coal from USA	Ruhr	Saar	Lorraine
1	0.10	0.09	0.10	0.15
2	0.08	0.08	0.08	0.11
3	0.07	0.07	0.07	0.10

Metallurgical coke is what is left in the coke after coke screenings are separated.

Processing one ton of coal requires the heat equivalent of 0.611 Kth of coke oven gas. One unit of blast furnace gas is equivalent to 0.927 Kth of coke oven gas.

The annual processing capacities of the three plants are 9×10^5 tons, 7×10^5 tons, and 3×10^5 tons of coal, respectively.

Saar coal cannot be used in plants 1 and 2. USA coal and Lorraine coal cannot be used in plant 3. The percentage of Lorraine coal in the coal used at plant 1 cannot exceed 30. The percentage of Lorraine coal in the coal used at plant 2 cannot exceed 35. The percentage of Saar coal in the coal used at plant 3 cannot exceed 40.

Coke oven gas can be bought or sold in any amounts at \$11/Kth. Coke screenings can be sold in any amounts at \$98/ton. Blast furnace gas can be purchased in any amounts at \$8/unit. The prices of coal are as follows:

Coal source	
USA	\$80.39/ton
Ruhr	\$80.90/ton at plant 1 and 2 \$83.93/ton at plant 3
Saar	\$80.14/ton
Lorraine	\$68.70/ton

All the coke screenings produced are sold. All the coke oven gas produced is either used up at the plants or sold.

It is required to produce a total of 10^6 tons of metallurgical coke at minimal cost.

—(*J. F. Collard*)

1.24 A farmer is planning his operations over a three-year period. At the beginning of the period he has two bushels of grain. At the beginning of years 1, 2, 3 he has to decide how much he will plant. A bushel of grain planted at the beginning of a year yields λ bushels by the end of that year. The profit per bushel of grain sold at the beginning of year i is expected to be p_i dollars for $i = 1, 2, 3, 4$. The farmer will sell all the grain he has available at the end of the third year and close down his farm. Determine an optimal selling-planting program that will maximize his total profit. Use the input-output approach.

1.25 A company manufactures a product, the demand for which varies from month to month. The raw material and labor availability exhibit seasonal variations.

Month	Cost of labor (dollars/ton of production) during: Regular time	Overtime	Limit on raw material availability (enough to make tons of product)	Demand (tons)	Selling price (dollars/ton)
1	\$4	\$6	600	400	18
2	during these		450	700	18
3	months		425	600	18
4	\$6	\$9	1200	900	25
5	during these		1300	900	25
6	months		1600	900	25
7			1600	800	25
8			1500	600	25
9			1300	800	25
10	\$4	\$6	500	1200	30
11			500	1100	30
12			500	1400	30

During the months 10, 11, 12, 1, 2, 3 the company can hire at most enough labor to produce 1200 and 600 tons per month during regular time and overtime, respectively. In months 4, 5, 6, 7, 8, 9 these labor capacities are 800 and 500 tons, respectively. The produce during a month can be sold anytime during the next month or later. Storage costs are \$1.00/ton from one month to the next for the product. Raw material cannot be stored. It has to be used up in the month in which it is obtained. Operations begin in month 1 with a stock of 50 tons of the product. At the end of month 12 the company should have a stock of at least 50 tons of the product. Determine an optimum production schedule.

1.26 An agency controls the operation of a system consisting of two water reservoirs with one hydroelectric power generation plant attached to each. The planning horizon for the system is a year, divided into six periods. Reservoir 1 has a capacity for holding 3500 kilo acre-ft of water and reservoir 2 has a capacity of 5500 kilo acre-ft. At any given instant of time, if the reservoir is at its full capacity, the additional inflowing water is spilled over a spillway. Spilled water does not produce any electricity.

During each period some specified minimal amount of water must be released from the reservoirs to meet the downstream requirements for recreation, irrigation, and navigation purposes. However, there is no upper limit on the amount of water that can be released from the reservoirs. Any unreleased water is stored (up to the capacity of the reservoir) and can be used for release in subsequent periods. All water released from the reservoirs (even though it is released for recreation and other purposes) produces electricity.

It can be assumed that during each period, the water inflows and releases occur at a constant rate. Also, on an average 1 acre-ft of water released from reservoir 1 produces 310 kWh of electricity and 1 acre-ft released from reservoir 2 produces 420 kWh.

At the beginning of the year reservoir 1 contains 1800 kilo acre-ft of water and reservoir 2 contains 2500 kilo-acre-ft of water. The same amounts of water must be left in the respective reservoirs at the end of the year.

The electricity produced can either be sold to a local firm (called a class I customer) or to class II customers.

A class I customer buys electricity on an annual basis; he requires that specified percentages of it should be supplied in the various periods. He pays $10.00/1000 kWh.

A class II customer buys electricity on a period by period basis. He will purchase any amount of electricity in any period at $5.00 per thousand kWh.

The data for the problem is given below.

Period	Inflows in kilo acre-ft into reservoir 1	Inflows in kilo acre-ft into reservoir 2	Minimum release from reservoir 1	Minimum release from reservoir 2	Percentage of annual energy sold to class I customer to be delivered in period
1	547	2616	200	304	10
2	1471	2335	200	578	12
3	982	1231	200	995	15
4	146	731	200	1495	32
5	32	411	200	558	21
6	159	497	200	392	10

Operate the system to maximize the total annual revenue from the sale of electricity.

—(*S. Parikh*)

1.27 A manufacturer has a permit to operate for five seasons. He can manufacture only during the first four seasons and in the fifth period he is only allowed to sell any leftover products.

He can manufacture two types of products. One unit of product 1 requires five man-hours in preparatory shop and three man-hours in finishing shop. Each unit of product 2 requires six man-hours in preparatory shop and one man-hour in finishing shop. During each season he has at most 12,000 man-hours in preparatory and 15,000 man-hours in the finishing shop (only during the first four seasons).

The product manufactured during some season can be sold anytime from the next season onward. However, selling requires some marketing effort and it is expected that 0.1 and 0.2 man-hours of marketing effort are required to sell 10 units of products 1 and 2, respectively. Man-hours for marketing effort can be hired at the following rates:

Season	Normal rate per man-hour	Maximum number of man-hour that can be hired at normal rate	Overtime rate per man-hour
2	$2	400	$20
3	$4	300	$20
4	$1	600	$20
5	$10	1000	$20

There is no limit on the number of man-hours that can be hired for marketing effort at the overtime rate.

If a unit of product is available for sale during a season, but is not sold in that season, the manufacturer has to pay carryover charges of \$2 per unit to put it up for sale again in the next season. The selling prices in the various seasons are expected to be:

Season	Expected selling price per unit of	
	Product 1	Product 2
2	\$20	45
3	25	40
4	30	40
5	15	30

How should the manufacturer operate in order to maximize his total profit?

1.28 Below is the composition of various foods used in making cereals.

Food	Percentage content (by weight)				Price (dollars/kg)	Availability per day (kg)
	Protein	Starch	Minerals, vitamins, etc.	Other material		
1	45	12	4	39	0.68	1500
2	7	38	1	54	0.27	500
3	12	25	2	61	0.31	1000
4	27	40	3	30	0.45	2000

The other material in each food is fiber, water, etc. The company blends these food materials and makes two kinds of cereals. In the process of blending 3% of protein, 5% of starch and 10% of minerals and vitamins are completely lost from the mix. For each 100 kg of foods added in the blend, the blending process adds 5 kg of other material (mainly water and fat).

Cereal type 1 sells for \$1.50/kg. It should contain at least 22% of protein, 2% of minerals and vitamins, and at most 30% of starch by weight. Cereal type 2 sells for \$1.00/kg. It should contain at least 30% starch by weight. What is the optimal product mix for the company?

1.29 A contractor is working on a project, work on which is expected to last for a period of T weeks. It is estimated that during the jth week, the contractor will need u_j man-hours of labor, $j = 1$ *to* T, for this project. The contractor can fulfill these requirements either by hiring laborers over the entire T week horizon (called *steady labor*) or by hiring laborers on a weekly basis each week (called *casual labor*) or by employing a combination of both. One man-hour of steady labor costs c_1 dollars; the cost is the same each week. However, the cost of casual labor may vary from week to week, and it is expected to be c_{2j} dollars/man-hour, during week j, $j = 1$ to T. How can he fulfill his labor requirements at minimal cost?

—(*S. Kedia and G. Ponce Compos*)

1.30 There are m refineries with the jth refinery having the capacity to supply a_i gallons of fuel. There are n cities with a demand for this fuel, and the demand in the jth city is b_j gallons. f_{ij} is the fraction of a gallon of fuel consumed in transportation when 1 gallon of fuel leaves refinery i to city j by a delivery truck. Find a feasible shipping schedule that minimizes the total amount of fuel consumed by the delivery trucks.

1.31 A farmer has three farms. He can grow three different crops on them. Data for the coming season is given below.

Farm	Usable acreage	Water available in acre-ft at \$25/acre-ft	Cost per acre-ft of water beyond amounts in previous column
1	400	1200	\$50
2	600	2200	\$60
3	450	1100	\$70

Crop	Maximum acreage farmer can plant	Water consumption (acre-ft/acre)	Yield units/acre	Selling price per unit
A	500	6	50	\$25/unit up to 10,000 units \$20/unit beyond 10,000 units
B	700	5	100	\$15/unit up to 40,000 units \$8/unit beyond 40,000 units
C	350	4	75	\$45/unit up to 20,000 units \$42/unit beyond 20,000 units

To maintain a uniform work load among the three farms, the farmer adopts the policy that the percentage of usable acreage planted must be the same at each of the three farms. However, any combination of the crops may be grown at any of the farms. The only expenses the farmer counts are for the water used at the various farms. How much acreage at each farm should be devoted to the various crops to maximize the farmer's *net profit*?

1.32 A farmer can lease land up to maximum of 1000 acres. He has to pay \$5 per acre per year if he leases up to 600 acres. For any land beyond 600 acres he can lease at \$8 per acre per year.

He grows corn on the land. He can grow corn at *normal* level or at an *intense* level (more fertilizer, frequent irrigation, etc.) Normal level yields 70 bushels/acre. Intense level yields 100 bushels/acre. Here are the requirements.

Requirements per acre per year	Normal level	Intense level
Labor (man-hours)	6	9
Materials (seed, fertilizer, water etc.)	\$20	\$35

Harvesting requires 0.5 man-hours of labor per bushel harvested. He can sell corn at the rate of \$2.50/bushel in the wholesale market. The farmer can also raise poultry. Poultry is measured in poultry units. To raise one poultry unit requires 25 bushels of corn, 20 man-hours of labor and 25 ft^2 of shed floor space. He can either use the corn that he has grown himself or buy corn from the retail market. He gets corn at the rate of \$3.50/bushel from the retail market. He can sell at the price of \$175 per poultry unit in the wholesale market up to 200 units. Any amount of poultry over 200 units sells for \$160 per unit. He has only one shed for raising poultry with 15,000 ft^2 of floor space. He and his family can contribute 4000 man-hours of labor per year at no cost. If he needs more labor he can hire it at \$3 per man-hour up to 3000 man-hours. For any amount of labor hired over 3000 man-hours, he has to pay \$6 per man-hour. Maximize his net profit.

1.33 A company makes products 1 and 2. Machines 1, 2, and 3 are required in the manufacture of these products. The relevant data is tabulated below:

Product	Machine-hours required to manufacture one unit of product on machine		
	1	2	3
1	3	9	2
2	7	4	5
Machine-hours available on the machine	350	970	1420

The manufactured products may be within specification or not. Product i, which is within specifications, is denoted by W_i, and the same product that is outside the specifications is denoted by R_i, $i = 1, 2$. Both W_i and R_i can be sold in the market. The market price of R_i is c_{2i}/unit for $i = 1, 2$. W_i can be sold either as W_i or it can also be sold as R_i. If W_i is sold as W_i, it sells for c_{1i}/unit and if it is sold as R_i it is sold at the same price as R_i. However, R_i cannot be sold as W_i.

When product i is manufactured, an average proportion, p_i, comes out as W_i, $i = 1, 2$. L_i, ℓ_i are the minimum demands for W_i and R_i, respectively. K_i and k_i are the maximum amount of W_i and R_i, respectively, that can be sold. Determine a production and marketing schedule for the company that maximizes its total sales revenue subject to all the constraints.

1.34 There are 14 food ingredients available. They can be classified into 7 groups. It is required to mix them into a minimum cost diet satisfying the following constraints.

Group	Food ingredient	Percent of nutrient content					Cost per ton
		Protein	Fat	Fiber	Ca	Ph	
1	1 Dried beet pulp	9	0.5	20	0.7	0.05	\$64
	2 Dried citrus pulp	6	3	16	2	0.1	35
2	3 Ground yellow corn	8.5	4	2.5	0.02	0.25	55
	4 Ground oats	12	4.5	12	0.1	0.4	54
3	5 Corn molasses	3.5			0.6	0.1	19
4	6 Wheat middlings	16	4	8	0.1	0.9	64
	7 Wheat bran	16	4	10.5	0.1	1.2	62
5	8 Distiller grains	26	8.5	9	0.15	0.6	77
	9 Corn gluten feed	24	2	8	0.3	0.65	66
6	10 Cotton seed meal	41	1.5	13	0.1	1.2	74
	11 Linseed oil meal	34	1	8	0.35	0.8	85
	12 Soybean oil meal	45	0.5	6.5	0.2	0.6	108
7	13 Ground lime stone				36	0.5	10
	14 Ground phosphate				32	14	66

Here are the constraints:

(i) Protein content should be at least 20%.

(ii) Fat content should be at least 3%.

(iii) Fiber content should be at most 12%.

(iv) Ca content should be between 1% and 2%.

(v) Ph content should be between 0.6% and 2%.

(vi) Ca content should be greater than or equal to the Ph content.

(vii) The following manufacturer's requirements (for reasons of taste, texture, and appearance) should be met.

Group	Food ingredient	Range of flexibility of percent of ingredient in diet mix	Range of flexibility of group percent in diet mix
1	1	4 to 20	5 to 20
	2	1 to 20	
2	3	1 to 25	20 to 35
	4	1 to 25	
3	5	5 to 14	5 to 14
4	6	5 to 30	10 to 30
	7	5 to 30	
5	8	1 to 15	2 to 25
	9	1 to 25	
6	10	1 to 35	3 to 35
	11	1 to 35	
	12	1 to 35	
7	13	0 to 2	
	14	1 to 2	

(*G. Brigham*)

1.35 A company manufactures products 1, 2, and 3 with labor and material as inputs. Here are the data.

Item	Items requirement to produce one ton of product 1	2	3	Amount of item available per month
Labor (man-hours)	7	2	4	46
Material (tons)	4	5	5	x_4
Profit ($/ton)	4	1	5	

The material required in this process is obtained from outside. There are two suppliers for this material. The amount of material available, called x_4 in the above tableau, depends on the supplier chosen. The small supplier can provide at most 25 tons of material per month. The big supplier will only supply if at least 35 tons of the material is taken per month. By company regulations, the company has to choose one of these two suppliers and get the material only from him. The company would like to adopt the policy that maximizes its monthly profit. Can this be found out by solving one LP? If not, discuss how you can solve this problem.

1.36 The following table gives some of the data collected during a ballistic evaluation of a new armor plate material. t is the thickness of the target armor plate in inches and v is the striking velocity of the projectile (in feet per second) required to penetrate the plate.

v	t
2300	4.087
2800	4.596
2850	4.646
2900	4.697
3200	5.002

v is the control parameter. Estimate t (the thickness of the plate that a projectile with striking velocity v would penetrate) as a function of v. Use the data to fit the following curve:

$$t = a_3 v + a_1 (a_3 v)^{\boldsymbol{a}_2}$$

where $a_1, \boldsymbol{a}_2, a_3$ are the coefficients to be estimated. $a_1, \boldsymbol{a}_2, a_3$ should be all nonnegative and $\boldsymbol{a}_2$ should lie between 0 and 1. Discuss a practical method for obtaining the values of $a_1, \boldsymbol{a}_2, a_3$ that give the best fit.

1.37 SEPARABLE CONVEX PROGRAMMING PROBLEM. Consider the following problem:

$$\begin{aligned} &\text{Minimize} \quad \textstyle\sum f_i(x_i) \\ &\text{Subject to} \quad \textstyle\sum x_i^2 = 1 \\ &k_i \geqq x_i \geqq \ell_i \qquad \text{for all } i \end{aligned}$$

where $f_i(x_i) = c_i x_i^{-\boldsymbol{a}_i}$ and c_i, $\boldsymbol{a}_i$, ℓ_i, k_i are given positive numbers, and $0 < \ell_i < k_i$ for all i. $f_i(x_i)$ is a nonlinear, convex function for all i. (See Chapter 7 for the definition of a convex function in the feasible region.) Transform this into a problem in which a convex nonlinear objective function has to be minimized subject to linear constraints on the variables. Discuss how an approximate solution of this problem can be obtained by using a piecewise linear approximation for the objective function. How can the precision of the solution obtained be improved?

2 the simplex method

2.1 INTRODUCTION

In an LP we have to optimize the objective function subject to linear equality and/or inequality constraints, and sign restrictions on the variables. If there are no sign restrictions on the variables and if there are no inequality constraints the LP will be of the form:

$$\text{Minimize} \quad z(x) = cx$$
$$\text{Subject to} \quad Ax = b$$

where A, b, c, are given matrices or vectors. It can be shown (see section 3.3.3) that in this case $z(x)$ is either a constant on the set of feasible solutions of the system, or it can be made to diverge to $-\infty$ on this set. Hence, such LPs are trivial, and not of practical interest. Hereafter, we will assume that the LPs being discussed consist of at least one inequality constraint or a variable restricted in sign.

As a naive approach for solving such problems it may be tempting to use the familiar technique in elementary calculus of trying to solve the system of equations obtained by putting the partial derivatives of the objective function equal to zero. This technique does not work here because of the constraints. Even the standard Lagrange multiplier technique does not apply directly to this problem because we have to deal with inequality constraints, and the Lagrange multiplier technique only handles equality constraints. We take recourse to the Karush-John-Kuhn-Tucker theories to handle inequality constraints. Unfortunately these theories provide only certain necessary conditions that every optimum solution must satisfy, and they do not provide any systematic method for solving these conditions. In any case it is clear that any method that can find an optimum solution of an LP must contain in itself a method for finding a feasible solution satisfying a system of linear equality and inequality constraints and sign restrictions on the variables. It can be shown that this problem of finding a feasible solution to start with can itself be posed as another LP. Solving such a linear programming formulation seems to be the only practicable approach for finding a feasible solution, satisfying all the constraints in the problem. Thus, none of the classical methods in calculus or linear algebra offer any approaches for solving LPs or for solving systems of linear inequalities. Special techniques had to be developed to tackle these problems. This chapter discusses these special techniques.

It would be convenient if the final answer to a problem can be obtained either as a simple formula, in terms of the data, or in some tabular form from which the answers can be looked up easily. It will be shown in Chapter 3 that this is generally impossible for LPs. LPs can only be solved by means of an "*algorithm*," which is a step-by-step procedure that leads to an answer to the problem in a finite number of steps. The most successful algorithm for solving LPs is the simplex algorithm developed by George B. Dantzig, or one of its many variants. The simplest form of the simplex algorithm will be discussed first to establish notation and develop ideas. The theory will be discussed in Chapter 3.

2.2 TRANSFORMING THE PROBLEM INTO THE STANDARD FORM

Here we discuss how to transform any LP into a standard form. It can be done by adopting the following steps.

In an LP model, sign restrictions on a variable would require that the variable can either be nonnegative or nonpositive. We will refer to such conditions as *restrictions* in the LP model. All the other conditions in an LP model (whether it involves one or many variables) will be referred to as the *constraints* of the model. For example, in the following system of conditions

$$\begin{aligned} x_1 - 3x_2 + 16x_3 &= 4 \\ -x_1 + 7x_2 \qquad\quad &\leqq 6 \\ x_2 - \quad 3x_3 &\geqq 0 \\ x_3 &\geqq -6 \\ x_3 &\leqq 18 \end{aligned} \tag{2.1}$$

and

$$\begin{aligned} x_1 &\geqq 0 \\ x_2 &\leqq 0 \end{aligned} \tag{2.2}$$

(2.2) are the restrictions and (2.1) are the constraints.

Lower Bound Constraints If there is any variable in the problem that has a required lower bound other than zero, for example, $x_1 \geqq \ell_1$ where ℓ_1 is some specified number, eliminate this variable by substituting $x_1 = y_1 + \ell_1$ wherever it occurs in the problem. Clearly the constraint $x_1 \geqq \ell_1$ is equivalent to requiring $x_1 = y_1 + \ell_1$ and $y_1 \geqq 0$. In this manner all lower bounded variables in the problem can be replaced by variables with lower bounds of zero.

Equality Constraints Convert all the remaining inequality constraints (including upper bound constraints on variables, if any) into equations by introducing appropriate slack variables. A constraint of the form,

$$\sum_{j=1}^{n} a_{1j}x_j \leqq b_1$$

is transformed into

$$\sum_{j=1}^{n} a_{1j}x_j + x_{n+1} = b_1, \qquad x_{n+1} \geqq 0$$

And a constraint of the form

$$\sum_{j=1}^{n} a_{2j}x_j \geqq b_2$$

is transformed into

$$\sum_{j=1}^{n} a_{2j}x_j - x_{n+2} = b_2, \qquad x_{n+2} \geqq 0$$

Here, x_{n+1}, x_{n+2} are the slack variables. The coefficient of the slack variable in the transformed equality constraint depends on whether the original inequality constraint was a "$\leqq$" or a "$\geqq$" inequality. Slack variables are always nonnegative, and represent unutilized capacities, etc. For example, in the above constraints, the slack variables are

$$x_{n+1} = b_1 - \sum_{j=1}^{n} a_{1j}x_j$$

$$x_{n+2} = \sum_{j=1}^{n} a_{2j}x_j - b_2$$

In some textbooks, the name *slack variable* is only used for variables like x_{n+1} above; and the variables like x_{n+2} are called *surplus variables.* Here we will use the term *slack variable* for both these kinds of variables. The slack variables are as much a part of the original problem as the variables used in modeling the constraints. Slack variables remain in the problem throughout, and their values in an optimal feasible solution of the problem give significant information.

Note Each inequality constraint in the original problem leads to a different slack variable.

Here is the reason for transforming the system of constraints into a system of equations. In any system of equations we can:

1. Multiply all the coefficients on both sides of an equation by any nonzero number.
2. Multiply an equation by a real number and add it to another equation.

These transformations do not change the set of feasible solutions of the system. For example, consider the system:

$$\begin{aligned} 3x_1 - 7x_2 &= -1 \\ x_1 + x_2 &= 13 \end{aligned}$$

Add two times the second equation to the first. Then multiply the second equation by four. This leads to the system

$$\begin{aligned} 5x_1 - 5x_2 &= 25 \\ 4x_1 + 4x_2 &= 52 \end{aligned}$$

It can be verified that both systems have the same solution.

When transformations of type 1 and 2 are carried out on a system of inequalities, the set of feasible solutions is normally altered. For example, consider the following system of inequalities for which the set of feasible solutions is indicated by the shaded region in Figure 2.1. Now transform the system by adding two times the second inequality to the first. The new system and its set of feasible solutions are given in Figure 2.2. Clearly the transformed system does not have the same set of feasible solutions as the original system.

However, when a system of inequality constraints is transformed into a system of equations by introducing the appropriate slack variables, then transformations of type 1 and 2 can be carried out on the resulting system of equations, and these will not change the set of feasible solutions of it.

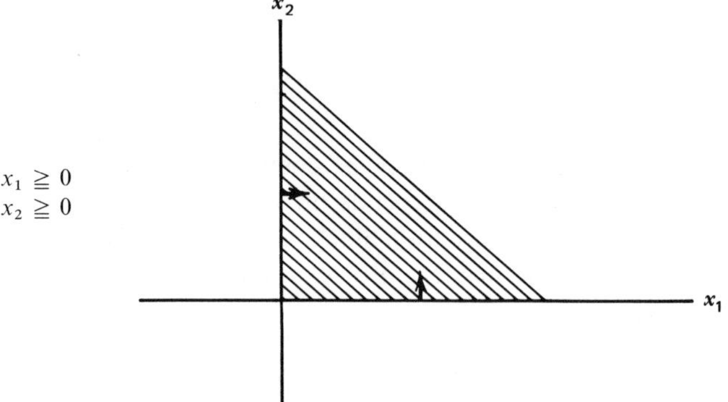

Figure 2.1 System of Inequalities

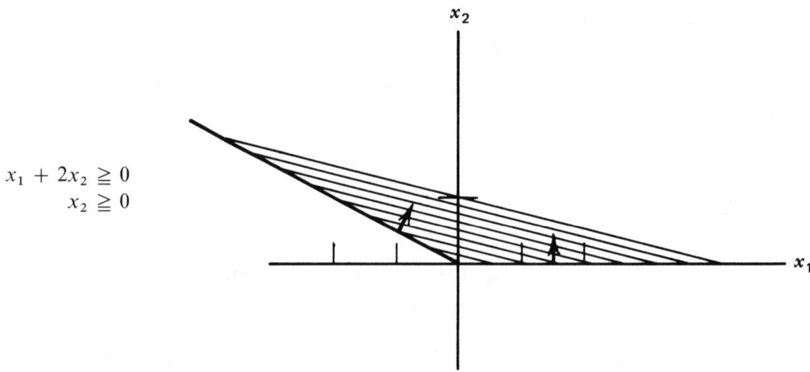

Figure 2.2 Transformed System

The simplex method uses transformations of type 1 and 2 on the system of constraints to find an optimum solution. Hence it is essential that the system of constraints of an LP be transformed into a system of equations by introducing appropriate slack variables before applying the simplex method on it.

Nonnegative Variables In the standard form we require all the variables to be nonnegative. If the original problem contains some unrestricted variables, they can be handled by two methods. In one method each unrestricted variable is transformed into the difference of two nonnegative variables as in section 1.1.7. For example, let the variable x_1 in the problem be unrestricted in sign. Any real variable can be expressed as a difference of two nonnegative variables. So we can substitute:

$$x_1 = x_1^+ - x_1^-$$

where

$$x_1^+ \geqq 0 \qquad \text{and} \qquad x_1^- \geqq 0 \tag{2.3}$$

If $x_1^+ - x_1^-$ is substituted for x_1 wherever it occurs in the original LP and the new variables x_1^+ and x_1^- are restricted to be nonnegative, the resulting LP is *equivalent* to the original one. From an optimum feasible solution of the transformed problem an optimum feasible solution of the original problem is obtained by using equation (2.3). Each unrestricted variable can be handled in a similar manner.

This method handles unrestricted variables conveniently, but it is not used much in practice, because it increases the number of variables in the problem and, hence, makes it more difficult to solve.

The other method for handling unrestricted variables takes advantage of their unrestricted nature and eliminates them from the problem. Suppose x_1 is an unrestricted variable in an LP. Since x_1 is a variable appearing in the problem, the coefficient of x_1 in at least one of the constraints must be nonzero. Suppose it is the ith constraint, which is

$$a_{i1}x_1 + a_{i2}x_2 + \cdots + a_{in}x_n = b_i$$

where $a_{i1} \neq 0$. This constraint is equivalent to

$$x_1 = (b_i - a_{i2}x_2 - \cdots - a_{in}x_n)/a_{i1}$$

Since there are no further restrictions on the variable x_1, once the optimal values of the other variables, $x_2, \ldots, x_n$, are obtained, the associated optimum value of x_1 can be obtained by substituting the values of x_2 to x_n in the above expression. Eliminate x_1 from the problem by storing this expression for x_1 somewhere, by substituting this expression for x_1 in all the other constraints and the objective function, and by eliminating the ith constraint from further consideration.

All the unrestricted variables in the problem can be eliminated from the problem in this manner one by one. Each unrestricted variable eliminated in this way reduces the number of variables in the LP by one and reduces the number of constraints by one; this makes the size of the problem smaller and hence easier to solve.

Problem

2.1 Explain why variables that must be nonnegative cannot be eliminated from the problem in the above manner.

If there is a variable in the model that must be nonpositive, say $x_1 \leqq 0$, let $x_1 = -y_1$, and eliminate x_1 from the model by using this equation. y_1 is a nonnegative variable.

After these transformations, all the variables in the transformed model are nonnegative variables. All restrictions in the model are nonnegativity restrictions on the variables.

Minimization Form Eliminate constant terms in the objective function. This does not change the set of optimum solutions for the problem. Then express the objective function in the *minimization form.* In the original statement of the problem, if the objective is

$$\text{Maximize} \quad z'(x) = \sum_{j=1}^{n} c'_j x_j$$

Change the problem with the new objective as

$$\text{Minimize} \quad z(x) = -z'(x) = \sum_{j=1}^{n} -c'_j x_j$$

This change does not affect the set of optimal feasible solutions of the problem, and the equation

$$\text{Maximum value of } z'(x) = -[\text{minimum value of } z(x)]$$

can be used to get the maximum value of the original objective function.

Nonnegative Right-Hand Side Constants Now the LP is transformed into the following form:

Minimize	$z(x) = \sum_{j=1}^{n} c_j x_j$		
Subject to	$\sum_{j=1}^{n} a_{ij} x_j = b_i$	$i = 1$ to m	These are the constraints in the problem
and	$x_j \geqq 0$	$j = 1$ to n	These are the sign restrictions on the variables.

Multiply the ith equation on both sides by -1, if b_i is negative. As a result of this, all the right-hand constants in the equality constraints become nonnegative.

Standard Form The problem is now in standard form. It may be written in detached coefficient tabular form.

Tableau 2.1. Original Tableau

x_1	$x_2 \ldots x_j \ldots x_n$	$-z$	b
a_{11}	$a_{12} \ldots a_{1j} \ldots a_{1n}$	0	b_1
$\vdots$	$\vdots \quad \vdots \quad \vdots$	$\vdots$	$\vdots$
a_{i1}	$a_{i2} \ldots a_{ij} \ldots a_{in}$	0	b_i
$\vdots$	$\vdots \quad \vdots \quad \vdots$	$\vdots$	$\vdots$
a_{m1}	$a_{m2} \ldots a_{mj} \ldots a_{mn}$	0	b_m
c_1	$c_2 \ldots c_j \ldots c_n$	1	0

Each row in the tableau (except the last row) represents a constraint in the problem. The ith row represents the constraint:

$$a_{i1}x_1 + a_{i2}x_2 + \cdots + a_{in}x_n = b_i$$

The last row in the tableau represents the equation

$$c_1x_1 + \cdots + c_nx_n - z = 0$$

This row is called the *cost row*, or the *objective row*. The c_j are known as the *original cost coefficients*. The numbers $b_1, \ldots, b_m$ are known as the *right-hand side constants*. The numbers a_{ij} are known as the *input-output coefficients*. The variables in the original tableau, $x_1, \ldots, x_n$, are all the variables in the model, including slack variables (excepting possibly some unrestricted variables that may have been eliminated from the problem). A *feasible solution* is any vector $x = (x_1, \ldots, x_n)$ that satisfies all the constraints and the sign restrictions.

2.3 CANONICAL TABLEAU

2.3.1 Nonsingular Linear Transformations

Each row of the tableau represents a constraint in the problem. From linear algebra it is well known that nonsingular linear transformations, that is transformations such as the following, transform the system into an *equivalent system*.

1. Multiplying all the entries in a row of the tableau by a nonzero constant, which is equivalent to multiplying both sides of the corresponding constraint by that nonzero constant.
2. Adding a constant multiple of a row of the tableau to another row, which is equivalent to adding a constant multiple of a constraint to another one.

Since they operate on the rows of the tableau, transformations of this type are known as *elementary row operations.* Given the equivalent system obtained after making several elementary row operations, a reverse series of elementary row operations can be performed on the equivalent system to yield the original system again.

Neither the set of feasible solutions nor the set of optimum solutions of the problem is changed by making such transformations. Hence we take the liberty to perform such transformations. The main approach of the simplex algorithm is to move from one basic feasible solution to another by making such transformations until an optimum solution is obtained.

Note Row operations are always performed element by element using corresponding elements in the respective rows. The following tableau illustrates how a row operation is performed using two rows, rows 1 and 2.

Row 1	a_{11}	$\dots a_{1j}$	$\dots a_{1n}$	b_1
Row 2	a_{21}	$\dots a_{2j}$	$\dots a_{2n}$	b_2
α(Row 1) + Row 2	$\alpha a_{11} + a_{21}$	$\dots \alpha a_{1j} + a_{2j}$	$\dots \alpha a_{1n} + a_{2n}$	$\alpha b_1 + b_2$
β(Row 2)	βa_{21}	$\dots \beta a_{2j}$	$\dots \beta a_{2n}$	βb_2

The third row in the above tableau gives α(row 1) + row 2, where α is any real number. The fourth row in the above tableau is β(row 2), where β is any nonzero real number. A numerical example is given below.

Row 1	1	0	3	−5	3
Row 2	0	1	2	4	1
−5(Row 1) +Row 2	−5	1	−13	29	−14

2.3.2 Canonical Form

The original tableau, Tableau 2.1, is in canonical form if there exists a unit matrix (see section 3.2) of order m among the first m rows of the original tableau (after the problem is written in standard form in which the right-hand side constants are nonnegative).

When this occurs it is easy to read out a feasible solution of the problem from the tableau. This is a special type of feasible solution known as a *basic feasible solution* (to be abbreviated as BFS). The significance of BFSs will be explained in Chapter 3, where it is shown that they correspond to *extreme or corner points* of the set of feasible solutions. A significant result of them is that if an LP in standard form has an optimum feasible solution, it has an optimum feasible solution that is a BFS. The simplex method for solving LPs draws on this result heavily, and it searches only among BFSs for an optimum feasible solution.

In defining a BFS for the problem, the variables are partitioned into two sets called *nonbasic* (or *independent*) variables and the *basic* (or *dependent*) variables. The nonbasic variables are all equal to zero in the BFS. The basic variables are such that the set of their column vectors in the tableau forms a column basis for the matrix of input-output coefficients.

If the problem is in canonical form, rearrange the numbers of the variables, if necessary, so that the column vectors corresponding to the variables $x_1, \ldots, x_m$ in the tableau form a unit matrix of order m among the first m rows. (This step of rearranging the numbers of the variables is really unnecessary. It is done purely for the convenience of referring to the basic variables as $x_1, \ldots, x_m$). Select the variables associated with the column vectors of the unit matrix as the *basic variables*. All the other variables in the problem become *nonbasic variables*. The column vectors in the tableau associated with the basic variables are called the present *basic column vectors*. All the other column vectors in the tableau are called the *nonbasic column vectors*.

The basic variable that corresponds to the first column vector of the unit matrix is called the *first basic variable* or the *basic variable in the first row*. (For convenience, we assume that this variable is called x_1.) In general the basic variable that corresponds to the rth column vector of the unit matrix is called the rth *basic variable* or the *basic variable in the* rth *row*. The *basic vector* is the vector of basic variables in their proper order. Record the basic variables in their order on the left-hand side of the tableau. When this is done, this tableau is of the form seen in Tableau 2.2.

Tableau 2.2.

Basic variables	$x_{m+1} \;\ldots\; x_n$	x_1	$x_2 \ldots x_m$	$-z$	b
x_1	$a_{1,m+1} \ldots a_{1n}$	1	$0 \ldots 0 \;\ldots$	0	b_1
x_2	$a_{2,m+1} \ldots a_{2n}$	0	$1 \ldots 0$	0	b_2
$\vdots$	$\vdots \qquad \vdots$	$\vdots$	$\vdots \quad \vdots$	$\vdots$	$\vdots$
x_m	$a_{m,m+1} \ldots a_{mn}$	0	$0 \ldots 1$	0	b_m
	$c_{m+1} \;\ldots\; c_n$	c_1	$c_2 \ldots c_m$	1	0

In putting the problem in standard form, the right-hand side constants have all been made nonnegative. The solution given by setting the ith basic variable equal to b_i for $i = 1$ to m, and all nonbasic variables equal to zero is a feasible solution of the problem. It satisfies all the constraints and is nonnegative as required.

2.3.3 Pricing Out Routine

Subtract suitable multiples of the first m rows from the cost row, so that the entries in it under all the basic column vectors become zero. This is called the *pricing out operation* or the *cost updating operation*. This gives Tableau 2.3, known as the *canonical tableau with respect to the present basic vector.* In the canonical tableau with respect to a given basic vector, the basic column vectors in their proper order should form the unit matrix, and they should all be priced out.

Tableau 2.3. Initial Canonical Tableau

Basic variables	$x_{m+1} \cdots x_j \cdots x_n$	$x_1 \ldots x_i \ldots x_m$	$-z$	b
x_1	$a_{1,m+1} \cdots a_{1j} \cdots a_{1n}$	$1 \ldots 0 \ldots 0$	0	b_1
$\vdots$	$\vdots \quad \vdots \quad \vdots$	$\vdots \quad \vdots \quad \vdots$	$\vdots$	$\vdots$
x_i	$a_{i,m+1} \cdots a_{ij} \cdots a_{in}$	$0 \ldots 1 \ldots 0$	0	b_i
$\vdots$	$\vdots \quad \vdots \quad \vdots$	$\vdots \quad \vdots \quad \vdots$	$\vdots$	$\vdots$
x_m	$a_{m,m+1} \cdots a_{mj} \cdots a_{mn}$	$0 \ldots 0 \ldots 1$	0	b_m
$-z$	$\overline{c}_{m+1} \cdots \overline{c}_j \cdots \overline{c}_n$	$0 \ldots 0 \ldots 0$	1	$-z_0$

Here is the initial BFS:

$$\begin{aligned} \text{All nonbasic variables} &= 0 \\ i\text{th basic variable} &= b_i \qquad i = 1 \text{ to } m \\ \text{Objective value} &= z_0 \end{aligned}$$

The simplex algorithm always requires a BFS like this to start with. The $\overline{c}_j$ are called the *relative cost coefficients with respect to the basic vector* $(x_1, \ldots, x_m)$, "relative" because their values depend on the choice of the basic vector. They are obtained by 'pricing out' all the basic column vectors. It will be explained later that these $\overline{c}_j$'s are the marginal rates of change. A unit change in x_j from its present value while retaining feasibility results in a change of $\overline{c}_j$ units in the objective value.

Words such as *basis* and *basic feasible solution*, have been used without being defined adequately. Rigorous definitions of these terms are given in Chapter 3.

In some problems the original input-output coefficient matrix may not contain a full unit matrix of order m as a submatrix. In such problems there is no way of directly obtaining a BFS from the original tableau. From the original tableau we may not be able to even decide whether a feasible solution exists. For such problems there is a method of introducing *artificial variables* and transforming the problem of finding a feasible solution into another LP. This augmented LP is formulated in such a way that it has an initial BFS (which is artificial) that can be obtained directly. Starting with this BFS, the augmented problem can be solved by applying the simplex algorithm. The augmented LP is known as the *Phase I problem.* An optimum solution of the Phase I problem either provides a BFS of the original problem or a proof that the original problem has no feasible solution at all. If a BFS of the original problem is obtained at the termination of the Phase I problem, the original problem is then solved by applying the simplex algorithm starting with it. This latter part of the computational effort is known as the *Phase II problem.*

To solve an LP, *the simplex algorithm* always requires a starting BFS. The *simplex method* for solving a general LP applies the simplex algorithm in two phases as discussed above. If the original tableau of the problem in standard form contains a full unit matrix as a submatrix, a BFS can be obtained directly without working on Phase I. Therefore, in this case we proceed directly to Phase II of the simplex method.

2.4 PHASE I WITH A FULL ARTIFICIAL BASIS

Before discussing the simplex algorithm we will discuss how to set up the phase I problem for obtaining an initial BFS of an LP if the original tableau corresponding to it (in standard form with nonnegative right-hand side constants) is not in canonical form.

Let the original tableau for the problem be Tableau 2.1 and suppose the first m rows of this tableau (the constraint rows) do not contain the unit matrix of order m as a submatrix. Augment the tableau with the *artificial variables* $x_{n+1}, \ldots, x_{n+m}$, whose column vectors form the unit matrix of order m. Let the cost coefficient of each of these artificial variables in the original objective row be equal to zero. Now the augmented problem has an initial BFS that is obtained directly:

$$\begin{aligned} x_i &= 0 \qquad i = 1, \ldots, n \\ x_{n+i} &= b_i \qquad i = 1, \ldots, m \end{aligned} \tag{2.4}$$

Therefore, the augmented problem is in canonical form with the artificial vector $(x_{n+1}, \ldots, x_{n+m})$ as the basic vector.

Any feasible solution of the augmented problem, $(\hat{x}_1, \ldots, \hat{x}_{n+1}, \ldots, \hat{x}_{n+m})$, in which all the artificial variables $\hat{x}_{n+1}, \ldots, \hat{x}_{n+m}$ are zero yields the feasible solution $(\hat{x}_1, \ldots, \hat{x}_n)$ of the original problem. This is the main idea exploited in Phase I of the simplex method. Starting with the initial BFS (2.4), try to get a feasible solution of the augmented problem in which all the artificial variables are zero. This may be achieved by restricting all the artificial variables to be nonnegative and minimizing their sum, that is,

$$\text{Minimize} \quad w = x_{n+1} + \cdots + x_{n+m}$$

w is known as the *Phase I objective function* or the *Phase I cost function* or the *infeasibility form.* Here is the augmented tableau.

Tableau 2.4. Augmented Tableau with a Full Artificial Basis

Basic variables	$x_1 \ldots x_n$	$x_{n+1} \ldots x_{n+m}$	$-z$	$-w$	b
x_{n+1}	$a_{11} \ldots a_{1n}$	$1 \quad \ldots 0$	0	0	b_1
$\vdots$	$\vdots \quad \vdots$	$\vdots \quad \vdots$	$\vdots$	$\vdots$	$\vdots$
x_{n+m}	$a_{m1} \ldots a_{mn}$	$0 \quad \ldots 1$	0	0	b_m
$-z$	$c_1 \ldots c_n$	$0 \quad \ldots 0$	1	0	0
$-w$	$0 \ldots 0$	$1 \quad \ldots 1$	0	1	0

The last row here is the *Phase I objective row.* As long as the simplex method is in Phase I, the objective is to minimize w. In this context the original objective row is known as the *Phase II objective row.* This problem of minimizing w, starting with a full artificial basis, is called the *Phase I problem.*

Price out the basic column vectors of the augmented tableau by subtracting suitable multiples of the first m rows from the Phase I objective row and the

Phase II objective row so that all the entries under the basic columns in these rows become zero. This gives the initial canonical tableau for the Phase I problem.

Tableau 2.5. Initial Canonical Tableau for the Phase I Problem

Basic variables	$x_1 \ldots x_n$	$x_{n+1} \ldots x_{n+m}$	$-z$	$-w$	b
x_{n+1}	$a_{11} \ldots a_{1n}$	$1 \;\ldots\; 0$	0	0	b_1
$\vdots$	$\vdots \quad \vdots$	$\vdots \quad \vdots$	$\vdots$	$\vdots$	$\vdots$
x_{n+m}	$a_{m1} \ldots a_{mn}$	$0 \;\ldots\; 1$	0	0	b_m
$-z$	$\bar{c}_1 \ldots \bar{c}_n$	$0 \;\ldots\; 0$	1	0	$-z_0$
$-w$	$\bar{d}_1 \ldots \bar{d}_n$	$0 \;\ldots\; 0$	0	1	$-w_0$

The Phase I problem is an LP in canonical form, with an initial BFS. Therefore, the simplex algorithm can be applied to solve it.

INTERPRETATION

The original constraints are

$$\sum_{j=1}^{n} a_{ij}x_j = b_i \qquad i = 1 \text{ to } m$$

$$x_j \geqq 0 \qquad \text{for all } j \tag{2.5}$$

The constraints of the corresponding Phase I problem are

$$\sum_{j=1}^{n} a_{ij}x_j + x_{n+i} = b_i \qquad i = 1 \text{ to } m \tag{2.6}$$

$$x_j \geqq 0 \qquad \text{for all } j \text{ and } x_{n+i} \geqq 0 \qquad \text{for } i = 1 \text{ to } m$$

The following conclusions can be drawn:

1. If $(\hat{x}_1, \ldots, \hat{x}_n)$ is a feasible solution of the original problem [i.e., it satisfies (2.5)], then $(\hat{x}_1, \ldots, \hat{x}_n, \hat{x}_{n+1}, \ldots, \hat{x}_{n+m})$ where $\hat{x}_{n+1}, \ldots, \hat{x}_{n+m}$ are all equal to zero is a feasible solution of the Phase I problem [i.e., it satisfies (2.6)].
2. Conversely, if $(\bar{x}_1, \ldots, \bar{x}_{n+1}, \ldots, \bar{x}_{n+m})$ is a feasible solution of the Phase I problem in which $\bar{x}_{n+1}, \ldots, \bar{x}_{n+m}$ all happen to be zero, $(\bar{x}_1, \ldots, \bar{x}_n)$ is a feasible solution to the original problem.
3. In (2.6), all the artificial variables $x_{n+1}, \ldots, x_{n+m}$ are restricted to be nonnegative. The sum of nonnegative numbers is always nonnegative. Therefore the Phase I objective value is greater than or equal to zero on the set of feasible solutions of the Phase I problem.
4. If the original problem has a feasible solution, by 1, a feasible solution to the Phase I problem can be constructed from it by setting all the artificial variables equal to zero. This feasible solution makes the value of w equal to zero. Since w is nonnegative on the set of feasible solutions of the Phase I problem, any feasible solution that makes w equal to zero must be an optimum solution. Therefore, if the original problem has a feasible solution, the minimum value of w in the Phase I problem is zero.

5. Conversely, if the minimum value of w in the Phase I problem is zero, there must exist a feasible solution $(\bar{x}_1, \ldots, \bar{x}_n; \bar{x}_{n+1}, \ldots, \bar{x}_{n+m})$ that makes w equal to zero. Therefore $\bar{x}_{n+1} + \cdots + \bar{x}_{n+m} = 0$. Since each of $\bar{x}_{n+1}, \ldots, \bar{x}_{n+m}$ is nonnegative, their sum can be zero only if each of them is zero. Hence $\bar{x}_{n+1}, \ldots, \bar{x}_{n+m}$ are all equal to zero. So $(\bar{x}_1, \ldots, \bar{x}_m)$ is a feasible solution of the original problem. Therefore, if the minimum value of w in the Phase I problem is zero, the original problem has a feasible solution.
6. From these arguments we conclude that the original problem has a feasible solution iff the minimum value of w in the Phase I problem is zero.
7. When the Phase I problem is solved, two things can happen.
   (a) The minimum value of w is greater than zero. By 6 this implies that the original problem cannot have a feasible solution. Therefore the original constraints and the sign restrictions together must be inconsistent. An example of such a system is

$$x_1 + x_2 = -1$$
$$x_1 \geqq 0 \qquad x_2 \geqq 0$$

   (b) The minimum value of w is zero. From an optimum solution the Phase I problem, obtain a feasible solution to the original problem as in 2.

Note If the original tableau contains some column vectors of the unit matrix of order m, the variables corresponding to these column vectors can be picked as basic variables in the initial basic vector. It is necessary to augment the tableau only with artificial variables whose column vectors are the column vectors of the unit matrix that do not appear in the original tableau. If this is done, the initial basic vector to the augmented problem consists of some original problem variables and some artificial variables. *The Phase I objective function is always the sum of the artificial variables that were introduced into the problem.* Hence the original Phase I cost coefficient of any original problem variable is zero, and of any artificial variable is one. The original Phase II cost coefficient of any artificial variable is zero. The initial canonical tableau for the Phase I problem is obtained by pricing out all the basic variables in both the objective rows.

2.5 THE SIMPLEX ALGORITHM

2.5.1 The Initial Stage

This is an algorithm that can be used to solve any LP in standard form for which a starting feasible basic vector is known. Suppose the problem is of the form:

$$\text{Minimize} \quad z(x) = \sum_{j=1}^{\mathrm{N}} c_j x_j$$

$$\text{Subject to} \quad \sum_{j=1}^{\mathrm{N}} a_{ij} x_j = b_i \qquad i = 1, \ldots, \mathrm{M}$$

$$x_j \geqq 0 \qquad j = 1, \ldots, \mathrm{N}$$

Suppose the initial basic vector is $(x_1, \ldots, x_{\mathrm{M}})$. Let the canonical tableau with respect to this basic vector be:

Tableau 2.6. Canonical Tableau

Basic variables	$x_{M+1} \dots x_N$	$x_1 \dots x_M$	$-z$	values
x_1	$\bar{a}_{1,M+1} \dots \bar{a}_{1,N}$	$1 \dots 0$	0	$\bar{b}_1$
$\vdots$	$\vdots \quad \vdots$	$\vdots \quad \vdots$	$\vdots$	$\vdots$
x_M	$\bar{a}_{M,M+1} \dots \bar{a}_{M,N}$	$0 \dots 1$	0	$\bar{b}_M$
$-z$	$\bar{c}_{M+1} \dots \bar{c}_N$	$0 \dots 0$	1	$-\bar{z}_0$

In every canonical tableau, the column vectors of the basic variables in their proper order form a unit matrix. Also, the cost row has zero entries under all the basic column vectors. The entries in the canonical tableau are called *updated entries* or *entries in the canonical tableau with respect to the present basic vector*, in order to distinguish them from the entries of the original tableau. The BFS corresponding to this basic vector is:

Nonbasic variables	$x_i = 0$	$i = \mathrm{M} + 1, \dots, \mathrm{N}$
Basic variables	$x_i = \bar{b}_i$	$i = 1, \dots, \mathrm{M}$
Objective value	$z = \bar{z}_0$	

The updated right-hand side constants in the canonical tableau will always represent the values of the current basic variables in the corresponding BFS.

2.5.2 The Fundamental Optimality Criterion

An *optimum feasible solution* for this problem is a feasible solution that gives the objective function its minimum value among all feasible solutions. Hence, a feasible solution $\tilde{x} = (\tilde{x}_1, \dots, \tilde{x}_N)$ is an optimum feasible solution if

$$z(\tilde{x}) \leqq z(x) \qquad \text{for all other feasible solutions } x.$$

This is the fundamental criterion for checking whether a particular feasible solution is optimal or not.

2.5.3 Termination Criteria

Now we check whether the current BFS is optimal to the problem. If it is, we terminate the algorithm. The criteria for determining it are known as *termination criteria.* We describe the termination criteria for any canonical tableau obtained during the course of the algorithm.

OPTIMALITY

The present BFS is optimal if all the relative cost coefficients with respect to this basic vector are nonnegative, that is,

$$\bar{c}_j \geqq 0 \qquad \text{for all } j$$

Of course, $\bar{c}_j$ is zero if x_j is a basic variable. The criterion requires that the relative cost coefficients of all nonbasic variables be nonnegative.

Note The optimality criterion requires that the relative cost coefficients $\bar{c}_j$ with respect to the present basic vector (and not the original cost coefficients c_j) should be nonnegative.

Discussion The present BFS of the problem makes the objective function equal to $\bar{z}_0$. Hence, if we can prove that $z(x) \geqq \bar{z}_0$ for every other feasible solution x, the present BFS must be optimal by the fundamental optimality criterion of section 2.5.2. From the last row of the present canonical tableau we see that

$$\sum_{j=M+1}^{N} \bar{c}_j x_j - z(x) = -\bar{z}_0$$

that is,

$$z(x) = \bar{z}_0 + \sum_{j=M+1}^{N} \bar{c}_j x_j$$

x_j must be nonnegative for all j in every feasible solution. Hence, if $\bar{c}_j$ is nonnegative for all j, $\sum_{j=M+1}^{N} \bar{c}_j x_j$ must be nonnegative. Hence, $z(x)$ must be greater than or equal to $\bar{z}_0$ at every feasible solution x. Thus, the present BFS must be optimal to the problem.

UNBOUNDEDNESS

The objective function is unbounded below (i.e., there exists a class of feasible solutions along which it diverges to $-\infty$), if there exists an s such that, in a canonical tableau with respect to some feasible basic vector,

$$\bar{c}_s < 0 \qquad \text{and } \bar{a}_{is} \leqq 0 \qquad \text{for all } i = 1, \ldots, M$$

where $\bar{c}_s$ is the relative cost coefficient of x_s with respect to this basic vector, and $\bar{a}_{is}$, $i = 1$ to M are the entries in the column vector corresponding to x_s in the canonical tableau with respect to this basic vector.

Discussion Consider the canonical tableau with respect to some feasible basic vector. For convenience in referring to it assume that the basic vector is $(x_1, \ldots, x_M)$. Suppose the column vector of x_s satisfies the unboundedness criterion in the present canonical tableau. Rearranging the column vectors, if necessary, let the canonical tableau be Tableau 2.7.

Tableau 2.7. Canonical Tableau

Basic variables	Other nonbasic variables	x_s	$x_1 \ldots x_M$	$-z$	b
x_1		$\bar{a}_{1s}$	$1 \ldots 0$	0	$\bar{b}_1$
$\vdots$		$\vdots$	$\vdots \quad \vdots$	$\vdots$	$\vdots$
x_M		$\bar{a}_{Ms}$	$0 \ldots 1$	0	$\bar{b}_M$
$-z$		$\bar{c}_s$	$0 \ldots 0$	1	$-\bar{z}$

The present BFS is:

$$i\text{th basic variable} \quad x_i = \bar{b}_i \qquad i = 1 \text{ to M}$$
$$\text{All nonbasic variables} = 0$$
$$\text{Objective function} \quad z = \bar{z}$$

Since this is a feasible solution, all the $\bar{b}_i$ are nonnegative. Denote this BFS by $\bar{x}$. In $\bar{x}$, all the current nonbasic variables, including x_s, are zero. Try to increase the value of x_s from zero to some nonnegative value, say λ, keeping all the remaining nonbasic variables fixed at zero. From the results in section 2.3.1, any vector that satisfies the system of constraints represented by the present canonical tableau also satisfies the original system of constraints and vice versa. Hence, if the equality constraints in the problem are to be satisfied, the values of the basic variables have to be changed such that the following equations hold.

$$\begin{array}{llllll} \bar{a}_{1s}x_s & + x_1 & & & = \bar{b}_1 \\ \bar{a}_{2s}x_s & & + x_2 & & = \bar{b}_2 \\ \vdots & & \ddots & & \vdots \\ \bar{a}_{Ms}x_s & & & + x_M & = \bar{b}_M \\ \bar{c}_s x_s & & & -z & = -\bar{z} \end{array} \tag{2.7}$$

This implies that the new solution is

$$\begin{array}{rl} \text{1st basic variable, } x_1 & = \bar{b}_1 - \bar{a}_{1s}\lambda \\ \text{2nd basic variable, } x_2 & = \bar{b}_2 - \bar{a}_{2s}\lambda \\ \vdots & \quad\vdots \\ i\text{th basic variable, } x_i & = \bar{b}_i - \bar{a}_{is}\lambda \\ \vdots & \quad\vdots \\ \text{Mth basic variable, } x_M & = \bar{b}_M - \bar{a}_{Ms}\lambda \\ x_s & = \lambda \\ \text{All other nonbasic variables} & = 0 \\ z & = \bar{z} + \bar{c}_s\lambda \end{array} \tag{2.8}$$

Since the column vector of x_s satisfied the unboundedness criterion, $\bar{a}_{is} \leqq 0$ for all i, and $\bar{c}_s < 0$. In the solution given in (2.8) the values of all the variables x_j remain nonnegative for every $\lambda \geqq 0$. Hence, it is a feasible solution for every $\lambda \geqq 0$. However, as λ is made larger and larger, the objective value corresponding to this solution, which is equal to $\bar{z} + \bar{c}_s\lambda$, decreases indefinitely, since $\bar{c}_s < 0$. By chosing a λ arbitrarily large, we get from (2.8) a feasible solution to the problem at which the value of $z(x)$ is less than any arbitrary real number. Hence in this case the objective function, $z(x)$, is unbounded below on the set of feasible solutions.

Note An LP in which the objective function $z(x)$ has to be minimized is said to be *unbounded below*, if $z(x)$ diverges to $-\infty$ on the set of feasible solutions of the problem. In a similar manner an LP in which an objective function $z'(x)$ has to be maximized, is said to be *unbounded above* if $z'(x)$ diverges to $+\infty$ on the set of feasible solutions.

2.5.4 Improving a Nonoptimal Basic Feasible Solution

If neither the optimality nor the unboundedness criteria are satisfied, the simplex algorithm moves to a better feasible solution [a feasible solution $\hat{x}$ is said to be better than the present feasible solution $\bar{x}$ if $z(\hat{x}) \leqq z(\bar{x})$] from the present one. The simplex algorithm does this by changing the value of one judiciously selected nonbasic variable from its present value of 0 to some nonnegative value. The important thing to remember is that *in each step, the simplex algorithm tries to change the value of only one nonbasic variable.* The nonbasic variable for this change is selected so that the objective value decreases as a result of this change. It can be shown that this will happen only if the nonbasic variable selected for the change has a negative relative cost coefficient in the present canonical tableau.

If x_s is the selected variable, it must satisfy

$$\bar{c}_s < 0$$

After some simple calculations, the nonnegative value that this nonbasic variable can take, denoted by θ, is determined. When x_s, is made equal to θ, the values of the present basic variables have to be recomputed so that the resulting solution satisfies all the constraints. The modified value of at least one of the basic variables turns out to be zero. One such basic variable is selected, dropped from the basic vector, and x_s is made a basic variable in its place. This gives a new basic vector, and by making the necessary transformations of the type discussed in section 2.3.1 the column vector of x_s is transformed into the appropriate column vector of the unit matrix, and this leads to the canonical tableau with respect to the new basic vector. This whole operation is called *bringing the nonbasic variable* x_s into the *basic vector.* x_s is known as the *entering variable* in this operation.

Any nonbasic variable with a strictly negative relative cost coefficient is a candidate for being the entering variable. The column vector of the entering variable in the present canonical tableau is called the *pivot column.* When there are several nonbasic variables x_j with $\bar{c}_j < 0$, the entering variable should be selected from among them, in such a way that the solution of the problem can be completed with the least computational effort. However, at present, no such selection method is known. One good rule, which is commonly used and which seems to work well in practice, is to select the nonbasic variable x_s as the entering variable where

$$\bar{c}_s = \text{minimum } \{\bar{c}_j, j = 1 \text{ to N}\} \tag{2.9}$$

If there is a tie in (2.9), break the tie arbitrarily. This rule has the rationale of choosing the variable with the largest per unit capability of reducing the value of z. (In this sense, it may be viewed as a *steepest descent rule.*) A rule like this to decide which nonbasic variable should be selected as the entering variable is known as a *pivot column choice rule.* There is no theoretical justification to believe that the pivot column choice rule of (2.9) solves every LP with the least computational effort. In fact, examples can be constructed to show that it may require the algorithm to go through more *iterations* before termination, than some other pivot column choice rule. However, on most practical problems it has given excellent results, and it is widely used.

When there are many variables in the problem, it is normally solved on a digital computer. In some commercially available programs for the simplex algorithm, all the relative cost coefficients are not even computed after each basic vector change. The realtive cost coefficients are computed one by one, and as soon as a negative relative cost coefficient turns up, that variable is selected as the entering variable and the remaining relative cost coefficients are not even computed.

Assume that the present basic vector is $(x_1, \ldots, x_M)$ and let the canonical tableau with respect to this basic vector be Tableau 2.6. All the nonbasic variables excepting the entering variable, x_s, remain equal to zero in the next solution. After we substitute the value zero for those variables, the remaining system of constraints is (2.7). Hence if x_s is given the value λ, the new solution [as in (2.8)] is

ith basic variable	$x_i = \bar{b}_i - \bar{a}_{is}\lambda$	$i = 1$ to M
	$x_s = \lambda$	
All other nonbasic variables	$= 0$	
Objective function	$z = \bar{z}_0 + \bar{c}_s\lambda$	

From these equations it is clear that if x_s is increased from its present value of 0 to the nonnegative value λ, the objective value decreases from its present value of $\bar{z}_0$, because $\bar{c}_s < 0$. *The change in the objective value is* $\bar{c}_s$ *units per unit change in the value of* x_s.

This is the reason for requiring that the relative cost coefficient of the entering variable should be negative. If the relative cost coefficient of the entering variable is zero, the objective value remains unchanged and if the relative cost coefficient of the entering variable is positive, the objective value increases.

Since $\bar{c}_s < 0$, the maximum decrease in objective value in this step can be achieved by giving λ the maximum possible value. However, the value of λ should be such that the new values of the present basic variables are all nonnegative. The new value of the ith basic variable is $\bar{b}_i - \bar{a}_{is}\lambda$, and it should be nonnegative, for all $i = 1$ to M. But we know that $\bar{b}_i \geqq 0$, because it is the value of the ith basic variable in the present BFS. Also, $\lambda \geqq 0$. Hence, if $\bar{a}_{is} \leqq 0$, the above restriction is satisfied automatically for any $\lambda \geqq 0$. On the other hand, if $\bar{a}_{is} > 0$, then $\bar{b}_i - \bar{a}_{is}\lambda \geqq 0$ implies $\lambda \leqq \bar{b}_i/\bar{a}_{is}$. Thus the maximum value that we can give to λ is

$$\begin{aligned} \theta &= \text{minimum}\left\{\frac{\bar{b}_i}{\bar{a}_{is}}: i \text{ such that } \bar{a}_{is} > 0\right\} \qquad \text{if at least one } \bar{a}_{is} > 0, \\ &= \infty, \text{ if all } \bar{a}_{is} \leqq 0 \qquad \text{for all } i = 1 \text{ to M} \end{aligned}$$

The second alternative, $\theta = \infty$, happens only when $\bar{a}_{is} \leqq 0$ for all i (see the unboundedness criterion, section 2.5.3). If the unboundedness criterion is not satisfied, then the maximum value that x_s can have is

$$\theta = \frac{\bar{b}_r}{\bar{a}_{rs}} = \text{minimum}\left\{\frac{\bar{b}_i}{\bar{a}_{is}}: i \text{ such that } \bar{a}_{is} > 0\right\}$$

This is known as the *minimum ratio* in this step. The operation of computing θ is known as the *minimum ratio test.* When x_s is given the value θ, x_r, the present rth basic variable becomes equal to zero. Replace the present rth basic variable by x_s. The rth row in the present canonical tableau is called the *pivot row.* The element in the pivot row and the pivot column, that is, $\bar{a}_{rs}$, is known as the *pivot element.* The present rth basic variable is the *leaving variable*, or the *dropping variable.*

All the basic vectors for this LP contain the same number of variables. Giving λ the maximum value that it can take (subject to nonnegativity restrictions on the variables) is also necessary to drive one of the present basic variables to zero and, thus, retain the same number of variables in the new basic vector.

2.5.5 Minimum Ratio Test

This test determines the leaving variable. Let $(\bar{b}_1, \ldots, \bar{b}_M)$ be the vector of values of the basic variables in the present solution. Let x_s be the entering variable and let $(\bar{a}_{1s}, \ldots, \bar{a}_{Ms})^T$ be the pivot column. Then the minimum ratio in this pivot step is

$$\theta = \frac{\bar{b}_r}{\bar{a}_{rs}} = \text{minimum}\left\{\frac{\bar{b}_i}{\bar{a}_{is}} : i \text{ such that } \bar{a}_{is} > 0\right\}$$

If there is a tie for r in this minimum, it is selected arbitrarily among the tied values of i. The leaving variable is the present rth basic variable in this pivot step.

The purpose of the minimum ratio test is to guarantee that the new basic solution obtained when x_s is brought into the basic vector satisfies the nonnegativity restrictions on the variables.

2.5.6 Pivot Operation

Bringing the nonbasic variable x_s into the basic vector as the rth basic variable produces a new canonical tableau with respect to the new basic vector by transforming the present column vector of x_s into the rth column vector of the unit matrix, using transformations of the type discussed in section 2.3.1. In linear algebra such a transformation is called a *pivot operation* with the column vector of x_s in the present canonical tableau as the *pivot column* and the present rth row as the *pivot row* and the entry contained in both of them, that is, $\bar{a}_{rs}$, as the *pivot element*. It transforms the pivot column into the column vector of the unit matrix with a "1" entry in the pivot row and zero entries in all the other rows. The following elementary row operations have to be performed to get the new canonical tableau.

1. For each $i \neq r$, subtract a suitable multiple of the pivot row from the ith row so that the entry in the pivot column in the ith row becomes zero.
2. Similarly, subtract a suitable multiple of the pivot row from the cost row so that the entry in the pivot column in the cost row becomes zero. This is the pricing out of the new basic column vector, or the *updating* of the objective row.
3. Divide the pivot row by the pivot element.

This completes the pivot operation, and we get a new canonical tableau in which x_s is the rth basic variable.

2.5.7 Change in Cost Due to Pivoting

The objective value corresponding to the new BFS is:

$$\text{The new objective value} = \text{the previous objective value} + \theta\bar{c}_s \qquad (2.10)$$

This follows from the objective row equation from the new tableau. Since $\theta \geqq 0$ and $\bar{c}_s < 0$, the new basic solution will be better than the present one.

2.5.8 The Iterative Process

Find out whether the optimality or the unboundedness termination criteria are satisfied in the new canonical tableau. If not, repeat the process until one of the termination criteria is satisfied. This will happen after a finite number of pivot operations. If the algorithm terminates by satisfying the optimality criterion, the canonical tableau corresponding to the final basic vector, which satisfies the optimality criterion, is known as an *optimal canonical tableau* for the problem. The BFS provided by the optimal canonical tableau is an optimum solution of the problem.

2.6 OUTLINE OF THE SIMPLEX METHOD FOR A GENERAL LINEAR PROGRAM

2.6.1 Case I

If the original tableau is in canonical form, an initial feasible basic vector associated with the unit matrix is chosen and the simplex algorithm applied directly from there. Termination may occur in two different ways.

1. If the final tableau satisfies the optimality criterion, the final BFS is an optimal feasible solution.
2. If the final tableau satisfies the unboundedness criterion, the objective function $z(x)$ is unbounded below, and there is no finite optimal solution (see section 2.5.3).

2.6.2 Case II

If the original tableau is not in canonical form, the simplex method involves two phases. If the original problem in standard form is the one in Tableau 2.1, the original Phase I tableau with a full artificial basis is Tableau 2.4, where $x_{n+1}, \ldots, x_{n+m}$ are the artificial variables and w, the Phase I objective function, is their sum. The choice of $x_{n+1}, \ldots, x_{n+m}$ as the initial basic vector for the phase I problem, and pricing out the basic column vectors in both the Phase I and II cost rows leads to the initial canonical tableau.

Suppose in some general stage of the Phase I problem, we find the canonical tableau to be as arranged in Tableau 2.8.

Tableau 2.8

Basic variables	$x_1 \ldots x_n$	$x_{n+1} \ldots x_{n+m}$	$-z$	$-w$	b
	$\bar{a}_{11} \ldots \bar{a}_{1n}$	$\beta_{11} \ldots \beta_{1m}$	0	0	$\bar{b}_1$
	$\vdots \quad \vdots$	$\vdots \quad \vdots$	$\vdots$	$\vdots$	$\vdots$
	$\bar{a}_{m1} \ldots \bar{a}_{mn}$	$\beta_{m1} \ldots \beta_{mm}$	0	0	$\bar{b}_m$
	$\bar{c}_1 \ldots \bar{c}_n$	$-\pi_1 \ldots -\pi_m$	1	0	$-\bar{z}$
	$\bar{d}_1 \ldots \bar{d}_n$	$-\sigma_1 \ldots -\sigma_m$	0	1	$-\bar{w}$

The $\bar{d}_j$ are the *Phase I relative cost coefficients.* Phase I terminates in this stage if $\bar{d}_j \geqq 0$, for $j = 1$ to n.

INFEASIBILITY CRITERION

At Phase I termination, if the minimum value of $w = \bar{w}$ is positive, the original problem has no feasible solution. See section 2.4.

SWITCHING OVER TO PHASE II

If $\bar{w} = 0$ at Phase I termination, a BFS of the original problem is obtained from the final tableau of the Phase I problem by suppressing the artificial variables. Having obtained a feasible solution of the original problem, the algorithm should be continued in such a way that all subsequent solutions obtained under the algorithm are also feasible to the original problem. This requires that the value of w must be kept equal to zero in all the subsequent steps. Thus any nonbasic artificial variable at this stage will never again be considered as a candidate to enter the basic vector. All such variables and their column vectors are deleted from the tableau. In normal computation, artificial variables are deleted from the tableau the moment they leave the basic vector during Phase I.

From the $m + 2$th row of the terminal Phase I tableau we get:

$$w = \bar{w} + \sum_{j=1}^{n} \bar{d}_j x_j + \sum_{j=1}^{m} (-\sigma_j) x_{n+j} \tag{2.11}$$

where $\bar{w}$ is 0 according to the assumption made here. Since we want to keep w equal to 0 in all subsequent iterations, any artificial variable x_{n+i} that is still in the tableau must remain equal to zero in all subsequent steps. Basic artificial variables can be kept in the basic vector until they are replaced by some original problem variables during Phase II iterations, but they must be equal to zero in all the subsequent BFSs obtained under the algorithm.

From (2.11) it is also clear that if any original problem variable, x_j, is such that $\bar{d}_j > 0$, that variable must be equal to zero in all subsequent steps, if w is to be kept equal to zero. Hence such variables will never be considered eligible to enter the basic vector during Phase II. Any variable x_j such that $\bar{d}_j > 0$ in the final Phase I tableau is permanently set equal to zero and its column vector is deleted from the tableau.

Thus, in all subsequent iterations, all the variables that are candidates for entering into the basic vector will be original problem variables, x_j, for which $\bar{d}_j = 0$. From (2.11) this automatically guarantees that the value of w remains equal to zero and hence that any artificial variable which is still in the basic vector will have a value of zero in all the subsequent BFSs obtained in the algorithm. The Phase I objective is not needed any more and it is deleted from the tableau.

Thus, switching over to Phase II requires the following steps:

1. All artificial variables and their column vectors are deleted from the tableau when they leave the basic vector during Phase I.
2. All original problem variables x_j for which the final Phase I relative cost coefficient, $\bar{d}_j$, is positive are set equal to zero and their column vectors are deleted from the tableau.
3. The Phase I objective row is deleted from the tableau.
4. Any artificial variable that is still in the basic vector can be left in it until it is replaced by some original problem variable during Phase II. However the values of all the artificial variables will be zero in all solutions obtained during Phase II.
5. During Phase II, the Phase II objective row is used for checking optimality, unboundedness, etc.

REDUNDANT EQUATIONS

The original system of constraints is

$$\sum_{j=1}^{n} a_{ij} x_j = b_j \qquad i = 1 \text{ to } m$$

If one of these equations can be obtained as a linear combination of the others, that equation is redundant and it can be deleted from the system. For example, consider this system:

$$\begin{aligned} x_1 + x_2 + x_3 + x_4 + x_5 &= 5 \\ x_1 + x_2 + 2x_3 + 2x_4 + 2x_5 &= 8 \\ x_1 + x_2 \qquad\qquad\qquad\qquad &= 2 \\ x_3 + x_4 + x_5 &= 3 \end{aligned}$$

In this system the first equation is the sum of equations 3 and 4. The second equation is the sum of equation 3 and two times equation 4. Hence the first two equations are redundant, and they may be deleted from the system.

If the system of equations is consistent, the number of nonredundant equations in the system is the rank of the coefficient matrix $A = (a_{ij})$. If A has rank m, there are no redundant equations in the system. The existence of redundant equations can be discovered during Phase I of the simplex method. Assume that all the nonbasic artificial column vectors have been deleted from the tableau. In the remaining final Phase I canonical tableau, if there is a row in which all the entries are zero excepting a single "1" entry in an artificial column vector, then that row corresponds to a redundant equation in the original system. All such rows and the corresponding artificial column vectors can be deleted from the tableau before entering Phase II.

Example
The following is the canonical tableau obtained at phase I termination of an LP with three constraints.

Basic variable	x_1	x_2	x_3	x_4	x_5	x_7	$-z$	$-w$	b
x_3	3	-1	1	1	0	0	0	0	6
x_7	0	0	0	0	0	1	0	0	0
x_5	2	1	0	2	1	0	0	0	5
$-z$	-2	-4	0	-3	0	0	1	0	-10
$-w$	0	0	0	0	0	0	0	1	0

x_1 to x_5 are original problem variables. x_7 is the only artificial variable left in the tableau at phase I termination. Notice that the second row in this tableau represents a redundant constraint. Hence delete it from the tableau. Thus initiate phase II with the following canonical tableau.

Basic variable	x_1	x_2	x_3	x_4	x_5	$-z$	b
x_3	3	-1	1	1	0	0	6
x_5	2	1	0	2	1	0	5
$-z$	-2	-4	0	-3	0	1	-10

ELIMINATING THE ARTIFICIAL VARIABLES AT THE END OF PHASE I

After eliminating the rows corresponding to redundant equations from the terminal Phase I canonical tableau, there may still be some artificial variables that are in the basic vector. Suppose the ith basic variable is an artificial variable. The updated right-hand side constant, in the ith row, $\bar{b}_i$, must be zero, since all the artificial variables are zero in the present BFS. Since this is not a redundant row there must be some j such that $\bar{a}_{ij} \neq 0$. Pick any j for which $\bar{a}_{ij} \neq 0$. The artificial variable can be replaced from the basic vector by x_j. To do this, a pivot operation is performed with the jth column vector as the pivot column, and the ith row as the pivot row. After the pivot operation the artificial column vector can be deleted from the tableau.

Each artificial variable in the basic vector can be deleted in this manner if desired. However, it is not necessary to remove artificial variables from the basic vector before going over to Phase II. They can be left in the basic vector until they are replaced from it during Phase II.

IMPROVING THE PHASE I OBJECTIVE FUNCTION

If the Phase I termination criterion is not satisfied by the present canonical tableau, the Phase I part of the simplex method is continued. An original problem variable, whose Phase I relative cost coefficient, $\bar{d}_j$, is negative, is selected as the entering variable. If several $\bar{d}_j$ are negative, a convenient pivot column choice rule is to select the variable x_s as the entering variable where s is such that

$$\bar{d}_s = \text{minimum } \{\bar{d}_j : j = 1 \text{ to } n\}$$

After the pivot, repeat the same process.

2.6.3 Alternate Optimum Solutions

Suppose an optimum solution is obtained when an LP is solved. If some of the nonbasic variables have zero relative cost coefficients in the final optimum canonical tableau, alternate optimum solutions for the problem can be obtained by bringing into the basic vector a nonbasic variable with zero relative cost coefficient. See section 3.10.

2.6.4 Numerical Examples

In these numerical examples the nonbasic artificial variables are retained in the tableau until Phase I is terminated, solely for the purpose of illustration. Normally artificial variables are deleted from the tableau once they leave the basic vector.

Example 1
Phase II Applied Directly

$$\begin{array}{lrl} \text{Minimize} & z = & -14G_1 - 18x_3 - 16x_4 - 80x_5 \\ \text{Subject to} & & 4.5G_1 + 8.5x_3 + 6x_4 + 20x_5 \leqq 6000 \\ & & G_1 + x_3 + 4x_4 + 40x_5 \leqq 4000 \\ & & x_3, \quad x_4, \quad x_5 \geqq 0 \\ & & G_1 \text{ unrestricted in sign} \end{array}$$

Replace the unrestricted variable G_1 by $x_1 - x_2$, where x_1, x_2 are two nonnegative variables. Convert the inequalities into equality constraints by adding the slack variables x_6, x_7.

Original Tableau

Basic variables	x_1	x_2	x_3	x_4	x_5	x_6	x_7	$-z$	Basic values	Ratio
x_6	4.5	−4.5	8.5	6	20	1	0	0	6000	300
x_7	1	−1	1	4	(40)	0	1	0	4000	100 min
$-z$	−14	14	−18	−16	−80	0	0	1	0	
					↑		↓			

This is in canonical form with respect to the basic vector (x_6, x_7). Start Phase II directly. Looking at $\bar{c}_j$'s, the most negative $\bar{c}_j$ is $\bar{c}_5 = -80$. Therefore bring x_5 into the basic vector. By the minimum ratio test x_7 is the leaving variable. The pivot element is 40, which has been circled. The pivot operation requires the following elementary row operations in the order listed:

$$\text{Row 1} - \left(\frac{20}{40}\right)(\text{Row 2})$$

$$\text{Row 3} + \left(\frac{80}{40}\right)(\text{Row 2})$$

$$\left(\frac{1}{40}\right)(\text{Row 2})$$

This gives the new canonical tableau:

Basic variables	x_1	x_2	x_3	x_4	x_5	x_6	x_7	$-z$	Basic values	Ratio
x_6	4	-4	⑧	4	0	1	$-\frac{1}{2}$	0	4000	500 min
x_5	$\frac{1}{40}$	$-\frac{1}{40}$	$\frac{1}{40}$	$\frac{1}{10}$	1	0	$\frac{1}{40}$	0	100	4000
$-z$	-12	12	-16	-8	0	0	2	1	8000	

Continuing we get

Basic variables	x_1	x_2	x_3	x_4	x_5	x_6	x_7	$-z$	Basic values	Ratio
x_3	$\frac{1}{2}$ (circled)	$-\frac{1}{2}$	1	$\frac{1}{2}$	0	$\frac{1}{8}$	$-\frac{1}{16}$	0	500	1000 min
x_5	$\frac{1}{80}$	$-\frac{1}{80}$	0	$\frac{7}{80}$	1	$-\frac{1}{320}$	$\frac{17}{640}$	0	$\frac{350}{4}$	7000
$-z$	-4	4	0	0	0	2	1	1	16,000	

Optimum Tableau

Basic variables	x_1	x_2	x_3	x_4	x_5	x_6	x_7	$-z$	Basic values
x_1	1	-1	2	1	0	$\frac{1}{4}$	$-\frac{1}{8}$	0	1000
x_5	0	0	$-\frac{1}{40}$	$\frac{3}{40}$	1	$-\frac{1}{160}$	$\frac{9}{320}$	0	75
$-z$	0	0	8	4	0	3	$\frac{1}{2}$	1	20,000

Now the optimality criterion is satisfied. Hence, the optimum solution is

$$x_1 = 1000 \qquad x_5 = 75$$
$$\text{and all other } x_j\text{'s} = 0$$
$$\text{the minimum objective value} = -20{,}000$$

In terms of the original problem variables, the optimum solution is

$$G_1 = 1000 \qquad x_5 = 75 \qquad x_3 = x_4 = 0$$

Example 2

Illustration of Phase I and Phase II

$$\begin{array}{lrcl}
\text{Maximize} & x_1 + 2x_2 + 3x_3 - x_4 & & \\
\text{Subject to} & x_1 + 2x_2 + 3x_3 & = & 15 \\
& -2x_1 - x_2 - 5x_3 & = & -20 \\
& x_1 + 2x_2 + x_3 + x_4 & = & 10 \\
& x_1, \quad x_2, \quad x_3, \quad x_4 & \geqq & 0
\end{array}$$

After putting the problem in standard form, we get the original tableau:

x_1	x_2	x_3	x_4	$-z$	b
1	2	3	0	0	15
2	1	5	0	0	20
1	2	1	1	0	10
−1	−2	−3	1	1	0

This tableau is not in canonical form. Artificial variables x_5, x_6, x_7 are introduced and the original Phase I tableau is:

Original Phase I Tableau with Full Artificial Basis

x_1	x_2	x_3	x_4	x_5	x_6	x_7	$-z$	$-w$	b
1	2	3	0	1	0	0	0	0	15
2	1	5	0	0	1	0	0	0	20
1	2	1	1	0	0	1	0	0	10
−1	−2	−3	1	0	0	0	1	0	0
0	0	0	0	1	1	1	0	1	0

(x_5, x_6, x_7) constitute an artificial basic vector.

Pricing out the basic column vectors in the Phase I objective row, we have the following tableau.

Phase I Initial Tableau

Basic variables	x_1	x_2	x_3	x_4	x_5	x_6	x_7	$-z$	$-w$	Basic values	Ratio
x_5	1	2	3	0	1	0	0	0	0	15	5
x_6	2	1	(5)	0	0	1	0	0	0	20	4 min
x_7	1	2	1	1	0	0	1	0	0	10	10
$-z$	−1	−2	−3	1	0	0	0	1	0	0	
$-w$	−4	−5	−9	−1	0	0	0	0	1	−45	

The most negative $\bar{d}_j$ is $\bar{d}_3$. Thus, x_3 is brought into the basic vector. By the minimum ratio test x_6 is the leaving variable. Here is the new canonical tableau.

Basic variables	x_1	x_2	x_3	x_4	x_5	x_6	x_7	$-z$	$-w$	Basic values	Ratio
x_5	$-\frac{1}{5}$	$\textcircled{\frac{7}{5}}$	0	0	1	$-\frac{3}{5}$	0	0	0	3	$\frac{15}{7}$ min
x_3	$\frac{2}{5}$	$\frac{1}{5}$	1	0	0	$\frac{1}{5}$	0	0	0	4	20
x_7	$\frac{3}{5}$	$\frac{9}{5}$	0	1	0	$-\frac{1}{5}$	1	0	0	6	$\frac{30}{9}$
$-z$	$\frac{1}{5}$	$-\frac{7}{5}$	0	1	0	$\frac{3}{5}$	0	1	0	12	
$-w$	$-\frac{2}{5}$	$-\frac{16}{5}$	0	-1	0	$\frac{9}{5}$	0	0	1	-9	

Continuing

Basic variables	x_1	x_2	x_3	x_4	x_5	x_6	x_7	$-z$	$-w$	Basic values	Ratio
x_2	$-\frac{1}{7}$	1	0	0	$\frac{5}{7}$	$-\frac{3}{7}$	0	0	0	$\frac{15}{7}$	
x_3	$\frac{3}{7}$	0	1	0	$-\frac{1}{7}$	$\frac{2}{7}$	0	0	0	$\frac{25}{7}$	
x_7	$\frac{6}{7}$	0	0	$\textcircled{1}$	$-\frac{9}{7}$	$\frac{4}{7}$	1	0	0	$\frac{15}{7}$	$\frac{15}{7}$
$-z$	0	0	0	1	1	0	0	1	0	15	
$-w$	$-\frac{30}{35}$	0	0	-1	$\frac{16}{7}$	$\frac{3}{7}$	0	0	1	$-\frac{15}{7}$	
x_2	$-\frac{1}{7}$	1	0	0	$\frac{5}{7}$	$-\frac{3}{7}$	0	0	0	$\frac{15}{7}$	
x_3	$\frac{3}{7}$	0	1	0	$-\frac{1}{7}$	$\frac{2}{7}$	0	0	0	$\frac{25}{7}$	
x_4	$\frac{6}{7}$	0	0	1	$-\frac{9}{7}$	$\frac{4}{7}$	1	0	0	$\frac{15}{7}$	
$-z$	$-\frac{6}{7}$	0	0	0	$\frac{16}{7}$	$-\frac{4}{7}$	-1	1	0	$\frac{90}{7}$	
$-w$	0	0	0	0	1	1	1	0	1	0	

Therefore at this point Phase I terminates, and the last artificial variable has left the basic vector. Now drop the Phase I objective row and the columns of the artificial variables x_5, x_6, x_7, and $-w$ and then proceed with Phase II.

Tableau at the End of Phase I

Basic variables	x_1	x_2	x_3	x_4	$-z$	Basic values	Ratios
x_2	$-\frac{1}{7}$	1	0	0	0	$\frac{15}{7}$	
x_3	$\frac{3}{7}$	0	1	0	0	$\frac{25}{7}$	$\frac{25}{3}$
x_4	$\frac{6}{7}$ (circled)	0	0	1	0	$\frac{15}{7}$	$\frac{15}{6}$ min
$-z$	$-\frac{6}{7}$	0	0	0	1	$\frac{90}{7}$	
x_2	0	1	0	$\frac{1}{6}$	0	$\frac{5}{2}$	
x_3	0	0	1	$-\frac{1}{2}$	0	$\frac{35}{14}$	
x_1	1	0	0	$\frac{7}{6}$	0	$\frac{15}{6}$	
$-z$	0	0	0	1	1	15	

This is an optimum tableau because all $\overline{c}_j$ are now nonnegative. Here is the optimum feasible solution.

$$x_1 = \frac{15}{6} \qquad x_2 = \frac{5}{2} \qquad x_3 = \frac{35}{14} \qquad x_4 = 0$$

The optimum objective value is -15. The optimum value of the original objective function is $-(\min z) = -(-15) = +15$.

Example 3

An Infeasible LP

$$\begin{aligned} \text{Minimize} \quad & x_1 + x_2 \\ \text{Subject to} \quad & x_1 + x_2 \leqq -1 \\ & x_1 - x_2 \geqq 0 \\ & x_1 \geqq 0 \quad x_2 \geqq 0 \end{aligned}$$

Introduce slack variables x_3, x_4.

x_1	x_2	x_3	x_4	$-z$	b
−1	−1	−1	0	0	1
1	−1	0	−1	0	0
1	1	0	0	1	0

After introducing the artificial variables x_5, x_6 the original Phase I tableau is:

x_1	x_2	x_3	x_4	x_5	x_6	$-z$	$-w$	b
−1	−1	−1	0	1	0	0	0	1
1	−1	0	−1	0	1	0	0	0
1	1	0	0	0	0	1	0	0
0	0	0	0	1	1	0	1	1

Initial Phase I Canonical Tableau

Basic variables	x_1	x_2	x_3	x_4	x_5	x_6	$-z$	$-w$	Basic values
x_5	-1	-1	-1	0	1	0	0	0	1
x_6	1	-1	0	-1	0	1	0	0	0
$-z$	1	1	0	0	0	0	1	0	0
$-w$	0	2	1	1	0	0	0	1	-1

Since $\bar{d} \geqq 0$, for all $j = 1$ to 4, Phase I terminates with this tableau and, since the value of $w = -(-1) = 1$, the original problem is infeasible. The constraints and sign restrictions of the original problem when represented on the x_1, x_2 Cartesian plane are as in Figure 2.3. The line $x_1 + x_2 = -1$ is indicated on the plane. Any point satisfying the constraint $x_1 + x_2 \leqq -1$ lies on the side of the arrow drawn on the line $x_1 + x_2 = -1$. In a similar manner all the constraints and sign restrictions are indicated by the corresponding lines and arrows pointing from them. Clearly, if there is a feasible solution to the problem, it must lie in the intersection of the dotted and shaded regions of the plane. Obviously no such point exists since the regions have an empty intersection.

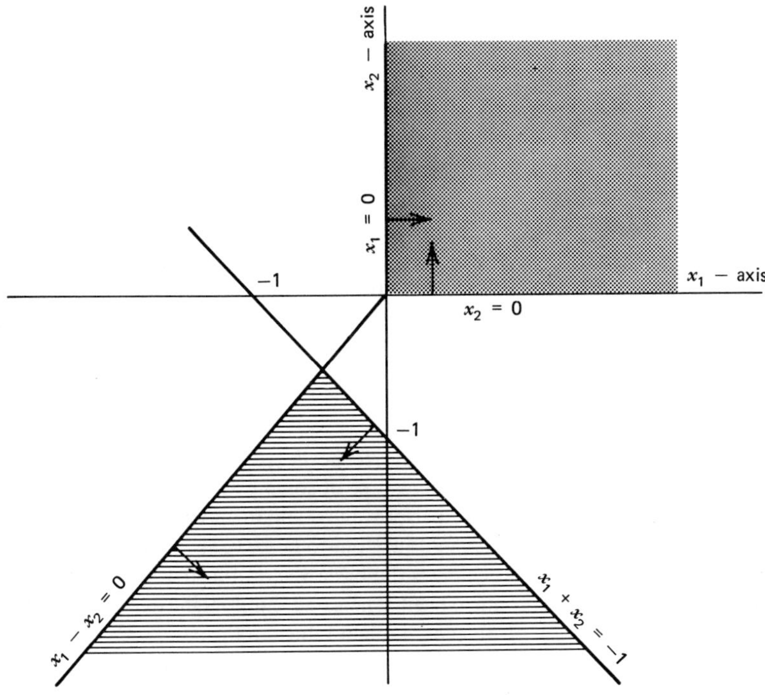

Figure 2 3

Example 4

Illustration of Unboundedness

$$\begin{array}{lrl} \text{Minimize} & -x_1 - & x_2 \\ \text{Subject to} & x_1 + & x_2 \geqq 1 \\ & x_1 - & x_2 \geqq 0 \\ & x_1 \geqq 0 & x_2 \geqq 0 \end{array}$$

After introducing the slack variables x_3, x_4 the original tableau is

x_1	x_2	x_3	x_4	$-z$	b
1	1	−1	0	0	1
1	−1	0	−1	0	0
−1	−1	0	0	1	0

After introducing artificial variables, x_5, x_6, we get the original Phase I tableau:

x_1	x_2	x_3	x_4	x_5	x_6	$-z$	$-w$	b
1	1	−1	0	1	0	0	0	1
1	−1	0	−1	0	1	0	0	0
−1	−1	0	0	0	0	1	0	0
0	0	0	0	1	1	0	1	0

Pricing out the artificial column vectors, we find that the initial Phase I canonical tableau is:

Basic variables	x_1	x_2	x_3	x_4	x_5	x_6	$-z$	$-w$	Basic values	Ratio
x_5	1	1	−1	0	1	0	0	0	1	1
x_6	(1)	−1	0	−1	0	1	0	0	0	0 min
$-z$	−1	−1	0	0	0	0	1	0	0	
$-w$	−2	0	1	1	0	0	0	1	−1	

DISCUSSION OF DEGENERACY

A BFS of an LP in standard form is said to be *degenerate* if the value of at least one of the basic variables is zero in it. In the course of solving an LP by the simplex algorithm, if we get a degenerate BFS and if we perform the standard pivot operation to improve the objective value, the minimum ratio could turn out to be zero, as it happened here. When this happens the pivot operation just changes the basic vector, but the BFS corresponding to it and the objective value remain unchanged. The same thing might happen in the next pivot step and, after a series of such pivot steps, it might return to the basic vector that started these degenerate pivot steps. In this case we say that the *simplex algorithm has cycled due to degeneracy* and unless something is done, the algorithm can cycle among this set of degenerate basic vectors indefinitely without ever moving out of the cycle and reaching a terminal basic vector.

In a nondegenerate BFS the values of all the basic variables are positive. Hence, the minimum ratio in a pivot step from a nondegenerate BFS is always positive and such a step in the course of simplex algorithm is guaranteed to result in a strict decrease in the objective value. Since the objective value never increases in the course of the algorithm, once a strict decrease in it occurs, we will never go back to the previous basic vector in subsequent steps. Thus cycling can never begin with a nondegenerate BFS.

There are special techniques for resolving the cycling problem under degeneracy. Among these, the lexico minimum ratio rule will be discussed later. (See Chapter 9.) However, in most practical applications of linear programming it has been found that these special techniques for resolving degeneracy are unnecessary. When a degenerate pivot has to be made, we continue as usual as if nothing has happened; after a small number of degenerate pivots the algorithm practically always resolves the degeneracy and continues on its way to a terminal basis. We now continue with our numerical example.

Basic variables	x_1	x_2	x_3	x_4	x_5	x_6	$-z$	$-w$	Basic values	Ratios
x_5	0	②	-1	1	1	-1	0	0	1	$\frac{1}{2}$ min
x_1	1	-1	0	-1	0	1	0	0	0	
$-z$	0	-2	0	-1	0	1	1	0	0	
$-w$	0	-2	1	-1	0	2	0	1	-1	
x_2	0	1	$-\frac{1}{2}$	$\frac{1}{2}$	$\frac{1}{2}$	$-\frac{1}{2}$	0	0	$\frac{1}{2}$	
x_1	1	0	$-\frac{1}{2}$	$-\frac{1}{2}$	$\frac{1}{2}$	$\frac{1}{2}$	0	0	$\frac{1}{2}$	
$-z$	0	0	-1	0	1	0	1	0	1	
$-w$	0	0	0	0	1	1	0	1	0	

Terminate Phase I and begin Phase II.

Initial Phase II Canonical Tableau

Basic variables	x_1	x_2	x_3	x_4	$-z$	Basic values
x_2	0	1	$-\frac{1}{2}$	$\frac{1}{2}$	0	$\frac{1}{2}$
x_1	1	0	$-\frac{1}{2}$	$-\frac{1}{2}$	0	$\frac{1}{2}$
$-z$	0	0	-1	0	1	1

The column vector of x_3 satisfies the unboundedness criterion in this canonical tableau. Actually the solution

$$\begin{aligned} x_2 &= \frac{1}{2} + \frac{\lambda}{2} \\ x_1 &= \frac{1}{2} + \frac{\lambda}{2} \\ \text{Slack variable,}\quad x_3 &= \lambda \\ \text{Slack variable,}\quad x_4 &= 0 \\ \text{Objective value}\quad &= -1 - \lambda \end{aligned} \tag{2.12}$$

is a feasible solution of the original problem for all $\lambda \geqq 0$. And as λ tends to $+\infty$, the objective value tends to $-\infty$. The set of feasible solutions is the shaded region in Figure 2.4. For each $\lambda \geqq 0$, the point represented by (2.12) is on the thick line and, as λ tends to $+\infty$, it moves on this line in the direction of the arrow.

This thick line is called an *extreme half-line* (to be defined in Chapter 3) of the set of feasible solutions and the objective function $z(x)$ diverges to $-\infty$ along this line.

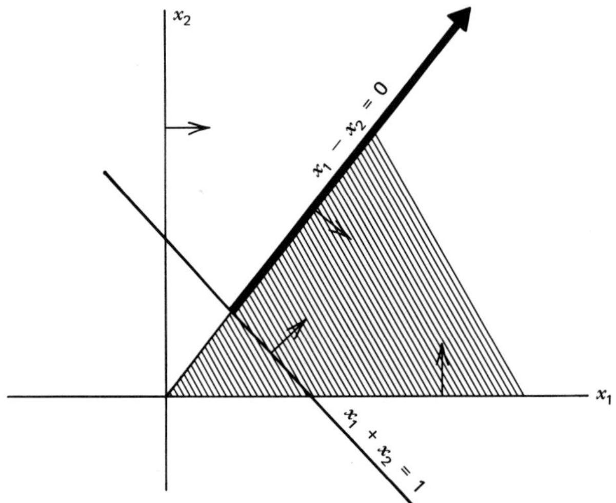

Figure 2.4

Example 5

Illustration of Eliminating Unrestricted Variables

Consider this LP

$$\begin{array}{lrcl}
\text{Maximize} & 2y_1 + y_2 + x_1 \qquad + 5x_3 - x_4 & & \\
\text{Subject to} & -2y_1 \qquad - 6x_1 + 2x_2 - 14x_3 \qquad & = & 0 \\
& 2y_1 + y_2 \qquad - 2x_2 + 2x_3 \qquad & = & 5 \\
& y_1 \qquad + 4x_1 + x_2 + 10x_3 \qquad & = & 15 \\
& x_1 + 2x_2 + x_3 + x_4 & = & 10 \\
& y_1 + y_2 - 5x_1 - 2x_2 - 10x_3 \qquad & = & -15 \\
& y_1 \text{ and } y_2 \text{ are unrestricted} & & \\
& x_1, x_2, x_3, x_4 \geqq 0 & &
\end{array}$$

From the first constraint we obtain

$$y_1 = -3x_1 + x_2 - 7x_3$$

Elimination of y_1 from the second constraint leads to

$$y_2 = 6x_1 + 12x_3 + 5$$

These are the only constraints that y_1 and y_2 have to satisfy. By using them, y_1 and y_2 can be eliminated from the problem. This reduces the problem to

$$\begin{array}{lrcl}
\text{Maximize} & x_1 + 2x_2 + 3x_3 - x_4 + 5 & & \\
\text{Subject to} & x_1 + 2x_2 + 3x_3 \qquad & = & 15 \\
& x_1 + 2x_2 + x_3 + x_4 & = & 10 \\
& -2x_1 - x_2 - 5x_3 \qquad & = & -20 \\
& x_j \geqq 0 \quad \text{for all } j & &
\end{array}$$

Maximizing $x_1 + 2x_2 + 3x_3 - x_4 + 5$ has the same set of optimum feasible solutions as maximizing $x_1 + 2x_2 + 3x_3 - x_4$. Hence the problem reduces to the problem in Example 2. The optimum solution obtained there is

$$x_1 = \frac{15}{6} \qquad x_2 = \frac{5}{2} \qquad x_3 = \frac{35}{14} \qquad x_4 = 0$$

Substituting in the expressions for y_1, y_2, the optimum solution for the original problem is

$$y_1 = -\frac{45}{2} \qquad y_2 = 50 \qquad x_1 = \frac{15}{6} \qquad x_2 = \frac{5}{2} \qquad x_3 = \frac{35}{14} \qquad x_4 = 0$$

and the maximum value of the original objective function is 20.

Comments The version of the simplex algorithm discussed here is the version developed originally. It is not computationally efficient. All the computations to be performed in the simplex algorithm can be carried out much more efficiently by using the inverse of the basis in each step. These improved methods require matrix theoretic manipulations. In Chapter 3 the necessary matrix algebra is reviewed and the improved methods of carrying out the simplex algorithm are discussed in Chapters 3 and 5. The study of the original simplex algorithm using canonical tableaus is very useful to introduce the basic concepts of linear programming and the simplex algorithm; for this reason it has been discussed here in great detail.

2.7 THE BIG-*M* METHOD

To solve a general LP the simplex algorithm is applied twice, first to a Phase I problem for obtaining an initial BFS and then to the Phase II problem that solves the original problem starting with the feasible basic vector obtained at the end of Phase I. There have been several attempts to combine these two phases into a single problem. *The Big-M method* is one of them. *In this section, the symbol "M" will be used to denote an arbitrarily large positive number.* Consider the LP in standard form:

$$\begin{aligned} &\text{Minimize} \quad z(x) = \textstyle\sum c_j x_j \\ &\text{Subject to} \quad \sum_{j=1}^{n} a_{ij} x_j = b_i \qquad i = 1 \text{ to } m \\ &\qquad x_j \geqq 0 \qquad \text{for all } j \end{aligned} \tag{2.13}$$

Suppose a full artificial basis is introduced. Let the artificial variables be called $t_1, \ldots, t_m$. The augmented problem is

$$\begin{aligned} &\text{Minimize} \quad Z(x, t) = \textstyle\sum c_j x_j + M(t_1 + \cdots + t_m) \\ &\text{Subject to} \quad \sum_{j=1}^{n} a_{ij} x_j + t_i = b_i \qquad i = 1 \text{ to } m \\ &\qquad x_j, t_i \geqq 0 \qquad \text{for all } i, j \end{aligned} \tag{2.14}$$

It is not necessary to give a specific value to M, but it is treated as a parameter that is very much larger than any number with which it is compared. The problem is now an LP with $(t_1, \ldots, t_m)$ as an initial feasible basic vector. Hence the simplex algorithm can be applied directly to solve it.

During the application of the simplex algorithm, the relative cost coefficient of x_j is of the form $\bar{c}_j + M\bar{c}_j^*$ where $\bar{c}_j$ is the constant term [coming from the original objective function, $z(x)$] and the $\bar{c}_j^*$ is the coefficient of the parameter M. If $\bar{c}_j^* < 0$, this relative cost coefficient is negative, whatever $\bar{c}_j$ may be (since M is very large). If $\bar{c}_j^* > 0$, this relative cost coefficient is positive. If $\bar{c}_j^* = 0$, the sign of this relative cost coefficient is the sign of $\bar{c}_j$. To make the computations easier, the $\bar{c}_j$ and the $\bar{c}_j^*$ are kept in two different rows of the tableau. The actual relative cost coefficient of x_j is $\bar{c}_j + \boldsymbol{M}\bar{c}_j^*$. If x_j is the entering variable, it should be priced out in both the $\bar{c}$ row and the $\bar{c}_j^*$ row.

When applied to this problem, the simplex algorithm might terminate in several ways.

1. If the simplex algorithm leads to an optimum solution $(\tilde{x}, \tilde{t})$ in which $\tilde{t} = 0$, then $\tilde{x}$ is an optimum feasible solution of the original problem.
2. If the simplex algorithm leads to an optimum solution $(\tilde{x}, \tilde{t})$ in which $\tilde{t} \neq 0$, the optimality criterion being satisfied independent of how large a value the parameter $\boldsymbol{M}$ has, the original problem has no feasible solution.
3. If the unboundedness criterion is satisfied independent of how large a value the parameter $\boldsymbol{M}$ has, then $Z(x, t)$ is unbounded below in (2.14) for all values of $\boldsymbol{M}$ sufficiently large. In this case, $z(x)$ is unbounded below in (2.13) if it is feasible. A proof of this is discussed in section 4.5.11. The Big-$\boldsymbol{M}$ method terminates here.

   In this case, the feasibility of (2.13) can only be determined by continuing the algorithm after changing the original objective function, $\sum c_j x_j$ to 0. With this change, the objective function of (2.14) becomes $Z^1(x, t) = \boldsymbol{M}(t_1 + \cdots + t_m)$. So $Z^1(x, t) \geqq 0$ at all feasible solutions (x, t) of (2.14); hence, it is bounded below.

   Making this change in the objective function is equivalent to making all the entries in the $\bar{c}$ row equal to zero while leaving the $\bar{c}^*$ row as it is in the present tableau. From now on all the relative cost coefficients are treated to be equal to $\boldsymbol{M}\bar{c}_j^*$ for all j and, hence, have the same sign as $\bar{c}_j^*$. The application of the simplex algorithm is continued starting again from the present basis.

   Let the optimum solution obtained at termination be $(\hat{x}, \hat{t})$. If $\hat{t} = 0$, (2.13) is feasible and $z(x)$ is unbounded below on it. If $\hat{t} \neq 0$, (2.13) has no feasible solution.

Therefore if (2.13) has an optimum feasible solution, the Big-$\boldsymbol{M}$ method will find it after the application of the simplex algorithm once. If (2.13) is infeasible or if $z(x)$ is unbounded below on it, we may have to apply the simplex algorithm twice before completely determining the status of the problem.

It is impossible to conclude theoretically whether the Big-$\boldsymbol{M}$ method solves LPs with less computational effort than the standard Phase I, II approach. Practical experience seems to indicate that both approaches take approximately the same amount of work.

Example

Consider the LP:

x_1	x_2	x_3	x_4	x_5	$-z$	b
1	−1	0	1	−2	0	1
0	1	0	−2	2	0	4
0	−1	1	2	1	0	6
3	−12	5	23	−9	1	0 (z minimize)

$x_j \geqq 0$ for all j.

Since the column vectors of x_1, x_3 are unit vectors, they can be included in the initial basic vector. Only one artificial variable need be introduced. As explained earlier the objective function for the Big-M problem is entered in two rows in the following tableau. The artificial variable is t.

Original Tableau for the Big-M Problem

	x_1	x_2	x_3	x_4	x_5	t	$-Z$	b
	1	−1	0	1	−2	0	0	1
	0	1	0	−2	2	1	0	4
	0	−1	1	2	1	0	0	6
$\bar{c}$ row	3	−12	5	23	−9	0	1	0
$\bar{c}^*$ row	0	0	0	0	0	1		0

Picking (x_1, t_1, x_3) as the initial basic vector and pricing out in both the cost rows leads to the initial canonical tableau.

Initial Canonical Tableau

Basic variables	x_1	x_2	x_3	x_4	x_5	t	$-Z$	b
x_1	1	−1	0	1	−2	0	0	1
t	0	(1)	0	−2	2	1	0	4
x_3	0	−1	1	2	1	0	0	6
$\bar{c}$ row	0	−4	0	10	−8	0	1	−33
$\bar{c}^*$ row	0	−1	0	2	−2	0		−4

Remember that the actual relative cost coefficient is the entry in $\bar{c}$ row plus M times the entry in $\bar{c}^*$ row. Thus the relative cost coefficient of x_2 is $-4-M$, and it is negative when M is large. Hence x_2 is a candidate to enter the basic vector. The value of $-Z$ in the present solution is $-33-4M$. Bringing x_2 into the basic vector leads to

Basic variables	x_1	x_2	x_3	x_4	x_5	t	$-Z$	b
x_1	1	0	0	−1	0	1	0	5
x_2	0	1	0	−2	2	1	0	4
x_3	0	0	1	0	3	1	0	10
$\bar{c}$ row	0	0	0	2	0	4	1	−17
$\bar{c}^*$ row	0	0	0	0	0	1		0

This is a terminal basis. Clearly the solution $(x_1, x_2, x_3, x_4, x_5) = (5, 4, 10, 0, 0)$ is optimal to the original problem with an objective value of $z(x) = 17$.

PROBLEMS

2.2 Consider the following LP:

x_1	x_2	x_3	x_4	x_5	x_6	x_7	$-z$	b
0	1	0	α	1	0	3	0	β
0	0	1	-2	2	Δ	-1	0	2
1	0	0	0	-1	2	1	0	3
0	0	0	δ	3	γ	ξ	1	0

$x_j \geqq 0$ for all j, $\quad z$ to be minimized.

The entries $\alpha, \beta, \gamma, \delta, \Delta, \xi$ in the tableau are parameters. *Each of the following questions is independent, and they all refer to this original problem.* Clearly state the range of values of the various parameters that will make the conclusions in the following questions true. B_1 is the basis for this problem corresponding to the basic vector

$$x_{B_1} = (x_2, x_3, x_1)$$

1. The present tableau is such that phase II of the simplex method can be applied using this as an initial tableau.
2. Row 1 in the present tableau indicates that the problem is infeasible.
3. B_1 is a feasible but nonoptimal basis for this problem.
4. B_1 is a feasible basis for the problem and the present tableau indicates that z is unbounded below.
5. B_1 is a feasible basis, x_6 is a candidate to enter the basic vector, and when x_6 is the entering variable, x_3 leaves the basic vector.
6. B_1 is a feasible basis, x_7 is a candidate to enter the basic vector, but when x_7 is the entering variable, the solution and the objective value remain unchanged.

2.3 Consider the LPs

$$\begin{aligned} \text{Minimize} \quad & z(x) = cx \\ \text{Subject to} \quad & Ax = b \\ & x \geqq 0 \end{aligned}$$

and

$$\begin{aligned} \text{Minimize} \quad & z(x) = (\mu c)x \\ \text{Subject to} \quad & Ax = (\lambda b) \\ & x \geqq 0 \end{aligned}$$

where λ and μ are strictly positive real numbers, and A, b, c are the same matrices or vectors in both the problems. Obtain a relationship between the optimum solutions of the two problems. Explain briefly why this relationship fails when either λ or μ is negative.

2.4 Solve the following LP:

$$\begin{aligned} \text{Minimize} \quad z(x) = \quad & x_1 - 2x_2 - 4x_3 + 4x_4 \\ \text{Subject to} \quad & \quad\; - x_2 + 2x_3 + x_4 \leqq 4 \\ & -2x_1 + x_2 + x_3 - 4x_4 \leqq 5 \\ & x_1 - x_2 \qquad\quad + 2x_4 \leqq 3 \\ & x_j \geqq 0 \quad \text{for all } j \end{aligned}$$

If $z(x)$ is unbounded below in this problem, find an extreme half line along which it diverges to $-\infty$. Find out a feasible solution on this half line that corresponds to an objective value of -415.

2.5 The following is a canonical tableau obtained in the course of solving an LP.

Basic variables	$x_1 \ldots x_m$	x_{m+1}	x_n	$-z$	b
x_1	$1 \ldots 0$	$\bar{a}_{1,m+1} \cdots$	$\bar{a}_{1,n}$	0	$\bar{b}_1$
$\vdots$	$\vdots \quad \vdots$	$\vdots$	$\vdots$	$\vdots$	$\vdots$
x_m	$0 \ldots 1$	$\bar{a}_{m,m+1} \cdots$	$\bar{a}_{m,n}$	0	$\bar{b}_m$
$-z$	$0 \ldots 0$	$\bar{c}_{m+1}$	$\ldots \bar{c}_n$	1	$-\bar{z}$

$x_j \geqq 0$, $\quad z$ to be minimized

In this problem it is known that every feasible solution satisfies $\sum x_j \leqq \alpha$, where α is a known positive number. Suppose the present canonical tableau does not satisfy the optimality criterion. Let $\bar{c}_s$ = minimum $\{\bar{c}_1, \ldots, \bar{c}_n\}$. Prove that $\bar{z} + \alpha\bar{c}_s$ is a lower bound for the minimum objective value in this Problem.

Solve LPs 2.6 to 2.22.

2.6

$$\begin{array}{ll} \text{Minimize} & 12x_1 - 10x_2 - 30x_3 \\ \text{Subject to} & -3x_1 + 2x_2 + 8x_3 \leqq 17 \\ & -x_1 + x_2 + 3x_3 \leqq 9 \\ & -2x_1 + x_2 + 8x_3 \leqq 16 \\ & x_j \geqq 0 \quad \text{for all } j \end{array}$$

2.7

$$\begin{array}{ll} \text{Minimize} & -x_1 - 4x_2 - 5x_3 \\ \text{Subject to} & x_1 + 2x_2 + 3x_3 \leqq 2 \\ & 3x_1 + x_2 + 2x_3 \leqq 2 \\ & 2x_1 + 3x_2 + x_3 \leqq 4 \\ & x_j \geqq 0 \quad \text{for all } j \end{array}$$

Use only one pivot step.

2.8

$$\begin{array}{ll} \text{Minimize} & -3x_1 - 11x_2 - 9x_3 + x_4 + 29x_5 \\ \text{Subject to} & x_2 + x_3 + x_4 - 2x_5 \leqq 4 \\ & x_1 - x_2 + x_3 + 2x_4 + x_5 \geqq 0 \\ & x_1 + x_2 + x_3 - 3x_5 \leqq 1 \\ & x_1 \text{ unrestricted}, \quad x_j \geqq 0 \text{ for all } j \neq 1 \end{array}$$

2.9

$$\begin{array}{ll} \text{Minimize} & -17x_1 + 13x_2 + 2x_3 - 8x_4 \\ \text{Subject to} & 2x_1 - x_2 + x_4 \leqq 4 \\ & -x_1 + x_3 - x_4 \geqq -3 \\ & x_1 + 3x_2 - 2x_3 \leqq 4 \\ & x_j \geqq 0 \quad \text{for all } j \end{array}$$

2.10

$$\begin{array}{ll} \text{Minimize} & x_6 \\ \text{Subject to} & 2x_2 + x_3 + x_4 - x_5 = 0 \\ & 3x_1 - 2x_2 - x_3 + 2x_4 \leqq 15 \\ & x_1 + x_3 \leqq 5 \\ & -9x_1 + 8x_2 + x_3 - 2x_4 - x_6 = 0 \\ & x_j \geqq 0 \quad \text{for all } j \neq 6 \\ & x_6 \text{ unrestricted} \end{array}$$

2.11

$$\begin{array}{ll} \text{Minimize} & x_1 \\ \text{Subject to} & x_1 + x_2 + x_3 + x_4 + x_5 \geqq 4 \\ & x_1 - 2x_2 - x_3 + 2x_4 \geqq -2 \\ & 5x_1 - 4x_2 - x_3 + 8x_4 + 2x_5 \leqq 1 \\ & x_1 \text{ unrestricted}, \\ & x_j \geqq 0 \quad \text{for all } j \neq 1 \end{array}$$

2.12

$$\begin{array}{ll} \text{Maximize} & x_5 \\ \text{Subject to} & -x_1 + 5x_2 + x_4 \geqq x_5 \\ & 2x_1 - 2x_3 + x_4 \geqq x_5 \\ & -2x_2 + 2x_3 + x_4 \geqq x_5 \\ & -x_1 - 2x_2 - x_3 - x_4 = -4 \\ & x_j \geqq 0 \quad \text{for all } j \end{array}$$

2.13

$$\begin{array}{ll} \text{Minimize} & x_1 + 3x_2 - x_3 + x_4 - x_5 + x_6 \\ \text{Subject to} & x_1 - x_3 + x_4 + 2x_5 - x_6 + x_7 = 6 \\ & x_1 + x_2 + x_5 + x_6 + 4x_7 = 14 \\ & 2x_1 - 2x_2 - 3x_3 + 3x_4 + 3x_5 - 4x_6 + 2x_7 = 22 \\ & x_j \geqq 0 \quad \text{for all } j \end{array}$$

2.14

$$\begin{array}{ll} \text{Minimize} & -5x_1 - 4x_3 - 13x_4 + 3x_5 + 11x_6 + x_7 \\ \text{Subject to} & -x_1 + 2x_2 + 5x_3 - x_4 + 5x_5 + 4x_6 + x_7 = 2 \\ & x_1 + x_2 + 4x_3 + 4x_4 + x_5 - x_6 - x_7 = 7 \\ & x_j \geqq 0 \quad \text{for all } j \end{array}$$

2.15

$$\begin{array}{ll} \text{Minimize} & -x_2 - 4x_3 + 2x_5 - 2x_7 + 10 \\ \text{Subject to} & 2x_1 - x_2 - x_3 + x_4 - 2x_5 + x_6 - x_7 = 4 \\ & 2x_1 + 2x_2 - 3x_3 + x_4 + 2x_5 + 2x_6 + x_7 = 6 \\ & 6x_1 + 2x_2 - 3x_3 + 4x_4 + x_5 + 2x_6 = 12 \\ & 2x_1 + x_2 + x_3 + 2x_4 + x_5 - x_6 = 2 \\ & x_j \geqq 0 \quad \text{for all } j \end{array}$$

2.16

$$\begin{array}{ll} \text{Minimize} & -x_1 + 2x_2 - x_3 - x_4 + 2x_5 - x_6 + 2x_7 + 3x_8 - 5x_9 \\ \text{Subject to} & x_1 + 5x_2 - 3x_3 + 4x_4 - 3x_5 + 2x_8 - 8x_9 = -3 \\ & -5x_2 - x_3 - 5x_4 + 4x_5 - x_6 + 2x_7 + x_9 = -4 \\ & x_1 + 2x_2 + 5x_4 + x_5 + x_6 + x_7 - 2x_8 + 3x_9 = 9 \\ & x_j \geqq 0 \quad \text{for all } j \end{array}$$

2.17

$$\begin{array}{ll} \text{Minimize} & -12x_1 - x_2 - 2x_3 - 5x_4 - 4x_5 + 6x_6 \\ \text{Subject to} & 8x_1 + x_2 + 3x_3 + 2x_4 + 3x_5 - 3x_6 = 17 \\ & -5x_1 - x_2 - x_3 - x_4 - 2x_5 + 2x_6 = -12 \\ & -5x_1 - x_3 - 2x_4 - x_5 + 4x_6 \leqq -8 \\ & x_j \geqq 0 \quad \text{for all } j \end{array}$$

2.18

$$\begin{array}{ll} \text{Minimize} & 34x_1 + 5x_2 + 19x_3 + 9x_4 \\ \text{Subject to} & -2x_1 - x_2 - x_3 - x_4 \leqq -9 \\ & 4x_1 - 2x_2 + 5x_3 + x_4 \leqq 8 \\ & -4x_1 + x_2 - 3x_3 - x_4 \leqq -5 \\ & x_j \geqq 0 \quad \text{for all } j \end{array}$$

2.19 Minimize $x_1 + x_2 + x_3 + x_4 + x_5 + x_6 + x_7$

Subject to

$$\begin{aligned} x_1 \quad + 2x_3 + 3x_4 - 2x_5 - x_6 + x_7 &= 1 \\ 3x_1 + 2x_2 + 2x_3 + 5x_4 - 2x_5 + x_6 + 3x_7 &= 5 \\ -2x_1 + x_2 \quad + x_4 - x_5 + x_6 - 2x_7 &= 4 \\ x_j \geqq 0 \quad \text{for all } j \end{aligned}$$

2.20 Minimize $2x_1$

Subject to

$$\begin{aligned} x_1 \quad - x_4 \quad &= 3 \\ x_1 - x_2 \quad - 2x_5 &= 1 \\ 2x_1 \quad + x_3 \quad + x_5 &= 7 \\ x_j \geqq 0 \quad \text{for all } j \end{aligned}$$

2.21 Maximize x_8

Subject to

$$\begin{aligned} -x_1 \quad - x_3 + x_4 + x_5 \quad &= -2 \\ x_1 \quad - 2x_3 - 3x_4 \quad + x_6 \quad - x_8 &= 4 \\ x_1 \quad - 2x_4 \quad + x_7 \quad &= 3 \\ 2x_1 + x_2 \quad - x_8 &= 6 \\ x_8 \text{ unrestricted}, x_j \geqq 0 \quad \text{for all } j \neq 8 \end{aligned}$$

Use at most one artificial variable.

2.22 (a) Minimize $-3x_1 + 5x_2 + 3x_3 + 9x_4 + 4x_5$

Subject to

$$\begin{aligned} -x_1 + x_2 + x_3 + 2x_4 + x_5 \quad &= 0 \\ x_3 + 3x_4 + 3x_5 + x_6 &= 5 \\ -2x_1 + 2x_2 + x_3 + x_4 - x_5 - x_6 &= -5 \\ x_j \geqq 0 \quad \text{for all } j \end{aligned}$$

(b) Minimize $-x_1$

Subject to

$$\begin{aligned} x_1 + x_2 - x_3 + x_4 - x_5 + 2x_6 &= 2 \\ 2x_1 - x_2 - x_3 - 2x_4 + x_5 - x_6 &= 3 \\ 3x_1 \quad - 2x_3 - x_4 \quad + x_6 &= 5 \\ x_j \geqq 0 \quad \text{for all } j \end{aligned}$$

2.23 Find the optimum solution or the extreme ray along which z diverges to $-\infty$ in each of the following LPs.

	x_1	x_2	x_3	x_4	x_5	x_6	$-z$	$= b$
	1	0	2	1	2	−1	0	3
(a)	2	1	2	0	1	−1	0	6
	−1	0	2	0	−3	0	1	10
	0	1	0	1	2	1	0	2
(b)	1	−3	−1	1	1	0	0	0
	0	−1	−1	2	2	0	1	−4
	1	0	−1	−2	−7	−9	0	5
(c)	0	1	−2	−1	−8	−10	0	4
	0	0	3	0	6	10	1	6

$x_j \geqq 0$ for all j in each problem

2.24 Solve the following LP:

$$\begin{array}{ll} \text{Minimize} & 3x_1 + 5x_2 + 7x_3 + 9x_4 + 11x_5 \\ \text{Subject to} & 9x_1 + 25x_2 + 28x_3 + 18x_4 + 66x_5 = 105 \\ & x_j \geqq 0 \quad \text{for all } j \end{array}$$

2.25 Consider the one constraint LP:

$$\begin{array}{ll} \text{Minimize} & \sum c_j x_j \\ \text{Subject to} & \sum a_j x_j = b \\ & x_j \geqq 0 \quad \text{for all } j \end{array}$$

(a) Develop a simple test for checking the feasibility of this problem.
(b) Develop a simple test for checking unboundedness.
(c) Develop a simple method for obtaining an optimum solution directly.

2.26 Find a feasible solution of each of the following systems:

(a)
$$\begin{aligned} x_1 + x_2 + x_3 + x_4 + x_5 &= 2 \\ -x_1 + 2x_2 + x_3 - 3x_4 + x_5 &= 1 \\ x_1 - 3x_2 - 2x_3 + 2x_4 - 2x_5 &= -4 \\ x_j \geqq 0 \quad \text{for all } j \end{aligned}$$

(b)
$$\begin{aligned} -4x_1 + x_2 + 4x_3 &= 3 \\ 2x_1 + x_2 + 5x_3 &= 2 \\ x_j \geqq 0 \quad \text{for all } j \end{aligned}$$

(c)
$$\begin{aligned} -5x_1 + x_2 - 3x_3 + 3x_4 + 8x_5 &\leqq -3 \\ 3x_1 - x_2 + 2x_3 - x_4 - 5x_5 &\leqq 2 \\ -2x_1 + x_2 - x_3 \qquad + 3x_5 &\leqq 2 \\ x_j \geqq 0 \quad \text{for all } j \end{aligned}$$

(d)
$$\begin{aligned} x_1 - 2x_2 + x_3 + 2x_4 - x_5 &= 5 \\ -x_1 - 3x_2 + 5x_3 + 6x_4 - x_5 &= 10 \\ -3x_1 + x_2 + 3x_3 + 2x_4 + x_5 &= 3 \\ x_j \geqq 0 \quad \text{for all } j \end{aligned}$$

2.27 Use a linear programming formulation to show that the constraints

$$\begin{aligned} 4x_1 + x_2 &\leqq 4 \\ 2x_1 - 3x_2 &\leqq 6 \\ x_1, \quad x_2 &\geqq 0 \end{aligned}$$

imply

$$x_1 + 2x_2 \leqq 8$$

2.28 Show that none of the feasible solutions of

$$\begin{aligned} 2x_1 - x_2 - x_3 + 2x_4 + x_5 &\leqq 3 \\ -3x_1 + x_2 + 4x_3 - 5x_4 - 2x_5 &\leqq -4 \\ x_j \geqq 0 \quad \text{for all } j \end{aligned}$$

satisfies

$$-6x_1 + 8x_2 + 7x_3 - 9x_4 - 5x_5 \leqq -18$$

2.29 Consider the LP

$$\begin{array}{ll} \text{Minimize} & z(x) = cx \\ \text{Subject to} & Ax = b \\ & x \geqq 0 \end{array}$$

where A is a matrix of order $m \times n$ and rank, m. It is required to transform this into an LP of the form:

$$\begin{aligned} \text{Minimize} \quad & z(x) = cx \\ \text{Subject to} \quad & Fx \geqq f \\ & x \geqq 0 \end{aligned}$$

where F is a matrix of order $(m + 1) \times n$. Show how this can be done.

2.30 Consider the LP

$$\begin{aligned} \text{Minimize} \quad & z(x) = cx \\ \text{Subject to} \quad & Ax = b \\ \text{And} \quad & l_j \leqq x_j \leqq k_j \qquad j = 1 \text{ to } n \end{aligned}$$

where A, b, c, are given $m \times n$, $m \times 1$, $1 \times n$ matrices respectively and l_j, k_j are given constants for each $j = 1$ to n. l_j and k_j are finite and $l_j \leqq k_j$ for all j. Some of the l_j and even k_j may be negative. Show how to transform this problem into an LP in n or less nonnegative variables, in which each variable is restricted to be less than or equal to 1.

2.31 Solve the following LPs by the Big-M method.

(a)

$$\begin{aligned} \text{Minimize} \quad & x_1 \\ \text{Subject to} \quad & x_1 + x_2 + x_3 + 2x_4 - x_5 - x_6 \geqq 2 \\ & -x_1 - 2x_3 + x_4 - x_5 - 2x_6 \geqq 3 \\ & x_2 - 3x_3 + 4x_4 - 3x_5 - 5x_6 \leqq 7 \\ & x_j \geqq 0 \quad \text{for all } j \end{aligned}$$

(b)

$$\begin{aligned} \text{Minimize} \quad & -2x_2 + 2x_3 + x_4 \\ \text{Subject to} \quad & x_1 + x_3 + x_4 - x_5 + x_6 + 2x_7 = 6 \\ & x_2 + x_4 - x_5 + x_6 = 5 \\ & -x_2 + x_3 - x_4 + x_5 + x_7 = -3 \\ & x_j \geqq 0 \quad \text{for all } j \end{aligned}$$

(c)

$$\begin{aligned} \text{Minimize} \quad & x_1 + x_2 + x_3 - 3x_4 + 6x_5 + 4x_6 \\ \text{Subject to} \quad & x_1 + x_2 + 3x_4 - x_5 + 2x_6 = 6 \\ & x_2 + x_3 - x_4 + 4x_5 + x_6 = 3 \\ & x_1 + x_3 - 2x_4 + x_5 + 5x_6 = 5 \\ & x_j \geqq 0 \quad \text{for all } j \end{aligned}$$

(d)

$$\begin{aligned} \text{Minimize} \quad & 3x_1 - 2x_3 + x_4 \\ \text{Subject to} \quad & x_1 + 2x_2 - 2x_3 - x_4 + x_5 \geqq 5 \\ & 5x_1 + 6x_2 - 3x_3 - x_4 - x_5 \leqq 10 \\ & 3x_1 + 2x_2 + x_3 + x_4 - 3x_5 \geqq 3 \\ & x_j \geqq 0 \quad \text{for all } j \end{aligned}$$

3 the geometry of the simplex method

3.1 EUCLIDEAN VECTOR SPACES: DEFINITIONS AND GEOMETRICAL CONCEPTS

Familiarity with the notion of the n-dimensional Euclidean space, $\mathbf{R}^n$, as a linear vector space is assumed. Every point $x \in \mathbf{R}^n$ is an *ordered vector* of the form $x = (x_1, \ldots, x_n)$ where each of the x_j is a real number. If the order changes, the vector changes. For example, the vectors (1, 2) and (2, 1) are different. Each vector can be viewed as the coordinate vector of a point in the Euclidean vector space of appropriate dimension. When all the x_j are equal to zero, the vector is known as the zero vector and it is also denoted by the symbol "0." It will always be clear from the context whether "0" refers to the real number 0 or the zero vector in some Euclidean space.

The vector $x \in \mathbf{R}^n$, is said to be *nonnegative* if each of its components, x_j, is greater than or equal to zero. Symbolically this is written as $x \geqq 0$. (Notice the two lines under the inequality sign.) Obviously $0 \geqq 0$.

The vector $x \in \mathbf{R}^n$ is said to be *semipositive* if each of the components x_j is greater than or equal to zero and at least one of them is strictly greater than zero. We denote this by $x \geq 0$. (Notice the single line under the inequality sign.) That is, $x \geq 0$ iff $x \geqq 0$ and $x \neq 0$ or, iff $x \geqq 0$ and $\sum x_j$ is strictly positive. Obviously $0 \ngeq 0$.

The vector x is said to be *positive* if each of its components, x_j, is strictly greater than zero. This is denoted by $x > 0$.

TRANSLATE

If $\mathbf{S} \subset \mathbf{R}^n$, $\hat{x} \in \mathbf{R}^n$, the *translate* of $\mathbf{S}$ to $\hat{x}$ is the set $\{x : x = \hat{x} + y, y \in \mathbf{S}\}$.

TRANSLATE LINEAR, AFFINE, AND CONVEX COMBINATIONS

Let $x = (x_1, \ldots, x_n)$ be a vector in $\mathbf{R}^n$ and α a real number. Then $\alpha x = (\alpha x_1, \ldots, \alpha x_n)$. Thus multiplication of a vector by a real number is the multiplication of each coordinate in the vector by that real number.

If $x^1 = (x_1^1, \ldots, x_n^1) \neq 0$, for $0 \leqq \alpha \leqq 1$, the vector αx^1 represents a point between 0 (the origin) and the point x^1, on the line segment joining them. For $\alpha > 1$, the point αx^1 is a point beyond x^1 on the straight line obtained by starting at the origin and joining it to x^1 and continuing in that direction. For $\alpha < 0$, the vector αx^1 is a point beyond the origin on the straight line obtained by joining x^1 to 0 and continuing in that direction. See Figure 3.2.

Let $x^1 = (x_1^1, \ldots, x_n^1)$ and $x^2 = (x_1^2, \ldots, x_n^2)$ be two vectors in $\mathbf{R}^n$. Their sum $x^1 + x^2$ is obtained by adding the corresponding coordinates in x^1 and x^2. Thus $x^1 + x^2 = (x_1^1 + x_1^2, \ldots, x_n^1 + x_n^2)$. If x^1, x^2 are distinct from 0 and not equal to each other, the three points x^1, x^2, 0 determine a unique two-dimensional subspace of $\mathbf{R}^n$. $x^1 + x^2$ is the fourth vertex of the parallelogram in this two-dimensional subspace, whose other vertices are $0, x^1, x^2$. This is known as the *parallelogram law of addition of vectors.* See Figure 3.1.

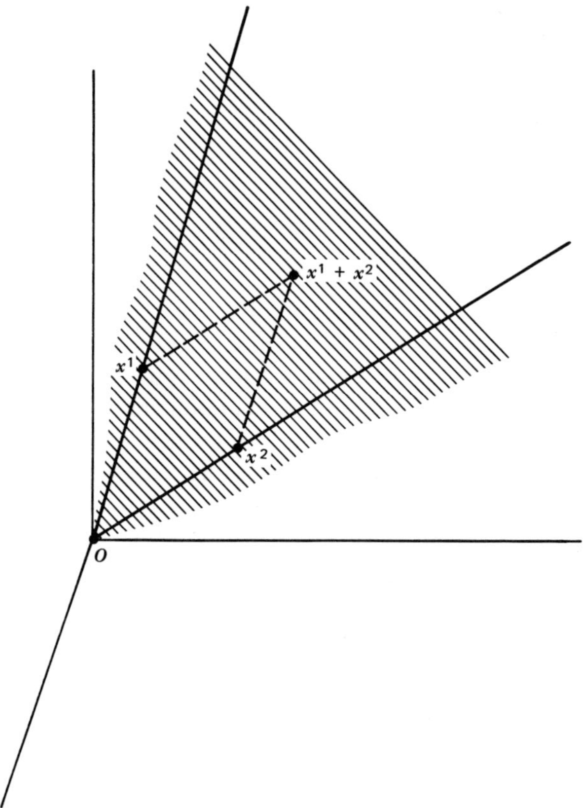

Figure 3.1 The linear hull of $\{x^1, x^2\}$ in $\mathbf{R}^3$ is the plane containing $0, x^1, x^2$.

Let $\{x^1, \ldots, x^k\}$ be a finite set of points in $\mathbf{R}^n$ where $x^r = (x_1^r, \ldots, x_n^r)$, for $r = 1$ to k. A *linear combination* of these points is any point x of the form: $x = \alpha_1 x^1 + \cdots + \alpha_k x^k$, where $\alpha_1, \ldots, \alpha_k$ are all real numbers. The set of all linear combinations of $x^1, \ldots, x^k$ is the subspace of $\mathbf{R}^n$ of smallest dimension containing $x^1, \ldots, x^k$; it is called the *linear hull* of $\{x^1, \ldots, x^k\}$. See Figures 3.1 and 3.2.

Example

Let $x^1 = (1, 0, -1)$, $x^2 = (-2, 3, 17)$. Any point of the form $x = (\alpha_1 - 2\alpha_2, 3\alpha_2, -\alpha_1 + 17\alpha_2)$ is a linear combination of x^1 and x^2.

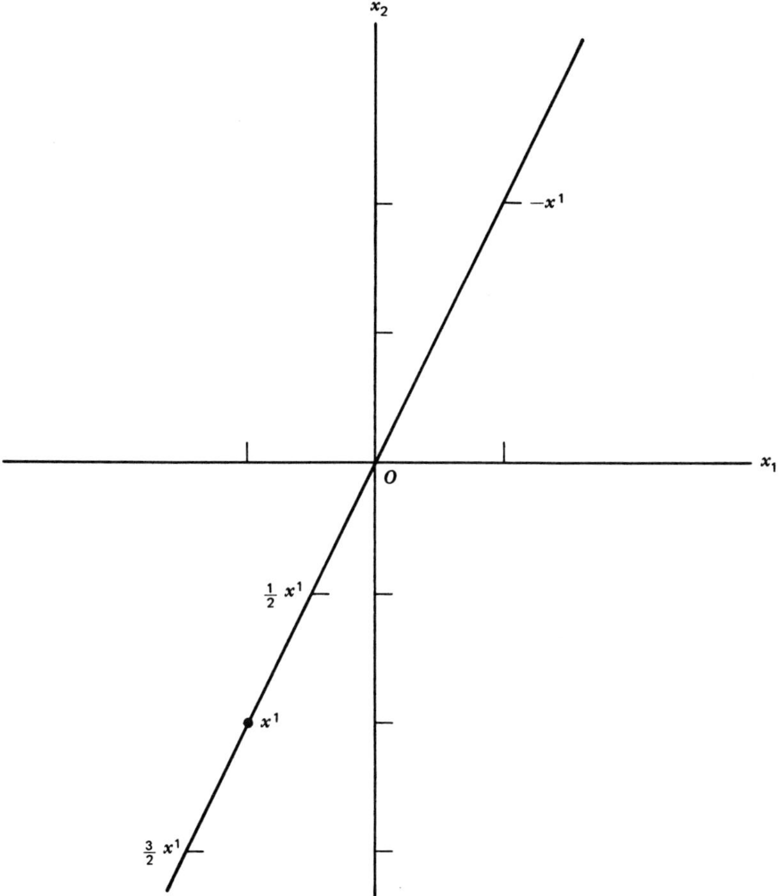

Figure 3.2 The linear hull of $\{x^1\}$ in $\mathbf{R}^2$.

Example

Let $x^1 = (-1, -2)$. The linear hull of $\{x^1\}$ is the set of all points in $\mathbf{R}^2$ that are scalar multiples of x^1. It is the set of all points on the straight line joining x^1 and the origin in Figure 3.2.

An *affine combination* of $x^1, \ldots, x^k$ is any point of the form

$$x = \alpha_1 x^1 + \cdots + \alpha_k x^k \tag{3.1}$$

where $\alpha_1, \ldots, \alpha_k$ are real numbers satisfying

$$\alpha_1 + \cdots + \alpha_k = 1 \tag{3.2}$$

The set of all affine combinations of $x^1, \ldots, x^k$ is known as the *affine hull* of $\{x^1, \ldots, x^k\}$. The affine hull of a set of points in $\mathbf{R}^n$ is a subset of its linear hull.

Clearly the point x in (3.1) can be expressed as

$$x = x^1 + (\alpha_1 - 1)x^1 + \alpha_2 x^2 + \cdots + \alpha_k x^k$$

and by (3.2) this is $x^1 + [\alpha_2(x^2 - x^1) + \cdots + \alpha_k(x^k - x^1)]$. If

$$y = \alpha_2(x^2 - x^1) + \cdots + \alpha_k(x^k - x^1),$$

y is a linear combination of $x^2 - x^1, \ldots, x^k - x^1$. As $\alpha_2, \ldots, \alpha_k$ assume all possible real values, y varies over the entire linear hull of $\{(x^2 - x^1), \ldots, (x^k - x^1)\}$.

Therefore the affine hull of $\{x^1, \ldots, x^k\}$ is the translate of the linear hull of $\{(x^2 - x^1), \ldots, (x^k - x^1)\}$ to x^1.

Example

Let $x^1 = (1, 0, 0)$, $x^2 = (0, 1, 0)$, $x^3 = (0, 0, 1)$. An affine combination of x^1, x^2, x^3 is a point of the form $x = (\alpha_1, \alpha_2, \alpha_3)$ where $\alpha_1 + \alpha_2 + \alpha_3 = 1$. Thus the affine hull of x^1, x^2, x^3 is the hyperplane of dimension 2 containing x^1, x^2, and x^3. See Figure 3.3.

Example

Let $x^1 = (3, 1)$, $x^2 = (4, 3)$. The affine hull of x^1, x^2 is the set of all points on the straight line joining x^1, x^2. Also it is the translate to x^1 of the linear hull of $\{(4, 3) - (3, 1) = (1, 2)\}$ in $\mathbf{R}^2$. See Figure 3.4.

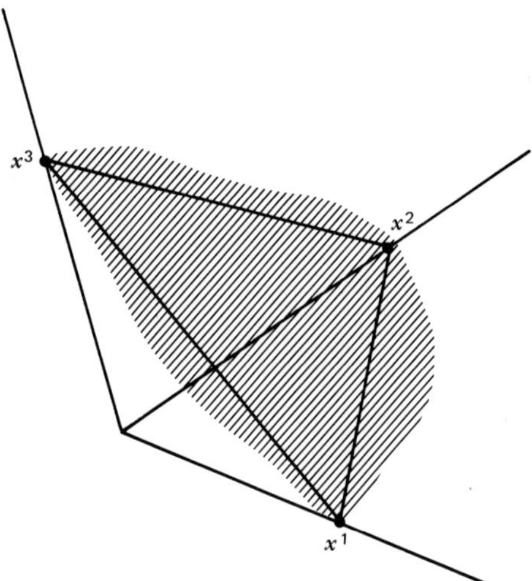

Figure 3.3 Affine hull of $\{x^1, x^2, x^3\}$ is the hyperplane containing x^1, x^2, and x^3.

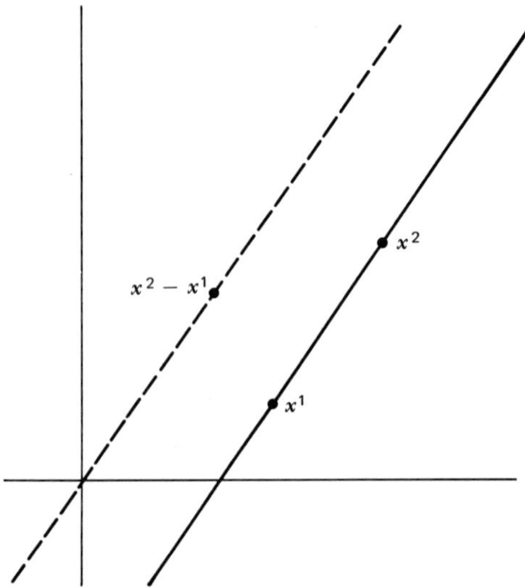

Figure 3.4 Affine hull of $\{x^1, x^2\}$ is the straight line through x^1 and x^2.

A *convex combination* of $x^1, \ldots, x^k$ is a point x of the form

$$x = \alpha_1 x^1 + \cdots + \alpha_k x^k$$

where $\alpha_1, \ldots, \alpha_k$ are real numbers satisfying:

$$\alpha_1 + \cdots + \alpha_k = 1$$

and

$$\alpha_1 \geqq 0, \ldots, \alpha_k \geqq 0$$

The set of all convex combinations of $x^1, \ldots, x^k$ is known as the *convex hull* of $\{x^1, \ldots, x^k\}$. It is a subset of the affine hull of the same set of points.

Example

If $x^1 = (1, 0)$, $x^2 = (0, 1)$ the convex hull of $\{x^1, x^2\}$ is the set of all points in $\mathbf{R}^2$ of the form (α_1, α_2) with $\alpha_1 \geqq 0$, $\alpha_2 \geqq 0$, $\alpha_1 + \alpha_2 = 1$. See Figures 3.5 and 3.6.

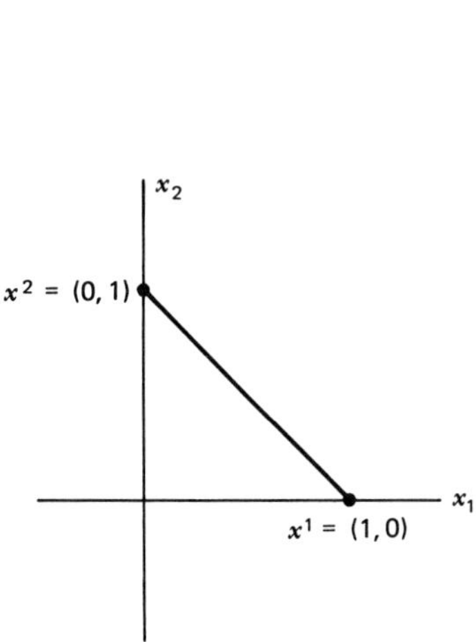

Figure 3.5 Convex hull of $\{x^1, x^2\}$.

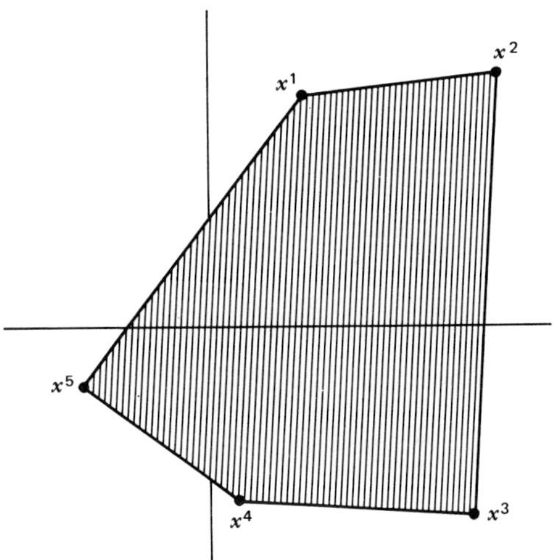

Figure 3.6 Convex hull of $\{x^1, \ldots, x^5\}$ in $\mathbf{R}^2$.

In general the convex hull of any two points in $\mathbf{R}^n$ is the set of all points on the *line segment* joining these two points.

STRAIGHT LINES, LINE SEGMENTS, RAYS AND HALF-LINES IN $\mathbf{R}^n$

A straight line in $\mathbf{R}^n$ is the locus of a point, all of whose coordinates are affine functions of a parameter θ,

$$x = \begin{pmatrix} x_1 \\ \vdots \\ x_n \end{pmatrix} = \begin{pmatrix} a_1 \\ \vdots \\ a_n \end{pmatrix} + \theta \begin{pmatrix} b_1 \\ \vdots \\ b_n \end{pmatrix}$$

as the parameter assumes all real values, where at least one of the b_i is nonzero [i.e., $(b_1, \ldots, b_n) \neq 0$]. A straight line is also the affine hull of any two points on it. If x^1 and x^2 are two distinct points in $\mathbf{R}^n$, a general point on the straight line joining them is $x(\theta) = x^1 + \theta(x^2 - x^1)$, where θ is a real-valued parameter. As θ takes all real values, $x(\theta)$ generates all the points on the straight line joining x^1 and x^2.

In $\mathbf{R}^2$, a straight line is the set of all points satisfying a single linear equation. However, for $n > 2$, the set of all points satisfying a single linear equation in $\mathbf{R}^n$ will be a hyperplane of dimension $n - 1$, and not a straight line. So this parametric representation is the most convenient method for representing straight lines in $\mathbf{R}^n$ for $n \geqq 2$.

Problem

3.1 As θ varies between 0 and 1 prove that $x(\theta) = x^1 + \theta(x^2 - x^1)$ generates all the points on the line segment joining x^1 and x^2.

Let $\tilde{x} \in \mathbf{R}^n$, $\tilde{x} \neq 0$. The *ray generated by* $\tilde{x}$ is the set $\{x: x = \lambda\tilde{x}, \lambda \geqq 0\}$. If $\hat{x} \in \mathbf{R}^n$, then the set $\{x: x = \hat{x} + \theta\tilde{x}, \theta \geqq 0\}$ is known as the *half-line* through $\hat{x}$ parallel to the ray generated by $\tilde{x}$. It starts at $\hat{x}$ (this point corresponds to $\theta = 0$) and as θ ranges from 0 to ∞, it traces the line through $\hat{x}$ parallel to the ray of $\tilde{x}$. See Figure 3.7. Every ray contains the origin, and a half-line is a translate of a ray.

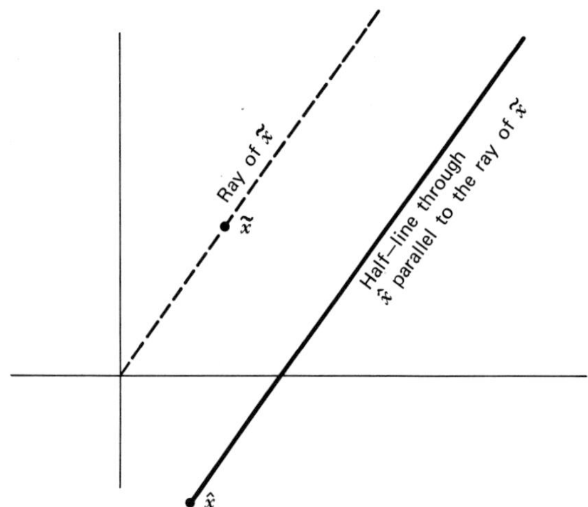

Figure 3.7

HYPERPLANE

A hyperplane in $\mathbf{R}^n$ is the set of all points $x = (x_1, \ldots, x_n) \in \mathbf{R}^n$, satisfying a single linear equation,

$$a_1x_1 + \cdots + a_nx_n = b$$

where $a_1, \ldots, a_n$, b are given numbers and at least one of the a_i is nonzero, that is, $(a_1, \ldots, a_n) \neq 0$.

HALF-SPACE

Consider an inequality constraint $a_{11}x_1 + \cdots + a_{1n}x_n \geqq b_1$, where

$$(a_{11}, \ldots, a_{1n}) \neq 0.$$

The set of all $x \in \mathbf{R}^n$ satisfying this inequality is the set of all points lying on one side of the hyperplane $a_{11}x_1 + \cdots + a_{1n}x_n = b_1$, and is known as a *half-space*. See Figure 3.8. Any equality constraint of the form

$$a_{11}x_1 + \cdots + a_{1n}x_n = b_1$$

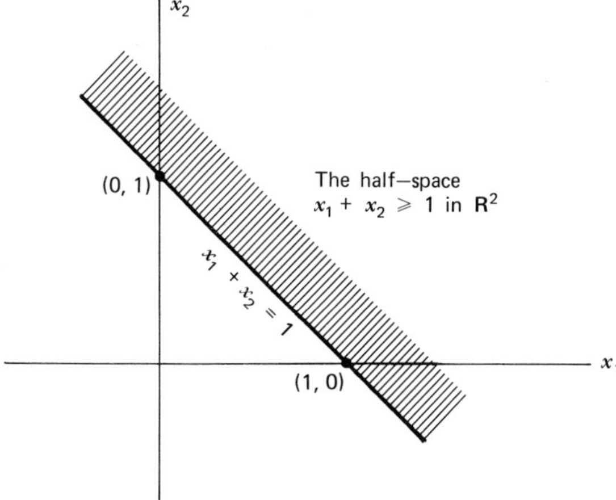

Figure 3.8

is equivalent to the pair of inequality constraints

$$a_{11}x_1 + \cdots + a_{1n}x_n \geqq b_1$$

and

$$a_{11}x_1 + \cdots + a_{1n}x_n \leqq b_1$$

both of which must hold.

In a general LP there may be some equality constraints, inequality constraints, and sign restrictions on the variables. Each sign restriction is actually an inequality constraint. Each equality constraint is equivalent to a pair of inequality constraints. Hence, all the constraints in any LP may be expressed as linear inequality constraints. The set of all points satisfying a linear inequality constraint is a half-space. Every feasible solution of the LP must satisfy all the constraints and, hence, it must be in each of the corresponding half-spaces. Hence in *linear programming problems, the set of feasible solutions is the intersection of a finite number of half-spaces.*

CONVEX SETS

A subset $\mathbf{K} \subset \mathbf{R}^n$ is said to be a *convex set* if every convex combination of any pair of points in $\mathbf{K}$ is also in $\mathbf{K}$. That is, if $\tilde{x} \in \mathbf{K}$, and $\hat{x} \in \mathbf{K}$, then $\alpha\tilde{x} + (1 - \alpha)\hat{x} \in \mathbf{K}$ for all $0 \leqq \alpha \leqq 1$. Hence, if $\mathbf{K}$ is a convex set, the line segment joining any pair of points in $\mathbf{K}$ lies entirely in $\mathbf{K}$.

Examples of convex sets in the plane are in Figure 3.9. Figure 3.10 contains examples of nonconvex sets.

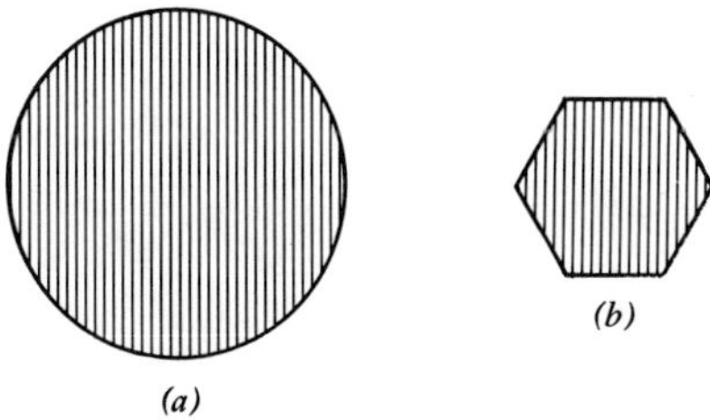

Figure 3.9 Convex sets (*a*) All points inside or on the circle. (*b*) All points inside or on the polygon.

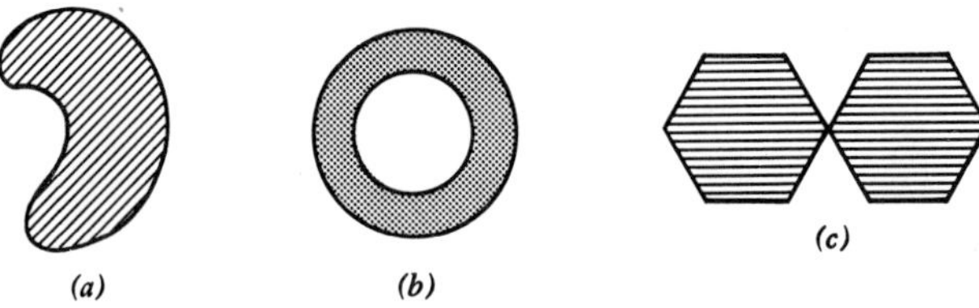

Figure 3.10 Non-convex sets (*a*) All points inside or on the cashew nut. (*b*) All points on or between two circles. (*c*) All points on at least one of the two polygons.

Problem

3.2 Prove that every half-space is a convex set.

3.3 Prove that if **K** is a convex set as defined above, and if $\{x^1,\dots,x^r\}$ is a finite set of points in **K**, then the convex hull of $\{x^1, \dots, x^r\}$ lies entirely in **K**.

3.4 Prove that the intersection of a family of convex sets is a convex set.

3.5 Prove that the set of feasible solutions of the system of constraints.

$$Ax + By = b$$
$$Ex + Fy \geqq d$$
$$x \geqq 0, \; y \text{ unrestricted}$$

is a convex set.

CONVEX POLYHEDRAL SETS AND CONVEX POLYTOPES

The intersection of a finite number of half-spaces is known as a *convex polyhedral set* or a *convex polyhedron*. From the discussion earlier, it is clear that the set of feasible solutions of an LP is a *convex polyhedron*. A convex polyhedron that is bounded is known as a *convex polytope* (Figure 3.11).

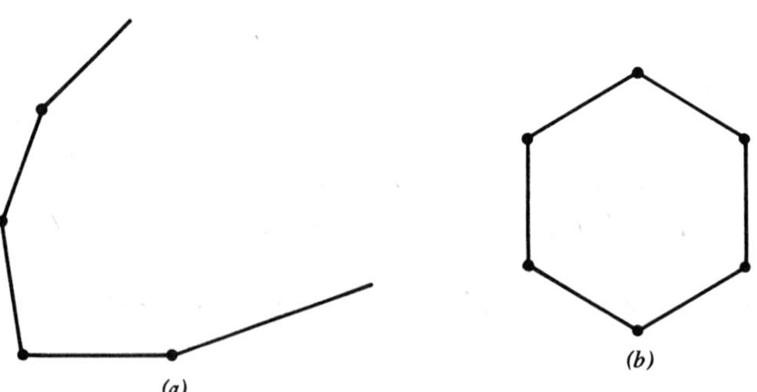

Figure 3.11 (*a*) A convex polyhedron which is not a polytope. (*b*) A convex polytope.

CONES, CONVEX CONES, CONVEX POLYHEDRAL CONES, AND POS CONES

A subset $\mathbf{S} \subset \mathbf{R}^n$ is called a *cone* iff $x \in \mathbf{S}$ implies that $\alpha x \in \mathbf{S}$ for all $\alpha \geqq 0$, that is, the ray generated by any point in a cone lies entirely in the cone.

A *convex cone* is a cone that is also a convex set. Hence if **S** is a convex cone and $x \in \mathbf{S}$, $y \in \mathbf{S}$, all convex combinations of x, y must lie in **S**. Also all nonnegative multiples of these convex combinations must also lie in **S**, since **S** is a cone.

Therefore **S** is a convex cone iff $x \in \mathbf{S}$, $y \in \mathbf{S}$ imples $\alpha x + \beta y \in \mathbf{S}$ for all $\alpha \geqq 0$, $\beta \geqq 0$.

Problem

3.6 Prove that **S** is a convex cone iff **S** contains every nonnegative linear combination of any finite number of points in **S**.

A convex cone **S** is a *convex polyhedral cone* if it is the intersection of a finite number of half-spaces. Thus every convex polyhedral cone is the set of feasible solutions of a system of constraints of the form "$Ax \geqq 0$" (Figures 3.12 to 3.14).

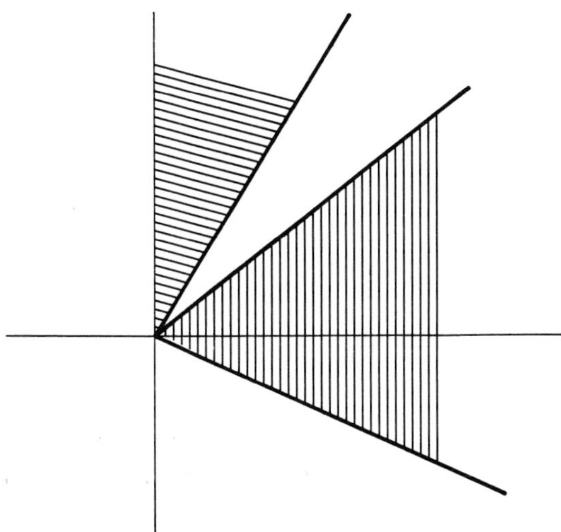

Figure 3.12 A cone in $\mathbf{R}^2$ that is not convex.

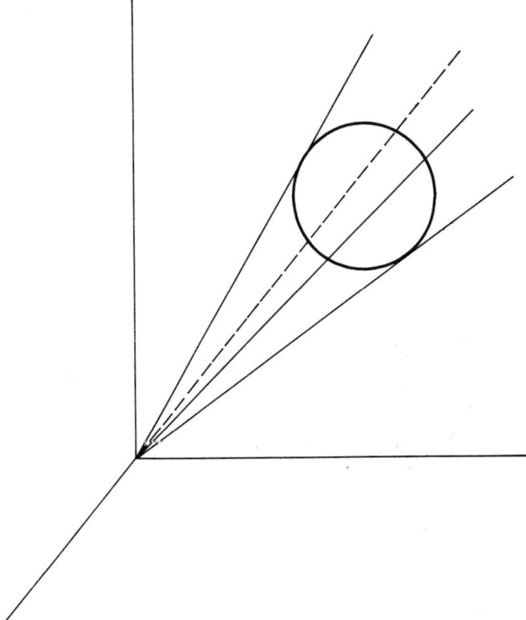

Figure 3.13 A convex cone in $\mathbf{R}^3$ that is not polyhedral.

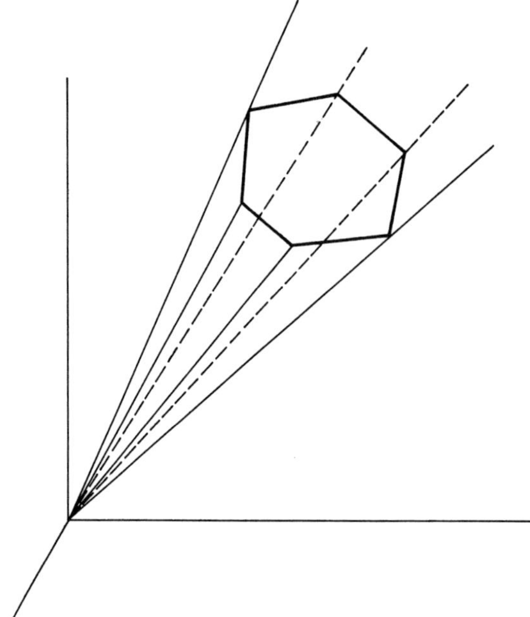

Figure 3.14 A convex polyhedral cone in $\mathbf{R}^3$.

Let $\{x^1, \ldots, x^r\}$ be a finite set of points in $\mathbf{R}^n$. The set $\{y: y = \alpha_1 x^1 + \cdots + \alpha_r x^r,$ $\alpha_1 \geqq 0, \ldots, \alpha_r \geqq 0\}$ is known as the *Pos cone of* $\{x^1, \ldots, x^r\}$, and denoted by pos $\{x^1, \ldots, x^r\}$.

3.2 MATRICES

A matrix is a rectangular array of numbers. Here is a matrix with m rows and n columns.

$$A = \begin{bmatrix} a_{11} & \cdots & a_{1j} & \cdots & a_{1n} \\ \vdots & & \vdots & & \vdots \\ a_{i1} & \cdots & a_{ij} & \cdots & a_{in} \\ \vdots & & \vdots & & \vdots \\ a_{m1} & \cdots & a_{mj} & \cdots & a_{mn} \end{bmatrix}$$

The entry in the ith row and the jth column of A is a_{ij} and it is known as the *(i, j)th entry* in A. The *order* of the matrix A is $m \times n$. When the order of A is understood, we will also denote A by (a_{ij}).

The coefficients of the variables in a system of linear equations is a matrix. For example, A is the matrix of coefficients of the variables in the system

$$\begin{array}{lcl} a_{11}x_1 + \cdots + a_{1n}x_n & = & b_1 \\ \vdots \qquad\qquad\quad \vdots & & \vdots \\ a_{m1}x_1 + \cdots + a_{mn}x_n & = & b_m \end{array}$$

If α is a real number, αA is the matrix obtained by multiplying each entry in A by α. Thus the (i, j)th entry in αA is αa_{ij}.

Two matrices $A = (a_{ij})$ and $B = (b_{ij})$ can only be added if they are of the same order. In this case $A + B = C$, where C is of the same order as A and B and the (i, j)th entry in C is $c_{ij} = a_{ij} + b_{ij}$. Thus matrix addition consists of adding the entries in corresponding positions.

The transpose of the matrix A, denoted by A^{T}, is obtained by recording each row vector of A as a column. Thus

$$A^{\mathrm{T}} = \begin{bmatrix} a_{11} & \cdots & a_{i1} & \cdots & a_{m1} \\ \vdots & & \vdots & & \vdots \\ a_{1j} & \cdots & a_{ij} & \cdots & a_{mj} \\ \vdots & & \vdots & & \vdots \\ a_{1n} & \cdots & a_{in} & \cdots & a_{mn} \end{bmatrix}$$

Hence, if A is of order $m \times n$, A^{T} will be of order $n \times m$.

A row vector in $\mathbf{R}^n$ can be treated as a matrix of order $1 \times n$. Likewise a column vector in $\mathbf{R}^n$ can be treated as a matrix of order $n \times 1$.

A square matrix is a matrix that has the same number of rows and columns. The matrix

$$D = \begin{pmatrix} d_{11} & \cdots & d_{1m} \\ \vdots & & \vdots \\ d_{m1} & \cdots & d_{mm} \end{pmatrix}$$

is a square matrix. D is said to be of *order m*. The entries $d_{11}, d_{22}, \ldots, d_{mm}$ are the *diagonal entries in D*. They constitute its *principal diagonal*. All the other entries in D are its off-diagonal entries.

A *unit matrix* or *identity matrix* is a square matrix in which all diagonal entries are equal to one, and all off-diagonal entries are zero. Thus

$$I = \begin{pmatrix} 1 & 0 & 0 \\ 0 & 1 & 0 \\ 0 & 0 & 1 \end{pmatrix}$$

is the unit matrix of order 3.

The rules for multiplication among matrices have evolved out of the study of linear transformation on systems of linear equations. Matrix multiplication is not commutative, that is, the order in which matrices are multiplied is very important. If A is a matrix of order $m \times n$ and B is a matrix of order $r \times s$, the product AB is defined only if $n = r$. In this case the product AB is a matrix C of order $m \times s$, where the (i, j)th entry in C is $c_{ij} = \sum_{t=1}^{n} a_{it}b_{tj}$. Hence the (i, j)th entry in AB is the sum of the products of the entries in the ith row in A, by the corresponding entry in the jth column of B. Therefore, matrix multiplication is known as *row by column multiplication.*

Example

Let

$$A = \begin{pmatrix} 1 & -1 & -5 \\ 2 & -3 & -6 \end{pmatrix} \qquad B = \begin{pmatrix} -2 & -4 \\ -6 & -4 \\ 3 & -2 \end{pmatrix}$$

Then,

$$AB = \begin{pmatrix} -11 & 10 \\ -4 & 16 \end{pmatrix}.$$

When the product AB exists, the product BA may not be defined, and even if it is defined, AB and BA may not be equal.

If A is a matrix of order $m \times n$, and π, x are vectors, the products πA and Ax exist only if π is a row vector in $\mathbf{R}^m$ and x is a column vector in $\mathbf{R}^n$.

3.3 LINEAR INDEPENDENCE OF A SET OF VECTORS AND SIMULTANEOUS LINEAR EQUATIONS

A set of vectors $\{A_{.1}, \ldots, A_{.k}\}$ in $\mathbf{R}^n$ is said to be *linearly dependent* iff it is possible to express the zero vector as a nonzero linear combination of vectors in the set; that is, iff there exist real numbers $\alpha_1, \ldots, \alpha_k$, not all zero such that

$$0 = \alpha_1 A_{.1} + \cdots + \alpha_k A_{.k}$$

The set $\{A_{.1}, \ldots, A_{.k}\}$ is said to be *linearly independent* if it is not linearly dependent.

Note From the definition it is clear that the empty set of vectors is linearly independent.

Thus a nonempty set of vectors $\{A_{.1}, \ldots, A_{.k}\}$ is linearly independent iff the system of simultaneous linear equations in $\alpha_1, \ldots, \alpha_k$,

$$\alpha_1 A_{.1} + \cdots + \alpha_k A_{.k} = 0$$

has the unique solution, $\alpha_1 = \alpha_2 = \cdots = \alpha_k = 0$.

Problem

3.7 Prove that any set of vectors containing the zero vector is linearly dependent.

3.3.1 An Algorithm for Testing Linear Independence

Let $\{A_{.1}, \ldots, A_{.k}\}$ be a set of vectors in $\mathbf{R}^m$ where $A_{.j} = (a_{1j}, \ldots, a_{mj})^{\mathrm{T}}$ for $j = 1$ to k. Suppose it is required to test whether this set is linearly independent. Write each vector in the set as a row vector in a tableau.

Tableau

Vector			
$A_{.1}$	a_{11}	$\cdots\cdots$	a_{m1}
$\vdots$	$\vdots$		$\vdots$
$A_{.k}$	a_{1k}	$\cdots\cdots$	a_{mk}

In the course of the algorithm the rows in the tableau will be altered. But at every step, each row in the tableau will be a nonzero linear combination of the vectors in the original set $\{A_{.1}, \ldots, A_{.k}\}$. The expression of each row in the current tableau as a nonzero linear combination of vectors in the set $\{A_{.1}, \ldots, A_{.k}\}$ will be indicated on the left-hand side of the tableau.

A convenient method for keeping track of the expression of each row of the tableau as a linear combination of $A_{.1}, \ldots, A_{.k}$, is to store this expression in detached coefficient form. To do this, the original tableau is recorded as below.

Original Tableau

$A_{.1}$	$A_{.2} \ldots A_{.k}$		
1	$0 \ldots 0$	a_{11}	$a_{21} \ldots a_{m1}$
0	$1 \ldots 0$	a_{12}	$a_{22} \ldots a_{m2}$
$\vdots$	$\vdots \quad \vdots$	$\vdots$	$\vdots \quad \vdots$
0	$0 \ldots 1$	a_{1k}	$a_{2k} \ldots a_{mk}$

The matrix on the left-hand side here is the identity matrix of order k. The ith row in this left-hand portion of the tableau contains the coefficients of $A_{.1}, \ldots, A_{.k}$ in the expression for the vector in this row on the right-hand portion of the tableau.

Whenever row operations are carried out, they are also performed on the left-hand portion of the tableau. In a general stage of the algorithm, if the entries in the ith row of the tableau are:

β_{i1}	$\beta_{i2} \ldots \beta_{ik}$	$\bar{a}_{1i}$	$\bar{a}_{2i} \ldots \bar{a}_{mi}$

this implies that the vector $(\bar{a}_{1i}, \ldots, \bar{a}_{mi})$ is $\sum_{j=1}^{k} \beta_{ij} A_{.j}$. The algorithm starts with the original tableau.

Step 1 Pick row 1 for examining.

Step 2 If all the entries in the right-hand portion of the tableau in the row being examined are zero, the left-hand portion in this row provides an expression for the zero vector as a nonzero linear combination of the vectors $A_{.1}, \ldots, A_{.k}$; hence, the set is linearly dependent. If not, go to Step 3. The row that is being examined becomes the *pivot row* in Step 3.

Step 3 If the pivot row is the last row in the tableau, conclude that the set $\{A_{.1}, \ldots, A_{.k}\}$ is linearly independent and terminate. Otherwise, go to Step 4.

Step 4 Pick a nonzero element in the right-hand portion of the tableau in the pivot row. Choose this as the pivot element, and perform a pivot in its column in the tableau, with that element as the pivot element. The column in the current tableau, in which the pivot element lies, is known as the *pivot column* for this stage.

Pivoting requires subtracting suitable multiples of the pivot row from the other rows in the tableau, so as to transform all the entries in the pivot column, except the pivot element, into zero, and then dividing the pivot row by the pivot element. The same operations are performed on the left-hand side of the tableau.

From the nature of the pivot operation it is clear that each of the rows in the new tableau is a nonzero linear combination of the original vectors $A_{.1}, \ldots, A_{.k}$. In the new tableau pick the next row for examining and go back to Step 2.

In at most k pivot steps the algorithm will either conclude that the set $\{A_{.1}, \ldots, A_{.k}\}$ is linearly independent, or it finds an expression for the zero vector as a nonzero linear combination of the given vectors $\{A_{.1}, \ldots, A_{.k}\}$.

Example

Find out whether the set of column vectors of the matrix A is a linearly independent set.

$$A = \begin{bmatrix} 0 & 1 & 1 & 2 \\ 1 & 0 & 1 & 2 \\ 2 & 1 & 1 & 4 \\ 0 & 1 & 1 & 2 \\ 1 & 0 & 1 & 2 \end{bmatrix}$$

Let $A_{.j}$ be the jth column vector of A. Here is the tableau.

Vector				Pivot column					
$A_{.1}$	$A_{.2}$	$A_{.3}$	$A_{.4}$						
1	0	0	0	0	(1)	2	0	1	Pivot row
0	1	0	0	1	0	1	1	0	
0	0	1	0	1	1	1	1	1	
0	0	0	1	2	2	4	2	2	

Pick the first row as the pivot row and the nonzero element "1" in the second column as the pivot element. The second column in the tableau is the pivot column for this step. In this case, the pivoting operation requires (a) subtracting the first row from the third row and (b) subtracting twice the first row from the fourth row on the tableau. This transforms the tableau as follows:

Vector				Pivot column					
$A_{.1}$	$A_{.2}$	$A_{.3}$	$A_{.4}$						
1	0	0	0	0	1	2	0	1	
0	1	0	0	(1)	0	1	1	0	Pivot row
−1	0	1	0	1	0	−1	1	0	
−2	0	0	1	2	0	0	2	0	

Continuing, we find that the tableau is again transformed into:

Vector						Pivot column			
$A_{.1}$	$A_{.2}$	$A_{.3}$	$A_{.4}$						
1	0	0	0	0	1	2	0	1	
0	1	0	0	1	0	1	1	0	
−1	−1	1	0	0	0	(−2)	0	0	Pivot row
−2	−2	0	1	0	0	−2	0	0	

Vector								
$A_{.1}$	$A_{.2}$	$A_{.3}$	$A_{.4}$					
0	-1	1	0	0	1	0	0	1
$-\frac{1}{2}$	$\frac{1}{2}$	$\frac{1}{2}$	0	1	0	0	1	0
$\frac{1}{2}$	$\frac{1}{2}$	$-\frac{1}{2}$	0	0	0	1	0	0
-1	-1	-1	1	0	0	0	0	0

The last row in the right-hand side of the tableau has become the zero vector. So

$$-A_{.1} \quad -A_{.2} \quad -A_{.3} \quad +A_{.4} = 0$$

which exhibits the linear dependence of the set of column vectors $\{A_{.1}, A_{.2}, A_{.3}, A_{.4}\}$

Example

From the above example it is clear that the set of column vectors $\{A_{.1}, A_{.2}, A_{.3}\}$ is linearly independent.

Problem

3.8 Is the set of column vectors of the following matrix D linearly independent?

$$D = \begin{bmatrix} 1 & 2 & 1 & 1 \\ 0 & 0 & 1 & 0 \\ 2 & 3 & 0 & 2 \\ 1 & 1 & 0 & 3 \\ 3 & 0 & 0 & 0 \end{bmatrix}$$

3.3.2 The Inverse of a Nonsingular Square Matrix

Inverse matrices are only defined for square matrices. If D is a square matrix of order m, its inverse is defined to be a square matrix E of order m satisfying $ED = DE = I$, where I is the unit matrix of order m. The inverse of D exists iff the set of row vectors of D is linearly independent, and such matrices are known as *nonsingular square matrices.* A square matrix in which the set of row vectors is linearly dependent is know as a *singular square matrix*. If D is a nonsingular square matrix, its inverse is normally denoted by D^{-1}.

A square matrix that can be transformed into the unit matrix by rearranging its row vectors is known as a *permutation matrix*. The following is a permutation matrix of order 4.

$$\begin{bmatrix} 0 & 0 & 0 & 1 \\ 1 & 0 & 0 & 0 \\ 0 & 1 & 0 & 0 \\ 0 & 0 & 1 & 0 \end{bmatrix}$$

If P is a permutation matrix of order m and A is any matrix of order $m \times n$, then PA is a matrix that is obtained by permuting the row vectors of A. That's why P is called a permutation matrix.

Problem

3.9 If P is a permutation matrix and P^T is its transpose, prove that $PP^T = P^TP = I$, the identity matrix. Hence, if P is a permutation matrix, $P^{-1} = P^T$.

Let B be a square matrix of order m, and b be a given column vector of order $m \times 1$. Consider the following problems:

1. To find B^{-1}.
2. To find the unique expression for b as a linear combination of the column vectors of B.

We discuss a method for solving these problems by performing row operations. Let I be the unit matrix of order m. Start with the following initial tableau.

Initial Tableau

B	I	b

Step 1 Pick the first row of the tableau to examine.

Step 2 The row being examined is known as the *Pivot row*.

If the pivot row consists of zero entries in all the first m columns of the current tableau, the set of row vectors of B is not linearly independent and B must be singular. Hence, B^{-1} does not exist. Terminate.

Otherwise, pick a nonzero entry in the pivot row among the first m columns of the tableau and make it the pivot element. The column containing the pivot element is known as the *pivot column*. By subtracting suitable multiples of the pivot row from all the other rows, transform all the entries in the pivot column, excepting the pivot element, into zero. Finally divide the pivot row by the pivot element.

Step 3 If the pivot row is the last row in the tableau, go to Step 4. Otherwise, pick the next row to examine and go back to Step 2.

Step 4 If all the rows in the tableau have been examined, the row operations performed so far would have transformed the matrix contained among the first m column vectors of the tableau (which was B originally) into a *permutation matrix*. Let this matrix be P.

Rearrange the rows of the tableau so that P becomes the identity matrix. This operation is equivalent to multiplying every column in the tableau on the left by P^T.

The square matrix contained among the $m + 1$ to $2m$ columns of the final tableau is B^{-1}. (Notice that the initial tableau contained the identity matrix in this same position.)

Let $B_{.j}$ denote the jth column vector of the original matrix, B. If the last column in the final tableau is $\bar{b} = (\bar{b}_1, \ldots, \bar{b}_m)^T$, then $b = \sum \bar{b}_j B_{.j}$; this gives the unique expression of b as a linear combination of the column vectors of B. $\bar{b}$ is known as the *representation* of b as a linear combination of the column vectors of B. Actually, $\bar{b} = B^{-1}b$.

Problem

3.10 Find the inverse of

$$\begin{bmatrix} 0 & 2 & 3 & 1 \\ 1 & 0 & 2 & 2 \\ 0 & -1 & 0 & -3 \\ 1 & 0 & -2 & 0 \end{bmatrix}$$

and find the unique expressions for $(1, 2, 3, 4)^T$ and $(-2, 1, 2, 1)^T$ as linear combinations of the column vectors of the above matrix.

3.11 Apply the inverse finding algorithm on this matrix:

$$B = \begin{bmatrix} 0 & 1 & -1 & -1 \\ 0 & 0 & 2 & 3 \\ 1 & 2 & 0 & 1 \\ -1 & 0 & 0 & 0 \end{bmatrix}$$

If $A = (a_{ij})$ is a matrix of order $m \times n$, we will denote by $A_{i.}$ the ith row vector of A, and by $A_{.j}$, the jth column vector of A. Thus

$$\begin{aligned} A_{i.} &= (a_{i1}, \ldots, a_{in}) \\ A_{.j} &= (a_{1j}, \ldots, a_{mj})^T \end{aligned}$$

Let $\{i_1, \ldots, i_r\} \subset \{1, \ldots, m\}$ and $\{j_1, \ldots, j_s\} \subset \{1, \ldots, n\}$. Then the matrix

$$\begin{pmatrix} a_{i_1, j_1} & a_{i_1, j_2} & \cdots & a_{i_1, j_s} \\ a_{i_2, j_1} & a_{i_2, j_2} & \cdots & a_{i_2, j_s} \\ a_{i_r, j_1} & a_{i_r, j_2} & \cdots & a_{i_r, j_s} \end{pmatrix}$$

is known as a *submatrix* of the matrix A generated by the subset $\{i_1, \ldots, i_r\}$ of rows and the subset $\{j_1, \ldots, j_s\}$ of columns.

Let $\mathbf{\Gamma}$ be a subset of row vectors of A. $\mathbf{\Gamma}$ is said to be a *maximal linearly independent subset of row vectors of A* if it satisfies the following properties: (1) it is linearly independent and (2) either $\mathbf{\Gamma}$ contains all the row vectors of A, or, including any other row vector of A that is not already in $\mathbf{\Gamma}$, into $\mathbf{\Gamma}$ makes it linearly dependent.

Clearly, if $\mathbf{\Gamma}$ is a maximal linearly independent subset of row vectors of A, every row vector of A can be expressed as a linear combination of row vectors in $\mathbf{\Gamma}$. Here is a fundamental result from linear algebra: given a matrix A, every maximal linearly independent subset of row vectors of A contains the same number of row vectors in it. This number (the number of row vectors in any maximal linearly independent subset of row vectors of A) is known as the *rank* of the matrix A.

Given a matrix A of order $m \times n$, its rank can be found by performing pivots on it. The following procedure not only finds the rank of A, it also finds a maximal linearly independent subset of rows of A and an expression for each of the remaining rows of A as a linear combination of rows from this subset.

Step 1 Enter the matrix in the form of a tableau, and put a unit matrix of order m on its left-hand side. Label the ith column vector of this unit matrix by $A_{i.}$

$A_{1.}$	$A_{2.} \ldots A_{i.} \ldots A_{m.}$		
1	$0 \ldots 0 \ldots 0$	a_{11}	$a_{12} \ldots a_{1n}$
0	$1 \ldots 0 \ldots 0$	a_{21}	$a_{22} \ldots a_{2n}$
$\vdots$	$\vdots \ \ddots \ \vdots \quad \vdots$	$\vdots$	$\vdots \quad \vdots$
0	$0 \ldots 1 \ldots 0$	a_{i1}	$a_{i2} \ldots a_{in}$
$\vdots$	$\vdots \quad \vdots \ \ddots \ \vdots$	$\vdots$	$\vdots \quad \vdots$
0	$0 \ldots \quad \ldots 1$	a_{m1}	$a_{m2} \ldots a_{mn}$

As in section 3.3.1, at any stage of the algorithm, the ith row on the left-hand side of the tableau contains the coefficients of $A_{1.}, \ldots, A_{m.}$, in the expression for the vector in this row on the right-hand portion of this tableau. That is, if the entries in the ith row of the tableau in a general stage of the algorithm are:

β_{i1}	$\beta_{i2} \ldots \beta_{im}$	$\bar{a}_{i1}$	$\bar{a}_{i2} \ldots \bar{a}_{in}$

then $(\bar{a}_{i1}, \bar{a}_{i2}, \ldots, \bar{a}_{in}) = \sum_{r=1}^{m} \beta_{ir} A_{r.}$ The algorithm starts with the original tableau. Select row 1 of the tableau for examining and go to Step 2.

Step 2 Let t be the number of the row being examined in the current tableau. If all the entries in the right-hand portion of the current tableau in row t are zero, go to Step 5 if $t = m$, or to Step 4 if $t < m$.

If there are some nonzero entries in the right-hand portion of the current tableau in row t, go to Step 5 if $t = m$, or select this row as the pivot row and go to Step 3 if $t < m$.

Step 3 Select a nonzero entry in the right-hand portion of the tableau in the pivot row and make it the pivot element. The column in which the pivot element lies is the pivot column.

Perform the pivot. This consists of subtracting suitable multiples of the pivot row from the other rows in the tableau so as to transform all the entries in the pivot column, excepting the pivot element, into zero, and then dividing the pivot row by the pivot element. These row operations are performed on both the right-hand and left-hand portions of the tableau. After the pivot go to Step 4.

Step 4 Select the next row for examining and go to Step 2.

Step 5 Identify the rows in the tableau at this stage that have nonzero entries in the right-hand portion of the tableau. Suppose these are rows $i_1, \ldots, i_r$. Then the rank of the matrix A is r, and $\{A_{i_1.}, A_{i_2.}, \ldots, A_{i_r.}\}$ is a maximal linearly independent subset of row vectors of A.

If row t in the tableau at this stage has all zero entries in the right-hand portion, and if $\beta_{t1}, \ldots, \beta_{tm}$ are the entries in the left-hand portion of the tableau at this stage in row t, then

$$\sum_{u=1}^{m} \beta_{tu} A_{u.} = 0$$

β_{tt} will be equal to 1 and

$$\sum_{\substack{u=1\\u\neq t}}^{m} (-\beta_{tu})A_{u.}$$

is an expression for $A_{t.}$, as a linear combination of row vectors from the maximal linearly independent subset obtained above.

Example

Find the rank of the matrix A given below, a maximal linearly independent subset of row vectors of it, and an expression for each of its other rows as a linear combination of rows from this maximal set.

$$A = \begin{bmatrix} 1 & 2 & -1 & 1 & 0 & -2 \\ 2 & 4 & -2 & 2 & 0 & -4 \\ 0 & 0 & 3 & 1 & 2 & 2 \\ 2 & 4 & 1 & 3 & 2 & -2 \\ 1 & -1 & -1 & 1 & -2 & 1 \end{bmatrix}$$

The original tableau is the first tableau given below. In each tableau, the pivot element is circled.

$A_{1.}$	$A_{2.}$	$A_{3.}$	$A_{4.}$	$A_{5.}$						
1	0	0	0	0	(1)	2	−1	1	0	−2
0	1	0	0	0	2	4	−2	2	0	−4
0	0	1	0	0	0	0	3	1	2	2
0	0	0	1	0	2	4	1	3	2	−2
0	0	0	0	1	1	−1	−1	1	−2	1
1	0	0	0	0	1	2	−1	1	0	−2
−2	1	0	0	0	0	0	0	0	0	0
0	0	1	0	0	0	0	3	(1)	2	2
−2	0	0	1	0	0	0	3	1	2	2
−1	0	0	0	1	0	−3	−2	0	−2	3
1	0	−1	0	0	1	2	−4	0	−2	−4
−2	1	0	0	0	0	0	0	0	0	0
0	0	1	0	0	0	0	3	1	2	2
−2	0	−1	1	0	0	0	0	0	0	0
−1	0	0	0	1	0	−3	−2	0	−2	3

The nonzero rows in the right-hand portion of the terminal tableau are rows 1, 3, 5. Hence $\{A_{1.}, A_{3.}, A_{5.}\}$ is a maximal linearly independent subset of row vectors of A, and the rank of A is 3. From rows 2 and 4 in the final tableau, we have

$$-2A_{1.} + A_{2.} = 0$$

and

$$-2A_{1.} - A_{3.} + A_{4.} = 0$$

These equations give, $A_{2.} = 2A_{1.}$ and $A_{4.} = 2A_{1.} + A_{3.}$, which are the required expressions.

In the above discussion we only used the row vectors of A and, hence, the rank defined there might appropriately be called the *row rank of the matrix A*. The *column rank of the matrix A* can be defined in a similar manner by replacing the word "row" in the above discussion by the word "column." This will be the same as the row rank of the matrix A^{T}. The *rank theorem* in linear algebra states that the row rank of any matrix is equal to its column rank. This number is therefore called the rank of the matrix. If A is of order $m \times n$, and rank r, clearly $r \leqq m$ and $r \leqq n$.

Let $\mathbf{F} = \{A_1, \ldots, A_k\}$ be a nonempty set of vectors in $\mathbf{R}^n$. A subset $\mathbf{E} \subset \mathbf{F}$, is said to be a *maximal linearly independent subset of* $\mathbf{F}$ if (i) $\mathbf{E}$ is linearly independent, and (ii) either $\mathbf{E} = \mathbf{F}$, or for every $A_j \in \mathbf{F} \backslash \mathbf{E}$, $\mathbf{E} \cup \{A_j\}$ is linearly dependent. A fundamental theorem in linear algebra states that all maximal linearly independent subsets of $\mathbf{F}$ have the same cardinality and this number is known as the *rank* of the set $\mathbf{F}$. See references [9 to 13].

3.3.3 Systems of Simultaneous Linear Equations

Let A be an $m \times n$ matrix and b an $m \times 1$ column vector. The system

$$Ax = b \tag{3.3}$$

is a system of m linear equations in n unknowns. The ith equation in this system is

$$A_{i.}x = b_i \qquad \text{for } i = 1 \text{ to } m$$

This system of equations can also be viewed as

$$\sum_{j=1}^{n} A_{.j}x_j = b$$

Hence, if x is a solution of this system, it is the vector of coefficients in an expression of b as a linear combination of the column vectors of A. Thus (3.3) has a solution iff b can be expressed as a linear combination of the column vectors of A. This leads to the following standard result of linear algebra.

Result (3.3) has a solution iff the rank of the augmented matrix $(A \vdots b)$ is equal to the rank of A. If rank $(A \vdots b) = 1 + \text{rank}(A)$, (3.3) has no solution.

Problem

3.12 If A is a square nonsingular matrix, prove that (3.3) has a solution and that this solution is unique.

The system of equations (3.3) is said to be *consistent* if it has a solution. Otherwise it is *inconsistent*. Let the rank of A = rank of $(A \vdots b) = r$. If $r = m$, all the equations in the system (3.3) are said to be linearly independent. In this case (3.3) is known as a *linearly independent system of equations*. This implies that there are no redundant equations in it. If $r < m$, system (3.3) has exactly $m - r$ redundant equations. By performing elementary row operations, (3.3) can be transformed into an equivalent system of r equations. Since r is the rank of the matrix A, $r \leqq m$ and $r \leqq n$. Hence, every consistent system of linear equations is equivalent to a system in which the number of equations is less than or equal to the number of unknowns.

The system of constraints (excluding the sign restrictions on variables) in an LP in standard form is a system of equations. Hence in discussing LPs in standard form, we can assume that the number of equality constraints is less than or equal to the number of variables.

The system of equations (3.3) can be solved by transforming it into *row echelon normal form*. The row echelon normal form represents an equivalent system of equations obtained by performing row operations on (3.3) and eliminating any redundant equations. Perform the following steps.

Step 1 Express (3.3) in detached coefficient form as in the tableau below.

$x_1 \ldots x_j \ldots x_n$	b
$a_{11} \ldots a_{1j} \ldots a_{1n} =$	b_1
$\vdots \quad \vdots \quad \vdots$	$\vdots$
$a_{i1} \ldots a_{ij} \ldots a_{in} =$	b_i
$\vdots \quad \vdots \quad \vdots$	$\vdots$
$a_{m1} \ldots a_{mj} \ldots a_{mn} =$	b_m

Pick the first row of the tableau for examining and go to Step 2.

Step 2 The row being examined in the present tableau is known as the *pivot row* for this stage. If all the entries in the pivot row on both sides of the tableau are zero, this row corresponds to a redundant equation in the system. Eliminate this row from the tableau and go to Step 3.

If all the entries in the pivot row in the present tableau on the left-hand side are zero, and the entry in the pivot row on the right-hand side is nonzero, this row represents the inconsistent equation, "0 = a nonzero number." Since this equation is obtained as a linear combination of equations from the original system, that system must be inconsistent. Hence, the system has no solution and we terminate.

If a nonzero entry exists in the pivot row on the left-hand side of the present tableau, pick any one of those entries as the pivot element. The column in which the pivot element lies is the *pivot column*. The variable corresponding to the pivot column is known as the *dependent variable* (*or basic variable* in linear programming terminology) *in the present pivot row.*

Perform the pivot. (In linear algebra, this operation is known as *Gauss-Jordan elimination.*) The pivot operation transforms the pivot column into the column vector of the unit matrix with a "1" entry in the pivot row and zero entries elsewhere, by using elementary row operations. Then go to Step 3.

Step 3 If there are some more rows in the tableau to be examined, pick the next row for examining and go to Step 2.

If all the rows have been examined and the system is not inconsistent, let the final tableau be as below.

Final Tableau

Dependent variable in the row	x_1	$x_2 \ldots x_n$	b
	$\bar{a}_{11}$	$\bar{a}_{12} \ldots \bar{a}_{1n}$	$\bar{b}_1$
	$\vdots$	$\vdots \quad \vdots$	$\vdots$
	$\bar{a}_{r1}$	$\bar{a}_{r2} \ldots \bar{a}_{rn}$	$\bar{b}_r$

The number of rows in this tableau, r, will be equal to m, if no redundant constraints were found in Step 2. Otherwise, $m - r$ is the number of redundant constraints eliminated from the system. The rank of the matrix A is r.

All the variables other than the dependent variables are known as *independent variables* (*nonbasic variables* in linear programming terminology).

The matrix consisting of the column vectors of the dependent variables in the final tableau is a permutation matrix. Hence a solution to the problem is obtained by giving arbitrary values to the independent variables, substituting these values in the final tableau, and then obtaining the values of the dependent variables. In particular, a solution is

$$\begin{aligned} \text{All independent variables} &= 0 \\ \text{Dependent variables in } i\text{th row} &= \bar{b}_i \qquad i = 1 \text{ to } r \end{aligned}$$

Note 1 This algorithm for solving a system of linear equations cannot guarantee that the solution obtained will be nonnegative. Hence this algorithm cannot be used directly to find a *feasible solution* (i.e., a nonnegative solution) of an LP in standard form.

Note 2 The algorithm (based on Gauss-Jordan elimination) discussed here for solving systems of linear equations may not perform very well from the numerical analysis point of view. Round-off errors can accumulate in this algorithm. The newer algorithms discussed in numerical analysis books reduce the round-off error accumulation. Most computer programs in use today are based on these newer algorithms. See Chapter 17, and also reference [12].

Problems

3.13 Solve the following systems:

(a)
$$\begin{aligned} 2x_1 - 3x_2 + 4x_3 &= 8 \\ 6x_1 + 5x_2 - 7x_3 &= 4 \end{aligned}$$

(b)
$$\begin{aligned} x_1 - x_2 + x_3 - x_4 + x_5 &= 3 \\ 2x_1 + 3x_3 + 2x_4 + 2x_5 &= -2 \\ 4x_1 - 2x_2 + 5x_3 + 4x_5 &= 2 \end{aligned}$$

(c)
$$\begin{aligned} x_1 + 2x_2 + 3x_3 &= 2 \\ -x_1 + x_2 - 2x_3 &= 1 \\ x_1 + 5x_2 + 4x_3 &= 5 \\ 3x_2 + x_3 &= 3 \\ -x_1 + 4x_2 - x_3 &= 4 \end{aligned}$$

3.14 Consider this LP in which there are no inequality constraints or sign restrictions.

$$\begin{aligned} \text{Minimize} \quad & z(x) = cx \\ \text{Subject to} \quad & Ax = b \end{aligned}$$

Prove that either $z(x)$ is a constant on the set of feasible solutions of this system

$$\left[\text{which happens when rank} \begin{pmatrix} A \\ \cdots \\ c \end{pmatrix} = \text{rank}\ (A)\right]$$

or that if there exist two feasible solutions that give different values to $z(x)$, then $z(x)$ is unbounded below on the set of feasible solutions of this system

$$\left[\text{which happens when rank} \begin{pmatrix} A \\ \cdots \\ c \end{pmatrix} = 1 + \text{rank}\ (A)\right].$$

3.4 BASIC FEASIBLE SOLUTIONS AND EXTREME POINTS OF CONVEX POLYHEDRA

3.4.1 Extreme Points

Let $\mathbf{\Gamma}$ be a convex subset of $\mathbf{R}^n$. A point $\bar{x} \in \mathbf{\Gamma}$ is said to be an *extreme point* of $\mathbf{\Gamma}$ if it is impossible to express it as a convex combination of two other *distinct points* in $\mathbf{\Gamma}$. That is, $\bar{x}$ is an extreme point of $\mathbf{\Gamma}$ iff $x^1 \in \mathbf{\Gamma}$, $x^2 \in \mathbf{\Gamma}$ and $0 < \alpha < 1$ such that:

$$\bar{x} = \alpha x^1 + (1 - \alpha)x^2 \text{ implies that } x^1 = x^2 = \bar{x}$$

Another suggestive name for extreme points is *corner points.* See Figure 3.15. Extreme points play a very important role in solving problems related to convex polyhedra.

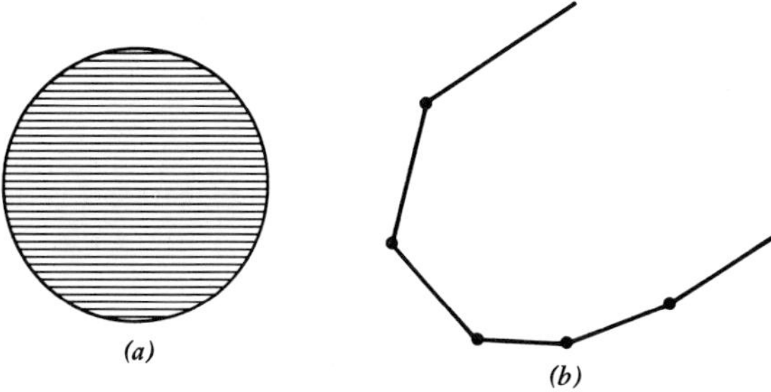

Figure 3.15 (*a*) For this convex set all boundary points are extreme points. (*b*) For this convex polyhedron all bold corner points are extreme points.

3.4.2 Basic Feasible Solutions of a System of Linear Equations in Nonnegative Variables

Consider the system of linear equations in nonnegative variables

$$\begin{aligned} Ax &= b \\ x &\geqq 0 \end{aligned} \tag{3.4}$$

where A, b are matrices of orders $m \times n$, $m \times 1$, respectively, and $x \in \mathbf{R}^n$.

Any vector x that satisfies $Ax = b$, (and may or may not satisfy $x \geqq 0$), is known as a *solution of (3.4)*. A solution of (3.4) that satisfies $x \geqq 0$, is known as a *feasible solution of (3.4)*

Let $\mathbf{K}$ be the set of feasible solutions of (3.4). If $\bar{x} \in \mathbf{K}$, the set of column vectors of A that $\bar{x}$ *uses* is $\{A_{.j}: j \text{ such that } \bar{x}_j > 0\}$. The feasible solution $\bar{x} \in \mathbf{K}$ is said to be a *basic feasible solution* for (3.4) iff the set of column vectors of A that $\bar{x}$ *uses* is a linearly independent set, that is, the set $\{A_{.j}: j \text{ such that } \bar{x}_j > 0\}$ is linearly independent.

Given a general system of linear constraints and restrictions, a *feasible solution* is a vector that satisfies all the constraints and restrictions in the system. A feasible solution is defined to be a basic feasible solution iff it corresponds to an extreme point of the set of feasible solutions. We will use the abbreviation "BFS" for "basic feasible solution." The definition of a BFS given in the paragraph above applies only to a system of linear equality constraints in nonnegative variables.

For a general system of constraints including some inequality constraints or unrestricted variables the definition of a BFS has to be modified so that every BFS is an extreme point and vice versa. Before giving the definition for a general system of constraints, we will prove that a BFS of a system of equality constraints in nonnegative variables is an extreme point of the set of feasible solutions and vice versa.

Theorem Let **K** be the set of feasible solutions of (3.4). $\tilde{x} \in \mathbf{K}$ is an extreme point of **K** iff it is a BFS.

Proof Suppose $\tilde{x}$ is not a BFS. We will prove that it cannot be an extreme point of **K**. In the vector $\tilde{x}$, some of the $\tilde{x}_j$ may be zero and some positive. For ease in referring to them, assume that $\tilde{x}_1, \ldots, \tilde{x}_r$ are all positive and $\tilde{x}_{r+1}, \ldots, \tilde{x}_n$ are all zero. Since $\tilde{x}$ is a feasible solution of (3.4),

$$\sum_{j=1}^{n} A_{.j}\tilde{x}_j = b$$

$$\therefore \quad \sum_{j=1}^{r} A_{.j}\tilde{x}_j = b \tag{3.5}$$

The set of column vectors of A that $\tilde{x}$ *uses* is $\{A_{.1}, \ldots, A_{.r}\}$. Since $\tilde{x}$ is not a BFS, this set must be linearly dependent. Hence, there exists $(\alpha_1, \ldots, \alpha_r) \neq 0$ such that

$$\sum_{j=1}^{r} A_{.j}\alpha_j = 0 \tag{3.6}$$

Combining (3.5) and (3.6) gives:

$$\sum_{j=1}^{r} A_{.j}(\tilde{x}_j + \theta\alpha_j) = b \tag{3.7}$$

From (3.7) we conclude that if

$$\begin{aligned} x_j(\theta) &= \tilde{x}_j + \theta\alpha_j \qquad && \text{for } j = 1 \text{ to } r \\ &= 0 && \text{for } j = r + 1 \text{ to } n \end{aligned}$$

then $x(\theta) = (x_1(\theta), \ldots, x_n(\theta))^{\mathrm{T}}$ is a solution of (3.4), for all real values of θ. Also since $(\alpha_1, \ldots, \alpha_r) \neq 0$, $x(\theta_1) \neq x(\theta_2)$ for $\theta_1 \neq \theta_2$.

Since $\tilde{x}_1, \tilde{x}_2, \ldots, \tilde{x}_r$ are all positive, when the absolute value of θ is sufficiently small, $x_j(\theta) > 0$ for all $j = 1$ to r; hence, $x(\theta)$ is a feasible solution of (3.4). Let ε be a small positive number such that $x(\theta)$ is a feasible solution of (3.4) for all θ whose absolute value is less than or equal to ε. Then clearly

$$\tilde{x} = \frac{x(-\varepsilon) + x(+\varepsilon)}{2}$$

Since $x(-\varepsilon)$, $x(+\varepsilon)$ are both feasible solutions of (3.4) and $x(-\varepsilon) \neq x(+\varepsilon)$, this implies that $\tilde{x}$ is not an extreme point of **K**.

We will now prove that if $\tilde{x}$ is a BFS, it is an extreme point of **K**. As before, let $\tilde{x}_j > 0$ for $j = 1$ to r and $\tilde{x}_j = 0$ for $j = r + 1$ to n. Since $\tilde{x}$ is a BFS, the set $\{A_{.1}, \ldots, A_{.r}\}$ is linearly independent. Suppose $\tilde{x}$ is not an extreme point of **K**. Then there must exist feasible solutions $\bar{x}$ and $\hat{x}$ such that

$$\tilde{x} = \alpha\bar{x} + (1 - \alpha)\hat{x}$$

where

$$0 < \alpha < 1 \qquad \text{and} \qquad \bar{x} \neq \hat{x}$$

So $\tilde{x}_j = \alpha\bar{x}_j + (1-\alpha)\hat{x}_j$ for $j = 1$ to n. Since $\tilde{x}_j = 0$ for $j = r+1$ to n, this implies that $\bar{x}_j = \hat{x}_j = 0$ for $j = r+1$ to n. Therefore,

$$\sum_{j=1}^{r} A_{.j}\bar{x}_j = b$$

$$\sum_{j=1}^{r} A_{.j}\hat{x}_j = b$$

Hence

$$\sum_{j=1}^{r} A_{.j}(\bar{x}_j - \hat{x}_j) = 0$$

However, since the set $\{A_{.1}, \ldots, A_{.r}\}$ is linearly independent, $\bar{x}_j - \hat{x}_j = 0$ for $j = 1$ to r. By earlier arguments $\bar{x}_j = \hat{x}_j = 0$ for $j = r+1$ to n. Therefore, $\bar{x} = \hat{x}$, a contradiction. Hence, $\tilde{x}$ must be an extreme point of **K**.

Example

Consider the system

x_1	x_2	x_3	x_4	x_5	b
1	0	−1	3	−1	1
0	1	1	4	2	4
0	0	0	−7	3	0

$x_j \geqq 0$ for all j

For this system $x = (2, 3, 1, 0, 0)^T$ is not a BFS because it *uses* the column vectors of x_1, x_2, and x_3 and this set is not linearly independent, because

$$(\text{column vector } x_1) - (\text{column vector } x_2) + (\text{column vector } x_3) = 0.$$

Example

Consider the system

x_1	x_2	x_3	x_4	x_5	x_6	b
1	−2	0	1	−1	−3	3
0	1	1	−1	2	1	1
1	−1	1	0	1	−2	4

$x_j \geqq 0$ for all j

The feasible solution $x = (5, 1, 0, 0, 0, 0)^T$ is clearly a BFS of this system.

3.4.3 Basic Feasible Solutions of a System of Equality and Inequality Constraints in Nonnegative Variables

The definition of a BFS for a system of linear constraints is always specified in such a way that a BFS is an extreme point and vice versa. While extreme points are defined in terms of geometric concepts, BFSs are defined in algebric terms using the concept of linear independence. For a BFS to correspond to an extreme point and vice versa, it is necessary to formulate the definition of a BFS taking into account the type of constraints in the system, the presence of inequality constraints, unrestricted variables, etc.

Consider the system consisting of some equality constraints and some inequality constraints in nonnegative variables.

$$\begin{aligned} Ax &= b \\ Dx &\geqq d \\ x &\geqq 0 \end{aligned} \tag{3.8}$$

Transform this into an equivalent system of equality constraints in nonnegative variables, by introducing slack variables corresponding to the inequality constraints. It becomes

$$\begin{aligned} Ax + Os &= b^* \\ Dx - Is &= d \\ x \geqq 0 \quad s &\geqq 0 \end{aligned} \tag{3.9}$$

$s = Dx - d$, and it is the vector of slack variables. A feasible solution x of (3.8) is said to be a *basic feasible solution* (BFS) of (3.8) iff the associated feasible solution $(x; s)$ (where $s = Dx - d$) of (3.9) is a BFS of (3.9) as defined in section 3.4.2.

This definition guarantees that every BFS of a system of constraints like (3.8) is an extreme point of the set of feasible solutions and vice versa.

Example

Consider the system of constraints

$$\begin{aligned} x_1 \qquad\qquad &\geqq 6 \\ x_2 + x_3 &\geqq 2 \\ x_1, x_2, \quad x_3 &\geqq 0 \end{aligned}$$

The point $\tilde{x} = (7, 2, 0)^T$ is a feasible solution of the system, and the set of column vectors corresponding to positive x_j in this system is $\{(1, 0)^T, (0, 1)^T\}$, which is linearly independent. However, $\tilde{x}$ is not a BFS according to the definition here. If we introduce the slack variables, the system becomes

$$\begin{aligned} x_1 \qquad\qquad - s_1 \qquad &= 6 \\ x_2 + x_3 \qquad - s_2 &= 2 \\ x_1, x_2, \quad x_3, s_1, \quad s_2 &\geqq 0 \end{aligned}$$

and $\tilde{x}$ corresponds to $(\tilde{x}, \tilde{s}) = (7, 2, 0, 1, 0)^T$ and the set of column vectors corresponding to x_1, x_2, s_1 in this system is $\{(1, 0)^T, (0, 1)^T, (-1, 0)^T\}$, which is linearly dependent. Actually,

$$\tilde{x} = \frac{1}{2}(6, 2, 0)^T + \frac{1}{2}(8, 2, 0)^T$$

and both $(6, 2, 0)^T$ and $(8, 2, 0)^T$ are feasible solutions of the original system of constraints. Hence $\tilde{x}$ is not an extreme point.

Problem

3.15 Prove that a BFS of (3.8), as defined here, is an extreme point of the set of feasible solution of (3.8) and vice versa.

3.4.4 Basic Feasible Solutions of a System of Equality Constraints

Consider the system of equality constraints in variables, some of which are restricted in sign and some of which are not:

$$\begin{aligned} Dx + Ey &= b \\ x \geqq 0, \qquad & y \text{ unrestricted} \end{aligned} \tag{3.10}$$

* Here O is the zero matrix of appropriate order.

where D is a matrix of order $m \times n_1$, and E is a matrix of order $m \times n_2$. Let the set of feasible solutions of this system be $\mathbf{F}$. Then a feasible solution $(\tilde{x}, \tilde{y})$ of (3.10) is called a *basic feasible solution* (BFS) of (3.10) iff the set of column vectors

$$\{D_{.j}: j \text{ such that } \tilde{x}_j > 0\} \cup \{E_{.j}: \text{for all } j\}$$

which contains all the column vectors of E and all the column vectors of D, corresponding to positive $\tilde{x}_j$, is linearly independent.

Problem

3.16 Prove that $(\tilde{x}, \tilde{y}) \in \mathbf{F}$ is an extreme point of $\mathbf{F}$ iff it is a BFS as defined above.

If $n_1 = 0$, that is, if (3.10) contains only the unrestricted variables y, then a solution is a BFS only if the column vectors of E form a linearly independent set. In this case, if (3.10) has a solution, it must be unique. So a system of equality constraints in unrestricted variables has a basic solution iff it has a unique solution. If the solution is not unique, the set of solutions of such a system is a translate of a subspace and it cannot have any BFSs or extreme points.

TRANSFORMING AN UNRESTRICTED VARIABLE INTO THE DIFFERENCE OF TWO NONNEGATIVE VARIABLES

As in Chapter 2, (3.10) can be transformed into a system of equality constraints in nonnegative variables by expressing each unrestricted variable as the difference of two nonnegative variables. Then the system becomes

$$\begin{gathered} Dx + Ey^+ - Ey^- = b \\ x \geqq 0, \qquad y^+ \geqq 0, \qquad y^- \geqq 0 \end{gathered} \tag{3.11}$$

The correspondence between feasible solutions of (3.10) and those of (3.11) is not one to one. For example, if $(\tilde{x}, \tilde{y}^+, \tilde{y}^-)$ is feasible to (3.11), then $(\tilde{x}, (\tilde{y}_i^+ + \delta), (\tilde{y}_i^- + \delta))$ obtained by adding a δ to each entry in $\tilde{y}^+$ and $\tilde{y}^-$ is also feasible to (3.11) for all $\delta \geqq 0$, and all these feasible solutions of (3.11) correspond to the feasible solution $(\tilde{x}, \tilde{y}^+ - \tilde{y}^-)$ of (3.10).

Every BFS of (3.11) need not correspond to a BFS of (3.10).

Problem

3.17 Consider the system

$$x_1 + x_2 = 1$$

Prove that this system has no BFSs and the set of feasible solutions of this system has no extreme points. Now transform the system by expressing each variable x_1, x_2 as the difference of two nonnegative variables. Find all the BFSs of the transformed system.

3.4.5 Basic Feasible Solutions of a General Linear System of Constraints

Consider a system of linear constraints in which there may be some equality constraints and some inequality constraints, some unrestricted variables and some variables restricted to be nonnegative. Transform this system into a system of equality constraints by introducing slack variables corresponding to the inequality constraints. A feasible solution of the original system is called a *basic feasible solution* (BFS), iff the corresponding solution (with the corresponding values of slack variables) is a BFS of the transformed system of equality constraints as defined in sections 3.4.2 to 3.4.4.

Problems

3.18 Let Γ be the set of feasible solutions of the system

$$\begin{aligned} Ax + Dy &= b \\ Ex + Fy &\geqq d \\ x \geqq 0, \qquad & y \text{ unrestricted} \end{aligned}$$

Prove that $(x, y) \in \Gamma$ is an extreme point of Γ iff it is a BFS as defined above.

3.19 Prove that 0 is the unique extreme point of the convex cone, which is the set of feasible solutions of

$$Ax \geqq 0$$

where A is a matrix of order $m \times n$ and rank n.

3.5 THE USEFULNESS OF BASIC FEASIBLE SOLUTIONS IN LINEAR PROGRAMMING

Every LP can be put in standard form by the transformations discussed in Chapter 2. The problem in standard form is an equivalent problem in which the constraints are all linear equations in nonnegative variables. Hence, for studying LPs, it is sufficient to restrict the discussion to systems of linear equations in nonnegative variables. In all subsequent sections of this chapter, we will only consider problems in which the constraints are linear equations in nonnegative variables.

3.5.1 Basis and its Primal Feasibility

Consider the system of constraints (3.4), where A is a matrix of order $m \times n$ and $x \in \mathbf{R}^n$. Suppose (3.4) does not contain any redundant constraints, that is, the rank of A is m.

Any solution, x, of (3.4) is the vector of the coefficients of the column vectors of A in an expression for b as a linear combination of the column vectors of A. *Hence in studying the system (3.4), the vector space of the columns of matrix A plays an important role.*

Any nonsingular square submatrix of A of order m is known as a *basis* for this system. Suppose B is a basis. The column vectors of A and the variables in the problem can be partitioned into the *basic* and the *nonbasic parts* with respect to this basis B. Each column vector of A, which is in the basis B, is known as a *basic column vector.* All the remaining column vectors of A are called the *nonbasic column vectors.*

Let x_B be the vector of the variables associated with the basic column vectors. The variables in x_B are known as the *basic variables* with respect to the basis B, and x_B is the *basic vector.*

Let x_D be the vector of the remaining variables, which are called the *nonbasic variables.* Let D be the matrix consisting of the column vectors associated with x_D. Rearranging the variables and column vectors, if necessary, (3.4) can be viewed as

$$\begin{aligned} Bx_B + Dx_D &= b \\ x_B \geqq 0, \qquad x_D &\geqq 0 \end{aligned}$$

The *basic solution of (3.4) corresponding to the basis B* is obtained by setting all the nonbasic variables equal to zero, and then solving the remaining system of equations for the values of the basic variables. Since B is nonsingular, this leads to the solution

$$x_D = 0$$
$$x_B = B^{-1}b$$

B is said to be a *primal feasible basis* for (3.4) iff the basic solution corresponding to it satisfies the nonnegativity restrictions, that is, iff $B^{-1}b \geqq 0$

3.5.2 Nondegeneracy

Suppose $\bar{x}$ is a BFS of (3.4). Let the rank of A be m. Since any set of $m + 1$ or more column vectors in $\mathbf{R}^m$ is linearly dependent, the number of positive variables $\bar{x}_j$ cannot exceed m. $\bar{x}$ is said to be a *nondegenerate BFS* of (3.4), if exactly m of the $\bar{x}_j$ are positive. If the number of positive $\bar{x}_j$ is less than m, $\bar{x}$ is known as a *degenerate BFS* of (3.4).

If $\bar{x}$ is a nondegenerate BFS of (3.4), the column vectors of A corresponding to positive $\bar{x}_j$ form a basis of (3.4) for which $\bar{x}$ is the associated basic solution.

If $\bar{x}$ is a degenerate BFS of (3.4), the set of column vectors of A corresponding to positive $\bar{x}_j$ is a linearly independent set, but does not consist of enough vectors to form a basis. It is possible to augment this set with some more column vectors of A and make it into a basis. However, there may be several ways of augmenting the set of column vectors of A *used* by $\bar{x}$ to make it into a basis. Every basis that includes all the column vectors of A corresponding to positive $\bar{x}_j$ is a basis that has $\bar{x}$ as its associated basic solution. The basic vectors corresponding to each of these bases differ in only the zero-valued basic variables.

In summary, every nondegenerate BFS, $\bar{x}$, is the basic solution associated with the basis consisting of the column vectors of A corresponding to positive $\bar{x}_j$. A degenerate BFS, $\bar{x}$, is the basic solution associated with any basis that contains all the column vectors of A corresponding to positive $\bar{x}_j$.

Hence a nondegenerate BFS has a unique basis associated with it. A degenerate BFS may have many bases associated with it.

Problem

3.20 Find all the bases of the system

x_1	x_2	x_3	x_4	x_5	b
1	0	−1	3	−1	1
0	1	1	4	2	4
0	0	0	−7	3	0

$x_j \geqq 0$ for all j

associated with the BFS $x = (1, 4, 0, 0, 0)^T$.

A primal feasible basis, B, of (3.4) is said to be a *nondegenerate basis* if $B^{-1}b > 0$. It is said to be a *degenerate basis* if at least one of the components in the vector $B^{-1}b$ is zero.

3.5.3 Finite Number of Basic Feasible Solutions

Each basis for (3.4) consists of m column vectors of A. Hence the total number of distinct bases for (3.4) is less than or equal to $\binom{n}{m}$. We know that each BFS is associated with one or more bases for (3.4). Hence the total number of distinct BFSs of (3.4) cannot exceed $\binom{n}{m}$ and, hence, is finite.

3.5.4 Existence of a Basic Feasible Solution

Theorem If the system of linear equality constraints in nonnegative variables, (3.4), has a feasible solution, it has a BFS.

Proof

Case 1 If $b = 0$, $x = 0$ is a BFS of (3.4) by definition.

Case 2 Let $b \neq 0$. In this case $x = 0$ is not feasible to (3.4). Therefore every feasible solution of (3.4) must be semipositive. Suppose $\tilde{x}$ is a feasible solution. Suppose

$$\begin{aligned} \tilde{x}_j &> 0 \qquad \text{for } j \in \{j_1, \ldots, j_r\} \\ &= 0 \qquad \text{otherwise} \end{aligned} \tag{3.12}$$

Then $\{A_{.j_1}, \ldots, A_{.j_r}\}$ is the set of column vectors of A that the feasible solution $\tilde{x}$ *uses*. If $\{A_{.j_1}, \ldots, A_{.j_r}\}$ is linearly independent, $\tilde{x}$ is a BFS of (3.4) and we have proved the theorem. If not, there must exist real numbers $\alpha_1, \ldots, \alpha_r$, not all zero, such that

$$\alpha_1 A_{.j_1} + \cdots + \alpha_r A_{.j_r} = 0$$

By feasibility of $\tilde{x}$ and (3.12)

$$\tilde{x}_{j_1} A_{.j_1} + \cdots + \tilde{x}_{j_r} A_{.j_r} = b$$

$$\therefore (\tilde{x}_{j_1} + \theta\alpha_1) A_{.j_1} + \cdots + (\tilde{x}_{j_r} + \theta\alpha_r) A_{.j_r} = b$$

for all real values of θ. Hence, if

$$\begin{aligned} \bar{x}_j(\theta) &= 0 \qquad && \text{for all } j \notin \{j_1, \ldots, j_r\} \\ &= \tilde{x}_{j_t} + \theta\alpha_t \qquad && \text{for } j = j_t,\ t = 1 \text{ to } r \end{aligned}$$

then $\bar{x}(\theta) = (\bar{x}_1(\theta), \ldots, \bar{x}_n(\theta))^{\mathrm{T}}$ is a solution of (3.4) for all real values of θ. By picking θ properly we can also guarantee that $\bar{x}(\theta)$ is a feasible solution of (3.4). For this we have to pick θ to satisfy

$$\tilde{x}_{j_t} + \theta\alpha_t \geqq 0 \qquad \text{for all } t = 1 \text{ to } r$$

that is,

$$\theta\alpha_t \geqq -\tilde{x}_{j_t} \qquad \text{for all } t = 1 \text{ to } r$$

Let

$$\begin{aligned} \theta_1 &= \max\left\{-\frac{\tilde{x}_{j_t}}{\alpha_t} : \quad t \text{ such that } \alpha_t > 0\right\} \\ &= -\infty \text{ if } \alpha_t \leqq 0 \qquad \text{for all } t \\ \theta_2 &= \min\left\{-\frac{\tilde{x}_{j_t}}{\alpha_t} : \quad t \text{ such that } \alpha_t < 0\right\} \\ &= +\infty \text{ if } \alpha_t \geqq 0 \qquad \text{for all } t \end{aligned}$$

Then for all values of θ satisfying $\theta_1 \leqq \theta \leqq \theta_2$, $\bar{x}(\theta)$ is a feasible solution of (3.4). Since $(\alpha_1, \ldots, \alpha_r) \neq 0$, at least one of the two number θ_1 or θ_2 must be finite. Let δ be either θ_1 or θ_2, which is finite. Clearly $\bar{x}(\delta)$ is a feasible solution in which at most $(r - 1)$ of the variables are positive.

Hence, starting with a feasible solution $\tilde{x}$ in which r variables are strictly positive, we have constructed another feasible solution $\bar{x}(\delta)$ in which at most $(r - 1)$ variables are strictly positive.

Either $\bar{x}(\delta)$ is a BFS, in which case we are done, or we can apply the same procedure on it and construct another feasible solution of (3.4) in which the number of strictly positive variables is at least one less than the corresponding number for

$\bar{x}(\delta)$. When this procedure is repeated, we are guaranteed to find a BFS of (3.4) after at most $(r - 1)$ applications of the procedure, since in this case every feasible solution of (3.4) is semipositive.

How to Find a Basic Feasible Solution Given a Feasible Solution

The above proof is constructive. It gives a procedure for constructing a BFS from a feasible solution. Consider the following system:

x_1	x_2	x_3	x_4	x_5	x_6	x_7	b
1	1	3	4	0	0	0	28
0	−1	−1	−2	1	0	0	−13
0	1	1	2	0	1	0	13
1	0	2	2	0	0	−1	15

$$x_j \geqq 0 \quad \text{for all } j$$

$\tilde{x} = (1, 2, 3, 4, 0, 0, 0)^T$ is a feasible solution of this system. Denote the column vector corresponding to x_j by $A_{.j}$. Then $\tilde{x}$ *uses* $\{A_{.1}, A_{.2}, A_{.3}, A_{.4}\}$. Using the algorithm of section 3.3.1, we find that this set is linearly dependent, and that

$$-2A_{.1} - A_{.2} + A_{.3} = 0$$

Hence $\tilde{x}$ is not a BFS, of the system. Since $\tilde{x}$ is feasible

$$A_{.1} + 2A_{.2} + 3A_{.3} + 4A_{.4} = b = (28, -13, 13, 15)^T$$

Using these two equations we get

$$(1 - 2\theta)A_{.1} + (2 - \theta)A_{.2} + (3 + \theta)A_{.3} + 4A_{.4} = b$$

To keep the coefficients of $A_{.1}, \ldots, A_{.4}$ nonnegative, θ must satisfy $\theta_1 = -3 \leqq \theta \leqq \frac{1}{2} = \theta_2$. Putting $\theta = \theta_1 = -3$ leads to

$$7A_{.1} + 5A_{.2} + 4A_{.4} = b \tag{3.13}$$

This implies that $\bar{x} = (7, 5, 0, 4, 0, 0, 0)^T$ is another feasible solution. $\bar{x}$ *uses* only the column vectors $\{A_{.1}, A_{.2}, A_{.4}\}$, which is a subset of the set of column vectors that $\tilde{x}$ *used*. Applying the test for linear independence to this set $\{A_{.1}, A_{.2}, A_{.4}\}$, we find that

$$-2A_{.1} - 2A_{.2} + A_{.4} = 0 \tag{3.14}$$

Multiplying equation (3.14) by θ and adding to (3.13) yields

$$(7 - 2\theta)A_{.1} + (5 - 2\theta)A_{.2} + (4 + \theta)A_{.4} = b \tag{3.15}$$

If the coefficients of $A_{.1}, A_{.2}, A_{.4}$ in (3.15) are to be nonnegative, we verify that θ must satisfy $-4 \leqq \theta \leqq 5/2$. Putting $\theta = 5/2$ in (3.15) leads to

$$2A_{.1} + \left(\frac{13}{2}\right)A_{.4} = b$$

which implies that $\hat{x} = (2, 0, 0, 13/2, 0, 0, 0)^T$ is another feasible solution of the system. It is easily verified that $\hat{x}$ is a BFS.

Unfortunately the theorem proved here is not necessarily true for a general linear systems of constraints. Consider

$$x_1 \geqq 6$$
$$x_2 \quad \text{unrestricted}$$

The set of all feasible solutions of this system is a half-space. Obviously this set has no extreme points or BFSs. See Figure 3.16.

We will now discuss the importance of BFSs in LPs, which are expressed in standard form, and the way in which these solutions are used by the simplex algorithm.

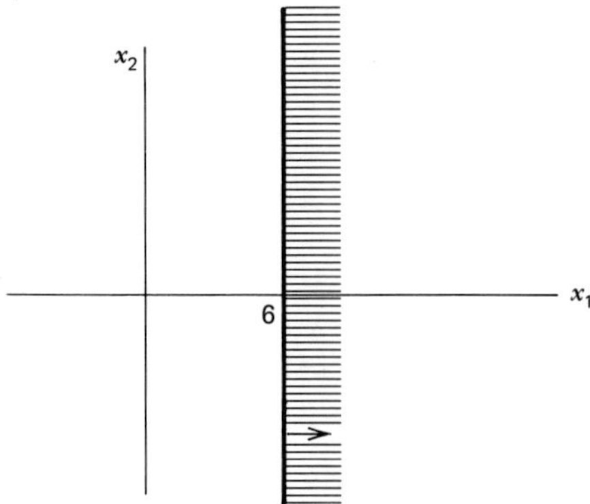

Figure 3.16

3.5.5 Existence of an Optimum Basic Feasible Solution

Theorem If an LP in standard form has an optimum feasible solution, then it has a basic optimum feasible solution.

Proof Consider the LP in standard form:

$$\begin{aligned} \text{Minimize} \quad & z(x) = cx \\ \text{Subject to} \quad & Ax = b \\ & x \geqq 0 \end{aligned} \tag{3.16}$$

Let $\tilde{x}$ be an optimum feasible solution of (3.16). Suppose

$$\begin{aligned} \tilde{x}_j &> 0 \qquad \text{for all } j \in \{j_1, \ldots, j_r\} \\ &= 0 \qquad \text{otherwise} \end{aligned} \tag{3.17}$$

$\tilde{x}$ *uses* the set of column vectors $\{A_{.j_1}, \ldots, A_{.j_r}\}$. If this set is linearly independent, $\tilde{x}$ is a BFS and since it is also optimal to (3.16), we have proved the theorem. Suppose the set $\{A_{.j_1}, \ldots, A_{.j_r}\}$ is not linearly independent. Hence, there exist numbers $\alpha_1, \ldots, \alpha_r$ not all zero, such that

$$\alpha_1 A_{.j_1} + \cdots + \alpha_r A_{.j_r} = 0 \tag{3.18}$$

We will now show that the assumption $\tilde{x}$ is optimal to (3.16) implies that any $\alpha = (\alpha_1, \ldots, \alpha_r)$ satisfying (3.18) must also satisfy

$$\alpha_1 c_{j_1} + \cdots + \alpha_r c_{j_r} = 0 \tag{3.19}$$

Suppose not. Let $\alpha_1 c_{j_1} + \cdots + \alpha_r c_{j_r} \neq 0$. From (3.17) and the feasibility of $\tilde{x}$, we have

$$\tilde{x}_{j_1} A_{.j_1} + \cdots + \tilde{x}_{j_r} A_{.j_r} = b$$

From this and (3.18)

$$(\tilde{x}_{j_1} + \theta\alpha_1) A_{.j_1} + \cdots + (\tilde{x}_{j_r} + \theta\alpha_r) A_{.j_r} = b \tag{3.20}$$

Let

$$\begin{aligned} \hat{x}_j(\theta) &= \tilde{x}_j + \theta\alpha_j \qquad \text{for } j \in \{j_1, \ldots, j_r\} \\ &= 0 \qquad \text{otherwise} \end{aligned} \tag{3.21}$$

and let $\hat{x}(\theta)$ be the column vector of $(\hat{x}_j(\theta))$. Then from (3.20)

$$A\hat{x}(\theta) = b \tag{3.22}$$

and from (3.21)

$$z(\hat{x}(\theta)) = z(\tilde{x}) + \theta(\alpha_1 c_{j_1} + \cdots + \alpha_r c_{j_r}) \tag{3.23}$$

Each of $\alpha_1, \ldots, \alpha_r$ are finite real numbers and $\tilde{x}_1, \ldots, \tilde{x}_r$ are all strictly positive. From (3.21) it is clear that a finite positive number, $\varepsilon > 0$, can be found such that for all θ satisfying $-\varepsilon \leqq \theta \leqq \varepsilon$, $\hat{x}(\theta)$ is also nonnegative. Thus, from (3.22) we see that $\hat{x}(\theta)$ is a feasible solution for all θ between $-\varepsilon$ and $+\varepsilon$.

If $\alpha_1 c_{j_1} + \cdots + \alpha_r c_{j_r} > 0$, let $\theta_1 = -\varepsilon$ and if $\alpha_1 c_{j_1} + \cdots + \alpha_r c_{j_r} < 0$, let $\theta_1 = +\varepsilon$. Then $\hat{x}(\theta_1)$ is a feasible solution of (3.16) and from (3.23), $z(\hat{x}(\theta_1)) < z(\tilde{x})$, which contradicts the assumption that $\tilde{x}$ is an optimum feasible solution of (3.16). Hence, (3.19) must hold. By using equation (3.18), we can obtain another feasible solution $\bar{x}$, such that the set of column vectors that $\bar{x}$ *uses* is a proper subset of the column vectors that $\tilde{x}$ uses, namely $\{A_{.j_1}, \ldots, A_{.j_r}\}$ as in section 3.5.4. By (3.19), any such feasible solution $\bar{x}$ that we obtain must also satisfy: $z(\bar{x}) = z(\tilde{x})$ and so $\bar{x}$ is also an optimum feasible solution. Hence, when this procedure is applied repeatedly, an optimal BFS of (3.16) will be obtained after at most r applications of the procedure.

Problem

3.21 For the LP

$$\begin{array}{lrl} \text{Minimize} & z(x) = & x_1 + x_2 + x_3 + 2x_4 + 2x_5 + 2x_6 \\ \text{Subject to} & & x_1 + x_2 + 2x_4 + x_5 + x_7 = 16 \\ & & x_2 + x_3 + x_4 + 2x_5 + 5x_7 = 19 \\ & & x_1 + x_3 + x_4 + x_5 + 2x_6 + 2x_7 = 13 \\ & & x_j \geqq 0 \quad \text{for all } j \end{array}$$

$x = (1, 2, 3, 4, 5, 0, 0)^T$ is known to be an optimum feasible solution. Obtain an optimum BFS from it.

Note The theorem discussed here is true for LPs in which the constraints are all equality constraints in nonnegative variables. It may not be true for a problem in general form.

3.5.6 Finiteness of the Simplex Algorithm

Consider again the LP in standard form, (3.16), where A is a matrix of order $m \times n$ and rank m. If the problem has an optimum feasible solution, then it has an optimum BFS by the results in section 3.5.5. Hence it is enough to search among the finite number of BFSs of (3.16) for an optimum feasible solution. This is what the simplex algorithm does. All the solutions generated in the course of the simplex algorithm are basic solutions.

The LP(3.16) is said to be a *nondegenerate LP* if every vector $\bar{x}$ that satisfies, $A\bar{x} = b$, has at least m nonzero $\bar{x}_j$ in it. This happens iff b cannot be expressed as a linear combination of any set of $m - 1$ or less column vectors of A, that is, iff $b \in \mathbf{R}^m$ does not lie in any subspace of $\mathbf{R}^m$ generated by a set of $m - 1$ or less column vectors of A.

If (3.16) is a nondegenerate LP, every BFS of it has exactly m positive components and, hence, has a unique associated basis. In this case when the simplex algorithm is applied to solve (3.16), the minimum ratio will turn out to be strictly positive in every pivot step. Therefore the objective value decreases strictly in each pivot step. Hence a basis that appears once in the course of the algorithm can never reappear. Also the total number of bases for (3.16) is finite. Hence the simplex algorithm must either terminate with an optimum basic vector or by satisfying the unboundedness criterion after a *finite number of pivot steps.*

Suppose (3.16) is a degenerate LP. While solving (3.16) by the simplex algorithm, a basic vector in which some basic variables are zero might be obtained. During the pivot operation on such a basis, the minimum ratio may turn out to be zero. If the minimum ratio is zero, the pivot operation just leads to a new basis associated with the same old BFS. In such a pivot step, the BFS and the objective value remain unchanged. Such a pivot step in the course of the simplex algorithm is known as a *degenerate pivot step.*

The next pivot step in the algorithm might also turn out to be a degenerate pivot step. Actually there might be a sequence of consecutive degenerate pivot steps. In all these steps the BFS of the problem and the objective value do not change at all. During these pivot steps, the algorithm is just moving along bases all of which are associated with the same BFS. After several of these degenerate pivot steps, the algorithm might return to the basis that started this sequence of degenerate pivot steps, completing a *cycle.* See Chapter 9 for an example of this cycling. The algorithm can now go through the same cycle again and again indefinitely without ever reaching a basis satisfying a termination criterion. This phenomenon is known as *cycling under degeneracy.*

Cycling does not seem to occur among the numerous degenerate LPs encountered in practical applications. However, the fact that it can occur has been established in specially constructed examples, one of which is discussed in Chapter 9.

Pivot row choice rules known as "*techniques for resolving degeneracy*" have been developed which guarantee that a basis once examined never reappears in the course of the simplex algorithm. The *lexicographic minimum ratio rule* discussed in Chapter 9 is one of them. When this is applied, the new basis is picked in such a way that there is a *strict lexico decrease* in the objective function. For the definitions of these terms and all other details, see Chapter 9. Since every pivot step results in a strict lexico decrease in the objective function, we conclude that a basis which appeared once in the course of the application of the simplex algorithm with the lexicographic pivot row choice rule, can never reappear.

Hence after a finite number of pivot steps, the algorithm must terminate with a basis, in which either the optimality criterion or the unboundedness criterion is satisfied. This leads to the following theorem:

Theorem Starting with a primal feasible basis, the simplex algorithm (with some technique for resolving degeneracy, if necessary) finds, after a finite number of pivot steps, one of the following.

(i) An optimal feasible basis, the canonical tableau with respect to which satisfies the optimality criterion.

(ii) A feasible basis, the canonical tableau with respect to which satisfies the unboundedness criterion.

Problem

3.22 Consider a general LP in which there is an unrestricted variable, x_1. Suppose this problem is put in standard form, during which the unrestricted variable, x_1, is expressed as the difference of two nonnegative variables as in $x_1 = x_1^+ - x_1^-$, $x_1^+ \geqq 0$, $x_1^- \geqq 0$. Let the transformed problem in standard form be called (P). Prove the following:

(i) While solving (P) by the simplex method at least one of the variables in the pair (x_1^+, x_1^-) remains equal to zero always.

(ii) If (P) has an optimum feasible solution, it has an optimum feasible solution that satisfies $x_1^+ x_1^- = 0$, and the simplex method finds only an optimum feasible solution of (P) that satisfies this property.

3.6 THE EDGE PATH TRACED BY THE SIMPLEX ALGORITHM

In this section we study how the simplex algorithm walks on the set of feasible solutions of an LP before a termination criterion is satisfied.

3.6.1 A System of *m* Linearly Independent Equations in *m* + 1 Nonnegative Variables

Consider the system of linear equations in nonnegative variables

$$\begin{aligned} Ey &= d \\ y &\geqq 0 \end{aligned} \tag{3.24}$$

where E is a matrix of order $m \times (m + 1)$, and rank m, and $y \in \mathbf{R}^{m+1}$. Since E is of rank m, there exists a square submatrix of E of order m that is nonsingular. Suppose it is $B = (E_{.1} \vdots \cdots \vdots E_{.m})$. Then the system of equations $Ey = d$ can be written as

$$B(y_1, \ldots, y_m)^{\mathrm{T}} + E_{.m+1} y_{m+1} = d$$

That is,

$$(y_1, \ldots, y_m)^{\mathrm{T}} = B^{-1}(d - E_{.m+1} y_{m+1})$$

Giving a value of θ to y_{m+1}, this is equivalent to

$$\begin{aligned} (y_1, \ldots, y_m)^{\mathrm{T}} &= B^{-1}d - \theta(B^{-1}E_{.m+1}) \\ y_{m+1} &= \theta \end{aligned}$$

As θ varies from $-\infty$ to $+\infty$ this generates a straight line in $\mathbf{R}^{m+1}$. The set of feasible solutions of (3.24) is the intersection of this straight line with the nonnegative orthant of $\mathbf{R}^{m+1}$. The following possibilities exist.

1. The straight line does not intersect the nonnegative orthant of $\mathbf{R}^{m+1}$. In this case the set of feasible solutions of (3.24) is empty. Figure 3.17 illustrates this situation in $\mathbf{R}^2$.
2. The intersection of this straight line and the nonnegative orthant of $\mathbf{R}^{m+1}$ is nonempty and bounded. In this case the set of feasible solutions of (3.24) is a line segment of the straight line. Hence it has exactly two extreme points, which are the end points of this line segment. See Figure 3.18.
3. The intersection of this straight line and the nonnegative orthant of $\mathbf{R}^{m+1}$ is nonempty and unbounded. In this case the set of feasible solutions of (3.24) is a half-line. Hence it has exactly one extreme point, which is the end point of this half-line. See Figure 3.19.

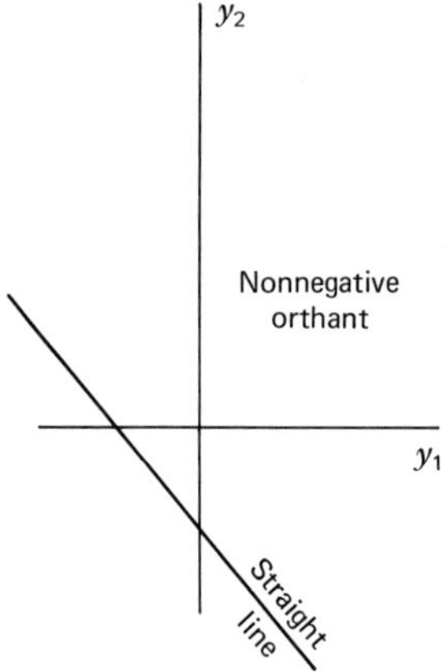

Figure 3.17

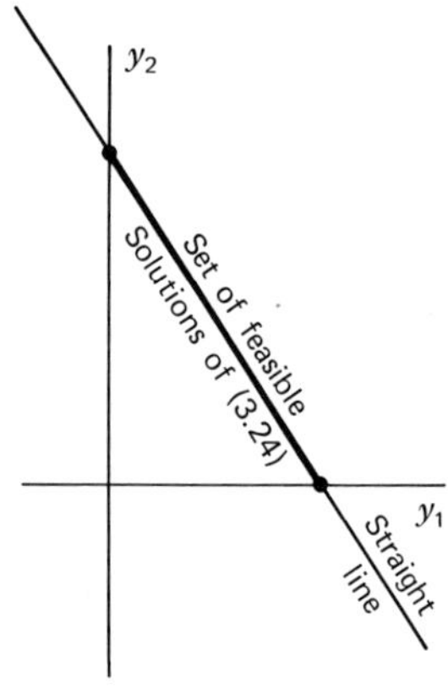

Figure 3.18

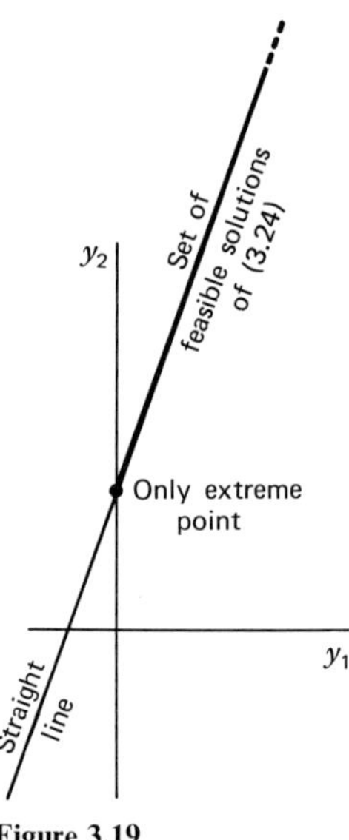

Figure 3.19

Hence, any linearly independent system of m equations in $m + 1$ nonnegative unknowns can have at most two BFSs. Notice that the linear independence of the system of constraints is essential for this result to hold.

Suppose we are solving the LP (3.16) where A is of order $m \times n$ and rank m, by the simplex algorithm. In each step, the simplex algorithm works with a basis, which is a nonsingular square submatrix of A consisting of m column vectors of A. The column vectors of A in the basis are the *basic column vectors* in this step of the algorithm. The algorithm then picks exactly one of the nonbasic column vectors, called the *pivot column*, and tries to bring it into the basis. During this step of the algorithm, all the other $n - m - 1$ nonbasic variables are set equal to zero. Hence in this step, the simplex algorithm is dealing with the subset of the set of feasible solutions, in which only the present basic variables and the entering variable are free to vary, while all the other variables are zero. This subset corresponds to the *restricted system* of (3.16) which is obtained by eliminating the $(n - m - 1)$ zero valued nonbasic variables and their column vectors from (3.16).

This restricted system is clearly of the form (3.24). Hence the results of this section are useful in determining what the simplex algorithm does in a pivot step. Before discussing this material, we will review some concepts related to convex polyhedra.

3.6.2 Adjacency, Bounded Edges

Consider the convex polyhedron represented in Figure 3.20. Clearly every point on the line segment joining the extreme points x^1 and x^2 cannot be expressed as a convex combination of any pair of points in the convex polyhedron that are not on this line segment. This, however, is not true of points on the line segment joining the extreme points x^1 and x^3. Extreme points like x^1 and x^2 are known as *adjacent extreme points* of the convex polyhedron. The extreme points x^1 and x^3 are not adjacent.

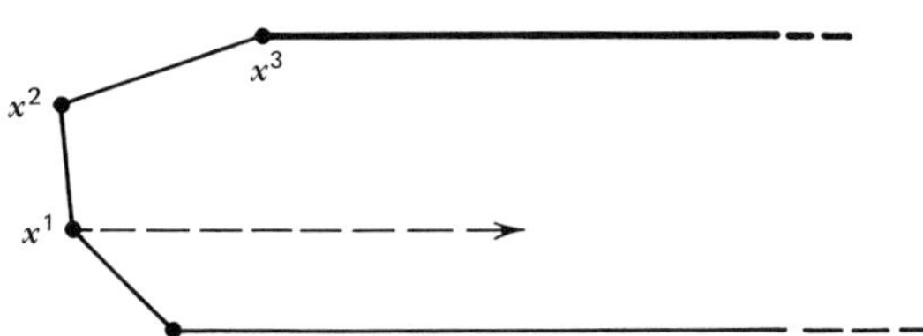

Figure 3.20

Two extreme points $\tilde{x}$ and $\hat{x}$ of a convex polyhedron, **K**, are said to be *adjacent* iff every point, $\bar{x}$, on the line segment joining them has the property that if

$$\bar{x} = \alpha x^1 + (1 - \alpha)x^2$$

where

$$0 < \alpha < 1 \qquad \text{and} \qquad x^1 \in \mathbf{K},\ x^2 \in \mathbf{K}$$

then both x^1 and x^2 must themselves be on the line segment joining $\tilde{x}$ and $\hat{x}$.

The line segment joining a pair of adjacent extreme points of a convex polyhedron is called a *bounded edge* or an *edge* of the convex polyhedron.

Consider the convex polyhedron **K**, which is the set of feasible solutions of (3.16). Here is an algebraic characterization of adjacency on **K**. Two distinct extreme points of **K**, $\tilde{x}$ and $\hat{x}$ are adjacent extreme points of **K** iff the set $\{A_{.j}$: j such that either $\tilde{x}_j$ or $\hat{x}_j$ or both are $>0\}$, is linearly dependent, and it contains a column vector such that when this column vector is deleted from the set, the remaining set is linearly independent. See reference [5] at the end of this chapter.

Problem

3.23 Prove that the definitions of adjacency given here are equivalent.

3.6.3 Adjacency in the Simplex Algorithm

Consider the LP (3.16) where A is a matrix of order $m \times n$ and rank m. Let **K** be the set of feasible solutions of (3.16). Suppose this problem is being solved by the simplex algorithm. Suppose B is the current basis and for convenience in referring to it assume that

$$B = (A_{.1} \vdots \cdots \vdots A_{.m})$$

Let $\tilde{x}$ be the BFS corresponding to the present basis B. Therefore

$$\tilde{x}_{m+1} = \tilde{x}_{m+2} = \cdots = \tilde{x}_n = 0$$

Suppose in this step the entering variable is x_{m+1}. The pivot column in this step is

$$\bar{A}_{.m+1} = B^{-1}A_{.m+1} = (\bar{a}_{1,m+1}, \ldots, \bar{a}_{m,m+1})^{\mathrm{T}}$$

We assume that $\bar{A}_{.m+1} \nleqq 0$. The case where $\bar{A}_{.m+1} \leqq 0$ corresponds to unboundedness; this is discussed in section 3.7.

During this pivot step, the remaining nonbasic variables $x_{m+2}, \ldots, x_n$ are fixed at 0. Only the nonbasic variable x_{m+1} is changed from its present value of 0 to a nonnegative λ and the values of the basic variables are reevaluated to satisfy the equality constraints. By the discussion in Chapter 2, this leads to the solution, $x(\lambda) = (\tilde{x}_1 - \bar{a}_{1,m+1}\lambda, \ldots, \tilde{x}_m - \bar{a}_{m,m+1}\lambda, \lambda, 0, \ldots, 0)^{\mathrm{T}}$.

The maximum value that λ can take is the *minimum ratio* in this pivot step, θ, and this is determined to keep $x(\lambda)$ nonnegative. After the pivot step the simplex algorithm moves to the BFS, $x(\theta)$. If the minimum ratio, θ, is equal to zero in this pivot step, this new BFS is $\tilde{x}$ itself. In this case the pivot step is a *degenerate pivot step*. Hence we have the following result.

1. In a degenerate pivot step, the simplex algorithm remains at the same BFS, but it obtains a new basis associated with it.

If the minimum ratio in this pivot step, θ, is strictly positive, the simplex algorithm moves to the new BFS, $x(\theta)$, which is distinct from $\tilde{x}$.

Let $\bar{x}$ be a point on the line segment joining $\tilde{x}$ and $x(\theta)$. Clearly the line segment joining $\tilde{x}$ and $x(\theta)$ is itself generated by varying λ in $x(\lambda)$ from 0 to θ. Hence $\bar{x}$ is $x(\lambda)$ for some value of λ between 0 and θ. Therefore, $\bar{x}_{m+2} = \cdots = \bar{x}_n = 0$. If $\bar{x}$ is expressed as a convex combination, $\bar{x} = \alpha x^1 + (1 - \alpha)x^2$ where $0 < \alpha < 1$ and x^1, x^2 are both distinct feasible solutions of (3.16), then

$$x^1_{m+2} = \cdots = x_n^1 = x^2_{m+2} = \cdots = x_n^2 = 0$$

Hence from the arguments in section 3.6.1, it is clear that both x^1, x^2 must be points of the form $x(\lambda)$ for some values of λ between 0 and θ. Therefore x^1, x^2 themselves lie on the line segment joining $\tilde{x}$ and $x(\theta)$. This implies that $\tilde{x}$ and $x(\theta)$ are themselves adjacent extreme points of $\mathbf{K}$. This leads to the following result.

2. During a nondegenerate pivot step, the simplex algorithm moves from one extreme point of the set of feasible solutions to an adjacent extreme point. The edge joining the extreme points is generated by giving the entering variable all possible values between 0 and the minimum ratio.

Example

Consider the following system of constraints:

Basic variables	x_1	x_2	x_3	x_4	x_5	x_6	b
x_1	1	0	0	2	−1	−1	0
x_2	0	1	0	−3	1	−2	2
x_3	0	0	1	−1	2	0	1

$x_j \geqq 0$ for all j

Let $A_{.j}$ denote the column vector corresponding to the variable x_j in this system for $j = 1$ to 6. Let $B = (A_{.1}, A_{.2}, A_{.3})$ be a basis. The BFS corresponding to this basis is

$$\tilde{x} = (0, 2, 1, 0, 0, 0)^{\mathrm{T}}$$

If the nonbasic variables x_4 is the entering variable, the minimum ratio turns out to be zero. This gives a new basis $(A_{.4}, A_{.2}, A_{.3})$ which is also associated with the same BFS $\tilde{x}$. This is a degenerate pivot step that results in a new basis, but no change in the actual BFS.

On the other hand, if the nonbasic variable x_5 is the entering variable, the minimum ratio is: $\theta = \text{minimum } \{2/1, 1/2\} = 1/2$ and, hence, x_5 enters the basic vector (x_1, x_2, x_3) replacing x_3 from it. Keeping x_4, x_6 equal to zero, giving x_5 a value of λ, and evaluating the values of x_1, x_2, x_3 leads to the solution:

$$x(\lambda) = (\lambda, 2 - \lambda, 1 - 2\lambda, 0, \lambda, 0)^{\mathrm{T}}$$

Therefore the new BFS is

$$\hat{x} = x\left(\frac{1}{2}\right) = \left(\frac{1}{2}, \frac{3}{2}, 0, 0, \frac{1}{2}, 0\right)^{\mathrm{T}}$$

and the new basis is $\hat{B} = (A_{.1}, A_{.2}, A_{.5})$. $\tilde{x}$ and $\hat{x}$ are adjacent BFSs. The line segment joining them is an edge of the set of feasible solutions of the system. Clearly,

$$\begin{aligned} x(\lambda) &= \frac{(\theta - \lambda)\tilde{x} + \lambda\hat{x}}{\theta} \\ &= \frac{(1/2 - \lambda)\tilde{x} + \lambda\hat{x}}{1/2} \end{aligned}$$

As λ varies between 0 and θ, $x(\lambda)$ generates all the points on the edge joining $\tilde{x}$ and $\hat{x}$. See Figure 3.21.

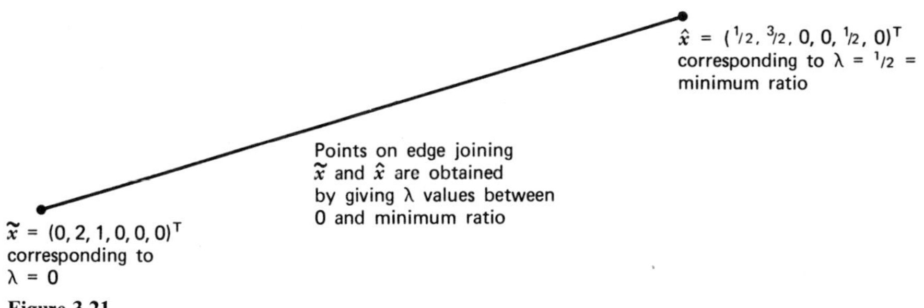

Figure 3.21

Problem

3.24 If $\tilde{x}$ and $\hat{x}$ are two distinct extreme points of **K**, develop a simple algorithm to check whether they are adjacent.

Hence, starting from a BFS, the simplex algorithm walks along the edges of the convex polyhedral set of feasible solutions (moving from an extreme point to an adjacent extreme point in every nondegenerate pivot step and sitting at the same extreme point in every degenerate pivot step) until termination occurs. Any path similar to that, which moves from an extreme point of a convex polyhedron **K** to another extreme point along the edges of **K** is known as an *edge path*, or *adjacent extreme point path*, or just *adjacency path*. The simplex algorithm moves along an adjacency path on the set of feasible solutions of the problem, along which the objective value steadily improves.

3.7 HOMOGENEOUS SOLUTIONS, UNBOUNDEDNESS, RESOLUTION THEOREMS

Consider the system of equations in nonnegative variables in (3.16). A *homogeneous solution corresponding to* (3.16) is a vector y satisfying

$$\begin{aligned} Ay &= 0 \\ y &\geqq 0 \end{aligned}$$

The set of all homogeneous solutions is a convex polyhedral cone. If $\tilde{x}$ is a feasible solution of (3.16), and $\hat{y}$ is a homogeneous solution corresponding to (3.16), then $\tilde{x} + \theta\,\hat{y}$ is also a feasible solution of (3.16) for any $\theta \geqq 0$. For example consider the problem in Example 4 of section 2.6.4.

x_1	x_2	x_3	x_4	b
1	1	-1	0	1
1	-1	0	-1	0

$x_j \geqq 0$ for all j

$\tilde{x} = (1/2, 1/2, 0, 0)^{\mathrm{T}}$ is a feasible solution and $\hat{y} = (1, 1, 2, 0)^{\mathrm{T}}$ is a homogeneous solution. For $\theta \geqq 0$, $\tilde{x} + \theta\hat{y}$ is a point on the thick half-line in the Figure 2.4 in section 2.6.4.

Resolution Theorem 1 Every feasible solution of (3.16) can be expressed as the sum of (i) a convex combination of the BFSs of (3.16), and (ii) a homogeneous solution corresponding to (3.16).

Proof Suppose $\tilde{x}$ is a feasible solution of (3.16). Suppose

$$\begin{aligned} \tilde{x}_j &> 0 \qquad \text{for} \quad j \in \{j_1, \ldots, j_k\} \\ &= 0 \qquad \text{otherwise} \end{aligned} \tag{3.25}$$

Hence the set of column vectors of A *used* by $\tilde{x}$ is $\{A_{.j_1}, \ldots, A_{.j_k}\}$. The proof of this theorem is based upon induction on the number k, the number of column vectors in this set.

Case 1 Suppose $k = 0$, This can only happen when $\tilde{x} = 0$ and since $\tilde{x}$ is assumed feasible, $b = 0$. In this case, $\tilde{x} = 0$, is itself a BFS of (3.16) and also a homogeneous solution corresponding to (3.16). Hence

$$\tilde{x} = \underbrace{0}_{\text{BFS}} + \underbrace{0}_{\substack{\text{Homogeneous} \\ \text{solution}}}$$

and hence the theorem holds in this case.

Case 2 $k > 1$.

Induction Hypothesis Suppose the theorem holds for every feasible solution that *uses* a set of $k - 1$ or less column vectors of A.

We will now show that under the induction hypothesis, the theorem also holds for $\tilde{x}$, which *uses* a set of k column vectors of A. If $\{A_{.j_1}, \ldots, A_{.j_k}\}$ is linearly independent, then $\tilde{x}$ is a BFS of (3.16) and in this case

$$\tilde{x} = \underbrace{\tilde{x}}_{\text{BFS}} + \underbrace{0}_{\substack{\text{Homogeneous}\\ \text{solution}}}$$

Hence the theorem holds for $\tilde{x}$. On the other hand, if $\{A_{.j_1}, \ldots, A_{.j_k}\}$ is not linearly independent, there exist real numbers $\alpha_1, \ldots, \alpha_k$ not all zero, such that

$$\alpha_1 A_{.j_1} + \cdots + \alpha_k A_{.j_k} = 0$$

By the feasibility of $\tilde{x}$ and (3.25)

$$\tilde{x}_{j_1} A_{.j_1} + \cdots + \tilde{x}_{j_k} A_{.j_k} = b$$

So

$$(\tilde{x}_{j_1} + \theta\alpha_1) A_{.j_1} + \cdots + (\tilde{x}_{j_k} + \theta\alpha_k) A_{.j_k} = b$$

Let

$$\begin{aligned} \theta_1 &= \text{maximum}\left\{\frac{-\tilde{x}_{j_t}}{\alpha_t} : t \text{ such that } \alpha_t > 0\right\} \\ &= -\infty \text{ if } \alpha_t \leqq 0 \text{ for all } t \end{aligned}$$

and

$$\begin{aligned} \theta_2 &= \text{minimum}\left\{\frac{-\tilde{x}_{j_t}}{\alpha_t} : t \text{ such that } \alpha_t < 0\right\} \\ &= +\infty \text{ if } \alpha_t \geqq 0 \text{ for all } t \end{aligned}$$

Since $\alpha = (\alpha_1, \ldots, \alpha_k) \neq 0$, at least one of the numbers θ_1, θ_2 must be finite. If θ_r is finite, define

$$\begin{aligned} \hat{x}_{j_t}(\theta_r) &= \tilde{x}_{j_t} + \theta_r \alpha_t \qquad \text{for } t = 1 \text{ to } k \\ \hat{x}_j(\theta_r) &= 0 \qquad \text{if } j \notin \{j_1, \ldots, j_k\} \end{aligned}$$

then $\hat{x}(\theta_r) = (\hat{x}_1(\theta_r), \ldots, \hat{x}_n(\theta_r))^{\mathrm{T}}$ is a feasible solution of (3.16) that *uses* a set of at most $k - 1$ column vectors of A. There are two possible subcases here:

(i) Suppose all the real numbers $\alpha_1, \ldots, \alpha_k$ are of the same sign. In this case only one of the numbers θ_1, θ_2 is finite and it is θ_r. If all $\alpha_t \geqq 0$ for $t = 1$ to k, then $\theta_1 < 0$ is finite and $\theta_2 = +\infty$. On the other hand, if $\alpha_t \leqq 0$ for $t = 1$ to k, then $\theta_1 = -\infty$ and $\theta_2 > 0$ is finite. Hence if

$$\begin{aligned} \hat{y}_{j_t} &= |\theta_r \alpha_t| \qquad \text{for } t = 1 \text{ to } k \\ \hat{y}_j &= 0 \qquad \text{for } j \notin \{j_1, \ldots, j_k\} \end{aligned}$$

then from the above equations $\hat{x}(\theta_r) = \tilde{x} - \hat{y}$, so $\tilde{x} = \hat{x}(\theta_r) + \hat{y}$. Obviously $\hat{y}$ is a homogeneous solution corresponding to (3.16). Since $\hat{x}(\theta_r)$ is a feasible solution of (3.16) that *uses* a set of $k - 1$ or less column vectors of A, the theorem holds for it by the induction hypothesis and, hence, $\hat{x}(\theta_r) = x^1 + y^1$ where x^1 is a convex combination of BFSs of (3.16) and y^1 is homogeneous solution corresponding to (3.16). Therefore,

$$\tilde{x} = \underbrace{x^1}_{\substack{\text{Convex combination}\\ \text{of BFSs}}} + \underbrace{(y^1 + \hat{y})}_{\substack{\text{Homogeneous}\\ \text{solution}}}$$

$y^1 + \hat{y}$ is a homogeneous solution corresponding to (3.16) because both y^1 and $\hat{y}$ are homogeneous solutions and the sum of two homogeneous solutions is also a homogeneous solution. Hence the theorem holds for $\tilde{x}$.

(ii) Suppose the real numbers $\alpha_1, \ldots, \alpha_k$ are not all of the same sign. In this case both θ_1, θ_2 are finite. θ_1 is strictly negative and θ_2 is strictly positive. Therefore

$$\begin{aligned}\tilde{x} &= \frac{\theta_2 \hat{x}(\theta_1) + (-\theta_1)\hat{x}(\theta_2)}{\theta_2 - \theta_1} \\ &= \beta \hat{x}(\theta_1) + (1 - \beta)\hat{x}(\theta_2)\end{aligned}$$

where

$$\beta = \frac{\theta_2}{\theta_2 - \theta_1}, \qquad 0 < \beta < 1$$

Since both $\hat{x}(\theta_1)$ and $\hat{x}(\theta_2)$ are feasible solutions of (3.16) *using* a set of $k - 1$ or less column vectors of A, the theorem holds for them by the induction hypothesis. Therefore

$$\begin{aligned}\hat{x}(\theta_1) &= x^1 + y^1 \\ \hat{x}(\theta_2) &= x^2 + y^2\end{aligned}$$

where x^1, x^2 are both convex combinations of BFSs of (3.16) and y^1, y^2 are both homogeneous solutions corresponding to (3.16).

$$\tilde{x} = \underbrace{(\beta x^1 + (1 - \beta)x^2)}_{\substack{\text{Convex combination} \\ \text{of BFSs}}} + \underbrace{(\beta y^1 + (1 - \beta)y^2)}_{\substack{\text{Homogeneous} \\ \text{solution}}}$$

Hence the theorem holds for $\tilde{x}$. By Case 1 and by induction, the theorem therefore holds for k. Therefore the theorem is true in general.

BOUNDEDNESS OF CONVEX POLYHEDRA

If the set of feasible solutions of

$$\begin{aligned}Ax &= b \\ x &\geqq 0\end{aligned}$$

is nonempty, by Resolution Theorem I, it is a convex polytope (i.e., is bounded) iff

$$\begin{aligned}Ay &= 0 \\ y &\geqq 0\end{aligned}$$

has a unique solution, namely, $y = 0$.

EXTREME HOMOGENEOUS SOLUTIONS

A homogeneous solution corresponding to (3.16) is called an *extreme homogeneous solution* iff it is a BFS of

$$\begin{aligned}Ay &= 0 \\ \sum_{j=1}^{n} y_j &= 1 \\ y_j &\geqq 0 \qquad \text{for all } j\end{aligned} \tag{3.26}$$

Thus there are only a finite number of distinct extreme homogeneous solutions. The constraint that the sum of all the variables should be equal to 1 eliminates 0 from consideration and normalizes the solution.

Lemma 1 Either 0 is the unique homogeneous solution, or every homogeneous solution corresponding to (3.16) can be expressed as a nonnegative linear combination of extreme homogeneous solutions.

Proof Suppose 0 is not the only homogeneous solution corresponding to (3.16). Every nonzero homogeneous solution must be semipositive and, hence, it must satisfy

$$Ay = 0$$
$$\sum_{j=1}^{n} y_j = \alpha > 0$$
$$y_j \geqq 0 \qquad \text{for each } j$$

Hence it is equal to a feasible solution of (3.26) multiplied by α. Thus every homogeneous solution corresponding to (3.16) including zero, is a nonnegative multiple of some feasible solution of (3.26). Apply Resolution Theorem I to the system of constraints (3.26). The homogeneous system corresponding to (3.26) is

$$Ay = 0$$
$$\sum_{j=1}^{n} y_j = 0$$
$$y_j \geqq 0 \qquad \text{for each } j$$

$\sum y_j = 0$ and $y \geqq 0$ together imply that $y = 0$. Hence the only homogeneous solution corresponding to (3.26) is $y = 0$. Therefore by Resolution Theorem I, every feasible solution of (3.26) can be expressed as a convex combination of BFSs of (3.26), that is, extreme homogeneous solutions corresponding to (3.16). Hence every homogeneous solution corresponding to (3.16) can be expressed as a nonnegative linear combination of extreme homogeneous solutions corresponding to (3.16).

Resolution Theorem II for Convex Polyhedra

Let **K** be the set of feasible solutions of (3.16).

(i) If 0 is the unique homogeneous solution corresponding to (3.16), every feasible solution of (3.16) can be expressed as a convex combinations of BFSs of (3.16).

(ii) If 0 is not the only homogeneous solution corresponding to (3.16), every feasible solution of (3.16) is the sum of
a convex combination of BFSs of (3.16), and
a nonnegative combination of extreme homogeneous solutions corresponding to (3.16).

Proof This theorem follows from Resolution Theorem I and Lemma 1.

Discussion Let $\mathbf{K}^{\Delta}$ be the convex hull of all the extreme points of **K**, that is, BFSs of (3.16). Let $\mathbf{K}^{\angle}$ be the cone of all homogeneous solutions corresponding to (3.16). If $\mathbf{\Gamma}_1$ and $\mathbf{\Gamma}_2$ are two subsets of $\mathbf{R}^n$, define

$$\mathbf{\Gamma}_1 + \mathbf{\Gamma}_2 = \{x + y : x \in \mathbf{\Gamma}_1, y \in \mathbf{\Gamma}_2\}$$

Then resolution Theorem II states that

$$\text{if} \quad \mathbf{K}^{\angle} = \{0\}, \qquad \mathbf{K} = \mathbf{K}^{\Delta}$$
$$\text{if} \quad \mathbf{K}^{\angle} \neq \{0\}, \qquad \mathbf{K} = \mathbf{K}^{\Delta} + \mathbf{K}^{\angle}$$

Let $x^1, \ldots, x^p$ be all the distinct BFSs of (3.16). If $\mathbf{K}^{\angle} = \{0\}$, then every feasible solution of (3.16) is of the form

$$x = \alpha_1 x^1 + \cdots + \alpha_p x^p$$
$$\text{where } \alpha_1 + \cdots + \alpha_p = 1 \text{ and } \alpha_t \geqq 0 \text{ for all } t$$

If $\mathbf{K}^{\angle} \neq \{0\}$, let $y^1, \ldots, y^L$ be all the extreme homogeneous solutions corresponding to (3.16). Then every feasible solution of (3.16) is of the form

$$x = \alpha_1 x^1 + \cdots + \alpha_p x^p + \beta_1 y^1 + \cdots + \beta_L y^L$$
$$\text{where} \quad \alpha_1 + \cdots + \alpha_p = 1$$
$$\text{and} \quad \alpha_t \geqq 0, \beta_r \geqq 0 \quad \text{for all } t, r$$

This is a very useful theorem in the theory of linear programming. The *decomposition principle*, which helps in decomposing very large LPs into several smaller ones is based primarily on this theorem.

Corollary 1 If the LP (3.16) has a feasible solution, it has an optimum feasible solution iff $z(y) \geqq 0$ for every homogeneous solution, y, corresponding to (3.16).

Corollary 2 Suppose there exists an extreme homogeneous solution corresponding to (3.16), y^r, such that $cy^r < 0$. In this case, if (3.16) is feasible, $z(x)$ is unbounded below on the set of feasible solutions of (3.16).

Proofs of Corollaries 1 and 2 Let $x^1, \ldots, x^p$ be all the BFSs of (3.16). Suppose 0 is the unique homogeneous solution corresponding to (3.16). Then only Corollary 1 can hold. Let x be any feasible solution of (3.16). By Resolution Theorem II,

$$x = \alpha_1 x^1 + \cdots + \alpha_p x^p$$
$$\text{where} \quad \alpha_1 + \cdots + \alpha_p = 1$$
$$\text{and} \quad \alpha_t \geqq 0 \quad \text{for all } t$$

Thus

$$cx = \alpha_1(cx^1) + \cdots + \alpha_p(cx^p)$$

and therefore

$$\underset{t = 1 \text{ to } p}{\text{Minimum}} (cx^t) \leqq cx \leqq \underset{t = 1 \text{ to } p}{\text{maximum}} (cx^t)$$

Hence, the BFS that minimizes cx among the finite number of BFSs of (3.16) must be optimal to (3.16).

Now consider the case where 0 is not the only homogeneous solution corresponding to (3.16). Let $y^1, \ldots, y^L$ be all the extreme homogeneous solutions corresponding to (3.16). If x is a feasible solution of (3.16), by Resolution Theorem II,

$$x = \sum_{t=1}^{p} \alpha_t x^t + \sum_{r=1}^{L} \beta_r y^r \tag{3.27}$$

where

$$\sum_{t=1}^{p} \alpha_t = 1 \text{ and } \alpha_t \geqq 0, \beta_r \geqq 0 \text{ for all } t, r.$$

Therefore,

$$cx = \sum_{t=1}^{p} \alpha_t(cx^t) + \sum_{r=1}^{L} \beta_r(cy^r)$$

By Lemma 1, $z(y) \geqq 0$ for all homogeneous solutions y iff $cy^r \geqq 0$ for all extreme homogeneous solutions y^r. In this case, from (3.27), we conclude that $z(x)$ is minimized by the BFS, which gives the least value to $z(x)$ among $x^1, \ldots, x^p$.

On the other hand, if there exists a y satisfying $Ay = 0$, $y \geqq 0$, $cy < 0$, then by Lemma 1, there must exist an extreme homogeneous solution y^r such that $cy^r < 0$. By making β_r sufficiently large in (3.27) we can obtain a feasible solution here that makes $z(x)$ less than any given real number. This proves Corollary 2.

Corollary 3 If $z(x)$ is unbounded below on the set of feasible solutions of the LP (3.16), it remains unbounded below even if the b vector is changed as long as the problem remains feasible.

Proof Since $z(x)$ is unbounded below on (3.16) by Corollaries 1 and 2, there exists a y^1 satisfying

$$\begin{aligned} Ay^1 &= 0 \\ cy^1 &< 0 \\ y^1 &\geqq 0 \end{aligned}$$

Suppose b is changed to b^1; let x^1 be a feasible solution of the modified LP.

$$\begin{array}{ll} \text{Minimize} & z(x) = cx \\ \text{Subject to} & Ax = b^1 \\ & x \geqq 0 \end{array}$$

Then $x^1 + \beta y^1$ is also a feasible solution of this LP for every $\beta \geqq 0$, and since $c(x^1 + \beta y^1) = cx^1 + \beta(cy^1)$ tends to $-\infty$ as β tends to $+\infty$ (because $cy^1 < 0$), $z(x)$ is unbounded below in this modified problem too.

UNBOUNDED EDGES OR EXTREME HALF-LINES OF CONVEX POLYHEDRA

Let $\mathbf{K}$ be the set of feasible solutions of (3.16). Suppose $\tilde{x}$ is a BFS of (3.16) and $\hat{y}$ an extreme homogeneous solution corresponding to (3.16). Then every point on the half-line

$$\{x\colon x = \tilde{x} + \theta\hat{y}, \theta \geqq 0\} \tag{3.28}$$

is a feasible solution for (3.16). This half-line is called an *unbounded edge* or an *extreme half-line* of $\mathbf{K}$ if every point $\bar{x}$ on it has the property that $\bar{x} = \alpha x^1 + (1 - \alpha)x^2$, for some $0 < \alpha < 1$. $x^1 \in \mathbf{K}$ and $x^2 \in \mathbf{K}$, imply that both x^1 and x^2 are also on this half-line.

Equivalently the half-line in (3.28) is an extreme half-line of $\mathbf{K}$ iff the set of column vectors $\{A_{.j}\colon j$ such that either $\tilde{x}_j$ or $\hat{y}_j$ or both are positive$\}$ is linearly dependent, and it contains a column vector such that when this column vector is deleted, the remaining set is linearly independent.

Problems

3.25 Prove that the two definitions of an extreme half-line given here are equivalent.

3.26 Let $\bar{x} \in \mathbf{K}$. If $\bar{x}$ is not an extreme point of $\mathbf{K}$, prove that it lies on an extreme half-line of $\mathbf{K}$ iff properties i and ii hold.

(i) Let $\mathbf{S} = \{j\colon j$ such that $\bar{x}_j > 0\}$. There exists numbers α_j such that

$$\begin{array}{ll} & \sum_{j \in \mathbf{S}} \alpha_j A_{.j} = 0 \\ \alpha_j \geqq 0 & \text{for all } j \in \mathbf{S} \\ \alpha_j > 0 & \text{for at least one } j \in \mathbf{S} \end{array}$$

(ii) Let $\mathbf{U} = \{j\colon \alpha_j > 0\} \subset \mathbf{S}$, then the set of column vectors obtained by deleting from $\{A_{.j}\colon j \in \mathbf{S}\}$ any one of the column vectors $A_{.j}$ for $j \in \mathbf{U}$ is linearly independent.

Example

Consider Example 4 in section 2.6.4.

x_1	x_2	x_3	x_4	b
1	1	−1	0	1
1	−1	0	−1	0

$x_j \geqq 0$ for all j

Let $\bar{x} = (1/2, 1/2, 0, 0)^T$ and $\hat{y} = (1/4, 1/4, 1/2, 0)^T$. Then $\bar{x}$ is a BFS of this system and $\hat{y}$ an extreme homogeneous solution corresponding to it. The half-line $\{x: x = \bar{x} + \hat{y}\theta, \theta \geqq 0\}$ is an extreme half-line of the set of feasible solutions. It is the thick half-line in Figure 2.4.

Another Example

In Figure 3.20, the thick half-line through x^3 is an extreme half-line. The parallel dashed half-line through x^1 is not an extreme half-line.

Example: An Illustration of Resolution Theorems

We have discussed resolution theorems for convex polyhedra that are the sets of feasible solutions of systems of linear equations in nonnegative variables. Similar theorems can be proved for general convex polyhedra. While considering the set of feasible solutions of a general system of linear equations and inequalities, the associated homogeneous system is obtained by changing the right-hand side constants in all the equations and inequalities to zero. For example, consider the general system:

$$\begin{aligned} Ax + Dy &= b \\ Ex + Fy &\geqq d \\ x \geqq 0, \quad & y \text{ unrestricted} \end{aligned}$$

The homogeneous system associated with it is:

$$\begin{aligned} Ax + Dy &= 0 \\ Ex + Fy &\geqq 0 \\ x \geqq 0, \quad & y \text{ unrestricted.} \end{aligned}$$

We will illustrate resolution theorems by considering a system of linear inequalities in two variables, because in this case all the relevant sets can be drawn conveniently on the two-dimensional Cartesian plane. Let **K** be the set of feasible solutions of the system of constraints:

$$\begin{aligned} x_1 - x_2 &\geqq -5 \\ x_1 - 3x_2 &\leqq -9 \\ x_1 &\geqq 1 \\ x_2 &\geqq 4 \\ x_1 + x_2 &\geqq 6 \end{aligned}$$

The homogeneous system associated with this is:

$$\begin{aligned} x_1 - x_2 &\geqq 0 \\ x_1 - 3x_2 &\leqq 0 \\ x_1 &\geqq 0 \\ x_2 &\geqq 0 \\ x_1 + x_2 &\geqq 0 \end{aligned}$$

The last constraint in this homogeneous system is automatically implied by the others and hence can be deleted. Let $\mathbf{K}^{\angle}$ be the set of feasible solutions of the homogeneous system.

$\mathbf{K}^{\angle}$ is the dotted cone in Figure 3.22. The extreme homogeneous solutions associated with this system are $x^5 = (1/2, 1/2)$, and $x^6 = (3/4, 1/4)$. The rays of x^5, x^6 are the extreme rays of $\mathbf{K}^{\angle}$.

The extreme points of $\mathbf{K}$ are x^1, x^2, x^3, x^4 marked in Figure 3.22. $\mathbf{K}^{\Delta}$ is their convex hull and it is the part marked with dashed lines in $\mathbf{K}$. Every vector in $\mathbf{K}$ can be expressed as the sum of a vector in $\mathbf{K}^{\Delta}$ and a vector in $\mathbf{K}^{\angle}$. This is resolution theorem I. Equivalently, resolution theorem II states that every vector in $\mathbf{K}$ can be expressed as

$$\alpha_1 x^1 + \alpha_2 x^2 + \alpha_3 x^3 + \alpha_4 x^4 + \beta_1 x^5 + \beta_2 x^6$$

$$\text{where} \quad \alpha_1 + \alpha_2 + \alpha_3 + \alpha_4 = 1, \quad \text{and} \quad \alpha_1, \alpha_2. \alpha_3, \alpha_4, \beta_1, \beta_2 \geqq 0.$$

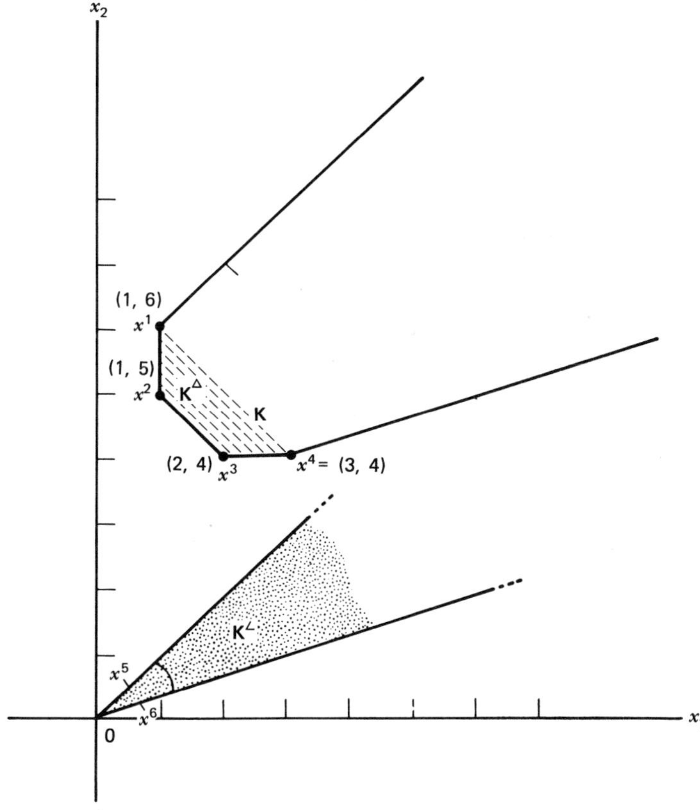

Figure 3.22

$\mathbf{K}$ has two extreme half-lines. The extreme half line of $\mathbf{K}$ through x^1 is parallel to the ray of x^5. It is $\{x: x = x^1 + \lambda x^5, \lambda \geqq 0\}$. The extreme half line of $\mathbf{K}$ through x^4 is parallel to the ray of x^6. It is $\{x: x = x^4 + \lambda x^6, \lambda \geqq 0\}$.

While considering a convex polyhedron that is the set of feasible solutions of a general system of linear equations and inequalities, the definition of an extreme homogeneous solution and the statements of the resolution theorems have to be modified appropriately. See references [3] and [6 to 8] at the end of this chapter. For our study of linear programming, it is adequate to consider systems of linear equations in nonnegative variables, and hence we have only discussed this case.

UNBOUNDEDNESS IN THE SIMPLEX ALGORITHM

Consider the LP (3.16). Suppose $z(x)$ is unbounded below on the set of feasible solutions of this problem. We will now show that at termination, the simplex algorithm provides an extreme half-line in this problem along which $z(x)$ diverges to $-\infty$. In solving this problem, the simplex algorithm will terminate with a canonical tableau in which the unboundedness criterion is satisfied. Suppose the terminal basis is the basis B. For convenience in referring to it, assume that the associated basic vector is $x_B = (x_1, \ldots, x_m)^T$. Suppose the final canonical tableau is:

Basic variables	$x_1 \ldots x_m$	$x_{m+1} \ \ldots \ x_n$	$-z$	b
x_1	$1 \ldots 0$	$\bar{a}_{1,m+1} \ldots \bar{a}_{1n}$	0	$\bar{b}_1$
$\vdots$	$\vdots \quad \vdots$	$\vdots \quad \vdots$	$\vdots$	$\vdots$
x_m	$0 \ldots 1$	$\bar{a}_{m,m+1} \ldots \bar{a}_{mn}$	0	$\bar{b}_m$
$-z$	$0 \ldots 0$	$\bar{c}_{m+1} \ \ldots \bar{c}_m$	1	$-\bar{z}$

In this canonical tableau, suppose the column vector of the nonbasic variable x_{m+1} satisfies the unboundedness criterion. Therefore

$$\bar{c}_{m+1} < 0$$

and

$$\bar{a}_{i,m+1} \leqq 0 \qquad \text{for all } i \tag{3.29}$$

$\bar{A}$ is obtained from A by doing a sequence of elementary row operations on A. Hence, if a vector $\bar{y}$ satisfies $\bar{A}\bar{y} = 0$, then $A\bar{y} = 0$ must also hold. Define $\bar{y}$ by

$$\begin{aligned} \bar{y}_i &= -\bar{a}_{i,m+1} && \text{for } i = 1 \text{ to } m \\ &= 1 && \text{for } i = m+1 \\ &= 0 && \text{for } i \geqq m+2 \end{aligned}$$

By (3.29), $\bar{y} \geq 0$. From the final canonical tableau, clearly $\bar{A}\bar{y} = 0$. So $A\bar{y} = 0$, also. Normalizing $\bar{y}$ so that the sum of all the components in it becomes equal to 1 leads to $\hat{y}$ where

$$\begin{aligned} \hat{y}_i &= \frac{-\bar{a}_{i,m+1}}{1 - \sum_{i=1}^{m} \bar{a}_{i,m+1}} && \text{for } i = 1 \text{ to } m \\ &= \frac{1}{1 - \sum_{i=1}^{m} \bar{a}_{i,m+1}} && \text{for } i = m+1 \\ &= 0 \quad \text{for all other } i \end{aligned}$$

Then $\hat{y} \geqq 0$ and $\sum_{j=1}^{n} \hat{y}_j = 1$ and by the above discussion $A\hat{y} = 0$. Clearly, $\hat{y}$ is an extreme homogeneous solution corresponding to (3.16). From the canonical tableau, we also verify that

$$z(\hat{y}) = c\hat{y} = \left(\frac{\bar{c}_{m+1}}{1 - \sum_{i=1}^{m} \bar{a}_{i,m+1}} \right) < 0$$

The BFS of (3.16) associated with the terminal basis is $\bar{x}$, where

$$\begin{aligned} \bar{x}_i &= \bar{b}_i && \text{for } i = 1 \text{ to } m \\ &= 0 && \text{for } i = m+1 \text{ to } n \end{aligned}$$

The half-line constructed in the simplex algorithm, along which $z(x)$ diverges to $-\infty$, is (see section 2.5.3)

$$\begin{aligned} x_i &= \bar{b}_i - \lambda\bar{a}_{i,\,m+1} && \text{for } i = 1 \text{ to } m \\ &= \lambda && \text{for } i = m+1 \\ &= 0 && \text{for } i \geqq m+2 \end{aligned}$$

for $\lambda \geqq 0$. This half-line is exactly the half-line

$$\{x\colon x = \bar{x} + \alpha\hat{y},\, \alpha \geqq 0\}$$

Every point on this half-line makes the variables $x_{m+2} = \cdots = x_n = 0$. Also every feasible solution of (3.16) in which the variables $x_{m+2} = \cdots = x_n = 0$ clearly lies on this line. If x^0 is any point on this line and $x^0 = \alpha x^1 + (1-\alpha)x^2$ for some $0 < \alpha < 1$ and feasible solutions x^1 and x^2, then $x^1_{m+2} = \cdots = x_n{}^1 = x^2_{m+2} = \cdots = x_n{}^2 = 0$. Hence x^1 and x^2 are on the half-line, too. Hence this half-line is an extreme half-line of the set of feasible solutions of (3.16).

Thus when the objective function in the LP being solved is unbounded, the simplex algorithm terminates by providing an extreme half-line of the set of feasible solutions along which the objective function diverges.

Example

Consider the LP of Example 4, section 2.6.4.

x_1	x_2	x_3	x_4	$-z$	b	
1	1	−1	0	0	1	
1	−1	0	−1	0	0	
−1	−1	0	0	1	0	minimize z

$x_j \geqq 0$ for all j

Consider the basic vector (x_1, x_2). This is a feasible basic vector. The associated BFS is $\tilde{x} = (1/2, 1/2, 0, 0)^{\mathrm{T}}$. Here is the canonical tableau with respect to this basic vector.

x_1	x_2	x_3	x_4	$-z$	b
1	0	$-\frac{1}{2}$	$-\frac{1}{2}$	0	$\frac{1}{2}$
0	1	$-\frac{1}{2}$	$\frac{1}{2}$	0	$\frac{1}{2}$
0	0	−1	0	1	1

From the updated column vector of x_3 we construct the extreme homogeneous solution:

$$\hat{y} = \left(\frac{1}{4}, \frac{1}{4}, \frac{1}{2}, 0\right)^{\mathrm{T}}$$

The extreme half-line along which $z(x)$ diverges to $-\infty$ is

$$\left\{x: x = \left(\frac{1}{2}, \frac{1}{2}, 0, 0\right)^{\mathrm{T}} + \lambda\left(\frac{1}{4}, \frac{1}{4}, \frac{1}{2}, 0\right)^{\mathrm{T}}, \lambda \geqq 0\right\}$$

which is the extreme half-line obtained in section 2.6.4, Example 4.

3.8 THE MAIN GEOMETRICAL ARGUMENT IN THE SIMPLEX ALGORITHM

From the results discussed in the previous sections, we derive the following result, which is the main mathematical result in the simplex algorithm: Let $\mathbf{K}$ be a convex polyhedron in $\mathbf{R}^n$, and $z(x)$ be a linear objective function defined on it.

1. If x^* is any extreme point of $\mathbf{K}$, either x^* minimizes $z(x)$ on $\mathbf{K}$ or there exists an edge (bounded or unbounded) of $\mathbf{K}$ through x^*, such that $z(x)$ decreases strictly as we move along this edge from x^*.
2. If $z(x)$ is bounded below on $\mathbf{K}$, and $\tilde{x}$ is any extreme point of $\mathbf{K}$, either $\tilde{x}$ minimizes $z(x)$ on $\mathbf{K}$, or there exists an adjacent extreme point of $\tilde{x}$ on $\mathbf{K}$, say $\hat{x}$, such that $z(\hat{x}) < z(\tilde{x})$.

ADJACENT VERTEX METHODS

For minimizing a linear objective function, $z(x)$, on a convex polyhedron $\mathbf{K}$, the simplex algorithm starts at an extreme point of $\mathbf{K}$. In each step it moves from the present extreme point of $\mathbf{K}$ along an edge of $\mathbf{K}$ incident at this extreme point, such that $z(x)$ decreases as the algorithm travels along this edge. If the edge is an unbounded edge, $z(x)$ must be unbounded below on $\mathbf{K}$. If the edge is bounded, the algorithm moves to the adjacent extreme point on the other side of this edge and continues in the same manner until either, (1) an extreme point is reached with the property that $z(x)$ does not decrease by moving from this extreme point along every edge of $\mathbf{K}$ containing this extreme point, or (2) an unbounded edge is found along which $z(x)$ diverges to $-\infty$.

Algorithms of this type are known as *adjacent vertex methods.* The class of optimization problems that can be solved by adjacent vertex methods includes LPs, fractional linear programming problems (those in which a function of the form $z(x) = (dx + \alpha)/(fx + \beta)$ has to be minimized subject to linear equality constraints in nonnegative variables) and some other class of problems in which a nonlinear objective function $\theta(x)$ satisfying certain monotonicity properties has to be minimized subject to linear equations in nonnegative variables. See reference [4] at the end of this chapter.

3.9 THE DIMENSION OF A CONVEX POLYHEDRON

A subset $\mathbf{S} \subset \mathbf{R}^n$ is said to be a *subspace of* $\mathbf{R}^n$ if: $x^1 \in \mathbf{S}$ and $x^2 \in \mathbf{S}$ implies $\alpha x^1 + \beta x^2 \in \mathbf{S}$ for all real values of α, β, that is, the linear hull of any pair of points in a subspace is in the subspace. All subspaces contain the origin O.

A subset $\{x^1, \ldots, x^k\}$ of points in a subspace $\mathbf{S}$ is said to be a *maximal linearly independent subset of points in* $\mathbf{S}$ if $\{x^1, \ldots, x^k\}$ is linearly independent and $\{x^1, \ldots, x^k, x\}$ is linearly dependent for every $x \in \mathbf{S}$.

It can be shown that all maximal linearly independent subsets of a given subspace of $\mathbf{R}^n$, contain the same number of points in them. This number is called the *dimension* of that subspace. Also, if $\{x^1, \ldots, x^k\}$ is a maximal linearly independent subset of points in a subspace $\mathbf{S}$, then $\mathbf{S}$ is the linear hull of $\{x^1, \ldots, x^k\}$.

A subset $\mathbf{\Gamma} \subset \mathbf{R}^n$ is said to be an *affine space* if: $x^1 \in \mathbf{\Gamma}$ and $x^2 \in \mathbf{\Gamma}$, implies $\alpha x^1 + (1 - \alpha)x^2 \in \mathbf{\Gamma}$ for all real values of α. Let $\mathbf{\Gamma}$ be an affine space and let $x^0 \in \mathbf{\Gamma}$. Then $\{x: x = y - x^0, y \in \mathbf{\Gamma}\}$ is a subspace of $\mathbf{R}^n$. The *dimension* of $\mathbf{\Gamma}$ is defined as the dimension of this subspace. Hence the dimension of $\mathbf{\Gamma}$ is the maximum value of r such that there exist points $x^1,\ldots,x^r$ *in* $\mathbf{\Gamma}$ satisfying: $\{x^1 - x^0,\ldots, x^r - x^0\}$ is a linearly independent set.

Let A be a matrix of order $m \times n$ and rank r. Then the set of solutions of

$$Ax = 0$$

is a subspace of $\mathbf{R}^n$ of dimension $n - r$. See references [2, 6 to 11] at the end of this chapter. The set of solutions of

$$Ax = b$$

is either the empty set, or an affine space of dimension $n - r$.

Let $\mathbf{K}$ be a convex polyhedral subset of $\mathbf{R}^n$. If $\mathbf{K} \neq \emptyset$, its dimension is defined to be the dimension of the smallest dimension affine space containing it. Equivalently, let x^0 be any point in $\mathbf{K}$. The dimension of $\mathbf{K}$ is the maximum number r such that there exist points $x^1, \ldots, x^r$ in $\mathbf{K}$ satisfying: $\{x^1 - x^0, \ldots, x^r - x^0\}$ is a linearly independent set.

The dimension of a convex polyhedron that is the set of feasible solutions of a system of linear equations in nonnegative variables is easily specified, under nondegeneracy. Let $\mathbf{K}$ be the set of feasible solutions of (3.4) where A is a matrix of order $m \times n$ and rank m. Let b be nondegenerate (i.e., every solution for the system has at least m nonzero coordinates, or equivalently, b does not lie in the linear hull of any set containing $m - 1$ or less columns of A). Then the dimension of $\mathbf{K}$ is $n - m$. If b is degenerate (i.e., not nondegenerate), the dimension of $\mathbf{K}$ is $n - m$ or less. See example below, and problems 3.51, 3.52.

Degenerate Example

Consider the system

$$\begin{aligned} x_1 - x_3 &= 0 \\ x_2 + x_3 &= 0 \\ x_1, x_2, x_3 &\geqq 0 \end{aligned}$$

Here $n = 3$, $m = 2$. Hence $n - m = 1$. However the system has the unique feasible solution $x = 0$, and since the set of feasible solutions contains a single point, its dimension is zero.

3.10 ALTERNATE OPTIMUM FEASIBLE SOLUTIONS AND FACES OF CONVEX POLYHEDRA

Consider the LP (3.16) where A is of order $m \times n$ and rank m. Suppose B_1 is an optimum feasible basis of this problem. Let the canonical tableau with respect to this basis be as follows.

Basic variables	x_B	$-z$	Updated nonbasic column vectors $x_{m+1} \ \cdots x_j \ \cdots x_n$	
x_B	I	0	$\bar{a}_{1,m+1} \cdots \bar{a}_{1j} \cdots \bar{a}_{1n}$ $\vdots \quad \vdots \quad \vdots$ $\bar{a}_{m,m+1} \cdots \bar{a}_{mj} \cdots \bar{a}_{mn}$	$B_1^{-1}b$
$-z$	0	1	$\bar{c}_{m+1} \ \cdots \bar{c}_j \ \cdots \bar{c}_n$	$-\bar{z}$

The optimum objective value in this problem is $\bar{z}$. Therefore, by the fundamental optimality criterion any feasible solution x that makes $z(x) = \bar{z}$ is an optimum feasible solution of this problem and vice versa. Hence the set of all optimum feasible solutions of (3.16) is the set of feasible solutions of

$$\begin{aligned} Ax &= b \\ z(x) &= cx = \bar{z} \\ x &\geqq 0 \end{aligned}$$

This leads to the following results.

1. The set of optimum feasible solutions of an LP is itself a convex polyhedral set. Consequently,
2. every convex combination of optimum feasible solutions is also optimal.

Since B_1 is an optimum basis, all the relative cost coefficients $\bar{c}_j$ must be nonnegative. The cost row in the canonical tableau leads to

$$z(x) = \bar{z} + \sum_{\substack{j \text{ such that} \\ x_j \text{ is nonbasic}}} \bar{c}_j x_j$$

From this equation it is clear that every feasible solution of (3.16), in which the variables x_j corresponding to strictly positive $\bar{c}_j$ are zero, is an optimum feasible solution. Hence, if $\Gamma = \{j : j \text{ such that } \bar{c}_j > 0\}$, the set of optimum feasible solutions of (3.16) is the set of feasible solutions of (3.16) in which $x_j = 0$ for all $j \in \Gamma$.

Thus, if x_s is a nonbasic variable in the final optimum canonical tableau, and $\bar{c}_s = 0$, and if the operation of bringing x_s into the basic vector involves a nondegenerate pivot step, this step yields an alternate optimum BFS. Alternate optimal bases are obtained by bringing into an optimal basic vector any nonbasic variable with a zero relative cost coefficient.

The LP (3.16) has a unique optimum feasible solution if there exists an optimum basis, such that in the canonical tableau corresponding to it, the relative cost coefficients of all the nonbasic variables are strictly positive.

Example

Consider the canonical tableau.

Basic variables	x_1	x_2	x_3	x_4	x_5	x_6	x_7	$-z$	b
x_1	1	0	0	0	1	1	0	0	3
x_2	0	1	0	1	1	-2	3	0	0
x_3	0	0	1	-2	1	1	7	0	2
$-z$	0	0	0	0	6	0	8	1	-10

An optimum basic vector of this problem is (x_1, x_2, x_3). The optimum solution corresponding to it is $x^{\mathrm{T}} = (3, 0, 2, 0, 0, 0, 0)^{\mathrm{T}}$ with objective value $= 10$. When x_6 is brought into this basic vector, it drives x_3 out of the basic vector and leads to the following new optimum canonical tableau.

Basic variables	x_1	x_2	x_3	x_4	x_5	x_6	x_7	$-z$	Basic values
x_1	1	0	−1	2	0	0	−7	0	1
x_2	0	1	2	−3	3	0	17	0	4
x_6	0	0	1	−2	1	1	7	0	2
$-z$	0	0	0	0	6	0	8	1	−10

and a new optimum BFS, $x^T = (1, 4, 0, 0, 0, 2, 0)$.

Problems

3.27 Obtain all the optimum BFSs to this problem.

3.28 In this problem prove that every feasible solution in which $x_5 = x_7 = 0$ is an optimum feasible solution and vice versa.

3.29 Write down a general expression that represents an optimum feasible solution to this problem.

FACES OF A CONVEX POLYHEDRON

Let $\mathbf{K} \subset \mathbf{R}^n$ be a convex polyhedron. Suppose $d \neq 0$ is a row vector of order $1 \times n$. Let $\mathbf{H}$ represent the hyperplane, defined by

$$\mathbf{H} = \{x: dx = d_0\}$$

for some real number d_0. If

$$dx \geqq d_0 \qquad \text{for all } x \in \mathbf{K}$$

the hyperplane $\mathbf{H}$ is said to have the convex polyhedron $\mathbf{K}$ on one of its sides. If $\mathbf{H}$ has $\mathbf{K}$ on one of its sides and, in addition, if there is a point $\tilde{x}$ such that $\tilde{x} \in \mathbf{K} \cap \mathbf{H}$, that is, if $\tilde{x} \in \mathbf{K}$ and $d\tilde{x} = d_0$, then $\mathbf{H}$ is said to be a *supporting hyperplane* for the convex polyhedron $\mathbf{K}$ at the point $\tilde{x} \in \mathbf{K}$. See Figure 3.23.

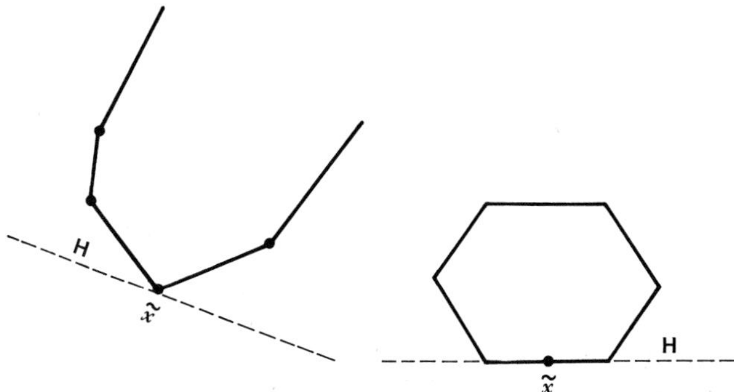

Figure 3.23 Supporting hyperplanes.

A *face* of a convex polyhedron $\mathbf{K}$ is either the convex polyhedron itself, the empty set, or the intersection of the convex polyhedron with a supporting hyperplane. A *facet* of a convex polyhedron is a face of it, whose dimension is one less than the dimension of the convex polyhedron (Figure 3.24).

Theorem The set of optimum feasible solutions of an LP is a face of its set of feasible solutions.

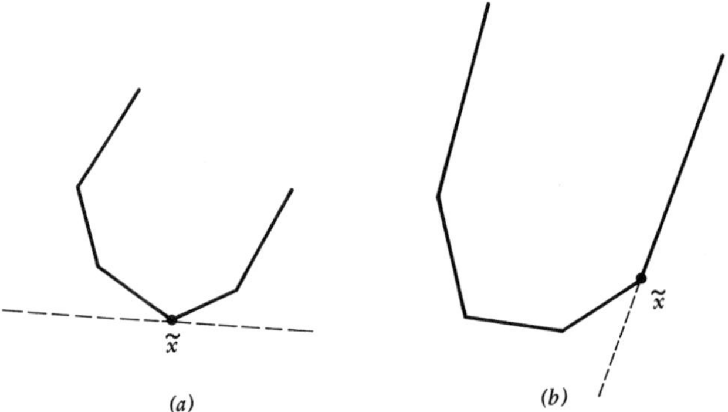

Figure 3.24 (*a*) Face $\{\bar{x}\}$. This is not a facet. (*b*) Face is the extreme half-line through $\bar{x}$, which is a facet.

Proof Consider the LP (3.16). Let **K** be the set of feasible solutions of this LP. If $z(x)$ is unbounded below on **K**, there are no optimum feasible solutions; hence, the set of optimum feasible solutions is the empty set, this is a face of **K** by definition. On the other hand, if the LP has an optimum feasible solution, let $\bar{z}$ be the optimum objective value. Then $z(x) = cx \geqq \bar{z}$ for all $x \in \mathbf{K}$, and the set of optimum feasible solutions is the set $\{x: x \in \mathbf{K} \text{ and } cx = \bar{z}\}$; this is a face of **K** by definition.

Problems

3.30 Let **K** be the set of feasible solutions of

$$\begin{aligned} Ax &= b \\ x &\geqq 0 \end{aligned}$$

Let $\mathbf{J} \subset \{1, 2, \ldots, n\}$. Prove that the set of feasible solutions of

$$\begin{aligned} Ax &= b \\ x &\geqq 0 \\ x_j &= 0 \qquad \text{for all } j \subset \mathbf{J} \end{aligned}$$

is a face of **K** and conversely that every face of **K** is of this form.

3.31 Let **K** be a convex polyhedron in the two-dimensional plane and suppose that $z(x)$ is a linear objective function defined on **K**. If the problem of minimizing $z(x)$ on **K** has three distinct extreme points of **K** optimal to it, then $z(x)$ is a constant on **K**.

3.11 PIVOT MATRICES

Consider the LP (3.16) where A is a matrix of order $m \times n$ and rank m. The main operations used in the simplex algorithm are the *pivot operations*. The pivot on the input-output coefficient matrix with the sth column as the *pivot column*, the rth row as the *pivot row*, and a_{rs} as the *pivot element* can only be performed if the pivot element is nonzero. It transforms the matrix from

$$A = \begin{bmatrix} a_{11} & \cdots & a_{is} & \cdots & a_{1n} \\ \vdots & & \vdots & & \vdots \\ a_{r1} & \cdots & \textcircled{a_{rs}} & \cdots & a_{rn} \\ \vdots & & \vdots & & \vdots \\ a_{m1} & \cdots & a_{ms} & \cdots & a_{mn} \end{bmatrix}$$

into

$$\bar{A} = \begin{bmatrix} \bar{a}_{11} & \cdots & 0 & \cdots & \bar{a}_{1n} \\ \vdots & & \vdots & & \vdots \\ \bar{a}_{r1} & \cdots & 1 & \cdots & \bar{a}_{rn} \\ \vdots & & \vdots & & \vdots \\ \bar{a}_{m1} & \cdots & 0 & \cdots & \bar{a}_{mn} \end{bmatrix}$$

by transforming the pivot column into the rth column vector of the unit matrix of order m. Clearly

$$\begin{aligned} \bar{a}_{ij} &= a_{ij} - \frac{a_{rj}a_{is}}{a_{rs}} && \text{for } i \neq r \\ &= \frac{a_{rj}}{a_{rs}}, && \text{for } i = r \end{aligned}$$

Let P be the square matrix of order m that differs from the unit matrix only in its rth column.

$$P = \begin{bmatrix} 1 & \cdots & 0 & -a_{1s}/a_{rs} & 0 & \cdots & 0 \\ \vdots & & \vdots & \vdots & \vdots & & \vdots \\ 0 & \cdots & 1 & -a_{r-1,\,s}/a_{rs} & 0 & \cdots & 0 \\ 0 & \cdots & 0 & 1/a_{rs} & 0 & \cdots & 0 \\ 0 & \cdots & 0 & -a_{r+1,\,s}/a_{rs} & 1 & \cdots & 0 \\ \vdots & & \vdots & \vdots & \vdots & & \vdots \\ 0 & \cdots & 0 & -a_{ms}/a_{rs} & 0 & \cdots & 1 \end{bmatrix}$$

It is easily verified that

1. The (r, r)th entry in P is the reciprocal of the pivot element; it is nonzero and hence, P is nonsingular.
2. $\bar{A} = PA$

For this reason the matrix P is known as the *pivot matrix* corresponding to this pivot step. The column vector in which P differs from the unit matrix depends on the position of the pivot row in the matrix A. If the pivot row is the rth row, P differs from the unit matrix only in its rth column. Pivot matrices belong to a class of matrices called *elementary matrices* in linear algebra. An elementary matrix is a square nonsingular matrix which is either a permutation matrix, or a matrix which differs from the unit matrix in just a row or a column.

In the kth stage of the algorithm, suppose the canonical tableau is:

$x_1 \ldots x_s \ldots x_n$	$-z$	b	
$\bar{a}_{11} \ldots \bar{a}_{1s} \ldots \bar{a}_{1n}$	0	$\bar{b}_1$	
$\vdots \quad \vdots \quad \vdots$	$\vdots$	$\vdots$	
$\bar{a}_{r1} \ldots (\bar{a}_{rs}) \ldots \bar{a}_{rn}$	0	$\bar{b}_r$	Pivot row
$\vdots \quad \vdots \quad \vdots$	$\vdots$	$\vdots$	
$\bar{a}_{m1} \ldots \bar{a}_{ms} \ldots \bar{a}_{mn}$	0	$\bar{b}_m$	
$\bar{c}_1 \ldots \bar{c}_s \ldots \bar{c}_n$	1	$-\bar{z}$	

Pivoting on this tableau with the sth column as the pivot column and the rth row as the pivot row is equivalent to multiplying every column in the tableau on the left by the pivot matrix:

$$P^k = \begin{bmatrix} 1 & \cdots & 0 & -\bar{a}_{1s}/\bar{a}_{rs} & 0 & \cdots & 0 \\ \vdots & & \vdots & \vdots & \vdots & & \vdots \\ 0 & \cdots & 1 & -\bar{a}_{r-1,s}/\bar{a}_{rs} & 0 & \cdots & 0 \\ 0 & \cdots & 0 & 1/\bar{a}_{rs} & 0 & \cdots & 0 \\ 0 & \cdots & 0 & -\bar{a}_{r+1,s}/\bar{a}_{rs} & 1 & \cdots & 0 \\ \vdots & & \vdots & \vdots & \vdots & & \vdots \\ 0 & \cdots & 0 & -\bar{c}_s/\bar{a}_{rs} & 0 & \cdots & 1 \end{bmatrix} \quad (3.30)$$

(the column containing $-\bar{a}_{1s}/\bar{a}_{rs}$ is labelled "rth-column")

which is of order $m + 1$, since each column in the tableau has $m + 1$ entries including the cost row.

Since a pivot matrix always differs from the unit matrix in just one column vector, it can be stored easily by storing this column vector and its position in the pivot matrix. The revised simplex method with the product form of the inverse discussed in Chapter 5 is based on the use of these pivot matrices.

3.12 CANONICAL TABLEAUS IN MATRIX NOTATION

Let B be a feasible basis of the LP (3.16). Let x_B be the basic vector and c_B the corresponding vector of cost coefficients. Suppose x_D is the vector of nonbasic variables and c_D the vector of their cost coefficients. After partitioning into the basic and nonbasic parts, (3.16) in tableau form is as follows.

x_B	$-z$	x_D	b
B	0	D	b
c_B	1	c_D	0

The canonical tableau with respect to the basis B is obtained by transforming the submatrix under the variables x_B, $-z$ into the unit matrix of order $m + 1$. This is achieved by multiplying each column in the tableau on the left by the matrix

$$T = \left(\begin{array}{c|c} B & 0 \\ \hline c_B & 1 \end{array}\right)^{-1} = \left(\begin{array}{c|c} B^{-1} & 0 \\ \hline -c_B B^{-1} & 1 \end{array}\right) \quad (3.31)$$

Hence this is the canonical tableau with respect to the basis B.

Basic variables	x_B	$-z$	x_D	Basic values
x_B	I	0	$B^{-1}D$	$B^{-1}b$
$-z$	0	1	$c_D - c_B B^{-1}D$	$-c_B B^{-1}b$

The column vector of x_j in the original tableau is $\begin{pmatrix} A_{.j} \\ \cdots \\ c_j \end{pmatrix}$.

The column vector of x_j in the canonical tableau is known as the *updated column vector of* x_j. It is $\begin{pmatrix} \bar{A}_{.j} \\ \cdots \\ \bar{c}_j \end{pmatrix}$ where

$$\bar{A}_{.j} = B^{-1}A_{.j}$$
$$\bar{c}_j = c_j - c_B B^{-1} A_{.j} = c_j - c_B \bar{A}_{.j}{}^*$$

The updated right-hand side constant vector is $\begin{pmatrix} \bar{b} \\ \cdots \\ -\bar{z} \end{pmatrix}$ where

$$\bar{b} = B^{-1}b$$
$$\bar{z} = c_B B^{-1} b$$

These are the values of the basic vector and the objective function in the basic solution of (3.16) corresponding to the basis B.

The matrix in (3.31) is known as the *inverse matrix* corresponding to this basis. The revised simplex method with the explicit form of the inverse discussed in Chapter 5 is based on the use of inverse matrices.

3.13 WHY IS AN ALGORITHM REQUIRED TO SOLVE LINEAR PROGRAMS?

Let $f(c, A, b)$ denote the optimum objective value in (3.16) as a function of the problem data c, A, and b.

$$\begin{aligned} f(c, A, b) &= \text{minimum value of } cx \\ \text{subject to} \quad Ax &= b \\ x &\geqq 0 \end{aligned} \tag{3.16}$$

Is it possible to obtain $f(c, A, b)$ by a compact formula without computing it by an algorithm for given c, A, b? Consider the simple case where A, c remain fixed, but b may vary. In this case let $g(b) = f(c, A, b)$. The set of $b \in \mathbf{R}^m$ for which (3.16) has a feasible solution is the convex cone $\{b : b = \sum A_{.j} x_j \text{ for some } x_1 \geqq 0, \ldots, x_n \geqq 0\}$. This cone is pos $\{A_{.1}, \ldots, A_{.n}\}$ = pos (A). A basis B of (3.16) is a feasible basis if $B^{-1}b \geqq 0$, that is, if b can be expressed as a nonnegative combination of column vectors of B. Hence, B is a feasible basis for (3.16) iff $b \in$ pos (B). B is an optimum basis for (3.16) if it is feasible and if the relative cost coefficients with respect to B are all nonnegative. Any basis B with respect to which the relative cost coefficients are all nonnegative is known as a *dual feasible basis* (to be discussed in detail in Chapter 4). Hence if B is a dual feasible basis for (3.16) and if $b \in$ pos (B), then an optimum solution of (3.16) is

$$\begin{aligned} x_B &= B^{-1}b \\ \text{All other } x_j &= 0 \\ \text{Minimum objective value} &= g(b) = c_B B^{-1} b \end{aligned}$$

Therefore as b varies in the cone pos (B), $g(b)$ is a linear function of b. The number of dual feasible bases for (3.16) grows very rapidly with $n - m$. Its growth rate is possibly of the order of $(n - m)!$ Hence, unless $n - m$ is a very small number, the total number of dual feasible bases for (3.16) is likely to be very large.

* In some books $c_B \bar{A}_{.j}$ is denoted by the symbol z_j and the optimality criterion stated as $c_j - z_j \geqq 0$ for all j, for the minimization problem.

Even if we are able to tabulate all these cones, pos (B), corresponding to the various dual feasible bases for (3.16) and the formula for $g(b)$ as a linear function of b in each one of them, the table will be very long because of the large number of these bases. To find out the value of $g(b)$ for any given b, we have to search the table for a cone pos (B) in which the vector b lies. This can be a very tedious job unless a systematic algorithm (or an iterative scheme) is developed for this search, because of the length of the table. This is exactly what the simplex algorithm does anyway. It is an iterative scheme for solving (3.16).

LPs can only be solved efficiently by using some algorithm that guarantees that the optimal solution will be obtained in a finite number of iterations. The simplex algorithm or one of its modern variants provide such an algorithm.

WHY IS IT CALLED THE SIMPLEX ALGORITHM

Let $\{x^1, \ldots, x^{n+1}\}$ be a set of points in $\mathbf{R}^n$ such that $\{x^2 - x^1, \ldots, x^{n+1} - x^1\}$ is a linearly independent set. Then the convex hull of $\{x^1, \ldots, x^{n+1}\}$ is known as an *n-simplex.*

If the set $\{x^1, \ldots, x^n\}$ of points in $\mathbf{R}^n$ is linearly independent, clearly the convex hull of $\{0, x^1, \ldots, x^n\}$ is an n-simplex. The cone pos $\{x^1, \ldots, x^n\}$ is known as a *simplicial cone in* $\mathbf{R}^n$.

If pos $\{x^1, \ldots, x^n\}$ is a simplicial cone, then for any i, pos $\{x^1, \ldots, x^{i-1}, x^{i+1}, \ldots, x^n\}$ is a facet of it. Verify this for $n = 3$ from Figure 3.25.

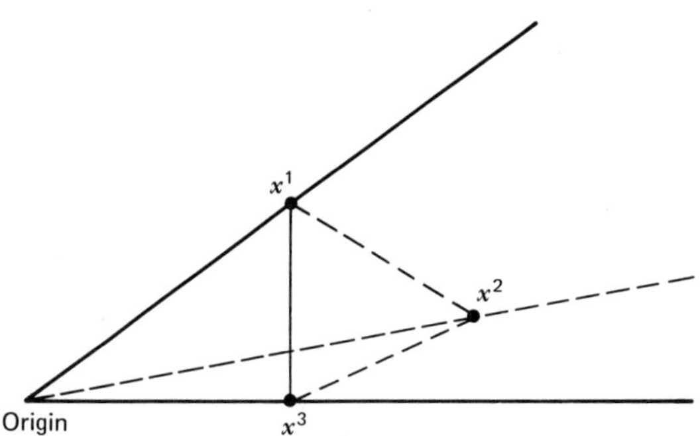

Figure 3.25 A simplical cone in $\mathbf{R}^3$.

Consider the LP

$$\begin{aligned} \text{Minimize} \quad & cx \\ \text{Subject to} \quad & Ax = b \\ & x \geqq 0 \end{aligned} \tag{3.32}$$

where A is a matrix of order $m \times n$ and rank m. The original column vector including the cost coefficient, associated with the variable x_j in this problem is $A_{.j} = (a_{1j}, \ldots, a_{mj}, c_j)^{\mathrm{T}}, j = 1$ to n.

Let $y = (y_1, \ldots, y_m, y_{m+1})$ be the coordinates of a general point in $\mathbf{R}^{m+1}$. Plot each of the points $A_{.j}, j = 1$ to n in $\mathbf{R}^{m+1}$. If $x_j \geqq 0$ for all j, the point $\sum_{j=1}^{n} x_j A_{.j}$ will be a point in the cone pos $\{A_{.1}, \ldots, A_{.n}\}$. If $\bar{x}$ is a feasible solution of (3.32) with an objective value of $\bar{z}$, then $\sum_{j=1}^{n} \bar{x}_j A_{.j} = (b_1, \ldots, b_m, \bar{z})^{\mathrm{T}}$, and this is a point in pos $\{A_{.1}, \ldots, A_{.n}\}$. Conversely, if $(b_1, \ldots, b_m, \alpha)^{\mathrm{T}}$ is a point in pos $\{A_{.1}, \ldots, A_{.n}\}$, by the definition of the pos cone, there must exist nonnegative numbers $\hat{x}_1, \ldots, \hat{x}_n$ such that $(b_1, \ldots, b_m, \alpha)^{\mathrm{T}} = \sum_{j=1}^{n} \hat{x}_j A_{.j}$.

Consider the straight line in $\mathbf{R}^{m+1}$, denoted by $\mathbf{L}$, defined by

$$\begin{aligned} y_1 &= b_1 \\ &\vdots \\ y_m &= b_m \\ y_{m+1} &= z \end{aligned} \tag{3.33}$$

where z is treated as a real-valued parameter here. This is a line in $\mathbf{R}^{m+1}$, parallel to the y_{m+1}-axis, through the point $(b_1, \ldots, b_m, 0)^{\mathrm{T}}$.

From the above argument it is clear that if $(b_1, \ldots, b_m, \alpha)^{\mathrm{T}}$ is a point on $\mathbf{L} \cap \text{pos}\,\{A_{.1}, \ldots, A_{.n}\}$, there exists a feasible solution of (3.32) that has an objective value of α (i.e., it makes $cx = \alpha$). Hence the problem of minimizing cx in (3.32) is equivalent to the problem of finding the minimum value of the parameter z such that the point (3.33) on $\mathbf{L}$ is in pos $\{A_{.1}, \ldots, A_{.n}\}$. This is the problem of finding the lowest point (in the direction of the y_{m+1}-axis) in $\mathbf{L} \cap \text{pos}\,\{A_{.1}, \ldots, A_{.n}\}$. Hence the straight line $\mathbf{L}$ is called the *requirement line* and we want to find the lowest point on it which lies in pos $\{A_{.1}, \ldots, A_{.n}\}$.

In the course of solving this problem by the simplex algorithm, consider a step in which the feasible basic vector is $(x_1, \ldots, x_m)$, say. Let the objective value of the present BFS be $\bar{z}$. In the present BFS, only the basic variables $x_1, \ldots, x_m$ can take positive values; all the nonbasic variables, $x_{m+1}, \ldots, x_n$ are zero. These facts imply that the point $(b_1, \ldots, b_m, \bar{z})^{\mathrm{T}}$ is in pos $\{A_{.1}, \ldots, A_{.m}\}$.

Let x_{m+1} be the entering variable in this step of the algorithm. Let $\bar{A}_{.m+1} = (\bar{a}_{1,m+1}, \ldots, \bar{a}_{m,m+1}, \bar{c}_{m+1})^{\mathrm{T}}$ be the updated column vector of x_{m+1} with respect to the present basic vector $(x_1, \ldots, x_m)$. By the entering variable choice rule in the algorithm, the relative cost coefficient of x_{m+1}, $\bar{c}_{m+1} < 0$. This implies that $\{A_{.1}, \ldots, A_{.m}, A_{.m+1}\}$ is linearly independent and, hence, that pos $\{A_{.1}, \ldots, A_{.m}, A_{.m+1}\}$ is a simplicial cone in $\mathbf{R}^{m+1}$. During this step, the algorithm is at the point $(b_1, \ldots, b_m, \bar{z})^{\mathrm{T}}$ of the requirement line, on the facet pos $\{A_{.1}, \ldots, A_{.m}\}$ of this simplicial cone. Two cases can happen now.

Case 1 $\bar{c}_{m+1} < 0$ and $(\bar{a}_{1,m+1}, \ldots, \bar{a}_{m,m+1}) \leqq 0$. In this case the half-line of the requirement line, obtained by letting the parameter $z \leqq \bar{z}$, lies entirely in the simplicial cone pos $\{A_{.1}, \ldots, A_{.m}, A_{.m+1}\}$. Since the simplicial cone pos $\{A_{.1}, \ldots, A_{.m+1}\}$ is a subset of pos $\{A_{.1}, \ldots, A_{.n}\}$, this implies that the part of the requirement line for $z \leqq \bar{z}$ is completely in pos $\{A_{.1}, \ldots, A_{.n}\}$. Hence there is no lowest point in $\mathbf{L} \cap \text{pos}\,\{A_{.1}, \ldots, A_{.n}\}$, and the objective function is unbounded below in (3.32).

Case 2 Suppose $\bar{a}_{is} > 0$ for at least one i. In this case there is a pivot element and when x_{m+1} enters the basic vector, it replaces one of the present basic variables, say x_1, from the basic vector. The new basic vector is $(x_2, \ldots, x_m, x_{m+1})$. Suppose the BFS corresponding to it has an objective value $\hat{z}$. By arguments similar to those used earlier, this implies that the point $(b_1, \ldots, b_m, \hat{z})$ on the requirement line is on pos $\{A_{.2}, \ldots, A_{.m}, A_{.m+1}\}$. Pos $\{A_{.1}, \ldots, A_{.m}\}$ and pos $\{A_{.2}, \ldots, A_{.m+1}\}$ are two facets of the simplicial cone pos $\{A_{.1}, \ldots, A_{.m+1}\}$. Hence the requirement line enters this simplicial cone at the point $(b_1, \ldots, b_m, \bar{z})$ on its facet pos $\{A_{.1}, \ldots, A_{.m}\}$, and leaves it at the point $(b_1, \ldots, b_m, \hat{z})$ on its facet pos $\{A_{.2}, \ldots, A_{.m+1}\}$.

This pivot step has the effect of moving from the point $(b_1, \ldots, b_m, \bar{z})$ on the facet pos $\{A_{.1}, \ldots, A_{.m}\}$, to the point $(b_1, \ldots, b_m, \hat{z})$ lying below on the requirement line and on the other facet pos $\{A_{.2}, \ldots, A_{.m+1}\}$ of the simplicial cone pos $\{A_{.1}, \ldots,$

$A_{.m+1}\}$. Hence each pivot step of the algorithm has the effect of moving down the requirement line from one facet of a simplicial cone to another. Since the algorithm consists of a sequence of such moves across these simplicial cones until the bottom point in $\mathbf{L} \cap \text{pos}\,\{A_{.1}, \ldots, A_{.n}\}$ is reached, it has been named the simplex algorithm.

Example

We will illustrate with an LP in which $m = 1$, $n = 5$. In this case the simplicial cones of the earlier discussion, will be simplicial cones in $\mathbf{R}^2$, and it is convenient to draw them on the two dimensional Cartesian plane. The LP is:

x_1	x_2	x_3	x_4	x_5		
1	3	4	1	-1	$= 2$	
3	2	1	-2	-2	$= z$	minimize

$$x_j \geqq 0 \qquad \text{for all } j$$

Since there is only one constraint in this LP, we have to deal with basic vectors of one variable only. The column vectors of x_1 to x_5 (denoted by $A_{.1}$ to $A_{.5}$) are plotted on the y_1, y_2-Cartesian plane as in the earlier discussion. Since the right-hand side constant $= b_1 = 2$, the requirement line is the line marked in Figure 3.26.

The canonical tableau with respect to (x_1) as the basic vector is

Basic variables	x_1	x_2	x_3	x_4	x_5	$-z$	
x_1	1	3	4	1	-1	0	2
$-z$	0	-7	-11	-5	1	1	-6

The present objective value is 6. Hence the point corresponding to this basic vector is the point P_1 on the requirement line in Figure 3.26. P_1 is the intersection of the requirement line and pos $\{A_{.1}\}$. In this canonical tableau x_3 has the most negative relative cost efficient, and we select it as the entering variable. x_3 replaces x_1 as the basic variable. The effect of this change is to move from the point P_1 on the facet pos $\{A_{.1}\}$ to the point P_2 on the facet pos $\{A_{.3}\}$ of the simplicial cone pos $\{A_{.1}, A_{.3}\}$. So during this step the algorithm walks across the simplicial cone pos $\{A_{.1}, A_{.3}\}$. The next canonical tableau is given below.

Basic variables	x_1	x_2	x_3	x_4	x_5	$-z$	b
x_3	$\frac{1}{4}$	$\frac{3}{4}$	1	$\frac{1}{4}$	$-\frac{1}{4}$	0	$\frac{1}{2}$
$-z$	$\frac{11}{4}$	$\frac{5}{4}$	0	$-\frac{9}{4}$	$-\frac{7}{4}$	1	$-\frac{1}{2}$

x_4 has the most negative relative cost coefficient in this tableau and, hence, it is selected as the next entering variable. It replaces x_3 as the basic variable. The effect is to move from the point P_2 on the facet pos $\{A_{.3}\}$ and the requirement line to the point P_3 on the facet pos $\{A_{.4}\}$ of the simplicial cone pos $\{A_{.3}, A_{.4}\}$. Here is the next canonical tableau.

Basic variables	x_1	x_2	x_3	x_4	x_5	$-z$	b
x_4	1	3	4	1	-1	0	2
$-z$	5	8	9	0	-4	1	4

x_5 is the only nonbasic variable with a negative relative cost coëfficient. The unboundedness criterion of the simplex algorithm is satisfied. We verify that the portion of the requirement line below the point P_3 lies entirely in the simplicial cone pos $\{A_{.4}, A_{.5}\}$. In Figure 3.26, the various simplicial cones that the algorithm walks through are marked by angle signs.

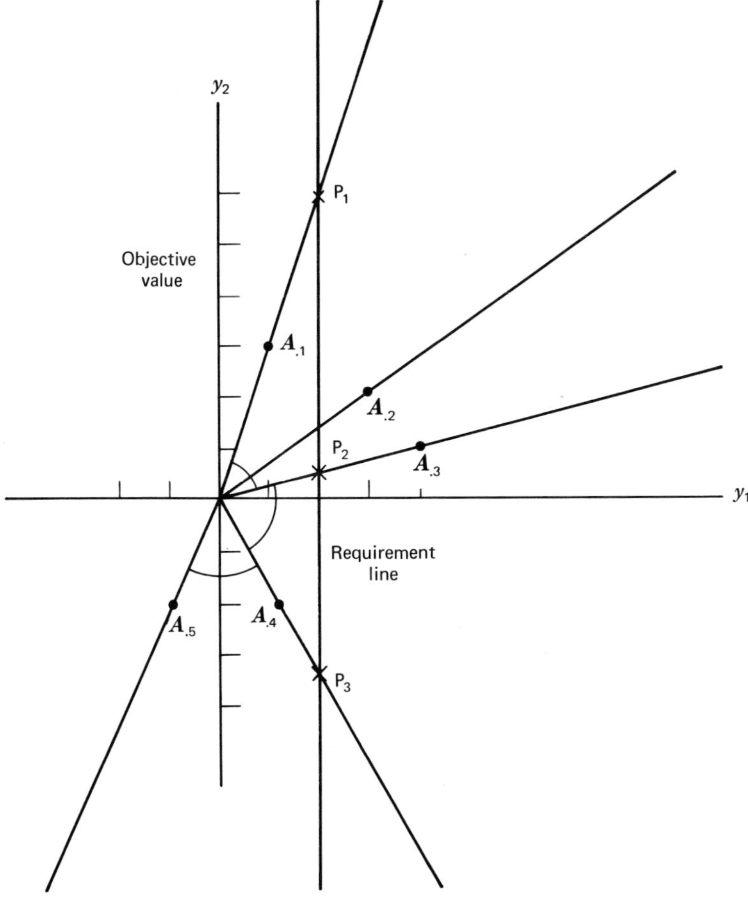

Figure 3.26

Note: Historically, the name originated in the geometry of an LP, which is (3.32) with the additional constraint $\Sigma x_j = 1$. For this LP, the set to be considered is $\boldsymbol{\Gamma}$ = convex hull of $\{A_{.1}, \dots, A_{.n}\}$. This LP is the problem of finding the lowest point in $\mathbf{L} \cap \boldsymbol{\Gamma}$. In solving this problem, the simplex algorithm goes through a sequence of moves across simplexes which are subsets of $\boldsymbol{\Gamma}$. See section 7.3 in Dantzig's book, reference [1] of Chapter 18.

3.14 PHASE I OF THE SIMPLEX METHOD USING ONE ARTIFICIAL VARIABLE

Consider the LP (3.16). The aim of Phase I of the simplex method is to find a feasible basis for this problem. The objective row is normally carried in the tableau during Phase I and updated whenever a pivot is made. In tableau form this is the problem:

x	$-z$	
A	0	b
c	1	0

As in section 3.3.3, transform the matrix A into row echelon normal form by row operations. If the system "$Ax = b$" is inconsistent, this will show up during this step as an equation of the form "0 = some nonzero number" and we terminate. If the system "$Ax = b$" is consistent, all the redundant constraints are eliminated in the process of putting the tableau in row echelon normal form. The column vectors of the basic variables in their proper order form a unit matrix in the final transformed tableau. For convenience in referring to them, we will assume that the basic variables are, $x_1, \ldots, x_m$, in that order. Let the final tableau be as follows.

Basic variables	$x_1 \ldots x_m$	$x_{m+1} \ldots x_n$	$-z$	b
x_1	$1 \ldots 0$	$\bar{a}_{1,m+1} \cdots \bar{a}_{1n}$	0	$\bar{b}_1$
$\vdots$	$\vdots \quad \vdots$	$\vdots \quad \vdots$	$\vdots$	$\vdots$
x_m	$0 \ldots 1$	$\bar{a}_{m,m+1} \cdots \bar{a}_{mn}$	0	$\bar{b}_m$
$-z$	$0 \ldots 0$	$\bar{c}_{m+1} \ldots \bar{c}_n$	1	$-\bar{z}$

If $\bar{b}_i \geqq 0$ for all i, the present basis is a feasible basis, and we are done. Otherwise $\bar{b}_i < 0$ for some i. In this case let

$$\bar{b}_r = \text{minimum } \{\bar{b}_i : i = 1 \text{ to } m\} \tag{3.34}$$

Augment the tableau with an artificial variable t whose column vector consists of all "-1"s. This is the Phase I tableau.

Basic variables	$x_1 \ldots x_m$	$x_{m+1} \ldots x_n$	t	$-z$	$-w$	b
x_1	$1 \ldots 0$	$\bar{a}_{1,m+1} \cdots \bar{a}_{1n}$	-1	0	0	$\bar{b}_1$
$\vdots$	$\vdots \quad \vdots$	$\vdots \quad \vdots$	$\vdots$	$\vdots$	$\vdots$	$\vdots$
x_m	$0 \ldots 1$	$\bar{a}_{m,m+1} \cdots \bar{a}_{mn}$	-1	0	0	$\bar{b}_m$
$-z$	$0 \ldots 0$	$\bar{c}_{m+1} \ldots \bar{c}_n$	0	1	0	$-\bar{z}$
$-w$	$0 \ldots 0$	$0 \ldots 0$	1	0	1	0

Replace the present rth basic variable x_r by the artificial variable t. The pivot row for this operation is the rth row and the pivot column is the column vector of t. The new basic vector is $(x_1, \ldots, x_{r-1}, t, x_{r+1}, \ldots, x_m)$, and by the definition of of r in equation (3.34), and the column vector of t, this is a feasible basic vector for the Phase I problem.

Example

Consider this LP:

x_1	x_2	x_3	x_4	x_5	x_6	$-z$	b
1	−1	0	1	−1	2	0	1
0	1	0	−1	1	1	0	−4
0	1	1	1	2	−2	0	−4
−30	4	2	−7	−8	−9	1	0 (z minimize)

$x_j \geqq 0$ for all j

The tableau is transformed into row echelon normal form by selecting (x_1, x_2, x_3) as the basic vector. This leads to the following tableau.

Basic variables	x_1	x_2	x_3	x_4	x_5	x_6	$-z$	b
x_1	1	0	0	0	0	3	0	−3
x_2	0	1	0	−1	1	1	0	−4
x_3	0	0	1	2	1	−3	0	0
$-z$	0	0	0	−7	−14	83	1	−74

In the basic solution corresponding to this basic vector, the most negative basic variable is x_2 with a value of −4, that is, $\bar{b}_2$ is the most negative $\bar{b}_i$. Hence, row 2 will be the pivot row to get a feasible basis for the Phase I problem. Introducing the artificial variable t and the Phase I objective function, w, the Phase I tableau is as follows.

Basic variables	x_1	x_2	x_3	x_4	x_5	x_6	t	$-z$	$-w$	b
x_1	1	0	0	0	0	3	−1	0	0	−3
x_2	0	1	0	−1	1	1	−1	0	0	−4
x_3	0	0	1	2	1	−3	−1	0	0	0
$-z$	0	0	0	−7	−14	83	0	1	0	−74
$-w$	0	0	0	0	0	0	1	0	1	0

Performing the pivot leads to the initial canonical tableau for the Phase I problem.

Basic variables	x_1	x_2	x_3	x_4	x_5	x_6	t	$-z$	$-w$	b
x_1	1	−1	0	1	−1	2	0	0	0	1
t	0	−1	0	1	−1	−1	1	0	0	4
x_3	0	−1	1	3	0	−4	0	0	0	4
$-z$	0	0	0	−7	−14	83	0	1	0	−74
$-w$	0	1	0	−1	1	1	0	0	1	−4

Starting with this feasible basis, the Phase I problem can be solved in the usual manner.

Problem

3.32 Complete the solution of this LP.

From a naive point of view, solving a Phase I problem with a single artificial variable, in which only that variable has to be eliminated from the basic vector, appears easier than solving the Phase I problem with a full artificial basis, in which m artificial variables have to be eliminated from the basic vector to get a feasible Phase II basic vector. There is no theoretical justification for this point of view. Practical experience indicates that both the Phase I models require approximately the same amount of computational effort.

PROBLEMS

3.33 Let $A = \{A_{.1}, \ldots, A_{.n}\}$ be a basis for $\mathbf{R}^n$. Let $B = \{B_{.1}, \ldots, B_{.k}\}$ be a linearly independent set of vectors in $\mathbf{R}^n$. Prove that a subset of $n - k$ vectors in A can be found that together with the vectors in B forms a basis for $\mathbf{R}^n$.

3.34 (a) Let $B_{.1} = (1, 1, 3)^T$, $B_{.2} = (1, -2, 1)^T$, $B_{.3}(1, -1, 4)^T$, $b = (1, -4, 5)^T$. Let $I_{.j}$ be the jth column vector of the unit matrix of order 3. Find a representation of b as a linear combination of (i) $\{B_{.1}, I_{.2}, I_{.3}\}$, (ii) $\{B_{.1}, I_{.2}, B_{.3}\}$, and (iii) $\{B_{.1}, B_{.2}, B_{.3}\}$. In each case check whether the representation is unique.

(b) Consider the system of equations:

$$\begin{aligned} 2x_1 - 4x_2 - 2x_3 &= -2 \\ 3x_1 - 2x_2 + x_3 &= 2 \\ -x_1 + 3x_2 + 2x_3 &= \alpha \end{aligned}$$

Find out the set of all α for which this system is (i) inconsistent, (ii) has at least one solution. In this case find all the solutions of the system.

3.35 For the system

$$\begin{aligned} -x_3 + x_4 &= 0 \\ x_2 + x_3 + x_4 &= 3 \\ -x_1 + x_2 + x_3 + x_4 &= 2 \\ x_j \geqq 0 \quad & \text{for all } j \end{aligned}$$

find the basic solution associated with the basic vector (x_1, x_3, x_4).

3.36 For the system

x_1	x_2	x_3	x_4	x_5	x_6	x_7	x_8	x_9	b
1	1	−2	0	3	1	0	0	0	9
0	−1	0	−1	−1	0	1	0	0	−6
0	0	1	1	−1	0	0	1	0	1
1	−1	2	2	−3	0	0	0	1	1
2	−1	1	2	−2	1	1	1	1	5

$$x_j \geqq 0 \quad \text{for all } j$$

is $(3, 2, 1, 2, 2, 0, 0, 0, 0)^T$ a BFS? If not construct a BFS from it.

3.37 For the system

x_1	x_2	x_3	x_4	x_5	x_6	x_7	b
1	−3	0	−8	1	1	0	−13
0	0	−2	−2	0	1	3	−4
−1	−4	0	−10	0	0	2	−19
0	0	3	3	0	1	0	6

$$x_j \geqq 0 \quad \text{for all } j$$

is $(1, 2, 1, 1, 0, 0, 0)^T$ a BFS? If not, construct a BFS from it.

3.38 For the system

x_1	x_2	x_3	x_4	x_5	x_6	
1	−1	2	2	2	−1	$\leqq$ 7
−1	2	−1	0	2	−1	= 0
1	2	2	1	2	−1	$\leqq$ 12
−1	2	2	2	−1	−1	$\leqq$ 11

$$x_j \geqq 0 \quad \text{for all } j$$

are 0, $(1, 2, 3, 0, 0, 0)^T$, $(1, 2, 3, 1, 0, 0)^T$ BFSs?

3.39 Find out all the BFSs of the system

$$-6 \leqq x_1 \leqq 2$$
$$-5 \leqq x_2 \leqq 3$$

3.40 Find all the adjacent extreme points of the BFSs associated with the basic vector (x_1, x_2, x_3) in the following system.

x_1	x_2	x_3	x_4	x_5	x_6	b
1	0	0	1	0	−1	0
0	1	0	−1	1	2	3
0	0	1	2	−1	1	2

$x_j \geqq 0$ for all j

3.41 For the system

$$\begin{aligned} x_1 + x_2 \quad\quad - 3x_4 + 3x_5 + x_6 &= 4 \\ x_1 + 2x_2 \quad\quad - 5x_4 + 5x_5 + 3x_6 &= 7 \\ - x_2 + x_3 + 2x_4 - 5x_5 + x_6 &= -3 \\ x_j \geqq 0 \quad \text{for all } j & \end{aligned}$$

is $(4, 9, 0, 3, 0, 0)^T$ an extreme point? Why? If not, is it on an edge? If it is, is the edge bounded or unbounded?

3.42 For the system

$$\begin{aligned} 4x_1 + x_2 &\geqq 4 \\ x_1 - x_2 &\leqq 2 \\ -2x_1 + x_2 &\leqq 2 \\ x_1 + x_2 &\geqq 2 \\ x_1 + 3x_2 &\geqq 3 \end{aligned}$$

find out all the extreme points, extreme homogeneous solutions, and extreme half-lines. Derive an expression for a general feasible solution of this system as in resolution theorem II.

3.43 For the system

x_1	x_2	x_3	x_4	x_5	x_6	b
1	0	0	1	−1	1	3
0	1	0	2	2	−1	4
0	0	1	−1	1	2	6

$x_j \geqq 0$ for all j

let $\bar{x}$ be the BFS associated with the basic vector (x_1, x_2, x_3). Find $\bar{x}$ and all the adjacent extreme points of $\bar{x}$. Use this information to check whether $\bar{x}$ minimizes $z(x) = 3x_1 - 2x_2 - 10x_3 - 11x_4 + 18x_5 - 10x_6$ on the set of feasible solutions of this system.

3.44 Let $\mathbf{K}$ be a convex polyhedron and let $\bar{x}$ be a point in it. Prove that $\bar{x}$ is an extreme point of $\mathbf{K}$ iff $\mathbf{K}\backslash\{\bar{x}\}$ is also a convex set.

3.45 Solve the following LPs using only one artificial variable.

(a) Minimize $-x_1 + 2x_2 - 3x_3 + x_4 + 3x_5 - x_6$

$$\begin{array}{lrrrrrrrrl} \text{Subject to} & x_1 & & & -2x_4 & + x_5 & + x_6 & = & 3 \\ & & x_2 & & -2x_4 & -2x_5 & + x_6 & = & -6 \\ & & & x_3 & + x_4 & -2x_5 & -2x_6 & = & -12 \end{array}$$

$$x_j \geqq 0 \quad \text{for all } j$$

(b) Maximize x_8

$$\begin{array}{lrrrrrrrrl} \text{Subject to} & -x_1 & & -x_3 & + x_4 & + x_5 & & & & = -2 \\ & x_1 & & -2x_3 & -3x_4 & & + x_6 & & - x_8 & = 4 \\ & x_1 & & & -2x_4 & & & + x_7 & & = 3 \\ & 2x_1 & + x_2 & & -5x_4 & & & & - x_8 & = 6 \end{array}$$

$$x_j \geqq 0 \quad \text{for all } j$$

3.46 Solve the following LPs. Determine the set of alternate optima in each case.

(a) Maximize $12x_1 + x_2 + 2x_3 + 5x_4 + 4x_5 - 6x_6$

$$\begin{array}{lrl} \text{Subject to} & 8x_1 + x_2 + 3x_3 + 2x_4 + 3x_5 - 3x_6 & = 17 \\ & -5x_1 - x_2 - x_3 - x_4 - 2x_5 + 2x_6 & = -12 \\ & -5x_1 - x_3 - 2x_4 - x_5 + 4x_6 & \leqq -8 \end{array}$$

$$x_j \geqq 0 \quad \text{for all } j$$

(b) Minimize $-2x_1 + x_2 + 18x_3 + 6x_4 - 2x_8$

$$\begin{array}{lrl} \text{Subject to} & x_1 + x_3 - x_4 - 2x_6 + x_7 - x_8 & = 6 \\ & x_2 - x_3 + x_4 - 2x_5 + x_6 + 2x_7 + x_8 & = 3 \\ & x_3 + 2x_4 - 3x_5 + x_6 - 2x_7 + x_8 & = 3 \end{array}$$

$$x_j \geqq 0 \quad \text{for all } j$$

3.47 Solve the following LP using (x_5, x_6, x_7) as the initial basic vector.

x_1	x_2	x_3	x_4	x_5	x_6	x_7	x_8	$-z$ =	b
1	1	0	0	−2	0	3	−3	0	7
−2	1	0	−3	1	3	−3	0	0	−5
0	1	1	−2	0	0	3	−4	0	9
1	1	1	3	4	2	7	−5	1	0

$x_j \geqq 0$ for all j, z to be minimized

Also show that the set of all optimum solution of this problem is an unbounded set.

3.48 Consider the LP

$$\begin{array}{ll} \text{Minimize} & z(x) = cx \\ \text{Subject to} & Ax = b \\ & x \geqq 0 \end{array}$$

If c_j is decreased, keeping all the other cost coefficients unaltered, prove that the value of x_j in an optimum solution increases.

3.49 EXTREME POINT RANKING Consider the LP

$$\begin{array}{ll} \text{Minimize} & z(x) = cx \\ \text{Subject to} & Ax = b \\ & x \geqq 0 \end{array}$$

Let **K** be the set of feasible solutions and x^1 an optimum BFS of this problem. Suppose it is required to rank all the BFSs of this problem in increasing order of the value of $z(x)$; that is, it is required to order the BFSs as a sequence $x^1, x^2, \ldots,$ with the property that

$$z(x^{r+1}) = \text{minimum } \{z(x)\text{: } x \text{ a BFS excluding } x^1, \ldots, x^r\}$$

(a) Prove that x^2 is determined by $z(x^2) = \text{minimum } \{z(x)\text{: } x \text{ an adjacent extreme point of } x^1 \text{ on } \mathbf{K}\}$

(b) In general, suppose $x^1, \ldots, x^r$ in the ranked sequence have been determined. Prove that x^{r+1} can be determined by

$$\begin{aligned} z(x^{r+1}) = \text{minimum } \{z(x)\text{: } & x \text{ an adjacent extreme point of} \\ & x^1 \text{ or } x^2, \ldots, \text{ or } x^r, \text{ excluding} \\ & x^1, x^2, \ldots, x^r\} \end{aligned}$$

(c) Using these results develop an algorithm for ranking the BFSs of the problem in increasing order of the value of $z(x)$, assuming that the problem is nondegenerate. What problems are posed by degeneracy?

3.50 FRACTIONAL PROGRAMMING It is required to solve the following problem (see reference [1])

$$\begin{aligned} \text{Minimize} \quad & f(x) = \frac{cx + \alpha}{dx + \beta} \\ \text{Subject to} \quad & Ax = b \\ & x \geqq 0 \end{aligned} \tag{1}$$

where α, β are real numbers. Let **K** be the set of feasible solutions of this problem. Assume that **K** is bounded. Consider the following LP:

$$\begin{aligned} \text{Minimize} \quad & \frac{cy + \alpha t}{\gamma} \\ \text{Subject to} \quad & Ay - bt = 0 \\ & dy + \beta t = \gamma \\ & y, t \geqq 0 \end{aligned} \tag{2}$$

where γ is a nonzero parameter and t is a nonnegative real-valued variable. Prove the following:

(a) If $dx + \beta > 0$ for all $x \in \mathbf{K}$, then (2) is infeasible when $\gamma = -1$. Put $\gamma = 1$ and let $(\tilde{y}, \tilde{t})$ be an optimum solution of (2). Then $\tilde{t} \neq 0$ and $\tilde{x} = (1/\tilde{t})\tilde{y}$ is an optimum solution of (1).

(b) If $dx + \beta < 0$ for all $x \in \mathbf{K}$, (1) is infeasible when $\gamma = 1$. Also, if $(\tilde{y}, \tilde{t})$ is an optimum solution of (2) in this case when $\gamma = -1$, then $\tilde{t} \neq 0$, and $\tilde{x} = (1/\tilde{t})\tilde{y}$ is an optimum solution of (1).

(c) If $dx + \beta$ is neither strictly positive, nor strictly negative on **K**, it must assume the value 0 for some $x \in \mathbf{K}$. In this case problem (1), is not well-defined. Explain what happens to problem (2) in this case.

—*A. Charnes and W. W. Cooper*

3.51 Let **K** be the set of feasible solutions of

$$\begin{aligned} Ax &= b \\ x &\geqq 0 \end{aligned}$$

where A, b are given matrices of orders $m \times n$ and $m \times 1$, respectively. Let I be the unit matrix of order n. If $\mathbf{K} \neq \emptyset$, prove that the dimension of **K** is $n - \alpha$, where $\alpha = \text{rank}\,(\{A_{i.}, i = 1 \text{ to } m\} \cup \{I_{t.}\text{: } t \text{ such that } x_t = 0 \text{ for all } x \in \mathbf{K}\})$.

3.52 Let A be a matrix of order $m \times n$, and b, a column vector in $\mathbf{R}^m$. Let $\mathbf{K}$ be the set of feasible solutions of

$$Ax \geqq b$$

Prove that if $\mathbf{K} \neq \emptyset$, its dimension is $n - \alpha$ where $\alpha = \text{rank } \{A_{i.}: i \text{ such that } A_{i.}x = b_i,$ for all $x \in \mathbf{K}\}$.

3.53 Read each of the following statements carefully and check whether it is *true* or *false*. Justify your answer by constructing a simple illustrative example, if possible, or by a simple proof. In all these statements, problem (1) is the LP.

$$\begin{aligned} \text{Minimize} \quad & z(x) = cx \\ \text{Subject to} \quad & Ax = b \qquad (1) \\ & x \geqq 0 \end{aligned}$$

where A is a matrix of order $m \times n$. $\mathbf{K}$ is the set of feasible solutions of this problem (1). $f(b)$ is the minimum objective value in this problem, as a function of the right-hand side constant vector, b.

1. The number of positive x_j in any BFS of (1) can never exceed the rank of A.
2. If rank of A is m, and (1) is nondegenerate, every feasible solution in which exactly m variables are positive, must be a BFS.
3. The unboundedness criterion will never be satisfied while solving the Phase I problem.
4. While solving the Phase I problem associated with (1), if the unboundedness criterion is satisfied, (1) must be infeasible.
5. If (1) is feasible, termination occurs in solving the Phase I problem associated with (1) only when all the artificial variables leave the basic vector.
6. In solving (1) by the primal simplex method, the pivot element in each pivot step must be a positive number.
7. If a tie occurs in the pivot row choice during a pivot step while solving (1) by the simplex algorithm, the BFS obtained after this pivot step is degenerate.
8. While solving (1) by the simplex algorithm, if the BFS in the beginning of a pivot step is degenerate, the objective value remains unchanged in this pivot step.
9. Simplex algorithm moves from an extreme point of $\mathbf{K}$ only in a nondegenerate pivot step.
10. In solving an LP by the simplex algorithm, a new feasible solution is generated after every pivot step.
11. A feasible basis for (1) can be found by transforming the system into row echelon normal form.
12. Let the rank of A in (1) be m. If (1) has an optimum solution, there must exist an optimum basis. Every basis consists of m column vectors and a pivot step changes one column vector in a basis. Hence starting from a feasible basis, the simplex algorithm finds an optimum basis in at most m pivot steps.
13. Every feasible solution of (1) in which m variables are positive is a BFS.
14. If $\bar{x}$ and $\tilde{x}$ are two adjacent BFSs of (1), the total number of variables that are positive in either $\bar{x}$ or $\tilde{x}$, or both, is at most $m + 1$.
15. If $\bar{x}$ and $\tilde{x}$ are two BFSs for (1) such that the total number of variables that are positive in either $\bar{x}$ or $\tilde{x}$ or both is at most $m + 1$, then $\bar{x}$ and $\tilde{x}$ must be adjacent on $\mathbf{K}$.
16. If $\bar{x}$ is a feasible solution of (1) and the rank of $\{A_{.j}: j \text{ such that } \bar{x}_j > 0\}$ is one less than its cardinality, $\bar{x}$ lies on an edge of $\mathbf{K}$.
17. No feasible solution of (1) in which $m + 2$ or more variables are positive can lie on an edge of $\mathbf{K}$.
18. Every nondegenerate BFS of (1) lies on exactly $n - m$ edges of $\mathbf{K}$.

19. Every convex set has an extreme point.
20. Every nonempty convex polytope has an extreme point.
21. The number of extreme points of any convex set is always finite.
22. If (1) has an optimum solution, **K** must be a bounded set.
23. In every optimum solution of (1), no more than m variables can be positive.
24. An LP can have more than one optimum solution iff it is degenerate.
25. The total number of optimum feasible solutions of any LP is always finite.
26. The total number of BFSs of any LP is always finite.
27. If an LP has more than one optimum solution, it must have an uncountable number of optimum solutions.
28. The set of optimum solutions of any LP is always a bounded set.
29. If (1) has an optimum solution in which $m + 1$ or more variables are positive, it must have an uncountable number of optimum solutions.
30. If (1) is feasible and there is no semipositive linear combination of columns of A that equals the 0-vector, $f(b)$ is finite.
31. Any consistent system of linear equations is equivalent to a system in which the number of constraints is less than or equal to the number of variables.
32. A system of linear equations has a unique solution iff the number of constraints in the system is equal to the number of variables.
33. A system of n independent linear equations in n unknowns has a unique solution. Likewise a system of n linear inequalities in n unknowns has a unique solution.
34. A straight line in $\mathbf{R}^2$ is the set of solutions of a single linear equation. Likewise a straight line in $\mathbf{R}^n$ is the set of solutions of a single linear equation.
35. If all the points on a straight line in $\mathbf{R}^n$ lie in a convex polyhedron $\mathbf{\Gamma}$, that straight line contains no extreme pts of $\mathbf{\Gamma}$.
36. Let $\mathbf{\Gamma}$ be the convex hull of $\{x^1, \ldots, x^r\}$. The set of extreme points of $\mathbf{\Gamma}$ is a subset of $\{x^1, \ldots, x^r\}$.
37. If $\mathbf{K} \neq \emptyset$, its dimension is $n - r$ or less, where r is the rank of A.
38. If every edge of **K** contains exactly two extreme points, **K** must be bounded.
39. If $n = m + 1$ (i.e. (1) is a system of m equations in $m + 1$ nonnegative variables), then the number of BFS's of (1) is at most 2.
40. The number of edges of **K** through an extreme point of **K** is always greater than or equal to the dimension of **K**.
41. If b is nondegenerate, the number of edges of **K** through any extreme point of **K** is exactly equal to the dimension of **K**.
42. If **K** is unbounded there must exist a semipositive homogeneous solution corresponding to (1).
43. The set of homogeneous solutions corresponding to (1) is a subset of **K**.
44. In a pivot step of the simplex algorithm if the minimum ratio is zero, there is no change in the objective value. In general, the minimum ratio in a pivot step is the amount by which the objective value changes in that step.
45. Consider a step in solving (1) by the simplex algorithm in which x_j is the entering variable. The minimum ratio in this step is the value of x_j in the solution that will be obtained after this pivot step.
46. Let the rank of A be m. In the course of solving (1) by the simplex method, let $\bar{A}_{.j}$ be the updated column vector of x_j in some stage. Then $\bar{A}_{.j}$ is the column vector of the coefficients in an expression of $A_{.j}$ as a linear combination of the columns of the basis in that stage.
47. In solving (1) by the simplex algorithm, if x_j is the entering variable during a pivot step, then the pivot matrix corresponding to the pivot in that step differs from the unit matrix only in the jth column.

REFERENCES

1. A. Charnes and W. W. Cooper, "Programming with Linear Fractional Functionals," *Naval Research Logistica Quarterly*, *9*, pp. 181–186, 1962.
2. U. Eckhardt, "Theorems on the Dimension of Convex Sets", *Linear Algebra and its Applications*, *12*, pp. 63–76, 1975.
3. A. J. Goldman, "Resolution and Separation Theorems for Polyhedral Convex Sets" in *Linear Inequalities and Related Systems*, H. W. Kuhn and A. W. Tucker (editors), Princeton University Press, 1956.
4. B. Martos, "The Direct Power of Adacent Vertex Programming Methods," *Management Science*, *12*, 3, pp. 241–252, 1965.
5. K. G. Murty, "Adjacency on Convex Polyhedra," *SIAM Review*, *13*, 3, pp. 377–386, July 1971.
6. J. Stoer and C. Witzgall, *Convexity and Optimization in Finite Dimensions I*, Springer Verlag, New York 1970.
7. D. W. Walkup and R. J.-B. Wets, "Lifting Projections of Convex Polyhedra", *Pacific Journal of Mathematics*, *28*, *2*, pp. 465–475, 1969.
8. R. J.-B. Wets and C. Witzgall, "Towards an Algebraic Characterization of Convex Polyhedral Cones, *Numerische Mathematik*, *12*, pp. 134–138, 1968.

Selected References on Linear Algebra

9. D. Gale, *The Theory of Linear Economic Models*, McGraw-Hill, New York, 1960.
10. P. R. Halmos, *Finite Dimensional Vector Spaces*, D. Van Nostrand, New York, 1969.
11. B. Noble, *Applied Linear Algebra*, Prentice-Hall, Englewood Cliffs, N.J., 1969.
12. D. I. Steinberg, "*Computational Matrix Algebra*", McGraw-Hill, 1974.
13. R. M. Thrall and L. Tornheim, *Vector Spaces and Matrices*, Wiley, New York, 1957.

4 duality in linear programming

4.1 INTRODUCTION

When an optimum solution of an LP is obtained by the simplex algorithm, the updated cost row in the final canonical tableau is nonnegative. This updated cost row is obtained by subtracting suitable multiples of the constraint rows from the original cost row. For the LP (3.16), if B is an optimum basis and $\bar{c}$ is the updated cost row with respect to it, it was shown in section 3.12 that $\bar{c} = c - (c_B B^{-1})A$. By treating these multiplying coefficients as variables, another LP known as the *dual problem* associated with the original problem (which in this context is called the *primal problem*) can be constructed. The variables in the dual problem are multipliers associated with the constraints in the primal problem.

Principles of duality appear in various branches of mathematics, physics, and statistics. In linear programming, duality theory turns out to be of great practical use. Also, duality in linear programming admits an elegant economic interpretation.

4.2 AN EXAMPLE OF A DUAL PROBLEM

The fact that the dual problem arises from economic considerations is illustrated by the following diet problem of the *cost minimizing housewife* and the *profit maximizing vitamin pill manufacturer*. A housewife is trying to make a *minimal cost diet* from six available primary foods (called 1, 2, 3, 4, 5, 6) so that the diet contains at least 9 units of vitamin A and 19 units of vitamin C. Here are the data on the foods.

Nutrient	Number of units of nutrients per kg of food 1	2	3	4	5	6	Minimum number of units of nutrient required in diet
Vitamin A	1	0	2	2	1	2	9
Vitamin C	0	1	3	1	3	2	19
Cost of food (cents/kg)	35	30	60	50	27	22	

To formulate the housewifes' problem suppose her diet consists of x_j kg of food j, $j = 1$ to 6. The constraints are

$$\begin{aligned} \text{Vitamin A content} &= x_1 \quad + 2x_3 + 2x_4 + x_5 + 2x_6 \geqq 9 \\ \text{Vitamin C content} &= x_2 + 3x_3 + x_4 + 3x_5 + 2x_6 \geqq 19 \\ \text{and} \quad & x_j \geqq 0 \quad \text{for all } j \end{aligned}$$

The objective function to be minimized is the cost of the diet $z(x) = 35x_1 + 30x_2 + 60x_3 + 50x_4 + 27x_5 + 22x_6$. In tabular form here is the housewife's problem.

x_1	x_2	x_3	x_4	x_5	x_6	
1	0	2	2	1	2	$\geqq 9$
0	1	3	1	3	2	$\geqq 19$
35	30	60	50	27	22	$= z(x)$ minimize

(4.1)

and $\quad x_j \geqq 0 \quad$ for all j

A manufacturer visualizes the scope for starting a new business in this situation. He proposes to manufacture synthetic pills of each nutrient and to sell them to the housewife. For his business to thrive he has to persuade the housewife to meet all her nutrient requirements by using his pills instead of the primary foods. However, since the housewife is cost conscious she will not use the pills unless the manufacturer can convince her that the prices of his pills are competitive in comparison with each of the available primary foods. This imposes several constraints on the prices he can charge for the pills. Let

$$\pi_1 = \text{price of vitamin A in pill form (cents/unit)}$$

$$\pi_2 = \text{price of vitamin C in pill form (cents/unit)}$$

Consider one primary food, say food 5. One kilogram of it contains one unit of vitamin A and three units of vitamin C. In terms of the manufacturer's prices, its intrinsic worth is therefore $\pi_1 + 3\pi_2$. One kilogram of this food costs 27 cents. The housewife will obviously compare these two quantities and unless

$$\pi_1 + 3\pi_2 \leqq 27$$

she will conclude that the manufacturer's prices are not competitive. Hence, to win her business, the manufacturer's price vector must satisfy the above constraint. A similar constraint on (π_1, π_2) arises from considering each primary food.

Since the manufacturer is trying to make money and not give it away, he requires that both π_1 and π_2 should be nonnegative. Also, since the housewife is cost conscious, if she decides to use the pills instead of the primary foods, she will buy just as many pills as are required to satisfy the minimal nutrient requirements exactly (buying more would cost extra money since the prices are nonnegative). Hence, the manufacturer's sales revenue will be $v(\pi) = 9\pi_1 + 19\pi_2$; and the manufacturer would like to see this maximized.

Thus the prices that the manufacturer can charge for his nutrient pills are obtained by solving the problem:

π_1	π_2	
1	$0 \leqq 35$	
0	$1 \leqq 30$	
2	$3 \leqq 60$	
2	$1 \leqq 50$	(4.2)
1	$3 \leqq 27$	
2	$2 \leqq 22$	
9	$19 = v(\pi)$, maximize	
and	$\pi_i \geqq 0$ for all i	

The following facts are clear:

1. The input-output coefficient tableau corresponding to the manufacturer's problem in (4.2) is just the transpose of the input-output tableau of the housewife's problem in (4.1) and vice versa.
2. The right-hand side constants in (4.1) are the coefficients of the objective function in (4.2) and vice versa.
3. Each column vector of (4.1) leads to a constraint in (4.2) and vice versa.
4. There is a variable in (4.2) corresponding to each constraint in (4.1), and (4.2) contains a constraint corresponding to each variable in (4.1).
5. (4.1) is a minimization problem in which the constraints are "$\geqq$" type, and (4.2) is a maximization problem, in which the constraints are "$\leqq$" type.
6. From the arguments of the housewife it is clear that she will not buy the pills unless they provide the required nutrients to her just as cheaply as the primary foods. This seems to suggest that

$$\text{The maximum value of } v(\pi) \leqq \text{the minimum value of } z(x) \tag{4.3}$$

This is in fact true. This follows from the weak duality theorem to be proved later. The LPs (4.1) and (4.2) are the *duals of each other*. The inequality (4.3) is one relationship between them.

4.3 DUAL VARIABLES ARE THE PRICES OF THE ITEMS

Given a general LP, the dual problem associated with it can be constructed using similar arguments. In this context, the original LP is known as the *primal problem*. The primal problem, and its dual, constitute a pair of LPs known as a *primal-dual pair of LPs*. It will be shown that the dual of the dual problem is an LP that is equivalent to the primal problem. Hence each problem in a primal-dual pair of LPs is the dual of the other.

In LPs, each constraint usually requires that the total amount of some item utilized is less than or equal to the total amount of this item available, or that the total number of units of some items produced should be greater than or equal to the known requirement for this item.

In the dual problem there will be a dual variable associated with this primal constraint, and it can be interpreted as the *price* of that item in the dual problem.* If the primal problem is to minimize the cost, the dual problem may be interpreted as that of determining the prices of the items to maximize the profit. In the dual problem the dual variables associated with inequality constraints of the primal problem will be restricted in sign.

In a general LP there may also be some equality constraints. In writing the corresponding dual problem, a dual variable is also associated with each primal equality constraint. However such dual variables will be unrestricted in sign in the dual problem. The justification for this is provided later. A negative-valued dual variable can also be interpreted as a price by treating negative prices as subsidies, for example, many cities would be happy to sell their garbage at $-\$5$ per ton, which is another way of saying that the city would be happy to pay \$5 per ton, to someone who offers to remove the garbage.

4.4 HOW TO WRITE THE DUAL OF A GENERAL LINEAR PROGRAM

Suppose the primal problem consists of some equality constraints, some inequality constraints, and sign restrictions on the variables. As in section 2.2 the word "*constraint*" is used here to refer to the equality or inequality conditions in the problem (i.e., the conditions that lead to a row in the original input-output tableau); the sign restrictions on the variables are considered separately.

By transforming the variables, if necessary, make all the sign restrictions in the problem into nonnegativity restrictions on the variables.

If there are any upper bound constraints on some of the variables, they should be treated as constraints in writing the dual problem.

The nonnegativity restriction on a variable is actually a lower bound restriction on it. Hence, if there is a nonnegativity restriction on a variable and a lower bound constraint, the implicit sign restriction becomes redundant and can be deleted. Thus any variable on which there is a lower bound constraint is treated as an unrestricted variable. All lower bound constraints on the variables (excepting nonnegativity restrictions) are treated as constraints in writing the dual problem.

It is not necessary to transform the problem into standard form before writing its dual.

Make sure that all the inequality constraints in the problem are of the *right type* before writing the dual. If the problem is a minimization problem, all the inequalities should be of the "$\geqq$" type (i.e., of the form $\sum a_i x_i \geqq b$). If the problem is a maximization problem all the inequalities should be of the "$\leqq$" type (i.e., of the form $\sum a_i x_i \leqq b$). Any inequality of the wrong type can be converted into an inequality of the right type by multiplying both sides of it by -1 and reversing the inequality sign. Also, all sign restrictions on the variables should be nonnegativity restrictions.

MNEMONIC FOR THE RIGHT TYPE OF INEQUALITIES

Imagine the problem of minimizing on the real line. Let $x \in \mathbf{R}^1$. Suppose the problem is to minimize x subject to a constraint of the form

$$x \leqq b \tag{4.4}$$

* Dual variables are also known as *shadow prices*, *marginal values* or *simplex multipliers*.

for some real number b. This minimization problem has no optimum solution, because the value of x can be decreased arbitrarily on the set of feasible solutions for (4.4). On the other hand, if the problem is of the form

$$\begin{aligned} &\text{Minimize} \quad x \\ &\text{Subject to} \quad x \geqq b \end{aligned} \tag{4.5}$$

there is an optimum solution, namely, $x = b$. This suggests that for a minimization problem, inequality constraints should be expressed in the "$\geqq$" form to be of the right type.

In a similar manner, for a maximization problem, inequality constraints should be expressed in the "$\leqq$" form to be of the right type.

DUAL PROBLEM

There will be one dual variable associated with each constraint in the primal (excluding nonnegativity restrictions on individual variables, if any). If the primal constraint is an inequality constraint of the right type, the associated dual variable is restricted to be nonnegative in the dual problem. If the primal constraint is an equality constraint, the associated dual variable is unrestricted in sign.

There is one dual constraint corresponding to each primal variable. Thus each column vector of the primal tableau leads to a constraint in the dual problem. Let π be the row vector of dual variables. Let $A_{.j}$ be the column vector of the coefficients of the primal variable x_j in the primal constraints, and c_j, the coefficient of x_j in the primal objective function. Then the dual constraint corresponding to x_j is $\pi A_{.j} = c_j$, if x_j is unrestricted in sign in the primal problem; or either the inequality constraint $\pi A_{.j} \leqq c_j$ or $\pi A_{.j} \geqq c_j$, whichever is of the *right type* for the dual problem, if x_j is restricted to be nonnegative in the primal problem.

If the primal is a minimization problem, the dual is a maximization problem and vice versa. The right-hand side constants of the primal problem are the coefficients of the dual objective function and vice versa.

Example 4.1

Let the primal problem be

$$\begin{array}{lrl} \text{Minimize} & z(x) = & 5x_1 - 6x_2 + 7x_3 + x_4 \\ \text{Subject to} & & x_1 + 2x_2 - x_3 - x_4 = -7 \\ & & 6x_1 - 3x_2 + x_3 - 7x_4 \geqq 14 \\ & & -2.8x_1 - 17x_2 + 4x_3 + 2x_4 \leqq -3 \\ & & x_1 \geqq 0, \quad x_2 \geqq 0, \\ & & x_3 \text{ and } x_4 \text{ unrestricted in sign} \end{array}$$

By converting the inequalities to the right type, the problem becomes

$$\begin{array}{clrl} \text{Associated dual variable} & \text{Minimize} & z(x) = & 5x_1 - 6x_2 + 7x_3 + x_4 \\ \pi_1 & & & x_1 + 2x_2 - x_3 - x_4 = -7 \\ \pi_2 & & & 6x_1 - 3x_2 + x_3 - 7x_4 \geqq 14 \\ \pi_3 & & & 2.8x_1 + 17x_2 - 4x_3 - 2x_4 \geqq 3 \\ & & & x_1, x_2 \geqq 0 \\ & & & x_3, x_4 \text{ unrestricted} \end{array}$$

The problem in detached coefficient form is given below.

Associated dual variable	Primal problem					
	x_1	x_2	x_3	x_4		
π_1	1	2	-1	$-1 =$	-7	
π_2	6	-3	1	$-7 \geqq$	14	
π_3	2.8	17	-4	$-2 \geqq$	3	
	5	-6	7	$1 =$	$z(x)$	Minimize
	$x_1 \geqq 0, x_2 \geqq 0, x_3$ and x_4 unrestricted					

Here is the dual problem in detached coefficient form.

Associated primal variable	π_1	π_2	π_3		
x_1	1	6	$2.8 \leqq$	5	
x_2	2	-3	$17 \leqq$	-6	
x_3	-1	1	$-4 =$	7	
x_4	-1	-7	$-2 =$	1	
	-7	14	$3 =$	$v(\pi)$	maximize
	π_1 unrestricted, $\pi_2 \geqq 0$, $\pi_3 \geqq 0$				

Also verify that the dual of the dual problem is the primal.

Example 4.2

Consider the primal problem:

$$\begin{aligned}
\text{Maximize} \quad z'(x) = \; & 3x_1 - 2x_2 - 5x_3 + 7x_4 + 8x_5 \\
\text{Subject to} \quad & x_2 - x_3 + 3x_4 - 4x_5 = -6 \\
& 2x_1 + 3x_2 - 3x_3 - x_4 \geqq 2 \\
& -x_1 + 2x_3 - 2x_4 \leqq -5 \\
& -2 \leqq x_1 \leqq 10 \\
& 5 \leqq x_2 \leqq 25 \\
& x_3, x_4 \geqq 0 \\
& x_5 \text{ unrestricted}
\end{aligned}$$

Transforming all the inequality constraints into the right type, here is the primal in detached coefficient form.

Associated dual variable	x_1	x_2	x_3	x_4	x_5		
π_1		1	-1	3	$-4 =$	-6	
π_2	-2	-3	3	1	$\leqq$	-2	
π_3	-1		2	-2	$\leqq$	-5	
π_4	1				$\leqq$	10	
π_5	-1				$\leqq$	2	
π_6		1			$\leqq$	25	
π_7		-1			$\leqq$	-5	
	3	-2	-5	7	$8 =$	$z'(x)$	maximize
	x_1, x_2, x_5 unrestricted; $x_3, x_4 \geqq 0$						

Therefore the dual problem in detached coefficient form is as below.

Associated primal variable	π_1	π_2	π_3	π_4	π_5	π_6	π_7			
x_1		-2	-1	1	-1			$=$	3	
x_2	1	-3				1	-1	$=$	-2	
x_3	-1	3	2					$\geqq$	-5	
x_4	3	1	-2					$\geqq$	7	
x_5	-4							$=$	8	
	-6	-2	-5	10	2	25	-5	$=$	$v(\pi)$	minimize
	π_1 unrestricted, π_2 to $\pi_7 \geqq 0$									

It can be verified again that the dual of the dual problem is the primal.

Example 4.3

Let A be a $m \times n$ matrix, b an $m \times 1$ column vector, c a $1 \times n$ row vector. Then, if the primal is

$$\begin{aligned} \text{Minimize} \quad z(x) &= cx \\ \text{Subject to} \quad Ax &\geqq b \\ x &\geqq 0 \end{aligned} \tag{4.6}$$

The dual is

$$\begin{aligned} \text{Maximize} \quad v(\pi) &= \pi b \\ \text{Subject to} \quad \pi A &\leqq c \\ \pi &\geqq 0 \end{aligned} \tag{4.7}$$

where π is the $1 \times m$ row vector of dual variables and x, the $n \times 1$ column vector of primal variables.

Example 4.4

If the primal is

$$\begin{aligned} \text{Minimize} \quad z(x) &= cx \\ \text{Subject to} \quad Ax &= b \\ x &\geqq 0 \end{aligned} \tag{4.8}$$

The dual is

$$\begin{aligned} \text{Maximize} \quad v(\pi) &= \pi b \\ \text{Subject to} \quad \pi A &\leqq c \\ \pi &\text{ unrestricted} \end{aligned} \tag{4.9}$$

Example 4.5

In addition, let E be an $r \times n$ matrix, and d an $r \times 1$ column vector. If the primal is:

$$\begin{aligned} \text{Minimize} \quad z(x) &= cx \\ \text{Subject to} \quad Ax &= b \\ Ex &\geqq d \\ x &\geqq 0 \end{aligned}$$

The dual is

$$\begin{aligned} \text{Maximize} \quad v(\pi, \mu) &= \pi b + \mu d \\ \text{Subject to} \quad \pi A + \mu E &\leqq c \\ \pi \text{ unrestricted}, \quad & \mu \geqq 0 \end{aligned}$$

where π is a row vector of order $1 \times m$ and μ, a row vector of order $1 \times r$.

Example 4.6
Consider the general LP

$$\begin{aligned} \text{Minimize} \quad & z(x, y) = cx + dy \\ \text{Subject to} \quad & Ax + By = b \\ & Ex + Fy \geqq g \\ x \geqq 0, \quad & y \text{ unrestricted in sign} \end{aligned} \tag{4.10}$$

where $A(m_1 \times n_1)$, $B(m_1 \times n_2)$, $b(m_1 \times 1)$, $E(m_2 \times n_1)$, $F(m_2 \times n_2)$, $g(m_2 \times 1)$, $c(1 \times n_1)$, and $d(1 \times n_2)$ are given matrices, and $x(n_1 \times 1)$, and $y(n_2 \times 1)$ are the variables.

Let $\pi(1 \times m_1)$ be the vector of dual variables associated with the equality constraints in (4.10) and $\mu(1 \times m_2)$ be the vector of dual variables associated with the inequality constraints. Then the dual problem is

$$\begin{aligned} \text{Maximize} \quad & v(\pi, \mu) = \pi b + \mu g \\ \text{Subject to} \quad & \pi A + \mu E \leqq c \\ & \pi B + \mu F = d \\ & \pi \text{ unrestricted}, \quad \mu \geqq 0 \end{aligned} \tag{4.11}$$

Example 4.7 Dual of the Transportation Problem
Suppose ore is available at m mines and is required at n plants. Let b_j (>0) denote the minimum amount of ore required at plant j, $j = 1$ to n and let a_i (>0) denote the maximum amount of ore that can be shipped from the ith mine, $i = 1$ to m. Let c_{ij} be the cost of shipping one unit of ore from the ith mine to the jth plant. The problem is to determine how much ore to ship from each mine to each plant in order to meet the requirements at the plants with minimum transportation cost.

Let x_{ij} denote the number of units of ore shipped from the ith mine to the jth plant. Then the total transportation cost, which is to be minimized, is $z(x) = \sum\sum x_{ij}c_{ij}$. Since a_i is the maximum number of units that can be shipped out of mine i, (x_{ij}) must satisfy

$$\text{Amount of ore shipped out of mine } i = \sum_j x_{ij} \leqq a_i \qquad i = 1 \text{ to } m$$

Similarly b_j is the minimum amount of ore required at the jth plant. This leads to

$$\text{Amount of ore reaching plant } j = \sum_i x_{ij} \geqq b_j \qquad j = 1 \text{ to } n$$

and all the variables x_{ij} have to be nonnegative. Transforming all the inequalities into the right type, the problem becomes:

$$\begin{aligned} \text{Minimize} \quad & z(x) = \sum\sum c_{ij}x_{ij} \\ \text{Subject to} \quad & \sum_j (-x_{ij}) \geqq -a_i \qquad i = 1 \text{ to } m \qquad (4.12) \\ & \sum_i x_{ij} \geqq b_j \qquad j = 1 \text{ to } n \qquad (4.13) \\ & x_{ij} \geqq 0 \qquad \text{for all } i, j \end{aligned}$$

Associate the dual variable u_i to the ith constraint in (4.12), $i = 1$ to m, and the dual variable v_j to the jth constraint in (4.13), $j = 1$ to n. The column vector corresponding to the variable x_{ij} consists of only two nonzero coefficients in the primal constraints, namely, a "-1" entry in the ith constraint in (4.12) and a "$+1$" entry in the jth constraint in (4.13). Thus the dual of this transportation problem is

$$\begin{aligned} \text{Maximize} \quad & w(u, v) = -\sum u_i a_i + \sum v_j b_j \\ \text{Subject to} \quad & -u_i + v_j \leqq c_{ij} \qquad \text{for all } i, j \\ & u_i \geqq 0, v_j \geqq 0 \qquad \text{for all } i, j \end{aligned}$$

Problems

4.1 Suppose a trucking company offers to take all the available ore at the mines and deliver the required amounts of ore at the plants. Let u_i be the price they agree to pay for one unit of ore at the ith mine, $i = 1$ to m, and let v_j be the price at which they agree to sell one unit of ore at the jth plant, $j = 1$ to n. Discuss an economic interpretation of the dual problem as that of maximizing the net revenue of the trucking company.

4.2 Consider the transportation problem in which the constraints are all stated as equations.

$$\begin{array}{lll} \text{Minimize} & z(x) = \sum\sum c_{ij}x_{ij} & \\ \text{Subject to} & \sum_j x_{ij} = a_i & i = 1 \text{ to } m \\ & \sum_i x_{ij} = b_j & j = 1 \text{ to } n \\ & x_{ij} \geqq 0 & \text{for all } i, j \end{array}$$

A transportation problem in this form will arise if b_j is the exact amount required in plant j and a_i is the exact amount to be shipped out of mine i. In this form the transportation problem has a feasible solution iff $\sum a_i = \sum b_j$. That's why a transportation problem in this form is known as a *balanced transportation problem.* Write down the dual of this problem. Prove that the dual problem always has a feasible solution.

4.3 Write down the dual for this form of the transportation problem

$$\begin{array}{lll} \text{Minimize} & z(x) = \sum\sum c_{ij}x_{ij} & \\ \text{Subject to} & \sum_j x_{ij} \leqq a_i & i = 1 \text{ to } m \\ & \sum_i x_{ij} = b_j & j = 1 \text{ to } n \\ & x_{ij} \geqq 0 & \text{for all } i, j \end{array}$$

4.4 Write down the dual of the assignment problem

$$\begin{array}{lll} \text{Minimize} & z(x) = \sum\sum c_{ij}x_{ij} & \\ \text{Subject to} & \sum_j x_{ij} = 1 & i = 1 \text{ to } n \\ & \sum_i x_{ij} = 1 & j = 1 \text{ to } n \\ & x_{ij} \geqq 0 & \text{for all } i, j \end{array}$$

4.5 DUALITY THEORY FOR LINEAR PROGRAMMING

The primal and the dual problems are generally two distinct LPs. (There is a small class of LPs called "*self-dual linear programs.*" Any LP in this class is equivalent to its dual. See problem 4.34.) The dual variables do not appear in the statement of the primal problem and vice versa.

4.5.1 Dual of the Dual Is Primal

Let us obtain the dual problem of the dual LP, (4.7), in Example 4.3. Problem (4.7) can be rewritten as

$$\begin{array}{ll} \text{Maximize} & v(\pi) = b^{\mathrm{T}}\pi^{\mathrm{T}} \\ \text{Subject to} & A^{\mathrm{T}}\pi^{\mathrm{T}} \leqq c^{\mathrm{T}} \\ & \pi^{\mathrm{T}} \geqq 0 \end{array}$$

Therefore, there are n constraints in this problem. Let x_j denote the dual variable associated with the jth constraint. Let $x^{\mathrm{T}} = (x_1, \ldots, x_n)$ be the dual vector. Then it can be verified that the dual problem of (4.7) is (4.6) of Example 4.3.

Problem

4.5 Verify that the dual of the dual is the primal for the problems in Examples 4.4, 4.5, and 4.6.

Thus in any primal dual pair of LPs, each is the dual of the other.

4.5.2 Duals of Equivalent Problems Are Equivalent

Let (P) refer to an LP and let (D) be its dual. Let ($\tilde{\text{P}}$) be an LP that is equivalent to (P). Let ($\tilde{\text{D}}$) be the dual of ($\tilde{\text{P}}$). Then ($\tilde{\text{D}}$) is equivalent to (D). For example, consider the primal (4.6) in Example 4.3. Including the slack vector s (a column vector of order $m \times 1$) the problem may be written in the equivalent form

$$\begin{aligned} \text{Minimize} \quad & z(x, s) = cx + 0s \\ \text{Subject to} \quad & Ax - Is = b \\ & x \geqq 0 \qquad s \geqq 0 \end{aligned}$$

This is a problem of the form discussed in Example 4.4. Denoting the dual vector by π (a row vector of order $1 \times m$), we find that the dual is

$$\begin{aligned} \text{Maximize} \quad & v(\pi) = \pi b \\ \text{Subject to} \quad & \pi A \leqq c \\ & \pi(-I) \leqq 0 \\ & \pi \text{ unrestricted} \end{aligned}$$

The constraint $\pi(-I) \leqq 0$ implies $\pi \geqq 0$. So this dual problem is equivalent to (4.7).

Problems

4.6 The LP (4.10) in Example 4.6 can be written in an equivalent form by expressing each unrestricted variable y_j as the difference of two nonnegative variables, $y_j^+ - y_j^-$. Show that the dual of this equivalent problem is equivalent to (4.11).

4.7 An equivalent LP to the one in Example 4.4 is obtained by replacing the constraints "$Ax = b$" by the equivalent system of inequality constraints:

$$\begin{aligned} Ax &\geqq b \\ Ax &\leqq b \end{aligned}$$

Show that the dual of this equivalent problem is equivalent to the dual found in Example 4.4.

This explains why the dual variables associated with primal equality constraints are unrestricted in sign.

4.8 Express the primal in Example 4.5 in an equivalent form by introducing slack variables corresponding to the inequality constraints. Show that the dual of this equivalent problem is equivalent to the dual obtained in Example 4.5.

4.5.3 Weak Duality Theorem

In a primal dual pair of LPs let x be a primal feasible solution and $z(x)$ the corresponding value of the primal objective function that is to be *minimized*. Let π be a dual feasible solution and $v(\pi)$ the corresponding dual objective function that is to *maximized*. Then,

$$z(x) \geqq v(\pi)$$

Proof We will prove it for the case where the primal and the dual are stated as in Example 4.3. The primal is (4.6) and the dual is (4.7). Then,

$$Ax \geqq b \text{ (because } x \text{ is primal feasible)}$$

$$\pi Ax \geqq \pi b \text{ (because } \pi \geqq 0) \tag{4.14}$$

$$\pi A \leqq c \text{ (because } \pi \text{ is dual feasible)}$$

$$\pi Ax \leqq cx \text{ (because } x \geqq 0) \tag{4.15}$$

Combining (4.14) and (4.15) we get

$$cx \geqq \pi Ax \geqq \pi b$$

that is,

$$z(x) \geqq v(\pi)$$

which proves the theorem when the primal and dual are stated in this form.

In general, every LP can be transformed into an equivalent problem in standard form by the transformations discussed in Chapter 2. Suppose the problem in standard form is

$$\begin{aligned} \text{Minimize} \quad & cx \\ \text{Subject to} \quad & Hx = h \\ & x \geqq 0 \end{aligned}$$

This same problem can be restated as

$$\begin{aligned} \text{Minimize} \quad & cx \\ \text{Subject to} \quad & Hx \geqq h \\ & -Hx \geqq -h \\ & x \geqq 0 \end{aligned} \tag{4.16}$$

which is in the same form as (4.6). So the proof given above applies to it. However, by section 4.5.2, the dual of (4.16) is equivalent to the dual of the original problem. Therefore the weak duality theorem must hold for any LP in general.

4.5.4 Note

One may be tempted to summarize the statement of the weak duality theorem as: "In any primal dual pair of LPs the primal objective value of any primal feasible solution is always greater than or equal to the dual objective value of any dual feasible solution." However, this is not true always. In a primal-dual pair, each problem is the dual of the other. One of the problems in the pair is a minimization problem and the other a maximization problem. The weak duality theorem states that the above statement is true *only when* the minimization problem in the pair is called the primal and the other called the dual.

Problem

4.9 Prove the weak duality theorem from first principles (i.e., using arguments similar to those used here) for the cases when the primal and dual are stated as in Examples 4.4, 4.5, and 4.6.

4.5.5 Corollaries of the Weak Duality Theorem

Considering any primal, dual pair of LPs, let the *primal* refer to the minimization problem in the pair and the dual refer to the maximization problem in the pair.

1. The primal objective value of any primal feasible solution is an upper bound to the maximum value of the dual objective in the dual problem.
2. The dual objective value of any dual feasible solution is a lower bound to the minimum value of the primal objective in the primal problem.
3. If the primal problem is feasible and its objective function is unbounded below on the primal feasible solution set, the dual problem cannot have a feasible solution.
4. If the dual problem is feasible and its objective function is unbounded above on the dual feasible solution set, the primal problem cannot have a feasible solution.
5. The converse of 3. is the following:
   "If the dual problem is infeasible and the primal problem is feasible, the primal objective function is unbounded below on the primal feasible solution set."
   Similarly, the converse of 4. is:
   "If the primal problem is infeasible and the dual problem is feasible, the dual objective function is unbounded above on the dual feasible solution set."
   Both these results are true and will be proved after a discussion of the fundamental duality theorem.

It is possible that both the primal and the dual problems in a primal, dual pair, have no feasible solutions. For example, consider the following:

Minimize $= 2x_1 - 4x_2$	Maximize $v(\pi) = \pi_1 + 2\pi_2$
Subject to $x_1 - x_2 = 1$	Subject to $\pi_1 - \pi_2 \leqq 2$
$-x_1 + x_2 = 2$	$-\pi_1 + \pi_2 \leqq -4$
$x_1 \geqq 0 \quad x_2 \geqq 0$	π_1, π_2 unrestricted

Clearly both problems in the pair are infeasible. Thus, even though the result in 5. is true, the fact that the dual problem is infeasible in a primal, dual pair of LPs does not imply that the primal objective function is unbounded on the primal feasible solution set, unless it is known that the primal is feasible.

4.5.6 Sufficient Optimality Criterion in Linear Programming

In a primal, dual pair of LPs let $z(x)$ be the primal objective function and $v(\pi)$, the dual objective function. If $\bar{x}, \bar{\pi}$ are a pair of primal and dual feasible solutions satisfying:

$$z(\bar{x}) = v(\bar{\pi})$$

then $\bar{x}$ is an optimal feasible solution of the primal and $\bar{\pi}$ is an optimum feasible solution of the dual.

Proof Suppose the primal denotes the minimization problem in the primal, dual pair. Let x be any primal feasible solution. By the weak duality theorem, because $\bar{\pi}$ is dual feasible, we have:

$$z(x) \geqq v(\bar{\pi}) \qquad \text{for all } x \text{ primal feasible}$$

But

$$z(\bar{x}) = v(\bar{\pi}) \qquad \text{by hypothesis}$$

So

$$z(x) \geqq z(\bar{x}) \qquad \text{for all } x \text{ primal feasible}$$

So $\bar{x}$ is optimal to the primal problem. Similarly we show that $\bar{\pi}$ is optimal to the dual problem.

Example

Considering the problems of the housewife and the pillmaker, let (4.1) be the primal and (4.2) the dual. We verify that

$$\bar{x}^{\mathrm{T}} = (\bar{x}_1, \bar{x}_2, \bar{x}_3, \bar{x}_4, \bar{x}_5, \bar{x}_6) = (0, 0, 0, 0, 5, 2)$$

is a primal feasible solution, and that

$$\bar{\pi} = (\bar{\pi}_1, \bar{\pi}_2) = (3, 8)$$

is a dual feasible solution. Also

$$z(\bar{x}) = v(\bar{\pi}) = 179 \text{ cents}$$

Therefore, by the sufficient optimality criterion $\bar{x}$ represents the optimal diet in the housewife's problem and $\bar{\pi}$ represents the optimal prices that the pillmaker can adopt to maximize his profit while keeping his nutrient pills economically competitive.

The converse of this sufficient optimality criterion is the fundamental duality theorem that is discussed in the next section.

4.5.7 The Fundamental Duality Theorem

In a primal, dual pair of LPs, if either the primal or the dual problem has an optimal feasible solution, then the other does also, and the two optimal objective values are equal.

Proof We will prove the fundamental duality theorem for the case where the primal and dual problems are stated as in Example 4.4. The primal problem is (4.8), where A is of order $m \times n$. Suppose the primal problem has an optimal feasible solution. In this case the primal problem has an optimal feasible basis by the results in Chapter 3. Suppose the basic vector $x_B = (x_1, \ldots, x_m)^{\mathrm{T}}$ is an optimal primal feasible basic vector. Let B be the basis associated with the basic vector x_B. Let D be the $m \times (n - m)$ matrix of the column vectors in A associated with the nonbasic variables. After partitioning into the basic and nonbasic parts as in section 3.12, the problem can be represented in tableau form as follows.

Original Tableau

Basic variables x_B	$(-z)$	Nonbasic variables x_D	Right-hand side constants
B	0	D	b
c_B	1	c_D	0

By the results in section 3.12, the canonical tableau with respect to the optimum basis B is as below.

Optimal Canonical Tableau

Basic variables	x_B	$-z$	x_D	Basic values
x_B	I	0	$B^{-1}D$	$B^{-1}b$
$-z$	0	1	$c_D - c_B B^{-1}D$	$-c_B B^{-1}b$

The last row in the optimal canonical tableau is $\overline{c} = c - c_B B^{-1} A$. By the optimality criterion, this basis is optimal only if

$$\overline{c} = c - c_B B^{-1} A \geqq 0$$

That is

$$(c_B B^{-1}) A \leqq c \tag{4.17}$$

From the tableau, it is clear that the optimal objective value is $(c_B B^{-1})b$. If we define $\pi = c_B B^{-1}$, by (4.17) we have $\pi A \leqq c$ and $\pi b = (c_B B^{-1})b =$ optimal primal objective value. By the sufficient optimality criterion, section 4.5.6, π is a dual optimal solution. This completes the proof when the primal and dual are as stated here.

In general, every LP can be transformed into an equivalent problem in standard form by the transformations discussed in Chapter 2. The equivalent problem in standard form is of the same type as the primal problem in Example 4.4 (i.e., a minimization problem involving equality constraints in nonnegative variables); hence, our proof of the fundamental duality theorem applies to it. However, by section 4.5.2, the dual of the equivalent problem in standard form is equivalent to the dual of the original problem. Thus, the fundamental duality theorem must hold for it too. This completes the general proof of the fundamental duality theorem.

Example

From section 4.5.6 verify that the fundamental duality theorem holds for the problems of the housewife and the pill manufacturer.

4.5.8 Corollaries of the Fundamental Duality Theorem

(i) ALTERNATIVE STATEMENT OF THE DUALITY THEOREM

An alternate statement of the fundamental duality theorem is: "If both the problems in a primal, dual pair of LPs have feasible solutions, then both have optimal feasible solutions, and the optimal objective values of both the problems are equal." This is easily proved by using the weak duality theorem and the fundamental duality theorem.

(ii) SEPARATION PROPERTY OF OBJECTIVE VALUES

Consider a primal, dual pair of LPs and suppose the minimization problem in the pair is the primal with the objective function $z(x)$. Suppose the dual objective function is $v(\pi)$. If both the problems have feasible solutions, then the values assumed by the two objective functions at feasible solutions of the respective problems are separated on the real line as in Figure 4.1.

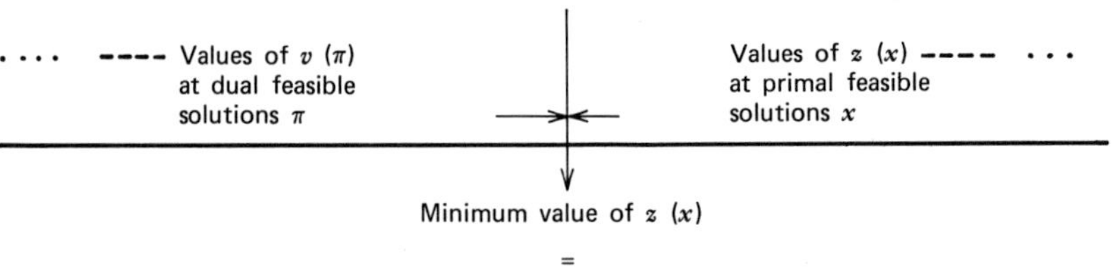

Figure 4.1

(iii) PRIMAL OBJECTIVE UNBOUNDED
If the primal is the minimization problem in a primal, dual pair, and if the primal is feasible and the dual is infeasible, then the primal cannot have an optimal feasible solution, that is, the primal objective function is unbounded below.

(iv) DUAL OBJECTIVE UNBOUNDED
If the dual is the maximization problem in a primal, dual pair, and if the dual is feasible and the primal infeasible, then the dual cannot have an optimal feasible solution, that is, the dual objective function is unbounded above.

Notes Parts (iii) and (iv) here are the converses of parts 3. and 4. of section 4.5.5, respectively, discussed in part 5. of section 4.5.5.

4.5.9 Dual Feasibility of a Basis

Consider the primal LP in standard form, (4.8), in Example 4.4. Let B be a basis for (4.8) and let x_B denote the corresponding basic vector. Let c_B be the row vector of the basic cost coefficients. From the arguments in section 4.5.7 the dual solution corresponding to the basis B is given by $\pi = c_B B^{-1}$. Partitioning the variables, column vectors, etc., into basic and nonbasic parts as in section 4.5.7, the primal problem (4.8) can be written as:

$$\begin{array}{lrl} \text{Minimize} & z(x) = & c_B x_B + c_D x_D \\ \text{Subject to} & & B x_B + D x_D = b \\ & & x_B \geqq 0 \quad x_D \geqq 0 \end{array}$$

Its dual can be written as:

$$\begin{array}{lrl} \text{Maximize} & v(\pi) = & \pi b \\ \text{Subject to} & & \pi B \leqq c_B \\ & & \pi D \leqq c_D \\ & & \pi \text{ unrestricted} \end{array}$$

The primal basic solution of (4.8) corresponding to the basis B is obtained by solving the system

$$\begin{array}{rl} & x_D = 0 \\ \text{and} & B x_B = b \end{array}$$

and the basis is *primal feasible* iff the solution satisfies the remaining nonnegativity constraints on the primal variables, that is, iff $x_B = B^{-1}b \geqq 0$. The dual solution corresponding to the basis B is obtained by solving

$$\pi B = c_B$$

and the basis B is *dual feasible* if this dual solution satisfies the remaining dual constraints, namely, $\pi D \leqq c_D$. Hence, if B is a dual feasible basis, the dual solution corresponding to it is $\pi = c_B B^{-1}$ and this will satisfy all the dual constraints if $\pi A \leqq c$. That is,

$$[c - (c_B B^{-1})A] \geqq 0 \tag{4.18}$$

If B is primal feasible, (4.18) is just the optimality criterion. Thus a basis B of (4.8) is optimal if it is both primal and dual feasible.

The simplex algorithm always deals with primal feasible bases. But until (4.18) is satisfied, it does not reach an optimal basis. Thus, all but the final optimal basis encountered in the simplex algorithm are dual infeasible. When dual feasibility is attained, the simplex algorithm terminates. The simplex algorithm starts with a primal feasible (but dual infeasible) basis, and it tries to attain dual feasibility, keeping primal feasibility throughout the algorithm. On the other hand, it is possible to develop an algorithm that starts with a dual feasible but primal infeasible basis and tries to attain primal feasibility, keeping dual feasibility throughout the algorithm. Such an algorithm is appropriately called the *dual simplex algorithm*, and it is discussed in Chapter 6.

4.5.10 Relative Cost Coefficients Are Dual Slack Variables

Consider the LP (4.8) in standard form again. Suppose the basis B, associated with the basic vector x_B and cost vector c_B, is an optimum basis for this problem. Then $\bar{\pi} = c_B B^{-1}$ is an optimal feasible solution of the dual of (4.8). The dual constraints are $\pi A_{.j} \leqq c_j$ for $j = 1$ to n. Hence, the dual slacks corresponding to the optimum dual feasible solution π are

$$\bar{c}_j = (c_j - \bar{\pi} A_{.j}) = (c_j - c_B B^{-1} A_{.j})$$

Thus, the relative cost coefficients of the simplex algorithm are the dual slack variables. If any of the relative cost coefficients with respect to the basis B are negative, since that slack variable is negative, the associated dual solution is dual infeasible and, hence, the basis B is dual infeasible. If all the relative cost coefficients are nonnegative, the associated dual solution is dual feasible and, hence, B is a dual feasible basis. Therefore, the optimality criterion of the simplex algorithm is just the dual feasibility criterion.*

Example
Consider the LP:

x_1	x_2	x_3	x_4	x_5	x_6	$-z$	b
1	0	0	−1	2	4	0	7
0	1	0	1	−2	6	0	19
0	0	1	2	3	−7	0	21
3	−4	−8	16	−17	19	1	0

(4.19)

$x_j \geqq 0$ for all j. z to be minimized

The dual problem is

$$\begin{array}{llr} \text{Maximize} & 7\pi_1 + 19\pi_2 + 21\pi_3 & \\ \text{Subject to} & \pi_1 \qquad\qquad\qquad \leqq & 3 \\ & \qquad\quad \pi_2 \qquad\quad \leqq & -4 \\ & \qquad\qquad\qquad \pi_3 \leqq & -8 \\ & -\pi_1 + \pi_2 + 2\pi_3 \leqq & 16 \\ & 2\pi_1 - 2\pi_2 + 3\pi_3 \leqq & -17 \\ & 4\pi_1 + 6\pi_2 - 7\pi_3 \leqq & 19 \\ & \pi_i \text{ unrestricted for all } i & \end{array} \qquad (4.20)$$

* The relative cost coefficients (i.e. those from an optimum tableau) are also known as the *shadow costs* associated with the activities in the primal LP model.

The dual constraints can be written in an equivalent manner as

$$
\begin{aligned}
3 - \pi_1 \qquad\qquad\qquad\quad & \geqq 0 \\
-4 \qquad\quad - \pi_2 \qquad\quad & \geqq 0 \\
-8 \qquad\qquad\qquad - \pi_3 & \geqq 0 \\
16 + \pi_1 - \pi_2 - 2\pi_3 & \geqq 0 \\
-17 - 2\pi_1 + 2\pi_2 - 3\pi_3 & \geqq 0 \\
19 - 4\pi_1 - 6\pi_2 + 7\pi_3 & \geqq 0
\end{aligned}
\tag{4.21}
$$

and the expressions on the left-hand side in (4.21) are obviously the slacks corresponding to the constraints in the dual problem.

Consider the basic vector (x_1, x_2, x_3) for (4.19). Here is the canonical tableau with respect to this basic vector.

x_1	x_2	x_3	x_4	x_5	x_6	$-z$	b
1	0	0	-1	2	4	0	7
0	1	0	1	-2	6	0	19
0	0	1	2	3	-7	0	21
0	0	0	39	-7	-25	1	223

Hence the relative cost coefficient vector with respect to the basic vector (x_1, x_2, x_3) is $\bar{c} = (0, 0, 0, 39, -7, -25)$.

The dual solution associated with the basis corresponding to the basic vector (x_1, x_2, x_3) is derived by solving the system obtained by treating the first three constraints in (4.20) as equations. It is $\pi = (3, -4, -8)$. Substituting this solution in (4.21) we verify that the dual slack vector is equal to the relative cost vector $\bar{c}$ obtained from the canonical tableau.

The most commonly used method for solving linear programming problems is the revised simplex method (see Chapter 5). It obtains and uses the dual solution at each stage, to compute the relative cost coefficients.

4.5.11 Unboundedness Criterion and Dual Infeasibility

Again consider the LP (4.8) in standard form. When this problem is solved by the simplex algorithm, suppose the unboundedness criterion is satisfied. Let the terminal basis obtained be B and, for convenience in referring to it, let us assume that the associated basic vector is $(x_1, \ldots, x_m)$. Since the unboundedness criterion is satisfied at this stage, there must exist a nonbasic variable, say x_{m+1}, such that the column vector of x_{m+1} in the canonical tableau at this stage satisfies the unboundedness criterion. Let the following be the canonical tableau with respect to this basis

Basic variables	$x_1 \ldots x_m$	$x_{m+1} \quad \cdots \quad x_n$	$-z$	Basic values
x_1	$1 \ldots 0$	$\bar{a}_{1,m+1} \cdots \bar{a}_{1,n}$	0	$\bar{b}_1$
$\vdots$	$\vdots \quad \vdots$	$\vdots \quad \vdots$	$\vdots$	$\vdots$
x_m	$0 \ldots 1$	$\bar{a}_{m,m+1} \cdots \bar{a}_{m,n}$	0	$\bar{b}_m$
$-z$	$0 \ldots 0$	$\bar{c}_{m+1} \quad \cdots \bar{c}_n$	1	$-\bar{z}$

Our assumptions are that

$$\bar{c}_{m+1} < 0$$

and (4.22)

$$\bar{a}_{i,\,m+1} \leqq 0 \qquad \text{for } i = 1 \text{ to } m$$

From the canonical tableau with respect to the basis B, the LP (4.8) is equivalent to

$$\begin{aligned} &\text{Minimize} \quad z(x) = \left(\sum_{j=m+1}^{n} \bar{c}_j x_j \right) + \bar{z} \\ &\text{Subject to} \quad x_i + \sum_{j=m+1}^{n} \bar{a}_{ij} x_j = \bar{b}_i \qquad i = 1 \text{ to } m \\ &\qquad\qquad\qquad x_j \geqq 0 \qquad \text{for all } j = 1 \text{ to } n \end{aligned} \tag{4.23}$$

Since $\bar{z}$ is a known constant, it can be ignored in the minimization process. The dual of (4.23) is

$$\begin{aligned} &\text{Maximize} \quad v(\pi) = \left(\sum_{i=1}^{m} \bar{b}_i \pi_i \right) + \bar{z} \\ &\text{Subject to} \qquad \pi_i \leqq 0 \qquad \text{for } i = 1 \text{ to } m \\ &\qquad\qquad \sum_{i=1}^{m} \bar{a}_{ij} \pi_i \leqq \bar{c}_j \qquad \text{for } j = m+1 \text{ to } n \end{aligned} \tag{4.24}$$

However, the constraints

$$\pi_i \leqq 0 \qquad \text{for } i = 1 \text{ to } m$$

and

$$\sum_{i=1}^{m} \bar{a}_{i,\,m+1} \pi_i \leqq \bar{c}_{m+1}$$

are inconsistent because of (4.22). Thus, the dual constraints (4.24) are inconsistent and, hence, the dual of (4.23) is infeasible. But (4.23) is equivalent to (4.8) and so the dual of (4.23) is equivalent to the dual of (4.8). So the dual of (4.8) must be infeasible too. This shows that the unboundedness criterion in the primal simplex algorithm just verifies dual inconsistency.

UNBOUNDEDNESS UNDER VARIABLE RIGHT-HAND SIDE CONSTANT VECTOR

Consider the LP (4.8) in standard form again. If (4.8) is feasible and unbounded below, that is, $z(x)$ can be made to diverge to $-\infty$ on the set of feasible solutions of (4.8), then the dual constraints in (4.9) must be inconsistent. But the dual constraints are independent of the primal right-hand side constant vector b and, hence, if they are inconsistent for some b, they remain inconsistent for any b. We therefore conclude that (4.8) remains unbounded even if we change b, as long as it remains feasible. Thus,

1. If the LP (4.8) is unbounded below for some b, then it is unbounded below for every b for which it is feasible.
2. If the LP (4.8) has an optimal feasible solution, then it will have an optimal feasible solution even after changing b, as long as it remains feasible.

These results will be useful when we consider sensitivity analysis and parametric linear programming.

UNBOUNDEDNESS IN THE BIG-*M* METHOD

Again consider the LP in standard form (4.8), where A is a matrix of order $m \times n$. Its dual is (4.9). To solve (4.8) by the big-M method (see section 2.7) the artificial variables $t_1, \ldots, t_m$ are introduced, and the problem modified to

$$\begin{aligned} \text{Minimize} \quad & z(x) = cx + \sum_{i=1}^{m} Mt_i \\ \text{Subject to} \quad & Ax + It = b \\ & x \geqq 0 \quad t \geqq 0 \end{aligned} \tag{4.25}$$

where M is an arbitrarily large positive number. The dual of (4.25) is

$$\begin{aligned} \text{Maximize} \quad & \pi b \\ \text{Subject to} \quad & \pi A \leqq c \\ & \pi_i \leqq M \qquad \text{for } i = 1 \text{ to } m \\ & \pi \text{ unrestricted in sign} \end{aligned} \tag{4.26}$$

The only constraints in (4.26) that (4.9) does not include are the restrictions "$\pi_i \leqq M$," for $i = 1$ to m. Clearly, if (4.26) is infeasible for arbitrarily large M, (4.9) must be infeasible too.

In the course of solving (4.25) by the simplex algorithm, if the unboundedness criterion is satisfied independent of how large a value M has, then its dual, (4.26) must be infeasible. So (4.9) must be infeasible too. Hence, in this case $z(x)$ is unbounded below if (4.8) is feasible. This proves the results discussed in section 2.7.

4.5.12 Complementary Slackness Property

Another important corollary of the duality theorem is known as the *complementary slackness theorem.* When problems in mathematical economics are modeled as LPs, this theorem can be interpreted as indicating an *equilibrium or economic stability* when optimality conditions are satisfied. That is why this theorem is also called the *equilibrium theorem* or the *price equilibrium theorem.* In this context any optimum feasible dual vector is called an *equilibrium price vector.*

Consider a pair of primal and dual LPs. By the manner in which the dual problem is defined, there is a dual variable corresponding to each primal constraint. This dual variable is sign restricted iff the corresponding primal constraint is an inequality. A primal slack variable is associated with this primal inequality constraint (the difference between the two sides of the inequality in any feasible solution). Thus, associated with each nonnegative dual variable, there is a primal slack variable. Hence, each sign restricted variable in one problem can be paired with its associated slack variable in the other problem. We now state the complementary slackness theorem.

Complementary Slackness Theorem (Equilibrium Theorem) A pair of primal and dual feasible solutions are optimal to the respective problems in a primal, dual pair of LPs, iff, whenever these feasible solutions make a slack variable in one problem strictly positive, the value (in these feasible solutions) of the associated nonnegative variable of the other problem is zero.

Proof We will prove this theorem for the case where the primal, dual pair of problems are stated as in (4.6) and (4.7) in Example 4.3 where A is of order $m \times n$. The dual variable π_i is associated with the primal constraint $A_{i.}x \geqq b_i$ and hence, its associated primal slack variable is $A_{i.}x - b_i = v_i$ for $i = 1$ to m.

The primal variable x_j is associated with the dual constraint $\pi A_{.j} \leqq c_j$ and, hence, its associated dual slack variable is $c_j - \pi A_{.j} = u_j$ for $j = 1$ to n. In this example, each primal and dual variable is restricted to be nonnegative; hence, each of them has an associated slack variable of the other problem. Let $\bar{x}$, $\bar{\pi}$ be a pair of primal and dual feasible solutions respectively. Then $\bar{v} = A\bar{x} - b$, and $\bar{u} = c - \bar{\pi}A$ are the vectors of the slack variables corresponding to these solutions. The theorem states that $\bar{x}$, $\bar{\pi}$ are optimal to the respective problems iff:

$$\text{Whenever} \quad \bar{v}_i = A_{i.}\bar{x} - b_i > 0 \quad \text{we have } \bar{\pi}_i = 0 \tag{4.27}$$

$$\text{And whenever} \quad \bar{u}_j = c_j - \bar{\pi}A_{.j} > 0 \quad \text{we have } \bar{x}_j = 0 \tag{4.28}$$

Another way of writing (4.27) and (4.28) is

$$\bar{v}_i\bar{\pi}_i = (A_{i.}\bar{x} - b_i)\bar{\pi}_i = 0 \qquad \text{for all } i = 1 \text{ to } m \tag{4.29}$$

$$\bar{u}_j\bar{x}_j = (c_j - \bar{\pi}A_{.j})\bar{x}_j = 0 \qquad \text{for all } j = 1 \text{ to } n \tag{4.30}$$

Note (i) Conditions (4.27) or (4.29) only require that if $\bar{v}_i > 0$, then $\bar{\pi}_i = 0$. They *do not require* that if $\bar{v}_i = 0$, then $\bar{\pi}_i$ be > 0; that is, both $\bar{v}_i$ and $\bar{\pi}_i$ could be equal to zero, and the conditions of the complementary slackness theorem would be satisfied.

(ii) Conditions (4.27) and (4.28) automatically imply that if $\bar{\pi}_i > 0$, then $\bar{v}_i = 0$ and that if $\bar{x}_j > 0$, then $\bar{u}_j = 0$.

We will prove the "if" portion of the theorem first. By primal feasibility

$$A_{i.}\bar{x} \geqq b_i \qquad \text{for all } i = 1 \text{ to } m$$

Multiplying both sides by $\bar{\pi}_i \geqq 0$ (by dual feasibility), summing over i and using (4.29),

$$\sum \bar{\pi}_i(A_{i.}\bar{x}) = \sum \bar{\pi}_i b_i \tag{4.31}$$

Similarly from dual feasibility, $\bar{\pi}A_{.j} \leqq c_j$ for all $j = 1$ to n. Multiplying both sides of this by $\bar{x}_j$ and summing over j and using (4.30) leads to

$$\sum \bar{x}_j(\bar{\pi}A_{.j}) = \sum \bar{x}_j c_j \tag{4.32}$$

However, $\sum \bar{\pi}_i(A_{i.}\bar{x}) = \sum \bar{x}_j(\bar{\pi}A_{.j}) = \bar{\pi}A\bar{x}$. From (4.31) and (4.32) $\sum \bar{x}_j c_j = \sum \bar{\pi}_i b_i$. Hence, from the sufficient optimality criterion $\bar{x}$, $\bar{\pi}$ are optimal feasible solutions of the respective problems.

To prove the "only if" portion of the theorem suppose $\bar{x}$ and $\bar{\pi}$ are a pair of optimal feasible solutions. We have to show that they satisfy the complementary slackness conditions (4.27) and (4.28). From the weak duality theorem we know that $c\bar{x} \geqq \bar{\pi}A\bar{x} \geqq \bar{\pi}b$. But from the duality theorem $c\bar{x} = \bar{\pi}b$. So $c\bar{x} = \bar{\pi}A\bar{x} = \bar{\pi}b$. So $c\bar{x} - \bar{\pi}A\bar{x} = 0$. This implies that

$$\sum (c_j - \bar{\pi}A_{.j})\bar{x}_j = 0 \tag{4.33}$$

However, $\bar{x}_j \geqq 0$ and $c_j - \sum_i \bar{\pi}_i a_{ij} \geqq 0$ by primal and dual feasibility. Therefore, the left side of (4.33) is a sum of nonnegative quantities. It is zero only if each term in the sum is zero. Thus (4.30) must hold. Similarly, $\bar{\pi}(A\bar{x} - b) = 0$ implies that (4.29) must hold. This completes the proof for the case where the primal and dual are as stated in Example 4.3.

Proof of complementary slackness theorem, when the primal and dual problems are stated in a different form, is exactly similar.

Note The complementary slackness theorem does not say anything about the values of unrestricted variables (corresponding to equality constraints in the other problem) in a pair of optimal feasible solutions of the primal and dual problems, respectively.

It is concerned only with nonnegative variables of one problem and the slack variables corresponding to the associated inequality constraints in the other problem of the pair of primal and dual LPs.

Corollary Consider a primal, dual pair of LPs. Let $\bar{x}$ be an optimal feasible solution of the primal LP. Then the following statements can be made about every dual optimum feasible solution.

(i) If x_j is a variable restricted to be nonnegative in the primal problem and if $\bar{x}_j > 0$, then the dual inequality constraint associated with the primal variable, x_j, is satisfied as an equation by *every* dual optimum feasible solution.

(ii) If the primal problem consists of any inequality constraints, let $\bar{v}$ represent the values of the corresponding slack variables, at the primal feasible solution $\bar{x}$. Then, if a slack variable $\bar{v}_i > 0$, the dual variable associated with it is equal to zero in *every* dual optimum feasible solution.

Given an optimal feasible solution of one of the problems in the primal, dual pair, the above results can be used to characterize the set of all optimal feasible solutions of the other problem.

Example

Consider the problems of the housewife and the pill manufacturer. The housewife's problem is (4.1). The slack variables in this problem are:

$$v_1 = (x_1 + 2x_3 + 2x_4 + x_5 + 2x_6 - 9)$$
$$v_2 = (x_2 + 3x_3 + x_4 + 3x_5 + 2x_6 - 19)$$

Suppose we are told that $\bar{x} = (0, 0, 0, 0, 5, 2)^{\mathrm{T}}$ is an optimal feasible solution to the housewife's problem. The dual problem is (4.2). Since x_5 and x_6 are positive, every dual optimum feasible solution must satisfy

$$\pi_1 + 3\pi_2 = 27$$
$$2\pi_1 + 2\pi_2 = 22$$

But this system of equations has a unique solution; $\bar{\pi} = (3, 8)$ and we verify that this is dual feasible. This also verifies the assertion that $\bar{x}$ is indeed optimal to the primal problem. Also, since $\bar{\pi}_1 > 0$ and $\bar{\pi}_2 > 0$, (v_1, v_2) must be equal to 0 at every primal optimum feasible solution. We verify that $\bar{v} = 0$. Thus, the dual optimum feasible solution is unique, and it is $\bar{\pi}$.

Consider the LP (4.8) in standard form and its dual (4.9). The complementary slackness conditions for optimality of this primal, dual pair of problems are

$$x_j(c_j - \pi A_{.j}) = 0 \qquad \text{for} \qquad j = 1 \text{ to } n \tag{4.34}$$

that is,

$$x_j\bar{c}_j = 0 \qquad \text{for} \qquad j = 1 \text{ to } n$$

Problems

4.10 Prove from first principles, as above, that $\bar{x}$ and $\bar{\pi}$ are primal and dual optimal, respectively, to (4.8) if they are feasible to the respective problems and satisfy (4.34).

4.11 Write down the complementary slackness conditions when the primal and dual are as stated in Examples 4.5 and 4.6 and prove the complementary slackness theorem for each case from first principles.

4.12 Write down the complementary slackness conditions for the transportation and assignments problems discussed in Example 4.7.

Suppose B is an optimum basis for the LP (4.8) in standard form. Let $\bar{c}$ be the vector of relative cost coefficients with respect to the basis B. If $\bar{c}_j > 0$, then $x_j = 0$ in every optimum feasible solution of (4.8) from the complementary slackness conditions for optimality. Also any feasible solution of (4.8) in which $x_j = 0$, for all j such that $\bar{c}_j > 0$ is an optimum feasible solution of (4.8). This shows that the set of optimum feasible solutions of (4.8) is characterized as in section 3.10.

ECONOMIC INTERPRETATION

Consider a diet problem. An example of a diet problem is the problem of the housewife (4.1) and its dual is the nutrient pill manufacturer's problem (4.2). The pill manufacturer has to set the prices of the nutrient pills in such a way that the total value (in terms of these prices) of the nutrients contained per unit of each food available to the housewife is less than or equal to the cost per unit of this food in the market.

In a general diet problem, suppose there are m nutrients, and n foods, and let a_{ij} be the number of units of nutrient i per unit of food j. Let c_j be the cost in the market per unit of food j. Let b_i be the minimal requirement for nutrient i. Let π_i denote the price charged per unit of nutrient i (in the form of a pill) by the pill manufacturer. The constraints on the prices are:

$$\sum_{i=1}^{m} \pi_i a_{ij} \leqq c_j \qquad \text{for all } j = 1 \text{ to } n$$

Given these prices, the housewife checks whether she can meet her nutrient requirements cheaper by buying the foods rather than the nutrient pills. Let x_j denote the number of units of food j that she buys. Since the cost per unit of food j is c_j, and she can get all the nutrition contained in one unit of food j from the pill manufacturer at a cost of $\sum_i \pi_i a_{ij}$, she would definitely not like to buy food j if

$$\sum_i \pi_i a_{ij} < c_j \tag{4.35}$$

Thus she would make $x_j = 0$ whenever (4.35) holds. Similarly, if the pill manufacturer associates a positive price per unit of nutrient i (i.e., $\pi_i > 0$), then she would try to meet the bare minimal requirements for this nutrient (since the restriction is only that the amount of nutrient i in the diet is greater than or equal to the minimal requirement, and getting any more than the minimal requirement costs money under this set-up). Thus, if $\pi_i > 0$, she will try to satisfy

$$\sum_j a_{ij} x_j = b_i$$

A similar interpretation can be given about the prices that the pill manufacturer will adopt if he knows the amounts of foods that the housewife is buying.

These are precisely the complementary slackness conditions for this primal, dual pair. When they are satisfied, there is no incentive for the housewife to change the levels of her food purchases or for the pill manufacturer to change his prices. The minimum cost incurred by the housewife is precisely the maximum revenue that the pill manufacturer can make. Therefore, these conditions may be considered *economic stability conditions or equilibrium conditions.*

4.5.13 Discussion on How Various Algorithms Solve the Problem

There are three sets of conditions to be satisfied by the solutions of a primal, dual pair of LPs in order to be optimal: (1) primal feasibility, (2) dual feasibility, and (3) complementary slackness property.

The simplex algorithm discussed in Chapter 2 obtains a sequence of solutions satisfying primal feasibility and complementary slackness properties. When dual feasibility is attained, it terminates.

The dual simplex algorithm (see Chapter 6) obtains a sequence of solutions satisfying complementary slackness and dual feasibility properties. Hence, that algorithm terminates when it attains primal feasibility. There are other algorithms for solving LPs, known as *out of kilter algorithm, complementary pivot algorithm,* etc. These algorithms generate a sequence of solutions of the primal-dual pair of LPs, satisfying primal and dual feasiblity properties. These algorithms terminate when the complementary slackness property is satisfied.

4.6 OTHER INTERPRETATIONS AND APPLICATIONS OF DUALITY

4.6.1 Dual Variables as Lagrange Multipliers

Problems in which an objective function has to be optimized subject to equality constraints can be handled by the Lagrange multiplier technique of classical calculus. However, when the constraints in the problem include some inequality constraints, the Lagrange multipliers associated with them have to satisfy sign restrictions and complementary slackness conditions. In this context, the Lagrange multipliers are known as the Kuhn-Tucker-Lagrange multipliers. For LPs, it turns out that the dual variables are the Kuhn-Tucker-Lagrange multipliers. Also the complementary slackness conditions in the *Kuhn-Tucker necessary conditions for optimality* (or the *Karush-Kuhn-Tucker necessary conditions for optimality*, see references [8, 9, 10] listed at the end of Chapter 16) in an LP are precisely the complementary slackness conditions for optimality in a primal-dual pair of LPs. Consider the LP (4.6). Associate the multiplier vector π to the constraints and the multiplier vector μ to the sign restrictions. The *Kuhn-Tucker-Lagrangian* corresponding to (4.6) is

$$L(x; \pi, \mu) = cx - \pi(Ax - b) - \mu x$$

The Kuhn-Tucker necessary conditions for optimality in (4.6) are

$$\begin{aligned} \frac{\partial L}{\partial x} = c - \pi A - \mu &= 0 \\ Ax &\geqq b \\ x &\geqq 0 \\ \pi(Ax - b) &= 0 \\ \mu x &= 0 \\ \pi \geqq 0, \qquad \mu &\geqq 0 \end{aligned}$$

Equivalently, these are

$$\begin{aligned} \pi A &\leqq c \\ \pi &\geqq 0 \\ Ax &\geqq b \\ x &\geqq 0 \\ \pi(Ax - b) = (c - \pi A)x &= 0 \end{aligned}$$

These are precisely the dual and primal feasibility and the complementary slackness conditions for optimality.

The nonlinear complementary slackness conditions $\pi(Ax - b) = (c - \pi A)x = 0$, and the inequality constraints in the Kuhn-Tucker necessary conditions make them hard to solve directly. The simplex method provides a systematic method for solving these Kuhn-Tucker necessary conditions.

4.6.2 The Saddle Point Property

Consider the primal LP (4.6) and its dual, (4.7). The Kuhn-Tucker-Lagrangian associated with (4.6) is $L(x; \pi, \mu) = cx - \pi(Ax - b) - \mu x$. The point $(\bar{x}, \bar{\pi}, \bar{\mu})$ is said to be a *Kuhn-Tucker saddle point* of $L(x; \pi, \mu)$ if $\bar{\pi} \geqq 0, \bar{\mu} \geqq 0$, and

$$L(x, \bar{\pi}, \bar{\mu}) \geqq L(\bar{x}; \bar{\pi}, \bar{\mu}) \geqq L(\bar{x}; \pi, \mu) \qquad \text{for all } \pi \geqq 0, \mu \geqq 0, \text{ and all } x$$

Problem

4.13 Prove that $(\bar{x}, \bar{\pi}, \bar{\mu})$ is a Kuhn-Tucker saddle point of $L(x; \pi, \mu)$ iff $\bar{x}$ is an optimum feasible solution of (4.6), $\bar{\pi}$ is an optimum feasible solution of (4.7), and $\bar{\mu} = c - \bar{\pi}A$.

4.6.3 Dual Variables as Partial Derivatives of the Optimal Objective Value

Consider the LP in standard form (4.8). Assume that A, c are fixed, and for a given b let $f(b)$ be the minimum value of $z(x)$ in the problem. The dual problem is (4.9). We wish to investigate how $f(b)$ behaves as b changes, but A, c remain fixed. Let $\bar{x}$, $\bar{\pi}$ be a pair of primal and dual optimal feasible solutions for fixed b. By the duality theorem $f(b) = \bar{\pi}b$. So in a crude sense, one could conclude that $\partial f/\partial b = \bar{\pi}$, that is, the optimal dual feasible solution is the partial derivative vector of the optimal objective value of the LP with respect to the right-hand side constants. When the dual optimum feasible solution $\bar{\pi}$ is unique, the $\bar{\pi}_i$ are the *marginal rates of change of* $f(b)$ for small perturbations in b_i.

Thus, the optimal dual feasible solution can be used in sensitivity analysis, when A, c remain fixed but when small perturbations occur in the right-hand side constants. $f(b)$ is not know explicitly as a function of b. However, for any given b vector, the value of $f(b)$ can be computed by solving (4.8), and the vector of partial derivatives of $f(b)$, that is, $\nabla f(b)$, is the optimum dual solution. If it is required to explore for an optimum b vector in some specified feasible region, this provides the necessary information for finding it by using some sub-gradient optimization method. Methods for solving parametric LPs discussed in chapter 7 provide very useful tools to explore for an optimum b vector.

Similarly, suppose A, b remain fixed, but the cost vector c, is likely to vary. Let $g(c)$ be the minimum objective value in the problem as a function of c. If $\bar{x}$ is the unique optimum feasible solution of this LP then, $g(c) = c\bar{x}$. Hence, in a crude sense, we could conclude that $\partial g/\partial c = \bar{x}$ and that $\bar{x}_j$ are the marginal rates of change in $g(c)$ when A, b remain fixed but when c_j is subject to small perturbations.

THE USEFULNESS OF THE OPTIMUM DUAL SOLUTION IN PRACTICAL APPLICATIONS

Consider the housewives problem (4.1), and its dual (4.2). From section 4.5.12, we know that the optimum dual solution is $\bar{\pi} = (3, 8)$, and the optimum objective value in the housewives problem is $9\bar{\pi} + 19\bar{\pi}_2 = 179$ cents. From the above discussion, we conclude that under the present conditions, the housewife can expect to spend 3 cents per each additional unit of vitamin A required, and 8 cents per each additional unit of vitamin C required. In the same way, she can expect to save 3 cents per each unit that the vitamin A requirement can be reduced, and 8 cents per each unit that the vitamin C requirement can be reduced, from their present levels. This kind of analysis is known as *marginal analysis*. One should of course remember that these rates 3 and 8 are only the rates of change for small changes in the requirements from their present levels and if large changes occur in the requirement levels, the values of these marginal rates might change.

Again consider the product mix problem of a company which can manufacture n different products using m different resources (raw materials, etc.). Let b_i units be the amount of resource i available, $i = 1$ to m. Let c_j \$/unit be the profit per unit of product j manufactured. Suppose the company wants to maximize its total profit. The product mix model for this company is a profit maximization LP, in n nonnegative variables, subject to m resource availability constraints. Let π_i be the dual variable associated with the constraint imposed by the ith resource availability. Let $\bar{\pi} = (\bar{\pi}_1, \ldots, \bar{\pi}_m)$ be the optimum dual solution. Then $\bar{\pi}_i$ \$ is the amount by which the company's profits go up, if the availability of resource i goes up by one unit from its present level of b_i units. If $\bar{\pi}_i > 0$, this implies that the company can increase its profits by increasing its supply of resource i. In most practical situations, the supplies of resources can be increased by spending some money. Suppose the company can acquire more units of resource i at the cost of α_i \$/unit. If $\bar{\pi}_i > \alpha_i$, the company can expect to increase its total net profit at the marginal rate of $(\bar{\pi}_i - \alpha_i)$\$ per each additional unit of resource i acquired at this cost. By comparing the marginal rates of change in the total profit corresponding to the various resources, the company can determine which resource is its most critical resource, etc.

Example

Consider the fertilizer problem (1.1), discussed in section 1.1.3. Let π_i be the dual variable associated with the material balance inequality of raw material i in this problem, $i = 1$ to 3. The resources of this company are raw materials 1, 2, 3. The optimum dual solution corresponding to this problem can be verified to be $\bar{\pi} = (5, 5, 0)$. This implies that the company's total sales revenue is not expected to change when small changes occur in the availability of raw material 3 (from its present availability level of 500 tons per month), because $\bar{\pi}_3 = 0$. Also, availability of one additional ton of either raw material 1 or 2 (from present levels), is expected to increase the company's total sales revenue by 5\$ (because $\bar{\pi}_1 = \bar{\pi}_2 = 5$). This implies that the company can improve its net gain by acquiring additional amounts of raw material 1 or 2, at a cost of at most 5\$ per unit (for either raw material).

In general the optimum dual solution corresponding to a linear programming model is the vector of the marginal rates of change in the objective value, per unit changes in the right hand side constants. In practical linear programming models, the right hand side constants may be either the limits on resource availability or targets for production levels of the various products. By comparing the marginal rates of change in the objective function corresponding to the various resource availabilities and production targets, the company can determine what is its most critical resource, or its most critical production target, etc. These marginal rates are the weights for assessing the relative importance (under present conditions) of the various resources in terms of their contribution to the objective function. They can be used to set up priorities for additional resource acquisition, and other such planning activities.

A linear programming model can be used to determine not only the present optimum solution, but, by utilizing the optimum dual solution, it will indicate changes that can be made in resource availabilities, etc. which will lead to further improvements in the objective function. When coupled with the techniques of sensitivity analysis (see Chapters 7, 8), this provides very useful quantitative information for planning to change the existing constraints to enable better performance.

4.6.4 Solving a Linear Program Is Equivalent to Solving a System of Equations in Nonnegative Variables

From the results in Chapters 2 and 3, every LP can be expressed in the form (4.6). Its dual is (4.7). From the sufficient optimality criterion of section 4.5.6, if x, π are respectively primal and dual feasible and satisfy

$$cx = \pi b$$

then x, π must be optimal to the respective problems. Rewriting all these constraints together, we have the system

$$\begin{aligned} Ax \qquad\quad &\geqq b \\ A^{\mathrm{T}}y &\leqq c^{\mathrm{T}} \\ cx - b^{\mathrm{T}}y &= 0 \\ x \geqq 0 \qquad y &\geqq 0 \end{aligned}$$

where $y = \pi^{\mathrm{T}}$. Introducing slack vectors u, v, this system is

$$\begin{aligned} Ax \qquad\qquad - v \qquad &= b \\ A^{\mathrm{T}}y \qquad + u &= c^{\mathrm{T}} \\ cx - b^{\mathrm{T}}y \qquad\qquad &= 0 \\ x \geqq 0, \quad y \geqq 0, \quad u \geqq 0, \quad v &\geqq 0 \end{aligned} \tag{4.36}$$

Thus, if $(\bar{x}, \bar{y}, \bar{u}, \bar{v})$ is a feasible solution of (4.36), then $\bar{x}$ is an optimum feasible solution of (4.6) and $\bar{\pi} = \bar{y}^{\mathrm{T}}$ is an optimum feasible solution of the dual problem.

Conversely, if $\bar{x}$ is any optimum feasible solution of (4.6), and $\bar{y} = \bar{\pi}^{\mathrm{T}}$ is any optimum feasible solution of the dual, and $\bar{v} = A\bar{x} - b$, $\bar{u} = c^{\mathrm{T}} - A^{\mathrm{T}}\bar{y}$, then $(\bar{x}, \bar{y}, \bar{u}, \bar{v})$ is feasible to (4.36).

Thus, solving the LP (4.6) is equivalent to solving the system of equations (4.36) in nonnegative variables. Hence any algorithm for solving systems of simultaneous linear equations in nonnegative variables can be used for solving LPs directly without a need for optimization. Conversely, any algorithm for solving LPs can be used to solve a system of simultaneous linear equations in nonnegative variables, by solving a Phase I type problem.

Numerous algorithms for solving systems of simultaneous linear equations in unrestricted variables (based on pivoting, elimination, reduction, triangularization, etc.) are discussed in classical linear algebra. However, the problem of solving a system of simultaneous linear equations in *nonnegative variables* is much harder, and this had to wait until the linear programming era.

4.6.5 Formulation of a Linear Program as a Linear Complementarity Problem

Consider the LP (4.6) and its dual (4.7). Introducing slack vectors, these systems can be written down together as

$$\begin{aligned} u &= \qquad -A^{\mathrm{T}}y + c^{\mathrm{T}} \\ v &= Ax \qquad\qquad - b \\ \begin{pmatrix} u \\ \cdots \\ v \end{pmatrix} &\geqq 0 \quad \begin{pmatrix} x \\ \cdots \\ y \end{pmatrix} \geqq 0 \end{aligned} \tag{4.37}$$

From the complementary slackness theorem, if the solution u, v, x, y of (4.37) satisfies

$$u^{\mathrm{T}}x + v^{\mathrm{T}}y = \begin{pmatrix} u \\ \cdots \\ v \end{pmatrix}^{\mathrm{T}} \begin{pmatrix} x \\ \cdots \\ y \end{pmatrix} = 0$$

then x is optimal to (4.6) and $\pi = y^{\mathrm{T}}$ is optimal to its dual (4.7). Let

$$\xi = \begin{pmatrix} u \\ \cdots \\ v \end{pmatrix}, \eta = \begin{pmatrix} x \\ \cdots \\ y \end{pmatrix}, M = \begin{pmatrix} 0 & \vdots & -A^{\mathrm{T}} \\ \cdots & \cdots & \cdots \\ A & \vdots & 0 \end{pmatrix}, v = \begin{pmatrix} c^{\mathrm{T}} \\ \cdots \\ -b \end{pmatrix}$$

Solving the LP (4.6) is equivalent to solving the system:

$$\begin{gathered} \xi - M\eta = v \\ \xi \geqq 0, \qquad \eta \geqq 0 \\ \xi^{\mathrm{T}}\eta = 0 \end{gathered} \tag{4.38}$$

The constraints $\xi \geqq 0, \eta \geqq 0, \xi^{\mathrm{T}}\eta = 0$ imply that $\xi_r \eta_r = 0$ for all r. Hence at most one variable in each pair (ξ_r, η_r) can take a positive value in any solution of (4.38). Hence each pair of variables (ξ_r, η_r) is known as a *complementary pair of variables* in (4.38). Problems similar to (4.38) are known as *linear complementarity problems.* Algorithms for solving them will be discussed in Chapter 16.

4.6.6 Applications in Two Person Zero-Sum Matrix Games

A *zero-sum two person game consists* of two players, each with a fixed finite set of choices in each play of the game. Suppose player I has m choices, namely, $1, 2, \ldots, m$. Suppose player II has n choices that are called $1, 2, \ldots, n$. If in a play of the game, player I chooses i and player II chooses j, player I gains a_{ij} dollars and player II gains $-a_{ij}$ dollars. In this case, a negative gain is considered as a loss and vice versa. Thus, the matrix $A = (a_{ij})$ is known as player I's payoff matrix and $-A$ is player II's payoff matrix.

In playing a game of this type, the players might pick their choices in a probabilistic manner just before each play. Let x_i be the probability with which player I picks his choice i, for $i = 1$ to m, and let y_j be the probability with which player II picks his choice j, for $j = 1$ to n.

The probability vector $x = (x_1, \ldots, x_m)^{\mathrm{T}}$ is known as player I's *mixed strategy*, and correspondingly the vector $y = (y_1, \ldots, y_n)^{\mathrm{T}}$ is player II's mixed strategy.

Suppose player I is interested in finding a mixed strategy x, which gives an expected payoff of at least α. The expected payoff of player I, if he uses mixed strategy x and player II chooses j, is: $x_1 a_{1j} + \cdots + x_m a_{mj}$. Since α is the guaranteed minimum expected payoff to player I under mixed strategy x, irrespective of what player II does, it must satisfy $x_1 a_{1j} + \cdots + x_m a_{mj} \geqq \alpha$ for all $j = 1$ to n. Hence, player I's problem is the LP

$$\begin{array}{ll} \text{Maximize} & 0x_1 + \cdots + 0x_m + \alpha \\ \text{Subject to} & a_{11}x_1 + \cdots + a_{m1}x_m - \alpha \geqq 0 \\ & \quad \vdots \\ & a_{1n}x_1 + \cdots + a_{mn}x_m - \alpha \geqq 0 \\ & x_1 + \cdots + x_m = 1 \\ & x_i \geqq 0 \quad \text{for all } i \\ & \alpha \text{ unrestricted in sign} \end{array} \tag{4.39}$$

Problem

4.14 Construct one feasible solution (x, α) to (4.39) in terms of (a_{ij}), thus proving that (4.39) is always feasible.

In a similar manner, let β denote the maximum expected payoff of player II, irrespective of what player I does. The problem of finding an optimal strategy y and payoff β of player II is the LP:

$$\begin{array}{lrcrcl} \text{Maximize} & 0y_1 + \cdots + & 0y_n & + \beta & & \\ & -a_{11}y_1 \quad \cdots - & a_{1n}y_n & - \beta & \geqq & 0 \\ & \vdots & \vdots & \vdots & & \vdots \\ & -a_{m1}y_1 - \cdots - & a_{mn}y_n & - \beta & \geqq & 0 \\ & y_1 + \cdots + & y_n & & = & 1 \\ & y_j \geqq 0 & \text{for all } j & & & \\ & \beta \text{ unrestricted in sign} & & & & \end{array} \tag{4.40}$$

Problem

4.15 As before, show that (4.40) is always feasible.

It is easily seen that problems (4.39) and (4.40) are a primal, dual pair of LPs. Since both problems are feasible, we conclude from the fundamental duality theorem that the maximum expected payoff of player I is equal to (−minimum expected loss of player II).

4.6.7 Applications in Proving Theorems of Alternatives for Linear Systems

An important theorem in the theory of convex sets is the *separating hyperplane theorem.* A consequence of this theorem is the following:

Let **K** be a closed convex subset of $\mathbf{R}^m$ and $b \in \mathbf{R}^m$. If $b \notin \mathbf{K}$, there exists a hyperplane in $\mathbf{R}^m$ that separates the point b from the subset **K**, that is, there exists $(\pi_1, \ldots, \pi_m) \neq 0$ and real number α, such that

$$\begin{aligned} \pi b &> \alpha \\ \pi y &\leqq \alpha \qquad \text{for all } y \in \mathbf{K} \end{aligned}$$

In this case, the hyperplane $\pi y = \alpha$ is known as a *separating hyperplane* separating the point b from the convex set **K** (Figure 4.2). Consider the special case of this theorem when **K** is a convex polyhedral cone. If **K** is a convex polyhedral cone, every point in it is a nonnegative linear combination of the extreme rays of the cone. Hence, there exists a finite set of points, say, $\{A_{.1}, \ldots, A_{.n}\}$ in $\mathbf{R}^m$ such that

$$\mathbf{K} = \{y\colon y = x_1 A_{.1} + \cdots + x_n A_{.n}, \qquad x_j \geqq 0 \text{ for all } j\} \tag{4.41}$$

Applying the separating hyperplane theorem to this case, we conclude that if b is not in the convex polyhedral cone **K**, there exists $\pi = (\pi_1, \ldots, \pi_m) \neq 0$, and a real number α such that

$$\begin{aligned} & \pi b > \alpha \\ \text{and} \qquad & \pi y \leqq \alpha \qquad \text{for all } y \in \mathbf{K} \end{aligned}$$

Since $0 \in \mathbf{K}$, we must have $\alpha \geqq 0$. Clearly $\pi y \leqq \alpha$ for all $y \in \mathbf{K}$ iff $\pi A_{.j} \leqq \alpha$ for all $j = 1$ to n, and $\alpha \geqq 0$. If there exists a $y \in \mathbf{K}$ such that $\pi y > 0$, since **K** is a cone, πy is unbounded above on **K**. These facts imply that α must be equal to zero.

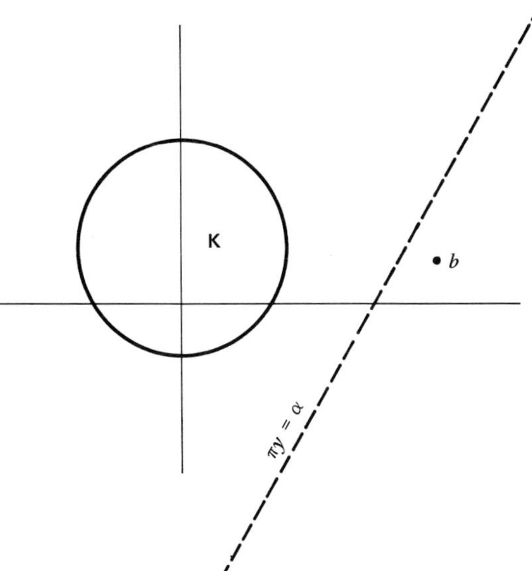

Figure 4.2

Combining these, we get the following result for this case: Either b is contained in the convex polyhedral cone **K** given by (4.41), or there exists a $\pi = (\pi_1, \ldots, \pi_m) \neq 0$ satisfying

$$\pi b > 0$$

$$\text{and} \qquad \pi A_{.j} \leqq 0 \qquad \text{for all } j = 1 \text{ to } n$$

Equivalently, *either the system*

$$\begin{aligned} Ax &= b \\ x &\geqq 0 \end{aligned} \tag{I}$$

has a solution x, or the system

$$\begin{aligned} \pi A &\leqq 0 \\ \pi b &> 0 \end{aligned} \tag{II}$$

has a solution π, *but not both.*

This result guarantees that exactly one of the two systems (I) or (II) has a feasible solution, and the other system is inconsistent. In mathematical programming literature results of this type are known as *theorems of the alternatives.* The theorem of the alternative that is deduced here from the separating hyperplane theorem of convex sets is known as *Farkas lemma.* It is a fundamental result that has found numerous applications in linear, nonlinear, and integer programming. We now show how Farkas lemma can be proved using the duality theorem.

Farkas Lemma Let A be a given $m \times n$ matrix and b a column vector of order $m \times 1$, then exactly one of the following systems has a feasible solution, and the other system is inconsistent.

$$\text{Either} \quad \begin{aligned} Ax &= b \\ x &\geqq 0 \\ x &\in \mathbf{R}^n \end{aligned} \tag{I}$$

$$\text{or} \quad \begin{aligned} \pi A &\leqq 0 \\ \pi b &> 0 \\ \pi &\in \mathbf{R}^m \end{aligned} \tag{II}$$

Proof Consider the primal LP

$$\begin{array}{ll} \text{Minimize} & z(x) = 0x \\ \text{Subject to} & Ax = b \\ & x \geqq 0 \end{array} \tag{4.42}$$

Its dual is

$$\begin{array}{ll} \text{Maximize} & v(\pi) = \pi b \\ \text{Subject to} & \pi A \leqq 0 \end{array}$$

If system (I) has a feasible solution, it is optimal to the primal problem, since $z(x) = 0$ for every x. So by the duality theorem the maximum value of $v(\pi)$ is zero in the dual problem. Hence, there cannot exist any dual feasible solution that makes $v(\pi) > 0$. Thus (II) is infeasible.

The dual problem here is always feasible, since $\pi = 0$ is a dual feasible solution. If system (I) is infeasible, the primal problem is infeasible, and by the fundamental duality theorem $v(\pi)$ is unbounded above on the dual feasible solution set. Thus, (II) must have a feasible solution.

Problem

4.16 Construct a proof of the fundamental duality theorem using Farkas lemma.

4.17 Using Farkas lemma, prove the following theorem of the alternative. If A is an $m \times n$ matrix and $b \in \mathbf{R}^m$, exactly one of the following two systems has a solution.

$$\text{Either} \quad Ax = b \tag{I}$$

$$\text{or} \quad \begin{array}{l} \pi A = 0 \\ \pi b = 1 \end{array} \tag{II}$$

Theorems of the alternatives have numerous applications in mathematical programming.

CONCLUDING REMARKS

The dual problem was formulated by J. von Neumann. Duality theorems were proved by D. Gale, H.W. Kuhn, and A.W. Tucker. The optimum strategies for the two-person zero sum matrix game discussed in Section 4.6.6 are known as *mini. max. strategies* (because they minimize the maximum loss that the player may incur). J. von Neumann first defined these strategies. The complementary slackness theorem is due to G. B. Dantzig and A. Orden.

PROBLEMS

4.18 Write the duals of the following problems. Verify that the dual of the dual is the original problem. Also, transform each problem into standard form and verify that the dual of the problem in standard form is equivalent to the dual of original problem.

(a)

$$\begin{array}{lrl} \text{Maximize} & z'(x) = & -17x_2 + 83x_4 - 8x_5 \\ \text{Subject to} & & -x_1 - 13x_2 + 45x_3 + 16x_5 - 7x_6 \geqq 107 \\ & & 3x_3 - 18x_4 + 30x_7 \leqq 81 \\ & & 4x_1 - 5x_3 + x_6 = -13 \\ & & -10 \leqq x_1 \leqq -2 \\ & & -3 \leqq x_2 \leqq 17 \\ & & 16 \leqq x_3 \\ & & x_4 \leqq 0 \\ & & x_5 \text{ unrestricted} \\ & & x_6, x_7 \geqq 0 \end{array}$$

(b)

$$\begin{array}{ll} \text{Minimize} & z(x) = 3x_1 - 7x_2 + 6x_4 + 5x_5 - x_6 \\ \text{Subject to} & 5x_1 + 8x_2 + 3x_3 + 3x_4 + 2x_5 + 11x_6 = 200 \\ & 5 \leqq x_j \leqq 20 \quad \text{for all } j \end{array}$$

(c)

$$\begin{array}{ll} \text{Maximize} & 4x_1 - 3x_2 + 8x_3 \\ \text{Subject to} & -2 \leqq x_1 \leqq 6 \\ & 4 \leqq x_2 \leqq 14 \\ & -12 \leqq x_3 \leqq -8 \end{array}$$

(d)

$$\begin{array}{ll} \text{Maximize} & z'(x) = \sum_{j=1}^{n} c'_j x_j \\ \text{Subject to} & \sum_{j=1}^{n} a_{1j} x_j = b_1 \\ & \sum_{j=1}^{n} a_{2j} x_j \geqq b_2 \\ & \sum_{j=1}^{n} a_{3j} x_j \leqq b_3 \\ & \ell_j \leqq x_j \leqq k_j \quad \text{for all } j \end{array}$$

where ℓ_j, k_j are specified lower and upper bounds for x_j.

(e)

$$\begin{array}{lll} \text{Minimize} & z = 3y_1 + 4y_2 & \\ \text{Subject to} & \sum_{j=1}^{n} a_{ij} x_j = b_{i1} y_1 + b_{i2} y_2 & \text{for } i = 1 \text{ to } r \\ & \sum_{j=1}^{n} a_{ij} x_j \geqq b_{i1} y_1 + b_{i2} y_2 & \text{for } i = r + 1 \text{ to } m \\ & \sum_{j=1}^{n} x_j = 4 & \\ & x_j \geqq 0 \quad \text{for all } j & \\ & y_1, y_2 \text{ unrestricted in sign} & \end{array}$$

(f) The network Problem 1.22.

(g) The curve fitting Problem 1.12. Also prove that the dual problem obtained is feasible, by actually constructing a dual feasible solution.

4.19 Write the complementary slackness conditions for optimality for each of the LPs in Problem 4.18.

4.20 Consider the LP

$$\begin{array}{ll} \text{Minimize} & z(x) = cx \\ \text{Subject to} & \ell_i \leqq A_{i.} x \leqq k_i \quad \text{for } i = 1 \text{ to } m \end{array}$$

where A is an $m \times n$ matrix and $\ell = (\ell_1, \ldots, \ell_m)^T$ and $k = (k_1, \ldots, k_m)^T$ are given vectors. Write the dual of this problem and the complementary slackness conditions for optimality for this primal, dual pair. What conditions on the A matrix would guarantee that the dual problem is feasible?

4.21 Consider the LP

$$\begin{array}{ll} \text{Minimize} & z(x) = cx \\ \text{Subject to} & Ax = b \end{array}$$

where A is an $m \times n$ matrix. Write the dual of this problem. Using duality theorem prove that if the problem is feasible, then the minimum value of $z(x)$ in this problem is finite iff c can be expressed as a linear combination of the row vectors of A. Using this prove that either $z(x)$ is a constant or unbounded below on the set of solutions of this problem.

4.22 Consider the LP in standard form

$$\begin{aligned} \text{Minimize} \quad & z(x) = cx \\ \text{Subject to} \quad & Ax = b \\ & x \geqq 0 \end{aligned}$$

Write the corresponding Phase I problem with a full artificial basis. Write the dual of this Phase I problem and the complementary slackness conditions for optimality in this primal, dual pair.

4.23 A, b, c, are given matrices of order $m \times n$, $m \times 1$, and $1 \times n$. Prove that the LP

$$\begin{aligned} \text{Minimize} \quad & cx - b^{\mathrm{T}}y \\ \text{Subject to} \quad & Ax \geqq b \\ & -A^{\mathrm{T}}y \geqq -c^{\mathrm{T}} \\ & x \geqq 0 \quad y \geqq 0 \end{aligned}$$

is either infeasible or has an optimum objective value of zero.

4.24 Consider the following LP

$$\begin{aligned} \text{Minimize} \quad & \sum_{j=m+1}^{n} c_j x_j \\ \text{Subject to} \quad & x_i + \sum_{j=m+1}^{n} a_{ij} x_j = b_i \qquad i = 1 \text{ to } m \\ & x_j \geqq 0 \qquad \text{for } j = 1 \text{ to } n \end{aligned}$$

where $b_i > 0$ for all i and $c_j \geqq 0$ for $j = m + 1$ to n. Write the dual of this problem and prove that the dual problem has a unique optimal solution. Find that dual optimal solution.

4.25 Consider the LP

$$\begin{aligned} \text{Minimize} \quad & z(x) = cx \\ \text{Subject to} \quad & Ax = b \\ & x \geqq 0 \end{aligned}$$

where A is of order $m \times n$ and rank m. This problem has an optimum nondegenerate feasible basis. Using the results in Problem 4.24 prove that the dual of this problem has a unique optimal solution.

4.26 For the LP

$$\begin{aligned} \text{Minimize} \quad & z(x) = -2x_1 + 13x_2 + 3x_3 - 2x_4 + 5x_5 + 5x_6 + 10x_7 \\ \text{Subject to} \quad & x_1 - x_2 + 4x_4 - x_5 + x_6 - 4x_7 = 5 \\ & x_1 + 7x_4 - 2x_5 + 3x_6 - 3x_7 \geqq -1 \\ & 5x_2 + x_3 - x_4 + 2x_5 - x_6 - 2x_7 \leqq 5 \\ & 3x_2 + x_3 + x_4 + x_5 + x_6 - x_7 = 2 \\ & x_j \geqq 0 \quad \text{for all } j \end{aligned}$$

prove that $x = (6, 0, 1, 0, 1, 0, 0)^{\mathrm{T}}$ is an optimum feasible solution by using the complementary slackness theorem.

4.27 For the LP

$$\text{Maximize} \quad z(x) = x_1 + 2x_2 + x_3 - 3x_4 + x_5 + x_6 - x_7$$

$$\begin{array}{llllllll}
\text{Subject to} & x_1 + x_2 & & - x_4 & & + 2x_6 - 2x_7 \leqq 6 \\
& x_2 & & - x_4 + x_5 & & - 2x_6 + 2x_7 \leqq 4 \\
& x_2 + x_3 & & & & + x_6 - x_7 \leqq 2 \\
& x_2 & & - x_4 & & - x_6 + x_7 \leqq 1 \\
& & x_j \geqq 0 & & \text{for all } j
\end{array}$$

Check whether $x = (7, 0, 5/2, 0, 3, 0, 1/2)^T$ is an optimum feasible solution using the complementary slackness conditions for optimality.

4.28 For the transportation problem

$$\text{Minimize} \quad z(x) = \sum_{i=1}^{3} \sum_{j=1}^{4} c_{ij}x_{ij}$$

$$\sum_{j=1}^{4} x_{ij} = a_i \qquad i = 1 \text{ to } 3$$

$$\sum_{i=1}^{3} x_{ij} = b_j \qquad j = 1 \text{ to } 4$$

$$x_{ij} \geqq 0 \qquad \text{for all } i, j$$

$$\text{where} \quad c = (c_{ij}) = \begin{pmatrix} 7 & 2 & -2 & 8 \\ 19 & 5 & -2 & 12 \\ 5 & 7.2 & -9 & 3 \end{pmatrix}$$

$$\text{and} \quad a = (a_i) = (3, 3, 4), b = (b_j) = (2, 3, 2, 3)$$

find an optimum feasible solution using the complementary slackness conditions, given that $u = (u_i) = (0, 3, -4)$, $v = (v_j) = (7, 2, -5, 7)$ is an optimum dual solution.

4.29 For the LP

$$\begin{array}{llllllllll}
\text{Minimize} & z(x) = -x_1 - 2x_2 + 4x_3 & & + 5x_5 \\
\text{Subject to} & x_1 & + x_3 - 2x_4 & - x_5 + 2x_6 & & = 3 \\
& x_1 + x_2 & - x_4 & - 3x_5 + 3x_6 + x_7 & = & 7 \\
& -x_1 - 2x_2 & - x_3 - x_4 & + 5x_5 - 2x_6 & & \leqq -4 \\
& & x_j \geqq 0 & \text{for all } j
\end{array}$$

Check whether $x = (3, 4, 0, 0, 0, 0, 0)^T$ is an optimum feasible solution by using the complementary slackness conditions. Using these conditions characterize the set of optimum solutions. What does the fact that

$$\text{Column vector of } x_5 = -(\text{column vector of } x_1) - 2(\text{column vector of } x_2)$$

imply about the shape of the set of optimum solutions of this problem?

4.30 Consider the LP

$$\begin{array}{lll}
\text{Minimize} & z(x) = \sum c_j x_j \\
\text{Subject to} & \sum a_j x_j = b \\
& \ell_j \leqq x_j \leqq k_j & \text{for all } j
\end{array}$$

Write the dual problem. If $\bar{x}$ is an optimum feasible solution of this problem, prove that there exists a real number $\bar{\pi}$ such that for all j

$$\bar{\pi}a_j < c_j \quad \text{implies} \quad \bar{x}_j = \ell_j$$

$$\bar{\pi}a_j > c_j \quad \text{implies} \quad \bar{x}_j = k_j$$

and

$$\ell_j < \bar{x}_j < k_j \quad \text{implies} \quad \bar{\pi}a_j = c_j$$

4.31 If the LP

$$\begin{aligned} \text{Minimize} \quad & z(x) = cx \\ \text{Subject to} \quad & Ax \geqq b \end{aligned}$$

has a feasible solution, prove that it has an optimum feasible solution iff c lies in the pos cone generated by the row vectors of A.

4.32 For the LP

x_1	x_2	x_3	x_4	x_5	x_6	x_7	$-z$	b
1	4	1	−2	2	5	1	0	11
0	−2	0	2	−1	−1	−1	0	−4
1	1	0	−1	1	3	2	0	4
2	7	1	−1	2	13	2	1	0

$x_j \geqq 0$ for all j. Minimize z.

Compute the primal and dual solutions associated with the basis corresponding to the basic vector (x_3, x_5, x_1). Is it an optimum basis? Characterize the set of optimum solutions of this problem using the complementary slackness theorem.
Do the same for the basic vector (x_5, x_2, x_3, x_8) for the LP:

$$\begin{aligned} \text{Minimize} \quad & 3x_1 - 2x_2 + x_3 - x_4 - 5x_5 + 4x_6 - 2x_7 - 3x_8 \\ \text{Subject to} \quad & x_1 + x_2 + x_3 + x_4 + x_5 + x_6 + x_7 + x_8 = 14 \\ & x_1 + x_2 + x_3 + x_5 + x_6 + x_7 = 9 \\ & x_1 + x_2 + x_5 + x_6 = 5 \\ & x_1 + x_5 = 2 \\ & x_j \geqq 0 \quad \text{for all } j \end{aligned}$$

4.33 Consider a primal, dual pair of LPs both of which are feasible. Prove that there exists a pair of primal and dual feasible optimum solutions satisfying the properties:

(a) Whenever a constraint in one problem is tight, the associated variable in the other problem is nonzero.

(b) Whenever a nonnegative variable in one problem is zero, the associated constraint in the other problem is slack.

4.34 Consider the LP

$$\begin{aligned} \text{Minimize} \quad & x_1 + x_2 + x_3 \\ \text{Subject to} \quad & -x_2 + x_3 \geqq -1 \\ & x_1 - x_3 \geqq -1 \\ & -x_1 + x_2 \geqq -1 \\ & x_j \geqq 0 \quad \text{for all } j \end{aligned}$$

Prove that this problem is equivalent to its dual. Such LPs are known as *self-dual linear programs.* Assuming that A is a square matrix, obtain sufficient conditions on c, A, b, under which the LP

$$\begin{aligned} \text{Minimize} \quad & cx \\ \text{Subject to} \quad & Ax \geqq b \\ & x \geqq 0 \end{aligned}$$

is self-dual.

4.35 Read each of the following statements carefully and check whether it is *true* or *false*. Justify your answer by constructing a simple illustrative example, if possible, or by a simple proof. In all these statements, problem (1) and the rest of the notation, is the same as that in Problem 3.53.

1. In the statement of an LP, if there is a constraint that a variable has to be greater than or equal to a specified lower bound, then in writing down the dual problem, that variable is treated as an unrestricted variable.
2. When a problem is formulated as an LP, the dual variables in the associated dual can always be interpreted as the equilibrium prices associated with the *items*, each of which corresponds to a constraint in the original problem.
3. The dual variables are the Lagrange multipliers associated with primal constraints.
4. The dual problem in a primal, dual pair of LPs is always the maximization problem in the pair.
5. Since the dual is a maximization problem and the primal is a minimization problem, the dual objective value is always greater than or equal to the primal objective value.
6. A basis for (1) is primal feasible if all the basic variables are nonnegative in the associated primal basic solution. Likewise, the basis is dual feasible, if all the dual variables are nonnegative in the associated dual basic solution.
7. If B is a dual feasible basis for (1), and b is a nonnegative linear combination of the columns in B, B must be an optimum basis for (1).
8. Let $\bar{x}$ be a BFS for (1). If all the bases for which $\bar{x}$ is the associated primal basic solution are dual infeasible, the optimum objective value in (1) must be less than $c\bar{x}$.
9. Let $\bar{\pi}$ be an optimum dual solution corresponding to (1). If $\bar{\pi}_1$ is positive, we can expect that $f(b)$ will increase if b_1 is increased slightly from its present value, while all the other data in the problem remains unchanged.
10. If (1) has a nondegenerate optimum BFS, the dual problem has a unique optimum solution.
11. In solving (1) by the simplex algorithm starting with a feasible basis, a terminal basis is either dual feasible or shows conclusively that the dual constraints are inconsistent.
12. In every primal, dual pair of LPs the primal objective value is always greater than or equal to the dual objective value.
13. If the dual problem in a primal, dual pair is infeasible, the primal objective function must be unbounded in the primal problem.
14. In a primal dual pair, if the dual objective function is unbounded in the dual problem, the primal problem must be infeasible.
15. If $f(b)$ is $-\infty$, it is possible to change the b vector to a b' such that $f(b')$ is finite.
16. Complementary slackness conditions for optimality exist only for inequality constraints or sign restrictions.
17. Since all the constraints in (1) are equations, there are no complementary slackness conditions for optimality for problem (1).
18. By the complementary slackness theorem, if x_j is positive in an optimum solution of (1), $c_j - \pi A_{.j}$ must be zero in all dual optimum solutions. Conversely, if $c_j - \pi A_{.j}$ is zero in every dual optimum solution, x_j must be positive in every primal optimum solution of (1).
19. If a method for solving systems of linear inequalities can be developed, that method can be used to solve any linear programming problem right away, without doing any optimization.
20. If $\mathbf{S}$ and $\mathbf{T}$ are two distinct subsets of $\mathbf{R}^n$, there exists a hyperplane separating them.
21. If a system of linear inequalities is inconsistent, there exists a nonnegative linear combination of the inequalities in the system, which is inconsistent by itself.

5 revised simplex method

5.1 INTRODUCTION

In the simplex algorithm using canonical tableaus discussed in Chapter 2, a lot of computations are performed at every pivot step. Every time a pivot is performed, it is carried out on every column of the tableau. This can be very time-consuming. In the revised simplex method, all the necessary computations are carried out by using the formulas discussed in sections 3.11 and 3.12 and restricting the pivot operations to the inverse matrix. The revised simplex method is purely an efficient computational scheme for applying the main ideas of the simplex method. To solve an LP by the revised simplex method, first transform it into the *standard form.* Suppose it is

$$\begin{aligned} \text{Minimize} \quad & z(x) = cx \\ \text{Subject to} \quad & Ax = b \\ & x \geqq 0 \end{aligned} \tag{5.1}$$

where $b \geqq 0$ and A is of order $m \times n$. There are two forms of the revised simplex method depending on how the inverse matrix is stored; one uses the explicit inverse and the other uses the product form of the inverse.

5.2 REVISED SIMPLEX ALGORITHM WITH THE EXPLICIT FORM OF INVERSE WHEN PHASE II CAN BEGIN DIRECTLY

See whether there exists a unit matrix of order m as a submatrix of the matrix A. If it exists, Phase II can begin directly. Choose an initial basic vector associated with a basis that is the unit matrix of order m. Introduce a unit matrix of order $m + 1$ by the side of the tableau. The *inverse tableau* will be obtained from these column vectors.

Original Tableau (for Starting Phase II Directly)

$x_1 \ldots x_n$	$-z$	Columns for the inverse tableau	b
$a_{11} \ldots a_{1n}$	0	$1 \ldots 0$	b_1
$\vdots \quad \vdots$	$\vdots$	$\vdots \quad \vdots$	$\vdots$
$a_{m1} \ldots a_{mn}$	0		b_m
$c_1 \ldots c_n$	1	$0 \ldots 1$	0
These column vectors will remain unchanged.		Elementary row operations will be performed only on these columns while pivoting.	

To obtain the initial inverse tableau perform the row operations required for pricing out the basic column vectors in the cost row. These row operations are performed only on the $m + 1$ column vectors introduced for obtaining the inverse tableau and on the column vector of the right-hand side constants. The inverse tableau will be of this form:

First Inverse Tableau

Basic variables	Inverse tableau	Basic values
	$1 \ldots 0 \quad 0$	b_1
	$\vdots \quad \vdots \quad \vdots$	$\vdots$
	$0 \ldots 1 \quad 0$	b_m
$-z$	$-\pi_1^0 \ldots -\pi_m^0 \quad 1$	$-z_0$

If B is the current basis, in the notation of section 3.10 the inverse tableau will be

$$T = \begin{pmatrix} B & \vdots & 0 \\ \cdots & \cdots & \cdots \\ c_B & \vdots & 1 \end{pmatrix}^{-1} = \begin{pmatrix} B^{-1} & \vdots & 0 \\ \cdots & \cdots & \cdots \\ -c_B B^{-1} & \vdots & 1 \end{pmatrix}$$

In a general stage of the simplex algorithm, say, at the end of stage k, let B be the basis under consideration. Let T, the inverse tableau at this stage, be

Inverse Tableau at the End of Stage k

Basic variables	Inverse tableau $= T$	Basic values
	$\beta_{11} \ldots \beta_{1m} \quad 0$	$\bar{b}_1$
	$\vdots \quad \vdots \quad \vdots$	$\vdots$
	$\beta_{m1} \ldots \beta_{mm} \quad 0$	$\bar{b}_m$
$-z$	$-\pi_1 \ldots -\pi_m \quad 1$	$-\bar{z}$

The matrix (β_{ij}) in the inverse tableau is B^{-1}. The $(-\pi)$ in the last row of the inverse tableau is $(-c_B B^{-1})$, and π is the dual solution of (5.1) corresponding to the present basis B. The vector of basic values at this stage is

$$\begin{pmatrix} x_B \\ \cdots \\ -z \end{pmatrix} = \begin{pmatrix} \bar{b} \\ \cdots \\ -\bar{z} \end{pmatrix} = T\begin{pmatrix} b \\ \cdots \\ 0 \end{pmatrix}$$

From the results in Chapters 3 and 4 the relative cost coefficients with respect to the current basis are

$$\begin{aligned} \bar{c}_j = c_j - \pi A_{.j} &= c_j + \sum_{i=1}^{m} (-\pi_i)a_{ij} \\ &= (-\pi \;\vdots\; 1)\begin{pmatrix} A_{.j} \\ \cdots \\ c_j \end{pmatrix} \end{aligned}$$

for $j = 1$ to n. The a_{ij}, c_j, b_i, etc., here are the entries in the original tableau. While the pivot operations are confined to the inverse matrix, the $\bar{c}_j$ have to be computed for all j using this formula in each step.

THE OPTIMALITY CRITERION

The present BFS is an optimum feasible solution if the relative cost coefficients are all nonnegative, that is, $\bar{c}_j \geqq 0$ for each $j = 1$ to n.

CHOOSING THE PIVOT COLUMN

If the optimality criterion is not satisfied at this stage, pick a nonbasic variable x_s associated with a negative relative cost coefficient as the entering variable. A convenient pivot column choice rule is to pick x_s associated with $\bar{c}_s$ = minimum $\{\bar{c}_j : j = 1 \text{ to } n\}$, as the entering variable, as in section 2.5.4. The pivot column is the updated column vector of x_s with respect to the current basis. From the results in Chapters 3 and 4, it is

$$\begin{pmatrix} \bar{A}_{.s} \\ \cdots \\ \bar{c}_s \end{pmatrix} = T\begin{pmatrix} A_{.s} \\ \cdots \\ c_s \end{pmatrix}$$

UNBOUNDEDNESS CRITERION

If the pivot column is such that

$$\bar{c}_s < 0, \qquad \bar{A}_{.s} \leqq 0$$

the objective function is unbounded below on the set of feasible solutions of the problem and we terminate the algorithm. The feasible solution $x(\theta)$ given by

$$\begin{aligned} i\text{th basic variable} &= \bar{b}_i - \theta\bar{a}_{is} \qquad i = 1 \text{ to } m \\ x_s &= \theta \\ \text{all other nonbasic variables} &= 0 \\ z &= \bar{z} + \theta\bar{c}_s \end{aligned}$$

is feasible for all $\theta \geqq 0$ and as θ tends to $+\infty$, $z(x(\theta))$ tends to $-\infty$.

MINIMUM RATIO TEST

If the unboundedness criterion is not satisfied, x_s is brought into the basic vector. To determine the pivot row compute $\bar{b}_r/\bar{a}_{rs}$ = minimum $\{\bar{b}_i/\bar{a}_{is} : i \text{ such that } \bar{a}_{is} > 0\}$, and the pivot row is the rth row. Ties here are broken arbitrarily. The present rth basic variable is the leaving variable.

OBTAINING THE NEW INVERSE TABLEAU BY PIVOTING

The pivot column is included at the end of the inverse tableau. Performing the pivot transforms the inverse tableau into the inverse tableau with respect to the new basis. Here again the row operations required for the pivot operation are performed only on the column vectors of the inverse tableau and on the updated right-hand side constant vector; they are not performed on the column vectors of the original tableau.

Old Inverse Tableau (at the End of Stage k)

Basic variables	Inverse tableau		Basic values	Pivot column (updated column of x_s)	
	$\beta_{11} \cdots \beta_{1m}$	0	$\bar{b}_1$	$\bar{a}_{1s}$	
	$\vdots \quad \vdots$		$\vdots$	$\vdots$	
x_r	$\beta_{r1} \cdots \beta_{rm}$	0	$\bar{b}_r$	$\bar{a}_{rs}$ (circled)	Pivot row
	$\vdots \quad \vdots$		$\vdots$	$\vdots$	
	$\beta_{m1} \cdots \beta_{mm}$	0	$\bar{b}_m$	$\bar{a}_{ms}$	
$-z$	$-\pi_1 \cdots -\pi_m$	1	$-\bar{z}$	$\bar{c}_s$	

After the pivot, the pivot column is transformed into $I_{.r}$ (where I is the unit matrix of order $m + 1$) and is dropped. The pivot updates the inverse tableau and the column vector of basic values. This pivot operation is also called the operation of *updating the inverse tableau.*

OBTAINING THE NEW INVERSE TABLEAU BY USING PIVOT MATRICES

As discussed in Chapter 3, the new inverse tableau can also be obtained by multiplying the present inverse tableau on the left by the *pivot matrix* corresponding to this pivot step. The pivot matrix for this operation is the matrix P^k in equation (3.30), and it differs from the unit matrix of order $(m + 1)$ in just the rth column. Indicating the entries in the new inverse tableau with the superscript*, we get

$$T^* = P^k T \qquad \text{and} \qquad \begin{pmatrix} b^* \\ \cdots \\ -z^* \end{pmatrix} = P^k \begin{pmatrix} \bar{b} \\ \cdots \\ -\bar{z} \end{pmatrix}$$

Having obtained the new inverse tableau, check whether the new basis is terminal. If not, repeat the iterations until termination occurs.

THE DUAL SOLUTION

The last row of the inverse tableau provides the negative of the dual solution corresponding to the present basis directly. If an optimum basis for the problem is obtained at termination, the dual solution obtained from the last row of the terminal inverse tableau is an optimum dual feasible solution corresponding to (5.1).

STARTING PHASE II FROM AN ARBITRARY PRIMAL FEASIBLE BASIS

Even if the original input-output coefficient matrix, A, does not contain a unit matrix of order m as a submatrix, Phase II can begin directly, if some primal feasible basis is known. The inverse tableau corresponding to it can be computed using the formulas discussed above. Continue the iterations from there until either the optimality or unboundedness criteria are satisfied.

5.3 REVISED SIMPLEX METHOD USING PHASE I AND PHASE II

If the original input-output coefficient matrix in (5.1) does not contain a unit matrix of order m as a submatrix, and if a primal feasible basis is not known, begin Phase I with a full artificial basis. Suppose the artificial variables are $x_{n+1}, \ldots, x_{n+m}$.

Original Tableau for Phase I Problem

$x_1 \quad x_n$	$x_{n+1} \quad x_{n+m}$	$-z$	$-w$	Columns for the inverse tableau	
$a_{11} \ldots a_{1n}$	$1 \ldots 0$	0	0	$1 \ldots 0$	b_1
$\vdots \quad \vdots$	$\vdots \quad \vdots$	$\vdots$	$\vdots$	$\vdots \quad \vdots$	
$a_{m1} \ldots a_{mn}$	$0 \ldots 1$	0	0	$\ldots$	b_m
$c_1 \ldots c_n$	$0 \ldots 0$	1	0	$0 \ldots 0$	0
$0 \ldots 0$	$1 \ldots 1$	0	1	$0 \ldots 1$	0
These column vectors will remain unchanged.				This is a unit matrix of order $(m + 2)$. Row operations will be performed only on these column vectors.	

The first inverse tableau during Phase I is obtained by carrying out the row operations required for pricing out of the initial basic columns in the Phases I and II cost rows. These row operations are carried out only on the column vectors introduced for getting the inverse tableau and on the right-hand side constant vector. During Phase I, inverse tableaus are of order $m + 2$. Let B be a basis and let c_B and d_B, be the Phases II and I basic cost vectors, respectively, corresponding to it. The inverse tableau corresponding to the basis B is

$$T = \begin{pmatrix} B & \vdots & 0 \\ \cdots & \cdots & \cdots \\ -c_B & \vdots & \begin{matrix} 1 & 0 \end{matrix} \\ -d_B & \vdots & \begin{matrix} 0 & 1 \end{matrix} \end{pmatrix}^{-1} = \begin{pmatrix} B^{-1} & \vdots & 0 \\ \cdots & \cdots & \cdots \\ -c_B B^{-1} & \vdots & \begin{matrix} 1 & 0 \end{matrix} \\ -d_B B^{-1} & \vdots & \begin{matrix} 0 & 1 \end{matrix} \end{pmatrix}$$

In a general stage of the algorithm during Phase I, say, at the end of stage k, let B be the basis under consideration. Let the present inverse tableau be

Inverse Tableau during Phase I at the End of Stage k

Current basic variables	Inverse tableau $= T$			Basic values
	$\beta_{11} \ldots \beta_{1m}$	0	0	$\bar{b}_1$
	$\vdots \quad \vdots$	$\vdots$	$\vdots$	$\vdots$
	$\beta_{m1} \ldots \beta_{mm}$	0	0	$\bar{b}_m$
$-z$	$-\pi_1 \ldots -\pi_m$	1	0	$-\bar{z}$
$-w$	$-\sigma_1 \ldots -\sigma_m$	0	1	$-\bar{w}$

The present Phase I objective value is $\overline{w} = \sigma b$. The Phase I relative cost coefficients with respect to the present basis are

$$\overline{d}_j = d_j - \sigma A_{.j} = (-\sigma, 0, 1)\begin{pmatrix} A_{.j} \\ \cdots \\ c_j \\ d_j \end{pmatrix}$$

for $j = 1$ to n. Hence $\overline{d}_j$ is obtained by multiplying the $m + 2$th row vector in the inverse tableau on the right by the column vector of x_j in the original Phase I tableau.

PHASE I TERMINATION CRITERION

As in Chapter 2, Phase I of the simplex method terminates when $\overline{d}_j \geqq 0$ for all $j = 1$ to n.

INFEASIBILITY CRITERION

When the Phase I termination criterion is satisfied, if the Phase I objective value, $\overline{w}$, is positive, the original problem has no feasible solution and we terminate. The σ_i's from the terminal Phase I inverse tableau provide the coefficients of a linear combination of the original system of constraints that leads to an impossible constraint. Multiply the ith equation in the original system (5.1) by σ_i and add over $i = 1$ to m. This leads to the constraint

$$\sum_{j=1}^{n} (-\overline{d}_j)x_j = \overline{w}$$

Since $\overline{d}_j \geqq 0$ for all $j = 1$ to n and $\overline{w} > 0$, this constraint can never be satisfied by any $x \geqq 0$.

SWITCHING OVER TO PHASE II

When Phase I termination criterion is satisfied, if $\overline{w} = 0$, the present BFS of the Phase I problem yields a BFS of the original problem when the zero values of the artificial variables are suppressed.

In all subsequent iterations, all the original problem variables x_j where j is such that $\overline{d}_j > 0$ are set equal to zero and such variables are never considered eligible to enter the basic vector. The $m + 2$th row in the tableau is deleted, as also the $m + 2$th column in the inverse tableau. The inverse tableaus during Phase I are of order $m + 2$, but they will be of order $m + 1$ during Phase II. The computations during Phase II are carried out as in section 5.2.

Any artificial variables in the basic vector at the end of Phase I will be equal to zero in the corresponding basic solution, and they are retained in the basic vector, unless they are replaced from the basic vector during the iterations in Phase II. In all the iterations during Phase II, all the artificial variables in the basic vector will be equal to zero.

PIVOT COLUMN AND PIVOT ROW CHOICE IN PHASE I

If the Phase I termination criterion is not satisfied, an original problem variable x_s where s is such that $\overline{d}_s < 0$ is selected as the entering variable. A convenient rule is to pick x_s associated with $\overline{d}_s =$ minimum $\{\overline{d}_j : j = 1$ to $n\}$, as the entering variable, as discussed in Chapter 2. The pivot column is the updated column vector of x_s and it is

$$\begin{pmatrix} \overline{A}_{.s} \\ \cdots \\ \overline{c}_s \\ \overline{d}_s \end{pmatrix} = T\begin{pmatrix} A_{.s} \\ \cdots \\ c_s \\ d_s \end{pmatrix}$$

The pivot row is determined by performing the minimum ratio test as in section 5.2.

UPDATING THE INVERSE TABLEAU DURING PHASE I

The pivot column is entered at the end of the present inverse tableau and the pivot performed. This transforms the present inverse tableau into the inverse tableau with respect to the new basis. The new inverse tableau can also be obtained by multiplying the present inverse tableau on the left by the pivot matrix.

$$P^k = \begin{bmatrix} 1\cdots 0 & -\bar{a}_{1s}/\bar{a}_{rs} & 0\cdots 0 \\ \vdots \quad \vdots & \vdots & \vdots \quad \vdots \\ 0\cdots 1 & -\bar{a}_{r-1,s}/\bar{a}_{rs} & 0\cdots 0 \\ 0\cdots 0 & 1/\bar{a}_{rs} & 0\cdots 0 \\ 0\cdots 0 & -\bar{a}_{r+1,s}/\bar{a}_{rs} & 1\cdots 0 \\ \vdots \quad \vdots & \vdots & \vdots \quad \vdots \\ 0\cdots 0 & -\bar{a}_{ms}/\bar{a}_{rs} & 0\cdots 0 \\ 0\cdots 0 & -\bar{c}_s/\bar{a}_{rs} & 0\cdots 0 \\ 0\cdots 0 & -\bar{d}_s/\bar{a}_{rs} & 0\cdots 1 \end{bmatrix}$$

(the middle column is the rth column)

Since the inverse matrix is of order $m + 2$, the pivot matrices during Phase I are of order $m + 2$.

Check whether the new basis satisfies the Phase I termination criterion. Otherwise continue the iterations.

INITIAL BASIC VECTOR IN PHASE I

If the original input-output coefficient matrix contains some column vectors of the unit matrix of order m, then it is only necessary to introduce as many artificial variables as necessary to complete a unit submatrix of order m in the augmented tableau. The initial basic vector for the Phase I problem could consist of some original problem variables and some artificial variables. The Phase I objective function is always the sum of the artificial variables that are introduced.

Example

Consider the LP

$$\begin{aligned} \text{Minimize} \quad & z(x) = 3x_1 - x_2 - 7x_3 + 3x_4 + x_5 \\ \text{Subject to} \quad & 5x_1 - 4x_2 + 13x_3 - 2x_4 + x_5 = 20 \\ & x_1 - x_2 + 5x_3 - x_4 + x_5 = 8 \\ & x_j \geqq 0 \quad \text{for } j = 1, \dots, 5 \end{aligned}$$

The artificial variables x_6, x_7 are introduced; here is the Phase I original tableau.

Original Phase I Tableau

x_1	x_2	x_3	x_4	x_5	x_6	x_7	$-z$	$-w$	Columns for inverse tableau				b
5	−4	13	−2	1	1	0	0	0	1	0	0	0	20
1	−1	5	−1	1	0	1	0	0	0	1	0	0	8
3	−1	−7	3	1	0	0	1	0	0	0	1	0	0
0	0	0	0	0	1	1	0	1	0	0	0	1	0

First Inverse Tableau (Phase I)

Basic variables	Inverse tableau				Basic values	$\bar{A}_{.3}$	Ratios
x_6	1	0	0	0	20	(13)	$\frac{20}{13}$ minimum ratio
x_7	0	1	0	0	8	5	$\frac{8}{5}$
$-z$	0	0	1	0	0	-7	
$-w$	-1	-1	0	1	-28	-18	

The Phase I relative cost coefficient vector is $(\bar{d}_1, \bar{d}_2, \bar{d}_3, \bar{d}_4, \bar{d}_5) = (-6, 5, -18, 3, -2)$. Hence, x_3 is brought into the basic vector. The pivot column is recorded on the tableau. The new inverse tableau is obtained either by pivoting, or by multiplying the present inverse tableau on the left by the pivot matrix:

$$P^1 = \begin{bmatrix} \frac{1}{13} & 0 & 0 & 0 \\ -\frac{5}{13} & 1 & 0 & 0 \\ \frac{7}{13} & 0 & 1 & 0 \\ \frac{18}{13} & 0 & 0 & 1 \end{bmatrix}$$

Note In this numerical example, the pivot matrices corresponding to all the pivot steps are given for illustration. The new inverse tableau is obtained *either* by performing the pivot operations on the rows of the present inverse tableau *or* by multiplying the present inverse tableau on the left by the pivot matrix.

Second Inverse Tableau (Phase I)

Basic variables	Inverse tableau				Basic values	$\bar{A}_{.5}$	Ratios
x_3	$\frac{1}{13}$	0	0	0	$\frac{20}{13}$	$\frac{1}{13}$	$\frac{20/13}{1/13}$
x_7	$-\frac{5}{13}$	1	0	0	$\frac{4}{13}$	$\left(\frac{8}{13}\right)$	$\frac{4/13}{8/13}$
$-z$	$\frac{7}{13}$	0	1	0	$\frac{140}{13}$	$\frac{20}{13}$	
$-w$	$\frac{5}{13}$	-1	0	1	$-\frac{4}{13}$	$-\frac{8}{13}$	

$\bar{d} = (\bar{d}_1, \bar{d}_2, \bar{d}_3, \bar{d}_4, \bar{d}_5) = (12/13, -7/13, 0, 3/13, -8/13)$. Therefore, bring x_5 into the basic vector. The second pivot matrix is

$$P^2 = \begin{bmatrix} 1 & -\frac{1}{8} & 0 & 0 \\ 0 & \frac{13}{8} & 0 & 0 \\ 0 & -\frac{20}{8} & 1 & 0 \\ 0 & 1 & 0 & 1 \end{bmatrix}$$

Third Inverse Tableau (Phase I)

Basic variables	Inverse tableau				Basic values
x_3	$\frac{1}{8}$	$-\frac{1}{8}$	0	0	$\frac{3}{2}$
x_5	$-\frac{5}{8}$	$\frac{13}{8}$	0	0	$\frac{1}{2}$
$-z$	$\frac{3}{2}$	$-\frac{5}{2}$	1	0	10
$-w$	0	0	0	1	0

The artificials have all left the basic vector. Also $\bar{d} = 0$ and $\bar{w} = 0$ in this stage. Strike off the last row and the fourth column from the inverse tableau and start Phase II.

Third Inverse Tableau (Beginning Phase II)

Basic variables	Inverse tableau			Basic values	$\bar{A}_{.2}$	Ratios
x_3	$\frac{1}{8}$	$-\frac{1}{8}$	0	$\frac{3}{2}$	$-\frac{3}{8}$	
x_5	$-\frac{5}{8}$	$\frac{13}{8}$	0	$\frac{1}{2}$	$\boxed{\frac{7}{8}}$	$\frac{1/2}{7/8}$
$-z$	$\frac{3}{2}$	$-\frac{5}{2}$	1	10	$-\frac{9}{2}$	

$\bar{c} = (8, -9/2, 0, 5/2, 0)$ and, hence, x_2 enters the basic vector. The pivot matrix is

$$P^3 = \begin{bmatrix} 1 & \frac{3}{7} & 0 \\ 0 & \frac{8}{7} & 0 \\ 0 & \frac{36}{7} & 1 \end{bmatrix}$$

Basic variables	Inverse tableau			Basic values
x_3	$-\frac{1}{7}$	$\frac{4}{7}$	0	$\frac{12}{7}$
x_2	$-\frac{5}{7}$	$\frac{13}{7}$	0	$\frac{4}{7}$
$-z$	$-\frac{12}{7}$	$\frac{41}{7}$	1	$\frac{88}{7}$

Now $\bar{c} = (2/7, 0, 0, 4/7, 36/7)$ and, hence, this is an optimum basis. The optimum solution is $(x_1, x_2, x_3, x_4, x_5) = (0, 4/7, 12/7, 0, 0)$, with a minimum optimal objective value of $-88/7$.

5.4 REVISED SIMPLEX METHOD USING THE PRODUCT FORM OF THE INVERSE

In this method, the inverse tableau is not stored, but the pivot matrices generated at each stage are stored individually. Since each pivot matrix differs from the unit matrix in just one column, it can be stored very conveniently by storing that column and its position in the unit matrix. Its storage in this manner occupies very little core space in the computer.

Suppose we start out in Phase I. The initial inverse tableau is a square matrix of order $m + 2$. Suppose it is denoted by P^0. It differs from the unit matrix of order $(m + 2)$ only in the last row. It can be stored easily by storing the last row, and its position in P^0.

Suppose the pivot matrices generated in the course of the algorithm are P^1, $P^2 \cdots$ in that order. Successive inverse tableaus can be obtained by multiplying P^0 on the left by the pivot matrices in the order in which they are generated. Suppose at the kth stage we are still in Phase I, and the column vector of basic values is $(\bar{b}_1, \ldots, \bar{b}_m, -\bar{z}, -\bar{w})^{\mathrm{T}}$.

Let P^k be the pivot matrix corresponding to this stage. Then the values of the basic variables and the objective value in the next stage are

$$(b_1^*, \ldots, b_m^*, -z^*, -w^*)^{\mathrm{T}} = P^k(\bar{b}_1, \ldots, \bar{b}_m, \bar{z}, -\bar{w})^{\mathrm{T}}.$$

Let σ^* be the Phase I dual solution and π^* the Phase II dual solution corresponding to the next stage. Then

$$(-\sigma^*, 0, 1) = (0, \ldots, 0, 1)P^k P^{k-1} \cdots P^1 P^0 \tag{5.2}$$

where $(0, \ldots, 0, 1)$ is a row vector of order $m + 2$ with all zeros excepting the "1" in the $(m + 2)$th place. Starting from the left, the multiplications in (5.2) can be carried out very quickly. Having obtained σ^*, the Phase I relative cost coefficients can be calculated as in section 5.3.

Check whether the Phase I termination criterion is satisfied. If it is not, pick the nonbasic variable x_s with the most negative Phase I relative cost coefficient as the the entering variable. The pivot column is the updated column vector of x_s, and it is

$$\begin{pmatrix} \bar{A}_{.s} \\ \cdots \\ \bar{c}_s \\ \bar{d}_s \end{pmatrix} = P^k P^{k-1} \cdots P^1 P^0 \begin{pmatrix} A_{.s} \\ \cdots \\ c_s \\ d_s \end{pmatrix} \tag{5.3}$$

The multiplication in (5.3) is easily carried out by starting from the right-hand side. The pivot element in the pivot column is determined by performing the usual minimum ratio test. Let the pivot matrix corresponding to this pivot step be P^{k+1}. The algorithm is now ready for the next stage.

On the other hand, if all the Phase I relative cost coefficients are nonnegative, then Phase I is terminated. If $w^* > 0$, the problem is infeasible and we terminate. If $w^* = 0$, we move to Phase II by deleting the last row and last column in each of the previous pivot matrices. For convenience we will continue to denote the pivot matrices [after the $(m + 2)$ row and column are struck off] by the same symbols $P^1, \ldots, P^k$ as before.

Suppose we are in the rth stage of Phase II. Let $(\bar{b}_1, \ldots, \bar{b}_m, -\bar{z})^{\mathrm{T}}$ be the column vector of present basic values. Let P^r be the pivot matrix corresponding to the pivot in this step. Suppose the superscript "*" identifies the quantities after the pivot. Then,

$$\begin{pmatrix} b^* \\ \cdots \\ -z^* \end{pmatrix} = P^r \begin{pmatrix} \bar{b} \\ \cdots \\ -\bar{z} \end{pmatrix}$$

$$(-\pi^*, 1) = (0, \ldots, 0, 1)P^r P^{r-1} \cdots P^1$$

and the updated column vector of any nonbasic variable, x_s is

$$\begin{pmatrix} A^*_{.s} \\ \cdots \\ c^*_s \end{pmatrix} = P^r \cdots P^1 \begin{pmatrix} A_{.s} \\ \vdots \\ c_s \end{pmatrix}$$

These multiplications can be carried out conveniently as before. Using these, the Phase II iterations are repeated until termination.

Numerical Example

$$\begin{array}{ll} \text{Minimize} & z = x_1 + x_2 + x_3 + 3x_4 + 4x_5 + 5x_6 + 2x_7 \\ \text{Subject to} & x_1 + x_2 + 2x_4 + x_5 + x_6 + x_7 = \frac{3}{2} \\ & x_2 - x_3 + x_4 - x_6 + 5x_7 = 1 \\ & 2x_1 - x_3 + 2x_4 - x_5 + x_6 + x_7 = 1 \\ & x_j \geqq 0 \quad \text{for all } j \end{array}$$

Introducing artificial variables t_1, t_2, t_3, we find that the original tableau for Phase I is as follows.

Original Tableau for Phase I

x_1	x_2	x_3	x_4	x_5	x_6	x_7	t_1	t_2	t_3	$-z$	$-w$	b
1	1	0	2	1	1	1	1	0	0	0	0	$\frac{3}{2}$
0	1	−1	1	0	−1	5	0	1	0	0	0	1
2	0	−1	2	−1	1	1	0	0	1	0	0	1
1	1	1	3	4	5	2	0	0	0	1	0	0
0	0	0	0	0	0	0	1	1	1	0	1	0

Initial Inverse Tableau

Basic variables	P^0					Basic values
t_1	1	0	0	0	0	$\frac{3}{2}$
t_2	0	1	0	0	0	1
t_3	0	0	1	0	0	1
$-z$	0	0	0	1	0	0
$-w$	-1	-1	-1	0	1	$-\frac{7}{2}$

At this stage we have $(t_1, t_2, t_3, -z, -w)^{\mathrm{T}} = P^0 b = (3/2, 1, 1, 0, -7/2)^{\mathrm{T}}$. $(-\sigma, 0, 1) = (0, 0, 0, 0, 1)P^0 = (-1, -1, -1, 0, 1)$ and $\bar{d} = (-3, -2, 2, -5, 0, 1, -7)$. x_4 has a negative Phase I relative cost coefficient, and it can be the entering variable. The updated column vector of x_4 is: P^0(original column of x_4) $= (2, 1, 2, 3, -5)^{\mathrm{T}}$. Perform the minimum ratio test:

Basic variables	Present basic values	Pivot column	Ratio
t_1	$\frac{3}{2}$	2	$\frac{3}{4}$
t_2	1	1	1
t_3	1	2	$\frac{1}{2}$ min

x_4 enters the basic vector replacing t_3 from it. The pivot matrix corresponding to this basis change is

$$P^1 = \begin{bmatrix} 1 & 0 & -\frac{2}{2} & 0 & 0 \\ 0 & 1 & -\frac{1}{2} & 0 & 0 \\ 0 & 0 & \frac{1}{2} & 0 & 0 \\ 0 & 0 & -\frac{3}{2} & 1 & 0 \\ 0 & 0 & \frac{5}{2} & 0 & 1 \end{bmatrix}$$

The new basic values vector is

$$\begin{aligned}(t_1, t_2, x_4, -z, -w)^{\mathrm{T}} &= P^1(3/2, 1, 1, 0, -7/2)^{\mathrm{T}} \\ &= (1/2, 1/2, 1/2, -3/2, -1)^{\mathrm{T}}.\end{aligned}$$

For this stage we have

$$(-\sigma, 0, 1) = (0, 0, 0, 0, 1)P^1 P^0 = (0, 0, 5/2, 0, 1)P^0 = (-1, -1, 3/2, 0, 1).$$

Therefore,

$$\bar{d} = \left(-1, -1, \frac{3}{2}, 0, 1\right) \times \text{(original tableau)}$$
$$= \left(2, -2, -\frac{1}{2}, 0, -\frac{5}{2}, \frac{3}{2}, -\frac{9}{2}\right)$$

So x_2 is a candidate to enter the basic vector. The pivot column is

$$P^1P^0 \text{ (original column of } x_2) = P^1P^0(1, 1, 0, 1, 0)^{\mathrm{T}}$$
$$= P^1(1, 1, 0, 1, -2)^{\mathrm{T}} = (1, 1, 0, 1, -2)^{\mathrm{T}}.$$

For the ratio test we have

Basic variables	Basic values	Pivot column	Ratio
t_1	$\frac{1}{2}$	1	$\frac{1}{2}$
t_2	$\frac{1}{2}$	1	$\frac{1}{2}$
x_4	$\frac{1}{2}$	0	

x_2 can replace either t_1 or t_2 from the basic vector. Suppose we decide to drop t_2 from the basic vector. The new pivot matrix is

$$P^2 = \begin{bmatrix} 1 & -1 & 0 & 0 & 0 \\ 0 & 1 & 0 & 0 & 0 \\ 0 & 0 & 1 & 0 & 0 \\ 0 & -1 & 0 & 1 & 0 \\ 0 & 2 & 0 & 0 & 1 \end{bmatrix}$$

The new basic values vector is $(t_1, x_2, x_4, -z, -w)^{\mathrm{T}} = P^2(1/2, 1/2, 1/2, -3/2, -1)^{\mathrm{T}} = (0, 1/2, 1/2, -2, 0)^{\mathrm{T}}$. The new Phase I dual vector is $(-\sigma, 0, 1) = (0, 0, 0, 0, 1)P^2P^1P^0 = (0, 2, 0, 0, 1)P^1P^0 = (0, 2, 3/2, 0, 1)P^0 = (-1, 1, 1/2, 0, 1)$. The new vector of Phase I relative cost coefficients is

$$\bar{d} = \left(-1, 1, \frac{1}{2}, 0, 1\right) \times \text{(original tableau)}$$
$$= \left(0, 0, -\frac{3}{2}, 0, -\frac{3}{2}, -\frac{3}{2}, \frac{9}{2}\right)$$

Since some of the $\bar{d}_j$ are negative, the Phase I optimality criterion is not yet satisfied.

Note The present basis associated with the basic vector (t_1, x_2, x_4) is degenerate, since the value of t_1 in the basic solution is 0. Actually the present value of the Phase I objective value is 0. Hence the present BFS of the Phase I problem is an optimum solution to the Phase I problem. However, this is not an optimal basis for the Phase I problem. This situation can only arise under degeneracy. When we continue to apply the algorithm, all subsequent pivots will be degenerate pivots. This will change the basis without any change in the solution, until a basis satisfying the optimality criterion is obtained.

Suppose we bring x_3 into the basic vector. The updated column vector of x_3 is $P^2P^1P^0(0, -1, -1, 1, 0)^T = P^2P^1(0, -1, -1, 1, 2)^T = P^2(1, -1/2, -1/2, 5/2, -1/2)^T = (3/2, -1/2, -1/2, 3, -3/2)^T$. The ratio test gives the following values.

Basic variables	Basic values	Pivot column	Ratios
t_1	0	$\frac{3}{2}$	0
x_2	$\frac{1}{2}$	$-\frac{1}{2}$	
x_4	$\frac{1}{2}$	$-\frac{1}{2}$	

x_3 comes into the basic vector replacing t_1, and this involves a degenerate pivot step. The pivot matrix is

$$P^3 = \begin{bmatrix} \frac{2}{3} & 0 & 0 & 0 & 0 \\ \frac{1}{3} & 1 & 0 & 0 & 0 \\ \frac{1}{3} & 0 & 1 & 0 & 0 \\ -2 & 0 & 0 & 1 & 0 \\ 1 & 0 & 0 & 0 & 1 \end{bmatrix}$$

Now all the artificial variables have left the basic vector. The present basic vector (x_3, x_2, x_4) is an optimum basic vector of the Phase I problem, and is a feasible basic vector of the original problem. To proceed to Phase II, delete the last row and column from each of the pivot matrices P^0, P^1, P^2, P^3, and for convenience in notation, refer to them by the same name. Drop the Phase I cost row and the w column vector from the original tableau. The present basic value vector is $(x_3, x_2, x_4, -z)^T = P^3(0, 1/2, 1/2, -2)^T = (0, 1/2, 1/2, -2)^T$. The corresponding Phase II dual vector is $(-\pi, 1) = (0, 0, 0, 1)P^3P^2P^1 = (-2, 0, 0, 1)P^2P^1 = (-2, 1, 0, 1)P^1 = (-2, 1, 0, 1)$. Hence, the Phase II relative cost coefficient vector is $\bar{c} = (-\pi, 1) \times (\text{original tableau}) = (-1, 0, 0, 0, 2, 2, 5)$.

x_1 enters the basic vector. The updated column vector of x_1 is $P^3P^2P^1(1, 0, 2, 1)^T = P^3P^2(-1, -1, 1, -2)^T = (0, -1, 1, -1)^T$.

Basic variables	Basic values	Pivot column	Ratios
x_3	0	0	
x_2	$\frac{1}{2}$	-1	
x_4	$\frac{1}{2}$	1	$\frac{1}{2}$

x_1 comes into the basic vector replacing x_4 from it. The pivot matrix is

$$P^4 = \begin{pmatrix} 1 & 0 & 0 & 0 \\ 0 & 1 & 1 & 0 \\ 0 & 0 & 1 & 0 \\ 0 & 0 & 1 & 1 \end{pmatrix}$$

The new basic value vector is $(x_3, x_2, x_1, -z)^T = P^4(0, 1/2, 1/2, -2) = (0, 1, 1/2, -3/2)^T$. The new dual vector is

$$\begin{aligned}(-\pi, 1) &= (0, 0, 0, 1)P^4P^3P^2P^1 \\ &= (0, 0, 1, 1)P^3P^2P^1 = \left(-\frac{5}{3}, 0, 1, 1\right)P^2P^1 \\ &= \left(-\frac{5}{3}, \frac{2}{3}, 1, 1\right)P^1 = \left(-\frac{5}{3}, \frac{2}{3}, \frac{1}{3}, 1\right).\end{aligned}$$

Hence the vector of relative cost coefficients is $\bar{c} = (-5/3, 2/3, 1/3, 1) \times$ (original tableau) $=$ $(0, 0, 0, 1, 2, 3, 4)$.

Hence, the present basis is an optimal basis. The optimum solution to the problem is

$$(x_1, x_2, x_3, x_4, x_5, x_6, x_7) = \left(\frac{1}{2}, 1, 0, 0, 0, 0, 0\right)$$

with objective value $z = 3/2$.

5.5 ADVANTAGES OF THE PRODUCT FORM IMPLEMENTATION OVER THE EXPLICIT FORM IMPLEMENTATION

Each pivot matrix differs from the unit matrix in just one column vector. This column vector is known as the *eta vector* in that pivot matrix. The pivot matrix can be stored in a compact fashion by storing the eta vector in it, and its position in the matrix.

For solving small LP's (in which the number of constraints, m, is small), the product form is cumbersome (when using hand computation, in particular) and the method using the explicit form of the inverse would probably be preferred. However, for solving large practical problems using a digital computer, the product form implementation has several advantages, which are discussed below.

A matrix is said to be a *sparse matrix*, if the proportion of nonzero entries in it is small. Likewise a vector is said to be a *sparse vector* if the proportion of nonzero entries in it is small. The *sparsity* of a matrix or vector is determined by the smallness of the proportion of nonzero entries in them. In most real life LP's of the type (5.1), the matrix A is usually sparse. If A is sparse, it usually happens that the eta vectors in the pivot matrices obtained while solving (5.1) by the simplex method, are themselves sparse. But the explicit inverse of any basis obtained while solving (5.1) by the simplex method is usually not sparse.

When the eta vector is sparse, it can be stored very compactly by storing the nonzero entries in it and their respective positions in the vector. Let A be of order $m \times n$ in (5.1). If each eta vector is not sparse, the space required for storing m pivot matrices, will be approximately equal to the space required for storing the explicit inverse of any basis for (5.1). However, if A is sparse (as in most LP's arising in practical applications), the tendency of the eta vectors to be sparse, leads to a considerable reduction in storage space requirements when the problem is solved using the product form of the inverse.

Also, when the eta vector in a pivot matrix is sparse, the computational effort involved in multiplying that pivot matrix by a vector, is small. When all the eta vectors tend to be sparse, the computational effort involved in carrying out the revised simplex algorithm with the product form of the inverse tends to be smaller than that in carrying out the same algorithm using explicit inverses which are usually not sparse. Product form of the inverse is also numerically more stable and has a lower error accumulation.

Because of these advantages most of the commercially available computer programs for solving LP's are based on the revised simplex method using the product form of the inverse.

PROBLEMS

5.1 Solve the following LPs by the revised simplex method:

(a)

$$\begin{array}{ll} \text{Maximize} & 5x_1 - x_2 + x_3 - 10x_4 + 7x_5 \\ \text{Subject to} & 3x_1 - x_2 - x_3 = 4 \\ & x_1 - x_2 + x_3 + x_4 = 1 \\ & 2x_1 + x_2 + 2x_3 + x_5 = 7 \\ & x_j \geqq 0 \quad \text{for all } j \end{array}$$

(b)

$$\begin{array}{ll} \text{Minimize} & -x_1 + 2x_2 \\ \text{Subject to} & 5x_1 - 2x_2 \leqq 3 \\ & x_1 + x_2 \geqq 1 \\ & -3x_1 + x_2 \leqq 3 \\ & -3x_1 - 3x_2 \leqq 2 \\ & x_j \geqq 0 \quad \text{for all } j \end{array}$$

In this problem, plot the path taken by the algorithm on a diagram.

(c)

x_1	x_2	x_3	x_4	x_5	x_6	$-z$	
1	2	0	1	0	−6	0	11
0	1	1	3	−2	−1	0	6
1	2	1	3	−1	−5	0	13
3	2	−3	−6	10	−5	1	0

$x_j \geqq 0$ for all j, minimize z.

(d)

x_1	x_2	x_3	x_4	x_5	x_6	x_7	$-z$	b
1	−1	1	2	1	−1	0	0	2
−1	1	1	1	2	1	−2	0	4
1	1	−1	1	−1	2	−4	0	6
2	4	−6	4	2	1	−11	1	0

$x_j \geqq 0$ for all j, minimize z.

(e)

x_1	x_2	x_3	x_4	x_5	x_6	x_7	x_8	$-z$	b
2	1	1	−2	−1	−1	0	0	0	5
6	5	−1	−3	−1	0	1	0	0	10
2	3	−3	1	1	0	0	−1	0	3
1	0	−2	3	0	0	0	0	1	0

$x_j \geqq 0$ for all j, minimize z.

(f)

x_1	x_2	x_3	x_4	x_5	x_6	$-z$	b
−1	1	2	1	0	1	0	0
0	1	3	0	1	3	0	5
2	−1	−1	−2	1	1	0	5
−3	3	9	5	0	4	1	0

$x_j \geqq 0$ for all j, minimize z.

5.2 Solve the following LP by the revised simplex algorithm using (x_1, x_2, x_6) as an initial basic vector.

$$\begin{array}{llllllll} \text{Minimize} & 14x_1 & - 19x_2 & & + 21x_4 & + 52x_5 & & \\ \text{Subject to} & x_1 & & & - x_4 & + x_5 & + x_6 & = 3 \\ & x_1 & + x_2 & & - x_4 & + 3x_5 & & = 4 \\ & x_1 & + x_2 & + x_3 & - 3x_4 & & + x_6 & = 6 \\ & & & x_j \geqq 0 & \text{for all } j & & & \end{array}$$

5.3 Solve the following LP using (x_3, x_5) as an initial basic vector by the revised simplex algorithm with the product form of the inverse.

$$\begin{array}{llllllll} \text{Minimize} & 2x_1 + 5x_2 & & + 7x_4 & + 15x_5 & + 14x_6 & \\ \text{Subject to} & x_1 + 2x_2 & - x_3 & + x_4 & + 4x_5 & + 5x_6 & = 10 \\ & x_1 + 3x_2 & - 2x_3 & + 2x_4 & + 5x_5 & + 7x_6 & = 12 \\ & & x_j \geqq 0 & \text{for all } j & & & \end{array}$$

5.4 Get an upper bound on the number of additions, subtractions, multiplications, divisions, and comparisons to be performed in a general step of the following.
(a) The revised simplex algorithm with the explicit form of the inverse.
(b) The simplex algorithm using canonical tableaus, in the course of solving the same given LP, as a function of m and n. m, n are the number of constraints and the number of variables, respectively, after the problem is expressed in standard form.

REFERENCES

1. D. M. Smith and W. Orchard-Hays, "Computational Efficiency in Product Form LP Codes," in "Recent Advances in Mathematical Programming," Edited by R. L. Graves and P. Wolfe, McGraw-Hill, 1963.

6 the dual simplex method

6.1 INTRODUCTION

Consider the *LP*

$$\begin{aligned} \text{Minimize} \quad & z(x) = cx \\ \text{Subject to} \quad & Ax = b \\ & x \geqq 0 \end{aligned} \tag{6.1}$$

where A is a given matrix of order $m \times n$ and rank m. Let B be a basis for (6.1). B is a *primal feasible basis*, if $B^{-1}b \geqq 0$ and a *dual feasible basis* if $c - c_B B^{-1} A \geqq 0$, and an *optimal basis* for (6.1) if it is both primal and dual feasible.

The *primal simplex algorithm** or its variants start with a primal feasible basis and move to a *terminal basis*, by walking through a sequence of primal feasible bases along the edges of the set of feasible solutions. All the bases with the possible exception of the terminal basis obtained in the primal algorithm are dual infeasible. At each pivot step, the primal algorithm makes an attempt to reduce dual infeasibility while retaining primal feasibility.

The *dual simplex algorithm* does just the opposite. It starts with a dual feasible, but primal infeasible basis and walks to a terminal basis by moving along adjacent dual feasible bases. At each pivot step, this algorithm tries to reduce primal infeasibility while retaining dual feasibility. If a feasible basis is reached, the dual simplex algorithm terminates by declaring it as an optimum basis.

6.2 DUAL SIMPLEX ALGORITHM WHEN A DUAL FEASIBLE BASIS IS KNOWN INITIALLY

Let B be a known dual feasible basis for (6.1). Let $x_B = (x_1, \ldots, x_m)$ be the associated basic vector. The dual simplex algorithm can be carried out either using the canonical tableaus or using the inverse tableaus. If the algorithm is to be carried out using the canonical tableaus, let the canonical tableau with respect to the basis B be as follows.

* The simplex algorithm discussed in Chapters 2, 5, is also called the *primal simplex algorithm*, to distinguish it from the dual simplex algorithm.

Canonical Tableau with Respect to the Basis B

Basic variables	$x_{m+1} \ldots x_s \ldots x_n$	$x_1 \ldots x_m$	$-z$	b
x_1	$\bar{a}_{1,m+1} \ldots \bar{a}_{1s} \ldots \bar{a}_{1n}$	$1 \ldots 0$	0	$\bar{b}_1$
$\vdots$	$\vdots \quad \vdots \quad \vdots$	$\vdots \quad \vdots$	$\vdots$	$\vdots$
x_m	$\bar{a}_{m,m+1} \ldots \bar{a}_{ms} \ldots \bar{a}_{mn}$	$0 \ldots 1$	0	$\bar{b}_m$
$-z$	$\bar{c}_{m+1} \ldots \bar{c}_s \ldots \bar{c}_n$	$0 \ldots 0$	1	$-\bar{z}$

Solving the problem using the inverse tableaus leads to the *revised dual simplex algorithm.* If the problem is to be solved using the revised dual simplex algorithm, let the inverse tableau corresponding to the basis B be T as follows.

Inverse Tableau Corresponding to the Basis B

Basic variables	Inverse tableau $= T$			Basic values
x_1	$\beta_{11} \ldots$	β_{1m}	0	$\bar{b}_1$
$\vdots$	$\vdots$	$\vdots$	$\vdots$	$\vdots$
x_m	$\beta_{m1} \ldots$	β_{mm}	0	$\bar{b}_m$
$-z$	$-\pi_1 \ldots$	$-\pi_m$	1	$-\bar{z}$

where

$$(\beta_{ij}) = B^{-1}, \qquad \pi = c_B B^{-1} \tag{6.2}$$

$$\begin{pmatrix} \bar{b} \\ \cdots \\ -\bar{z} \end{pmatrix} = T \begin{pmatrix} b \\ \cdots \\ 0 \end{pmatrix} \tag{6.3}$$

$$\bar{c}_j = c_j + \sum_{i=1}^{m} (-\pi_i) a_{ij} \qquad \text{for } j = 1 \text{ to } n \tag{6.4}$$

$$\begin{pmatrix} \bar{A}_{.s} \\ \cdots \\ \bar{c}_s \end{pmatrix} = T \begin{pmatrix} A_{.s} \\ \cdots \\ c_s \end{pmatrix} \tag{6.5}$$

In general, if B is any basis with its inverse given by (6.2) and if $\bar{a}_{rs}$ is the entry in the canonical tableau of (6.1) with respect to the basis B, then for any $r = 1$ to m, and $s = 1$ to n,

$$\bar{a}_{rs} = (\beta_{r1}, \ldots, \beta_{rm}) A_{.s} \tag{6.6}$$

All the entries in the rth row of the canonical tableau with respect to the basis B can be conveniently obtained from the original tableau and the inverse tableau corresponding to the basis B by using (6.6).

Since the starting basis B is known to be dual feasible, $\bar{c}_j \geqq 0$ for all $j = 1$ to n.

OPTIMALITY CRITERION IN THE DUAL SIMPLEX ALGORITHM

The present basis is optimal if $\bar{b}_i \geqq 0$ for all $i = 1$ to m. If the optimality criterion is satisfied, the current basic solution is primal feasible and optimal; hence we terminate.

CHOOSING THE PIVOT ROW

If the optimality criterion is not satisfied, pick a row i such that $\bar{b}_i < 0$ and choose the pivot element from it. In the present basic solution, the ith basic variable is equal to $\bar{b}_i < 0$. The purpose of pivoting in this row is to obtain a new basic vector such that the value of the ith basic variable in the basic solution corresponding to that basic vector becomes positive. For this the pivot element, $\bar{a}_{ij}$, has to be chosen among the negative entries in the updated pivot row. *Hence, in the dual simplex algorithm, the pivot elements in all pivots will be negative.* This is just contrary to the primal simplex algorithm, where the pivot elements have to be positive in order to retain primal feasibility.

The updated right-hand side constant in the pivot row becomes positive as a result of the pivot step. The pivot step as it is carried out in the dual simplex algorithm will be called a *dual simplex pivot step.*

Where there are several i such that $\bar{b}_i$ is negative, choose any one of them for the pivot row. The most commonly used pivot row choice rule is to pick the pivot row as the rth row where $\bar{b}_r$ = minimum $\{\bar{b}_i : i = 1 \text{ to } m\}$. If there is a tie for r in this equation, break it arbitrarily.

PRIMAL INFEASIBILITY CRITERION

The original problem (6.1) is infeasible if in a canonical tableau, with respect to a basis, there is an i such that

$$\bar{b}_i < 0 \qquad \text{and} \qquad \bar{a}_{ij} \geqq 0 \qquad \text{for each } j = 1 \text{ to } n$$

Discussion Suppose this criterion is satisfied in the ith row of the canonical tableau. This row corresponds to the constraint

$$\sum_{j=1}^{n} \bar{a}_{ij} x_j = \bar{b}_i \tag{6.7}$$

Since this comes from a canonical tableau, it is a linear combination of the original constraints in (6.1). Hence, every feasible solution of (6.1) must satisfy (6.7). However, since $\bar{b}_i < 0$ and $\bar{a}_{ij} \geqq 0$ for all j, (6.7) cannot be satisfied by any $x \geqq 0$.

PIVOT ELEMENT

If the primal infeasibility criterion is not satisfied, perform the pivot and obtain a new basis. For this, the pivot row has already been picked. Let it be the rth row in the tableau. The pivot column has to be determined in such a way that it leads to a new basis, which is also dual feasible. If the pivot element is $\bar{a}_{rs}$ the relative cost coefficients in the next basis will be $\bar{c}_j - \bar{c}_s \bar{a}_{rj}/\bar{a}_{rs}$, $j = 1$ to n. Hence the next basis will remain dual feasible if

$$\bar{c}_j - \frac{\bar{c}_s \bar{a}_{rj}}{\bar{a}_{rs}} \geqq 0 \qquad \text{for all } j \tag{6.8}$$

Since the present basis is dual feasible, $\bar{c}_j \geqq 0$ for all $j = 1$ to n. In order to satisfy (6.8), the pivot column is picked as the sth column, where s is such that

$$\frac{\bar{c}_s}{-\bar{a}_{rs}} = \text{minimum} \left\{ \frac{\bar{c}_j}{-\bar{a}_{rj}} : \qquad j \text{ such that } \bar{a}_{rj} < 0 \right\} \tag{6.9}$$

Any ties for s in (6.9) are broken arbitrarily. If the computations are done using the inverse tableaus, the relative cost-coefficients $\bar{c}_j$ may be computed using equation (6.4), and the pivot row (in this case it is the *updated* rth *row*) can be computed using (6.6).

PIVOTING

If the canonical tableaus are being computed, the new canonical tableau is obtained by pivoting on the column vector of the variable x_s with $\bar{a}_{rs}$ as the pivot element. By elementary row operations, this column vector is transformed into the rth unit vector and priced out in the cost row. This gives the new canonical tableau.

If the inverse tableaus are being used to perform the computations, $\bar{A}_{.s}$, the pivot column (the updated column vector of x_s), is found by using equation (6.5). This pivot column is then entered by the side of the previous inverse tableau.

Previous Inverse Tableau

Basic variables	Inverse tableau			Basic values	Pivot column	
	$\beta_{11} \cdots$	β_{1m}	0	$\bar{b}_1$	$\bar{a}_{1s}$	
	$\vdots$	$\vdots$	$\vdots$	$\vdots$	$\vdots$	
	$\beta_{r1} \cdots$	β_{rm}	0	$\bar{b}_r$	$-\bar{a}_{rs}$ (circled)	Pivot element
	$\vdots$	$\vdots$	$\vdots$	$\vdots$	$\vdots$	
	$\beta_{m1} \cdots$	β_{mm}	0	$\bar{b}_m$	$\bar{a}_{ms}$	
	$-\pi_1 \cdots$	$-\pi_m$	1	$-\bar{z}$	$\bar{c}_s$	

The new inverse tableau is obtained by performing the pivot. After the pivot is complete, test whether the new basis satisfies either the optimality or the primal infeasibility criteria. Repeat the iterations until a basis that satisfies either the optimality or the primal infeasibility criterion is obtained.

Note about Unboundedness Since the dual problem is known to be feasible, the primal problem cannot be unbounded, by the weak duality theorem. The algorithm discussed here will terminate with a basis that satisfies either the optimality criterion or the infeasibility criterion.

Example Using Canonical Tableaus

Solve the following problem by the dual simplex algorithm:

Original Tableau

x_1	x_2	x_3	x_4	x_5	x_6	$-z$	b	
1	0	0	4	−5	7	0	8	
0	1	0	−2	4	−2	0	−2	
0	0	1	1	−3	2	0	2	
0	0	0	1	3	2	1	0	(minimize z)

$x_j \geqq 0 \quad$ for all j

(x_1, x_2, x_3) constitutes a dual feasible basic vector. Here is the canonical tableau with respect to this basic vector.

First Canonical Tableau

Basic variables	x_1	x_2	x_3	x_4	x_5	x_6	$-z$	Basic values	
x_1	1	0	0	4	−5	7	0	8	
x_2	0	1	0	(−2)	4	−2	0	−2	Pivot row
x_3	0	0	1	1	−3	2	0	2	
$-z$	0	0	0	1	3	2	1	0	
$-\frac{\bar{c}_j}{\bar{a}_{2j}}$ for $\bar{a}_{2j} < 0.$				$\frac{1}{2}$		$\frac{2}{2}$			

Since $\bar{b}_2$ is -2, the second row is picked as the pivot row. The minimum ratio $(-\bar{c}_j/\bar{a}_{2j})$ for j such that $\bar{a}_{2j} < 0$ occurs in the column of the nonbasic variable x_4. So x_4 is the entering variable. It replaces x_2 from the basic vector. The pivot element has been circled. The new canonical tableau is

Basic variables	x_1	x_2	x_3	x_4	x_5	x_6	$-z$	Basic values
x_1	1	2	0	0	3	3	0	4
x_4	0	$-\frac{1}{2}$	0	1	−2	1	0	1
x_3	0	$\frac{1}{2}$	1	0	−1	1	0	1
$-z$	0	$\frac{1}{2}$	0	0	5	1	1	−1

Since all the basic variables are positive, this basis is now optimal. So the optimal feasible solution is

$$(x_1, x_2, x_3, x_4, x_5, x_6) = (4, 0, 1, 1, 0, 0)$$
$$z(x) = 1$$

Example Using Inverse Tableaus

Consider the LP

$$\begin{aligned} \text{Minimize} \quad & z(x) = 8x_1 + 8x_2 + 9x_3 \\ \text{Subject to} \quad & x_1 + x_2 + x_3 \leqq 1 \\ & 2x_1 + 4x_2 + x_3 \geqq 8 \\ & x_1 - x_2 - x_3 \geqq 2 \\ & x_1 \geqq 0, \quad x_2 \geqq 0, \quad x_3 \geqq 0 \end{aligned}$$

Slack variables x_4, x_5, x_6 are introduced. Here is the original tableau (the initial right-hand side constants are not required to be nonnegative here)

Original Tableau

x_1	x_2	x_3	x_4	x_5	x_6	$-z$	b
1	1	1	1	0	0	0	1
−2	−4	−1	0	1	0	0	−8
−1	1	1	0	0	1	0	−2
8	8	9	0	0	0	1	0 (minimize z)

$x_j \geqq 0$ for all j

(x_4, x_5, x_6) constitutes a dual feasible basic vector here. The inverse tableau with respect to this basic vector is

Basic variables	Inverse tableau				Basic values	Pivot column
x_4	1	0	0	0	1	1
x_5	0	1	0	0	−8	(−4) Pivot element
x_6	0	0	1	0	−2	1
$-z$	0	0	0	1	0	8

Pick the second row as the pivot row. Below are the values of the relative cost coefficients and the updated second row.

		x_1	x_2	x_3	x_4	x_5	x_6
$\bar{a}_{2j}$	=	−2	−4	−1	0	1	0
$\bar{c}_j$	=	8	8	9	0	0	0
$-\dfrac{\bar{c}_j}{\bar{a}_{2j}}$	for $\bar{a}_{2j} < 0$ =	$\frac{8}{2}$	$\frac{8}{4}$	9			

x_2 enters the basic vector. The pivot column and the pivot element are recorded on the inverse tableau.

Second Inverse Tableau

Basic variables	Inverse tableau				Basic values	Pivot column x_1
x_4	1	$\frac{1}{4}$	0	0	−1	$\frac{1}{2}$
x_2	0	$-\frac{1}{4}$	0	0	2	$\frac{1}{2}$
x_6	0	$\frac{1}{4}$	1	0	−4	$\left(-\frac{3}{2}\right)$ Pivot element
$-z$	0	2	0	1	−16	4

Since x_6 is the most negative basic variable, the third row is picked as the next pivot row.

		x_1	x_2	x_3	x_4	x_5	x_6
$\bar{a}_{3j}$	$=$	$-\frac{3}{2}$	0	$\frac{3}{4}$	0	$\frac{1}{4}$	1
$\bar{c}_j$	$=$	4	0	7	0	2	0
$-\frac{\bar{c}_j}{\bar{a}_{3j}}$	for $\bar{a}_{3j} < 0 =$	$\frac{8}{3}$					

x_1 is the entering variable. The updated column vector of x_1 is computed and recorded on the inverse tableau.

Third Inverse Tableau

Basic variables	Inverse tableau				Basic values
x_4	1	$\frac{1}{3}$	$\frac{1}{3}$	0	$-\frac{7}{3}$
x_2	0	$-\frac{1}{6}$	$\frac{1}{3}$	0	$\frac{2}{3}$
x_1	0	$-\frac{1}{6}$	$-\frac{2}{3}$	0	$\frac{8}{3}$
$-z$	0	$\frac{8}{3}$	$\frac{8}{3}$	1	$-\frac{80}{3}$

The next pivot row is the first row. The updated first row is

$$(\bar{a}_{11}, \bar{a}_{12}, \bar{a}_{13}, \bar{a}_{14}, \bar{a}_{15}, \bar{a}_{16}) = \left(0, 0, 1, 1, \frac{1}{3}, \frac{1}{3}\right)$$

Since this contains no negative entries, the infeasibility criterion is satisfied. The row in the final inverse tableau corresponding to the pivot row is (1, 1/3, 1/3, 0). Hence the row obtained by multiplying row 1 of the original tableau by 1 and adding to it row 2 multiplied by 1/3 and row 3 multiplied by 1/3 yields the inconsistent equation

$$0x_1 + 0x_2 + x_3 + x_4 + \frac{1}{3}x_5 + \frac{1}{3}x_6 = -\frac{7}{3}$$

Since all the variables in the problem are restricted to be nonnegative, the left-hand side of this equation will always be nonnegative and, hence, the above equation can never be satisfied by any $x \geqq 0$.

Problems

6.1 Show that in the dual simplex algorithm the objective value steadily increases from one iteration to the next. Compare this with what happened in the primal simplex algorithm where the value of the objective function steadily decreased from one iteration to the next.

6.2 Show that the dual simplex algorithm is precisely the primal simplex algorithm applied to the dual problem.

6.3 DISADVANTAGES AND ADVANTAGES OF THE DUAL SIMPLEX ALGORITHM

In the dual simplex algorithm all but the final basic solution are primal infeasible. If we run out of computer time in the middle of solving an LP by the dual simplex algorithm, all the effort spent is wasted since we do not even have a feasible solution to the problem until the algorithm is carried out to the end. This is a disadvantage.

In the primal simplex algorithm all the basic solutions obtained are primal feasible. Thus, if it becomes necessary to terminate the computation before the optimality criterion is satisfied, we can at least be content with the best among the feasible solutions obtained. Even though it may not be optimal to the problem, it is feasible. That's why, if a primal feasible basis and another dual feasible basis are available for an LP, it is preferable to solve it by the primal simplex algorithm starting with the known primal feasible basis, than to approach it by the dual simplex algorithm starting with the known dual feasible basis.

However, the dual simplex algorithm is very useful in doing sensitivity analysis (i.e., postoptimality analysis). Suppose we have an optimal basis for an LP. After the optimal basis is obtained, it may become necessary to change the right-hand constant vector b. When the b vector is changed, the previous optimal basis may no longer be primal feasible. However, it is dual feasible, assuming that the cost coefficients remain unchanged. Starting from that basis the dual simplex algorithm can be applied to solve the problem for the new b vector. Similarly, the dual simplex algorithm is used in solving LPs with parametric right-hand side vector (Chapter 7).

In practical applications of linear programming, it may become necessary to introduce a new constraint into the LP model after it has been solved. The dual simplex algorithm can be used to find an optimum solution of the augmented model starting from an optimum solution of the present model (Chapter 8). The same property makes the dual simplex algorithm very useful in cutting plane approaches for solving integer programs (Chapter 14).

6.4 DUAL SIMPLEX METHOD WHEN A DUAL FEASIBLE BASIS IS NOT KNOWN AT THE START

Suppose a dual feasible basis for (6.1) is not known. By transforming the system of constraints "$Ax = b$" into row echelon normal form obtain an arbitrary basis for (6.1). Let B denote the basis. Let $x_B = (x_1, \ldots, x_m)$ be the basic vector. If this basis turns out to be dual feasible, it can be used as a starting dual feasible basis for the dual simplex algorithm.

However, the basis B may be neither primal nor dual feasible. If B is not dual feasible, let the canonical tableau with respect to the basis B be as follows.

Canonical Tableau

Basic variables	$x_1 \ldots x_m$	$x_{m+1} \ \ldots \ x_n$	$-z$	Basic values
x_1	$1 \ldots 0$	$\bar{a}_{1,m+1} \ldots \bar{a}_{1n}$	0	$\bar{b}_1$
$\vdots$	$\vdots \quad \vdots$	$\vdots \quad \vdots$	$\vdots$	$\vdots$
x_m	$0 \ldots 1$	$\bar{a}_{m,m+1} \ldots \bar{a}_{mn}$	0	$\bar{b}_m$
$-z$	$0 \ldots 0$	$\bar{c}_{m+1} \ldots \bar{c}_n$	1	$-\bar{z}$

Add an artificial variable, x_0 with cost coefficient zero, and an artificial constraint:

$$x_0 + x_{m+1} + \cdots + x_n = M \tag{6.10}$$

where $(x_{m+1}, \ldots, x_n)$ is the nonbasic vector and M is a very large positive number. It is not necessary to give a specific numerical value to M; it should always be considered to be strictly larger than any other number with which it is compared. The original problem modified by the addition of the constraint (6.10) will be called the *augmented problem.*

Tableau for the Augmented Problem

Basic variables	x_0	$x_1 \ldots x_m$	$x_{m+1} \;\ldots\; x_s \;\ldots\; x_n$	$-z$	Basic values
x_0	1	$0 \ldots 0$	$1 \;\ldots 1 \;\ldots 1$	0	M
x_1	0	$1 \ldots 0$	$\bar{a}_{1,m+1} \ldots \bar{a}_{1s} \ldots \bar{a}_{1n}$	0	$\bar{b}_1$
$\vdots$	$\vdots$	$\vdots \quad \vdots$	$\vdots \quad \vdots \quad \vdots$	$\vdots$	$\vdots$
x_m	0	$0 \ldots 1$	$\bar{a}_{m,m+1} \ldots \bar{a}_{ms} \ldots \bar{a}_{mn}$	0	$\bar{b}_m$
$-z$	0	$0 \ldots 0$	$\bar{c}_{m+1} \;\ldots \bar{c}_s \;\ldots \bar{c}_n$	1	$-\bar{z}$

Let $\bar{c}_s$ = minimum $\{\bar{c}_j: j = 1 \text{ to } n\}$. Since B is not a dual feasible basis of the original problem, $\bar{c}_s < 0$ is the most negative $\bar{c}_j$. Bring x_s into the basic vector, replacing x_0 from it. By the definition of $\bar{c}_s$, this pivot will transform all the entries in the cost row of the tableau into nonnegative numbers. Thus we get a dual feasible basis for the augmented problem. The basic vector $(x_s, x_1, \ldots, x_m)$ is a dual feasible basic vector for the augmented problem. Starting with this dual feasible basic vector apply the dual simplex algorithm and solve the augmented problem. Termination of this algorithm may occur in any of the three following ways:

Augmented Problem Primal Infeasible In this case the original problem is also infeasible because if the original problem has a feasible solution $\bar{x} = (\bar{x}_1, \ldots, \bar{x}_n)$, then $(\bar{x}_0, \bar{x}_1, \ldots, \bar{x}_n)$ where $\bar{x}_0 = M - \sum_{j=m+1}^{n} \bar{x}_j$ is a feasible solution of the augmented problem.

Augmented Problem Has an Optimum Basic Vector Containing x_0 Let $(\bar{x}_0, \bar{x}_1, \ldots, \bar{x}_n)$ be an optimal BFS of the augmented problem. In this case the optimum objective value of the augmented problem is obviously independent of M, as long as M is very large. Hence, $(\bar{x}_1, \ldots, \bar{x}_n)$ is an optimum BFS of the original problem (6.1). An optimal basis for (6.1) is obtained by suppressing the row and the column corresponding to x_0 from the optimum basis for the augmented problem.

Augmented Problem Has an Optimum Basic Vector Not Containing x_0 In this case the values of the basic variables in the final optimum BFS of the augmented problem depend on M. If the optimal objective value for the augmented problem depends on M, it must tend to $-\infty$ as M tends to $+\infty$. In this case the original problem is feasible and unbounded.

If the optimal objective value for the augmented problem is independent of M, the original problem is feasible and has an optimal feasible solution, which can be obtained from the optimal solution of the augmented problem by suppressing x_0 (which is equal to zero in this solution). In this case, an optimal BFS of the original problem may be obtained by decreasing the value of M until one of the basic variables in the final optimum BFS of the augmented problem vanishes.

Example 1

Original Tableau

x_1	x_2	x_3	x_4	x_5	$-z$	b	
1	2	0	1	-2	0	-3	
0	1	0	1	-3	0	1	
0	2	1	0	-1	0	1	
0	0	0	-1	1	1	0	(z minimize)

$$x_j \geqq 0 \qquad \text{for all } j$$

(x_1, x_2, x_3) forms a basic vector for this problem. Here is the canonical tableau that is neither primal nor dual feasible.

Canonical Tableau

x_1	x_2	x_3	x_4	x_5	$-z$	b
1	0	0	-1	4	0	-5
0	1	0	1	-3	0	1
0	0	1	-2	5	0	-1
0	0	0	-1	1	1	0

Below is the augmented problem.

x_0	x_1	x_2	x_3	x_4	x_5	$-z$	b
1	0	0	0	(1)	1	0	M
0	1	0	0	-1	4	0	-5
0	0	1	0	1	-3	0	1
0	0	0	1	-2	5	0	-1
0	0	0	0	-1	1	1	0

To obtain a dual feasible tableau, we replace x_0 in the basic vector by x_4.

Basic variables	x_0	x_1	x_2	x_3	x_4	x_5	$-z$	b	
x_4	1	0	0	0	1	1	0	M	
x_1	1	1	0	0	0	5	0	$M - 5$	
x_2	-1	0	1	0	0	(-4)	0	$1 - M$	Pivot row
x_3	2	0	0	1	0	7	0	$2M - 1$	
$-z$	1	0	0	0	0	2	1	M	
$-\dfrac{\bar{c}_j}{\bar{a}_{3j}}$ for $\bar{a}_{3j} < 0$	$\frac{1}{1}$					$\frac{2}{4}$			

Since M is an arbitrarily large positive number, the pivot row is the third row. By the dual simplex minimum ratio test, x_5 comes into the basic vector replacing x_2 from it.

Basic variables	x_0	x_1	x_2	x_3	x_4	x_5	$-z$	b
x_4	$\frac{3}{4}$	0	$\frac{1}{4}$	0	1	0	0	$\frac{3M+1}{4}$
x_1	$\boxed{-\frac{1}{4}}$	1	$\frac{5}{4}$	0	0	0	0	$\frac{-15-M}{4}$
x_5	$\frac{1}{4}$	0	$-\frac{1}{4}$	0	0	1	0	$\frac{M-1}{4}$
x_3	$\frac{1}{4}$	0	$\frac{7}{4}$	1	0	0	0	$\frac{M+3}{4}$
$-z$	$\frac{1}{2}$	0	$\frac{1}{2}$	0	0	0	1	$\frac{M+1}{2}$
x_4	0	3	4	0	1	0	0	-11
x_0	1	-4	-5	0	0	0	0	$15+M$
x_5	0	1	1	0	0	1	0	-4
x_3	0	1	3	1	0	0	0	-3
$-z$	0	2	3	0	0	0	1	-7

From the first row in this tableau it is clear that the problem is infeasible.

Example 2

Original Tableau

x_1	x_2	x_3	x_4	x_5	$-z$	b
1	0	0	1	-1	0	2
0	1	0	-1	-1	0	-4
0	0	1	-2	2	0	-3
2	1	0	0	0	1	0 (z minimize)

$$x_j \geqq 0 \text{ for all } j$$

Select (x_1, x_2, x_3) as the initial basic vector. Pricing out and introducing the artificial constraint, we get the original tableau for the augmented problem.

Original Tableau for the Augmented Problem

x_0	x_1	x_2	x_3	x_4	x_5	$-z$	b
1	0	0	0	$\boxed{1}$	1	0	M
0	1	0	0	1	-1	0	2
0	0	1	0	-1	-1	0	-4
0	0	0	1	-2	2	0	-3
0	0	0	0	-1	3	1	0

Bringing x_4 into the basic vector leads to a dual feasible tableau. Here is the corresponding inverse tableau.

Basic variables	First dual feasible inverse tableau					Basic values	Pivot column x_0
x_4	1	0	0	0	0	M	1
x_1	−1	1	0	0	0	$2 - M$	(−1)
x_2	1	0	1	0	0	$M - 4$	1
x_3	2	0	0	1	0	$2M - 3$	2
$-z$	1	0	0	0	1	M	1

The pivot row is the second row. Below are the updated second and the cost rows.

Updated	x_0	x_1	x_2	x_3	x_4	x_5
Pivot row	−1	1	0	0	0	−2
Cost row	1	0	0	0	0	4
Ratios	1 minimum					2

Hence, x_0 enters the basic vector. Its updated column is the pivot column and is entered on the right-hand side of the tableau.

Basic variables	Inverse tableau					Basic values	Pivot column (x_5)
x_4	0	1	0	0	0	2	−1
x_0	1	−1	0	0	0	$M - 2$	2
x_2	0	1	1	0	0	−2	(−2) Pivot row
x_3	0	2	0	1	0	1	0
$-z$	0	1	0	0	1	2	2

The third row is the pivot row.

Updated	x_0	x_1	x_2	x_3	x_4	x_5
Pivot row	0	1	1	0	0	−2
Cost row	0	1	0	0	0	2
Ratios						1

x_5 enters the basic vector replacing x_2 from it. Its updated column vector is the pivot column, and is entered on the right-hand side of the tableau.

Basic variables	Inverse tableau					Basic values
x_4	0	$\frac{1}{2}$	$-\frac{1}{2}$	0	0	3
x_0	1	0	1	0	0	$M - 4$
x_5	0	$-\frac{1}{2}$	$-\frac{1}{2}$	0	0	1
x_3	0	1	-1	1	0	1
$-z$	0	2	1	0	1	0

When M is large this tableau is primal feasible and, hence, optimal. $(x_0, x_1, \ldots, x_5)^T = (M - 4, 0, 0, 1, 3, 1)^T$ is optimal to the augmented problem with the optimum objective value being zero, Hence, $(x_1, \ldots, x_5)^T = (0, 0, 1, 3, 1)^T$ is optimal to the original problem with the optimal objective value equal to zero.

Example 3

x_1	x_2	x_3	x_4	x_5	x_6	$-z$	b
1	0	0	1	-3	7	0	-5
0	1	0	-1	1	-1	0	1
0	0	1	3	1	-10	0	8
1	3	0	0	0	-2	1	0 (z minimize)

$x_j \geqq 0$ for all j

Take (x_1, x_2, x_3) as the initial basic vector. Pricing out and introducing the artificial constraint leads to the following.

Basic variables	x_0	x_1	x_2	x_3	x_4	x_5	x_6	$-z$	b
x_0	1	0	0	0	1	1	(1)	0	M
x_1	0	1	0	0	1	-3	7	0	-5
x_2	0	0	1	0	-1	1	-1	0	1
x_3	0	0	0	1	3	1	-10	0	8
$-z$	0	0	0	0	2	0	-6	1	2

Bringing x_6 into the basic vector leads to a dual feasible tableau.

Basic variables	x_0	x_1	x_2	x_3	x_4	x_5	x_6	$-z$	b	
x_6	1	0	0	0	1	1	1	0	$\boldsymbol{M}$	
x_1	-7	1	0	0	-6	(-10)	0	0	$-5-7\boldsymbol{M}$	Pivot row
x_2	1	0	1	0	0	2	0	0	$1+\boldsymbol{M}$	
x_3	10	0	0	1	13	11	0	0	$8+10\boldsymbol{M}$	
$-z$	6	0	0	0	8	6	0	1	$2+6\boldsymbol{M}$	
Ratios	$\frac{6}{7}$				$\frac{8}{6}$	$\frac{6}{10}$				

Basic variables	x_0	x_1	x_2	x_3	x_4	x_5	x_6	$-z$	b
x_6	$\frac{3}{10}$	$\frac{1}{10}$	0	0	$\frac{4}{10}$	0	1	0	$\frac{3\boldsymbol{M}-5}{10}$
x_5	$\frac{7}{10}$	$-\frac{1}{10}$	0	0	$\frac{6}{10}$	1	0	0	$\frac{5+7\boldsymbol{M}}{10}$
x_2	$-\frac{4}{10}$	$\frac{2}{10}$	1	0	($-\frac{12}{10}$)	0	0	0	$-\frac{4\boldsymbol{M}}{10}$
x_3	$\frac{23}{10}$	$\frac{11}{10}$	0	1	$\frac{64}{10}$	0	0	0	$\frac{23\boldsymbol{M}+25}{10}$
$-z$	$\frac{18}{10}$	$\frac{6}{10}$	0	0	$\frac{44}{12}$	0	0	0	$\frac{18\boldsymbol{M}-10}{10}$
Ratios	$\frac{18}{4}$				$\frac{44}{12}$				
x_6	$\frac{5}{30}$	$\frac{5}{30}$	$\frac{1}{3}$	0	0	0	1	0	$\frac{\boldsymbol{M}-3}{6}$
x_5	$\frac{5}{10}$	0	$\frac{1}{2}$	0	0	1	0	0	$\frac{5+5\boldsymbol{M}}{10}$
x_4	$\frac{4}{12}$	$-\frac{2}{12}$	$-\frac{10}{12}$	0	1	0	0	0	$\frac{\boldsymbol{M}}{3}$
x_3	$\frac{5}{30}$	$\frac{13}{6}$	$\frac{64}{12}$	1	0	0	0	0	$\frac{\boldsymbol{M}+15}{6}$
$-z$	$\frac{10}{30}$	$\frac{8}{6}$	$\frac{44}{12}$	0	0	0	0	1	$\frac{\boldsymbol{M}-3}{3}$

This is an optimum tableau for the augmented problem for large $\boldsymbol{M}$. The optimum objective value is $-\frac{\boldsymbol{M}-3}{3}$ and this diverges to $-\infty$ as $\boldsymbol{M}$ moves toward $+\infty$. Hence, the original problem is feasible and unbounded below. This feasible solution

$$x^{\mathrm{T}} = \left(0, 0, \frac{M + 15}{6}, \frac{M}{3}, \frac{1 + M}{2}, \frac{M - 3}{6}\right)$$

$$\text{Objective value} = -\frac{M - 3}{3}$$

is feasible for all M sufficiently large and as M moves toward $+\infty$, the objective value decreases indefinitely.

6.5 COMPARISON OF THE PRIMAL AND THE DUAL SIMPLEX METHODS

Feature of the Primal Simplex Method	Comparative Feature of the Dual Simplex Method
1. The primal simplex algorithm needs a primal feasible basis to start with.	1. The dual simplex algorithm needs a dual feasible basis to start with.
2. Starting with a primal feasible basis, the primal simplex algorithm tries to attain dual feasibility, while maintaining primal feasibility throughout.	2. Starting with a dual feasible basis, the dual simplex algorithm tries to attain primal feasibility, while maintaining dual feasibility throughout.
3. In the primal simplex algorithm the optimality criterion is the dual feasibility criterion.	3. In the dual simplex algorithm the optimality criterion is the primal feasibility criterion.
4. Starting with a primal feasible basis, the primal simplex algorithm terminates either with an optimum basis or by establishing that the primal objective function is unbounded (this is the dual infeasibility criterion).	4. Starting with a dual feasible basis, the dual simplex algorithm terminates either with an optimum basis or by establishing primal infeasibility (in this case the dual objective function is unbounded).
5. Suppose the problem is being solved by the revised primal simplex algorithm starting with a known primal feasible basis. All the primal solutions obtained during the algorithm are primal feasible. The dual solutions obtained during the algorithm (excepting the final one when the optimality criterion is satisfied) are dual infeasible.	5. Suppose the problem is being solved by the revised dual simplex algorithm starting with a known dual feasible basis. All the dual solutions obtained during the algorithm are dual feasible. The primal solutions obtained in the algorithm (excepting the final one when the optimality criterion is satisfied) are primal infeasible.
6. Pivot elements are positive in all pivot steps, to maintain primal feasibility.	6. Pivot elements are negative in all pivot steps, to move closer to primal feasibility.
7. The pivot column is first selected by the problem solver, among	7. The pivot row is first selected by the problem solver, among those

Feature of the Primal Simplex Method	Comparative Feature of the Dual Simplex Method
those columns which correspond to a negative relative cost coefficient (i.e., negative dual slack variable). The pivot row is then determined by the algorithm using the primal simplex minimum ratio test, the purpose of which is to guarantee that primal feasibility is maintained in the next basis.	rows which have a negative updated right hand side constant (i.e. a negative valued primal basic variable). The pivot column is then determined, by the algorithm using the dual simplex minimum ratio test, the purpose of which is to guarantee that dual feasibility is maintained in the next basis.
8. The primal simplex minimum ratio test uses the updated right hand side constants and positive entries in the pivot column.	8. The dual simplex minimum ratio test uses the updated cost coefficients (i.e. the relative cost coefficients) and negative entries in the pivot row.
9. Dual infeasibility is established in the primal simplex algorithm, if the updated column vector of the entering variable (i.e., the pivot column) has no positive entries.	9. Primal infeasibility is established in the dual simplex algorithm, if the pivot row has no negative entries.
10. The primal solution in an intermediate stage of the primal simplex algorithm, is primal feasible but may not be optimal. Hence the corresponding objective value is an upper bound for the minimum objective value in the problem. Also the objective value steadily decreases as the algorithm progresses.	10. The primal solution in an intermediate stage of the dual simplex algorithm is primal infeasible, because some basic variables have negative values in it, violating the nonnegativity restrictions. It is an optimum solution of the problem obtained by relaxing the nonnegativity restrictions on the primal basic variables. Hence the corresponding objective value is a lower bound for the minimum objective value in the original problem. Also the objective value steadily increases as the algorithm progresses.
11. If a primal feasible basis for the original problem is not known, an augmented problem is created by introducing nonnegative artificial variables, so that the augmented problem has a readily available primal feasible basis.	11. If a dual feasible basis for the original problem is not known, an augmented problem is created by introducing an artificial inequality constraint (which can be written as an equation by including an appropriate nonnegative slack variable), so that the augmented problem has a readily available dual feasible basis.

Remarks The dual simplex algorithm has been developed by C. E. Lemke. See reference [1].

PROBLEMS

6.3 Solve the following LPs by the dual simplex method.

(a)

x_1	x_2	x_3	x_4	x_5	x_6	x_7	$-z$	b
1	2	0	−3	0	2	−3	0	−4
0	−1	0	2	1	−1	2	0	1
0	2	1	−3	0	1	−2	0	1
0	4	0	5	0	9	9	1	0

$x_j \geqq 0$ for all j; minimize z.

(b)

$$\begin{array}{ll} \text{Minimize} & 3x_1 + 4x_2 + 2x_3 + x_4 + 5x_5 \\ \text{Subject to} & x_1 - 2x_2 - x_3 + x_4 + x_5 \leqq -3 \\ & -x_1 - x_2 - x_3 + x_4 + x_5 \leqq -2 \\ & x_1 + x_2 - 2x_3 + 2x_4 - 3x_5 \leqq 4 \\ & x_j \geqq 0 \quad \text{for all } j. \end{array}$$

(c)

$$\begin{array}{ll} \text{Minimize} & -2x_1 - x_2 - 8x_4 + 2x_5 - 3x_7 \\ \text{Subject to} & 3x_1 + x_3 + 16x_4 - 2x_5 + 5x_6 + 4x_7 = 18 \\ & 2x_1 + x_3 + 11x_4 - x_5 + 3x_6 + 3x_7 = 11 \\ & x_1 - x_2 + x_3 + 7x_4 - 2x_5 + 2x_6 + x_7 = 6 \\ & x_j \geqq 0 \quad \text{for all } j \end{array}$$

Use (x_1, x_2, x_3) as the initial basic vector.

(d)

$$\begin{array}{ll} \text{Minimize} & -2x_1 + 4x_2 + x_3 - x_4 + 6x_5 + 8x_6 - 9x_7 - 5x_8 \\ \text{Subject to} & x_1 + x_4 - 2x_5 + x_6 + x_7 - 2x_8 = -3 \\ & x_2 - x_4 + x_5 + x_6 - 3x_7 - x_8 = -14 \\ & x_3 + x_4 - x_5 - 2x_6 - x_7 + x_8 = -5 \\ & x_j \geqq 0 \quad \text{for all } j \end{array}$$

(e)

$$\begin{array}{ll} \text{Minimize} & -2x_1 + 4x_2 + 2x_3 + x_4 - 4x_5 - 10x_6 \\ \text{Subject to} & 5x_2 + 2x_3 + x_4 - 3x_5 - 9x_6 - 4x_7 = -8 \\ & x_1 - 3x_2 - x_3 - x_4 + 2x_5 + 8x_6 = 7 \\ & -2x_1 - x_3 + x_4 - 5x_6 + 6x_7 = -3 \\ & x_j \geqq 0 \quad \text{for all } j \end{array}$$

6.4 Solve the following system using the dual simplex algorithm:

$$\begin{array}{l} 5x_1 + 4x_2 - 7x_3 \leqq 1 \\ -x_1 + 2x_2 - x_3 \leqq -4 \\ -3x_1 - 2x_2 + 4x_3 \leqq 3 \\ x_j \geqq 0 \quad \text{for all } j \end{array}$$

6.5 Show that the method of generating an artificially dual feasible basis by introducing the artificial constraint (6.10), is the dual analogue of generating an artificially primal feasible basis by introducing an artificial variable.

6.6 Consider the LP (6.1). Suppose a dual feasible basis for it is known, and suppose the LP is being solved by the dual simplex algorithm. At the kth stage of this algorithm, suppose the primal infeasibility criterion is satisfied. Using the data in the canonical tableau or the inverse tableau at this stage, explain how to construct a half-line of dual feasible solutions, along which the dual objective function diverges to $+\infty$.

REFERENCE

1. C. E. Lemke, "The Dual Method of Solving the Linear Programming Problem," *Naval Research Logistics Quarterly*, *1*, pp. 36–47, 1954.

7 parametric linear programs

7.1 INTRODUCTION

In practical applications of linear programming, it often happens that the coefficients of the objective function or the right-hand side constants are not precisely known at the time the problem is being solved. For example, in modeling an automobile manufacturer's problem, the coefficients of the objective function may depend on several parameters such as the price of steel. The automobile manufacturer may have no control over the values of these parameters, and they may vary from time to time. Each variation in the values of the parameters changes his objective function, and thus, creates a new problem. In order to plan his strategy well in advance, the manufacturer clearly requires the optimum solution for the model as a function of the parameters for all possible values of the parameters beforehand.

Example

Consider the fertilizer product mix problem discussed in section 1.1. The availability of raw materials 1, 2, 3 at present is $b = (b_1, b_2, b_3)^T = (1500, 1200, 500)^T$. Suppose the company has opened new quarries. When the quarries are in full production, they are expected to provide additional supplies of raw materials 1, 2, 3 in the amounts $b^* = (500, 200, 300)^T$. For several years before they come into full production, the quarries may work only at a fraction of their full production capacity. If the fraction is λ, the availability of raw materials, 1, 2, 3 will be $b + \lambda b^* = (1500 + 500\lambda, 1200 + 200\lambda, 500 + 300\lambda)^T$, $0 \leqq \lambda \leqq 1$. With this the fertilizer product mix problem becomes a parametric right-hand side problem. Optimum solutions of this problem are required for all values of λ between 0 and 1.

Since only linear models are being considered here, it is assumed that coefficients of the objective function or the right-hand side constant vector vary linearly with the parameters. First, consider the case where there is only one parameter, λ. Problems in which the data depends on the values of two or more parameters are much harder to tackle. The problems to be solved are of the following form:

$$\begin{aligned} \text{Minimize} \quad & z_\lambda(x) = (c + \lambda c^*)x \\ \text{Subject to} \quad & Ax = b \\ & x \geqq 0 \end{aligned} \tag{7.1}$$

and

$$\begin{aligned} \text{Minimize} \quad & z(x) = cx \\ \text{Subject to} \quad & Ax = b + \lambda b^* \\ & x \geqq 0 \end{aligned} \tag{7.2}$$

where A is a matrix of order $m \times n$, b, b^*, c, c^* are vectors of appropriate dimensions and λ is a parameter that takes real values.

The LP (7.1) is known as the *parametric cost problem* and (7.2) is known as the *parametric right-hand side problem.* LPs in which both cost coefficients and the right-hand side constants simultaneously depend on the value of a parameter are considered later on.

Problem

7.1 Show that every parametric right-hand side problem is the dual of a parametric cost problem and vice versa.

In (7.1) let $z(x) = cx$, $z^*(x) = c^*x$. Hence, $z_\lambda(x) = z(x) + \lambda z^*(x)$. For any given value of λ, (7.1) is a standard LP and, hence, can be solved by the simplex method or one of its variants. If B is a feasible basis for (7.1), it is an optimum basis for a specified value of λ if all the relative cost coefficients with respect to it are nonnegative for that value of λ. The updating operation can be carried out by treating each cost coefficient as a function of λ, without giving any specific value for it. All the relative cost coefficients will turn out to be functions of λ, and B is an optimum basis for all values of λ that keep all the relative cost coefficients nonnegative. Computing all the relative cost coefficients as functions of λ is made easier by keeping, the parametric objective function in two rows of the tableau. One row contains the constant terms of the objective function, namely, the c_j's and this is the row corresponding to $z(x)$. The second row contains all the coefficients of λ in the objective function, namely, the c_j^*'s, and this is the row corresponding to $z^*(x)$. Treating a cost coefficient as a function of λ and pricing it out is equivalent to pricing out in both the cost rows. If we represent the objective function in this manner, the original tableau for (7.1) is as follows.

Tableau 7.1. Original Tableau for the Parametric Cost Problem

$x_1 \ldots x_j \ldots x_n$	$-z$	$-z^*$	b
$a_{11} \ldots a_{1j} \ldots a_{1n}$	0	0	b_1
$\vdots \quad \vdots \quad \vdots$	$\vdots$	$\vdots$	$\vdots$
$a_{m1} \ldots a_{mj} \ldots a_{mn}$	0	0	b_m
$c_1 \ldots c_j \ldots c_n$	1	0	0
$c_1^* \ldots c_j^* \ldots c_n^*$	0	1	0

In a similar manner here is the original tableau for the parametric right-hand side problem (Tableau 7.2).

Tableau 7.2. Original Tableau for the Parametric Right-Hand Side Problem

$x_1 \cdots x_j \cdots x_n$	$-z$	b	b^*
$a_{11} \cdots a_{1j} \cdots a_{1n}$	0	b_1	b_1^*
$\vdots \quad \vdots \quad \vdots$	$\vdots$	$\vdots$	$\vdots$
$a_{m1} \cdots a_{mj} \cdots a_{mn}$	0	b_m	b_m^*
$c_1 \cdots c_j \cdots c_n$	1	0	0

The question of feasibility of (7.1) does not depend on λ, since λ affects only the objective function in (7.1). Thus, if (7.1) is infeasible for some λ, it is infeasible for all values of λ and we terminate.

If $z(x)$ is unbounded below on the set of feasible solutions of (7.2) for some specified value of λ, it remains unbounded below for every value of λ for which (7.2) is feasible (see section 4.5.11). Hence in this case the only remaining question is to find the set of values of λ for which (7.2) is feasible. This question does not depend on the objective function. Therefore, we can replace the original objective function by the zero objective function $z'(x) = 0$, in which all the cost coefficients are zero. $z'(x)$ is always zero and, hence, the modified problem cannot be unbounded. In solving the modified problem, the algorithm will determine the range of values of λ for which (7.2) is feasible.

7.2 ANALYSIS OF THE PARAMETRIC COST PROBLEM GIVEN AN OPTIMUM BASIS FOR SOME λ

Let $x_B = (x_1, \ldots, x_m)$, associated with the basis B, be an optimum basic vector for (7.1) for some specified value of λ, say, λ_0. If the problem is being solved using the canonical tableaus, let the canonical tableau with respect to the basis B, be as in Tableau 7.3.

Tableau 7.3

Basic variables	$x_1 \quad x_m$	$x_{m+1} \cdots x_j \cdots x_n$	$-z$	$-z^*$	Basic values
x_1	$1 \ldots 0$	$\bar{a}_{1,m+1} \cdots \bar{a}_{ij} \cdots \bar{a}_{1n}$	0	0	$\bar{b}_1$
$\vdots$	$\vdots \quad \vdots$	$\vdots \quad \vdots \quad \vdots$	$\vdots$	$\vdots$	$\vdots$
x_m	$0 \ldots 1$	$\bar{a}_{m,m+1} \cdots \bar{a}_{mj} \cdots \bar{a}_{mn}$	0	0	$\bar{b}_m$
$-z$	$0 \ldots 0$	$\bar{c}_{m+1} \cdots \bar{c}_j \cdots \bar{c}_n$	1	0	$-\bar{z}$
$-z^*$	$0 \ldots 0$	$\bar{c}^*_{m+1} \cdots \bar{c}^*_j \cdots \bar{c}^*_n$	0	1	$-\bar{z}^*$

If the problem is being solved using the inverse tableaus, let the inverse tableau corresponding to the basis B be as in Tableau 7.4.

Tableau 7.4

Basic variables	Inverse tableau			Basic values
x_1	$\beta_{11} \dots \beta_{1m}$	0	0	$\bar{b}_1$
$\vdots$	$\vdots \quad \vdots$	$\vdots$	$\vdots$	$\vdots$
x_m	$\beta_{m1} \dots \beta_{mm}$	0	0	$\bar{b}_m$
$-z$	$-\pi_1 \dots -\pi_m$	1	0	$-\bar{z}$
$-z^*$	$-\pi_1^* \dots -\pi_m^*$	0	1	$-\bar{z}^*$

Here $(\beta_{ij}) = B^{-1}$. If c_B, c_B^* are the cost vectors associated with the basic vector x_B, $\pi = c_B B^{-1}$ and $\pi^* = c_B^* B^{-1}$. The relative cost coefficient of x_j as a function of the parameter λ is $\bar{c}_j + \lambda \bar{c}_j^*$ where

$$\bar{c}_j = c_j - \pi A_{.j}$$
$$\bar{c}_j^* = c_j^* - \pi^* A_{.j}$$

7.2.1 Optimality Interval or the Characteristic Interval of the Basis B

The basis B is an optimum basis for all values of the parameter λ satisfying

$$\bar{c}_j + \lambda \bar{c}_j^* \geqq 0 \qquad \text{for all } j \tag{7.3}$$

Hence a necessary condition for B to be an optimum basis is

$$\bar{c}_j \geqq 0 \qquad \text{for all } j \text{ such that } \bar{c}_j^* = 0 \tag{7.4}$$

If (7.4) is violated, B can never be an optimum basis for (7.1). For B to be an optimum basis for a value of λ it is not necessary for either $\bar{c}_j$ or $\bar{c}_j^*$ to be nonnegative. Condition (7.3) is the only optimality criterion to be satisfied for B to be optimal. If $\bar{c}_j^* > 0$, (7.3) is equivalent to $\lambda \geqq -\bar{c}_j/\bar{c}_j^*$. If $\bar{c}_j^* < 0$, (7.3) is equivalent to $\lambda \leqq -\bar{c}_j/\bar{c}_j^*$. Let

$$\bar{\lambda}_B = \text{minimum} \left\{ \frac{-\bar{c}_j}{\bar{c}_j^*} : j \text{ such that } \bar{c}_j^* < 0 \right\} \tag{7.5}$$
$$= +\infty \qquad \text{if} \qquad \bar{c}_j^* \geqq 0 \qquad \text{for all } j$$

$$\underline{\lambda}_B = \text{maximum} \left\{ \frac{-\bar{c}_j}{\bar{c}_j^*} : j \text{ such that } \bar{c}_j^* > 0 \right\} \tag{7.6}$$
$$= -\infty \qquad \text{if} \qquad \bar{c}_j^* \leqq 0 \qquad \text{for all } j$$

$\underline{\lambda}_B$ is known as the *lower characteristic value* (or the *lower break point*) of the basis B, and $\bar{\lambda}_B$ is known as the *upper characteristic value* (or the *upper breakpoint*) of the basis B. The closed interval $\{\lambda: \underline{\lambda}_B \leqq \lambda \leqq \bar{\lambda}_B\} = [\underline{\lambda}_B, \bar{\lambda}_B]$ is known as the *characteristic interval* or the *optimality interval* of the basis B. For all values of the parameter λ in its characteristic interval, B is an optimum basis for the parametric cost problem (7.1).

If $\underline{\lambda}_B > \bar{\lambda}_B$, the optimality interval of the basis B is empty and B can never be an optimal basis. If $\underline{\lambda}_B = \bar{\lambda}_B$, the optimality interval of the basis B contains only a single point. For all values of λ outside the optimality interval of B, B is nonoptimal for (7.1). Hence in exploring for optimal solutions for all λ, once we leave its optimality interval, a basis will never reappear as an optimal basis.

If the optimality interval of the basis B is nonempty, for λ satisfying $\underline{\lambda}_B \leqq \lambda \leqq \overline{\lambda}_B$, an optimal BFS of (7.1) is:

$$\text{The basic vector: } x_B = \overline{b} = B^{-1}b$$
$$\text{All nonbasic variables} = 0$$
$$\text{Optimum objective value} = \overline{z} + \lambda\overline{z}^*$$

Hence in the interval $\underline{\lambda}_B \leqq \lambda \leqq \overline{\lambda}_B$, the optimal feasible solution is the same, but the optimal objective value is a linear function of λ. An optimum dual solution corresponding to (7.1) for values of λ in this interval is $\pi + \lambda\pi^*$.

7.2.2 Solving the Problem for $\lambda > \overline{\lambda}_B$

Suppose $\overline{\lambda}_B$ is finite. If it is required to solve (7.1) for $\lambda > \overline{\lambda}_B$, identify j that attains the minimum in (7.5). Suppose this is $j = s$. Then $\overline{\lambda}_B = -\overline{c}_s/\overline{c}_s^*$. When λ exceeds $\overline{\lambda}_B$, the relative cost coefficient of x_s becomes negative. Hence to solve the problem for $\lambda > \overline{\lambda}_B$, bring x_s into the present basic vector. When this is done, two cases might occur.

Case 1 The unboundedness criterion may be satisfied. This happens if the updated column vector of x_s is such that $\overline{a}_{is} \leqq 0$ for all i. In this case, the parametric objective function, $z_\lambda(x)$ is unbounded below on the set of feasible solutions of (7.1) whenever $\lambda > \overline{\lambda}_B$.

Case 2 If the unboundedness criterion is not satisfied, bring x_s into the basic vector as in the simplex algorithm or the revised simplex algorithm. Suppose the rth basic variable, x_r, drops out of the basic vector and is replaced by x_s. Remember to price out the pivot column in both the cost rows. Let the new basis be $\tilde{B}$.

Problems

7.2 Prove that the relative cost coefficients of any variable x_j with respect to the bases B and $\tilde{B}$ are both equal when $\lambda = \overline{\lambda}_B$. Hence show that B, $\tilde{B}$ are two alternate optimal bases for (7.1) when $\lambda = \overline{\lambda}_B$.

7.3 Let $\overline{c}_j$, $\overline{c}_j^*$, $\overline{a}_{ij}$ denote the updated entries with respect to the basis B and $\tilde{c}_j$, $\tilde{c}_j^*$, $\tilde{a}_{ij}$ denote the updated entries with respect to the basis $\tilde{B}$. Prove that the relative cost coefficient of the dropping variable x_r with respect to the new basis $\tilde{B}$ is

$$\tilde{c}_r + \lambda\tilde{c}_r^* = -(\overline{c}_s + \lambda\overline{c}_s^*)/\overline{a}_{rs}.$$

Hence show that $\tilde{B}$ is not an optimum basis for any $\lambda < \overline{\lambda}_B$.

7.4 By using the results in Problems 7.2 and 7.3, prove that the lower characteristic point of the new basis $\tilde{B}$ is $\overline{\lambda}_B$.

The optimality interval of $\tilde{B}$ is another closed interval that has its left end point coinciding with the right end point ($= \overline{\lambda}_B$) of the optimality interval of the previous basis B. The same procedure can be repeated with the basis $\tilde{B}$, and optimum solutions of (7.1) determined as λ increases further.

7.2.3 Solving the Problem for $\lambda < \underline{\lambda}_B$

Returning to the original basis B, suppose $\underline{\lambda}_B$ is finite. If it is required to solve (7.1) for $\lambda < \underline{\lambda}_B$, identify a j that attains the maximum in (7.6). Suppose this is $j = t$.

Then bring x_t into the basic vector x_B. If the unboundedness criterion is satisfied, $z_\lambda(x)$ is unbounded below on the set of feasible solutions of (7.1) for all $\lambda < \underline{\lambda}_B$. Otherwise, bring x_t into the basic vector and suppose the new basis is $\hat{B}$. From results similar to those in section 7.2.2, it is clear that B, $\hat{B}$ are alternate optimal bases for (7.1) when $\lambda = \underline{\lambda}_B$. The optimality interval of $\hat{B}$ is another closed interval with its right end point equal to the left end point $(=\underline{\lambda}_B)$ of the optimality interval of B. Repeating the same procedure, optimum solutions of (7.1) can be obtained as λ decreases further.

7.3 TO FIND AN OPTIMUM BASIS FOR THE PARAMETRIC COST PROBLEM

Solve (7.1) for some value of λ, say, $\lambda = \lambda_1$ (λ_1 could be 0) first, either by the simplex method or a variant of it. If (7.1) has an optimum feasible solution for λ_1, this will terminate with an optimum feasible basis for $\lambda = \lambda_1$. However, it may happen that $z_\lambda(x)$ is unbounded below on the set of feasible solutions of (7.1) when $\lambda = \lambda_1$. In this case, the simplex method terminates with a feasible basis, B, that satisfies the unboundedness criterion when $\lambda = \lambda_1$. Let $\bar{c}_j, \bar{c}_j^*, \bar{a}_{ij}$ denote the updated entries with respect to the basis B. Hence, there exists a nonbasic variable, x_s, such that $\bar{c}_s + \lambda_1\bar{c}_s^* < 0$ and $\bar{a}_{is} \leqq 0$ for all i. Clearly (7.1) remains unbounded below for all λ satisfying $\bar{c}_s + \lambda\bar{c}_s^* < 0$. Therefore, if $\bar{c}_s^* = 0$, and $\bar{c}_s < 0$, (7.1) is unbounded below for all λ and we terminate. Otherwise, let $\lambda_2 = -\bar{c}_s/\bar{c}_s^*$. If $\bar{c}_s^* > 0$,(7.1) is unbounded below for all $\lambda < \lambda_2$. If $\bar{c}_s^* < 0$, (7.1) is unbounded below for all $\lambda > \lambda_2$.

In either case, change the value of the parameter to λ_2 and try to solve the problem again. This is done by continuing the simplex algorithm starting from the basis B, in which the relative cost coefficients are given by $\bar{c}_j + \lambda_2\bar{c}_j^*$.

If the unboundedness criterion is satisfied again, change the value of λ again and continue the iterations of the simplex (or the revised simplex) algorithm with the new value of λ. In a finite number of iterations we will either find a value of λ and an optimal feasible basis corresponding to it, or conclude that the problem is unbounded for all values of λ.

7.4 SUMMARY OF THE RESULTS ON THE PARAMETRIC COST PROBLEM

1. The characteristic intervals of the optimal bases obtained in the course of the algorithm do not overlap (except at end points). The characteristic intervals and the intervals in which the objective function is unbounded form a partition of the real line.
2. All values of λ for which the objective function $z_\lambda(x)$ is unbounded below on the set of feasible solutions of (7.1) are either in open intervals of the form $(-\infty, \lambda)$ for some λ, or of the form $(\lambda, +\infty)$,
3. In exploring for optimum solutions of (7.1) for all λ, once λ leaves the characteristic interval of some optimum basis, that basis never reappears as an optimum basis. The total number of bases is finite. These two together imply that the total number of characteristic intervals obtained in the course of the algorithm will be finite. In a finite number of steps the parametric cost problem can be solved for all real values of λ.
4. In each characteristic interval, the optimum feasible solution is fixed, but the optimum objective value varies linearly with λ. Also the optimum objec-

tive values at the common end point of two consecutive characteristic intervals are equal. Hence, the optimal objective value of a parametric cost problem is a piecewise linear continuous function in the parameter λ.

7.5 ANALYSIS OF THE PARAMETRIC RIGHT-HAND SIDE PROBLEM GIVEN AN OPTIMAL BASIS FOR SOME λ

Let x_B be an optimum basic vector for (7.2) for some specified value of λ, say λ_0, and let B be the associated basis. Let c_B be the corresponding basic cost vector. Let

$$\bar{b} = B^{-1}b, \qquad \bar{b}^* = B^{-1}b^*, \qquad \bar{z} = c_B B^{-1}b, \qquad \bar{z}^* = c_B B^{-1}b^*$$

The BFS corresponding to the basis B as a function of λ is tabulated as in Tableau 7.5.

Tableau 7.5. Basic Values as Functions of λ

Basic variables	Updated right-hand side vectors		Basic values
	$\bar{b}_1$	$\bar{b}_1^*$	$\bar{b}_1 + \lambda\bar{b}_1^*$
	$\vdots$	$\vdots$	$\vdots$
x_B	$\bar{b}_i$	$\bar{b}_i^*$	$\bar{b}_i + \lambda\bar{b}_i^*$
	$\vdots$	$\vdots$	$\vdots$
	$\bar{b}_m$	$\bar{b}_m^*$	$\bar{b}_m + \lambda\bar{b}_m^*$
$-z$	$-\bar{z}^*$	$-\bar{z}^*$	$-\bar{z} - \lambda\bar{z}^*$

Since B is an optimum basis for (7.2) when $\lambda = \lambda_0$, B must be dual feasible. The parameter λ appears only in the right-hand side constant vector and, hence, dual feasibility of a basis for (7.2) is a property that is independent of λ. Hence, B is an optimum basis for (7.2) for all values of λ for which it is primal feasible, that is, for all λ satisfying $\bar{b}_i + \lambda\bar{b}_i^* \geqq 0$ for all i. Since B is primal feasible when $\lambda = \lambda_0$, we must have $\bar{b}_i + \lambda_0\bar{b}_i^* \geqq 0$ for all i. This implies that

$$\bar{b}_i \geqq 0 \qquad \text{for all} \qquad i \qquad \text{such that} \qquad \bar{b}_i^* = 0 \tag{7.7}$$

If (7.7) is not satisfied, B is not even primal feasible and it can never be optimal. $\bar{b}_i + \lambda\bar{b}_i^* \geqq 0$ for all i implies that $\lambda \geqq -\bar{b}_i/\bar{b}_i^*$ for all i such that $\bar{b}_i^* > 0$, and that $\lambda \leqq -\bar{b}_i/\bar{b}_i^*$ for all i such that $\bar{b}_i^* < 0$. Let

$$\begin{aligned} \bar{\lambda}_B &= \text{minimum } \{-\bar{b}_i/\bar{b}_i^* : i \text{ such that } \bar{b}_i^* < 0\} \\ &= +\infty, \text{ if } \bar{b}_i^* \geqq 0 \qquad \text{for all } i \end{aligned} \tag{7.8}$$

$$\begin{aligned} \underline{\lambda}_B &= \text{maximum } \{-\bar{b}_i/\bar{b}_i^* : i \text{ such that } \bar{b}_i^* > 0\} \\ &= -\infty, \text{ if } \bar{b}_i^* \leqq 0 \qquad \text{for all } i \end{aligned} \tag{7.9}$$

If $\underline{\lambda}_B > \bar{\lambda}_B$, B is never primal feasible for any λ. If $\underline{\lambda}_B \leqq \bar{\lambda}_B$, then the closed interval $\underline{\lambda}_B \leqq \lambda \leqq \bar{\lambda}_B$, that is, $[\underline{\lambda}_B, \bar{\lambda}_B]$ is the *characteristics interval or the optimality interval* of the basis B. For all values of λ in this optimality interval, the BFS listed in the tableau 7.5 is an optimum feasible solution of (7.2). Hence, in this interval, the optimum objective value and the values of the basic variables in the optimum BFS vary linearly with λ.

To Solve the Problem for $\lambda > \overline{\lambda}_B$

If $\overline{\lambda}_B$ is finite and if it is required to solve (7.2) for values of $\lambda > \overline{\lambda}_B$, identify the value of i that attains the minimum in (7.8). Suppose it is $i = r$. Then $\overline{\lambda}_B = -\overline{b}_r/\overline{b}_r^*$ and for $\lambda > \overline{\lambda}_B$, $\overline{b}_r + \lambda\overline{b}_r^* < 0$.

Let $\overline{A}_{r.} = (\overline{a}_{r1}, \ldots, \overline{a}_{rn})$, be the updated rth row vector with respect to the basis B. If $\overline{A}_{r.} \geqq 0$, that is, $\overline{a}_{rj} \geqq 0$ for all j, the problem is clearly infeasible whenever $\lambda > \overline{\lambda}_B$. This is the primal infeasibility criterion.

If the primal infeasibility criterion is not satisfied, make a dual simplex pivot step in the rth row. Suppose it gives a consecutive optimal basis $\tilde{B}$.

Problem

7.5 Prove that both B, $\tilde{B}$ are alternate optimal bases when $\lambda = \overline{\lambda}_B$. Also prove that $\underline{\lambda}_{\tilde{B}} = \overline{\lambda}_B$.

The next optimal basis $\tilde{B}$ has its characteristic interval to the right of $\overline{\lambda}_B$. The same procedure is repeated with the new basis $\tilde{B}$ until optimal feasible solutions are obtained for all required $\lambda > \overline{\lambda}_B$.

To Solve the Problem for $\lambda < \underline{\lambda}_B$

Returning to the original basis B, suppose $\underline{\lambda}_B$ is finite and you have to solve the problem for $\lambda < \underline{\lambda}_B$. Identify the i that attains the maximum in (7.9). Suppose it is $i = t$. If the updated tth row, $\overline{A}_{t.} \geqq 0$, (7.2) is primal infeasible for all $\lambda < \underline{\lambda}_B$. Otherwise a dual simplex pivot step is made in the tth row, and the procedure continued in a similar manner.

7.6 TO FIND AN INITIAL OPTIMUM BASIS FOR THE PARAMETRIC RIGHT-HAND SIDE PROBLEM GIVEN A FEASIBLE BASIS FOR SOME λ

Let B be a feasible basis for (7.2) for some specified value of λ, say, λ_1. Fix $\lambda = \lambda_1$ (i.e., treat the right-hand constant vector as equal to $b + \lambda_1 b^*$). Starting with the feasible basis B, minimize $z(x)$ by applying either the simplex (or the revised simplex) algorithm. Two possibilities may occur.

1. An optimal feasible solution exists when $\lambda = \lambda_1$. Then the simplex (or the revised simplex) algorithm will find an optimal basis for $\lambda = \lambda_1$.
2. The simplex algorithm may terminate by satisfying the unboundedness criterion for $\lambda = \lambda_1$.

In case 2, $z(x)$ is unbounded below on the set of feasible solutions of (7.2) for all values of the parameter λ for which (7.2) is feasible. To determine the range of values for which (7.2) is feasible, replace all the cost coefficients by zero. From now on, all the relative cost coefficients will be equal to zero. For this zero objective function, the final basis obtained under the simplex algorithm is optimal.

7.7 TO FIND A FEASIBLE BASIS FOR THE PARAMETRIC RIGHT-HAND SIDE PROBLEM

Pick an arbitrary value for λ, say λ_1 (λ_1 could be 0) and try to find a feasible basis. In this case (7.2) is an LP with the right-hand side constant vector $b + \lambda_1 b^*$. Write down all the constraints in (7.2) in such a way that the right-hand side constant vector is nonnegative when $\lambda = \lambda_1$, that is, if $b_i + \lambda_1 b_i^* < 0$, multiply both sides of the ith constraint in (7.2) by -1, for each i.

If the original tableau is canonical, a feasible unit basis is available for the case when $\lambda = \lambda_1$. If the tableau is not canonical, augment the tableau with a full artificial basis and formulate the corresponding Phase I problem as below.

$$\begin{aligned} \text{Minimize} \quad & w(x, y) = \sum_{i=1}^{m} y_i \\ \text{Subject to} \quad & Ax + Iy = b + \lambda b^* \\ & x \geqq 0, \quad y \geqq 0 \end{aligned} \tag{7.10}$$

Starting with the artificial basic vector solve the Phase I parametric problem by the procedures in sections 7.6, 7.5. There are two possibilities here.

1. It may turn out that the minimal objective value of the parametric Phase I problem is positive for all λ. In this case the original problem (7.2), is infeasible for all λ.
2. A value of the parameter may be found, say, λ_2 such that when $\lambda = \lambda_2$, the Phase I objective function has a minimum value of 0. In this case, the optimal basis for (7.10) when $\lambda = \lambda_2$ provides a feasible basis for (7.2).

7.8 SUMMARY OF THE RESULTS ON THE PARAMETRIC RIGHT-HAND SIDE PROBLEM

1. The characteristic intervals of the optimal bases obtained in the course of the algorithm do not overlap (except at end points).
2. All the values of λ for which (7.2) is infeasible are either in open intervals of the form $(-\infty, \lambda)$ for some λ or of the form $(\lambda, +\infty)$.
3. The characteristic intervals of the optimal bases obtained in the course of the algorithm and the open intervals in which (7.2) is infeasible, form a partition of the real line.
4. In exploring for optimum solutions of (7.2) for all λ, once λ leaves the characteristic interval of some optimum basis, that basis never reappears as an optimum basis (since it will not even be primal feasible). Since the total number of possible bases is finite, the total number of characteristic intervals obtained in the course of the algorithm will be finite.
5. In each interval, the optimal feasible solution and the optimum objective value vary linearly in the parameter λ, but the optimum dual solution is fixed. The optimal objective value is a piecewise linear continuous function in the parameter, λ.

7.9 CONVEX AND CONCAVE FUNCTIONS ON THE REAL LINE

Let $f(\lambda)$ be a real-valued function defined on the real line. It is said to be a *convex function* if:

$$f[\alpha\lambda_1 + (1 - \alpha)\lambda_2] \leqq \alpha f(\lambda_1) + (1 - \alpha)f(\lambda_2)$$

for all λ_1, λ_2, and $0 \leqq \alpha \leqq 1$. See Figures 7.1 and 7.2.

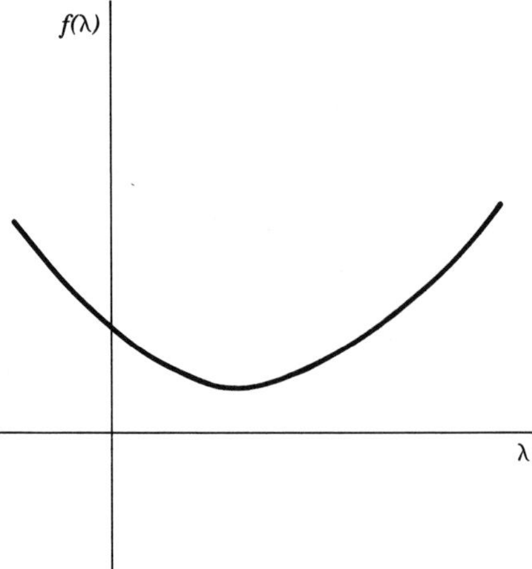

Figure 7.1 A convex function defined on the real line.

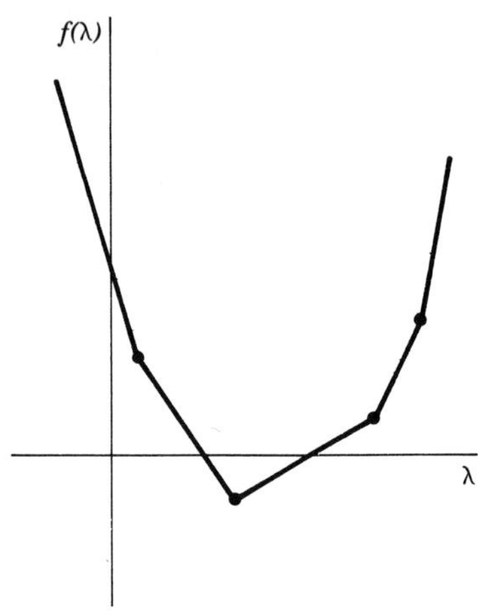

Figure 7.2 A piecewise linear convex function defined on the real line.

$f(\lambda)$ is said to be a *concave function* if $-f(\lambda)$ is a convex function, that is, if

$$f[\alpha\lambda_1 + (1 - \alpha)\lambda_2] \geqq \alpha f(\lambda_1) + (1 - \alpha)f(\lambda_2)$$

for all λ_1, λ_2, and $0 \leqq \alpha \leqq 1$. See Figures 7.3 and 7.4.

Problems

7.6 Let $f(\lambda)$ = minimum value of $z_\lambda(x)$ in (7.1). Prove that $f(\lambda)$ is a piecewise linear concave function.

7.7 Let $g(\lambda)$ = minimum value of $z(x)$ in (7.2). Prove that $g(\lambda)$ is a convex piecewise linear function in λ.

7.8 Consider the multiparameter parametric cost problem

$$\begin{aligned} \text{Minimize} \quad & z(x) = (c + \lambda_1 c^1 + \lambda_2 c^2 + \cdots + \lambda_t c^t)x \\ \text{Subject to} \quad & Ax = b \\ & x \geqq 0 \end{aligned}$$

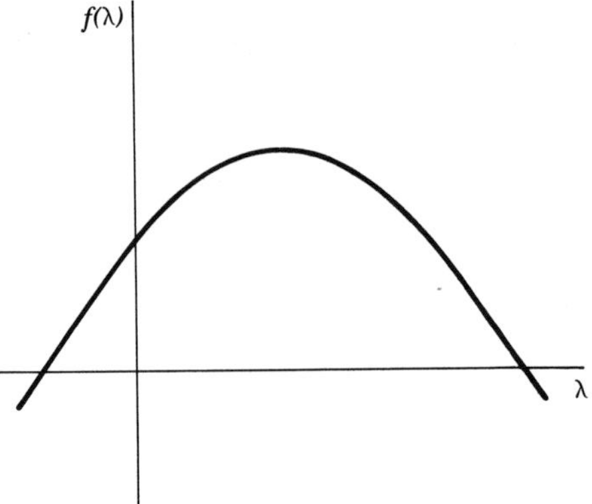

Figure 7.3 A concave function defined on the real line.

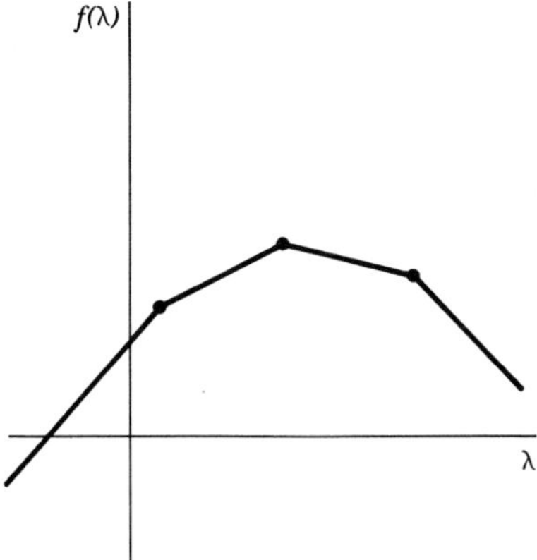

Figure 7.4 A piecewise linear concave function defined on the real line.

where t, the number of parameters is greater than or equal to 2. Let $\lambda = (\lambda_1, \dots, \lambda_t)$ be the parameter vector. If B is an optimum basis for this problem when the parameter vector is $\bar{\lambda}$, prove that the region of the parameter vector λ for which B remains an optimum basis is a convex polyhedral set in $\mathbf{R}^t$. Find this set.

This explains the difficulty of tackling parametric cost problems when there are two or more parameters. With only one parameter, the parameter space is the real line and convex sets in the real line are just intervals. Hence for a one parameter problem, the parameter space can be covered easily by passing from one interval (the characteristic interval of some optimal basis) to another. On the other hand, convex polyhedral sets in two or higher dimensional spaces can have several facets and, if the parametric cost problem involves two or more parameters, it becomes very complicated to cover the entire parameter space by convex polyhedral sets, which are the optimality regions of the various bases for the problem.

Problem

7.9 Just as in the parametric cost problem, show that the parametric right-hand side problem becomes very complicated when the right-hand side constant vector depends on two or more parameters.

Example: Parametric Cost Problem Using Canonical Tableaus

Original Tableau

x_1	x_2	x_3	x_4	x_5	x_6	x_7	$-z$	$-z^*$	b
1	0	0	−1	1	−1	2	0	0	1
0	1	0	0	1	−2	1	0	0	2
0	0	1	−3	2	1	−1	0	0	3
0	0	0	−6	12	30	−50	1	0	0
0	0	0	1	−2	−3	10	0	1	0

$x_j \geqq 0$ for all j, minimize $z + \lambda z^*$ for all λ

First solve the problem when λ is 0. (x_1, x_2, x_3) forms an initial feasible basic vector and the original tableau is actually the canonical tableau with respect to this basic vector. When λ is 0, the relative cost coefficient of x_4 is -6 and the updated column vector of x_4 is nonpositive. Hence, $z_\lambda(x)$ is unbounded below when λ is 0. It is unbounded below for all λ satisfying $-6 + \lambda < 0$, that is, $\lambda < 6$.

Now let $\lambda = 6$. It is verified that the basic vector (x_1, x_2, x_3) is optimal. The original tableau is an optimum canonical tableau when λ is 6. By using (7.5) and (7.6), it is verified that both the lower and the upper characteristic values corresponding to the basic vector (x_1, x_2, x_3) are equal to 6. Hence, the basic vector (x_1, x_2, x_3) is optimal when λ is in the singleton closed interval $[6, 6]$. To explore for optimum solutions for $\lambda > 6$, bring x_5 into the basic vector. This leads to

Basic variables	x_1	x_2	x_3	x_4	x_5	x_6	x_7	$-z$	$-z^*$	b
x_5	1	0	0	-1	1	-1	2	0	0	1
x_2	-1	1	0	(1)	0	-1	-1	0	0	1
x_3	-2	0	1	-1	0	3	-5	0	0	1
$-z$	-12	0	0	6	0	42	-74	1	0	-12
$-z^*$	2	0	0	-1	0	-5	14	0	1	2
$-\frac{\bar{c}_j}{\bar{c}_j^*}$ for $\bar{c}_j^* < 0$				6		$\frac{42}{5}$				Minimum $\bar{\lambda} = 6$
$-\frac{\bar{c}_j}{\bar{c}_j^*}$ for $\bar{c}_j^* > 0$	6						$\frac{74}{14}$			Maximum $\underline{\lambda} = 6$

This new BFS, $x = (0, 1, 1, 0, 1, 0, 0)^T$, is optimal again only for the single value $\lambda = 6$. The next entering variable is x_4.

Basic variable	x_1	x_2	x_3	x_4	x_5	x_6	x_7	$-z$	$-z^*$	b
x_5	0	1	0	0	1	-2	1	0	0	2
x_4	-1	1	0	1	0	-1	-1	0	0	1
x_3	-3	1	1	0	0	(2)	-6	0	0	2
$-z$	-6	-6	0	0	0	48	-68	1	0	-18
$-z^*$	1	1	0	0	0	-6	13	0	1	3
$-\frac{\bar{c}_j}{\bar{c}_j^*}$ for $\bar{c}_j^* < 0$						8				Minimum $\bar{\lambda} = 8$
$-\frac{\bar{c}_j}{\bar{c}_j^*}$ for $\bar{c}_j^* > 0$	6	6					$\frac{68}{13}$			Maximum $\underline{\lambda} = 6$

The feasible solution $x = (0, 0, 2, 1, 2, 0, 0,)^T$, $z = -3\lambda + 18$, is optimal for all $6 \leqq \lambda \leqq 8$. The next entering variable is x_6.

Basic variables	x_1	x_2	x_3	x_4	x_5	x_6	x_7	$-z$	$-z^*$	b
x_5	-3	2	1	0	1	0	-5	0	0	4
x_4	$-\frac{5}{2}$	$\frac{3}{2}$	$\frac{1}{2}$	1	0	0	-4	0	0	2
x_6	$-\frac{3}{2}$	$\frac{1}{2}$	$\frac{1}{2}$	0	0	1	-3	0	0	1
$-z$	66	-30	-24	0	0	0	76	1	0	-66
$-z^*$	-8	4	3	0	0	0	-5	0	1	9
$-\frac{\bar{c}_j}{\bar{c}_j^*}$ for $\bar{c}_j^* < 0$	$\frac{66}{8}$						$\frac{76}{5}$			Minimum $66/8 = \bar{\lambda}$
$-\frac{\bar{c}_j}{\bar{c}_j^*}$ for $\bar{c}_j^* > 0$		$\frac{30}{4}$	8							Maximum $= 8 = \underline{\lambda}$

Hence the BFS $x = (0, 0, 0, 2, 4, 1)^T$, $z = 66 - 9\lambda$ is optimal in the range $8 \leqq \lambda \leqq 33/4$. From the column vector of x_1 in this tableau, it is clear that if $\lambda > 33/4$, $z_\lambda(x)$ is unbounded below on the set of feasible solutions. This completes the solution of the problem for all real values of λ. The diagram of optimal values of $z_\lambda(x)$ is in Figure 7.5.

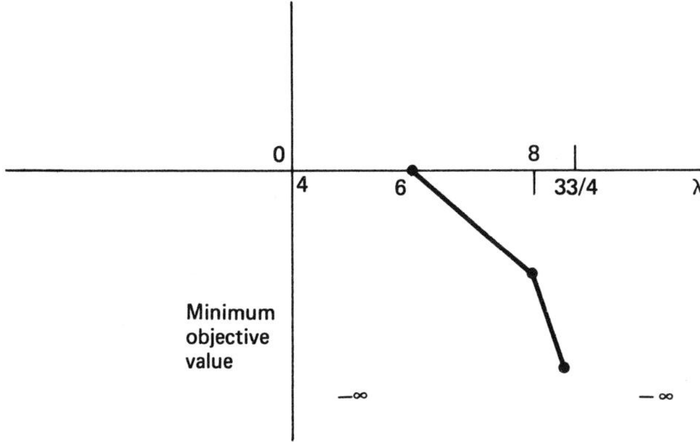

Figure 7.5

Example: Parametric Cost Problem Using Inverse Tableaus

Original Tableau

x_1	x_2	x_3	x_4	x_5	x_6	x_7	z	z^*	b
1	0	0	1	-2	1	1	0	0	1
0	1	0	1	1	-2	1	0	0	3
0	0	1	3	1	1	-2	0	0	7
0	0	0	4	0	1	3	1	0	0
0	0	0	-1	-2	2	1	0	1	0

$x_j \geqq 0$ for all j minimize $z + \lambda z^*$ for all λ.

(x_1, x_2, x_3) forms an optimal basic vector when $\lambda = 0$. Below is the inverse tableau corresponding to this basic vector.

Basic variables	Inverse tableau					Basic values	Pivot column (x_5)
x_1	1	0	0	0	0	1	-2
x_2	0	1	0	0	0	3	(1)
x_3	0	0	1	0	0	7	1
$-z$	0	0	0	1	0	0	0
$-z^*$	0	0	0	0	1	0	-2

The updated cost rows corresponding to this basis are:

			x_1	x_2	x_3	x_4	x_5	x_6	x_7	
$\bar{c}$			0	0	0	4	0	1	3	
$\bar{c}^*$			0	0	0	-1	-2	2	1	
$-\dfrac{\bar{c}_j}{\bar{c}_j^*}$	for	$\bar{c}_j^* < 0$				4	0			Minimum $\bar{\lambda} = 0$
$-\dfrac{\bar{c}_j}{\bar{c}_j^*}$	for	$\bar{c}_j^* > 0$						$-\dfrac{1}{2}$	-3	Maximum $= -\dfrac{1}{2} = \underline{\lambda}$

Hence, the basic vector (x_1, x_2, x_3) is optimal for all $-1/2 \leqq \lambda \leqq 0$. Hence, the optimum feasible solution in this interval is $x = (1, 3, 7, 0, 0, 0, 0)^{\mathrm{T}}$, $z = 0$. To investigate for $\lambda > 0$, bring x_5 into the basic vector. The updated column vector of x_5 is the pivot column, and this is entered on the right-hand side of the first inverse tableau.

Basic variables	Inverse tableau					Basic values	Pivot column (x_6)
x_1	1	2	0	0	0	7	-3
x_5	0	1	0	0	0	3	-2
x_3	0	-1	1	0	0	4	(3)
$-z$	0	0	0	1	0	0	1
$-z^*$	0	2	0	0	1	6	-2

Here are the updated cost vectors.

			x_1	x_2	x_3	x_4	x_5	x_6	x_7	
$\bar{c}$			0	0	0	4	0	1	3	
$\bar{c}^*$			0	2	0	1	0	-2	3	
$-\dfrac{\bar{c}_j}{\bar{c}_j^*}$	for	$\bar{c}_j^* < 0$						$\dfrac{1}{2}$		Minimum $= \bar{\lambda} = \dfrac{1}{2}$
$-\dfrac{\bar{c}_j}{\bar{c}_j^*}$	for	$\bar{c}_j^* > 0$		0		-4			-1	Maximum $= \underline{\lambda} = 0$

The BFS, $x = (7, 0, 4, 0, 3, 0, 0)^T$ and $z = -6\lambda$, is therefore optimal for all $0 \leqq \lambda \leqq 1/2$. The next variable to enter the basic vector is x_6, whose updated column vector is already entered on the right side of the inverse tableau.

Basic variables	Inverse tableau					Basic values
x_1	1	1	1	0	0	11
x_5	0	$\frac{1}{3}$	$\frac{2}{3}$	0	0	$\frac{17}{3}$
x_6	0	$-\frac{1}{3}$	$\frac{1}{3}$	0	0	$\frac{4}{3}$
$-z$	0	$\frac{1}{3}$	$-\frac{1}{3}$	1	0	$-\frac{4}{3}$
$-z^*$	0	$\frac{4}{3}$	$\frac{2}{3}$	0	1	$\frac{26}{3}$

The updated cost rows are:

			x_1	x_2	x_3	x_4	x_5	x_6	x_7	
$\bar{c}$			0	$\frac{1}{3}$	$-\frac{1}{3}$	$\frac{10}{3}$	0	0	4	
$\bar{c}^*$			0	$\frac{4}{3}$	$\frac{2}{3}$	$\frac{7}{3}$	0	0	1	
$-\frac{\bar{c}_j}{\bar{c}_j^*}$	for	$\bar{c}_j^* < 0$								Minimum $= \bar{\lambda} = +\infty$
$-\frac{\bar{c}_j}{\bar{c}_j^*}$	for	$\bar{c}_j^* > 0$		$-\frac{1}{4}$	$\frac{1}{2}$	$-\frac{10}{7}$			-4	Maximum $= \underline{\lambda} = \frac{1}{2}$

The BFS, $x = (11, 0, 0, 0, 17/3, 4/3, 0)^T$ and $z = 4/3 - 26\lambda/3$, is therefore optimal for all $1/2 \leqq \lambda$. To find optimum solutions for $\lambda < -1/2$ return to the tableau corresponding to the basic vector (x_1, x_2, x_3). The pivot column is the updated column of x_6. This leads to the inverse tableau:

Basic variables	Inverse tableau					Basic values	Pivot column (x_5)
x_6	1	0	0	0	0	1	-2
x_2	2	1	0	0	0	5	-3
x_3	-1	0	1	0	0	6	③
$-z$	-1	0	0	1	0	-1	2
$-z^*$	-2	0	0	0	1	-2	2

The updated cost rows are:

			x_1	x_2	x_3	x_4	x_5	x_6	x_7	
$\bar{c}$			-1	0	0	3	2	0	2	
$\bar{c}^*$			-2	0	0	-3	2	0	-1	
$-\dfrac{\bar{c}_j}{\bar{c}_j^*}$	for	$\bar{c}_j < 0$	$-\dfrac{1}{2}$			1			2	Minimum $= \bar{\lambda} = -\dfrac{1}{2}$
$-\dfrac{\bar{c}_j}{\bar{c}_j^*}$	for	$\bar{c}_j^* > 0$					-1			Maximum $= \underline{\lambda} = -1$

The BFS, $x = (0, 5, 6, 0, 0, 1, 0)^T$ and $z = 1 + 2\lambda$, is therefore optimal for $-1 \leqq \lambda \leqq -1/2$. The next pivot column is the updated column vector of x_5.

Basic variables	Inverse tableau					Basic values
x_6	$\dfrac{1}{3}$	0	$\dfrac{2}{3}$	0	0	5
x_2	1	1	1	0	0	11
x_5	$-\dfrac{1}{3}$	0	$\dfrac{1}{3}$	0	0	2
$-z$	$-\dfrac{1}{3}$	0	$-\dfrac{2}{3}$	1	0	-5
$-z^*$	$-\dfrac{4}{3}$	0	$-\dfrac{2}{3}$	0	1	-6

The updated cost rows are:

			x_1	x_2	x_3	x_4	x_5	x_6	x_7	
$\bar{c}$			$-\dfrac{1}{3}$	0	$-\dfrac{2}{3}$	$-\dfrac{5}{3}$	0	0	4	
$\bar{c}^*$			$-\dfrac{4}{3}$	0	$-\dfrac{2}{3}$	$-\dfrac{13}{3}$	0	0	1	
$-\dfrac{\bar{c}_j}{\bar{c}_j^*}$	for	$\bar{c}_j^* < 0$	$-\dfrac{1}{4}$		-1	$\dfrac{5}{13}$				Minimum $= \bar{\lambda} = -1$
$-\dfrac{\bar{c}_j}{\bar{c}_j^*}$	for	$\bar{c}_j^* > 0$							-4	Maximum $= \underline{\lambda} = -4$

The BFS, $x = (0, 11, 0, 0, 2, 5, 0)$ and $z = 5 + 6\lambda$, is optimal for all $-4 \leqq \lambda \leqq -1$. The next pivot column is the updated column vector of x_7, which is $(-1, 0, -1, 4, 1)^T$. Since $\bar{A}_{.7} \leqq 0$, $z_\lambda(x)$ is unbounded below for all $\lambda < -4$. The optimum objective value as a function of λ is drawn in Figure 7.6.

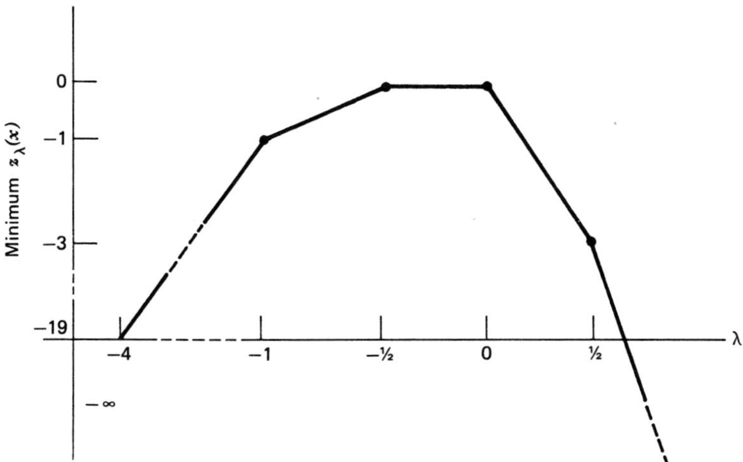

Figure 7.6

Example: Parametric Right-Hand Side Problem

x_1	x_2	x_3	x_4	x_5	x_6	x_7	$-z$	b	b^*
1	0	0	2	1	0	−1	0	−1	3
0	1	0	1	0	1	1	0	−2	1
0	0	1	0	1	1	2	0	−3	2
1	3	2	3	−5	1	3	1	0	0

$x_j \geqq 0$, $\quad z$ to be minimized

To find a feasible basis, try to solve the Phase I problem for some value of λ, say, 0.

Original Tableau for the Phase I Problem

x_1	x_2	x_3	x_4	x_5	x_6	x_7	x_8	x_9	x_{10}	$-z$	$-w$	b	b^*
−1	0	0	−2	−1	0	1	1	0	0	0	0	1	3
0	−1	0	−1	0	−1	−1	0	1	0	0	0	2	−1
0	0	−1	0	−1	−1	−2	0	0	1	0	0	3	−2
1	3	2	3	−5	1	3	0	0	0	1	0	0	0
0	0	0	0	0	0	0	1	1	1	0	1	0	0

x_8, x_9, x_{10} are the artificial variables and w is the Phase I objective function.

Phase I Initial Tableau

Basic variables	Inverse tableau					$\bar{b}$	$\bar{b}^*$	Ratios $-\bar{b}_i/\bar{b}_i^*$ for $\bar{b}_i^* < 0$	Ratios $-\bar{b}_i/\bar{b}_i^*$ for $\bar{b}_i^* > 0$	Pivot column (x_1)
x_8	1	0	0	0	0	1	−3	$\frac{1}{3}$		(−1)
x_9	0	1	0	0	0	2	−1	2		0
x_{10}	0	0	1	0	0	3	−2	$\frac{3}{2}$		0
$-z$	0	0	0	1	0	0	0			1
$-w$	−1	−1	−1	0	1	−6	6			1
								Minimum $= \bar{\lambda} = \frac{1}{3}$	Maximum $= \underline{\lambda} = -\infty$	

This is a feasible tableau for the Phase I problem when $\lambda = 0$. Computing the Phase I relative cost coefficients, we have, $\bar{d} = (1, 1, 1, 3, 2, 2, 2)$. Since $\bar{d} > 0$, this is an optimum basis for the Phase I problem. The value of $w = 6 - 6\lambda = 6$ when $\lambda = 0$. Hence, the original problem is infeasible when $\lambda = 0$. The basic vector (x_8, x_9, x_{10}) is feasible to the Phase I problem for all $-\infty \leqq \lambda \leqq 1/3$ and in this range, the optimum Phase I objective value is $6 - 6\lambda > 0$. Hence, the original problem is infeasible for all $\lambda \leqq 1/3$. Now try to find optimum solutions for the Phase I problem for $\lambda > 1/3$. For this, make a dual simplex pivot in the first row. The pivot row is generated using the inverse tableau. This leads to the following updated rows.

	x_1	x_2	x_3	x_4	x_5	x_6	x_7
$\bar{A}_{1.}$	−1	0	0	−2	−1	0	1
$\bar{d}$	1	1	1	3	2	2	2
$-\frac{\bar{d}_j}{\bar{a}_{1j}}$ for $\bar{a}_{1j} < 0$	1			$\frac{3}{2}$	2		

Hence, the entering variable is x_1. Its updated column vector is the pivot column and this is entered on the right-hand side of the inverse tableau. Here is the next inverse tableau.

Basic variables	Phase I Inverse tableau					$\bar{b}$	$\bar{b}^*$	Ratio $-\bar{b}_i/\bar{b}_i^*$ for $\bar{b}_i^* < 0$	Ratio $-\bar{b}_i/\bar{b}_i^*$ for $\bar{b}_i^* > 0$	Pivot column x_5
x_1	−1	0	0	0	0	−1	3		$\frac{1}{3}$	1
x_9	0	1	0	0	0	2	−1	2		0
x_{10}	0	0	1	0	0	3	−2	$\frac{3}{2}$		(−1)
$-z$	1	0	0	1	0	1	−3			−6
$-w$	0	−1	−1	0	1	−5	3			1
								Minimum $= \bar{\lambda} = \frac{3}{2}$	Maximum $= \underline{\lambda} = \frac{1}{3}$	

The new basic vector (x_1, x_9, x_{10}) is optimal for all $1/3 \leqq \lambda \leqq 3/2$ and in this region, minimum $w = 5 - 3\lambda$. Hence, for all λ in this region, the original problem is still infeasible. To obtain optimum solutions to the Phase I problem when $\lambda > 3/2$, a Phase I dual simplex pivot should be made on the third row. The pivot row and the Phase I relative cost coefficients are

	x_1	x_2	x_3	x_4	x_5	x_6	x_7
$\bar{A}_{3.}$	0	0	-1	0	-1	-1	-2
$\bar{d}$	0	1	1	1	1	2	3
$-\frac{\bar{d}_j}{\bar{a}_{3j}}$ for $\bar{a}_{3j} < 0$			1		1	2	$\frac{3}{2}$

The entering variable can be either x_3 or x_5, and we pick x_5. Its updated column is the pivot column.

Basic variables	Phase I Inverse tableau					$\bar{b}$	$\bar{b}^*$	Ratios $-\frac{\bar{b}_i}{\bar{b}_i^*}$: $\bar{b}_i^* < 0$	Ratios $-\frac{\bar{b}_i}{\bar{b}_i^*}$: $\bar{b}_i^* > 0$	Pivot column x_4
x_1	-1	0	1	0	0	2	1		-2	2
x_9	0	1	0	0	0	2	-1	2		$\textcircled{-1}$
x_5	0	0	-1	0	0	-3	2		$\frac{3}{2}$	0
$-z$	1	0	-6	1	0	-17	9			1
$-w$	0	-1	0	0	1	-2	1			1
								Minimum $= \bar{\lambda} = 2$	Maximum $= \underline{\lambda} = \frac{3}{2}$	

The basic vector (x_1, x_9, x_5) is optimal to the Phase I problem when $3/2 \leqq \lambda \leqq 2$, and in this range the minimum value of w is $2 - \lambda$. Thus, the Phase I objective attains the value zero for the first time when $\lambda = 2$. Hence, when $\lambda = 2$, a feasible solution of the original problem is $x_1 = 2 + 2$, $x_5 = -3 + 4$; all other $x_j = 0$. The Phase II relative cost vector corresponding to this basis is $\bar{c} = (0, 3, 8, 1, 0, 7, 16) \geqq 0$. Hence, the above feasible solution is optimal to the original problem when $\lambda = 2$. To get optimal solutions for $\lambda > 2$, make a Phase II dual simplex pivot in the second row.

	x_1	x_2	x_3	x_4	x_5	x_6	x_7
$\bar{A}_{2.}$	0	-1	0	-1	0	-1	-1
$\bar{c}$	0	3	8	1	0	7	16
$-\frac{\bar{c}_j}{\bar{a}_{2j}}$ for $\bar{a}_{2j} < 0$		3		1		7	16

The pivot column is the updated column of x_4, and x_4 replaces the artificial variable x_9, which is in the present basic vector with zero value when $\lambda = 2$. With this pivot all the artificial variables have left the basis, and we drop the w column and the Phase I relative cost row and go to Phase II.

Basic variables	Phase II Inverse tableau				$\bar{b}$	$\bar{b}^*$	Ratio $-\frac{\bar{b}_i}{\bar{b}_i^*}$: $\bar{b}_i^* < 0$	Ratio $-\frac{\bar{b}_i}{\bar{b}_i^*}$: $\bar{b}_i^* > 0$	Pivot column x_2
x_1	−1	2	1	0	6	−1	6		(−2)
x_4	0	−1	0	0	−2	1		2	1
x_5	0	0	−1	0	−3	2		$\frac{3}{2}$	0
$-z$	1	1	−6	1	−15	8			2
							Minimum = 6 = $\bar{\lambda}$	Maximum = $\underline{\lambda}$ = 2	

Thus $x_1 = 6 - \lambda$, $x_4 = -2 + \lambda$, $x_5 = -3 + 2\lambda$, all other $x_j = 0$, $z = 15 - 8\lambda$ is an optimum feasible solution for all $2 \leqq \lambda \leqq 6$. To get optimum solutions for $\lambda > 6$ make a dual simplex pivot in the first row.

	x_1	x_2	x_3	x_4	x_5	x_6	x_7
$\bar{A}_{1.}$	1	−2	−1	0	0	−3	−5
$\bar{c}$	0	2	8	0	0	6	15
$-\frac{\bar{c}_j}{\bar{a}_{1j}}$ for $\bar{a}_{1j} < 0$		1	$\frac{8}{1}$			$\frac{6}{3}$	$\frac{15}{5}$

The pivot column is the updated column of x_2.

Basic variables	Phase II inverse tableau				$\bar{b}$	$\bar{b}^*$	Ratios $-\bar{b}_i/\bar{b}_i^*$: $\bar{b}_i^* < 0$	Ratios $-\bar{b}_i/\bar{b}_i^*$: $\bar{b}_i^* > 0$
x_2	$\frac{1}{2}$	−1	$-\frac{1}{2}$	0	−3	$\frac{1}{2}$		6
x_4	$-\frac{1}{2}$	0	$\frac{1}{2}$	0	1	$\frac{1}{2}$		−2
x_5	0	2	−1	0	−3	2		$\frac{3}{2}$
$-z$	0	3	−5	1	−9	7		
							Minimum = $\bar{\lambda}$ = ∞	Maximum = $\underline{\lambda}$ = 6

Hence, the feasible solution, $x_2 = -3 + \lambda/2$, $x_4 = 1 + \lambda/2$, $x_5 = -3 + 2\lambda$, all other $x_j = 0$, $z = 9 - 7\lambda$, is optimal for all $6 \leqq \lambda$. See Figure 7.7.

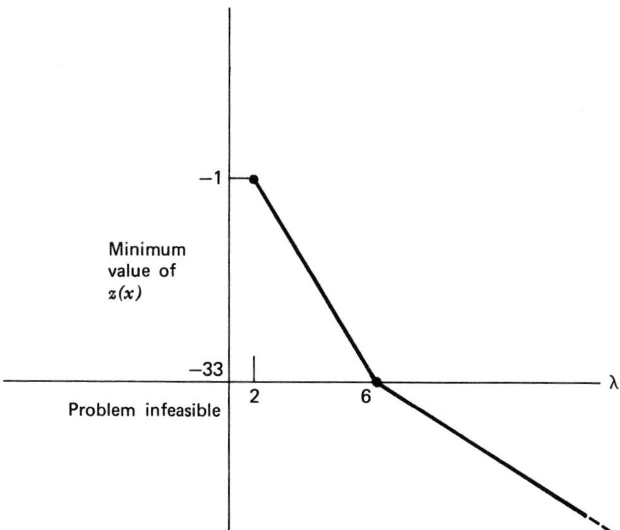

Figure 7.7 Optimum objective value as a function of the parameter

Problems

7.10 COMBINED PARAMETRIC COST AND RIGHT-HAND SIDE PROBLEMS. In practical applications, we encounter parametric LPs of the form:

$$\begin{aligned} \text{Minimize} \quad & (c + \lambda c^*)x \\ \text{Subject to} \quad & Ax = b + \lambda b^* \\ & x \geqq 0 \end{aligned}$$

where A, b, b^*, c, c^* are all given. Combine the ideas in sections 7.2, 7.3, 7.5, 7.6, and 7.7 and describe an algorithm for solving this problem for all real values of λ. Explain clearly when primal simplex or dual simplex pivots have to be made during the algorithm.

7.11 For the combined parametric problem in Problem 7.10 show that the optimum objective value may not be piecewise linear in λ and possibly neither convex nor concave. Construct illustrative examples.

7.12 Consider the LP:

$$\begin{aligned} \text{Minimize} \quad & cx \\ \text{Subject to} \quad & Ax = b \\ & x \geqq 0 \end{aligned}$$

Let B be an arbitrary basis to this problem. B may turn out to be neither primal nor dual feasible. However, it can be made both primal and dual feasible to a modified problem obtained by adding a parameter θ to all the negative updated right-hand side values and all the negative relative cost coefficients, and making the value of this parameter θ sufficiently large, say $\theta = \theta_1$. Discuss an application of the algorithm developed in Problem 7.10 to solve this LP by starting with $\theta = \theta_1$ and obtaining optimal solutions of the modified problem as θ decreases from θ_1 to 0. Solve the following problem using this algorithm.

x_1	x_2	x_3	x_4	x_5	x_6	x_7	$-z$	b
1	0	0	−1	1	−1	1	0	1
0	1	0	−2	−1	1	1	0	−2
0	0	1	1	−2	−3	2	0	−4
0	0	0	−4	6	3	−7	1	0 (z min)

$x_j \geqq 0 \quad$ for all j

PROBLEMS

7.13 Solve the following parametric cost problems for all real values of the parameter.

(a)
$$\begin{array}{llllllll} \text{Minimize} & & & (-4+4\lambda)x_4 & +(3+2\lambda)x_5 & +(1-2\lambda)x_6 & & \\ \text{Subject to} & x_1 & & -2x_4 & +x_5 & +x_6 & = & 5 \\ & & x_2 & -2x_4 & +x_5 & +2x_6 & = & 4 \\ & & x_3 & & +x_5 & -x_6 & = & 6 \\ & & & x_j \geqq 0 & \text{for all } j & & & \end{array}$$

(b)
$$\begin{array}{ll} \text{Minimize} & \lambda x_1 - x_2 \\ \text{Subject to} & 3x_1 - x_2 \leqq 5 \\ & 2x_1 + x_2 \leqq 3 \\ & x_1, x_2 \text{ unrestricted} \end{array}$$

(c)
$$\begin{array}{ll} \text{Minimize} & 2\lambda x_1 + (1-\lambda)x_2 - 3x_3 + \lambda x_4 + 2x_5 - 3\lambda x_6 \\ \text{Subject to} & x_1 + 3x_2 - x_3 + 2x_5 = 7 \\ & -2x_2 + 4x_3 + x_4 = 12 \\ & -4x_2 + 3x_3 + 8x_5 + x_6 = 10 \\ & x_j \geqq 0 \quad \text{for all } j \end{array}$$

(d) Minimize

$$\lambda x_1 + x_2 + x_3 + (9\lambda - 41)x_4 + (15-\lambda)x_5 + (20-4\lambda)x_6 + (9-\lambda)x_7 + (\lambda+2)x_8$$

Subject to

$$\begin{array}{l} x_1 + 2x_2 - 3x_4 + 3x_5 + 3x_6 + 5x_7 + 3x_8 = 22 \\ x_2 + x_3 - x_4 + 3x_5 + 3x_6 + x_7 + 2x_8 = 12 \\ 2x_2 + x_3 - 2x_4 + 4x_5 + 5x_6 + 3x_7 + 3x_8 = 21 \\ x_j \geqq 0 \quad \text{for all } j \end{array}$$

(e) Minimize

$$\lambda x_1 + 2\lambda x_2 + x_3 + (\lambda - 9)x_4 + (5\lambda + 6)x_5 + (9 - 4\lambda)x_6 + (4\lambda - 3)x_7$$

Subject to

$$\begin{array}{l} x_1 + x_3 + x_4 + 2x_5 - 2x_6 + 3x_7 = 5 \\ x_3 - x_4 + x_5 - x_6 + 2x_7 = 3 \\ x_1 + x_2 + x_4 + 3x_5 - 2x_6 + 2x_7 = 3 \\ x_j \geqq 0 \quad \text{for all } j \end{array}$$

(f) Minimize

$$(1-\lambda)(x_1 + x_2 + x_3) + (6 - 4\lambda)x_4 + (5 - 3\lambda)x_5 + (4 + \lambda)x_6 + (2 + 5\lambda)x_7$$

Subject to

$$\begin{array}{l} x_1 + x_2 + x_3 + 4x_4 + 2x_5 - 3x_7 = 15 \\ -x_1 - x_2 - 3x_4 - x_5 - 2x_6 + 2x_7 = -12 \\ x_2 + x_3 + 2x_4 + 3x_5 - x_6 - 2x_7 = 13 \\ x_j \geqq 0 \quad \text{for all } j \end{array}$$

7.14 Consider the parametric cost LP:

$$\begin{array}{ll} \text{Minimize} & 3x_1 + (4 + 8\lambda)x_2 + 7x_3 - x_4 + (1 - 3\lambda)x_5 \\ \text{Subject to} & 5x_1 - 4x_2 + 14x_3 - 2x_4 + x_5 = 20 \\ & x_1 - x_2 + 5x_3 - x_4 + x_5 = 8 \\ & x_j \geqq 0 \quad \text{for all } j \end{array}$$

For what range of values of λ is the basic vector (x_2, x_3) optimal for this problem? Determine optimum feasible solutions for all $\lambda \geqq 1$.

7.15 Consider the parametric cost LP:

$$\begin{aligned} \text{Minimize} \quad & (c + \lambda c^*)x \\ \text{Subject to} \quad & Ax = b \\ & x \geqq 0 \end{aligned}$$

The objective function is known to be unbounded below on the set of feasible solution of this problem when λ is λ_0. Prove that the objective function must be unbounded below on the set of feasible solutions of this problem either in the interval $\lambda \geqq \lambda_0$ or in the interval $\lambda \leqq \lambda_0$ or both.

7.16 Solve

$$\begin{aligned} \text{Minimize} \quad & x_1 + \lambda x_2 + \mu x_3 + x_4 + x_5 + x_6 + x_7 + \lambda x_8 + \mu x_9 \\ \text{Subject to} \quad & x_1 - x_4 - 2x_6 + 4x_7 + 2x_8 + x_9 = 4 \\ & x_2 + 2x_4 - 2x_5 + x_6 + 2x_7 + 4x_8 = 2 \\ & x_3 + x_5 + x_6 + x_7 + 2x_9 = 1 \\ & x_j \geqq 0 \quad \text{for all } j \end{aligned}$$

for all real values of the parameters λ, μ.

7.17 Solve the following parametric right-hand side problems for all real values of the parameter:

(a)

$$\begin{aligned} \text{Minimize} \quad & x_1 + 2x_2 - x_3 - 4x_4 - x_5 + 2x_6 + x_7 \\ \text{Subject to} \quad & x_1 + x_2 + x_3 + x_5 = 9 - 2\lambda \\ & x_2 + x_3 + x_4 - x_5 + 2x_6 - x_7 = 4 - \lambda \\ & x_1 + x_2 + 2x_3 + 2x_4 + 2x_5 + x_6 + x_7 = 5 - \lambda \\ & x_j \geqq 0 \quad \text{for all } j \end{aligned}$$

(b)

$$\begin{aligned} \text{Minimize} \quad & 2x_1 + 3x_2 + 4x_3 + 5x_4 \\ \text{Subject to} \quad & 2x_1 - x_2 + x_3 - x_4 \leqq 2 - \lambda \\ & x_1 + 2x_2 + x_3 - x_4 \leqq 10 - 2\lambda \\ & x_1 + x_2 - 2x_3 - x_4 \leqq 3 + \lambda \\ & x_j \geqq 0 \quad \text{for all } j \end{aligned}$$

7.18 Solve the following LP

$$\begin{aligned} \text{Minimize} \quad & x_1 + x_2 + 7x_3 + 3x_4 + x_5 + 2x_6 + x_7 \\ \text{Subject to} \quad & x_1 + 2x_2 - x_3 - x_4 + x_5 + 2x_6 + x_7 = 16 \\ & x_2 - 3x_4 - x_5 + 3x_6 - x_7 = 2 \\ & -x_1 - 3x_3 + 3x_4 + x_5 - x_7 = -4 \\ & x_j \geqq 0 \quad \text{for all } j \end{aligned}$$

In this problem, x_7, is the amount (in tons) of some product manufactured. It is rumored that the federal government will legislate that the company should manufacture some specified amount (which is unknown) of this product in the national interest. To prepare for this possibility, x_7 must be considered as a parameter and the problem solved for all nonnegative values of x_7. Work out the solution.

8 sensitivity analysis

8.1 INTRODUCTION

Suppose we take a practical problem, formulate it as the LP

$$\begin{aligned} \text{Minimize} \quad z(x) &= cx \\ \text{Subject to} \quad Ax &= b \\ x &\geqq 0 \end{aligned} \tag{8.1}$$

where A is a matrix of order $m \times n$ and rank m, and obtain an optimum feasible solution for it. In most real-life problems, the coefficients in A, c, and b are estimated from practical considerations. After the final optimal feasible solution has been obtained we may discover that some of the entries in b, c, or A have to be changed or that extra constraints or variables have to be introduced into the model. Solving the modified problem from scratch will be wasteful. *Sensitivity analysis* (also called *postoptimality analysis*) deals with the problem of obtaining an optimum feasible solution of the modified problem starting with the optimum feasible solution of the old problem.

We will consider the problem of introducing only one change at a time (e.g., how to get the new optimum feasible solution if the value of only one c_j has to be changed or if only one new constraint has to be added, etc.). If several changes have to be made, make them one at a time, or extend the methods discussed here in an obvious manner to take care of several simultaneous changes.

In Chapter 7 we have discussed some types of postoptimality analysis where the right-hand side vector or the cost coefficient vector vary linearly as a function of a parameter as it ranges over the real line. Here we discuss various other types of postoptimality analyses. Some of the postoptimality analyses discussed hcrc, for example, changes in cost coefficients or the right-hand side constants, can be viewed as specializations of the parametric analysis applied to those cases.

Let $\mathbf{K}$ denote the set of feasible solutions of (8.1). Suppose the optimal basis obtained for (8.1) is $\tilde{B}$, associated with the basic vector $x_{\tilde{B}} = (x_1, \ldots, x_m)$. $c_{\tilde{B}} = (c_1, \ldots, c_m)$. Let $\tilde{\pi} = c_{\tilde{B}}\tilde{B}^{-1}$ and let $\tilde{x}$ be the BFS of (8.1) corresponding to the basis $\tilde{B}$. For illustration we will use the following problem

Tableau 8.1. Original Problem

x_1	x_2	x_3	x_4	x_5	x_6	$-z$	b
1	2	0	1	0	−6	0	11
0	1	1	3	−2	−1	0	6
1	2	1	3	−1	−5	0	13
3	2	−3	−6	10	−5	1	0

$x_j \geqq 0$ for all j, z to be minimized

$x_{\tilde{B}} = (x_1, x_2, x_3)$ is an optimum basic vector for this problem.

Tableau 8.2. Optimum Inverse Tableau for the Problem in Tableau 8.1

Basic variables	Inverse tableau				Basic values
x_1	−1	−2	2	0	3
x_2	1	1	−1	0	4
x_3	−1	0	1	0	2
$-z$	−2	4	−1	1	−11

Therefore, $\tilde{x} = (3, 4, 2, 0, 0, 0)^{\mathrm{T}}$ is an optimum feasible solution of this problem and the minimum objective value is $z(\tilde{x}) = 11$.

8.2 INTRODUCING A NEW ACTIVITY

Suppose a new variable, called x_{n+1}, has to be introduced into the model, (8.1). Let $A_{.n+1}$, c_{n+1} be the input-output coefficient vector and the cost coefficient, respectively, of this new variable. Let $X^{\mathrm{T}} = (x^{\mathrm{T}}, x_{n+1})$. The augmented problem is

$$\begin{aligned} \text{Minimize} \quad & Z(X) = cx + c_{n+1}x_{n+1} \\ \text{Subject to} \quad & (A \;\vdots\; A_{.n+1})X = b \\ & X \geqq 0 \end{aligned} \tag{8.2}$$

$\tilde{B}$ remains a feasible basis to the augmented problem, and the BFS of (8.2) corresponding to it is $\tilde{X}^{\mathrm{T}} = (\tilde{x}^{\mathrm{T}}, 0)$. From the optimality criterion $\tilde{X}$ remains an optimum feasible solution of (8.2) if the relative cost coefficient of the new variable, x_{n+1}, with respect to the basis $\tilde{B}$, is nonnegative; that is, if $\bar{c}_{n+1} = c_{n+1} - \tilde{\pi}A_{.n+1} \geqq 0$. On the other hand, if $\bar{c}_{n+1} < 0$, solve (8.2) by using the inverse tableau corresponding to the basis $\tilde{B}$ as an initial tableau. Bring x_{n+1} into the basic vector and complete the solution of (8.2) according to the revised simplex algorithm.

It often happens that one pivot (that of bringing the new variable x_{n+1} into the basic vector $x_{\tilde{B}}$) is all that is necessary to solve (8.2). However, there is no theoretical guarantee that this will be the general case. In some problems, it may be necessary to perform several pivots before the algorithm reaches a terminal basis for (8.2).

Problems

8.1 Prove that the optimum objective value in (8.2) is less than or equal to the optimum objective value in (8.1). Construct a numerical example to show that the objective function in (8.2) may be unbounded below even though (8.1) has an optimum feasible solution.

8.2 How is the set of feasible solutions of the augmented problem related to **K**?

Example

Consider the LP in Tableau 8.1. Suppose a new variable x_7 corresponding to the column vector $A_{.7} = (1, 2, -3)^{\mathrm{T}}$ and cost coefficient $c_7 = -7$ has to be introduced. The relative cost coefficient of x_7 with respect to the basis in Tableau 8.2 is

$$\bar{c}_7 = c_7 + (-\tilde{\pi})A_{.7} = -7 + (-2, 4, -1)(1, 2, -3)^{\mathrm{T}} = 2$$

Hence, the previous optimum solution with $x_7 = 0$ is still optimal. The optimum feasible solution of the new problem is $\tilde{X} = (3, 4, 2, 0, 0, 0, 0)^{\mathrm{T}}$ with the optimal objective value $=$ $\tilde{Z} = 11$. Thus, if x_7 is the level of some new activity that has become available, we conclude that it is optimal to perform the new activity at zero level.

Example

Consider the LP in Tableau 8.1 again. Suppose we have to include a new variable x_7 with $A_{.7} = (3, -1, 1)^{\mathrm{T}}$ and $c_7 = 4$. The relative cost coefficient of x_7 with respect to the basis in Tableau 8.2 is $\bar{c}_7 = 4 + (-2, 4, -1)(3, -1, 1)^{\mathrm{T}} = -7 < 0$. We conclude that it is optimal to include the new activity in the basic vector. The updated column vector of x_7 with respect to the basis in Tableau 8.2 is

$$\begin{pmatrix} \bar{A}_{.7} \\ \cdots \\ \bar{c}_7 \end{pmatrix} = \begin{pmatrix} -1 & -2 & 2 & 0 \\ 1 & 1 & -1 & 0 \\ -1 & 0 & 1 & 0 \\ -2 & 4 & -1 & 1 \end{pmatrix} \begin{pmatrix} 3 \\ -1 \\ 1 \\ 4 \end{pmatrix} = \begin{pmatrix} 1 \\ 1 \\ -2 \\ -7 \end{pmatrix}$$

This is the pivot column. The minimum ratio test indicates that x_1 drops out of the basic vector when x_7 enters.

Basic variables	Inverse tableau				Basic values
x_7	-1	-2	2	0	3
x_2	2	3	-3	0	1
x_3	-3	-4	5	0	8
$-z$	-9	-10	13	1	10

Since the relative cost coefficient of x_6 here is -6, this basis is again not optimal to the augmented problem. Bring x_6 into the basic vector and continue the algorithm until a terminal basis for the augmented problem is obtained.

Problem

8.3 Complete the solution of this numerical problem.

8.3 INTRODUCING AN ADDITIONAL INEQUALITY CONSTRAINT

Consider the LP (8.1) and the optimum basis $\tilde{B}$ for it again. Suppose the additional constraint

$$A_{m+1.}x \leqq b_{m+1} \tag{8.3}$$

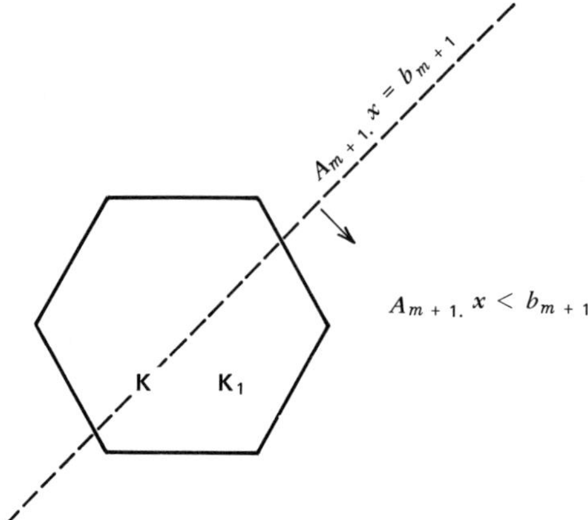

Figure 8.1 Introducing a new inequality constraint.

has to be introduced. Including this additional inequality constraint may make some of the points of **K** infeasible. Hence, it makes the set of feasible solutions smaller. Let $\mathbf{K}_1$ denote the set of feasible solutions of the augmented problem. $\mathbf{K}_1 \subset \mathbf{K}$. See Figure 8.1.

Hence, if the original problem has an optimum feasible solution, the augmented problem is either infeasible or it has an optimum feasible solution. Also $z(\tilde{x}) =$ minimum $\{z(x)\colon x \in \mathbf{K}\} \leqq$ minimum $\{z(x)\colon x \in \mathbf{K}_1\}$. Hence, if $\tilde{x} \in \mathbf{K}_1$, that is, $\tilde{x}$ satisfies (8.3), $\tilde{x}$ remains optimal to the augmented problem. On the other hand, if $\tilde{x}$ does not satisfy (8.3), then

$$\tilde{x}_{n+1} = -A_{m+1.}\tilde{x} + b_{m+1} < 0 \tag{8.4}$$

where x_{n+1} is the slack variable corresponding to (8.3). The augmented problem is

$$\begin{array}{ll} \text{Minimize} & Z(X) = \sum_{j=1}^{n} c_j x_j + 0x_{n+1} \\ \text{Subject to} & \begin{pmatrix} A & \vdots & 0 \\ \cdots & \cdots & \cdots \\ A_{m+1.} & \vdots & 1 \end{pmatrix} \begin{pmatrix} x \\ \cdots \\ x_{n+1} \end{pmatrix} = \begin{pmatrix} b \\ \cdots \\ b_{m+1} \end{pmatrix} \\ & x_j \geqq 0 \qquad \text{for all } j \end{array} \tag{8.5}$$

Let $\boldsymbol{A}$ be the input-output coefficient matrix of order $(m + 1) \times (n + 1)$ in (8.5), and $X^{\mathrm{T}} = (x^{\mathrm{T}}, x_{n+1})$. By including x_{n+1} as an additional basic variable, we can enlarge $\tilde{B}$ into a basis for (8.5), denoted by $\tilde{\boldsymbol{B}}$. $X_{\tilde{\boldsymbol{B}}} = (x_1, \ldots, x_m, x_{n+1})^{\mathrm{T}}$. The basic solution of (8.5) corresponding to the basis $\tilde{\boldsymbol{B}}$ is $\tilde{X}^{\mathrm{T}} = (\tilde{x}^{\mathrm{T}}, \tilde{x}_{n+1})$. By (8.4), $\tilde{\boldsymbol{B}}$ is an infeasible basis for (8.5). Since $c_{n+1} = 0$, the relative cost coefficient of x_j with respect to the basis $\tilde{\boldsymbol{B}}$ in (8.5) is equal to the relative cost coefficient of x_j with respect to the basis $\tilde{\boldsymbol{B}}$ in (8.1), which is $\bar{c}_j \geqq 0$ for all $j = 1$ to n. Thus, $\tilde{\boldsymbol{B}}$ is a dual feasible but primal infeasible basis for (8.5). Using $\tilde{\boldsymbol{B}}$, as an initial basis, we can solve (8.5) by using the dual simplex algorithm. We now discuss how to obtain the inverse tableau corresponding to the basis $\tilde{\boldsymbol{B}}$ for (8.5).

$$\tilde{B} = \begin{pmatrix} a_{11} & \cdots & a_{1m} \\ \vdots & & \vdots \\ a_{m1} & \cdots & a_{mm} \end{pmatrix}, \qquad \tilde{\boldsymbol{B}} = \begin{bmatrix} a_{11} & \cdots & a_{1m} & 0 \\ \vdots & & \vdots & \vdots \\ a_{m1} & \cdots & a_{mm} & 0 \\ a_{m+1,1} & \cdots & a_{m+1,m} & 1 \end{bmatrix}$$

Therefore

$$\tilde{\boldsymbol{B}}^{-1} = \begin{pmatrix} \tilde{B}^{-1} & \vdots & 0 \\ \cdots\cdots\cdots & \cdot & \cdots \\ -(a_{m+1,1}, \ldots, a_{m+1,m})\tilde{B}^{-1} & \vdots & 1 \end{pmatrix}$$

Thus $\tilde{\boldsymbol{B}}^{-1}$ can easily be obtained from $\tilde{B}^{-1}$. Also obviously $\tilde{\Pi}$ = dual solution of (8.5) corresponding to $\tilde{\boldsymbol{B}}$ is equal to $(\tilde{\pi}, 0)$.

Thus the inverse tableau for (8.5) corresponding to the basis $\tilde{\boldsymbol{B}}$ is easily obtained from the inverse tableau for (8.1) corresponding to the basis $\tilde{B}$. On the other hand, if (8.1) was solved by computing the canonical tableaus after each pivot step, to obtain the canonical tableau of (8.5) with respect to the basis $\tilde{\boldsymbol{B}}$, introduce the $(m+1)$th constraint row at the bottom of the canonical tableau of (8.1) with respect to the basis $\tilde{B}$, include x_{n+1} as the basic variable in that row, and then price out all the basic column vectors of $\tilde{B}$ in it.

Example

Returning to the LP in Tableau 8.1, suppose the new constraint, $x_1 - x_2 + 3x_3 \leqq -7$, has to be imposed. The optimum solution $\tilde{x} = (3, 4, 2, 0, 0, 0)^T$ violates this new constraint. Let x_7 be the slack variable. $x_7 = -x_1 + x_2 - 3x_3 - 7$. Here is the inverse tableau for the augmented problem corresponding to the basic vector (x_1, x_2, x_3, x_7).

Basic variables	Inverse Tableau					Basic values	Pivot column (x_4)
x_1	−1	−2	2	0	0	3	−1
x_2	1	1	−1	0	0	4	1
x_3	−1	0	1	0	0	2	2
x_7	5	3	−6	1	0	−12	(−4)
$-z$	−2	4	−1	0	1	−11	1

x_7 is the only negative-valued basic variable here. The updated row vector in which x_7 is the basic variable is:

	x_1	x_2	x_3	x_4	x_5	x_6	x_7	$-z$	b
Updated 4th row	0	0	0	−4	0	−3	1	0	−12
Updated cost row	0	0	0	1	3	8	0	1	−11
$-\dfrac{\bar{c}_j}{\bar{a}_{4j}}$ for $\bar{a}_{4j} < 0$				$\dfrac{1}{4}$		$\dfrac{8}{3}$			

The pivot column is the updated column vector of x_4, which is entered as the last column in the inverse tableau above. The pivot is performed and the dual simplex algorithm applied in a similar manner until termination.

Problems

8.4 Complete the solution of this augmented problem.

8.5 Show that the type of sensitivity analysis discussed here (introducing an additional inequality constraint) is the dual of the one discussed in section 8.2 (introducing a new sign restricted variable).

8.6 Prove that the set of extreme points of $\mathbf{K}_1$ consists of
(a) All extreme points of $\mathbf{K}$ that satisfy (8.3).

(b) Points of intersection of edges of **K** that do not completely lie on the hyperplane **H**, with **H**, where $\mathbf{H} = \{x: A_{m+1.}x = b_{m+1}\}$.

8.7 If every optimum feasible solution of (8.1) violates (8.3), and if $\mathbf{K}_1 \neq \emptyset$, prove that every optimum feasible solution of the augmented problem satisfies (8.3) as an equation, that is, lies on the hyperplane **H** defined in Problem 8.6.

8.4 INTRODUCING AN ADDITIONAL EQUALITY CONSTRAINT

Returning to the LP (8.1), suppose an additional equality constraint has to be introduced.

$$A_{m+1.}x = b_{m+1}$$

Let $\mathbf{H} = \{x: A_{m+1.}x = b_{m+1}\}$. The set of feasible solutions of the augmented problem is the intersection of **K** with the hyperplane **H**, which is denoted by $\mathbf{K}_2 = \mathbf{K} \cap \mathbf{H}$. See Figure 8.2 for an example. If $\bar{x} \in \mathbf{H}$, obviously $\bar{x}$ is optimal to the augmented problem. Suppose $\bar{x} \notin \mathbf{H}$. Then $A_{m+1.}\bar{x} \neq b_{m+1}$. Suppose $A_{m+1.}\bar{x} > b_{m+1}$. Then change the original problem by adding the constraints $A_{m+1.}x - x_{n+1} = b_{m+1}$, $x_{n+1} \geqq 0$. (If, on the other hand, it turned out that $A_{m+1.}\bar{x} < b_{m+1}$, then the coefficient of x_{n+1} should be $+1$ instead.) Here is the new problem,

$$\begin{aligned} \text{Minimize} \quad & Z(X) = cx + Mx_{n+1} \\ \text{Subject to} \quad & \begin{pmatrix} A & \vdots & 0 \\ \cdots & \cdots & \cdots \\ A_{m+1.} & \vdots & -1 \end{pmatrix} \begin{pmatrix} x \\ \cdots \\ x_{n+1} \end{pmatrix} = \begin{pmatrix} b \\ \cdots \\ b_{m+1} \end{pmatrix} \\ & X = \begin{pmatrix} x \\ \cdots \\ x_{n+1} \end{pmatrix} \geqq 0 \end{aligned} \tag{8.6}$$

where M is an arbitrarily large positive number. $X_{\tilde{B}} = (x_1, \ldots, x_m, x_{n+1})$ forms a feasible basic vector for (8.6). The corresponding BFS is $\bar{X}^{\mathrm{T}} = (\bar{x}^{\mathrm{T}}, \bar{x}_{n+1})$ where $\bar{x}_{n+1} = A_{m+1.}\bar{x} - b_{m+1}$. Also,

$$\tilde{B} = \begin{bmatrix} a_{11} & \cdots & a_{1m} & 0 \\ \vdots & & \vdots & \vdots \\ a_{m1} & \cdots & a_{mm} & 0 \\ a_{m+1,1} & \cdots & a_{m+1,m} & -1 \end{bmatrix}$$

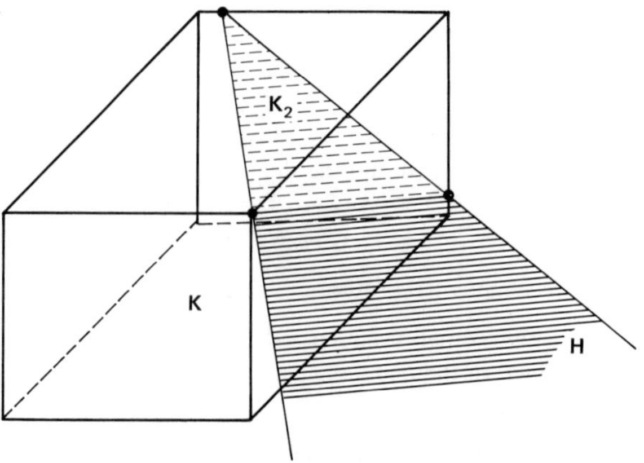

Figure 8.2 Introducing a new equality constraint.

Hence,

$$\tilde{\boldsymbol{B}}^{-1} = \begin{pmatrix} \tilde{B}^{-1} & \vdots & 0 \\ \cdots\cdots & & \cdots \\ (a_{m+1,1}, \ldots, a_{m+1,m})\tilde{B}^{-1} & \vdots & -1 \end{pmatrix}$$

Starting with the feasible basis $\tilde{\boldsymbol{B}}$, the augmented problem (8.6) is solved until optimality is reached. Suppose $\hat{X}$ is the final optimum solution obtained for (8.6). If $\hat{x}_{n+1} > 0$, then $\mathbf{K}_2 = \emptyset$. On the other hand, if $\hat{x}_{n+1} = 0$, then $\hat{x}$ is an optimum feasible solution of the augmented problem.

Problems

8.8 Returning to the LP in Tableau 8.1, suppose the additional constraint

$$x_1 + x_2 + x_3 + x_4 + x_5 + x_6 = 12$$

has to be imposed. Obtain an optimum feasible solution of the augmented problem.

8.9 Discuss an approach for solving the augmented problem with the additional equality constraint, using the dual simplex algorithm instead of the Big-$\boldsymbol{M}$ approach described here.

8.10 Prove that the set of extreme points of $\mathbf{K}_2$ consists of
(a) All extreme points of $\mathbf{K}$ that are in $\mathbf{H}$.
(b) Points of intersection of edges of $\mathbf{K}$ that do not completely lie in $\mathbf{H}$ with $\mathbf{H}$.

8.11 Let $\mathbf{H}$, $\mathbf{K}_2$ be defined as in the beginning of this section. Let

$$\mathbf{K}_1 = \mathbf{K} \cap \{x: \mathbf{A}_{m+1.}x \leqq b_{m+1}\}$$
$$\mathbf{K}_3 = \mathbf{K} \cap \{x: \mathbf{A}_{m+1.}x \geqq b_{m+1}\}$$

If $\mathbf{K}_2 \neq \emptyset$, prove that $\mathbf{K}_2$ contains an optimum solution of at least one of the following two problems.
(a) Minimize $z(x) = cx$, $x \in \mathbf{K}_1$
(b) Minimize $z(x) = cx$, $x \in \mathbf{K}_3$

8.5 COST RANGING OF A NONBASIC COST COEFFICIENT

Consider again the LP (8.1). Suppose x_r is a variable, which is not in the optimal basic vector $x_{\tilde{B}}$. Assuming that all the other cost coefficients except c_r remain fixed at their specified values, determine the range of values of c_r within which $\tilde{B}$ remains an optimal basis.

For this, treat c_r as a parameter. $\tilde{B}$ remains an optimal basis as long as

$$\bar{c}_r = c_r - \tilde{\pi}A_{.r} \geqq 0, \qquad \text{that is} \qquad c_r \geqq \tilde{\pi}A_{.r}.$$

The relative cost coefficients of all the other variables are independent of the value of c_r and, hence, they remain nonnegative.

If the new value of c_r is not in the closed interval $[\tilde{\pi}A_{.r}, \infty]$, x_r is the only nonbasic variable that has a negative relative cost coefficient with respect to the basis $\tilde{B}$ in the modified problem. Bring x_r into the basic vector and continue the application of the simplex algorithm until a terminal basis for the modified problem is obtained.

Example

Consider the LP in Tableau 8.1. The range of values of c_5 (whose present value is 10) for which the basic vector $x_{\tilde{B}} = (x_1, x_2, x_3)$ remains optimal is determined by

$$c_5 + (-2, 4, -1)(0, -2, -1)^{\mathrm{T}} \geqq 0$$

that is,

$$c_5 \geqq 7$$

Suppose the value of c_5 has to be modified to 6. When c_5 is changed from 10 to 6, the updated column vector of x_5 becomes $(2, -1, -1, -1)^{\mathrm{T}}$. With this as the pivot column, bring x_5 into the basic vector. This leads to the following tableau.

Basic variables	Inverse tableau				Basic values
x_5	$-\frac{1}{2}$	-1	1	0	$\frac{3}{2}$
x_2	$\frac{1}{2}$	0	0	0	$\frac{11}{2}$
x_3	$-\frac{3}{2}$	-1	2	0	$\frac{7}{2}$
$-z$	$-\frac{5}{2}$	3	0	1	$-\frac{19}{2}$

It can be verified that this tableau displays an optimum tableau for the modified problem.

8.6 COST RANGING OF A BASIC COST COEFFICIENT

Consider the LP (8.1). Suppose all the cost coefficients except c_1, which is a basic cost coefficient, remain fixed at their present values. Treating c_1 as a parameter determine the range of values of c_1 for which $\tilde{B}$ remains an optimal basis. Since c_1 is a basic cost coefficient, any change in c_1 changes the dual solution corresponding to the basis $\tilde{B}$, and all the relative cost coefficients. For simplicity let γ_1 denote the parameter and c_1 its present value. Let $\pi(\gamma_1)$, $\bar{c}_j(\gamma_1)$, etc., denote the dual solution, and the relative cost coefficient of x_j corresponding to the basis $\tilde{B}$ as functions of the parameter γ_1. Then

$$\pi(\gamma_1) = (\gamma_1, c_2, \ldots, c_m)\tilde{B}^{-1}$$

Hence, the dual vector is an affine function of γ_1. If we use this dual solution, the relative cost coefficient of x_j is

$$\bar{c}_j(\gamma_1) = c_j - \pi(\gamma_1)A_{.j} \qquad \text{for } j = 1 \text{ to } n$$

Hence, each $\bar{c}_j(\gamma_1)$ is itself an affine function of γ_1. The range of values of γ_1 for which $\tilde{B}$ remains an optimum basis is the range of γ_1 within which every $\bar{c}_j(\gamma_1)$ remains nonnegative. This range will turn out to be a nonempty closed interval. If it is necessary to modify the cost coefficient of x_1 to some value c_1' outside this closed interval, fix $\gamma_1 = c_1'$ and compute the relative cost coefficients $\bar{c}_j(c_1')$ using the formulas already developed. Bring one of the variables for which $\bar{c}_j(c_1') < 0$ into the basic vector and continue the applications of the simplex algorithm until a terminal basis for the modified problem is obtained.

Example

In the LP in Tableau 8.1, let γ_1 represent the cost coefficient of x_1 whose present value is 3. The dual solution corresponding to the basic vector (x_1, x_2, x_3) as a function of γ_1 is

$$\pi(\gamma_1) = (\gamma_1, 2, -3)\tilde{B}^{-1} = (-\gamma_1 + 5, -2\gamma_1 + 2, 2\gamma_1 - 5)$$

Using $\bar{c}_j(\gamma_1) = c_j - \pi(\gamma_1)A_{.j}$, the row of relative coefficients is $\bar{c}(\gamma_1) = (0, 0, 0, \gamma_1 - 2, -2\gamma_1 + 9, 2\gamma_1 + 2) \geqq 0$ iff $2 \leqq \gamma_1 \leqq 9/2$. So $\tilde{B}$ is an optimum basis whenever the cost coefficient of x_1 is in the interval $[2, 9/2]$, assuming that all the other cost coefficients remain at their present values.

Suppose γ_1 has to be changed to 5. Changing the cost coefficient of x_1 from the present value of 3 to 5, the modified dual solution is $\pi(\gamma_1 = 5) = (0, -8, 5)$. With this change, the inverse tableau becomes

Basic variables	Modified inverse tableau for basis B				Basic values	Pivot column (x_5)
x_1	-1	-2	2	0	3	(2) Pivot row
x_2	1	1	-1	0	4	-1
x_3	-1	0	1	0	2	-1
$-z$	0	8	-5	1	-17	-1

The modified relative cost coefficient of x_5 is -1. Hence, bring x_5 into the basic vector. The updated column vector of x_5 is entered as the pivot column in the inverse tableau. The new inverse tableau is:

Basic variables	Inverse tableau				Basic values
x_5	$-\frac{1}{2}$	-1	1	0	$\frac{3}{2}$
x_2	$\frac{1}{2}$	0	0	0	$\frac{11}{2}$
x_3	$-\frac{3}{2}$	-1	2	0	$\frac{7}{2}$
$-z$	$-\frac{1}{2}$	7	-4	1	$-\frac{31}{2}$

Remember that the cost coefficient of x_1 is now 5, and verify whether the new basis satisfies the termination criterion. Continue the algorithm if necessary.

Problems

8.12 Construct a counter example to the following: "If B is an optimum basis for (8.1), then it remains an optimum basis when some of the entries in c_B are decreased."

8.13 Is the following true? "x_1 is a basic variable in an optimum basic vector for (8.1). If c_1 is decreased and if the modified problem has an optimum feasible solution, then it has an optimum basic vector containing x_1."

8.7 RIGHT-HAND SIDE RANGING

In the LP (8.1) suppose we wish to determine the range of values of one of the right-hand side constants, say b_1, for which the basis $\tilde{B}$ remains optimal. Treat this right-hand side constant as a parameter and denote it by β_1; b_1 is the present value of β_1. We assume that all the other right-hand side constants stay fixed at their present values. $\tilde{B}$ is an optimum basis when $\beta_1 = b_1$. Hence, $\tilde{B}$ is dual feasible.

So $\tilde{B}$ is optimal for all values of β_1 for which it is primal feasible. The values of the basic variables $x_{\tilde{B}}$ are all functions of the parameter β_1, and $\tilde{B}$ is primal feasible as long as all these values are nonnegative.

$$x_{\tilde{B}}(\beta_1) = \tilde{B}^{-1}(\beta_1, b_2, \ldots, b_m)^{\mathrm{T}} \geqq 0$$

Since all these inequalities are linear in β_1, this will determine a closed interval of the form $\underline{\lambda} \leqq \beta_1 \leqq \bar{\lambda}$. For all values of β_1 in this closed interval, $\tilde{B}$ remains an optimal basis.

If it is necessary to modify the value of β_1 from its present value b_1, to a value b_1' outside the closed interval $[\underline{\lambda}, \bar{\lambda}]$, fix β_1 at b_1', and obtain the modified values of the basic variables. $\tilde{B}$ is still dual feasible, but primal infeasible. Starting with $\tilde{B}$, apply the dual simplex routine until a new terminal basis is obtained.

Example

In the LP in Tableau 8.1 the values of the basic variables in the basic vector (x_1, x_2, x_3) as functions of β_1 are

$$\tilde{B}^{-1}\begin{pmatrix}\beta_1\\6\\13\end{pmatrix} = \begin{pmatrix}-1 & -2 & 2\\1 & 1 & -1\\-1 & 0 & 1\end{pmatrix}\begin{pmatrix}\beta_1\\6\\13\end{pmatrix} = \begin{pmatrix}-\beta_1 + 14\\\beta_1 - 7\\-\beta_1 + 13\end{pmatrix}$$

All these are nonnegative iff $7 \leqq \beta_1 \leqq 13$. Hence, this is the interval within which the basic vector (x_1, x_2, x_3) remains optimal.

Problems

8.14 Obtain the new optimum feasible solution of the LP in Tableau 8.1 after b_1 is changed from its present value of 11 to 6.

8.8 CHANGES IN THE INPUT-OUTPUT COEFFICIENTS IN A NONBASIC COLUMN VECTOR

Let x_j be a variable that is not in the optimum basic vector $x_{\tilde{B}}$ of (8.1). a_{ij} is one of the input-output coefficients in the column vector of x_j. If all the other coefficients in the problem except a_{ij} remain fixed at their present values, what is the range of values of a_{ij} within which $\tilde{B}$ remains an optimal basis?

Treat a_{ij} as a parameter. To avoid confusion, call this parameter, α_{ij}, and let a_{ij} be its present value. Since x_j is a nonbasic variable, a change in α_{ij} does not affect the primal feasibility of $\tilde{B}$. A change in α_{ij} can only change the relative cost coefficient of x_j and this is obtained by

$$\bar{c}_j(\alpha_{ij}) = c_j + \left(\sum_{r \neq i} (-\tilde{\pi}_r)a_{rj}\right) - \tilde{\pi}_i\alpha_{ij}$$

As long as $\bar{c}_j(\alpha_{ij}) \geqq 0$, $\tilde{B}$ remains an optimal basis. This determines a closed interval for α_{ij}, and as long as α_{ij} is in this interval $\tilde{B}$ remains an optimal basis. If α_{ij} has to be changed to a value a_{ij}' outside this interval, make the change, bring x_j into the basic vector, and continue with the application of the simplex algorithm until a new terminal basis is obtained.

Example

In the LP in Tableau 8.1, the present value of α_{25} in the column vector of x_5 is -2. $\bar{c}_5(\alpha_{25}) = 10 + (-2, 4, -1)(0, \alpha_{25} - 1)^{\mathrm{T}} = 11 + 4\alpha_{25} \geqq 0$ iff $\alpha_{25} \geqq -11/4$. Thus $\tilde{B}$ remains an optimal basis for $\alpha_{25} \geqq -11/4$.

Suppose we change α_{25} from its present value of -2 to -3. The updated column vector of x_5 changes to $(4, -2, -1, -1)^{\mathrm{T}}$. With this pivot column bring x_5 into the basic vector and continue until termination is reached.

8.9 CHANGE IN A BASIC INPUT-OUTPUT COEFFICIENT

Let x_1 be a basic variable in the optimum basic vector $x_{\tilde{B}}$ for (8.1). Suppose we have to modify one input-output coefficient, say a_{11}, in the column vector of x_1 to a'_{11}. The modified column vector of x_1 will be $A'_{.1} = (a'_{11}, a_{21}, \ldots, a_{m1})^T$. Let x'_1 indicate the level of this activity corresponding to the new column vector $A'_{.1}$. The previous column vector $A_{.1}$ is no longer a part of the problem and it should be eliminated. Physically x'_1 replaces x_1 in the original tableau. We will refer to the altered problem by the name *modified problem.*

Construct a new problem by augmenting the present original tableau with the new variable x'_1 with its column vector $A'_{.1}$ and cost coefficient c_1, and changing the cost coefficient of x_1 to $\boldsymbol{M}$, where $\boldsymbol{M}$ is a very large positive number. This leads to the problem, which we call the "*new problem.*" In this problem x_1 plays the role of an artificial variable.

x_1	$x_2 \ldots x_m \ldots x_n$	x'_1	$-Z$	
$A_{.1}$	$A_{.2} \ldots A_{.m} \ldots A_{.n}$	$A'_{.1}$	0	b
M	$c_2 \quad c_m \quad c_n$	c_1	1	0

$\tilde{B}$ is still a feasible basis to this new problem. However, in the inverse tableau corresponding to the basis $\tilde{B}$, the dual vector has to be recomputed, since the cost coefficient of x_1 has been changed to $\boldsymbol{M}$. The new dual solution corresponding to $\tilde{B}$ is $\pi(\boldsymbol{M}) = (\boldsymbol{M}, c_2, \ldots, c_m)\tilde{B}^{-1}$. Hence the new inverse tableau is

Basic variables	Inverse tableau		Basic values
x_1 $\vdots$ x_m	$\tilde{B}^{-1}$	0	
$-Z$	$-\pi(M)$	1	

Starting with this feasible inverse tableau the primal simplex algorithm is applied to the new problem. The normal termination conclusions of the Big-$\boldsymbol{M}$ method apply.

8.10 PRACTICAL APPLICATIONS OF SENSITIVITY ANALYSIS

When studying a system using a linear programming model, techniques of sensitivity analysis can be used to evaluate new products (as in section 8.2); to determine the breakeven selling price of a new product, at which point it becomes competitive with the existing list of products in terms of profitability (as in sections 8.2 and 8.5); to evaluate new technologies or processes for making products (as in sections 8.8, 8.9); to assess how profitable it is to acquire additional resources and to determine which resources to acquire in what quantities (using the ideas discussed in sections 4.6.3, 8.7 and chapter 7); to evaluate the effects of changes in the costs (as in sections 8.5, 8.6, and chapter 7); and to determine optimal policies for handling new constraints that might arise. When used in this manner, the linear programming model not only determines an optimum solution to implement, but becomes a tool for optimal planning. Examples of some of these uses are discussed in the problems that follow.

Optimization models other than linear programming models (e.g. integer programming models to be discussed later on in this book, and nonlinear programming models) do not lend themselves that readily to a marginal analysis or sensitivity analysis as the linear programming model. That's why if a practical problem can be approximated reasonably closely by a linear programming model, it is so much easier to study it than otherwise.

PROBLEMS

8.15 Consider the housewifes' diet problem discussed in Chapter 4.

Original Tableau for the Housewifes' Diet Problem

x_1	x_2	x_3	x_4	x_5	x_6	x_7	x_8	$-z$	b
1	0	2	2	1	2	-1	0	0	9
0	1	3	1	3	2	0	-1	0	19
35	30	60	50	27	22	0	0	1	0 (minimize z)

$x_j \geqq 0$ for all j

Here x_1 to x_6 are the kilograms of the primary foods 1 to 6 in the housewifes' diet, and x_7, x_8 are the slack variables representing the excess amounts of the nutrients, vitamins A and C, in the diet over the minimum requirements. The basis B_1, associated with the basic vector (x_5, x_6) is optimal to this problem. The optimum inverse tableau is

Basic variables	Inverse tableau			Basic values
x_5	$-\frac{1}{2}$	$\frac{1}{2}$	0	5
x_6	$\frac{3}{4}$	$-\frac{1}{4}$	0	2
$-z$	-3	-8	1	-179

Answer each of the following questions with respect to this original problem:

(a) Suppose a new primary food, food 7, is available in the market at 88 cents/kg. One kilogram of this food contains two units of vitamin A and four units of vitamin C. Should the housewife include this new food in her diet? If not how much should the cost of this food decrease before the housewife can include it in her diet?

At this breakeven cost show that there is an optimum diet that includes food 7 and another that does not.

What is the optimum diet if food 7 is actually available at 32 cents/kg?

(b) Consider the original housewifes' diet problem again. The housewife has just read an article in a health magazine which says that another nutrient, vitamin E, is very important for the family's health. The vitamin E content of the six primary foods are 2, 3, 5, 2, 1, and 1 units/kg, respectively. The minimum requirement of vitamin E is 10 units. The housewife wants to include this additional constraint in the problem. How does this change her optimal diet?

(c) Referring to the original problem, in addition to meeting the minimum requirements on vitamins A and C, suppose the housewife decides to have the diet consist of a total of 2000 calories exactly. The calorie contents of the six primary foods are, 160, 20, 500, 280, 300, and 360 per kilogram, respectively. How does this additional requirement change her optimal diet?

(d) What is the marginal effect of increasing the minimum requirement of vitamin C on the cost of the optimal diet in the original housewife's problem? How much can the minimal requirement of vitamin C increase before the basis B_1 becomes non-optimal to the problem? What is an optimal diet when this requirement is 39 units?

(e) In the original housewives' problem, suppose that each additional unit of vitamin A in the diet is expected to bring an average savings of 10 cents in medical expenses. How does this alter the optimal diet?

(f) For what range of cost per kilogram of primary foods 4 and 5 does the basis B_1 remain optimal to the original housewives' problem?

(g) What happens to the optimal diet in the original housewifes' problem if the vitamin C content of food 4 changes? For what range of values of this quantity does the basis B_1 remain optimal? Suppose a richer version of food 4 containing $1 + \alpha$ units of vitamin C per kilogram is available at a cost of $50 + 4\alpha$ cents/kg for any $\alpha \geqq 0$. What is the minimum value of α at which it becomes attractive for the housewife to include it in her diet?

8.16 Data on different solvents is given below

	Solvent 1	Solvent 2	Solvent 3	Solvent 4	Chemical requirement in blend per kg
Chemical 1 content (units/kg)	180	120	90	60	$\geqq 90$
Chemical 2 content (units/kg)	3	2	6	5	$\leqq 4$
Cost (cent/kg)	16	12	10	11	

Let x_j be the proportion of solvent type j in the blend, $j = 1$ to 4. x_5 and x_6 are the slack variables associated with chemicals 1 and 2 requirements, respectively. The original tableau for the problem is

x_1	x_2	x_3	x_4	x_5	x_6	$-z$	b
1	1	1	1	0	0	0	1
180	120	90	60	-1	0	0	90
3	2	6	5	0	1	0	4
16	12	10	11	0	0	-1	0 (minimize z)

$$x_j \geqq 0 \quad \text{for all } j$$

The basis B_1 associated with the basic vector (x_2, x_3, x_5) is an optimum basis for the problem.

$$B_1^{-1} = \begin{bmatrix} \frac{3}{2} & 0 & -\frac{1}{4} \\ -\frac{1}{2} & 0 & \frac{1}{4} \\ 135 & -1 & -\frac{15}{2} \end{bmatrix}$$

Each part of the question below is independent of the others and they all refer to this original problem

(a) Write down the dual problem and the complementary slackness conditions for optimality. Obtain an optimum feasible solution for the dual problem from the above information.

(b) How much does the optimum objective value change if the minimal chemical 1 requirement is changed to 88 units? Why?

(c) Let β_1 be the minimal chemical 1 requirement. Its present value is 90. For what range of values of β_1 does B_1 remain optimal to the problem? When β_1 is 114 what is an optimum feasible solution to the problem?

(d) How much can the cost per kilogram of solvent 3 change before B_1 becomes non-optimal to the problem? When the cost per kilogram of solvent 3 becomes 11 cents, what is an optimum solution?

8.17 Consider the following LP

$$\begin{array}{llr} \text{Minimize} & z(x) = -2x_1 - 4x_2 - x_3 - x_4 & \\ \text{Subject to} & x_1 + 3x_2 + x_4 \leqq 8 & = \text{available amount of raw material 1} \\ & 2x_1 + x_2 \leqq 6 & = \text{available amount of raw material 2} \\ & x_2 + 4x_3 + x_4 \leqq 6 & = \text{available amount of raw material 3} \\ & x_j \geqq 0 \quad \text{for all } j & \end{array}$$

x_5, x_6, x_7 are the slack variables corresponding to the various inequalities. The basis B_1 corresponding to the basic vector (x_1, x_3, x_2) is optimal to the problem.

$$B_1^{-1} = \begin{bmatrix} -\frac{1}{5} & \frac{3}{5} & 0 \\ -\frac{1}{10} & \frac{1}{20} & \frac{1}{4} \\ \frac{2}{5} & -\frac{1}{5} & 0 \end{bmatrix}$$

(a) If the availability of only one of the raw materials can be marginally increased, which one should be picked? Why?

(b) For what range of values of b_1 (the amount of raw material 1 available) does the basis B_1 remain optimal? What is an optimum solution to the problem if $b_1 = 20$?

(c) If seven more units of raw material 1 can be made available (over the present 8 units), what is the maximum you can afford to pay for it? Why?

(d) The company has an option to produce a new product. Let x_8 be the number of units of this product manufactured. The input-output vector of x_8 will be

$$\left(10, 20, 24 - 3\lambda, -25 + \frac{\lambda}{4}\right)^{\mathrm{T}}$$

where λ is a parameter that can be set anywhere from 0 to 6. What is the minimum value of λ at which it becomes profitable to produce the product? What is an optimum solution when $\lambda = 4$?

8.18 Consider the following diet problem

Nutrient	Nutrient content in foods available (units/kg)			Minimum daily requirement (MDR) for nutrient
	Greens	Potatoes	Corn	
Vitamin A	10	1	9	5
C	10	10	10	50
D	10	11	11	10
Cost (cent/kg)	50	100	51	

Let x_1, x_2, and x_3 be the amounts of greens, potatoes, and corn included in the diet respectively. Let x_4, x_5, and x_6 be the slack variables representing the excess of vitamins A, C, and D, respectively, in the diet. The basis B_1 associated with the basic vector (x_4, x_1, x_6) is optimal to this problem

$$B_1^{-1} = \begin{pmatrix} -1 & 1 & 0 \\ 0 & 0.1 & 0 \\ 0 & 1 & -1 \end{pmatrix}$$

(a) Find the optimum primal and dual solutions associated with the basis B_1.

(b) A new food (milk) has become available. One liter of milk contains 0, 10, and 20 units, respectively, of vitamins A, C, and D and costs 40 cents. Would you recommend including it in the diet? Why? What is the highest price of milk at which it is still attractive to include it in the diet?

(c) Consider the original problem again. A nutrition specialist claims that the actual MDRs for vitamins A, C, and D should really be 5, $50 + 10\lambda$, and $10 + 15\lambda$, where λ is a nonnegative parameter; he proposes an experiment to estimate λ. Assuming that he is correct, find $\bar{\lambda}$, the maximum value of λ at which B_1 is still an optimum basis to the problem. What is an optimum solution if $\lambda = \bar{\lambda} + 1$?

8.19 Consider the LP (8.1), and its dual. Discuss what effects the following have on the primal and dual feasible solution sets and the respective optimal objective values.

(i) Introducing a new nonnegative primal variable.
(ii) Introducing a new inequality constraint in the primal problem.
(iii) Introducing a new equality constraint in the primal problem.

9 degeneracy in linear programming

9.1 GEOMETRY OF DEGENERACY

Consider the LP

$$\begin{array}{lrl} \text{Minimize} & z(x) = & cx \\ \text{Subject to} & Ax = & b \\ & x \geqq & 0 \end{array} \tag{9.1}$$

where A is a matrix of order $m \times n$ and rank m. If B is a basis for (9.1), it is a *nondegenerate basis* iff all the basic variables are nonzero in the basic solution of (9.1) corresponding to it, that is, iff all components in the vector $B^{-1}b$ are nonzero. A basis of (9.1) that is not nondegenerate is said to be a *degenerate basis*. The LP (9.1) is said to be *totally (primal) nondegenerate* iff every basis for (9.1) is nondegenerate.

Problem

9.1 Prove that (9.1) is totally nondegenerate iff every solution, x, of $Ax = b$, has at least m nonzero components.

Equivalently, (9.1) is totally nondegenerate iff b cannot be expressed as a linear combination of any set of $m - 1$ or less column vectors of A, that is, iff b does not lie in any subspace that is the linear hull of a set of $m - 1$ or less column vectors of the matrix A.

Examples

Since the objective function does not play any role in determining primal degeneracy or nondegeneracy, we omit it in these examples

$$A = \begin{pmatrix} 1 & 0 & 0 & 1 \\ 0 & 1 & 0 & 0 \\ 0 & 0 & 1 & -1 \end{pmatrix} \qquad b = \begin{pmatrix} -1 \\ 1 \\ 0 \end{pmatrix}$$

In this case, $m = 3$, and b can be expressed as a linear combination of $A_{.1}$, $A_{.2}$. Hence, this problem is primal degenerate. Also any basis for this problem that includes $A_{.1}$, $A_{.2}$ as basic vectors is a degenerate basis. Clearly the problem remains degenerate even if b is changed, as

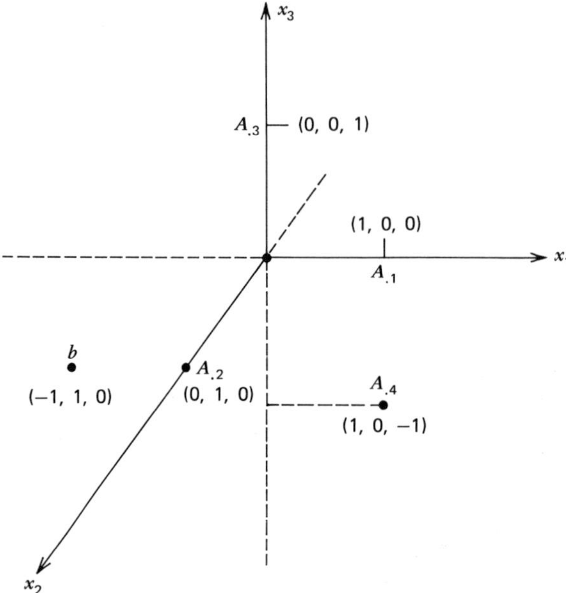

Figure 9.1

long as b lies on any of the coordinate planes, or on the linear hull of $\{A_{.2}, A_{.4}\}$. The problem becomes nondegenerate iff b does not lie on any of these four planes. See Figure 9.1.

Clearly if $b = (1, 2, 3)^T$, and A remains the same, the problem is nondegenerate.

THE b VECTOR IN DEGENERATE PROBLEMS

The LP (9.1) is degenerate iff b is in a subspace of $\mathbf{R}^m$ that is the linear hull of a set of $m - 1$ or less column vectors of A. There are at most $\sum_{r=1}^{m-1} \binom{n}{r}$ subsets of column vectors of A with $m - 1$ or less column vectors in them. The linear hull of each of these subsets is a subspace of $\mathbf{R}^m$ of dimension $m - 1$ or less. The LP (9.1) is degenerate iff b lies in one of these subspaces. Hence, when A is fixed, the set of all $b \in \mathbf{R}^m$ that make (9.1) degenerate is the set of all b that lie in a finite number of subspaces of dimension $m - 1$ or less. Thus in a statistical sense *almost all b* will make (9.1) nondegenerate. Also if (9.1) is degenerate, b can be perturbed to a point in its neighborhood that is not in any of the finite number of subspaces discussed earlier, and this would make (9.1) nondegenerate. One such perturbation that makes (9.1) nondegenerate is to replace, b by

$$b(\varepsilon) = b + (\varepsilon, \varepsilon^{\mathbf{2}}, \ldots, \varepsilon^{\mathbf{m}})^T \tag{9.2}$$

where ε is an arbitrarily small positive number, and in the symbol $\varepsilon^{\mathbf{r}}$, $\mathbf{r}$ is the exponent, that is, $\varepsilon^{\mathbf{r}} = \varepsilon \times \varepsilon \times \cdots \times \varepsilon$ with $\mathbf{r}$ ε's in this product. Notice the difference in type style between exponents and superscripts.

Theorem Given any $b \in \mathbf{R}^m$ there exists a positive number $\varepsilon_1 > 0$ such that whenever $0 < \varepsilon < \varepsilon_1$

$$\begin{aligned} Ax &= b(\varepsilon) \\ x &\geqq 0 \end{aligned} \tag{9.3}$$

is nondegenerate.

Proof Let $B_1, \ldots, B_L$ be all the bases for (9.3). Consider one of these bases B_r. Let $B_r^{-\mathbf{1}} = (\beta_{ij}^r)$. Since $B_r^{-\mathbf{1}}$ is nonsingular $(\beta_{i1}^r, \ldots, \beta_{im}^r) \neq 0$ for each $i = 1$ to m.

Let $\bar{b}^r = B_r^{-1}b$. Then

$$B_r^{-1}b(\varepsilon) = \begin{pmatrix} \bar{b}_1{}^r + \beta^r_{11}\varepsilon + \cdots + \beta^r_{1m}\varepsilon^{\mathbf{m}} \\ \vdots \\ \bar{b}_m{}^r + \beta^r_{m1}\varepsilon + \cdots + \beta^r_{mm}\varepsilon^{\mathbf{m}} \end{pmatrix}.$$

For each $i = 1$ to m, and $r = 1$ to L, the polynomial, $\bar{b}_i^r + \beta^r_{i1}\theta + \cdots + \beta^r_{im}\theta^{\mathbf{m}}$ is a nonzero polynomial in θ. Therefore, it has at most m roots. Let the roots be $\theta_{r,i,1}, \ldots, \theta_{r,i,k}$. If ε is not equal to one of these roots, then

$$\bar{b}_i^r + \beta^r_{i1}\varepsilon + \cdots + \beta^r_{im}\varepsilon^{\mathbf{m}} \neq 0.$$

Let $\mathbf{\Gamma}$ be the collection of the roots of all these polynomials in θ for $i = 1$ to m and $r = 1$ to L. $\mathbf{\Gamma}$ is a finite set consisting of at most $\mathrm{L}m^2$ real numbers. Hence, it is possible to pick a positive number $\varepsilon_1 > 0$ such that $\mathbf{\Gamma}$ contains no element belonging to the open interval $0 < \varepsilon < \varepsilon_1$. By the construction of the set $\mathbf{\Gamma}$ and the positive number ε_1, it is clear that in any basic solution of (9.3), all the basic variables are nonzero and, hence, that (9.3) is nondegenerate whenever $0 < \varepsilon < \varepsilon_1$.

ADJACENT BASES

Let $\mathbf{K}$ be the set of feasible solutions of (9.1). Every basis for (9.1) is a nonsingular square matrix of order m. Two feasible bases B_1 and B_2 are said to be *adjacent bases* if the associated basic vectors x_{B_1} and x_{B_2} contain $m - 1$ common basic variables. Thus two consecutive bases obtained in the course of the simplex algorithm are always *adjacent bases.*

ADJACENT EXTREME POINTS OF **K**

It was shown in Chapter 3 that if B_1 and B_2 are two feasible adjacent bases for (9.1), and if the BFSs corresponding to these two bases are *distinct* (i.e., not equal), then these two BFSs are adjacent extreme points of $\mathbf{K}$, and the line segment joining them is an *edge* of $\mathbf{K}$. The BFSs corresponding to two feasible adjacent bases B_1 and B_2 are the same iff (1) both B_1 and B_2 are degenerate bases, and (2) B_2 is obtained from B_1 by making a degenerate pivot step and vice versa.

Every extreme point of $\mathbf{K}$ is the BFS corresponding to some feasible basis. Let $\tilde{x}$ and $\hat{x}$ be two distinct extreme points of $\mathbf{K}$. Suppose $\tilde{x}$ is the BFS corresponding to a basis $\tilde{B}$ and $\hat{x}$ is the BFS corresponding to a basis $\hat{B}$. $\tilde{x}$ and $\hat{x}$ are adjacent extreme points of $\mathbf{K}$ if either (1) $\tilde{B}$ and $\hat{B}$ are adjacent bases, or (2) there exists a sequence of feasible adjacent bases.

$$\tilde{B}, B_1, B_2, \ldots, B_r; B_{r+1}, \ldots, B_t, \hat{B}$$

such that every pair of consecutive bases in the sequence are adjacent, and the BFSs corresponding to all the bases $B_1, B_2, \ldots, B_r$ are equal to $\tilde{x}$, and the BFSs corresponding to all the bases $B_{r+1}, \ldots, B_t$ are equal to $\hat{x}$.

ALL THE BASES CORRESPONDING TO AN EXTREME POINT

Suppose $\tilde{x}$ is an extreme point of $\mathbf{K}$. This implies that $\tilde{x}$ is a BFS of (9.1). Therefore the set $\mathbf{A}(\tilde{x}) = \{A_{.j}: j \text{ such that } \tilde{x}_j > 0\}$ is linearly independent.

Case 1 If the set $\mathbf{A}(\tilde{x})$ has exactly m column vector in it, it defines a basis for (9.1) and it is the unique basis corresponding to $\tilde{x}$. This will always be the case with every extreme point of $\mathbf{K}$ if (9.1) is nondegenerate.

Case 2 Suppose $\mathbf{A}(\tilde{x})$ has less than m column vectors in it. This can only happen if (9.1) is degenerate and $\tilde{x}$ is a degenerate BFS of it. In this case, treat $\mathbf{A}(\tilde{x})$ as a partial basis. Obtain the set of all bases for (9.1) that can be obtained by making this partial basis into a complete basis. This is the set of bases for (9.1) that contain all the column vectors in $\mathbf{A}(\tilde{x})$ as basic column vectors. It is the set of bases corresponding to $\tilde{x}$.

ADJACENT BASIS PATHS

Theorem 1 Let $\tilde{x}$ be an extreme point of $\mathbf{K}$. Let $\tilde{B}$ and $\bar{B}$ be two distinct bases for (9.1) corresponding to both of which $\tilde{x}$ is the BFS. Then there exists a sequence of bases

$$\tilde{B}, B_1, \ldots, \bar{B}$$

such that consecutive bases in the sequence are adjacent and all the bases in the sequence correspond to $\tilde{x}$.

Proof This follows from standard results in linear algebra, using the fact that both $\tilde{B}$ and $\bar{B}$ are bases obtained by completing a partial basis as above.

Theorem 2 If $\tilde{B}$ and $\hat{B}$ are a pair of feasible bases for (9.1), then there exists a sequence of bases

$$\tilde{B}, B_1, B_2, \ldots, B_\ell, \bar{B}$$

such that all the bases in the sequence are feasible and every pair of consecutive bases in the sequence are adjacent.

Proof Suppose $\tilde{x}$, $\hat{x}$ are the BFSs corresponding to the bases $\tilde{B}$ and $\hat{B}$, respectively. If $\tilde{x} = \hat{x}$ the result follows from the previous theorem. If $\tilde{x} \neq \hat{x}$, let $z^1(x) = \sum d_j x_j$ where $d_j = 1$ if $\hat{x}_j = 0$ and $d_j = 0$, otherwise. Since $\hat{x}$ is a BFS, it is the unique feasible solution of (9.1) that makes $z^1(x) = 0$. Also $z^1(x) \geqq 0$ for all $x \in \mathbf{K}$. Hence, $\hat{x}$ is the unique optimum feasible solution for the problem of minimizing $z^1(x)$ on $\mathbf{K}$. Starting with the feasible basis $\tilde{B}$, apply the simplex algorithm to the problem of minimizing $z^1(x)$ on $\mathbf{K}$. The simplex algorithm walks through a sequence of adjacent bases until it terminates with an optimal basis, the BFS corresponding to which must be $\hat{x}$. Combining this with the previous theorem the result follows.

EDGES OF **K** CONTAINING A SPECIFIED EXTREME POINT

Let $\tilde{x}$ be an extreme point of $\mathbf{K}$, and $\tilde{B}$ a basis corresponding to $\tilde{x}$. For convenience, assume $x_{\tilde{B}} = (x_1, \ldots, x_m)$. Let x_s be a nonbasic variable. Then $\tilde{x}_s = 0$. Keeping all the other nonbasic variables excepting x_s fixed at value zero, make $x_s = \alpha \geqq 0$ and recompute the values of the basic variables so as to satisfy the contraints. If $\bar{A}_{.s} = B^{-1}A_{.s}$, the new solution obtained is

$$x(\alpha) = (\tilde{x}_1 - \alpha\bar{a}_{1s}, \ldots, \tilde{x}_m - \alpha\bar{a}_{ms}, 0, \ldots, 0, \alpha, 0, \ldots, 0)^{\mathrm{T}}.$$

where the "α" is in the sth position in the vector on the right hand side of the above equation.

Let θ be the minimum ratio when x_s enters the basic vector $x_{\tilde{B}}$. If θ is 0, this is a degenerate pivot step. In this case, x_s enters the basic vector $x_{\tilde{B}}$ to give a new basic vector corresponding to $\tilde{x}$. If $\theta > 0$, $x(\alpha)$ is a feasible solution for all $0 \leqq \alpha \leqq \theta$. As α varies from 0 to θ, the point $x(\alpha)$ traces an edge of the convex polyhedron $\mathbf{K}$. If θ is finite, the edge traced by $x(\alpha)$ is a bounded edge of $\mathbf{K}$ and $x(\theta)$ is an adjacent extreme point of $\tilde{x}$. The new basis obtained when x_s is brought into the basic vector $x_{\tilde{B}}$ is a basis corresponding to the new extreme point $x(\theta)$. If $\theta = +\infty$, the edge traced by $x(\alpha)$ as α ranges from 0 to ∞ is an *unbounded edge* (*or an extreme half-line*) of $\mathbf{K}$. All the edges of $\mathbf{K}$ that contain the extreme point $\tilde{x}$ are generated in this manner using some basic vector corresponding to $\tilde{x}$ and a variable not in that basic vector.

If (9.1) is nondegenerate, the minimum ratio corresponding to every pivot step is strictly positive. Also, if $\tilde{x}$ is a BFS, there is a unique basis corresponding to it in this case. Hence there are exactly $n - m$ edges of **K** that contain the extreme point $\tilde{x}$.

A polyhedron that has the same number (equal to its dimension) of edges going out of each of its extreme points is known as a *simple convex polyhedron.* Hence, if (9.1) is nondegenerate, **K** is a simplex convex polyhedron. A bounded convex polyhedron is known as a *convex polytope.* Hence, if (9.1) is nondegenerate and **K** is bounded, then **K** is a *simple convex polytope.* In this case, each edge of **K** is a bounded edge. Hence, every extreme point of **K** has exactly $n - m$ adjacent extreme points.

Even if (9.1) is degenerate, if $\tilde{x}$ happens to be a nondegenerate BFS of (9.1), there are exactly $n - m$ edges of **K** through $\tilde{x}$. However, if $\tilde{x}$ is a degenerate BFS of (9.1), there are several bases corresponding to $\tilde{x}$. Each of these bases leads to some edges of **K** through $\tilde{x}$. In this case, to get all the edges of **K** through $\tilde{x}$, examine each basic vector corresponding to $\tilde{x}$ and generate the edges from it obtained by choosing one of the nonbasic variables corresponding to positive minimum ratio as the entering variable. In this case, the number of edges of **K** through $\tilde{x}$ and the number of adjacent extreme points of $\tilde{x}$ is likely to be much larger than $n - m$. And these numbers may be different for different extreme points of **K**.

Problem

9.2 Show that the following system is degenerate.

$$\sum_{j=1}^{N} x_{ij} = 1 \qquad \text{for } i = 1 \text{ to N}$$

$$\sum_{i=1}^{N} x_{ij} = 1 \qquad \text{for } j = 1 \text{ to N}$$

$$x_{ij} \geqq 0$$

Prove that $\tilde{x} = I$ = unit matrix of order N, is a BFS of this system. Let **K** be the set of feasible solutions of this system. Using the theory of the transportation problem (see Chapter 11) show that the number of adjacent extreme of points of $\tilde{x}$ in **K** is

$$\sum_{r=2}^{N} \binom{N}{r} (r - 1)!$$

Compare and see that this number is much larger than the dimension of this polytope, $N^2 - (2N - 1)$, the number of variables minus the number of independent equality constraints in this problem.

In general, a degenerate BFS has a much larger number of adjacent BFSs than a nondegenerate BFS.

9.2 RESOLUTION OF CYCLING UNDER DEGENERACY

9.2.1 Simplex Algorithm under Nondegeneracy

Theorem Suppose (9.1) is nondegenerate and is being solved by the simplex algorithm starting with a feasible basis:

(i) The minimum ratio in every pivot step will be positive.

(ii) The pivot row is always uniquely determined in every pivot step during the course of the algorithm, that is, there will be no ties in determining the minimum ratio in every pivot step.

(iii) The value of the objective function makes a strict decrease after every pivot step.
(iv) A basis that appeared in the course of the algorithm will never appear later on, that is, there can be no cycling.
(v) The algorithm reaches a terminal basis in a finite number of pivot steps.

Proof By nondegeneracy, the value of every basic variable is nonzero in any basic solution of (9.1). Hence, all the basic values will be positive in every BFS. This implies (i).

To prove (ii), let B be the current basis associated with the basic vector $x_B = (x_1, \dots, x_m)$ and suppose the nonbasic variable x_s is the entering variable. The current vector of basic values is $B^{-1}b = \bar{b}$. The pivot column is $B^{-1}A_{.s} = \bar{A}_{.s}$. The minimum ratio here is minimum $\{\bar{b}_i/\bar{a}_{is}: i \text{ such that } \bar{a}_{is} > 0\}$. If there is a tie for the i that attains the minimum here, suppose both $i = 1, 2$ are such that $\bar{b}_1/\bar{a}_{1s} = \bar{b}_2/\bar{a}_{2s}$ = minimum ratio here. Hence x_s can enter the basic vector replacing either x_1 or x_2 from it. If x_1 is the dropping variable, the value of the basic variable x_2 is zero in the BFS obtained after the pivot. This contradicts the hypothesis of nondegeneracy of (9.1). Hence, in each pivot step, there can be no ties in the minimum ratio test for determining the pivot row.

If x_s is the entering variable in a pivot step of the algorithm, $\bar{c}_s < 0$. If θ is the minimum ratio in this pivot step, $\theta > 0$ by nondegeneracy. The change in the objective value as a result of this pivot step is $\theta\bar{c}_s < 0$. This implies (iii).

Part (iv) follows from (iii).

Part (v) follows from (iv) and from the fact that the total number of bases is finite.

Hence, if (9.1) is nondegenerate, except for the choice of the pivot column at each stage, the simplex algorithm is fully automatic and the algorithm walks to a terminal basis without ever returning to a basis it has seen already. Under nondegeneracy, there cannot be any problem of cycling. The same claims cannot be made if (9.1) is degenerate. (See Chapters 2 and 3.) Parts (i) to (iv) of the theorem proved here may not hold. Starting from a degenerate feasible basis, the algorithm could go through a sequence of degenerate pivots, and return to the basis that started this sequence, and the sequence will be repeated again and again forever. This is the *cycling of the simplex algorithm under degeneracy*. If this happens, part (v) of the theorem cannot be guaranteed. See the example at the end of this chapter.

9.2.2 Perturbed Problem

Suppose (9.1) is degenerate. The general strategy for avoiding cycling is as follows. Replace b by $b(\varepsilon)$ from equation (9.2), and get the modified problem

$$\begin{aligned} \text{Minimize} \quad & z(x) = cx \\ \text{Subject to} \quad & Ax = b(\varepsilon) \\ & x \geqq 0 \end{aligned} \tag{9.4}$$

From section 9.1, we know that there exists a positive number ε_1 such that for all $0 < \varepsilon < \varepsilon_1$, (9.4) is nondegenerate. Thus for all ε that are positive but arbitrarily small, (9.4) can be solved by applying the simplex algorithm with no possibility of cycling occurring. It is not necessary to give ε any specific value as long as it is treated as *arbitrarily small*, which will mean that it is a positive number, strictly smaller than any other positive number with which it may be compared during the course of the algorithm.

9.2.3 Feasibility

Theorem If B is a feasible basis for (9.4) when ε is *arbitrarily small*, then B is a feasible basis for (9.1).

Proof Let $B^{-1} = (\beta_{ij})$ and $\bar{b} = B^{-1}b$. Then $B^{-1}b(\varepsilon) = (\bar{b}_1 + \beta_{11}\varepsilon + \cdots + \beta_{1m}\varepsilon^m, \ldots, \bar{b}_m + \beta_{m1}\varepsilon + \cdots + \beta_{mm}\varepsilon^m)^T$. If for some i, $\bar{b}_i < 0$, then $\bar{b}_i + \beta_{i1}\varepsilon + \cdots + \beta_{im}\varepsilon^m < 0$ when ε is arbitrarily small, which is a contradiction. Hence, $\bar{b}_i \geqq 0$ for all i. Thus B must also be a feasible basis for (9.1).

When (9.4) is solved by the simplex algorithm, treating ε as arbitrarily small, a basis that appeared once cannot reappear by nondegeneracy. If (9.4) is feasible, the terminal basis either satisfies the optimality or the unboundedness criteria. These termination criteria are independent of the right-hand side vector. Hence, the terminal basis obtained for (9.4) is also a terminal basis for (9.1). The final optimal solution of (9.1) is obtained from the corresponding thing of (9.4) by eliminating all the terms involving ε or its positive powers (in effect substituting 0 for ε).

Definition: Lexico Positive Let $\gamma = (\gamma_1, \ldots, \gamma_r)$. It is said to be *lexico positive* if $\gamma \neq 0$ and if the first nonzero component of γ is positive, irrespective of what values the subsequent components may have. We will indicate this by writing $\gamma \succ 0$. The symbol "$\succ$" stands for "lexicographically greater than." Thus, if $\gamma \succ 0$, exactly one of the following holds:

$$\text{Either } \gamma_1 > 0; \text{ or } \gamma_1 = 0, \gamma_2 > 0; \text{ or } \gamma_1 = 0, \gamma_2 = 0, \\ \gamma_3 > 0, \ldots, \text{or } \gamma_1 = 0, \gamma_2 = 0, \ldots, \gamma_{r-1} = 0, \gamma_r > 0$$

Given two vectors $\xi = (\xi_1, \ldots, \xi_r)$ and $\eta = (\eta_1, \ldots, \eta_r)$, we say that $\xi \succ \eta$ iff $\xi - \eta \succ 0$. ξ is said to be *lexico negative* (denoted by $\xi \prec 0$) if $-\xi \succ 0$.

PRIMAL FEASIBILITY FOR ARBITRARILY SMALL ε

Let B be a basis for (9.4). Let $B^{-1} = (\beta_{ij}), \bar{b} = B^{-1}b$.

Theorem B is a feasible basis for (9.4) for *arbitrarily small* ε iff $(\bar{b}_i, \beta_{i1}, \ldots, \beta_{im}) \succ 0$ for all i.

Proof Since B^{-1} is nonsingular, obviously $(\beta_{i1}, \ldots, \beta_{im}) \neq 0$ for all $i = 1$ to m. So $(\bar{b}_i, \beta_{i1}, \ldots, \beta_{im})$ is nonzero for all i. When ε is positive and arbitrarily small, the sign of $\bar{b}_i + \beta_{i1}\varepsilon + \cdots + \beta_{im}\varepsilon^m$ is obviously the same as the sign of the first nonzero coefficient in this polynomial, that is, the first nonzero component in $(\bar{b}_i, \beta_{i1}, \ldots, \beta_{im})$. So for arbitrarily small ε, $\bar{b}_i + \beta_{i1}\varepsilon + \cdots + \beta_{im}\varepsilon^m$ is positive iff the first nonzero component in the row vector $(\bar{b}_i, \beta_{i1}, \ldots, \beta_{im})$ is positive. Applying this for each i, we have proved the theorem.

This theorem makes it easy to test the feasibility of a basis for (9.4) for arbitrarily small ε. There is no need to give any specific value to ε. Given the updated right-hand side constant vector, and the inverse of the basis, testing the basis for feasibility in (9.4) requires the testing of lexico positiveness of m row vectors, which is done without ever mentioning the perturbation variable ε.

9.2.4 Pivot Row Choice for Arbitrarily Small ε

Let B be a feasible basis for (9.4) with $B^{-1} = (\beta_{ij})$, and suppose the nonbasic variable x_s has to be brought into the basic vector. $\bar{A}_{.s} = B^{-1}A_{.s}$ is the pivot column. If $(1/\bar{a}_{ps})(\bar{b}_p, \beta_{p1}, \ldots, \beta_{pm}) = (1/\bar{a}_{qs})(\bar{b}_q, \beta_{q1}, \ldots, \beta_{qm})$ for some $p \neq q$, then $(\beta_{p1}, \ldots, \beta_{pm}) = (\bar{a}_{ps}/\bar{a}_{qs}) \times (\beta_{q1}, \ldots, \beta_{qm})$, which is not possible since both $(\beta_{p1}, \ldots, \beta_{pm})$ and $(\beta_{q1}, \ldots, \beta_{qm})$ are row vectors of the nonsingular matrix B^{-1}. Hence, if $\bar{A}_{.s} \nleq 0$, there is a unique lexico minimum among the nonempty set of vectors. $\{(\bar{b}_i, \beta_{i1}, \ldots, \beta_{im})/\bar{a}_{is}: i \text{ such that } \bar{a}_{is} > 0\}$. The minimum ratio in this pivot step in solving (9.4) is as follows

$$\text{Minimum } \{(b_i + \beta_{i1}\varepsilon + \cdots + \beta_{im}\varepsilon^{\mathbf{m}})/\bar{a}_{is}: i \text{ such that } \bar{a}_{is} > 0\}$$

and clearly when ε is arbitrarily small, this minimum is attained by the same i that corresponds to the unique lexico minimum in the set of vectors $\{(\bar{b}_i, \beta_{i1}, \ldots, \beta_{im})/\bar{a}_{is}: i \text{ such that } \bar{a}_{is} > 0\}$.

Thus when ε is arbitrarily small, irrespective of its actual value, the pivot row is uniquely determined according to the following steps.

Step 1 Find out the i that attains the minimum of $(\bar{b}_i/\bar{a}_{is})$ over all i satisfying $\bar{a}_{is} > 0$. If the i attaining this minimum is unique, it determines the pivot row, and we stop. Otherwise go to Step 2.

Step 2 If there is a tie for the i attaining the minimum in Step 1, find the i that attains the minimum of $(\beta_{i1}/\bar{a}_{is})$ over all i that tied for the minimum in Step 1. If the minimum here is attained by a unique i, that determines the pivot row and we stop. Otherwise go to the next Step.

The general step is:

Step *t* If the pivot row is not determined unambiguously in Step $t-1$, find out the i that attains the minimum of $(\beta_{i,t-1}/\bar{a}_{is})$ over all i that tied for the minimum in Step $t-1$. If the minimum here is attained by a unique i, it defines the pivot row and we stop. Otherwise go to the next step.

During or before Step m the pivot row will be determined uniquely.

CHANGE IN OBJECTIVE VALUE

Let B be a feasible basis for (9.4) for arbitrarily small ε and let $B^{-1} = (\beta_{ij})$. Let $\bar{b} = B^{-1}b$ and c_B be the row vector of basic cost coefficients. The corresponding dual solution is $\pi = c_B B^{-1}$ and let $\bar{z} = c_B\bar{b}$. The BFS of (9.4) corresponding to the basis B and the objective value, $\bar{z}_0$, at this solution are*

$$\text{Basic vector } x_B = \begin{pmatrix} \bar{b}_1 + \beta_{11}\varepsilon + \cdots + \beta_{1M}\varepsilon^{\mathbf{m}} \\ \vdots \\ \bar{b}_m + \beta_{m1}\varepsilon + \cdots + \beta_{mm}\varepsilon^{\mathbf{m}} \end{pmatrix}$$

$$\bar{z}_0 = \bar{z} + \pi_1\varepsilon + \cdots + \pi_m\varepsilon^{\mathbf{m}}$$

Suppose the nonbasic variable x_s (with relative cost coefficient $\bar{c}_s$) is the entering variable. Suppose the pivot row is the rth row for this pivot step. Let z_1 be the objective value after the pivot. Then

$$\begin{aligned} z_1 &= \bar{z}_0 + \bar{c}_s(\bar{b}_r + \varepsilon\beta_{r1} + \cdots + \varepsilon^{\mathbf{m}}\beta_{rm})/\bar{a}_{rs} \\ &= \bar{z} + \bar{c}_s\frac{\bar{b}_r}{\bar{a}_{rs}} + \sum_{t=1}^{m} \varepsilon^{t}(\pi_t + \bar{c}_s\beta_{rt}/\bar{a}_{rs}) \end{aligned}$$

For arbitrarily small ε, $z_1 < \bar{z}_0$. This implies that $(\bar{z} + \bar{c}_s\bar{b}_r/\bar{a}_{rs}, \pi_1 + \bar{c}_s\beta_{r1}/\bar{a}_{rs}, \ldots, \pi_m + \bar{c}_s\beta_{rm}/\bar{a}_{rs}) \prec (\bar{z}, \pi_1, \ldots, \pi_m)$.

As long as we do not wish to give a specific value to ε, we can treat the vector $(\bar{b}_i, \beta_{i1}, \ldots, \beta_{im})$ as defining the value of the ith basic variable in the BFS corresponding to the basis B. Given this vector, the actual value of the ith basic

* As usual, all the nonbasic variables have zero value in this BFS.

variable for any particular value of ε is $\bar{b}_i + \varepsilon\beta_{i1} + \cdots + \varepsilon^m\beta_{im}$. Similarly the vector $(\bar{z}, \pi_1, \ldots, \pi_m)$ may be treated as defining the value of the objective function corresponding to the feasible basis B. Treating the values of the basic variables and the objective value as these vectors we have the following:

1. A basis is feasible to (9.4) iff the values of all the basic variables are lexico positive. When this happens, we say that the basis is *lexico feasible.*
2. The algorithm starts with a lexico feasible basis and all the bases obtained in the course of the algorithm will be lexico feasible.
3. The termination criteria and the pivot column choice are the same as under the regular simplex method.
4. Once a pivot column has been selected, the pivot row is selected by the rules discussed above. The new pivot row choice rule is known as the *lexico minimum ratio rule.* It determines the pivot row (and, hence, the pivot element) uniquely without any ambiguity.
5. When the simplex algorithm using these rules, terminates, the terminal tableau is a terminal tableau for (9.4) for all ε arbitrarily small and also for $\varepsilon = 0$.
6. The terminal basis for (9.4) obtained by the simplex algorithm using these rules is also a terminal basis for (9.1).

9.3 SUMMARY OF THE SIMPLEX ALGORITHM USING THE LEXICO MINIMUM RATIO RULE

Consider the LP (9.1). We assume that it is in standard form (i.e., that $b \geqq 0$).

Case 1 If A contains a units matrix of order m as a submatrix, pick it as the initial feasible basis. Since b is nonnegative and the basis is the unit matrix, it is obviously lexico feasible. If we start with this basis, the problem is solved using the revised simplex algorithm with the explicit form of the inverse. However, instead of using the regular minimum ratio rule, the lexico minimum ratio rule discussed in section 9.2.4 is used in each pivot step to determine the pivot row. In the revised simplex algorithm, the inverse tableau is computed and this contains all the data needed to perform the lexico minimum ratio rule instead of the usual minimum ratio rule. It only requires some additional work in identifying the pivot element when a tie occurs in the usual minimum ratio test. This guarantees that there will be strict lexico decrease in the vector $(\bar{z}, \pi_1, \ldots, \pi_m)$ after each pivot step. Hence, a basis that appeared once can never reappear. *There cannot be any cycling.* After a finite number of pivot steps, the algorithm obtains a terminal basis.

Case 2 If A does not contain a unit matrix of order m as a submatrix, provide a full artificial basis and formulate the Phase I problem.

$$\text{Minimize} \qquad w(y) = \sum_{i=1}^{m} y_i$$

$$\begin{aligned} \text{Subject to} \quad & Ax + Iy = b \\ & x \geqq 0, \quad y \geqq 0 \end{aligned}$$

For this Phase I problem, the artificial basic vector is an initial lexico feasible basic vector. Starting with this lexico feasible basic vector, solve this Phase I problem as under Case 1, using the revised simplex algorithm with the explicit form of the inverse using the lexico minimum ratio rule. When Phase I is completed, if the original problem turns out to be feasible, go over to Phase II as usual under the simplex algorithm and continue with the revised simplex routine using the lexico minimum ratio rule.

9.4 DO COMPUTER CODES USE THE LEXICO MINIMUM RATIO RULE?

If the LP is solved by the revised simplex method with the explicit form of the inverse, the algorithm in section 9.3, which guarantees that cycling cannot occur, only requires the replacement of the standard minimum ratio rule by the lexico minimum ratio rule for pivot row choice. It requires very little additional work. However, most computer programs in existence for solving LPs do not include this routine for the following reasons.

1. Cycling does not seem to occur in the course of solving most practical problems even though they are degenerate. Therefore for practical applications it seems unnecessary to include routines for avoiding cycling.
2. Most LP codes use the revised simplex algorithm with the product form of the inverse to conserve core space. Under this routine the actual basis inverse is never actually computed explicitly. It makes it possible for them to solve large problems. Consequently, it is very difficult to adopt the lexico minimum ratio rule, as it requires the explicit basis inverse in each pivot step.

Example

Consider the following problem by K. T. Marshall and J. W. Suurballe (reference [2]).

x_1	x_2	x_3	x_4	x_5	x_6	x_7	$-z$	b
1	0	0	1	1	1	1	0	1
0	1	0	$\frac{1}{2}$	$-\frac{11}{2}$	$-\frac{5}{2}$	9	0	0
0	0	1	$\frac{1}{2}$	$-\frac{3}{2}$	$-\frac{1}{2}$	1	0	0
0	0	0	-1	7	1	2	1	0

$x_j \geqq 0$ for all j, z to be minimized

Use (x_1, x_2, x_3) as the initial basic vector. Verify that the following sequence of pivot steps is such that the incoming variable is eligible to enter the basic vector according to the objective improvement routine and that the outgoing basic variable is determined by the usual minimum ratio test.

Pivot step number	Entering variable	Leaving variable
1	x_4	x_2
2	x_5	x_3
3	x_6	x_4
4	x_7	x_5
5	x_2	x_6
6	x_3	x_7

All these are degenerate pivot steps and at the end of pivot step 6, the initial basic vector (x_1, x_2, x_3) is obtained again.

Verify that the lexico minimum ratio rule selects the leaving variable as x_3 in the first pivot step when x_4 is the entering variable into the basic vector (x_1, x_2, x_3). After this pivot we are led to an optimum basis.

PROBLEMS

9.3 Consider the following LP

x_1	x_2	x_3	x_4	x_5	x_6	x_7	$-z$	b
1	0	0	0	$\frac{3}{5}$	$-\frac{32}{5}$	$\frac{24}{5}$	0	0
0	1	0	0	$\frac{1}{5}$	$-\frac{9}{5}$	$\frac{3}{5}$	0	0
0	0	1	0	$\frac{2}{5}$	$-\frac{8}{5}$	$\frac{1}{5}$	0	0
0	0	0	1	0	1	0	0	1
0	0	0	0	$-\frac{2}{5}$	$-\frac{2}{5}$	$\frac{9}{5}$	1	0

$x_j \geqq 0$ for all j, $\quad z$ to be minimized

In solving this problem by the simplex algorithm starting with the feasible basic vector (x_1, x_2, x_3, x_4), in each step select the entering variable as x_s where

$$s = \text{minimum } \{j: j \text{ such that } \bar{c}_j < 0\}$$

and pick the pivot row as the rth row where r = minimum $\{i: i$ such that row i is eligible to be the pivot row by the usual minimum ratio test$\}$. Using these rules show that cycling occurs. Verify that the use of the lexico minimum ratio rule avoids the cycling.

—*K. T. Marshall and J. W. Suuraballe*

9.4 Consider the following LP

x_1	x_2	x_3	x_4	x_5	x_6	x_7	$-z$	b
1	0	0	0	$\frac{1}{3}$	$-\frac{32}{9}$	$\frac{20}{9}$	0	0
0	1	0	0	$\frac{1}{6}$	$-\frac{13}{9}$	$\frac{5}{18}$	0	0
0	0	1	0	$\frac{2}{3}$	$-\frac{16}{9}$	$\frac{1}{9}$	0	0
0	0	0	1	0	0	0	0	1
0	0	0	0	-7	-4	15	1	0

$x_j \geqq 0$ for all j, $\quad z$ to be minimized

If the simplex algorithm is applied on this problem with (x_1, x_2, x_3, x_4) as the initial feasible basic vector, using the selection rules described in Problem 9.3, show that cycling occurs, even though the objective function is unbounded below on the set of feasible solutions of this problem.

—*J. W. Suurballe*

9.5 Consider the following LP:

x_1	x_2	x_3	x_4	x_5	x_6	$-z$	b
1	0	0	$-\frac{1}{3}$	$-\frac{1}{6}$	$-\frac{2}{3}$	0	$-\frac{1}{3}$
0	1	0	$\frac{32}{9}$	$\frac{13}{9}$	$\frac{16}{9}$	0	$\frac{8}{9}$
0	0	1	$-\frac{20}{9}$	$-\frac{5}{18}$	$-\frac{1}{9}$	0	$\frac{4}{9}$
1	1	1	1	1	1	1	0

$x_j \geqq 0$ for all j, $\quad z$ to be minimized

Verify that (x_1, x_2, x_3) is a dual feasible (but primal infeasible) basic vector. Starting with this basic vector, apply the dual simplex algorithm using the following choice rules.

(a) The pivot row is the row corresponding to the most negative updated right-hand side constant and also the topmost row among such rows, in case of a tie.

(b) The pivot column is the leftmost of the eligible columns as determined by the usual dual simplex minimum ratio test.

Verify that cycling occurs.

—*J. W. Suurballe*

9.6 Develop a method (similar to the lexico minimum ratio rule) for pivot column choice in the dual simplex algorithm that guarantees cycling cannot occur. (Use canonical tableaus if necessary.)

9.7 Consider the LP:

$$\begin{aligned} \text{Minimize} \quad & cx \\ \text{Subject to} \quad & Ax = b \\ & x \geqq 0 \end{aligned}$$

Let x_{B_1} be a feasible basic vector satisfying the property that exactly one basic variable is zero in the BFS corresponding to it (i.e., if B_1 is the associated basis, then $B_1^{-1}b \geqq 0$, and exactly one component of the vector $B_1^{-1}b$ is zero). Prove that x_{B_1} cannot be in a cycle of degenerate basic vectors for this problem.

REFERENCES

1. A. Charnes, "Optimality and Degeneracy in Linear Programming," *Econometrica*, *20*, 2, pp. 160–170, April 1952.

2. K. T. Marshall and J. W. Suurballe, "A Note on Cycling in the Simplex Method," *Naval Research Logistics Quarterly*, *16*, 1, pp. 121–137, March 1969.

3. Also see Dantzig's book, reference [1] in Chapter 18.

10 bounded variable linear programs

10.1 INTRODUCTION

Here we consider LPs in which some or all of the variables are restricted to lie within specified finite lower and upper bounds, known as *bounded variable linear programs*. Let p be the number of bound constraints, and let m be the number of other constraints in the problem. This problem can of course be solved by including all the bound restrictions as constraints, but the increased size of the problem is undesirable computationally. Here we develop special techniques that make it possible to solve this problem using bases of order m only.

10.2 BOUNDED VARIABLE PROBLEMS

10.2.1 Standard Form for Bounded Variable Linear Programs

We consider LPs of the form

$$\begin{array}{lll} \text{Minimize} & cy & \\ \text{Subject to} & Ay = b' & \\ & \ell_j \leqq y_j \leqq k_j & \text{for } j \in \mathbf{J} = \{1, \ldots, n_1\} \\ & y_j \geqq 0 & \text{for } j \in \bar{\mathbf{J}} = \{n_1 + 1, \ldots, n\} \end{array} \tag{10.1}$$

where A is a matrix of order $m \times n$. If $n_1 = n$, then $\bar{\mathbf{J}} = \emptyset$, and in this case all the variable in the problem are bounded. ℓ_j and k_j are finite and $\ell_j \leqq k_j$ for all $j \in \mathbf{J}$.

Substitute $y_j = x_j + \ell_j$ for all $j \in \mathbf{J}$ and $y_j = x_j$ for all $j \in \bar{\mathbf{J}}$ and eliminate the y_j from the problem. Let $b = b' - \sum_{j \in \mathbf{J}} A_{.j} \ell_j$. Transferring all the constant terms to the right-hand side and eliminating the constant terms from the objective function, the LP (10.1) is transformed into the equivalent problem:

$$\begin{aligned} \text{Minimize} \quad & z(x) = cx \\ \text{Subject to} \quad & Ax = b \\ & x_j \leqq U_j = k_j - \ell_j \qquad \text{for } j \in \mathbf{J} = \{1, \ldots, n_1\} \\ & x_j \geqq 0 \qquad \text{for all } j \end{aligned} \tag{10.2}$$

As usual, assume that all b_i are nonnegative. If some b_i is negative, multiply both sides of the ith equation in (10.2) by -1. If $U_j = 0$ for some $j \in \mathbf{J}$, the corresponding variable x_j should be equal to zero in all feasible solutions of (10.2), and it can be eliminated from further consideration. From now on we will assume that $U_j > 0$ for all $j \in \mathbf{J}$.

Now the problem is in standard form for bounded variable LPs.

10.2.2 Basic Feasible Solutions

From the definition in section 3.4.3. it can be verified that a feasible solution, $\bar{x}$, of (10.2) is a BFS iff the set $\{A_{.j}: j \in \mathbf{J}, j \text{ such that } 0 < \bar{x}_j < U_j\} \cup \{A_{.j}: j \in \bar{\mathbf{J}}, j \text{ such that } \bar{x}_j > 0\}$ is linearly independent.

For the LP (10.2), we define a *working basis* to be a square nonsingular submatrix of A of order m. Given a working basis for (10.2), a variable associated with a column vector of the working basis will be called a *basic variable* when referring to this working basis. All other variables will be called *nonbasic variables.*

From the previous discussion, it is clear that a feasible solution, $\bar{x}$, is a BFS (and, hence, an extreme point of the set of feasible solutions) iff there exists a working basis with the property that all nonbasic variables with respect to this working basis are either equal to their lower bound (0 here) or their upper bound (provided this upper bound is finite) in the solution $\bar{x}$. All basic variables have to satisfy the bound restrictions on them, but they can have any value within their respective bounds.

Numerical Example

Consider the system of constraints in Tableau 10.1.

Tableau 10.1

x_1	x_2	x_3	x_4	x_5	x_6	x_7	b
1	0	1	-1	1	2	1	6
0	1	0	1	-2	1	-2	4
0	0	1	-1	0	2	1	1

$$0 \leqq x_j \leqq 6 \quad \text{for } j = 1, 2$$
$$0 \leqq x_4 \leqq 4, 0 \leqq x_5 \leqq 2, 0 \leqq x_6 \leqq 10$$
$$\text{and } x_3, x_7 \geqq 0$$

Let $\bar{x}$ be $(3, 4, 5, 4, 2, 0, 0)^T$. Let B_1 be the working basis associated with the basic vector (x_1, x_2, x_3). $\bar{x}$ is a feasible solution. The nonbasic variables x_4, x_5 are at their upper bounds and the nonbasic variables x_6, x_7 are at their lower bounds in the feasible solution $\bar{x}$. The basic variables x_1, x_2, x_3 are strictly within their respective bounds in $\bar{x}$. Hence $\bar{x}$ is a BFS corresponding to the working basis B_1.

Consider the feasible solution $\tilde{x} = (5, 2, 0, 1, 0, 1, 0)^{\mathrm{T}}$ of this system. In this solution, the variables x_1, x_2, x_4, and x_6 are all strictly within their respective bounds. Since the set of four column vectors associated with these four variables is linearly dependent, $\tilde{x}$ is not a BFS.

Note For a general LP in standard form, given a basis, the basic solution associated with this basis is uniquely defined as in section 3.5.1. However, for the bounded variable LP (10.2), given a working basis, we have the freedom to set the nonbasic variables either at their lower bounds or upper bounds. Hence, there may be several basic solutions of (10.2) associated with a given working basis.

10.2.3 Optimality Criterion

Assigning the dual variables $\pi_1, \ldots, \pi_m; \mu_1, \ldots, \mu_{n_1}$, to the constraints in that order, the dual problem of (10.2) is

$$\begin{array}{lll} \text{Maximize} & \pi b - \mu \mathrm{U} & \\ \text{Subject to} & \pi A_{.j} - \mu_j \leqq c_j & \text{for } j \in \mathbf{J} \\ & \pi A_{.j} \qquad \leqq c_j & \text{for } j \in \bar{\mathbf{J}} \\ & \pi \text{ unrestricted} & \mu \geqq 0 \end{array}$$

The complementary slackness conditions for optimality in this primal-dual pair are

$$\begin{array}{rl} x_j(c_j - \pi A_{.j} + \mu_j) = 0 & \text{for } j \in \mathbf{J} \\ x_j(c_j - \pi A_{.j}) = 0 & \text{for } j \in \bar{\mathbf{J}} \\ \mu_j(\mathrm{U}_j - x_j) = 0 & \text{for } j \in \mathbf{J} \end{array}$$

Define $\bar{c}_j = c_j - \pi A_{.j}$. Dual feasibility requires $\bar{c}_j + \mu_j \geqq 0$ for all $j \in \mathbf{J}$ and $\bar{c}_j \geqq 0$ for all $j \in \bar{\mathbf{J}}$. For some $j \in \mathbf{J}$, if $\bar{c}_j < 0$, dual feasibility automatically requires $\mu_j > 0$ (because $\bar{c}_j + \mu_j \geqq 0$ for $j \in \mathbf{J}$) and the complementary slackness conditions imply that $x_j = \mathrm{U}_j$. For $j \in \mathbf{J}$ if $\bar{c}_j > 0$ then $\bar{c}_j + \mu_j > 0$ (because μ_j is a nonegative dual variable) and, in this case, the complementary slackness conditions imply that $x_j = 0$. Again for $j \in \mathbf{J}$, if x_j is strictly within its bounds, the complementary slackness property will only be satisfied if $\bar{c}_j = 0$ and $\mu_j = 0$.

For $j \in \bar{\mathbf{J}}$, $\bar{c}_j \geqq 0$ for dual feasibility and $\bar{c}_j > 0$ implies that $x_j = 0$ by the complementary slackness property.

Thus, given a feasible solution $\bar{x}$ of (10.2), it is optimum iff there exists a vector π such that when $\bar{c} = c - \pi A$, the variable x_j is at its lower bound 0 whenever the corresponding $\bar{c}_j > 0$, and at its finite upper bound U_j, whenever the corresponding $\bar{c}_j < 0$, in the solution $\bar{x}$.

Let $\bar{x}$ be a BFS of (10.2) associated with a working basis B. If x_j is a basic variable, $\bar{x}_j$ can lie strictly within its respective bounds, and by the complementary slackness property this arrangement requires that $\bar{c}_j$ is 0. Hence, the π vector associated with the working basis B can be obtained by solving $c_j - \pi A_{.j} = 0$ for all j such that $A_{.j}$ is a column vector in the working basis B. As usual, let c_B represent the basic cost coefficient vector. Then π is obtained by solving "$c_B = \pi B$." Therefore, $\pi = c_B B^{-1}$, which is the same as in the general LP model. Having computed π,

$\bar{c}$ is obtained from $\bar{c} = c - \pi A$. The optimality criteria are

$$\begin{array}{llll} \text{for} & j \in \mathbf{J}, \bar{c}_j < 0 & \text{implies} & \bar{x}_j = \mathrm{U}_j \\ & \bar{c}_j > 0 & \text{implies} & \bar{x}_j = 0 \\ \text{and} & j \in \bar{\mathbf{J}}, \bar{c}_j \geqq 0 & & \end{array}$$

Numerical Example

Let $z(x) = -3x_1 + 4x_2 + 2x_3 - 2x_4 - 14x_5 + 11x_6 - 5x_7$. Consider the problem of minimizing $z(x)$ on the set of feasible solutions of the numerical example in Tableau 10.1. Let B_1 be the working basis associated with the basic vector (x_1, x_2, x_3). It is required to check whether the feasible solution $\bar{x} = (3, 4, 5, 4, 2, 0, 0)^{\mathrm{T}}$ corresponding to the working basis B_1 is optimal to this problem.

$$\pi = c_{B_1} B_1^{-1} = (-3, 4, 2) \begin{pmatrix} 1 & 0 & 1 \\ 0 & 1 & 0 \\ 0 & 0 & 1 \end{pmatrix}^{-1} = (-3, 4, 5)$$

$$\bar{c} = c - \pi A = (0, 0, 0, -4, -3, 3, 1).$$

$\bar{c}_4 = -4$ and $\bar{x}_4$ is 4, the upper bound for x_4. $\bar{c}_5 = -3$ and $\bar{x}_5$ is 2, the upper bound for x_5. $\bar{c}_6, \bar{c}_7$ are both positive and both $\bar{x}_6, \bar{x}_7$ are equal to zero, the lower bound for the variables x_6, x_7. Hence, all the optimality conditions are satisfied and $\bar{x}$ is an optimum solution of this problem.

From this it is clear that all the computations required for solving the problem can be performed using the working bases only. This is described in the next section.

10.3 SIMPLEX METHOD USING WORKING BASES

10.3.1 Phase I

Consider the bounded variable LP (10.2), where A is a matrix of order $m \times n$ and b is a nonnegative vector. Phase I of the simplex method finds an initial BFS to this problem. Introduce the artificial variables $x_{n+1}, \ldots, x_{n+m}$, as in Tableau 10.2.

Table 10.2. Original Tableau for the Phase I Problem

$x_1 \ldots x_n$	$x_{n+1} \ldots x_{n+m}$	$-z$	$-w$	b
$a_{11} \ldots a_{1n}$	$1 \ \ldots \ 0$	0	0	b_1
$\vdots \quad \vdots$	$\vdots \quad \vdots$	$\vdots$	$\vdots$	$\vdots$
$a_{m1} \ldots a_{mn}$	$0 \ \ldots \ 1$	0	0	b_m
$c_1 \ldots c_n$	$0 \ \ldots \ 0$	1	0	0
$0 \ \ldots 0$	$1 \ \ldots \ 1$	0	1	0

$x_j \leqq \mathrm{U}_j$ for $j \in \{1, \ldots, n_1\} = \mathbf{J}$

$x_j \geqq 0$ for all j. Artificial variables are $x_{n+1}, \ldots, x_{n+m}$.

In the BFSs obtained during the algorithm, the nonbasic variable x_j for $j \in \mathbf{J}$ is either equal to zero or equal to its upper bound U_j. However, nonbasic x_j for $j \in \bar{\mathbf{J}}$ are always equal to zero in the solution. *In each stage of this algorithm, the present value of each nonbasic variable has to be clearly recorded.*

During Phase I, try to minimize w, the sum of all the artificial variable introduced. Tableau 10.3 contains the initial inverse tableau.

Tableau 10.3

Basic variables	Inverse tableau				Basic values
x_{n+1}	$1 \dots$	0	0	0	b_1
$\vdots$	$\vdots$	$\vdots$	$\vdots$	$\vdots$	$\vdots$
x_{n+m}	$0 \dots$	1	0	0	b_m
$-z$	$0 \dots$	0	1	0	0
$-w$	$-1 \dots$	-1	0	1	$-w_0 = -\sum_{i=1}^{m} b_i$

All nonbasic variables = 0.

To discuss the Phase I termination criterion, assume that during Phase I, in a general step, say step k, the inverse tableau is as found in Tableau 10.4.

Tableau 10.4

Basic variables	Inverse tableau				Basic values
x_{i_1}	$\beta_{11} \dots$	β_{1m}	0	0	$\bar{b}_1$
$\vdots$	$\vdots$	$\vdots$	$\vdots$	$\vdots$	$\vdots$
x_{i_m}	$\beta_{m1} \dots$	β_{mm}	0	0	$\bar{b}_m$
$-z$	$-\pi_1 \dots$	$-\pi_m$	1	0	$-\bar{z}$
$-w$	$-\sigma_1 \dots$	$-\sigma_m$	0	1	$-\bar{w}$

Values of nonbasic variables $x_j = \bar{x}_j$ for nonbasic x_j.

The working basis B in this step is the submatrix of A consisting of the column vectors corresponding to the present basic variables, $x_{i_1}, \dots, x_{i_m}$. Let c_B, d_B be the row vectors of the Phase II and Phase I cost coefficients of the basic variables in this step, as usual. Let $\mathbf{N} = \{j\colon j \text{ such that } x_j \text{ is a nonbasic variable at present}\}$. Then the entries in the inverse tableau are

$$(\beta_{ij}) = B^{-1}, \pi = c_B B^{-1}, \sigma = d_B B^{-1}$$

$$\bar{b} = \text{vector of basic values} = B^{-1}\Big(b - \sum_{j \in \mathbf{N} \cap \mathbf{J}} \bar{x}_j A_{.j}\Big)$$

$$\bar{z} = c_B \bar{b} + \sum_{j \in \mathbf{N} \cap \mathbf{J}} c_j \bar{x}_j$$

$$\bar{w} = \text{sum of the values of basic artificials}$$

$\bar{d}_j = d_j - \sigma A_{.j} = -\sigma A_{.j}$ for each $j = 1$ to n. The updated column vector of x_j with respect to the present working basis is $\bar{A}_{.j} = B^{-1}A_{.j}$, as usual. Phase I termination criteria are obtained by applying the optimality criteria of section 10.2.3 to the Phase I problem. These are discussed in the next section.

10.3.2 Phase I Termination Criteria

Phase I terminates if in the present BFS the value of

$$\begin{aligned} \bar{x}_j &= \mathrm{U}_j && \text{for all } j \text{ such that } j \in \mathbf{J} \text{ and } \bar{d}_j < 0 \\ &= 0 && \text{for all } j \text{ such that } j \in \mathbf{J} \text{ and } \bar{d}_j > 0 \\ \text{and} \quad \bar{d}_j &\geqq 0 && \text{for all } j \in \bar{\mathbf{J}} \end{aligned}$$

10.3.3 Choice of Nonbasic Variable for Change of Status during Phase I

Let Tableau 10.4 be the present inverse tableau and let the solution given there be the present solution. If Phase I termination criteria are not satisfied, select a nonbasic variable, x_s, which has led to a violation of the termination criteria, for changing its status. This variable will be called the *entering variable* as in Chapter 2. It belongs to one of the following types.

(i) $s \in \mathbf{J}, \bar{d}_s < 0,$ and $\bar{x}_s = 0$

or

(ii) $s \in \mathbf{J}, \bar{d}_s > 0,$ and $\bar{x}_s = \mathrm{U}_s$

or

(iii) $s \in \bar{\mathbf{J}}, \bar{d}_s < 0$

Changing the status of x_s to decrease the Phase I objective value further is discussed in the following sections.

10.3.4 Changing the Status of x_s Where $s \in \mathbf{J}$ and $\bar{d}_s < 0$

Here the entering variable, x_s, is such that its present value is zero and $\bar{d}_s < 0$. The fact that $\bar{d}_s < 0$ implies that increasing the value of this variable from its present value of 0 would reduce the Phase I objective value. As is usual in the simplex method, leave the values of all nonbasic variables other than x_s unchanged, change the value of x_s from 0 to a parameter λ, and reevaluate the values of all the present basic variables as functions of λ.

Clearly the new solution is given by:

$$\begin{aligned} x_j &= \bar{x}_j && \text{for all } j \in \mathbf{N}, j \neq s \\ x_s &= \lambda \\ x_{i_r} &= \bar{b}_r - \bar{a}_{rs}\lambda && r = 1 \text{ to } m \qquad (10.3) \\ z &= \bar{z} + \bar{c}_s\lambda \\ w &= \bar{w} + \bar{d}_s\lambda \end{aligned}$$

where $\mathbf{N} = \{j : j \text{ such that } x_j \text{ is a nonbasic variable}\}$. For the new solution to remain feasible, the parameter λ has to be chosen so that the nonnegativity restrictions hold on the values of all the variables and the upper bound restrictions hold on all variables x_j for $j \in \mathbf{J}$. The restriction that all the present basic variables should be nonnegative implies that

$$\bar{b}_r - \bar{a}_{rs}\lambda \geqq 0$$

that is,

$$\lambda \leqq \frac{\bar{b}_r}{\bar{a}_{rs}} \qquad \text{for all } r \text{ such that } \bar{a}_{rs} > 0$$

The restriction that all the present basic variables x_{i_r}, $i_r \in \mathbf{J}$ cannot exceed the corresponding upper bounds U_{i_r} implies that

$$\bar{b}_r - \bar{a}_{rs}\lambda \leqq U_{i_r}$$

that is,

$$\lambda \leqq \frac{\bar{b}_r - U_{i_r}}{\bar{a}_{rs}} \qquad \text{for all } r \text{ such that } i_r \in \mathbf{J} \text{ and } \bar{a}_{rs} < 0$$

Thus the maximum value that λ can have is $\theta = \min\,\{\bar{b}_r/\bar{a}_{rs}$ for all r such that $\bar{a}_{rs} > 0, (\bar{b}_r - U_{i_r})/\bar{a}_{rs}$ for all r such that $i_r \in \mathbf{J}$ and $\bar{a}_{rs} < 0, U_s\}$.

Case 1 If $\theta = U_s$, x_s becomes a nonbasic variable, whose value is equal to its upper bound in the next step. Hence, x_s moves from being a nonbasic variable with a value of 0 to one with a value equal to its upper bound. There is no change in the working basis. The only things that change in the inverse tableau are the values of the basic variables that are computed from (10.3) by substituting $\lambda = \theta = U_s$, and the value of the nonbasic variable x_s that is changed from 0 to U_s. Everything else remains the same and the algorithm moves to the next step.

Example

Consider the case where (x_1, x_2, x_3) is the present basic vector, while solving a bounded variable LP. Let x_4 whose value in the present solution, $\bar{x}_4 = 0$, which has $\bar{d}_4 < 0$, be the entering variable. Let U_1, U_2, U_3, U_4; the upper bounds on x_1, x_2, x_3, x_4; be 12, 30, 35, 3, respectively. Let the vector of basic values in the present BFS be $(x_1, x_2, x_3) = \bar{b} = (10, 20, 30)$.

Let $\bar{A}_{.4}$, the updated column vector of x_4, be $(1, -2, 0)^{\mathrm{T}}$. Then the new solution obtained after changing the value of x_4 from 0 to λ is

$$\begin{aligned} x_1 &= 10 - \lambda \\ x_2 &= 20 + 2\lambda \\ x_3 &= 30 \\ x_4 &= \lambda \\ x_j &= \bar{x}_j, \quad \text{its present value, for all other } j \end{aligned} \tag{10.4}$$

The maximum value that λ can have is 3. Hence in this case x_4 moves from being a zero-valued nonbasic variable to a nonbasic variable at upper bound. The solution after the change is obtained by substituting 3 for λ in (10.4). With the new solution, but the same inverse tableau and the same relative cost coefficients, the algorithm moves to the next step.

Case 2 Suppose $\theta < U_s$ and suppose $\theta = \bar{b}_t/\bar{a}_{ts}$ for some t such that $\bar{a}_{ts} > 0$. In this case, θ is strictly less than the upper bound for x_s, but if λ exceeds θ, the present tth basic variable x_{i_t} becomes negative. When $\lambda = \theta$, the basic variable x_{i_t} attains its lower bound 0. In this case, the basic variable x_{i_t} is dropped from the basic vector, and x_s becomes the tth basic variable in its place. The updated column vector of x_s is the pivot column and the new inverse tableau is obtained by performing the pivot with $\bar{a}_{ts}$ as the pivot element.

The new inverse tableau could also have been obtained by multiplying the present inverse tableau on the left by the appropriate pivot matrix corresponding to this pivot operation, instead of performing the pivot.

The vector of new basic values is obtained from (10.3) by substituting $\lambda = \theta$. The variable x_s is removed from the list of nonbasic variables and x_{i_t} is listed as a new nonbasic variable whose value in the new solution is zero. This completes this step and the algorithm moves to the next step.

Example

Consider the problem in which the data is the same as in Case 1, with the exception that $\bar{A}_{.4} = (1, -1, 15)^T$. In this case, the new solution obtained after changing the value of x_4 from 0 to λ is

$$\begin{aligned} x_1 &= 10 - \lambda \\ x_2 &= 20 + \lambda \\ x_3 &= 30 - 15\lambda \\ x_4 &= \lambda \\ x_j &= \bar{x}_j \qquad \text{its present value, for all other } j \end{aligned} \tag{10.5}$$

The maximum value that λ can have here is 2. When $\lambda = 2$, x_3 becomes equal to zero. Hence, x_4 replaces x_3 from the basic vector. The new inverse tableau is obtained as in Chapter 5, by performing a pivot with the updated column vector of x_4 as the pivot column and the row in which x_3 is the basic variable as the pivot row. The new solution however, is obtained by substituting 2 for λ in (10.5).

Case 3 Suppose $\theta < U_s$ and suppose $\theta = (\bar{b}_t - U_{i_t})/\bar{a}_{ts}$ for some t such that $\bar{a}_{ts} < 0$ and $i_t \in \mathbf{J}$. In this case when λ exceeds θ, the value of the present tth basic variable, x_{i_t} exceeds its specified upper bound U_{i_t} in the new solution. In this case, the variable x_{i_t} is made into a nonbasic variable whose value is equal to its upper bound in the new solution. The variable x_s becomes the tth basic variable in its place. The new inverse tableau is computed as in Case 2 above. The values of the basic variables in the new solution are obtained from (10.3) by substituting $\lambda = \theta$. The algorithm then moves to the next step.

Example

Consider the problem in which the data is the same as in Case 1, with the exception that $\bar{A}_{.4} = (-1, 1, -1)^T$. The new solution obtained after changing the value of x_4 from 0 to λ is

$$\begin{aligned} x_1 &= 10 + \lambda \\ x_2 &= 20 - \lambda \\ x_3 &= 30 + \lambda \\ x_4 &= \lambda \\ x_j &= \bar{x}_j, \qquad \text{its present value for all other } j \end{aligned} \tag{10.6}$$

In this case the maximum value that λ can have is 2. When $\lambda = 2$, x_1 reaches its upper bound of 12. Hence x_4 replaces x_1 from the basic vector. x_1 becomes a nonbasic variable at its upper bound. The new solution is obtained by substituting $\lambda = 2$ in (10.6) and the new inverse tableau is obtained by performing a pivot with the updated column of x_4 as the pivot column and the row in which x_1 is basic as the pivot row (as in Chapter 5).

10.3.5 Changing the Status of x_s Where $s \in \mathbf{J}$ and $\bar{d}_s > 0$

Here the entering variable x_s is such that its present value is equal to its upper bound and $\bar{d}_s > 0$. This implies that decreasing x_s from its present value U_s, leads to a reduction in the Phase I objective function. Leave the values of all the nonbasic variables other than x_s unchanged, change the value of x_s from U_s to a parameter λ, and obtain the values of the present basic variables as functions of λ. The new solution is

$$\begin{aligned} x_j &= \bar{x}_j, \text{ its present value for all } j \neq s, j \in \mathbf{N} \\ x_s &= \lambda \\ x_{i_r} &= \text{the present } r\text{th basic variable} \\ &= \bar{b}_r + \bar{a}_{rs}U_s - \bar{a}_{rs}\lambda \qquad r = 1 \text{ to } m \\ z &= \bar{z} - \bar{c}_s U_s + \bar{c}_s\lambda \\ w &= \bar{w} - \bar{d}_s U_s + \bar{d}_s\lambda \end{aligned} \tag{10.7}$$

Since $\bar{d}_s > 0$, the maximum decrease in the value of w can be achieved by making λ as low as possible consistent with the requirements for the feasibility of the new solution in (10.7). The restriction that all the present basic variables should be nonnegative implies that

$$\bar{b}_r + \bar{a}_{rs}U_s - \bar{a}_{rs}\lambda \geqq 0$$

that is,

$$\lambda \geqq (\bar{b}_r + \bar{a}_{rs}U_s)/\bar{a}_{rs} \qquad \text{for all } r \text{ such that } \bar{a}_{rs} < 0$$

The restriction that all the present basic variables x_{i_r}, $i_r \in \mathbf{J}$ should be less than or equal to their corresponding upper bounds, U_{i_r}, implies that $\bar{b}_r + \bar{a}_{rs}U_s - \bar{a}_{rs}\lambda \leqq U_{i_r}$, that is,

$$\lambda \geqq (\bar{b}_r + \bar{a}_{rs}U_s - U_{i_r})/\bar{a}_{rs} \qquad \text{for all } r \text{ such that } i_r \in \mathbf{J} \text{ and } \bar{a}_{rs} > 0.$$

Thus, the minimum value that λ can have is $\theta = \max\,\{(\bar{b}_r + \bar{a}_{rs}U_s)/\bar{a}_{rs}$ for all r such that $\bar{a}_{rs} < 0$, $(\bar{b}_r + \bar{a}_{rs}U_s - U_{i_r})/\bar{a}_{rs}$ for all r such that $i_r \in \mathbf{J}$ and $\bar{a}_{rs} > 0, 0\}$. There are several cases to consider here.

Case 1 Suppose $\theta = 0$. Then x_s moves from a nonbasic variable whose value is presently equal to its upper bound into a nonbasic variable whose value is equal to 0 in the next solution. There is no change in the working basis. The only changes in the inverse tableau occur in the values of the basic variables that are computed from (10.7) by substituting $\lambda = 0$. The algorithm then moves to the next step.

Example

Consider a problem in which the data is the same as in the example under Case 1 of section 10.3.4, with the exception that the entering variable, x_4, is presently a nonbasic variable at its upper bound, 3, with $\bar{d}_4 > 0$, and its updated column vector, $\bar{A}_{.4} = (-1, 2, -3)^{\mathrm{T}}$. Hence, x_4 should be decreased from its present value of 3. If the value of x_4 is changed to λ, the new solution is

$$\begin{aligned} x_1 &= 10 + 3(-1) - \lambda(-1) \\ x_2 &= 20 + 3(2) - \lambda(2) \\ x_3 &= 30 + 3(-3) - \lambda(-3) \\ x_4 &= \lambda \\ x_j &= \bar{x}_j, \qquad \text{its present value, for all other } j \end{aligned} \tag{10.8}$$

The minimum value that λ can have in (10.8) is zero. Hence x_4 is made into a zero-valued nonbasic variable in the next step. The new solution is obtained by substituting 0 for λ in (10.8). There is no change in the inverse tableau. The algorithm then moves to the next step.

Case 2 Suppose $\theta > 0$, and is equal to $(\bar{b}_t + \bar{a}_{ts}U_s)/\bar{a}_{ts}$ for some t such that $\bar{a}_{ts} < 0$. When $\lambda = \theta$ in (10.7), the present tth basic variable x_{i_t} assumes a value of 0 and it becomes a nonbasic variable whose value is 0 in the new solution. x_s becomes the tth basic variable in its place. The new inverse tableau is obtained as under Case 2 of section 10.3.4. The basic values in the new solution are obtained from (10.7) by substituting $\lambda = \theta$.

Example

Consider the problem in which the data is the same as in the example under Case 1 above (in this section) with the exception that the updated column vector of x_4 is $\bar{A}_{.4} = (-5, -6, -2)^T$. If the value of x_4 is changed to λ, the new solution here is

$$
\begin{aligned}
x_1 &= 10 + 3(-5) - \lambda(-5) \\
x_2 &= 20 + 3(-6) - \lambda(-6) \\
x_3 &= 30 + 3(-2) - \lambda(-2) \\
x_4 &= \lambda \\
x_j &= \bar{x}_j, \quad \text{its present value, for all other } j
\end{aligned}
\tag{10.9}
$$

The minimum value that λ can have in (10.9) is 1. When $\lambda = 1$, x_1 becomes equal to zero. So x_4 replaces x_1 from the basic vector. x_1 becomes a zero-valued nonbasic variable. The new inverse tableau is obtained as in Chapter 5 by pivoting with the updated column of x_4 as the pivot column and the row in which x_1 is the basic variable, as the pivot row. The new solution is obtained by substituting $\lambda = 1$ in (10.9). The algorithm then moves to the next step.

Case 3 Suppose $\theta > 0$ and is equal to $(\bar{b}_{ts} + \bar{a}_{ts}U_s - U_{i_t})/\bar{a}_{ts}$ for some t such that $i_t \in \mathbf{J}$ and $\bar{a}_{ts} > 0$. In this case when $\lambda = \theta$ in (10.7), the basic variable x_{i_t} is equal to its upper bound and it becomes a nonbasic variable whose value is equal to its upper bound in the next solution. x_s becomes the tth basic variable in its place. The new inverse tableau is obtained as in Case 2 of section 10.3.4 and the values of the basic variables in the new solution are computed by substituting $\lambda = \theta$ in (10.7). Go now to the next step.

Example

Consider the problem in which the data is the same as in the example under Case 1 of this section, with the exception that the updated column vector of the entering variable, x_4, is $\bar{A}_{.4} = (-1, 5, 0)^T$. If the value of x_4 is changed to λ, the new solution in this case is

$$
\begin{aligned}
x_1 &= 10 + 3(-1) - \lambda(-1) \\
x_2 &= 20 + 3(5) - \lambda(5) \\
x_3 &= 30 \\
x_4 &= \lambda \\
x_j &= \bar{x}_j, \quad \text{its present value, for all other } j
\end{aligned}
\tag{10.10}
$$

The minimum value that λ can have in (10.10) is 1. When $\lambda = 1$, x_2 becomes equal to 30, its upper bound. So x_4 replaces x_2 from the basic vector. x_2 becomes a nonbasic variable at its upper bound. The new inverse tableau is obtained by performing a pivot with the updated column of x_4 as the pivot column, and the row in which x_2 is basic as the pivot row. The new solution is obtained by substituting $\lambda = 1$ in (10.10). The algorithm then moves to the next step.

10.3.6 Changing the Status of x_s Where $s \in \bar{\mathbf{J}}$

Here the entering variable x_s is such that $s \in \bar{\mathbf{J}}$. Clearly $\bar{d}_s$ must be strictly negative. Here, there is no upper bound restriction on the entering variable x_s and increasing its value from its present value of 0 leads to a decrease in the Phase I objective function. Keep the values of all the nonbasic variables other than x_s unchanged. Making $x_s = \lambda$, compute the values of the basic variables as functions of λ, so that the new solution satisfies the equality constraints in the problem. The new solution as a function of λ is the same as in (10.3).

Since the Phase I objective function decreases as the value of x_s is increased, λ should be made as large as possible consistent with feasibility. Using arguments as in section 10.3.4, it can be seen that the maximum value that λ can have is $\theta = \min \{\bar{b}_r/\bar{a}_{rs}$ for all r such that $\bar{a}_{rs} > 0$, $(\bar{b}_r - \mathrm{U}_{i_r})/\bar{a}_{rs}$ for all r such that $i_r \in \mathbf{J}$ and $\bar{a}_{rs} < 0, \infty\}$. There are several cases to consider here.

Case 1 If $\theta = \bar{b}_t/\bar{a}_{ts}$ for some t such that $\bar{a}_{ts} > 0$, then x_{i_t} drops from the basic vector and becomes a nonbasic variable with value 0 in the next step, and x_s becomes the tth basic variable in its place. The new inverse tableau and the new feasible solution are all obtained as in Case 2 of section 10.3.4.

Case 2 If $\theta = (\bar{b}_t - \mathrm{U}_{i_t})/\bar{a}_{ts}$ for some t such that $i_t \in \mathbf{J}$ and $\bar{a}_{ts} < 0$, then x_{i_t} drops from the basic vector and becomes a nonbasic variable whose value is equal to its upper bound in the next step. The variable x_s becomes the tth basic variable in its place. The new solution and the new inverse tableau are obtained as in Case 3 of section 10.3.4.

The Phase I objective function is bounded below by zero. As the value of λ increases from 0 the Phase I objective value decreases linearly. Hence, during Phase I, the value of θ will never be equal to ∞, and one of the cases discussed above should occur in each step before Phase I termination.

Note Notice that the pivot elements in the pivot steps to be performed in Case 3 of section 10.3.4, Case 2 of section 10.3.5 and Case 2 of this section, are all negative.

10.3.7 Infeasibility

The algorithm is continued until the Phase I termination criteria are satisfied, which will occur in a finite number of steps. When the Phase I termination criteria are satisfied, if the value of w is strictly positive, the original problem must be infeasible and the algorithm terminates.

10.3.8 Switching over to Phase II

When the Phase I termination criteria are satisfied, if $\bar{w} = 0$, the present solution is a feasible solution of the original problem and we switch over to Phase II.

In the terminal Phase I step, if $\bar{d}_j < 0$ for some j, then $j \in \mathbf{J}$ and the value of x_j must be U_j in the present solution. Decreasing the value of x_j from its present U_j will increase the Phase I objective function from its present value of 0 and, hence, leads to an infeasible solution of the original problem. Hence, in all feasible solutions of the original problem, the variable x_j must be equal to U_j. Therefore, this variable x_j is permanently set equal to U_j, and all such variables are prevented from being chosen for a change in status during Phase II iterations.

Similarly, if $\bar{d}_j > 0$ for some j in the Phase I terminal step, the associated variable x_j must be equal to zero in all feasible solutions of the original problem. Hence, all such variables are permanently set equal to zero, and these variables are never again considered for change of status during Phase II iterations.

Hence all the remaining variables that are eligible for change in status during Phase II iterations must have $\bar{d}_j$ equal to 0 in the terminal Phase I step. Any changes made in the values of these variables will not alter the Phase I objective value from 0; hence, all subsequent solutions obtained in the algorithm will be feasible solutions of the original problem. If any artificial variable is in the basic vector at this stage, its value in the present solution must be equal to zero and it can be left as a basic variable until it is removed from the basic vector during some Phase II iteration. The Phase I objective row and the last unit column vector from the inverse tableau are deleted. (If the problem is being solved by the revised simplex method using the product form of the inverse, this is equivalent to deleting the last row and the last column from each pivot matrix obtained so far.) The Phase I objective row, the column vector of $-w$, and the column vectors of all nonbasic artificial variables are also deleted from the original tableau. Then the algorithm moves to Phase II.

PHASE II TERMINATION CRITERIA

As in the usual simplex algorithm, define $\bar{c}_j = c_j - \pi A_{.j}$ for each j such that x_j is eligible for change of status during Phase II. The optimality criteria for the Phase II problem are discussed in section 10.2.3.

10.3.9 Choice of the Entering Variable during Phase II

If Phase II optimality criteria are not satisfied, select a nonbasic variables, x_s, which has led to a violation of the optimality criteria, and make it the entering variable. It can belong to one of the following types.

(i) $s \in \mathbf{J}$, $\quad \bar{c}_s < 0 \quad$ and the present value of x_s is 0

In this case, increasing the value of x_s from its present value of 0 leads to a reduction in the value of $z(x)$. The necessary computations for obtaining the next solution are carried out as in section 10.3.4.

(ii) $s \in \mathbf{J} \quad \bar{c}_s > 0 \quad$ and the present value of x_s is U_s

In this case, the value of $z(x)$ will decrease if the value of x_s is decreased from its present value of U_s. The computation for obtaining the next solution are performed as in section 10.3.5.

(iii) $s \in \bar{\mathbf{J}} \quad$ and $\quad \bar{c}_s < 0$

In this case, the value of $z(x)$ decreases as the value of x_s is increased from its present value of 0. Making x_s equal to λ, the new solution as a function of λ is the same as in (10.3). As in section 10.3.6 we conclude that the maximum value that λ can have is $\theta = \min \{\bar{b}_r/\bar{a}_{rs}$ for all r such that $\bar{a}_{rs} > 0$, $(\bar{b}_r - U_{i_r})/\bar{a}_{rs}$ for all r such that $i_r \in \mathbf{J}$ and $\bar{a}_{rs} < 0, \infty\}$. There are several cases to be considered here.

Case 1 If $\theta = \infty$, $z(x)$ is unbounded below in this problem. As λ increases from 0, we obtain a feasible solution, the value of $z(x)$ corresponding to which diverges to $-\infty$. Hence if this happens, the algorithm terminates.

Case 2 If $\theta = \bar{b}_t/\bar{a}_{ts}$ for some t such that $\bar{a}_{ts} > 0$, the computations for obtaining the next solution are performed as in Case 1 of section 10.3.6.

Case 3 If $\theta = (\bar{b}_t - U_{i_t})/\bar{a}_{ts}$ for some t, the computations for obtaining the next solution are performed as under Case 2 of section 10.3.6.

Numerical Example

Consider the following bounded variable LP.

Original Tableau

x_1	x_2	x_3	x_4	x_5	x_6	x_7	x_8	$-z$	b
1	0	0	-1	1	2	2	$-\frac{1}{5}$	0	$\frac{16}{5}$
0	1	0	1	-2	1	-2	$\frac{2}{5}$	0	$\frac{28}{5}$
0	0	1	-1	0	2	1	$\frac{1}{5}$	0	$\frac{39}{5}$
-1	2	-2	4	-4	-5	1	$\frac{18}{5}$	1	0

z to be minimized, $x_j \geqq 0$ for all j and $x_1, x_2, x_3, x_4, x_5, x_6$, and x_8 all $\leqq 4$

In this problem **J** is $\{1, 2, 3, 4, 5, 6, 8\}$ and U_j is 4 for all $j \in$ **J**. Because of the upperbound restrictions, the usual solution $x_1 = 16/5$, $x_2 = 28/5$, $x_3 = 39/5$, and all other $x_j = 0$ is not feasible to this problem. (It violates the upperbound restrictions on x_2, x_3.) The original Phase I tableau is as follows.

x_1	x_2	x_3	x_4	x_5	x_6	x_7	x_8	x_9	x_{10}	x_{11}	$-z$	$-w$	b
1	0	0	-1	1	2	2	$-\frac{1}{5}$	1	0	0	0	0	$\frac{16}{5}$
0	1	0	1	-2	1	-2	$\frac{2}{5}$	0	1	0	0	0	$\frac{28}{5}$
0	0	1	-1	0	2	1	$\frac{1}{5}$	0	0	1	0	0	$\frac{39}{5}$
-1	2	-2	4	-4	-5	1	$\frac{18}{5}$	0	0	0	1	0	0
0	0	0	0	0	0	0	0	1	1	1	0	1	0

$x_j \geqq 0$ for all j and $x_j \leqq 4$ for all $j \in \{1, 2, 3, 4, 5, 6, 8\}$

x_9, x_{10}, x_{11} are the artificial variables. In Phase I, $w = x_9 + x_{10} + x_{11}$ is to be minimized. Here is the initial inverse tableau.

Initial Inverse Tableau (Phase I)

Basic variables	Inverse Tableau					Basic values	Basic values after $x_8 = 4$	Pivot column x_7
x_9	1	0	0	0	0	$\frac{16}{5}$	4	②
x_{10}	0	1	0	0	0	$\frac{28}{5}$	4	-2
x_{11}	0	0	1	0	0	$\frac{39}{5}$	7	1
$-z$	0	0	0	1	0	0	$-\frac{72}{5}$	1
$-w$	-1	-1	-1	0	1	$-\frac{83}{5}$	-15	-1
All nonbasics = 0.								

$\bar{d}$ is $(-1, -1, -1, 1, 1, -5, -1, -2/5)$. x_8 is entering. It updated column is

$$\left(-\frac{1}{5}, \frac{2}{5}, \frac{1}{5}, \frac{18}{5}, -\frac{2}{5}\right)^{\mathrm{T}}.$$

It becomes a nonbasic at the upper bound. The new vector of basic values is entered by the side of the inverse tableau. Next, x_7 is entering. Its updated column is $(2, -2, 1, 1, -1)^{\mathrm{T}}$. x_7 becomes a basic variable at value 2, replacing x_9 from the basic vector. The pivot column is entered by the side of the inverse tableau and the pivot element is circled. The next inverse tableau is

Basic variables	Inverse Tableau					Basic values	Basic values after $x_3 = 4$	Pivot column x_6
x_7	$\frac{1}{2}$	0	0	0	0	2	2	1
x_{10}	1	1	0	0	0	8	8	3
x_{11}	$-\frac{1}{2}$	0	1	0	0	5	1	①
$-z$	$-\frac{1}{2}$	0	0	1	0	$-\frac{82}{5}$	$-\frac{42}{5}$	-6
$-w$	$-\frac{1}{2}$	-1	-1	0	1	-13	-9	-4
$x_8 = 4$. All other nonbasics = 0.								

The new $\bar{d}$ is $(-1/2, -1, -1, 1/2, 3/2, -4, 0, -1/2)$. x_3 is entering. Its updated column is $(0, 0, 1, -2, -1)^T$. It becomes a nonbasic variable at upper bound. The new vector of basic values is entered by the side of the inverse tableau. Next x_6 is entering. It enters the basic vector at value 1, replacing x_{11} from the basic vector. The pivot column is already entered by the side of the inverse tableau and the pivot element is circled. The next inverse tableau is

Basic variables	Inverse Tableau					Basic values	Pivot column x_3	
x_7	1	0	-1	0	0	1	-1 (circled)	Pivot row
x_{10}	$\frac{5}{2}$	1	-3	0	0	5	-3	
x_6	$-\frac{1}{2}$	0	1	0	0	1	1	
$-z$	$-\frac{7}{2}$	0	6	1	0	$-\frac{12}{5}$	4	
$-w$	$-\frac{5}{2}$	-1	3	0	1	-5	3	

$x_3 = x_8 = 4$. All other nonbasics $= 0$.

The new $\bar{d}$ is $(-5/2, -1, 3, -3/2, -1/2, 0, 0, 7/10)$. x_3 has to be decreased from its present value of 4. Its updated column vector is $(-1, -3, 1, 4, 3)^T$. It becomes a basic variable at value 3, replacing x_7 from the basic vector. x_7 becomes a zero-valued nonbasic variable. The pivot column is entered by the side of the inverse tableau and the pivot element is circled. The new vector of basic values is obtained as in Case 2, section 10.3.5. The new inverse tableau is

Basic variables	Inverse Tableau					Basic values	Pivot column x_4	
x_3	-1	0	1	0	0	3	0	
x_{10}	$-\frac{1}{2}$	1	0	0	0	2	$\frac{3}{2}$ (circled)	Pivot row
x_6	$\frac{1}{2}$	0	0	0	0	2	$-\frac{1}{2}$	
$-z$	$\frac{1}{2}$	0	2	1	0	$\frac{8}{5}$	$\frac{3}{2}$	
$-w$	$\frac{1}{2}$	-1	0	0	1	-2	$-\frac{3}{2}$	

$x_8 = 4$. All other nonbasics $= 0$.

The new $\bar{d}$ is $(1/2, -1, 0, -3/2, 5/2, 0, 3, -1/2)$. x_4 is entering. It becomes a basic variable replacing x_{10} from the basic vector. The pivot column is entered on the inverse tableau and the pivot element is circled. The next inverse tableau is

Basic variables	Inverse Tableau					Basic values	Pivot column x_5
x_3	$-\frac{2}{3}$	$-\frac{2}{3}$	1	0	0	$\frac{5}{3}$	$\frac{2}{3}$
x_4	$-\frac{1}{3}$	$\frac{2}{3}$	0	0	0	$\frac{4}{3}$	$\boxed{-\frac{5}{3}}$
x_6	0	1	0	0	0	4	0
$-z$	1	-1	2	1	0	$-\frac{18}{5}$	-1
$-w$	0	0	0	0	1	0	

$x_8 = 4$. All other nonbasics $= 0$.

Now all the artificial variables have become nonbasic variables with values equal to zero. $\bar{w} = 0$ and $\bar{d} = 0$. So Phase I terminates. $\bar{c}$ is $(0, 1, 0, 0, -1, 0, 7, 17/5)$. x_5 is entering. Its updated column is $(2/3, -5/3, 0, -1)^{\mathrm{T}}$. It becomes a basic variable at value 8/5, replacing x_4 from the basic vector. x_4 becomes a nonbasic variable at its upper bound. The pivot column is entered by the side of the inverse tableau and the pivot element is circled. The new vector of basic values is obtained as in Case 3 of section 10.3.5. The new inverse tableau is

Basic variables	Inverse Tableau				Basic values
x_3	$-\frac{4}{5}$	$-\frac{2}{5}$	1	0	$\frac{3}{5}$
x_5	$\frac{1}{5}$	$-\frac{2}{5}$	0	0	$\frac{8}{5}$
x_6	0	1	0	0	4
$-z$	$\frac{6}{5}$	$-\frac{7}{5}$	2	1	$-\frac{14}{5}$

$x_8 = x_4 = 4$. All other nonbasics $= 0$.

The new $\bar{c}$ is $(1/5, 3/5, 0, -3/5, 0, 0, 41/5, 0)$. It is clear that the present solution is optimal.

PROBLEMS

10.1 By using upper bounding techniques, find a feasible solution of the following.

x_1	x_2	x_3	x_4	x_5	b
1	2	0	-1	2	2
2	-1	2	-5	2	0

$0 \leqq x_j \leqq 1$ for all j

10.2 Solve

$$\begin{array}{ll} \text{Minimize} & -12x_1 - 12x_2 - 9x_3 - 15x_4 - 90x_5 - 26x_6 - 112x_7 \\ \text{Subject to} & 3x_1 + 4x_2 + 3x_3 + 3x_4 + 15x_5 + 13x_6 + 16x_7 \leqq 35 \\ & 0 \leqq x_j \leqq 1 \quad \text{for all } j \end{array}$$

10.3 Solve the following bounded variable LPs.

(a)

x_1	x_2	x_3	x_4	x_5	x_6	x_7	$-z$	b
1	0	0	1	2	−1	1	0	13
0	1	0	−1	1	1	2	0	9
0	0	1	2	2	2	−1	0	5
3	4	−2	−5	3	2	−1	1	0

$0 \leqq x_j \leqq 5$ for all j, z to be minimized.

(b)

x_1	x_2	x_3	x_4	x_5	x_6	x_7	$-z$	b
1	0	0	2	−2	1	−8	0	0
0	1	0	1	1	−1	1	0	11
0	0	1	3	−1	−2	2	0	6
1	2	1	−1	2	1	−1	1	0

$0 \leqq x_j \leqq 4$ for all j, z to be minimized.

10.4 Discuss how to perform marginal analysis for the right-hand side constants b_i and the bounds l_j, k_j in the LP (10.1) when it is solved using working bases as in this chapter.

Also discuss how to perform the various types of sensitivity analysis considered in Chapter 8 and the parametric analysis considered in Chapter 7, on the LP (10.1), with the output obtained when it is solved using working bases.

11 primal algorithm for the transportation problem

11.1 THE BALANCED TRANSPORTATION PROBLEM: THEORY

We consider the transportation problem in which all the constraints are equality constraints.

$$\text{Minimize} \quad z(x) = \sum_{i=1}^{m} \sum_{j=1}^{n} c_{ij}x_{ij}$$

$$\text{Subject to} \quad \sum_{j=1}^{n} x_{ij} = a_i \qquad i = 1 \text{ to m} \tag{11.1}$$

$$\sum_{i=1}^{m} x_{ij} = b_j \qquad j = 1 \text{ to n}$$

$$x_{ij} \geqq 0 \qquad \text{for all } i, j$$

where $a_i > 0$, $b_j > 0$ for all i, j and $\sum a_i = \sum b_j$; and c_{ij} may be arbitrary real numbers.

Until now we have used the symbols m, n, to denote the number of constraints and the number of variables, respectively, in an LP model. In this chapter we will use the symbols m, n, to denote the number of sources and demand centers respectively, in a transportation model. Notice that the number of constraints in this transportation model is m + n and the number of variables is mn.

The nice structure of this problem makes it possible to compute an initial BFS by very simple techniques without the need to solve any complicated Phase I problem; to compute the dual solution and the relative cost coefficients corresponding to any basis very easily without the need to compute the basis inverse; and to compute the new BFS when a nonbasic variable is brought into the basic vector by making adjustments along a loop without doing any complicated pivoting. This simplified version of the primal revised simplex algorithm for the balanced transportation problem is discussed here. First, we review the main results on this problem.

Obviously $\sum a_i = \sum b_j$ is a necessary condition for feasibility of (11.1). From the results in section 11.2 of this chapter, it will be clear that this condition is sufficient for feasibility of (11.1)

The transportation problem (11.1) can be written down in the equivalent form

$$\text{Minimize} \quad z(x) = \sum_{i=1}^{m} \sum_{j=1}^{n} c_{ij} x_{ij}$$

$$\text{Subject to} \quad \sum_{j=1}^{n} -x_{ij} = -a_i \qquad i = 1 \text{ to m} \tag{11.2}$$

$$\sum_{i=1}^{m} x_{ij} = b_j \qquad j = 1 \text{ to n} \tag{11.3}$$

$$x_{ij} \geqq 0 \qquad \text{for all } i, j$$

11.1.1 Redundancy in the Constraints

Theorem There is exactly one redundant equality constraint in (11.2) and (11.3). When any one of the constraints in (11.2) or (11.3) is dropped, the remaining is a linearly independent system of constraints.

Proof Let $x = (x_{11}, \ldots, x_{1n}, x_{21}, \ldots, x_{mn})^T$ be the column vector of all the mn variables in this problem. Let $\boldsymbol{E}$ denote the $(m + n) \times mn$ coefficient matrix corresponding to the constraints in (11.2) and (11.3). The column vector in $\boldsymbol{E}$ associated with the variable x_{ij} contains only two nonzero entries: "-1" in row i and a "$+1$" in row $m + j$. Since the equation "$0 = 0$" is obtained as the sum of all the constraints in (11.2) and (11.3), *any one of the constraints in them may be treated as a redundant constraint.*

Let $\bar{\boldsymbol{E}}$ be the matrix obtained by striking off some row, say, the last row, from $\boldsymbol{E}$. $\bar{\boldsymbol{E}}$ is an $(m + n - 1) \times mn$ matrix. It is the coefficient matrix of the system of constraints:

$$\sum_{j=1}^{n} -x_{ij} = -a_i \qquad i = 1 \text{ to m} \tag{11.4}$$

$$\sum_{i=1}^{m} x_{ij} = b_j \qquad j = 1 \text{ to } n - 1 \tag{11.5}$$

Suppose the rank of $\bar{\boldsymbol{E}}$ is strictly less than $m + n - 1$. Then, there must exist a nonzero linear combination of the equations in (11.4) and (11.5) in which all the coefficients of the variables on the left-hand side are zero. Suppose this is obtained by multiplying the equations in (11.4) by $\alpha_1, \ldots, \alpha_m$ and the equations in (11.5) by $\alpha_{m+1}, \ldots, \alpha_{m+n-1}$, respectively, and summing. Since this is a nonzero linear combination $(\alpha_1, \ldots, \alpha_m, \ldots, \alpha_{m+n-1}) \neq 0$.

The variable x_{in} for any i from 1 to m occurs in exactly one equation among (11.4) and (11.5) with a nonzero coefficient, namely, the ith equation in (11.4). Since the coefficient of each x_{in} is zero in the linear combination discussed above, $\alpha_1 = \cdots = \alpha_m = 0$.

However, the variable x_{ij} for any j from 1 to $n - 1$ occurs with a nonzero coefficient in exactly one equation in (11.5), namely, the jth equation. Thus, if

$\alpha_1 = \cdots = \alpha_m = 0$ and if the linear combination discussed above were to produce an equation in which the coefficients of all the variables on the left-hand side are zero, we must have $\alpha_{m+1} = \cdots = \alpha_{m+n-1} = 0$. This contradicts the assumption that $(\alpha_1, \ldots, \alpha_{m+n-1}) \neq 0$. Hence, the rank of $\bar{E}$ must be $m + n - 1$.

Hence, when any one of the constraints in (11.2) and (11.3) is dropped, the remaining constraint is a linearly independent system of constraints.

Corollary A basic vector for the transportation problem (11.1) consists of $(m + n - 1)$basic variables.

11.1.2 Triangular Matrix

Let $D = (d_{ij})$ be a nonsingular square matrix of order N. If $d_{ij} = 0$ for $j \geqq i + 1$, D is said to be a *lower triangular matrix*. See Figure 11.1. A square matrix D is said to be an *upper triangular matrix*, if D^T is a lower triangular matrix.

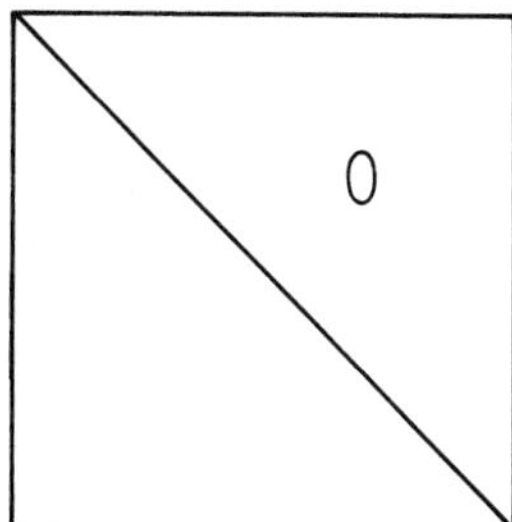

Figure 11.1 A lower triangular matrix.

If D is a nonsingular square matrix, which becomes a lower triangular matrix after a permutation of its columns and rows, D is said to be a *triangular matrix*. A triangular matrix satisfies the following properties:

(i) The matrix has a row that contains only a single nonzero entry.

(ii) The submatrix obtained from the matrix by striking off the row containing a single nonzero entry and the column in which this nonzero entry lies, also satisfies property (i). The same process can be repeated until all the rows and columns in the matrix are struck off.

Conversely, any matrix satisfying these properties is a triangular matrix.

SOLVING BY BACK SUBSTITUTION

If D is a lower triangular matrix, the system of equations "$Dy = d$" is very easily solved. The first equation

$$d_{11}y_1 = d_1$$

gives the value of y_1. Substituting the value of y_1 in the second equation

$$d_{21}y_1 + d_{22}y_2 = d_2$$

we obtain the value of y_2. In a similar way, the value of y_{r+1} is obtained from the $r + 1$th equation by substituting the values of $y_1, \ldots, y_r$ obtained from the earlier equations for $r = 1, \ldots, N - 1$. This method of solving a triangular system of equations is known as *back substitution.*

If D is any triangular matrix, the system "$Dy = d$," can still be solved by back substitution. Identify the equation containing a single nonzero entry on the left-hand side; solve that equation for the value of the variable associated with the nonzero coefficient in that equation; substitute the value of this variable in all the remaining equations and continue in the same manner with the remaining equations.

Note Let D be a lower triangular matrix of order N. Let $y = (y_1, \ldots, y_N)^T$ and $\pi = (\pi_1, \ldots, \pi_N)$. In solving the system of equations

$$Dy = d$$

the method discussed above computes $y_1, \ldots, y_N$ in that order. To compute y_{r+1}, the method uses the values of $y_1, \ldots, y_r$ obtained earlier, for each $r = 1$ to $N - 1$. Hence in some textbooks, the name *forward substitution* is used to describe this method when it is applied to solve the system of equations "$Dy = d$."

When this method is applied to solve the system of equations $\pi D = h$, it computes $\pi_N, \pi_{N-1}, \ldots, \pi_1$, in that order. To compute the value of π_r, it uses the values of $\pi_N, \ldots, \pi_{r+1}$, obtained earlier, for $r = N - 1$ to 1. Hence in some textbooks, the name *back substitution* is used for this method when it is applied to solve the system of equations "$\pi D = h$."

Here, we will use the term *back substitution*, to describe the application of this method for solving either a system of equations of the form "$Dy = d$" or a system of equations of the form "$\pi D = h$," whenever D is a triangular matrix,

11.1.3 Triangularity of the Basis in the Transportation Problem

Theorem Every basis for the transportation problem is triangular.

Proof The transportation problem (11.1) can be written in the form

$$\begin{aligned}
&\text{Minimize} && z(x) \\
&\text{Subject to} && \sum_{j=1}^{n} - x_{ij} = -a_i && i = 1 \text{ to } m \\
& && \sum_{i=1}^{m} x_{ij} = b_j && j = 1 \text{ to } n - 1 \\
& && x_{ij} \geqq 0 && \text{for all } i, j
\end{aligned} \tag{11.6}$$

The input-output coefficient matrix in these constraints is $\bar{E}$. Let B be a nonsingular square submatrix of $\bar{E}$ of order $(m + n - 1)$. Every column vector in $\bar{E}$ contains at most two nonzero entries: a "-1" and a "$+1$." If the total number of nonzero entries in B exceeds $2(m + n - 1)$, then every column in B must contain two nonzero entries, and since they are a "-1" and a "$+1$," the sum of all the rows in B must be zero, which is a contradiction to the nonsingularity of B. Thus, the total number of nonzero entries in B must be $< 2(m + n - 1)$. Since B is of order $(m + n - 1)$, there must exist a row of B with a single nonzero entry. By a similar argument, the submatrix of B obtained by striking off this row and the column containing this single nonzero entry must again have the same property, that is, it contains a row with a single nonzero entry. The same argument can be repeated again and again. Hence, B is triangular.

Problems

11.1 If B is any basis for (11.6), prove that the determinant of B is either $+1$ or -1.

11.2 Let F be any square submatrix of $\bar{E}$. By using an argument similar to that in the above proof, prove that the determinant of F is either 0, $+1$, or -1.

Given a basis for (11.6), since it is triangular, the basic solution of (11.6) corresponding to it can be computed by back substitution. Also the corresponding dual solution can be computed by back substitution. This makes it unnecessary to keep the basis inverse while solving the transportation problem. Also, since the nonzero entries in $\bar{E}$ are all equal to "-1" or "$+1$," finding the primal and dual solutions corresponding to a given basis by back substitution involves only additions and subtractions. No multiplications or divisions have to be performed. This leads to the following corollary.

Corollary In any basic solution of (11.6) the values of the basic variables are of the form

$$\sum_{i=1}^{m} \gamma_i a_i + \sum_{j=1}^{n-1} \psi_j b_j$$

where all the γ_i, ψ_j are equal to 0 or $+1$ or -1.

Integer Property If all the a_i and b_j are positive integers, then every basic solution of (11.1) is an integer vector. Hence, if all a_i, b_j are positive integers, and if the transportation problem is feasible, it has an optimum solution (x_{ij}) that is an integer vector.

11.1.4 Transportation Array

Clearly, the variables (x_{ij}) in the transportation problem can be arranged in an m × n array (known as the m × n *transportation array*) such that the constraints specify the sums of the variables in each row and column of the array. The variable x_{ij} corresponds to the cell (i, j) in such an array and vice versa. With the value of each variable x_{ij} entered in the appropriate cell, the array is denoted by x.

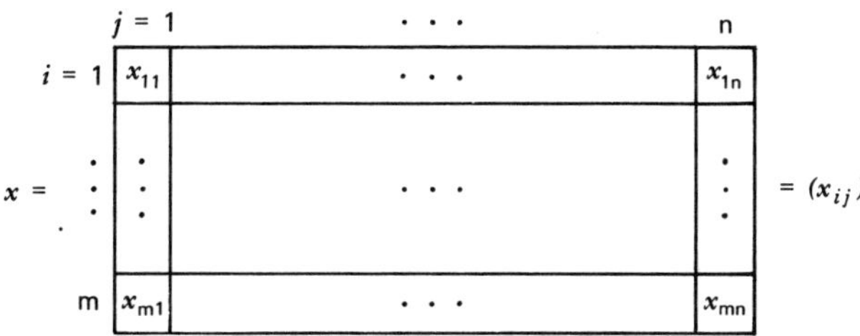

If (x_{ij}) satisfies all the constraints, then x represents a feasible solution of the transportation problem in array form.

The transportation array is not the same as the *simplex tableau* (the tableau of input-output coefficients used in the standard simplex algorithm, which will consist of m + n − 1 rows and mn columns for the transportation problem). While the simplex tableau displays the coefficients of the constraints, the transportation array is just a convenient method for displaying a feasible solution of the transportation problem. All the computations required for solving the transporttation problem can be organized using the transportation array only without ever requiring the simplex tableau.

11.1.5 Linear Independence of a Subset of Cells of the Array

Definition A subset of cells of the transportation array is said to be *linearly independent*, if the set of column vectors in (11.6) corresponding to variables associated with cells in the subset is linearly independent.

Definition A subset of $m + n - 1$ cells of the $m \times n$ transportation array is a *basic set* if it is linearly independent.

Test for a Basic Set of Cells.

Suppose a subset $\mathscr{B}$ of $m + n - 1$ cells of the $m \times n$ transportation array is given. Let B be the square matrix of order $m + n - 1$ whose column vectors are the column vectors in (11.6) corresponding to variables associated with cells in the subset $\mathscr{B}$. $\mathscr{B}$ is a basic set iff B is nonsingular. From section 11.1.3, if B is nonsingular, it is a triangular matrix. Hence, there exists a row vector of B with a single nonzero element. Each row in (11.6) represents a constraint that the sum of all the variables in a row or column of the transportation array is a specified number. Hence, there exists a row vector of B with a single nonzero element, iff there exists a row or column of the array that contains exactly one cell from $\mathscr{B}$. Also, B is a triangular matrix iff the same property again holds for the submatrix of B obtained by striking off the row of B containing a single nonzero element, and the column of B containing that nonzero element. This leads to the following simple test to check whether $\mathscr{B}$ forms a basic set.

Initially, none of the cells of the array are struck off. In each stage of the test all the cells in a selected row, or column of the array will be struck off. At any stage of the test, a row or column of the array that has some cells not yet struck off is known as a *remaining row or column.* $\mathscr{B}$ forms a basic set iff the following properties hold.

(i) At each stage, there exists a remaining row or column of the array that contains exactly one cell of $\mathscr{B}$ which has not yet been struck off.

(ii) Identify the remaining row or column of the array that contains exactly one cell of $\mathscr{B}$ which has not yet been struck off. Strike off all the cells in that row or column. The remaining part of the array must again satisfy property (i). Repeat until all the cells in the array are struck off.

Problem

11.3 Do the subsets of marked cells in the following arrays constitute basic sets?

y	y				
	y	y			
		y	y		
			y	y	y

					y
y			y	y	
y	y				
	y	y	y		

Example To Compute the Basic Solution
Consider the transportation problem with a_i, b_j as given below:

	$j = 1$	2	3	4	5	6	a_i
$i = 1$		□				□	7
2	□				□		17
3		□	□				5
4	□			□		□	24
b_j	15	10	9	3	8	8	

Compute the basic solution corresponding to the basic set marked by an "□" in the array. Every cell that is not marked by an "□" in the array is a nonbasic cell and, hence, the value of x_{ij} corresponding to all such cells is zero in the basic solution.

There must exist either a row or a column of the transportation array that contains exactly one basic cell. In this example it is column 4, which contains only the basic cell (4, 4). The sum of all the variables corresponding to cells in this column must equal $b_4 = 3$. This implies that the value of the basic variable $x_{44} = 3$ in the basic solution. Likewise, $x_{25} = 8$. A similar argument shows that $x_{32} = 10$. This and the fact that (3, 3) is the only other basic cell in the third row imply that $x_{33} = 5 - x_{32} = -5$. Consequently, $x_{13} = 9 - x_{33} = 14$. Hence, $x_{16} = 7 - x_{13} = -7$. Thus $x_{46} = 8 - x_{16} = 15$. And $x_{41} = 24 - x_{44} - x_{46} = 6$, implying that $x_{21} = 15 - x_{41} = 9$. Thus, the basic solution corresponding to this basic set is

	$j = 1$	2	3	4	5	6
$i = 1$			14			−7
2	9				8	
3		10	−5			
4	6			3		15

The values of all the nonbasic variables in the basic solution are 0, and these are not entered. This is an infeasible basis to the problem, since some of the basic values violate the nonnegativity restrictions.

1.1.6 θ Loops

A θ-loop is a subset of cells of the transportation array with the following properties.

(i) It is nonempty.

(ii) It contains either 0 or 2 cells from each row and column of the array.

(iii) No proper subset of it satisfies properties (i) and (ii).

If a subset of cells form a θ-loop, entries of $+\theta$ and $-\theta$ can be put alternately in the cells of the subset, such that if a row or column of the array contains a cell from the subset with a $+\theta$ entry, then it also contains one other cell from the subset, and it has a $-\theta$ entry. When this is done, every cell in the subset will end up having either a $+\theta$ entry or a $-\theta$ entry.

Example

θ				$-\theta$
	θ	$-\theta$		
$-\theta$		θ		
	$-\theta$			θ

A θ-loop with $+\theta$, $-\theta$ entries. Cells not in the θ-loop have no entries marked in them.

Example

		y	y		
		y	y		
				y	y
				y	y

The subset of cells marked with an y in the above tableau satisfies properties (i) and (ii), but not property (iii). Hence this subset is not a θ-loop.

Theorem 1 Linear Dependence of a θ-Loop A subset of cells of the transportation array that forms a θ-loop is linearly dependent.

Proof Put entries of $+\theta$ and $-\theta$ alternately among the cells in the θ-loop, and entries of 0 in all the cells of the array, not in the θ-loop. When $\theta = 1$, the sum of all the entries in each row and column of the array is zero. This implies that if we take the column vectors in (11.6) corresponding to the cells in the θ-loop and multiply them by $+1$ if the cell had a $+\theta$ entry, and by -1 if the cell had a $-\theta$ entry, and sum, we get the 0 vector. Hence, the set of these column vectors is linearly dependent.

Theorem 2 Triangularity of a Collection of Cells without a θ-Loop A nonempty collection of cells, Δ, of the transportation array that contains no θ-loop satisfies the following properties.

1. There exists a row or column of the array that contains exactly one cell from Δ.

2. Every nonempty subset of such a collection of cells, Δ, satisfies property 1.

Proof Suppose Δ does not satisfy property 1. Pick any cell from Δ, say (i_1, j_1), and put a $+\theta$ entry in it. By assumption, the i_1th row of the tableau must contain at least one more cell from Δ. Pick one of those, say (i_1, j_2), and put a $-\theta$ entry in it. This completes our work in row i_1 of the tableau and we will never again put any more $+\theta$ or $-\theta$ entries in it. We picked $(i_1, j_2) \in \Delta$ from column j_2. By assumption, this column must contain at least one more cell from Δ, say, (i_2, j_2). Put a $+\theta$ entry in (i_2, j_2) and drop column j_2 of the array from further consideration.

Repeat this process until we return to column j_1, which was the column we started with. When this happens, all the cells from Δ that have either a $+\theta$ or a $-\theta$ in them, form a θ-loop. Hence, Δ contains a θ-loop as a subset, contradicting the hypothesis. Hence, property 1. must hold. Since Δ contains no θ-loop, every nonempty subset of Δ cannot contain a θ-loop itself and, hence, it must satisfy property 1. too.

Theorem 3 Let Δ be a subset of cells of the transportation array. If Δ does not contain a θ-loop as a subset, Δ is linearly independent.

Proof If Δ is empty, it is obviously linearly independent. Suppose Δ is nonempty. If it is linearly dependent, there must exist a nonzero linear combination of the column vectors in (11.6) corresponding to the cells in Δ, which is equal to 0.

Enter the multipliers of the column vectors in this linear combination, in the corresponding cells in the $m \times n$ transportation array and enter 0 in all the cells of the array, not in Δ. Since the linear combination of the column vectors is the zero vector, the sum of all these entries in each row and column of the transportation array must be 0.

The subset of all the cells with nonzero entries is a nonempty subset of Δ, and since the sum of these entries in each row and column of the transportation array is 0, this subset of cells cannot satisfy property 1. of Theorem 2. This is a contradiction to the hypothesis that Δ contains no θ-loop.

TEST FOR LINEAR INDEPENDENCE

From these results it is clear that a subset of cells, Δ, of the transportation array is linearly independent iff there is no subset of Δ that is a θ-loop. Equivalently, if Δ is nonempty, it is linearly independent iff Δ and every nonempty subset of Δ satisfies the property that there exists a row or column of the array that contains exactly one cell of the subset.

This leads to the following simple test to check whether a subset of cells of the transportion array, Δ, is linearly independent.

Strike off all the cells in rows and columns of the transportation array that contain no cells of Δ. Δ is linearly independent iff the following properties hold.

(a) At each stage, there exists a row or column of the remaining transportation array that contains exactly one cell of Δ that is not yet struck off.

(b) Identify the remaining row or column of the transportation array that contains exactly one cell of Δ that is not yet struck off. Strike off all the cells in that row or column. The remaining part of the transportation array must again satisfy property (a). Repeat until all the cells are struck off.

Clearly this test is similar to the test for checking whether a given subset of cells in the transportation array is a basic set.

A θ-LOOP IS A MINIMAL LINEARLY DEPENDENT SET

A collection of cells of the transportation array is said to be a *minimal linearly dependent set* iff (1) the collection is linearly dependent, and (2) no proper subset of the collection is linearly dependent.

From the definition of a θ-loop and from the results discussed here, it is clear that a collection of cells from the transportation array is a minimal linearly dependent set iff the collection is a θ-loop.

11.1.7 θ-Loop In $\mathscr{B} \cup \{(p,q)\}$

Theorem Suppose $\mathscr{B}$ is a basic set of m + n − 1 cells from the m × n transportation array and (p, q) is a nonbasic cell. Then the collection of cells $\mathscr{B} \cup \{(p, q)\}$ contains exactly one θ-loop and this θ-loop contains the nonbasic cell (p, q).

Proof Since $\mathscr{B}$ forms a basic set, obviously $\mathscr{B}$ cannot contain any θ-loop. Thus, if there is a θ-loop in $\mathscr{B} \cup \{(p, q)\}$, that θ-loop must contain the cell (p, q). Since the rank of $\bar{E}$ is m + n − 1, no subset of m + n cells from the m × n transportation array can be linearly independent. So $\mathscr{B} \cup \{(p, q)\}$ must be linearly dependent, and by section 11.1.6, it must contain at least one θ-loop.

It is a standard result in linear algebra that a set consisting of a basis and exactly one nonbasic column vector contains a unique minimal linearly dependent set. Applying this result to the set of column vectors in (11.6) corresponding to the cells in $\mathscr{B} \cup \{(p, q)\}$ we conclude that $\mathscr{B} \cup \{(p, q)\}$ contains exactly one minimal linearly dependent set. Thus, $\mathscr{B} \cup \{(p, q)\}$ contains exactly one θ-loop.

HOW TO FIND THE θ-LOOP IN $\mathscr{B} \cup \{(p,q)\}$

Place an entry of $+\theta$ in the nonbasic cell (p, q). Make alternately entries of $-\theta$ and $+\theta$ among the basic cells such that in the end each row and column of the array contains either a $-\theta$ and $+\theta$ entries, or none at all. The set of all the cells marked with $-\theta$ and $+\theta$ entries at the end is the unique θ-loop in $\mathscr{B} \cup \{(p, q)\}$.*

There is another method for finding the θ-loop in $\mathscr{B} \cup \{(p, q)\}$. At the outset all the m + n cells in $\mathscr{B} \cup \{(p, q)\}$ are uncrossed. Look for a row or a column of the transportation array with a single uncrossed cell from the set $\mathscr{B} \cup \{(p, q)\}$. If such a row or column exists, cross out the single cell that it contains from the set $\mathscr{B} \cup \{(p, q)\}$. Repeat this process again and again until there is no row or column in the transportation array that contains exactly one uncrossed cell from the set $\mathscr{B} \cup \{(p, q)\}$. When this happens, all the remaining uncrossed cells from the set $\mathscr{B} \cup \{(p, q)\}$ form the θ-loop. Starting with a $+\theta$ entry in the cell (p, q) put alternately $-\theta$ and $+\theta$ entries in the remaining uncrossed cells from the set $\mathscr{B} \cup \{(p, q)\}$, until each $+\theta$ entry in any row or column is cancelled by a $-\theta$ entry.

There are other efficient labeling methods for finding the θ-loop in $\mathscr{B} \cup \{(p, q)\}$. They are discussed in section 11.4.

Problem

11.4 Let $\mathscr{B} = \{(i_1, j_1), \ldots, (i_{m+n-1}, j_{m+n-1})\}$ be a basic set of cells for the m × n transportation array and let (p, q) be a nonbasic cell. Prove that:

$$\begin{pmatrix}\text{The column vector}\\ \text{corresponding to}\\ x_{pq} \text{ in (11.6)}\end{pmatrix} = \sum_{t=1}^{m+n-1} \alpha_t \begin{pmatrix}\text{Column corresponding}\\ \text{to basic variable}\\ x_{i_t j_t} \text{ in (11.6)}\end{pmatrix}$$

where $\alpha_t = 0$ if basic cell (i_t, j_t) is not in the θ-loop in $\mathscr{B} \cup \{(p, q)\}$.

$= 1$ if basic cell (i_t, j_t) is in θ-loop of $\mathscr{B} \cup \{(p, q)\}$ and has an entry of $-\theta$ when this θ-loop is determined as in this section.

$= -1$ as above, and if the entry in this basic cell is $+\theta$ in the θ-loop.

* When the $+\theta$, $-\theta$ entries are entered in the cells of the θ-loop in this manner, the cells with the $+\theta$ entry are known as *recipient cells*, and the cells with the $-\theta$ entry are known as *donor cells*, in the θ-loop.

DANTZIG PROPERTY

From Problem 11.4 we know that the *representation* of any nonbasic column vector of the system of transportation constraints (11.6), as a linear combination of a basic set of column vectors for (11.6), has only entries of 0, +1, or −1 in it. Thus, in the simplex canonical tableau for (11.6) with respect to any basis, the entries in all the updated column vectors among the constraint rows will be either 0, +1, or −1. This property is known as the *Dantzig property.*

11.2 REVISED PRIMAL SIMPLEX ALGORITHM FOR THE BALANCED TRANSPORTATION PROBLEM

11.2.1 To Obtain an Initial Basic Feasible Solution

Initially all the cells in the transportation array are *admissible.* As this method progresses, all the cells in a selected row or a column of the transportation array will be made *inadmissible* in each stage. Also the row and column totals will be modified after each step. The modified row and column totals will be denoted by a_i' and b_j' respectively. Initially a_i' is the original a_i and b_j' is the original b_j for all i, j. There are several steps in each stage of this procedure.

Step 1 Pick any admissible cell at this stage. It may be worthwhile to pick a cell (r, s) of the form $c_{rs}=\text{minimum}\,\{c_{i,j}$: over all the cells (i, j) that are admissible at this stage$\}$. This may help in producing a BFS that is nearly optimal, even though there is no theoretical guarantee that this will occur.

Step 2 Make the cell (r, s) picked in Step 1 a *basic cell.* Make the value of the basic variable x_{rs} equal to

$$x_{rs} = \text{minimum}\,\{a_r', b_s'\} \tag{11.7}$$

where a_r', b_s' are the current values of the modified row and column totals. It is possible that the value on the right-hand side of (11.7) may turn out to be zero. Then, x_{rs} is still treated as a *basic variable* whose value in the BFS is zero.

Step 3 In Step 2, one of the following cases may occur.

Case 1 $a_r' < b_s'$. Make all the cells in the rth row inadmissible for all subsequent stages. Modify b_s' to $(b_s' - a_r')$. Keep all the other row and column totals unchanged. This completes the stage. Go to the next stage by returning to Step 1.

Case 2 $a_r' > b_s'$. In this case, make all the cells in the sth column inadmissible for all subsequent stages. Modify a_r' to $(a_r' - b_s')$ and go to the next stage by returning to Step 1.

Case 3 $a_r' = b_s'$ and the set of admissibie cells at this stage lie in two or more rows. Then make all the cells in the rth row inadmissible for all subsequent stages and modify b_s' to zero. Go to the next stage by returning to Step 1.

Case 4 $a_r' = b_s'$ and the set of admissible cells at this stage lie in a single row (namely, the rth row) but in two or more columns. Make all the cells in the sth column inadmissible for all subsequent stages. Modify a_r' to zero and go to the next stage by returning to Step 1.

Case 5 $a'_r = b'_s$ and at this stage (r, s) is the only admissible cell left. In this case terminate.

Discussion

1. The procedure in this section can only terminate when Case 5 occurs.
2. In each stage before termination, all the cells in either a row or a column of the array are made inadmissible. In the final stage during which termination occurs, only one admissible row and one admissible column are left.
3. In each stage exactly one basic cell is picked.
4. From (2) and (3) it is clear that the number of basic cells picked is exactly $\mathrm{m} + \mathrm{n} - 1$.
5. From the manner in which the basic cells are picked and the admissible cells are defined for the next stage, it is clear that the set of $(\mathrm{m} + \mathrm{n} - 1)$ cells picked cannot contain a θ-loop. Hence by the results in section 11.1, the set of $(\mathrm{m} + \mathrm{n} - 1)$ cells picked constitute a basic set of cells for the transportation array.
6. From the way the row and column totals are modified in each stage, it is clear that a'_i and b'_j always remain nonnegative for all i, j. Thus, the basic solution obtained is a nonnegative solution and, hence, is a BFS. Some of the variables associated with basic cells may have value equal to zero in the BFS, but they are clearly recorded as basic cells in order to distinguish them from nonbasic cells.

Example

Find an initial BFS of the transportation problem in which the row and column totals are

$$a = (a_1, a_2, a_3, a_4) = (6, 25, 20, 13)$$
$$b = (b_1, b_2, b_3, b_4, b_5, b_6) = (4, 16, 10, 9, 7, 18)$$

In this numerical example, we only wish to illustrate how to get an initial BFS for the transportation problem. Hence, to keep the illustration simple, the objective function is not provided here. Thus, whenever Step 1 has to be executed here, an admissible cell is picked arbitrarily without any regard to the objective function. Suppose (4, 3) is picked as the initial basic cell, the value of x_{43} should be set equal to minimum $\{13, 10\} = 10$. Then modify a'_4 to 3 and make all cells in column 3 inadmissible for the next stage. In the arrays that follow, we indicate the basic cells by putting a small square in the cell with the value of basic variable entered in the square. All inadmissible cells have a dotted line drawn through them. Continuing, we have the following arrays.

	$j = 1$	2	3	4	5	6	a'_i
$i = 1$							6
2							25
3							20
4			10				3
b'_j	4	16		9	7	18	

Next, the admissible cell (2, 5) is picked as a basic cell. x_{25} = minimum $\{25, 7\}$ = 7.

	$j = 1$	2	3	4	5	6	a_i'
$i = 1$							6
2					7		18
3							20
4			10				3
b_j'	4	16		9		18	

Picking (4, 4) for the next basic cell leads to:

	$j = 1$	2	3	4	5	6	a_i'
$i = 1$							6
2					7		18
3							20
4			10	3			
b_j'	4	16		6		18	

Pick (3, 2) for the next basic cell.

	$j = 1$	2	3	4	5	6	a_i'
$i = 1$							6
2					7		18
3		16					4
4			10	3			
b_j'	4			6		18	

Pick (3, 1) for the next basic cell.

	$j = 1$	2	3	4	5	6	a_i'
$i = 1$							6
2					7		18
3	4	16					
4			10	3			
b_j'	0			6		18	

Picking (2, 1) for the next basic cell leads to:

	$j = 1$	2	3	4	5	6	a_i'
$i = 1$							6
2	0				7		18
3	4	16					
4			10	3			
b_j'				6		18	

Pick (2, 6) as the next basic cell.

	$j = 1$	2	3	4	5	6	a_i'
$i = 1$							6
2	0				7	18	
3	4	16					
4			10	3			
b_j'				6		0	

Pick (1, 4) for the next basic cell.

	$j = 1$	2	3	4	5	6	a_i'
$i = 1$				6			0
2	0				7	18	
3	4	16					
4			10	3			
b_j'						0	

The only remaining cell is (1, 6), and we pick it as the last basic cell. This leads to the final BFS:

	$j = 1$	2	3	4	5	6
$i = 1$				6		0
2	0				7	18
3	4	16				
4			10	3		

All the cells without an entry are nonbasic cells and the values of the corresponding variables in the BFS are all zero.

11.2.2 Finding the Dual Solution

In this chapter we will use the symbols u_i, v_j to denote the dual variables. The dual of the transportation problem (11.1) is:

$$\begin{array}{lll} \text{Maximize} & w(u, v) = \sum a_i u_i + \sum b_j v_j & \\ \text{Subject to} & u_i + v_j \leqq c_{ij} \quad \text{for all } i, j & (11.8) \\ u_i, v_j & \text{unrestricted in sign for all } i, j & \end{array}$$

If $\mathscr{B}$ is a basic set of cells for the m $\times$ n transportation array, the dual solution corresponding to this basis is obtained by solving

$$u_r + v_s = c_{rs} \qquad \text{for all basic cells } (r, s) \tag{11.9}$$

The *relative cost coefficient* of any variable x_{ij} in the problem is the associated dual slack variable, namely, $c_{ij} - (u_i + v_j)$.

The transportation problem (11.1) contains exactly one redundant equation. Anyone of the constraints in (11.1) can be deleted, and the remaining is an equivalent linearly independent system of constraints. The dual of this equivalent problem is obtained by setting the dual variable corresponding to the deleted constraint equal to zero in (11.8).

To simplify the discussion, assume that the constraint corresponding to $j = \text{n}$ in (11.1) is deleted. This is equivalent to setting $v_\text{n} = 0$ in (11.8). Given a basis consisting of (m + n − 1) basic cells, the corresponding dual solution satisfies (11.9), which is a system of m + n − 1 equality constraints in m + n variables, and it, together with the constraint, $v_\text{n} = 0$, uniquely determines the dual solution. From the results in section 11.1, it is clear that the dual basic solution can be found by back substitution. There will be at least one basic cell in column n of the transportation array. Suppose (t, n) is a basic cell. Then "$v_\text{n} = 0$" and "$u_t + v_\text{n} = c_{t,\text{n}}$" together provide the value of the dual variable u_t in the dual basic solution. Substitute the value of u_t in each equation in (11.9) in which it occurs. By the triangularity of the basis this will help in determining the value of another dual variable, and so on.

After all the u_i and v_j are computed, the relative cost coefficients are calculated from $\bar{c}_{ij} = c_{ij} - u_i - v_j$.

If (u, v) is the dual basic solution corresponding to the basic set $\mathscr{B}$, it is a dual feasible solution iff $u_i + v_j \leqq c_{ij}$, that is, $\bar{c}_{ij} = c_{ij} - u_i - v_j \geqq 0$ for all i and j.

Example

Consider the transportation problem with the following cost matrix.

	$j = 1$	2	3	4	5	6
$i = 1$	$c_{11} = 10$	4	12	□ 1	12	□ 8
2	□ −3	2	6	2	□ 3	□ −1
3	□ 2	□ 4	2	10	2	1
4	10	−2	□ 5	□ −5	7	5

The c_{ij} are entered in the lower right corner in the cell. It is required to find the dual solution corresponding to the basic set consisting of all the cells with a square marked in the cell. In determining the dual solution, only the cost coefficients of the basic cells are used. We set $v_n = 0$ and find the rest of the dual solution by back substitution. Setting $v_6 = 0$ and looking at the equations (11.9) corresponding to the basic cells (2, 6) and (1, 6) leads to

$$u_2 + v_6 = -1, \quad \text{so} \quad u_2 = -1; \qquad u_1 + v_6 = 8, \quad \text{so} \quad u_1 = 8$$

Now using the values of u_1, u_2 in the equations (11.9) corresponding to basic cells (2, 1), (1, 4), and (2, 5), we get $v_1 = -2$, $v_4 = -7$, and $v_5 = 4$. In a similar way, we proceed and obtain the dual solution indicated on the following array. *The relative cost coefficients $\bar{c}_{ij}$ are entered in the upper left corner of the cell.*

(Each cell: upper left entry / lower right entry; □ marks a basic cell.)

	$j = 1$	2	3	4	5	6	u_i
$i = 1$	$4=\bar{c}_{11}$ / $c_{11}=10$	−4 / 4	1 / 12	□ / 1	0 / 12	□ / 8	8
2	□ / −3	3 / 2	4 / 6	10 / 2	□ / 3	□ / −1	−1
3	□ / 2	□ / 4	−5 / 2	13 / 10	−6 / 2	−3 / 1	4
4	10 / 10	−4 / −2	□ / 5	□ / −5	1 / 7	3 / 5	2
v_j	−2	0	3	−7	4	0	

In each basic cell, the relative cost coefficient is zero, and this is not entered in the above array.

Problem

11.5 We discussed a very simple algorithm for finding a primal feasible basis and the corresponding primal BFS for the transportation problem using the triangularity property. Develop a similar simple algorithm for finding a dual feasible basis to the transportation problem (11.1).

OPTIMALITY CRITERION

The present BFS is optimal if the relative cost coefficients $\bar{c}_{ij}$ are nonnegative for all i, j.

11.2.3 Improving a Nonoptimal Basic Feasible Solution

If the optimality criterion is not satisfied, pick some nonbasic cell whose relative cost coefficient is negative, and try to bring it into the basic set. Suppose the nonbasic cell to be brought into the basic set is (p, q), $\bar{c}_{pq} < 0$.

Find the θ-loop in the set of cells consisting of the cell (p, q) and the basic cells. Place an entry of $+\theta$ in the cell (p, q) and entries of $-\theta$ and $+\theta$ alternately among the cells in the θ-loop. Let $\tilde{x}$ be the present BFS. Let

$$\begin{aligned} x_{ij}(\theta) &= \tilde{x}_{ij} && \text{if } (i, j) \text{ is not in the } \theta\text{-loop} \\ &= \tilde{x}_{ij} + \theta && \text{if } (i, j) \text{ has a } +\theta \text{ entry} \\ &= \tilde{x}_{ij} - \theta && \text{if } (i, j) \text{ has a } -\theta \text{ entry} \end{aligned}$$

Then $x(\theta) = [x_{ij}(\theta)]$ is the solution obtained by giving a value θ to x_{pq}, retaining all the other nonbasic variables at zero value, and reevaluating the values of the basic variables. The maximum value of θ that keeps $x(\theta)$ nonnegative is

$$\theta_1 = \min \{x_{rs} : (r, s) \text{ a basic cell with } -\theta \text{ entry}\} \tag{11.10}$$

When $\theta = \theta_1$, $x_{rs}(\theta)$ becomes equal to zero for (r, s), which attains the minimum in (11.10).* This cell is dropped from the basic set and (p, q) is included as the basic cell replacing it. The new BFS is $x(\theta_1)$ with an objective value of $z(\tilde{x}) + \bar{c}_{pq}\theta_1$. The procedure is repeated with the new basic set.

Example

Solve the transportation problem, with the cost matrix given in section 11.2.2 and the a, b vectors given in section 11.2.1. We pick the BFS to this problem obtained in section 11.2.1. From sections 11.2.1 and 11.2.2, the relevant data is:

	$j = 1$	2	3	4	5	6
$i = 1$	$4=\bar{c}_{11}$; $c_{11}=10$	-4; 4	1; 12	$x_{14} = \boxed{6} + \theta$; 1	0; 12	$\boxed{0} - \theta$; 8
2	$\boxed{0} - \theta$; -3	3; 2	4; 6	10; 2	$\boxed{7}$; 3	$\boxed{18} + \theta$; -1
3	$\boxed{4} + \theta$; 2	$\boxed{16} - \theta$; 4	-5; 2	13; 10	-6; 2	-3; 1
4	10; 10	-4; θ; -2	$\boxed{10}$; 5	$\boxed{3} - \theta$; -5	1; 7	3; 5

The optimality criterion is not satisfied, since in particular $\bar{c}_{42} = -4 < 0$. Therefore, bring (4, 2) into the basic set. This leads to the θ-loop entered in the array. The basic cells with a "$+\theta$" entry are (1, 4), (2, 6), and (3, 1). The basic cells with a "$-\theta$" entry are (1, 6), (2, 1), (3, 2), and (4, 4). The new solution is therefore $(x_{14}, x_{16}, x_{21}, x_{25}, x_{26}, x_{31}, x_{32}, x_{42}, x_{43}, x_{44}) = (6 + \theta, 0 - \theta, 0 - \theta, 7, 18 + \theta, 4 + \theta, 16 - \theta, \theta, 10, 3 - \theta)$ and all other $x_{ij} = 0$. The maximum value that θ can have is $\theta_1 = \text{minimum } \{16, 0, 0, 3\} = 0$. So x_{42} enters the basic vector with a value $\theta_1 = 0$ and it can replace either x_{21} or x_{16} from the basic vector. Suppose we drop x_{21} from the basic vector. This is a degenerate pivot step and the new basis corresponds to the same BFS.

Compute the dual solution and the relative cost coefficients corresponding to the new basis as before. This leads to the next array.

	$j = 1$	2	3	4	5	6	u_i
$i = 1$	$8=\bar{c}_{11}$; $c_{11}=10$	0; 4	1; 12	$\boxed{6}$; 1	0; 12	$\boxed{0}$; 8	8
2	4; -3	7; 2	4; 6	10; 2	$\boxed{7}$; 3	$\boxed{18}$; -1	-1
3	$\boxed{4}$; 2	$\boxed{16} - \theta$; 4	-9; θ; 2	9; 10	-10; 2	-7; 1	8
4	14; 10	$\boxed{0} + \theta$; -2	$\boxed{10} - \theta$; 5	$\boxed{3}$; -5	1; 7	3; 5	2
v_j	-6	-4	3	-7	4	0	

* If there is a tie from the basic cell (r, s) that attains the minimum in (11.10), any one of the tied cells can be dropped from the basic set.

Bring the cell (3, 3) with relative cost coefficient $\bar{c}_{33} = -9$ into the basic set. The θ-loop generated is indicated in the array. The maximum possible value for θ is minimum $\{16, 10\} = 10$. When $\theta = 10$, the basic variable x_{43} becomes equal to zero and is dropped from the basic vector. The new BFS is:

	$j = 1$	2	3	4	5	6
$i = 1$				6		0
2					7	18
3	4	6	10			
4		10		3		

Compute the dual solution corresponding to the new basis and the new relative cost coefficients, and test whether the new basis satisfies the optimality criterion. Otherwise continue the algorithm until the optimality criterion is satisfied.

Problem

11.6 Complete the solution of this transportation problem: Also solve the following transportation problem:

								a_i
	3	18	5	3	9	6	−2	51
$c =$	4	0	4	6	6	8	18	16
	2	8	3	17	−4	18	16	17
	−6	7	19	9	3	0	12	11
b_j	9	12	25	12	10	20	7	

11.7 Solve the following transportation problem.

							a_i
	2	0	3	5	6	3	10
$c =$	1	3	4	0	9	7	24
	3	7	9	1	8	1	9
	4	8	7	3	6	8	36
b_j	24	4	1	19	23	8	

Start with $\{(1, 2), (1, 5), (2, 1), (2, 4), (3, 2), (3, 3), (3, 6), (4, 4), (4, 5)\}$ as the initial basic set.

11.8 In the balanced transportation problem (11.1), prove that the set of optimum feasible solutions is unaffected, if the cost matrix is modified by adding or subtracting a constant to all elements in a row or column of (c_{ij}). What happens to the set of optimum feasible solutions if all the elements in a row or column of the cost matrix are multiplied by a positive constant?

11.9 Consider the transportation problem:

							a_i
$c =$	11	−10	1	− 6	1	− 7	17
	11	11	10	12	12	10	7
	2	0	7	8	7	6	9
	− 4	− 2	2	3	4	2	12
b_j	10	12	7	3	8	5	

If $\tilde{u} = (-7, 8, 3, 0)$ and $\tilde{v} = (-4, -3, 2, 1, 4, 0)$ is $(\tilde{u}, \tilde{v})$ an optimum feasible solution of the dual problem? Test by using the complementary slackness theorem. At the same time, obtain all the optimal BFSs of this transportation problem.

11.10 Consider the balanced transportation problem and the solution $(\tilde{x}_{ij})$ tabulated within small circles inside the cells.

	$j = 1$	2	3	4	5	6	a_i
$i = 1$	$c_{11} = 2$	7	(4) 6	(5) 7	13	15	9
2	(3) 0	10	(6) 5	15	15	(7) 8	16
3	5	(11) 2	9	(3) 5	(10) 5	(8) 7	32
4	0	(1) 1	(2) 3	8	5	(9) 6	12
b_j	3	12	12	8	10	24	

(a) Is $(\tilde{x}_{ij})$ a BFS for this transportation problem? If not, using θ-loops, obtain a BFS from it.
(b) Is $(\tilde{x}_{ij})$ an optimum feasible solution?

11.3 THE UNIMODULARITY PROPERTY

Consider the LP

$$\begin{aligned} \text{Minimize} \quad & cx \\ \text{Subject to} \quad & Ax = b \\ & x \geqq 0 \end{aligned} \tag{11.11}$$

The matrix A is said to have the *unimodularity property* if the determinant of every square submatrix of A is either 0, +1, or −1. This automatically implies that all the entries in A itself are either 0, +1, or −1. From *Cramer's rule* (see Appendix 3) for solving a square nonsingular system of simultaneous linear equations, it is clear that if A has the unimodularity property and if b is an integer vector, every basic solution of (11.11) will be an integer vector. Hence, if A is unimodular and if it is required to solve the integer program,

$$\begin{aligned} \text{Minimize} \quad & cx \\ \text{Subject to} \quad & Ax = b \\ & x \geqq 0 \\ & x \text{ an integer vector} \end{aligned} \tag{11.12}$$

we can neglect the integer constraints and just solve (11.11) instead, as an ordinary LP using the simplex algorithm. The optimum BFS of (11.11) will actually be an optimum solution of the integer program (11.12).

It was proved in section 11.1, that the input-output coefficient matrix of the transportation problem is unimodular. Hence, as long as a_i, b_j are integers for all i, j, the optimum solution of (11.6) obtained by the simplex algorithm will be an integer solution. This is what makes integer programs of the transportation problem type so trivial.

Problems

11.11 Consider the assignment problem:

$$\begin{aligned} \text{Minimize} \quad z(x) = & \sum_{i=1}^{n} \sum_{j=1}^{n} c_{ij} x_{ij} \\ \text{Subject to} \quad & \sum_{i=1}^{n} x_{ij} = 1 \qquad \text{for all } j = 1 \text{ to n} \\ & \sum_{j=1}^{n} x_{ij} = 1 \qquad \text{for all } i = 1 \text{ to n} \\ & x_{ij} \geqq 0 \end{aligned}$$

Show that this can be solved by the algorithm discussed in section 11.2. Prove that this algorithm leads to an optimum solution to the assignment problem in which all the x_{ij} are either 0 or 1.

11.12 The problem of determining how much fraction of his life each man should spend with each woman in a group consisting of an equal number of men and women in order to maximize the total happiness of the group was discussed in Chapter 1. Show that this is an assignment problem and has an optimum solution in which every man spends all his life with only one woman and vice versa.

DISCUSSION ON THE MARRIAGE PROBLEM

Problem 11.12 is known as the *marriage problem*. The fact that there is an optimum solution of this problem in which every man spends all his life with a woman has been interpreted by sociologits, as supplying a proof to the controversial statement "there exists an optimum marriage policy in which everyone remains monogamous, and there is no divorce in society." However, this proof is valid only under the linearity assumption that the happiness derived by the ith man and the jth woman by spending a fraction x_{ij} of their lives together is $c_{ij}x_{ij}$ for any $x_{ij} \geqq 0$. In practice this linearity assumption may fail to hold over long periods of time.

Problems

11.13 Prove the following result by A. J. Hoffman and D. Gale by an inductive argument. Let A be an $m \times$ n matrix whose rows can be partitioned into two disjoint sets $\mathbf{S}$ and $\bar{\mathbf{S}}$ with the following properties:

(i) Every column vector of A contains at most two nonzero entries.

(ii) Every entry in A is 0, $+1$, or -1.

(iii) If two nonzero entries in a column vector of A have the same sign, then one of these nonzero entries is in **S** and the other is in $\bar{\mathbf{S}}$.

(iv) If two nonzero entries in a column vector of A have opposite signs, then both the rows in which these entries lie are in either **S** or $\bar{\mathbf{S}}$.

Then A is a unimodular matrix.

11.14 Using the result in Problem 11.13, prove that the input-output coefficient matrix of the transportation problem (11.2) and (11.3) has the unimodularity property.

11.4 LABELING METHODS IN THE PRIMAL TRANSPORTATION ALGORITHM

In sections 11.1 and 11.2, we discussed only a trial and error method for finding the θ-loop in $\mathscr{B} \cup \{(p, q)\}$. Unless the problem being solved is very small, this trial and error method is likely to be time consuming. Besides, it is very difficult to program it for solving the problem on a computer. The labeling method discussed in this section provides an efficient and systematic method for finding the θ-loop in $\mathscr{B} \cup \{(p, q)\}$. Most of the commercially available computer programs for solving transportation problems use this labeling method. The development of this labeling method has resulted in tremendous improvements in the efficiency with which large transportation problems can be solved using computers.

11.4.1 Review of Graph Theory Concepts

A *graph* is a pair of sets. One of them is a finite set, $\mathscr{N}$, of points (also called *nodes or vertices*). The other set is a set of lines, $\mathscr{A}$ called *edges*, each edge joining a pair of distinct points in $\mathscr{N}$. If $i \in \mathscr{N}$, $j \in \mathscr{N}$, $i \neq j$, the edge joining i and j will be denoted by $(i; j)$ or equivalently $(j; i)$, and this edge is said to be *incident* at points i and j. The graph itself is $\mathrm{G} = (\mathscr{N}, \mathscr{A})$. There is at most one edge between any pair of points, and every edge contains exactly two points of $\mathscr{N}$.

Given two nodes i_0 and i_* in $\mathscr{N}$, a path from i_0 to i_* in the graph G is a connected sequence of nodes and edges of the form

$$i_0, (i_0; i_1), i_1(i_1; i_2), i_2, \ldots, i_{k-1}, (i_{k-1}; i_*), i_*$$

such that each edge in the sequence is in $\mathscr{A}$. For brevity, we will represent the above path by the ordered set of edges $[(i_0; i_1), \ldots, (i_{k-1}; i_*)]$. The length of a path is defined to be equal to the number of edges in the sequence. Thus the length of the path from i_0 to i_* discussed above is k, even through some edges may appear more than once in the sequence.

A path is called a *simple path* if every node along the path appears in the sequence only once. A *cycle* is a path from a node to itself that contains at least two edges. A cycle is a *simple cycle* if it is a simple path from a node to itself.

A graph is a *connected graph* if there exists a path in the graph between every pair of points in the graph. Otherwise it is a *disconnected graph.* A connected graph that contains no cycles is called a *tree.* Given a set of points $\mathscr{N}$, the graph $\mathrm{G} = (\mathscr{N}, \mathscr{A})$ is said to be a *spanning tree* if G is a tree and if it contains an edge incident at every point in $\mathcal{N}$. If $G = (\mathcal{N}, \mathcal{A})$ is a tree, a point in $\mathcal{N}$ is said to be a *terminal node* (also called *end node or pendant node*) if there is exactly one edge in $\mathcal{A}$ incident at it. See Figure 11.2.

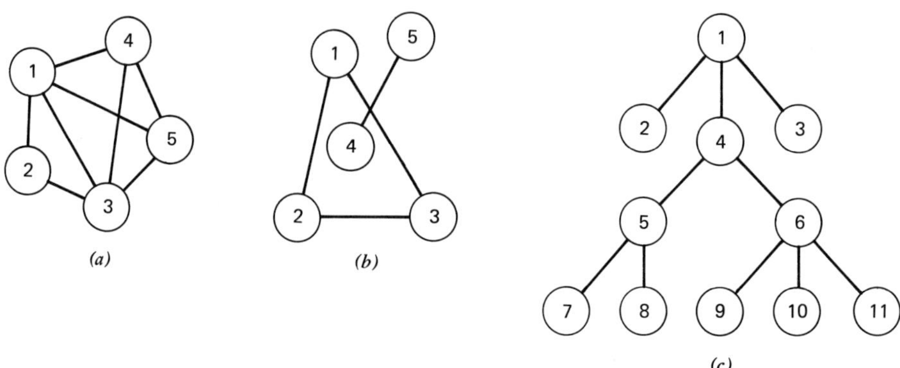

Figure 11.2 (*a*) A connected graph with 5 points and 8 edges. (*b*) A discounted graph. (*c*) A tree with 11 points. Points 2, 3, 7, 8, 9, 10, 11 are terminal nodes.

Theorem Let i_0, i_* be a pair of points on a tree G. Then there exists a unique simple path in G from i_0 to i_*.

Proof Since G is a tree, it is connected and, hence, there exists at least one path from i_0 to i_*. If there exist two distinct simple paths from i_0 to i_* in G, these paths together constitute a cycle from i_0 to i_0 contradicting the hypothesis that G is a tree.

Problems

11.15 Prove that every tree has at least one terminal node.

11.16 If $\mathcal{N} = \{1, \ldots, r\}$ and $G = (\mathcal{N}, \mathcal{A})$ is a spanning tree, prove that the number of edges in $\mathcal{A}$ is $r - 1$.

11.4.2 Applications to the Transportation Problem

Consider a balanced transportation problem with m sources and n markets. The ith row of the transportation array is associated with the ith source, $i = 1$ to m, and we will represent it by a point whose *serial number* is i. The jth column of the transportation array is associated with the jth market, $j = 1$ to n, and we will represent it by a point whose serial number is $m + j$. Let $\mathcal{N} = \{1, \ldots, m, m + 1, \ldots, m + n\}$. Let Δ be a subset of cells of the transportation array. Determine the corresponding set of edges $\mathcal{A}_\Delta = \{(i; m + j)$: the cell (i, j) is in $\Delta\}$.* Then $G_\Delta = (\mathcal{N}, \mathcal{A}_\Delta)$ is the graph associated with the subset Δ.

Theorem **Simple cycles and θ-Loops** Every θ-loop in Δ corresponds to a simple cycle in G_Δ and vice versa.

Proof This follows from the definitions of θ-loops and simple cycles.

Hence Δ contains a θ-loop iff there is a simple cycle in G_Δ. From this, it is clear that Δ forms a basic set of cells for the $m \times n$ transportation array iff $G_\Delta = (\mathcal{N}, \mathcal{A}_\Delta)$ is a spanning tree.

* Edges in the graph are denoted by unordered pairs $(i; j)$ with a semicolon in the middle and here i, j are serial numbers of points in the graph. Cells in the transportation array are denoted by ordered pairs (i, j) with a comma in the middle, this being the cell in row i and column j of the array.

TO DRAW $G_{\mathscr{B}}$ AS A ROOTED TREE

Suppose $\mathscr{B}$ is a basic set of cells for the m $\times$ n transportation array. Draw the graph $G_{\mathscr{B}} = (\mathcal{N}, \mathscr{A}_{\mathscr{B}})$ as follows: Pick m + n as the initial node. When drawn in the manner discussed below, the tree $G_{\mathscr{B}}$ is called a *rooted tree* with m + n as the *root*. Any other point in $\mathcal{N} = \{1, \ldots, m, m+1, \ldots, m+n\}$ could have been picked as the root, but to keep the discussion here compatible with section 11.2, we pick m + n as the root. The *predecessor index*, P(m + n), of the root node m + n is always $= \emptyset$, the empty set. This completes *stage* 0. We will now discuss what happens in a general stage, stage k, where $k \geqq 1$.

Stage k In this stage include in the tree, all points $i \in \mathcal{N}$ that have not yet been included, satisfying $(i; j) \in \mathscr{A}_{\mathscr{B}}$ for some j with the property that j is a point included in the graph in stage $k - 1$. The predecessor index, $P(i)$, of such a node i is defined to be j. Draw the edge $(i; j)$.

If the tree now contains all the points in $\mathcal{N}$, terminate. Otherwise go to stage $k + 1$.

Example

Consider the basic set, $\mathscr{B}$, of cells of the transportation array marked by a "T."

	$j = 1$	2	3	4	5	6	Associated point serial number	Predecessor index	Successor index	Younger brother index	Elder brother index
$i = 1$			T	T			1	8	7	$\emptyset$	$\emptyset$
2	T					T	2	10	5	4	$\emptyset$
3		T			T		3	6	9	$\emptyset$	$\emptyset$
4		T		T		T	4	10	6	$\emptyset$	2
Associated point serial number	5	6	7	8	9	10					
Predecessor index	2	4	1	4	3	$\emptyset$					
Successor index	$\emptyset$	3	$\emptyset$	1	$\emptyset$	2					
Younger brother index	$\emptyset$	8	$\emptyset$	$\emptyset$	$\emptyset$	$\emptyset$					
Elder brother index	$\emptyset$	$\emptyset$	$\emptyset$	6	$\emptyset$	$\emptyset$					

Column j of the transportation array corresponds to the point $4 + j$ in the graph. Row i corresponds to point i. The rooted tree corresponding to $\mathscr{B}$ is the graph in Figure 11.3 excluding the dashed line (1; 9).

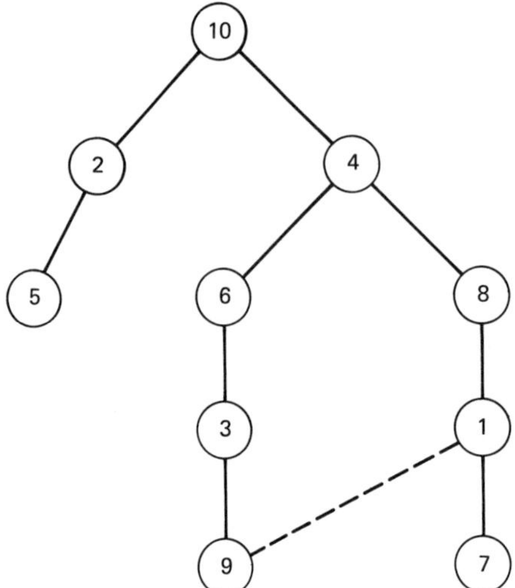

Figure 11.3

The predecessor indices are entered on the transportation array itself. The tree is drawn above in such a way that if points i and j are connected by an edge and j is below i, then $P(j) = i$. If $\Delta = \mathscr{B} \cup \{(1, 5)\}$, the graph G_Δ is the graph in Figure 11.3 including the dashed line $(1; 9)$, and it contains the simple cycle corresponding to the θ-loop in $\mathscr{B} \cup \{(1, 5)\}$.

PROPERTIES OF THE ROOTED TREE

The rooted tree corresponding to a basic set of cells of the transportation array has the following properties.

(i) All points included in the tree during a stage of even number in the procedure discussed above are associated with columns of the transportation array and all points included in the tree during a stage of odd number in the procedure are associated with rows of the transportation array.

(ii) A cell of the transportation array always corresponds to an edge in the graph between two points that are included in the tree at different stages in the procedure.

(iii) All points have a predecessor index except the root node. The predecessor index of a point associated with a row of the array is always the serial number of a point associated with a column of the array and vice versa.

11.4.3 The Successor and Brother Indices

Let i be any point in the rooted tree $G_{\mathscr{B}}$. The set of all points in this graph of the form $\{j: P(j) = i\}$ is known as the set of *immediate successors* of the point i. Clearly the edge $(i; j)$ is the last edge in the unique simple path from the root node to j for all points j in this set.

If i is a point in G, the set of points $\{j: j \neq i$ and j is such that i is a point on the unique simple path from the root node to $j\}$ is known as the set of all *descendants* of i. Clearly the set of all descendants of i is the union of the set of immediate successors of i and the sets of all descendants of j as j ranges over the set of immediate successors of i.

If i is not the root node and is a terminal node of $G_{\mathscr{B}}$, its set of immediate successors and its set of all descendants are empty.

In the course of the algorithm, it is necessary to compute the set of immediate successors and the set of descendants. For doing this computation, the successor and brother indices of points are defined. For all terminal nodes not equal to the root node, the *successor index* is defined to be the empty set, since these points have no successors. If the point i has a unique immediate successor j, the successor index of i, $S(i)$, is defined to be j. If a point, i, has two or more immediate successors, these successors are all *brothers* of each other. Thus, j and k are brothers iff they have the same predecessor index. These successors are arranged in some order as a sequence from left to right. This order can be chosen arbitrarily. If j and k are two points in this sequence and j appears to the left of k, then k is known as a *younger brother* of j and j is an *elder brother* of k. The leftmost of all these brothers is the *eldest son* of i. The *successor index*, $S(i)$, is the serial number of the eldest son of i.

The *younger brother index* of a point j, denoted by $YB(j)$, is the serial number of the eldest among the set of younger brothers of j (i.e., the leftmost of the younger brothers of j in the ordering chosen for the brothers of j), if the set of younger brothers of j is nonempty. $YB(j)$ is the empty set if j has no younger brothers.

The *elder brother index* of a point j, $EB(j)$, is the serial number of the youngest (i.e., the rightmost of the elder brothers of j in the sequence of brothers) of the set of elder brothers of j, if this set is nonempty. $EB(j)$ is the empty set if j has no elder brothers.

Example 1

For a transportation problem with $m = 6$, $n = 11$, the rooted tree corresponding to a basis is given in Figure 11.4. The root node is 17. Assuming that the brothers are ordered from left to right as they are recorded in the diagram above, the various indices are tabulated as below.

Point	1	2	3	4	5	6	7	8	9	10	11	12	13	14	15	16	17
Predecessor	13	7	17	15	17	11	5	4	2	3	2	4	6	1	3	3	∅
Successor	14	9	10	8	7	13	2	∅	∅	∅	6	∅	1	∅	4	∅	3
Younger brother	∅	∅	5	∅	∅	∅	∅	12	11	15	∅	∅	∅	∅	16	∅	∅
Elder brother	∅	∅	∅	∅	3	∅	∅	∅	∅	∅	9	8	∅	∅	10	15	∅

Example 2

For the example in section 11.4.2, the successor and brother indices are recorded on the transportation array.

TO OBTAIN THE SETS OF IMMEDIATE SUCCESSORS, YOUNGER BROTHERS AND DESCENDANTS

When the various indices are defined as in the previous sections, the set of younger brothers of a point j is empty if $YB(j) = \emptyset$. If $YB(j) \neq \emptyset$, the set of younger brothers of j is the union of $\{YB(j)\}$ and the set of younger brothers of $YB(j)$. Using this recursively, the set of younger brothers of any point in the graph can be found by using only the younger brother indices.

The set of immediate successors of a point, i, is empty if $S(i) = \emptyset$. If $S(i) \neq \emptyset$, the set of immediate successors of i is the union of $\{S(i)\}$ and the set of younger brothers of $S(i)$.

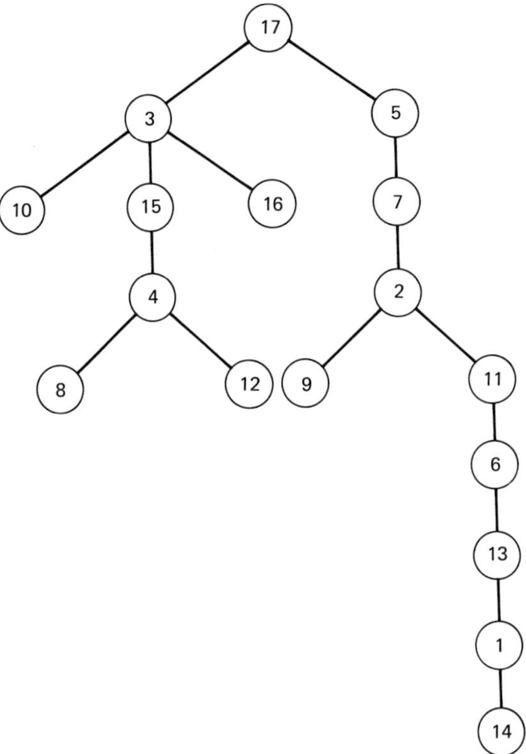

Figure 11.4

The set of descendants of a point i is empty if the successor index of i is empty. Otherwise it is the union of the set of immediate successors of i and the sets of descendants of each of the immediate successors of i.

11.4.4 To Obtain the Simple Path between a Point and the Root Node

Suppose i is a point in the rooted tree $G_{\mathscr{B}}$, which is not the root node. The following procedure finds the unique simple path between it and the root node.

Step 1 The edge $(i; P(i))$ is the first edge in the path. If $P(i)$ is the root node terminate. Otherwise pick $P(i)$ as the *current point* and go to Step 2.

Step 2 Suppose the current point is j. $(j, P(j))$ is the next edge in the path. Go to Step 3.

Step 3 If $P(j)$ is the root node, terminate. Otherwise, change the current point to $P(j)$ and go back to Step 2.

When the path obtained here is written in the reverse order, it becomes the simple path from the root node to i in the rooted tree $G_{\mathscr{B}}$.

Example

Consider the basis for the 4 × 6 transportation problem discussed in section 11.4.2. The root node in the rooted tree corresponding to this basis is 10. Suppose it is required to find the simple path from 9 to the root node. Point 9 corresponds to column 5 in the transportation array and its predecessor index is 3. So the edge (9; 3) is in the path. The predecessor index of row 3 is 6. So the edge (3; 6) is in the path. The predecessor index of point 6 (column 2) is 4, which puts the edge (6; 4) in the path. The predecessor index of point 4 (row 4) is 10, which puts (4; 10) in the path. Since 10 is the root node, the procedure terminates here. Hence the path is ((9; 3), (3; 6), (6; 4), (4; 10)). Verify this from Figure 11.3.

11.4.5 To Determine the θ-Loop in $\mathscr{B} \cup \{(p, q)\}$

Suppose $\mathscr{B}$ is a basic set of cells for the m $\times$ n transportation array and (p, q) is a nonbasic cell. The unique θ-loop in $\mathscr{B} \cup \{(p, q)\}$ can be determined easily using the predecessor indices.

The column q of the m $\times$ n transportation array corresponds to the point m + q in the rooted tree $G_{\mathscr{B}}$. Hence the cell (p, q) corresponds to the edge, $(p;$ m $+ q)$ in $G_{\mathscr{B}}$. The θ-loop in $\mathscr{B} \cup \{(p, q)\}$ corresponds to the unique simple cycle in the graph $(\mathscr{N}, \mathscr{A}_{\mathscr{B}} \cup \{(p; \text{m} + q)\})$.

Find the unique simple path from m + q to the root node. Also find the unique simple path from p to the root node. Eliminate all common edges in these two paths. Suppose the remaining portions of these paths are $((p; i_1), (i_1; i_2), \ldots, (i_k; i_*))$ and $((\text{m} + q; j_1), (j_1; j_2), \ldots, (j_t; i_*))$. The common point, i_*, in these two, is known as the *apex* of the simple cycle. The simple cycle itself is $((p; \text{m} + q), (\text{m} + q; j_1), \ldots, (j_t; i_*), (i_*; i_k), \ldots, (i_2; i_1), (i_1; p))$.

An edge $(i; j)$ of the rooted tree where i is a point corresponding to a row of the transportation array and j is a point corresponding to a column of the array corresponds to the cell $(i, j - \text{m})$ in the array. The cells in the θ-loop are the cells of the transportation array corresponding to the edges in the simple cycle. When all the cells in the θ-loop are determined, entries of $+\theta$ and $-\theta$ can be put alternately in them starting with a $+\theta$ entry in the cell (p, q).

In the primal algorithm for solving the transportation problem, the present basic cell that drops off from the basic set when (p, q) is the entering cell is determined as in section 11.2.3, once the θ-loop corresponding to (p, q) is determined. Suppose the cell (p, q) replaces the cell (r, s) from the basic set. Let the new basic set be $\mathscr{B}'$.

The graph $G_{\mathscr{B}'}$ corresponding to the new basic set is obtained from the graph $G_{\mathscr{B}}$ by deleting the edge $(r;$ m $+ s)$ and adding the edge $(p;$ m $+ q)$.

Example 1

Consider the basis for the 6 $\times$ 11 transportation problem discussed in section 11.4.3, Example 1. Suppose the θ-loop corresponding to the nonbasic cell (4, 5) has to be found. The edge corresponding to cell (4, 5) is (4; 11). The simple path from the point 4 to the root node is ((4; 15), (15; 3), (3; 17)) and the simple path from the point 11 to the root node is ((11; 2), (2; 7), (7; 5), (5; 17)). These two paths have no common edges and 17 is the common point on these two paths. Hence 17 is the apex of this simple cycle and the simple cycle is ((4; 11), (11; 2), (2; 7), (7; 5), (5; 17), (17; 3), (3; 15), (15; 4)).

So the θ-loop consists of the cells (4, 5), (2, 5), (2, 1), (5, 1), (5, 11), (3, 11), (3, 9), (4, 9). The cells with $+\theta$ entries are (4, 5), (2, 1), (5, 11), (3, 9) and the cells with $-\theta$ entries are (2, 5), (5, 1), (3, 11), (4, 9).

Example 2

For the same 6 $\times$ 11 transportation problem and the basic set $\mathscr{B}$ discussed in Example 1, find the θ-loop corresponding to the nonbasic cell (1, 3).

The cell (1, 3) corresponds to the edge (1; 9) in the rooted tree. The simple path from the point 1 to the root node is ((1; 13), (13; 6), (6; 11), (11; 2), (2; 7), (7; 5), (5; 17)), and the path from the point 9 to the root node is ((9; 2), (2; 7), (7; 5), (5; 17)). Eliminating the common edges between these two paths, it is seen that the simple cycle in this problem is ((9; 1), (1; 13), (13; 6), (6; 11), (11; 2), (2; 9)) and the apex is 2. This corresponds to the θ-loop consisting of the cells ((1, 3), (1, 7), (6, 7), (6, 5), (2, 5), (2, 3)). The cells with $+\theta$ entries are (1, 3), (6, 7), (2, 5) and the cells with $-\theta$ entries are (1, 7), (6, 5), (2, 3).

11.4.6 Updating the Rooted Tree and the Various Indices

Suppose $\mathscr{B}$ is a basic set of cells for the m $\times$ n transportation array and let $G_{\mathscr{B}}$ be the corresponding rooted tree. Suppose the rooted tree $G_{\mathscr{B}'}$ of an adjacent basic set $\mathscr{B}'$ is obtained by dropping the edge $(i_2; j_*)$ from $G_{\mathscr{B}}$ and adding the edge $(i_1; j_1)$. Arrange the points in the edge $(i_2; j_*)$ in such a way that $P(j_*) = i_2$.

Suppose i_0 is the apex of the simple cycle in $(\mathscr{N}, \mathscr{A}_{\mathscr{B}} \cup \{(i_1; j_1)\})$. The simple cycle consists of the edge $(i_1; j_1)$, a path from i_0 to j_1 in reverse order, and a path from i_0 to i_1. The edge $(i_2: j_*)$ being deleted will be contained on exactly one of these two paths. Suppose the points in the edge $(i_1; j_1)$ are arranged in such a way that the edge $(i_2: j_*)$ is on the path from i_0 to j_1 as in Figure 11.5.

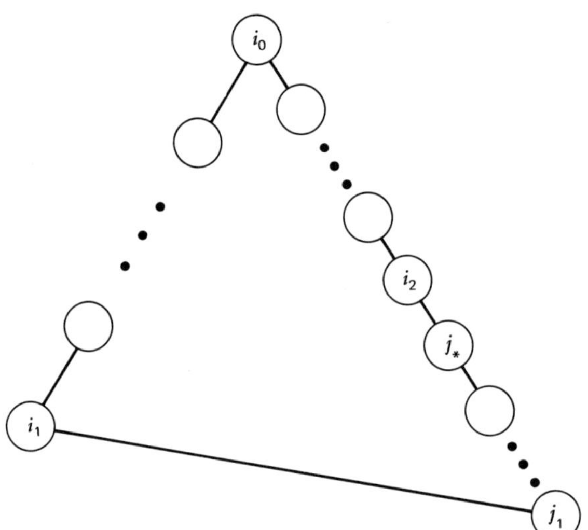

Figure 11.5 The simple cycle.

The path from j_1 to i_0 on this simple cycle contains the path from j_1 to j_*. Suppose this path is $((j_1; j_2), (j_2; j_3), \ldots, (j_t; j_*))$ where the edges in this path are written in such a way that $P(j_u) = j_{u+1}$ for $u = 1$ to $t - 1$ and $P(j_t) = j_*$. We will refer to j_* as j_{t+1}. See Figure 11.6.

Compute the set of all descendants of j_* in the present rooted tree $G_{\mathscr{B}}$ using the procedure in section 11.4.3. Let **H** be the set containing j_* and the set of all descendants of j_*. We will use the symbols P(i), S(i), YB(i), EB(i) to indicate present indices, and P′(i), S′(i), YB′(i), EB′(i) to indicate the updated indices (i.e., the new indices after the basis change). It is convenient to do the updating in the order indicated below.

UPDATING THE PREDECESSOR INDICES

The predecessor indices of all the points except $j_1, \ldots, j_t, j_{t+1}$ remain unchanged. Set $P'(j_1) = i_1$, $P'(j_u) = j_{u-1}$, for $u = 2$ to $t + 1$.

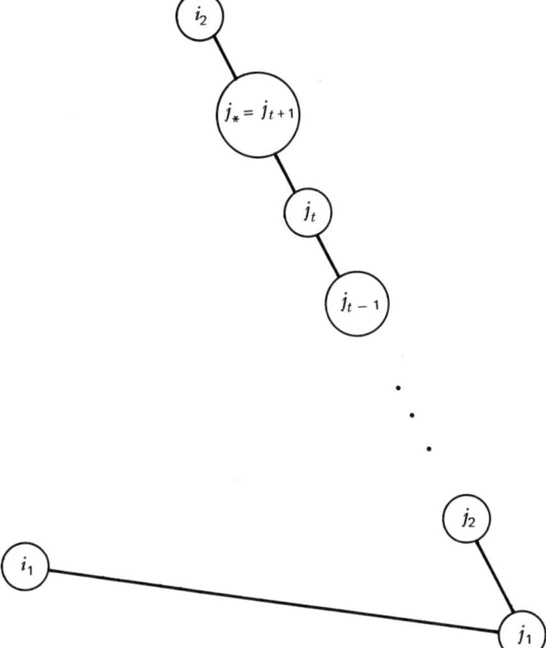

Figure 11.6

UPDATING THE SUCCESSOR INDICES

The successor indices for all points other than $i_1, j_1, \ldots, j_t, j_{t+1}, i_2$ remain unchanged. Set $S'(i_1) = j_1, S'(j_u) = j_{u+1}$ for $u = 1$ to t. If $S(i_2) = j_*$ set $S'(i_2) = YB(j_*)$. If $S(i_2) \neq j_*$, make $S'(i_2) = S(i_2)$. If $S(j_*) = j_t$, set $S'(j_*) = YB(j_t)$. If $S(j_*) \neq j_t$, make $S'(j_*) = S(j_*)$.

UPDATING THE BROTHER INDICES

To update the brother relationships, we will make the assumption that any new immediate successor of a point joins the previous immediate successors of this point as their eldest brother (i.e., any new immediate successor joins always at the left end of the sequence of brothers). Also, if a point is to be removed from the set of immediate successors, we will assume that the elder-younger brotherly relationship among the remaining points in this set remain unchanged. This is compatible with what has been done in updating the successor indices.

The brother indices may change only among the set **H** (the set containing j_* and the descendants of j_*), the point $S(i_1)$, and $YB(j_*)$, $EB(j_*)$, if these are not empty. For all other points, the brother indices remain unchanged.

Set $YB'(j_1) = S(i_1)$. For each $u = 1$ to $t + 1$, if $EB(j_u) \neq \varnothing$, set $YB'(EB(j_u)) = YB(j_u)$; and if $YB(j_u) \neq \varnothing$ set $EB'(YB(j_u)) = EB(j_u)$. If $S(i_1) \neq \varnothing$, set $EB'(S(i_1)) = j_1$. For each $u = 1$ to $t + 1$ set $EB'(j_u) = \varnothing$, because these points join as the eldest among their new set of brothers. For each $j = 2$ to $t + 1$, set $YB'(j_u) = S(j_{u-1})$. If $YB(j_*) \neq \varnothing$, $EB'(YB(j_*)) = EB(j_*)$. If $EB(j_*) \neq \varnothing$, $YB'(EB(j_*)) = YB(j_*)$. Leave all other brother indices unchanged.

Example

Consider the basis for the 6 × 11 transportation problem discussed in section 11.4.3, Example 1. Suppose the edge (3; 17) is being deleted and the edge (4; 11) is being added. The rooted tree corresponding to the new basis is shown in Figure 11.7.

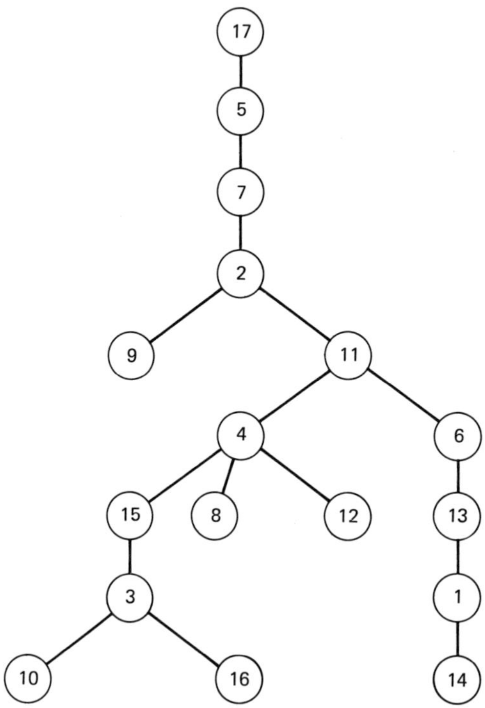

Figure 11.7

The new indices are obtained from the previous indices tabulated in Example 1 of section 11.4.3, using the updating procedure. They are tabulated as below.

Point	1	2	3	4	5	6	7	8	9	10	11	12	13	14	15	16	17
Predecessor	13	7	15	11	17	11	5	4	2	3	2	4	6	1	4	3	Ø
Successor	14	9	10	15	7	13	2	Ø	Ø	Ø	4	Ø	1	Ø	3	Ø	5
Younger brother	Ø	Ø	Ø	6	Ø	Ø	Ø	12	11	16	Ø	Ø	Ø	Ø	8	Ø	Ø
Elder brother	Ø	Ø	Ø	Ø	Ø	4	Ø	15	Ø	Ø	9	8	Ø	Ø	Ø	10	Ø

11.4.7 Updating the Dual Vector and the Relative Cost Coefficients

Suppose the basic set $\mathscr{B}'$ for the m × n transportation problem is obtained by bringing the cell (p, q) into the basic set $\mathscr{B}$. Suppose $(u, v, \bar{c})$ refer to the dual solution and the relative cost coefficients with respect to the basic set $\mathscr{B}$. Let $(u', v', \bar{c}')$ refer to the dual solution and the relative cost coefficients with respect to the new basic set $\mathscr{B}'$.

Suppose that the cell being dropped from the basic set when (p, q) is inserted corresponds to the edge $(i_2; j_*)$ in $G_{\mathscr{B}}$, where $P(j_*) = i_2$. As before, let **H** be the set containing j_* and all the descendants of j_* in the rooted tree $G_{\mathscr{B}}$. Let the entering cell (p, q) correspond to the edge $(i_1; j_1)$ in the graph, where i_1 and j_1 are such that the dropping edge $(i_2; j_*)$ is on the path from the apex to j_1. Let the path from j_1 to the apex be as in section 11.4.6. Here is the new dual solution.

$$
\begin{aligned}
u'_i &= u_i - \alpha\bar{c}_{pq} && \text{for } i \in \mathbf{H} \\
&= u_i && \text{for } i \notin \mathbf{H} \\
v'_j &= v_j + \alpha\bar{c}_{pq} && \text{for m} + j \in \mathbf{H} \\
&= v_j && \text{for m} + j \notin \mathbf{H} \\
\bar{c}'_{ij} &= \bar{c}_{ij} + \alpha\bar{c}_{pq} && \text{for } i \in \mathbf{H} \quad \text{and} \quad \text{m} + j \notin \mathbf{H} \\
&= \bar{c}_{ij} - \alpha\bar{c}_{pq} && \text{for } i \notin \mathbf{H} \quad \text{and} \quad \text{m} + j \in \mathbf{H} \\
&= \bar{c}_{ij} && \text{if } i \text{ and m} + j \text{ are both in } \mathbf{H} \text{ or not in } \mathbf{H}
\end{aligned}
$$

where $\alpha = 1$ if i_1 is the node corresponding to a row of the transportation array, and $= -1$, otherwise.

This follows from the fact that the new dual solution satisfies $u'_i + v'_j = c_{ij}$ for all basic cells (i, j) in the basic set $\mathscr{B}'$ and $\bar{c}_{ij} = c_{ij} - (u_i + v_j)$ for all i, j.

ORGANIZING THE COMPUTATIONS

Clearly all the computations can be performed using the various indices. There is no need to have a picture of the rooted tree for doing the computations. The indices can be recorded on the transportation array itself and all the computations performed using them.

Numerical Example

Consider the following balanced transportation problem. In this problem m = 5 and n = 8. So rows 1 to 5 correspond to points 1 to 5, respectively, in the graph and columns 1 to 8 correspond to points 6 to 13, respectively, in the graph. All the indices are serial numbers of the points.

An initial BFS for the problem is given in page 320. Every cell with a small square in the middle is a basic cell and the value of that basic variable is entered in the middle of that square. The dual solution corresponding to this basis, u_i and v_j for $i = 1$ to m, $j = 1$ to n, is tabulated. The original cost coefficient c_{ij} is recorded in the lower right corner of the cell (i, j) and the relative cost coefficient of a nonbasic cell is entered in the upper left corner. The predecessor indices, P(i), the successor indices, S(i), the younger brother indices, YB(i), and the elder brother indices, EB(i), of the various points corresponding to the rows and columns of the transportation array are entered. In determining the initial brother indices, it is assumed that all the brothers are arranged from left to right in increasing order of their serial number (i.e., initially, we assume that if points i, j are brothers, i is elder brother of j if $i < j$).

A diagram of the rooted tree corresponding to this basis is given below (Figure 11.8) to aid in understanding how the various indices are obtained. But all the indices could be directly read out from the array itself starting with the root node, which is the point corresponding to column 8. Since some of the $\bar{c}_{ij} < 0$, the present basis is not optimal. Suppose we decide to bring the nonbasic cell (3, 6) into the basic set. This entering cell (3, 6) corresponds to the edge (3; 11).

From the predecessor indices, the path from point 3 to the root node is ((3; 12), (12; 5), (5; 13)) and the path from point 11 to the root node is ((11; 1), (1; 10), (10; 5), (5; 13)). Elim-

	$j = 1$	2	3	4	5	6	7	8	a_i	u_i	P(i)	S(i)	YB(i)	EB(i)
$i = 1$	−6, 8	5, 10	1, 7	−5, 2	[4], 8	[5], 11	0, 2	−1, 8	9	9	10	11	∅	∅
2	1, 10	[3], 0	0, 1	6, 8	7, 10	−1, 5	4, 1	0, 4	3	4	7	∅	∅	∅
3	[1], 20	3, 14	3, 15	[12], 13	6, 20	−7, 10	[9], 8	−10, 5	22	15	12	6	4	∅
4	4, 25	18, 30	2, 15	1, 15	0, 15	2, 20	[17], 9	4, 20	17	16	12	∅	∅	3
5	9, 25	[8], 7	[19], 8	11, 20	[2], 10	2, 15	[8], 4	[6], 11	24	11	13	7	∅	∅
b_j	1	11	19	12	6	5	34	6						
v_j	5	−4	−3	−2	−1	2	−7	0						
P(m + j)	3	5	5	3	5	1	5	∅						
S(m + j)	∅	2	∅	∅	1	∅	3	5						
YB(m + j)	9	8	10	∅	12	∅	∅	∅						
EB(m + j)	∅	∅	7	6	8	∅	10	∅						

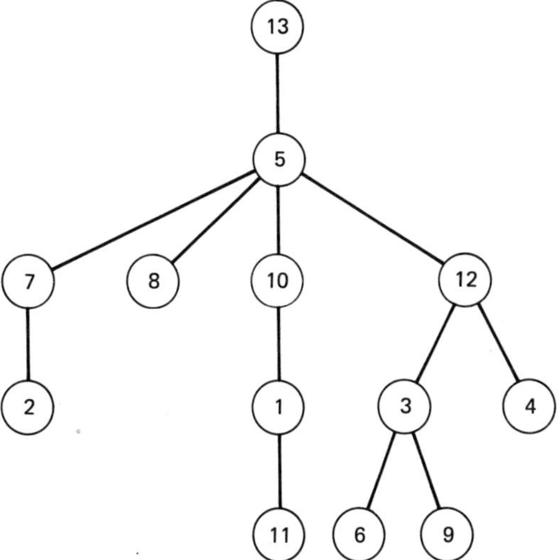

Figure 11.8 Rooted tree corresponding to the present basis.

inating the common edges of these two paths, notice that the apex is the point 5 and the simple cycle is ((3; 11), (11; 1), (1; 10), (10; 5), (5; 12), (12; 3). The simple cycle corresponds to the θ-loop consisting of the cells ((3, 6), (1, 6), (1, 5), (5, 5), (5, 7), (3, 7)). The cells (3, 6), (1, 5), and (5, 7) have a $+\theta$ entry in the θ-loop and the cells (1, 6), (5, 5), (3, 7) have a $-\theta$ entry. The minimum basic value (in the present BFS) in the cells with a $-\theta$ entry is minimum $\{5, 2, 9\} = 2$, and it occurs in cell (5, 5). So the cell (5, 5) leaves the basic set.

The leaving cell (5, 5) corresponds to the edge (10; 5). This edge appears on the path from point 11 to the apex. So the path corresponding to $(j_1; j_2), \dots, (j_t; j_*)$ of section 11.4.6 in this example is (11; 1), (1; 10). **H**, the set containing 10 and all the descendants of 10, is $\{10, 1, 11\}$. Since the entering cell is (3, 6), the $\bar{c}_{pq}$, corresponding to the discussion above is $\bar{c}_{36} = -7$.

The BFS, the various indices, the dual solution, and the relative cost coefficients are updated using the procedures discussed earlier. The new transportation array is on page 322. The new basis is again nonoptimal since some of the relative cost coefficients are still negative. The algorithm can be continued in a similar manner until optimality is attained.

Problem

11.17 Complete the solution of this problem.

CONCLUDING REMARKS

Methods for labeling trees were first discussed by E. Johnson (see reference [3]). F. Glover, D. Karney, and D. Klingman (reference [1]) developed the application of Johnson's tree labeling method to the transportation problem and called it the *augmented predecessor index method* (the method discussed in this section is a minor variant of it). See references [2 and 4] for other variants of this method.

11.5 SENSITIVITY ANALYSIS IN THE TRANSPORTATION PROBLEM

Consider the transportation problem (11.1). Let $\mathscr{B}$ be an optimum basic set of cells for it. Let $\tilde{x}$ be the corresponding optimum solution and let $\tilde{u}$, $\tilde{v}$, $\tilde{\bar{c}}$ be the corresponding dual solution and the relative cost vector for it.*

* ALTERNATE OPTIMUM SOLUTIONS: $\tilde{x}$ is the unique optimum solution if $\tilde{\bar{c}}_{ij} > 0$ for all nonbasic cells (i, j). Alternate basic optimum solutions can be obtained by bringing into $\mathscr{B}$ any nonbasic cell (i, j) with $\tilde{\bar{c}}_{ij} = 0$. Every convex combination of optimum solutions is also optimal. See sections 2.6, 3.10, 4.5, and 4.5.

	$j = 1$	2	3	4	5	6	7	8	u_i	P(i)	S(i)	YB(i)	EB(i)
$i = 1$	−13, 8	−2, 10	−6, 7	−12, 2	[6], 8	[3], 11	−7, 2	−8, 8	16	11	10	∅	∅
2	1, 10	[3], 0	0, 1	6, 8	14, 10	6, 5	4, 1	0, 4	4	7	∅	∅	∅
3	[1], 20	3, 14	3, 15	[12], 13	13, 20	[2], 10	[7], 8	−10, 5	15	12	11	4	∅
4	4, 25	18, 30	2, 15	1, 15	7, 15	9, 20	[17], 9	4, 20	16	12	∅	∅	3
5	9, 25	[8], 7	[19], 8	11, 20	7, 10	9, 15	[10], 4	[6], 11	11	13	7	∅	∅
v_j	5	−4	−3	−2	−8	−5	−7	0					
P(m + j)	3	5	5	3	1	3	5	∅					
S(m + j)	∅	2	∅	∅	∅	1	3	5					
YB(m + j)	9	8	12	∅	∅	6	∅	∅					
EB(m + j)	11	∅	7	6	∅	∅	8	∅					

11.5.1 Change in a Nonbasic Cost Coefficient

Let (i, j) be a cell that is not in $\mathscr{B}$. Suppose the cost coefficient c_{ij} is subject to change, while all the remaining data in the problem remains unchanged. Since (i, j) is a nonbasic cell, a change in c_{ij} changes only the relative cost coefficient of the cell (i, j). Hence $\mathscr{B}$ remains an optimum basic set for all c_{ij} satisfying: $c_{ij} - (\tilde{u}_i + \tilde{v}_j) \geqq 0$, that is, $c_{ij} \geqq \tilde{u}_i + \tilde{v}_j$. If the new value of c_{ij} violates this constraint, bring (i, j) into the basic set and continue the application of the algorithm until optimality is reached again.

11.5.2 Change in a Basic Cost Coefficient

Let (r, s) be a cell in $\mathscr{B}$ and suppose c_{rs} is subject to change. Denote the cost coefficient in the cell (r, s) by γ_{rs} and let c_{rs} be its present value. Since (r, s) is a basic cell, a change in γ_{rs} changes the dual solution. Treating γ_{rs} as a parameter, recompute the dual solution corresponding to the basic set $\mathscr{B}$, as a function of γ_{rs}. Let $u(\gamma_{rs})$, $v(\gamma_{rs})$ be the new dual solution. The new relative cost coefficients are obtained from $\bar{c}_{ij}(\gamma_{rs}) = c_{ij} - u_i(\gamma_{rs}) - v_j(\gamma_{rs})$ for all i, j. All $\bar{c}_{ij}(\gamma_{rs})$ are affine functions of γ_{rs}. $\mathscr{B}$ remains an optimum basic set for all values of γ_{rs} for which $\bar{c}_{ij}(\gamma_{rs}) \geqq 0$ for all i, j. This defines the *optimality interval* of the basic set $\mathscr{B}$. The solution of (11.1) corresponding to the basic set $\mathscr{B}$ does not change as a result of a change in γ_{rs} and, hence, $\tilde{x}$ remains an optimum solution of (11.1) for all values of γ_{rs} in the optimality interval.

If the new value of γ_{rs} lies outside the optimality interval of $\mathscr{B}$, replace c_{rs} by the new value of γ_{rs}, and compute all the relative cost coefficients $\bar{c}_{ij}(\gamma_{rs})$. Bring a nonbasic cell with a negative $\bar{c}_{ij}(\gamma_{rs})$ into the basic set and continue the application of the simplex algorithm until optimality is reached again.

Numerical Example

Consider the following transportation problem. An optimum BFS is entered in the array. Basic cells are marked with a "□" in the middle and the values of the basic variables are entered inside it. The original cost coefficients, c_{ij}, are entered in the lower right hand corner of each cell, and the relative cost coefficients of nonbasic cells are entered in the upper left corner of the cell. The u_i and v_j obtained as in section 11.2.2 are also entered.

$j =$	1	2	3	4	u_i
$i = 1$	$11 = \bar{c}_{11}$; $c_{11} = 15$	8; 11	[3]; 0	[4]; 1	1
2	9; 14	[9]; 4	[5]; 1	6; 8	2
3	[7]; 6	[6]; 5	5; 7	7; 10	3
v_j	3	2	−1	0	

Suppose c_{14} is subject to change. If we denote it by γ_{14}, the new dual solution and the relative cost coefficients are as follows.

$j =$	1	2	3	4	u_i
$i = 1$	11 15	8 11	[3] 0	[4] γ_{14}	γ_{14}
2	9 14	[9] 4	[5] 1	$7-\gamma_{14}$ 8	$1+\gamma_{14}$
3	[7] 6	[6] 5	5 7	$8-\gamma_{14}$ 10	$2+\gamma_{14}$
v_j	$4-\gamma_{14}$	$3-\gamma_{14}$	$-\gamma_{14}$	0	

All the relative cost coefficients are nonnegative iff $\gamma_{14} \leqq 7$. Hence for all values of $\gamma_{14} \leqq 7$, the present solution is optimal.

11.5.3 Parametric Cost Transportation Problem

Consider the transportation problem of the form (11.1) with the objective function $z(x) = \sum_i \sum_j (c_{ij} + \lambda c^*_{ij})x_{ij}$, where λ is a parameter; suppose you have to solve the problem for all real values of the parameter λ.

Fix λ at 0 and solve the problem. Let $\mathscr{B}$ be the optimum basic set obtained. Let (u_i), (v_j) be the dual solution corresponding to $\mathscr{B}$ when (c_{ij}) is the cost matrix. Similarly let (u^*_i), (v^*_j) be the dual solution corresponding to $\mathscr{B}$ when (c^*_{ij}) is the cost matrix. Enter u_i, u^*_i in separate columns on the array. Similarly enter v_j, v^*_j in separate rows on the array. The dual solution corresponding to $\mathscr{B}$ as a function of λ is $(u_i + \lambda u^*_i)$, $(v_j + \lambda v^*_j)$. The relative cost coefficient of the cell (i, j) as a function of λ is $(c_{ij} - u_i - v_j) + \lambda(c^*_{ij} - u^*_i - v^*_j) = \bar{c}_{ij} + \lambda \bar{c}^*_{ij}$. Enter $\bar{c}_{ij}$ and $\bar{c}^*_{ij}$ separately in the cell. The basic set $\mathscr{B}$ is optimum for all values of λ satisfying $\bar{c}_{ij} + \lambda \bar{c}^*_{ij} \geqq 0$. Thus, if

$$\underline{\lambda} = \text{maximum} \left\{ -\frac{\bar{c}_{ij}}{\bar{c}^*_{ij}} : \quad (i, j) \text{ such that } \bar{c}^*_{ij} > 0 \right\} \qquad (11.13)$$

$$= -\infty \qquad \text{if all } \bar{c}^*_{ij} \leqq 0$$

and

$$\bar{\lambda} = \text{minimum} \left\{ -\frac{\bar{c}_{ij}}{\bar{c}^*_{ij}} : \quad (i, j) \text{ such that } \bar{c}^*_{ij} < 0 \right\} \qquad (11.14)$$

$$= +\infty \qquad \text{if all } \bar{c}^*_{ij} \geqq 0$$

then $\mathscr{B}$ is an optimum basic set for all λ in the interval $\underline{\lambda} \leqq \lambda \leqq \bar{\lambda}$.

If you have to find an optimum solution of the problem for $\lambda > \bar{\lambda}$, identify the cell (i, j), which attains the minimum in (11.14) bring it into the basic set; repeat the analysis with the new basis.

Problem

11.18 Solve the following parametric cost transportation problem for all real values of λ.

	Cost Matrix						a_i
	$3 - \lambda$	$2 + \lambda$	$-1 + 2\lambda$	3	7λ	-8	14
	4	6	5	-2λ	4	-2	10
	0	$3 + 8\lambda$	-3	$-7 - 2\lambda$	1	5	5
	$2 + \lambda$	λ	0	$8 + \lambda$	4	3	5
b_j	6	9	4	3	7	5	

11.5.4 Changes in Material Requirements

After the transportation problem (11.1) is solved, suppose changes have to be made in one of the a_i's and one of the b_j's, while all the remaining data is unaltered. Suppose a_p has to be changed to $a_p + \delta$ and, likewise, b_q has to be changed to $b_q + \delta$. Find out the range of value of δ for which the present basic set $\mathscr{B}$ remains optimal.

The pth row corresponds to the point p and the qth column corresponds to the point $\mathrm{m} + q$ in the rooted tree $G_{\mathscr{B}}$. Find the unique simple path from p to $\mathrm{m} + q$ in $G_{\mathscr{B}}$ as in section 11.4.5. Suppose the edges in the path correspond to the cells $(p, j_1), (i_1, j_1), (i_1, j_2), \ldots, (i_t, q)$ in that order. Add amounts of $+\delta$ and $-\delta$, alternately [starting with $+\delta$ in (p, j_1)] to the present BFS. Let the new BFS be denoted by $\tilde{x}(\delta)$ as a function of δ. $\mathscr{B}$ is an optimum basic set, and $\tilde{x}(\delta)$ is the corresponding BFS for all values of δ for which $\tilde{x}(\delta)$ remains nonnegative.

Example

Consider the following optimum BFS for a transportation problem.

$j =$	1	2	3	4	5	6
$i = 1$	3		9			
2		6		7	4	
3				5		3
4	8				13	

In this problem suppose a_3 has to be changed to $8 + \delta$ and b_1 has to be changed to $11 + \delta$. The new BFS as a function of δ is as follows. This solution is optimal for all δ satisfying $-4 \leqq \delta \leqq 7$.

$j =$	1	2	3	4	5	6
$i = 1$	3		9			
2		6		$7 - \delta$	$4 + \delta$	
3				$5 + \delta$		3
4	$8 + \delta$				$13 - \delta$	

11.6 TRANSPORTATION PROBLEMS WITH INEQUALITY CONSTRAINTS

11.6.1 Transportation Model with Demand Constraints as Equations

$$\text{Minimize} \quad z(x) = \sum_{i=1}^{\mathrm{m}} \sum_{j=1}^{\mathrm{n}} c_{ij} x_{ij}$$

$$\text{Subject to} \quad \sum_{j=1}^{\mathrm{n}} x_{ij} \leqq a_i \qquad i = 1 \text{ to m}$$

$$\sum_{i=1}^{\mathrm{m}} x_{ij} = b_j \qquad j = 1 \text{ to n}$$

$$x_{ij} \geqq 0 \qquad \text{for all } i, j$$

This transportation problem is feasible iff $\sum a_i \geqq \sum b_j$. If we introduce slack variables, the problem may be rewritten as

$$\begin{aligned} \text{Minimize} \quad & z(x) = \sum_{i=1}^{m} \sum_{j=1}^{n+1} c_{ij} x_{ij} \\ \text{Subject to} \quad & \sum_{j=1}^{n+1} x_{ij} = a_i \qquad i = 1 \text{ to m} \\ & \sum_{i=1}^{m} x_{ij} = b_j \qquad j = 1 \text{ to n} + 1 \\ & x_{ij} \geqq 0 \qquad \text{for all } i, j \end{aligned}$$

Here $b_{n+1} = \sum a_i - \sum b_j$, $c_{i,n+1} = 0$ for all i, and the slack variable $x_{i,n+1}$ represents the amount of material left unutilized at the ith source. The problem with the slack variables is a balanced transportation problem and, hence, can be solved by the algorithm discussed earlier.

11.6.2 The Unbalanced Transportation Problem

Consider this transportation problem:

$$\begin{aligned} \text{Minimize} \quad & z(x) = \sum_{i=1}^{m} \sum_{j=1}^{n} c_{ij} x_{ij} \\ \text{Subject to} \quad & \sum_{j=1}^{n} x_{ij} \leqq a_i \qquad i = 1 \text{ to m} \qquad (11.15) \\ & \sum_{i=1}^{m} x_{ij} \geqq b_j \qquad j = 1 \text{ to n} \qquad (11.16) \\ & x_{ij} \geqq 0 \qquad \text{for all } i, j \end{aligned}$$

We will refer to this problem as Transportation Problem I. It is feasible iff $\sum a_i \geqq \sum b_j$.

Case 1 If $\sum a_i = \sum b_j$ every feasible solution of Problem I satisfies all the inequality constraints as equations. Hence, in this case, replace all the inequalities in (11.15) and (11.16) by equations and solve it by the algorithm discussed earlier.

Case 2 Suppose $\sum a_i > \sum b_j$. In this case Problem I is known as an *unbalanced transportation problem*. Also, suppose $c_{ij} \geqq 0$ for all i, j.

Problem

11.19 In this case, prove that there exists an optimum feasible solution $x = (x_{ij})$ of the transportation problem I that satisfies all the constraints in (11.16) as equations. Also, if $c_{ij} > 0$ for all i, j, prove that every optimum solution of this problem must satisfy (11.16) as equations.

Hence, in this case, an optimum feasible solution of Problem I can be obtained by replacing all the inequalities in (11.16) by equations and solving the remaining problem as in section 11.6.1.

Case 3 Suppose $\sum a_i > \sum b_j$ and some c_{ij} are negative. In this case, it may be profitable to oversupply (beyond the minimal demand b_j) the markets j for which some c_{ij} are negative. For each $i = 1$ to m, let

$$c_{i,\,n+1} = \text{minimum}\ \{0, c_{i1}, \ldots, c_{in}\}$$

For each i such that $c_{i,\,n+1} < 0$, let $j = r_i$ satisfy

$$c_{i,\,n+1} = c_{i,\,r_i} = \text{minimum}\ \{c_{i1}, \ldots, c_{in}\}$$

If there is a tie for r_i, break it arbitrarily. Also, let $b_{n+1} = \sum a_i - \sum b_j$. Consider the balanced transportation problem:

$$\begin{aligned} \text{Minimize} \quad & \sum_{i=1}^{m} \sum_{j=1}^{n+1} c_{ij} x_{ij} \\ \text{Subject to} \quad & \sum_{j=1}^{n+1} x_{ij} = a_i \qquad i = 1 \text{ to m} \\ & \sum_{i=1}^{m} x_{ij} = b_j \qquad j = 1 \text{ to n} + 1 \\ & x_{ij} \geqq 0 \qquad \text{for all } i, j \end{aligned} \tag{11.17}$$

Suppose $\tilde{x} = (\tilde{x}_{ij})$ is an optimum feasible solution of (11.17). For $j = 1$ to n, let

$$\begin{aligned} \hat{x}_{ij} &= \tilde{x}_{ij} && \text{for all } i \text{ such that } c_{i,\,n+1} \geqq 0, \\ &= \tilde{x}_{ij} && \text{for all } i \text{ such that } c_{i,\,n+1} < 0, j \neq r_i \\ &= \tilde{x}_{ij} + \tilde{x}_{i,\,n+1} && \text{for all } i \text{ such that } c_{i,\,n+1} < 0, j = r_i \end{aligned}$$

Then $\hat{x} = (\hat{x}_{ij})$ is an optimum feasible solution of the Transportation Problem I in this case.

Problem

11.20 Prove that $(\hat{x}_{ij})$ is an optimum feasible solution of the transportation problem I in this case.

—O. Merrill and R. Tobin

Interpretation An optimum solution for Problem I meets the minimal demand requirements b_j, $j = 1$ to n at minimal cost. Then, for each $i = 1$ to m any excess supply left over at the ith source is shipped to the destination j with the most negative cost coefficient c_{ij}, or just left unutilized if all $c_{ij} \geqq 0$ for $j = 1$ to n.

Problem

11.21 If all $c_{ij} \leqq 0$ prove that there is an optimum feasible solution for Problem I that satisfies all the inequalities (11.15) as equations. And if $c_{ij} < 0$ for all i, j, every optimum solution must satisfy (11.15) as equations.

Consider the transportation problem II:

$$\begin{aligned} \text{Minimize} \quad & \sum_{i=1}^{m} \sum_{j=1}^{n} c_{ij} x_{ij} \\ \text{Subject to} \quad & \sum_{j=1}^{n} x_{ij} = a_i \qquad i = 1 \text{ to m} && (11.18) \\ & \sum_{i=1}^{m} x_{ij} \geqq b_j \qquad j = 1 \text{ to n} && (11.19) \\ & x_{ij} \geqq 0 \qquad \text{for all } i, j \end{aligned}$$

It is feasible iff $\sum a_i \geqq \sum b_j$. If $\sum b_j = \sum a_i$, every feasible solution of the problem must satisfy all the inequalities in (11.19) as equations. In this case, replace all inequality signs in (11.19) by equation signs and solve the problem as a balanced transportation problem. If $\sum a_i > \sum b_j$, let r_i be defined by c_{i,r_i} = minimum $\{c_{i1}, \ldots, c_{in}\}$ for each i = 1 to m. Break ties for r_i in the above equation arbitrarily. Let $c_{i,\text{n}+1} = c_{i,r_i}$ for all i = 1 to m, and $b_{\text{n}+1} = \sum a_i - \sum b_j$. Consider the balanced transportation problem:

$$\begin{array}{lll} \text{Minimize} & \sum_{i=1}^{\text{m}} \sum_{j=1}^{\text{n}+1} c_{ij} x_{ij} & \\ \text{Subject to} & \sum_{j=1}^{\text{n}+1} x_{ij} = a_i & i = 1 \text{ to m} \\ & \sum_{i=1}^{\text{m}} x_{ij} = b_j & j = 1 \text{ to n} + 1 \\ & x_{ij} \geqq 0 & \text{for all } i, j \end{array}$$

Suppose $\tilde{x} = (\tilde{x}_{ij})$ is an optimum solution of this blanced problem. Let

$$\begin{aligned} \hat{x}_{ij} &= \tilde{x}_{ij} && \text{for all } i = 1 \text{ to m} \\ & && j = 1 \text{ to n}, j \neq r_i \\ &= \tilde{x}_{ij} + \tilde{x}_{i,\text{n}+1} && \text{if } j = r_i,\ i = 1 \text{ to m} \end{aligned}$$

Then $\hat{x} = (\hat{x}_{ij})$ is an optimum solution, of the transportation problem II.

Problem

11.22 Solve the following transportation problems.

(a)

							Maximum amount available (a_i)
	6	−4	10	3	11	12	8
$c =$	1	10	3	1	2	−7	11
	−3	−11	8	5	0	8	7
	8	2	7	19	3	4	23
Exact requirement (b_j)	4	7	3	9	2	10	

(b)

							Maximum amount available (a_i)
	3	2	1	1	4	3	19
$c =$	6	−3	−7	0	11	2	23
	3	0	4	0	1	3	17
	−4	−8	−8	−3	−2	0	29
Minimal requirement (b_j)	8	15	11	3	13	9	

(c)

							Exact supply to be drawn (a_i)
	13	9	7	4	4	4	39
$c =$	4	3	0	0	6	7	14
	−1	−3	−3	−3	2	7	17
	8	0	−9	0	−9	7	21
Minimal requirement (b_j)	18	8	5	9	13	19	

11.7 BOUNDED VARIABLE TRANSPORTATION PROBLEMS

Consider this transportation problem:

$$\begin{aligned}
\text{Minimize} \quad & z(x) = \sum\sum c_{ij}x_{ij} \\
\text{Subject to} \quad & \sum_{j=1}^{n} x_{ij} = a_i \qquad i = 1 \text{ to m} \\
& \sum_{i=1}^{m} x_{ij} = b_j \qquad j = 1 \text{ to n} \\
& x_{ij} \geqq 0 \qquad \text{for all } i, j \\
& x_{ij} \leqq \mathrm{U}_{ij} \qquad \text{for } (i, j) \in \mathbf{J}
\end{aligned} \tag{11.20}$$

Here $\mathbf{J}$ is the set of all cells (i, j) in the transportation array for which there is an upper bound restriction (or capacity restriction) on the amount that can be shipped from supply point i to demand point j. $\bar{\mathbf{J}}$ is the set of all the remaining cells in the transportation array. $a_i > 0, b_j > 0$ for all i, j. A necessary condition for feasibility is that $\sum a_i = \sum b_j$ and let this quantity be denoted by s.

A *working basis* for this problem is a basis for the m × n transportation problem as discussed in section 11.1. A feasible solution of (11.20) is a BFS if it is associated with a working basis with the property that all nonbasic variables x_{ij} for $(i, j) \in \mathbf{J}$ are either equal to 0 or U_{ij} in the solution.

Applying the results of section 10.2.3 to this problem, the optimality criteria can be derived. Let $\bar{x}$ be a BFS of (11.20) associated with a working basic set, $\mathscr{B}$. Compute $u = (u_1, \ldots, u_m)$, $v = (v_1, \ldots, v_n)$ (as in section 11.2) from

$$\begin{aligned}
u_i + v_j &= c_{ij} \qquad \text{for all } (i, j) \text{ in the basic set } \mathscr{B} \\
v_n &= 0
\end{aligned}$$

Compute $\bar{c}_{ij} = c_{ij} - u_i - v_j$ for all i, j. $\bar{x}$ is an optimum solution of (11.20) if

$$\begin{aligned}
\bar{c}_{ij} > 0 \quad & \text{implies} \quad \bar{x}_{ij} = 0 \\
\bar{c}_{ij} < 0 \quad & \text{implies} \quad (i, j) \in \mathbf{J} \quad \text{and} \quad \bar{x}_{ij} = \mathrm{U}_{ij}
\end{aligned}$$

11.7.1 To Obtain an Initial Basic Feasible Solution

For the uncapacitated transportation problem, it is very easy to obtain an initial BFS directly by the methods discussed in section 11.2. However, for the capacitated transportation problem, there is no such direct method available, and it may be necessary to solve a Phase I problem. One such method is discussed here. Consider the Phase I problem:

$$\begin{aligned}
\text{Minimize} \quad & w(x) = \sum_{i=1}^{m+1} \sum_{j=1}^{n+1} d_{ij}x_{ij} \\
\text{Subject to} \quad & \sum_{j=1}^{n+1} x_{ij} = a_i \qquad \text{for } i = 1 \text{ to m} + 1 \\
& \sum_{i=1}^{m+1} x_{ij} = b_j \qquad \text{for } j = 1 \text{ to n} + 1 \\
& x_{ij} \geqq 0 \qquad \text{for all } i, j \\
& x_{ij} \leqq \mathrm{U}_{ij} \qquad \text{for all } (i, j) \in \mathbf{J}
\end{aligned} \tag{11.21}$$

where $a_{m+1} = b_{n+1} = s$, d_{ij} is 0 for all $1 \leqq i \leqq m$ and $1 \leqq j \leqq n$; $d_{m+1, j} = d_{i, n+1} = 1$ for all $1 \leqq i \leqq m$ and $1 \leqq j \leqq n$ and $d_{m+1, n+1} = 0$. $x_{m+1, j}$ and $x_{i, n+1}$ are the artificial variables in this problem.

An initial feasible basis for (11.21) is marked in the transportation array below.

Initial Basic Feasible Solution of (11.21)

	$j = 1$	2	...	n	n + 1
$i = 1$					$\boxed{a_1}$
2 ⋮	Nonbasic cells All equal to zero				$\boxed{a_2}$ ⋮
m					$\boxed{a_m}$
m + 1	$\boxed{b_1}$	$\boxed{b_2}$	...	$\boxed{b_n}$	$\boxed{0}$

Since (11.21) is a transportation problem of order $(m + 1) \times (n + 1)$, a working basic set for it consists of $m + n + 1$ basic cells. Let $\bar{x}$ be the present BFS of (11.21) associated with the present working basic set $\mathscr{B}$. Compute the dual vector $\sigma = (\sigma_1, \ldots, \sigma_{n+1})$, $\mu = (\mu_1, \ldots, \mu_{m+1})$ from

$$\mu_i + \sigma_j = d_{ij} \qquad \text{for all cells } (i, j) \text{ in } \mathscr{B}$$

$$\sigma_{n+1} = 0$$

and compute $\bar{d}_{ij} = d_{ij} - \mu_i - \sigma_j$ for all i, j. $\bar{x}$ is an optimum solution of (11.21) if

$$\bar{d}_{ij} > 0 \quad \text{implies} \quad \bar{x}_{ij} = 0$$

$$\bar{d}_{ij} < 0 \quad \text{implies} \quad (i, j) \in \mathbf{J} \quad \text{and} \quad \bar{x}_{ij} = U_{ij}$$

CHOICE OF NONBASIC CELL FOR CHANGE OF STATUS DURING PHASE I

Let $\bar{x}$ be the present BFS of (11.21). If Phase I termination criteria are not satisfied, pick a nonbasic cell (p, q) that has led to a violation of the termination criteria, for change of status. It may belong to one of the following types.

1. $(p, q) \in \mathbf{J}, \bar{d}_{pq} < 0$ and $\bar{x}_{pq} = 0$
2. $(p, q) \in \mathbf{J}, \bar{d}_{pq} > 0$ and $\bar{x}_{pq} = U_{pq}$
3. $(p, q) \in \bar{\mathbf{J}}$ and $\bar{d}_{pq} < 0$

Find out the θ-loop associated with (p, q) with respect to the present working basic set $\mathscr{B}$ by starting with a "$+\theta$" entry in the cell (p, q).

11.7.2 Changing the Status of (p, q) Such That $\bar{d}_{pq} < 0$

In this case the value of the Phase I objective function can be decreased by increasing the value of x_{pq} from its present value of 0. The maximum value that can be given to x_{pq} is min $\{\bar{x}_{ij}$ for all basic cells (i, j) such that they are in the θ-loop with a $-\theta$ entry; $U_{ij} - \bar{x}_{ij}$ for all basic cells (i, j) such that $(i, j) \in \mathbf{J}$ and (i, j) is in the θ-loop with a $+\theta$ entry; U_{pq}, if $(p, q) \in \mathbf{J}\}$. There are several cases to consider here.

Case 1 If $(p, q) \in \mathbf{J}$ and the maximum value that x_{pq} can assume turns out to be U_{pq}, make x_{pq} into a nonbasic variable whose value is equal to its upper bound in the next step. Putting θ equal to U_{pq}, revise the values of all the basic variables in the θ-loop and then erase the θ-loop. Keep the same working basis. Move to the next step.

Case 2 If the maximum value that x_{pq} can assume turns out to be some $\bar{x}_{i_1, j_1}$ where (i_1, j_1) is a basic cell with a $-\theta$ entry in the θ-loop, change the working basic set by dropping the cell (i_1, j_1) from it and making the cell (p, q) a basic cell. Make the value of $x_{pq} = \bar{x}_{i_1, j_1}$ and revise the values of all the other basic variables in the θ-loop by substituting $\bar{x}_{i_1, j_1}$ for θ. Make (i_1, j_1) a zero-valued nonbasic cell.

Case 3 If the maximum value that x_{pq} can take turns out to be $U_{i_1, j_1} - \bar{x}_{i_1, j_1}$ for a basic cell (i_1, j_1) such that $(i_1, j_1) \in \mathbf{J}$ and (i_1, j_1) is in the θ-loop with a $+\theta$ entry, drop (i_1, j_1) from the working basic set and make it into a nonbasic cell at its upper bound in the next solution. Make x_{pq} equal to $U_{i_1, j_1} - \bar{x}_{i_1, j_1}$, and revise the values of all the other basic variables in the θ-loop by substituting the same value for θ. Make (p, q) a basic cell and go to the next step.

11.7.3 Changing the Status of (p, q) Such That $\bar{d}_{pq} > 0$

In this case (p, q) must be in $\mathbf{J}$ and the present value of x_{pq} must be U_{pq}. The Phase I objective value can be decreased by decreasing the value of x_{pq} from its present value of U_{pq}. Since the new value of x_{pq} will be $U_{pq} + \theta$, θ should be negative and as small as possible. The smallest value that θ can have is max $\{-U_{pq}; -\bar{x}_{ij}$ for all basic cells (i, j) with a $+\theta$ entry in the θ-loop; $\bar{x}_{ij} - U_{ij}$ for all basic cells (i, j) such that $(i, j) \in \mathbf{J}$ and (i, j) appears with a $-\theta$ entry in the θ-loop$\}$. There are several cases to consider here.

Case 1 If the smallest value that θ can take is $-U_{pq}$, then x_{pq} becomes a nonbasic variable whose value is 0 in the next step. There is no change in the working basis. Revise the values of all the basic variables in the θ-loop by substituting $-U_{pq}$ for θ and go to the next step.

Case 2 If the smallest value that θ can take happens to be $-\bar{x}_{i_1, j_1}$ for some basic (i_1, j_1) in the θ-loop, then (i_1, j_1) is dropped from the working basic set and made into a nonbasic cell with zero value. x_{pq} becomes a basic variable, whose value is $U_{pq} - \bar{x}_{i_1, j_1}$ in the next solution. Revise the values of all the other basic variables in the θ-loop by substituting the value $-\bar{x}_{i_1, j_1}$ for θ and go to the next step.

Case 3 If the smallest value that θ can take happens to be $\bar{x}_{i_1, j_1} - U_{i_1, j_1}$ for some basic cell (i_1, j_1) in the θ-loop with $(i_1, j_1) \in \mathbf{J}$, then x_{i_1, j_1} is made a nonbasic variable whose value is equal to its upperbound in the next solution. x_{pq} becomes a basic variable in its place with its value equal to $U_{pq} + \theta$ in the next solution. The values of all the other basic variables in the θ-loop are revised by substituting $\bar{x}_{i_1, j_1} - U_{i_1, j_1}$ for θ, and the algorithm then goes to the next step.

11.7.4 Phase I Termination

After a finite number of steps, the Phase I termination criteria will be satisfied. At phase I termination if any of $x_{m+1, j}$ for $1 \leqq j \leqq n$ or $x_{i, n+1}$ for $1 \leqq i \leqq m$ are strictly positive, the original problem (11.20) must be infeasible and the algorithm terminates.

If $x_{m+1, j}$ for $1 \leqq j \leqq n$ and $x_{i, n+1}$ for $1 \leqq i \leqq m$ are all zero, the original problem is feasible. In this case, if the present working basic set has $m + n - 1$ basic cells (i, j) among those in the range $1 \leqq i \leqq m$ and $1 \leqq j \leqq n$, these cells form a working basic set for the original problem (11.20). Therefore, erase the $(m + n)$th row and the $(n + 1)$th column, which are all cells corresponding to the artificial variables. Go over to Phase II.

On the other hand, suppose the working basic set at Phase I termination has less than $m + n - 1$, say $(m + n - 1) - r$, basic cells (i, j) among those in the range $1 \leqq i \leqq m$ and $1 \leqq j \leqq n$. In this case, pick r nonbasic cells that together with the present (m + n − 1) − r basic cells provide a working basic set and make these r cells into basic cells. Go over to Phase II.

11.7.5 Phase II

Given a BFS $\bar{x}$ associated with a working basis for (11.20), compute the corresponding u, v, $\bar{c}$, and check whether optimality criterion is satisfied. If not, for change of status pick a nonbasic variable, x_{pq}, which has led to a violation of the optimality criterion. It can belong to the following types.

1. $\bar{c}_{pq} < 0$ and the present value of x_{pq} is 0. In this case, the objective value can be decreased by increasing the value of x_{pq} from its present value of 0. These computations are carried out as in section 11.7.2.
2. $(p, q) \in \mathbf{J}, \bar{c}_{pq} > 0$ and the present value of x_{pq} is U_{pq}. In this case, the objective value can be decreased by decreasing the value of x_{pq} from its present value of U_{pq}. These computations are carried out as in section 11.7.3.

The algorithm will terminate in a finite number of steps with an optimum solution of the problem.

Numerical Example

Consider the following transportation problem.

					a_i
	1	1	2	15	21
$c =$	9	8	17	5	11
	4	2	12	2	28
b_j	8	16	25	11	

The amounts shipped from any supply point to any demand point should be less than or equal to 15.

In the following transportation arrays basic cells are always marked with a small square in the middle, with the value of the basic variable entered inside the small square. Nonbasic cells at upper bound are marked with a small circle in the middle, with the value of that variable entered in the middle of the circle. Below is the initial Phase I array.

$j =$	1	2	3	4	5	μ_i
$i = 1$	$-2 = \bar{d}_{11}$ $d_{11} = 0$	−2 0	−2 0	−2 0	[21] 1	1
2	−2 0	−2 0	−2 0	−2 0	[11] 1	1
3	−2 0	−2 0	−2 0	−2 0	[28] 1	1
4	[8] 1	[16] 1	[25] 1	[11] 1	[0] 0	0
σ_j	1	1	1	1	0	

x_{33} is entering. It becomes a nonbasic at upperbound. Next x_{24} is entering. It replaces x_{25} from the basic vector. Here is the transportation array after these changes.

$j =$	1	2	3	4	5	μ_i
$i = 1$	−2 ; 0	−2 ; 0	−2 ; 0	−2 ; 0	[21] ; 1	1
2	0 ; 0	0 ; 0	0 ; 0	[11] ; 0	1	−1
3	−2 ; 0	−2 ; 0	−2 ; (15) ; 0	−2 ; 0	[13] ; 1	1
4	[8] ; 1	[16] ; 1	[10] ; 1	[0] ; 1	[26] ; 0	0
σ_j	1	1	1	1	0	

Next the following changes are made in the order listed. Notice that the dual solution and the relative cost coefficients have to be recomputed after every basis change.

Entering variable	Leaving variable	Status of entering variable	Status of leaving variable
x_{31}	x_{41}	Basic at value 8	Zero-valued nonbasic
x_{32}	x_{35}	Basic at value 5	Zero-valued nonbasic
x_{12}	x_{42}	Basic at value 11	Zero-valued nonbasic
x_{13}	x_{43}	Basic at value 10	Zero-valued nonbasic
x_{34}	x_{44}	Basic at value 0	Zero-valued nonbasic

The transportation array at the end of these changes is:

$j =$	1	2	3	4	5
$i = 1$		[11]	[10]		[0]
2				[11]	
3	[8]	[5]	(15)	[0]	
4					[60]

Clearly Phase I terminates in this stage and the basic cells among the cells of the original transportation array constitute a working basic set for it. Thus the initial Phase II array is:

$j =$	1	2	3	4	u_i
$i = 1$	$-2 = c_{11}$; $c_{11} = 1$	[11] $+ \theta$; 1	[10] $- \theta$; 2	14 ; 15	1
2	2 ; 9	3 ; 8	11 ; 17	[11] ; 5	5
3	[8] ; 4	[5] $- \theta$; 2	9 ; (15) $+ \theta$; 12	[0] ; 2	2
v_j	2	0	1	0	

Since $\bar{c}_{33} = 9$, x_{33} should be reduced from its present value of 15. The θ-loop with an entry of $+\theta$ in the cell (3, 3) is entered. When θ is -5, x_{13} reaches its upperbound and leaves the basic vector to become a nonbasic at upperbound. Thus the new array is:

$j =$	1	2	3	4	u_i
$i = 1$	−2, θ, 1	$\boxed{6} - \theta$, 1	−9, (15), 2	14, 15	1
2	2, 9	3, 8	2, 17	$\boxed{11}$, 5	5
3	$\boxed{8} - \theta$, 4	$\boxed{10} + \theta$, 2	$\boxed{10}$, 12	$\boxed{0}$, 2	2
v_j	2	0	10	0	

x_{11} is entering. The θ loop is entered. When θ is 5, x_{32} reaches its upperbound. Thus the new array is:

$j =$	1	2	3	4	u_i
$i = 1$	$\boxed{5}$, 1	$\boxed{1}$, 1	−7, (15), 2	16, 15	−1
2	2, 9	1, 8	2, 17	$\boxed{11}$, 5	5
3	$\boxed{3}$, 4	−2, (15), 2	$\boxed{10}$, 12	$\boxed{0}$, 2	2
v_j	2	2	10	0	

The optimality criterion is now satisfied and hence the solution in this transportation array is optimal.

PROBLEMS

In all these problems, the entries inside the tableau are the unit transportation cost coefficients, c_{ij}.

11.23 There are three refineries that have gas available and four markets that require gas. The data is given below.

Market	1	2	3	4	Available gas (units of 10^6 gallons)
Refinery 1	4	7	9	10	8
2	6	4	3	6	10
3	9	6	4	8	6
Requirement (in units of 10^6 gallons)	5	3	8	4	

Determine an optimum transportation policy.

11.24 In the following transportation model, the total requirements at the market exceeds the amounts of material available at the plants. The deficits have to be made up by imports. All the markets have the same priority standing. Determine how much of the demand in each market should be fulfilled by each plant to utilize the existing available material at minimal transportation cost.

Market	1	2	3	4	5	Available at plant (tons)
Plant 1	3	9	5	6	7	10
2	2	1	8	10	13	25
3	.3	12	6	5	2	13
4	1	9	14	3	2	33
Required at markets (tons)	30	22	17	19	12	

11.25 For the following transportation problem is $\{(1, 1), (1, 4), (2, 2), (2, 5), (3, 3), (3, 4), (3, 5)\}$ a basic set?

$j =$	1	2	3	4	5	a_i
$i = 1$	9	16	4	11	19	8
2	8	6	8	12	8	10
3	1	12	3	23	6	30
b_j	5	4	9	8	22	

If so, find the primal and dual solutions corresponding to it. Starting with this solution find all the alternate optimum solutions of the problem.

If a_3 changes to $30 + \delta$ and b_4 changes to $8 + \delta$ find the range of values of δ for which the first optimal basic set obtained remains optimal.

Returning to the original problem (with $\delta = 0$), find out the range of values of c_{12} for which the first optimal BFS obtained remains optimal. What is an optimum solution when c_{12} changes to 2?

Returning to the original problem (with $\delta = 0$ and $c_{12} = 16$) find out the range of values of c_{35} for which the first optimal BFS obtained remains optimal. What is an optimum solution when c_{35} changes to 20?

11.26 Consider the following transportation problem.

$j =$	1	2	3	a_i
$i = 1$	1	3	2	10
2	2	1	3	15
3	3	2	1	$5 + \delta$
b_j	5	$10 + \delta$	15	

Let $f(\delta)$ be the minimum objective value in this problem as a function of δ. Find all the alternate optimum solutions of the problem when $\delta = 0$. Draw $f(\delta)$ in the range $\delta \geqq 0$, as a curve. At what values of δ does $f(\delta)$ change slope? Explain the slopes and the changes in it in terms of the optimum dual solution as a function of δ.

11.27 Solve the following transportation problem for $\lambda = 0$ and get two distinct optimum solutions.

$j =$	1	2	3	a_i
$i = 1$	5	$14 - 2\lambda$	7	15
2	6	7	8	20
3	$14 + \lambda$	$24 - 3\lambda$	9	4
b_j	7	11	21	

Solve it also for all $\lambda \geqq 0$.

11.28 Use $\{(1, 4), (2, 3), (2, 4), (3, 2), (3, 5), (4, 1), (4, 2), (4, 3)\}$ as an initial basic set for $\delta = 0$, and solve the problem.

$j =$	1	2	3	4	5	a_i
$i = 1$	6	11	14	11	4	$3 + \delta$
2	7	17	9	10	5	15
3	9	5	7	18	0	14
4	9	13	13	15	7	10
b_j	3	$9 + \delta$	14	7	9	

For what range of values of δ does the optimum basic set obtained remain optimal?

11.29 Show that the basic set $\{(1, 2), (1, 4), (2, 1), (2, 3), (3, 2), (3, 5), (4, 1), (4, 4)\}$ is an optimum basic set for the following problem when $\delta = 0$.

$j =$	1	2	3	4	5	a_i
$i = 1$	4	4	8	6	10	$14 - \delta$
2	3	8	2	16	14	$12 + 2\delta$
3	9	7	9	18	12	7
4	8	18	11	13	18	18
b_j	$17 + 4\delta$	$11 - \delta$	4	$17 - 2\delta$	2	

Using it as an initial basis, solve the problem for all $0 \leqq \delta \leqq 8$.

11.30 BOTTLENECK TRANSPORTATION PROBLEM There are m plants manufacturing a product and the ith plant has a_i units available. There are n markets where there is a demand for this and the demand in the jth market is b_j units. $\sum a_i = \sum b_j$. The time needed to make a delivery from the ith plant to the jth market is t_{ij} days. $i = 1$ to m, $j = 1$ to n. The activity of shipping from i to j is done independently for each i, j. If x_{ij} is the number of items shipped from i to j, and $x = (x_{ij})$ is a feasible solution of this problem, show that the time needed to implement the solution x is:

$$\theta(x) = \text{maximum } \{t_{ij}: (i, j) \text{ such that } x_{ij} > 0\}$$

Develop a method for minimizing $\theta(x)$. Solve the problem with the following data:

t_{ij}

$j =$	1	2	3	4	5	6	a_i
$i = 1$	13	18	7	23	47	5	23
2	39	55	3	45	12	29	9
3	23	9	21	32	23	9	17
4	72	11	16	14	35	19	11
b_j	7	8	19	5	13	8	

11.31 Solve the following bounded variable balanced transportation problems:

(a)

							a_i
	31	23	20	22	17	14	48
$c =$	17	18	19	5	16	7	43
	16	15	17	29	9	14	41
	22	9	3	22	25	2	46
b_j	19	37	26	30	30	36	

$0 \leqq x_{1j} \leqq 10$, $0 \leqq x_{ij} \leqq 20$ for $i = 2, 3$ and for all j and $x_{4j} \geqq 0$ for all j.

(b)

							a_i
	3	4	11	9	8	13	32
$c =$	2	7	6	14	5	2	102
	1	3	7	1	13	17	22
	5	9	4	2	3	18	20
b_j	36	3	17	36	34	50	

$0 \leqq x_{ij} \leqq 20$ for all i, j

11.32 Given a dual feasible but primal infeasible basis for (11.1), discuss how to solve it using the dual simplex algorithm starting with the given basis. In each step clearly explain how to select the entering variable and to update the array efficiently.

11.33 Consider the balanced transportation problem (11.1), when all the a_i and b_j are affine function of a real-valued parameter λ, that is, $a_i = \bar{a}_i + \lambda a_i^*$, $b_j = \bar{b}_j + \lambda b_j^*$, where the data $\bar{a}_i$, $\bar{b}_j$, a_i^*, b_j^* satisfies $\sum \bar{a}_i = \sum \bar{b}_j$ and $\sum a_i^* = \sum b_j^*$. Develop the parametric algorithm for solving this problem for all values of λ in the interval over which all the a_i and b_j remain positive.

11.34 $a_1, a_2, \ldots, a_m$; $b_1, b_2, \ldots, b_n$ are given positive integers. $\alpha = \sum a_i$, $\beta = \sum b_j$, $v =$ minimum $\{\alpha, \beta\}$. δ is a positive integer between 1 and v. Let $\mathbf{K}$ be the set of feasible solutions of the problem

$$\begin{aligned} \text{Minimize} \quad & \sum\sum c_{ij}x_{ij} \\ \text{Subject to} \quad & \sum_j x_{ij} \leqq a_i, \quad i = 1 \text{ to m} \\ & \sum_i x_{ij} \leqq b_j, \quad j = 1 \text{ to n} \\ & \sum_i \sum_j x_{ij} = \delta \\ & x_{ij} \geqq 0 \quad \text{for all } i, j \end{aligned}$$

(a) Prove that all the extreme points of **K** are integer points.
(b) Show that the above problem can be transformed into a transportation problem. Develop an algorithm for solving it, as δ varies from 1 to v.

11.35 Consider the following transportation model, where a_i, b_j are real numbers (some of them may be negative) satisfying

$$\sum_{i=1}^{m} a_i = \sum_{j=1}^{n} b_j,$$

and **J** is a specified subset of cells of the m $\times$ n transportation array.

$$\begin{aligned} \text{minimize} \quad & z(x) = \sum_{i=1}^{m} \sum_{j=1}^{n} c_{ij} x_{ij} \\ \text{subject to} \quad & \sum_{j=1}^{n} x_{ij} = a_i \quad \text{for } i = 1 \text{ to m} \\ & \sum_{i=1}^{m} x_{ij} = b_j \quad \text{for } j = 1 \text{ to n} \\ & x_{ij} \geqq 0 \quad \text{for } (i, j) \in \mathbf{J} \\ & x_{ij} \leqq 0 \quad \text{for } (i, j) \notin \mathbf{J} \end{aligned}$$

Develop a primal algorithm for solving it using the ideas discussed in section 11.7.

11.36 THE TRANSPORTATION PARADOX: Let $a = (a_1, \ldots, a_m)$, $b = (b_1, \ldots, b_n)$ and let $f(a, b)$ be the minimum value of $z(x)$ in the balanced transportation problem (11.1). Suppose $\tilde{a}, \tilde{b}, \hat{a}, \hat{b}$ are all positive vectors satisfying

$$\sum_{i=1}^{m} \tilde{a}_i = \sum_{j=1}^{n} \tilde{b}_j \quad \text{and} \quad \sum_{i=1}^{m} \hat{a}_i = \sum_{j=1}^{n} \hat{b}_j,$$

and $\tilde{a} \geqq \hat{a}$, $\tilde{b} \geqq \hat{b}$. $f(\tilde{a}, \tilde{b})$ is the minimum transportation cost when the supplies at sources are $\tilde{a}_i$ and the demands at the markets are $\tilde{b}_j$. Likewise $f(\hat{a}, \hat{b})$ is the minimum transportation cost when the supplies at the sources are $\hat{a}_i$ and the demands at the markets are $\hat{b}_j$. Suppose $c_{ij} > 0$ for all i, j. It seems reasonable to expect that the total cost of transportation will increase if the supplies available at the sources, and the demands at the markets both go up. Thus in this case, since $\tilde{a}_i \geqq \hat{a}_i$ for all i and $\tilde{b}_j \geqq \hat{b}_j$ for all j, one intuitively expects that $f(\tilde{a}, \tilde{b}) \geqq f(\hat{a}, \hat{b})$. However this may not be true, as the following example illustrates. This is known as the *transportation paradox*.
(i) Consider the balanced transportation problem with the following data.

	Unit transportation costs ($ ton)						available at source i (tons)
$j =$	1	2	3	4	5	6	a_i
$i = 1$	30	11	5	35	8	29	30
2	2	5	2	5	1	9	$10 + \delta$
3	35	20	6	40	8	33	45
4	19	2	4	30	10	25	30
Required at market j (tons) b_j	25	$20 + \delta$	6	7	22	35	

Compute $g(\delta)$, the minimum objective value in this problem, as a function of δ, in the range $0 \leqq \delta \leqq 22$. Show that $g(\delta)$ strictly decreases as δ increases in this range.

(ii) Give an explanation of the transportation paradox using the optimum dual solution. In particular prove the following. Let

$$a(\delta) = (a_1, \ldots, a_{i-1}, a_i + \delta, a_{i+1}, \ldots, a_m),$$
$$b(\delta) = (b_1, \ldots, b_{j-1}, b_j + \delta, b_{j+1}, \ldots, b_n).$$

Let $\tilde{\mathscr{B}}$ be an optimum basic set for the balanced transportation problem (11.1) with $a = a(0)$, $b = b(0)$, and let $(\tilde{u}, \tilde{v})$ be the dual solution corresponding to the basic set $\tilde{\mathscr{B}}$ obtained as in section 11.2. When a is changed to $a(\delta)$ and b is changed to $b(\delta)$, suppose the basic set $\tilde{\mathscr{B}}$ remains feasible to (11.1) in the range $0 \leqq \delta \leqq \delta_1$, where δ_1 is a positive number. Prove that the minimum objective value $f(a(\delta), b(\delta))$, is linear with slope $\tilde{u}_i + \tilde{v}_j$ in the interval $0 \leqq \delta \leqq \delta_1$, and that it is strict monotone decreasing in this interval iff $\tilde{u}_i + \tilde{v}_j < 0$.

(iii) The transportation paradox indicates that in a practical transportation model, one may be able to decrease the total transportation cost by increasing the supplies at some sources and making corresponding increases in the demands at some markets, from present levels. To take advantage of this phenomenon, we should examine the transportation model of the following form.

$$\begin{aligned} \text{minimize} \quad & \sum_{i=1}^{m} \sum_{j=1}^{n} c_{ij} x_{ij} \\ \text{subject to} \quad & \sum_{j=1}^{n} x_{ij} \geqq a_i, \quad i = 1 \text{ to m} \\ & \sum_{i=1}^{m} x_{ij} \geqq b_j, \quad j = 1 \text{ to n} \\ & x_{ij} \geqq 0, \quad \text{for all } i, j. \end{aligned}$$

where a_i, b_j are given positive numbers. Develop an efficient algorithm for solving this transportation model. (Use the algorithm developed in problem 11.35.) Discuss its uses in practical applications. (see reference [5]).

11.37 Discuss how to perform marginal analysis in the transportation models discussed in sections 11.1, 11.6.1, 11.6.2, and 11.7, and its usefulness in practical applications.

11.38 Consider the balanced transportation problem (11.1). Let $\mathscr{B}$ be a basic set of cells in the m × n transportation array for it. In section 11.2.2, the dual solution corresponding to this basic set has been computed by solving

$$u_i + v_j = c_{ij} \qquad \text{for all basic cells } (i, j) \text{ in } \mathscr{B} \tag{1}$$
$$v_n = 0 \tag{2}$$

Let $(\tilde{u}, \tilde{v})$ denote the dual solution obtained by solving (1), (2). Suppose we replace (2) by the equation

$$v_n = \alpha \tag{3}$$

where α is an arbitrary real number. Let $(\hat{u}, \hat{v})$ be the solution obtained by solving (1) and (3). Show that

$$\begin{aligned} \hat{v}_j &= \tilde{v}_j + \alpha && \text{for all } j \\ \hat{u}_i &= \tilde{u}_i - \alpha && \text{for all } i \\ \hat{u}_i + \hat{v}_j &= \tilde{u}_i + \tilde{v}_j && \text{for all } i, j. \end{aligned}$$

Hence $c_{ij} - (\tilde{u}_i + \tilde{v}_j) = c_{ij} - (\hat{u}_i + \hat{v}_j)$ for all i, j. Therefore the values of the relative cost coefficients are unaffected by replacing (2) by (3) in computing the dual solution.

If $\mathscr{B}$ is a dual feasible basic set, both $(\tilde{u}, \tilde{v})$ and $(\hat{u}, \hat{v})$ are dual feasible solutions, that is, they are feasible to (11.8), and yield the same values for the vector of relative cost coefficients. Also if $\mathscr{B}$ is an optimum basic set for (11.1), the conclusion drawn from a marginal analysis (see problem 11.37) are the same whether $(\tilde{u}, \tilde{v})$ or $(\hat{u}, \hat{v})$ is used, because $\tilde{u}_i + \tilde{v}_j = \hat{u}_i + \hat{v}_j$, for all i, j.

From these arguments, show that the dual solution of (11.1), corresponding to any basic set of cells $\mathscr{B}$, can be computed by selecting any one of the dual variables in (11.8), giving it an arbitrary real value, and then using (1) to compute the values of the other variables in the corresponding dual solution.

REFERENCES

1. F. Glover, D. Karney, and D. Klingman, "The Augmented Predecessor Index Method for Locating Stepping Stone Paths and Assigning Dual Prices in Distribution Problems," *Transportation Science*, *6*, 1, pp. 171–180, 1972.
2. F. Glover, D. Karney, D. Klingman, and A. Napier, "A Computation study on Start Procedures, Basis Change Criteria and Solution Algorithms for Transportation Problems," *Management Science*, *20*, 5, pp. 793–813, January 1974.
3. E. Johnson, "Networks and Basic solutions," *Operations Research*, 14, 4, pp. 89–95, 1966.
4. V. Srinivasan and G. L. Thompson, "Benefit Cost Analysis of Coding Techniques for the Primal Transportation Algorithm," *Journal of the ACM*, *20*, pp. 194–213, 1973.
5. W. Szwarc, "The Transportation Paradox," *Naval Research Logistics Quarterly*, *18*, 2 pp. 185–202, June 1971.

12 network algorithms

12.1 INTRODUCTION: NETWORKS

A *network* is a pair of sets, $(\mathcal{N}, \mathcal{A})$ where $\mathcal{N}$ is a set of *points* (also called *vertices* or *nodes*) and $\mathcal{A}$ is a set of *lines*, each line joining a pair of distinct points. A line joining points i and j is called an *arc* if it can only be used in one specified direction, say, from i to j; in this case it is denoted by the *ordered pair* (i, j). The arc (i, j) is *incident into point* j and *incident out of point* i. A line joining two points x and y that can be used either in the direction from x to y, or from y to x, is called an *edge* and it is denoted by the *unordered pair* (x; y). The edge (x; y) is said to be *incident* at points x, y. There can be more than one line joining points i, j. They will then be denoted by $(i, j)_1$, $(i, j)_2$, etc. However, we assume that there are *no self loops*, which are lines joining a point with itself. Figure 12.1 is the diagram of a network with 8 points, 16 arcs, and 7 edges. As it is drawn in the diagram the arcs (3, 5) and (2, 6) intersect, but their point of intersection is not a point in the network.

The network is a *directed network* if all the lines in it are arcs. It is an *undirected network* if all the lines in it are edges. If it contains both arcs and edges, it is a *mixed network.*

Many practical problems can be posed as network problems. For example, problems in the study of highway systems for vehicular traffic are problems on a network, in which the points are traffic centers and the lines (arcs or edges) are the streets joining pairs of centers. In a similar manner, problems in the study of airline systems, railroad systems, shipping systems, and in general all transportation planning problems are network problems on appropriately defined networks. Problems in natural gas, crude oil or other fluid flows are problems in the appropriate pipeline networks. The telephone network is the network for studying the flow of calls. Economic models can be treated as network models by representing factories, warehouses, and markets as points and by treating highways, railroads, waterways, and other transportation channels as lines. Thus network algorithms find a vast number of applications in communication, transportation, distribution, traffic analysis, and in solving various practical problems that can be posed as flow problems or combinatorial optimization problems. Some of the basic network algorithms are discussed in this chapter briefly.

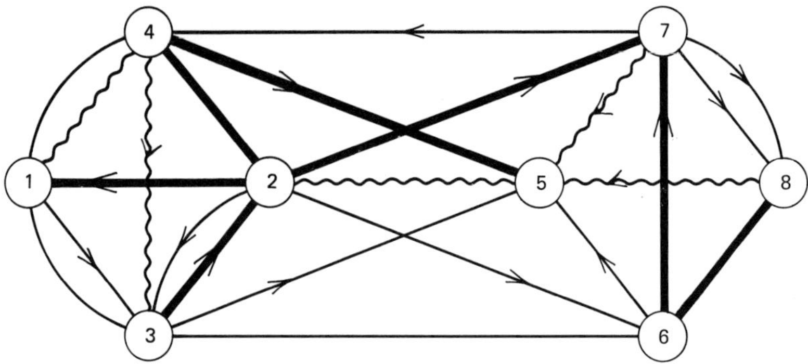

Figure 12.1

12.2 NOTATION

In this section G will denote a network $(\mathcal{N}, \mathcal{A})$.

DEGREE

The *degree* of a point in a network is the number of lines incident at that point.

CHAINS AND CIRCUITS

A chain from a point x_1 to a point x_k in G is a sequence of points and lines, x_1, $\mathbf{e}_1$, x_2, $\mathbf{e}_2, \ldots, \mathbf{e}_{k-1}$, x_k, where $\mathbf{e}_r$ is either the arc (x_r, x_{r+1}) or the edge $(x_r; x_{r+1})$ with its orientation selected to be from x_r to x_{r+1}, which we will treat as the arc (x_r, x_{r+1}) for $r = 1$ to $k - 1$. Hence all the lines in this chain are arcs directed toward the point x_k. A chain is said to be a *simple chain* if no point is repeated on it. A chain from x_1 to x_1 (with two or more arcs in it) is called a *circuit* containing x_1. See Figures 12.2 and 12.3.

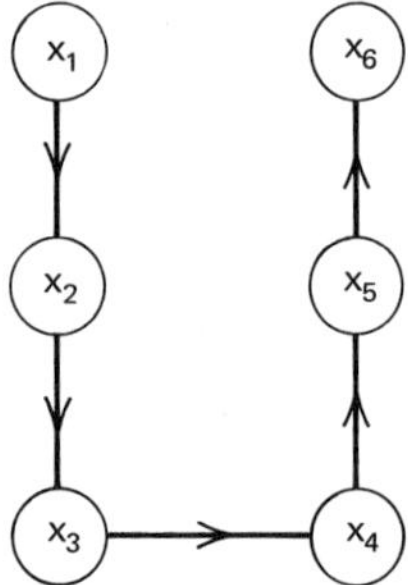

Figure 12.2 A simple chain.

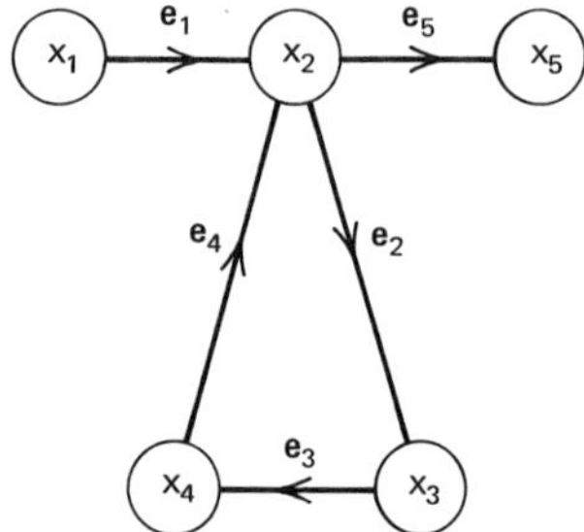

Figure 12.3 A chain which is not simple.

PATHS AND CYCLES

A *path* from x_1 to x_k is a sequence of points and lines, x_1, $\mathbf{e}_1$, x_2, $\mathbf{e}_2, \ldots, \mathbf{e}_{k-1}$, x_k, such that for each $r = 1$ to $k - 1$, $\mathbf{e}_r$ is either the arc (x_r, x_{r+1}) or the arc (x_{r+1}, x_r) or the edge $(x_r; x_{r+1})$ with some orientation selected for it, which we will again treat as an arc. It is a sequence of lines connecting the points x_1 and x_k, but the lines need not all be directed toward x_k. An arc whose orientation coincides with the direction of travel from x_1 to x_k is called a *forward arc of the path* and an arc whose orientation is opposite to the direction of travel from x_1 to x_k is called a *reverse arc of the path.* A path is a *simple path* if no point is repeated in it. See Figure 12.4. A path from x_1 to x_1 containing at least two arcs is a *cycle* containing x_1. A *simple cycle* is a cycle that does not contain another cycle as a proper subset.

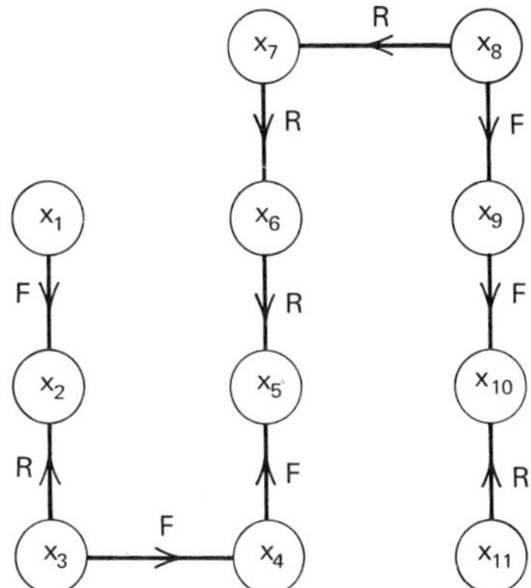

Figure 12.4 A path from x_1 to x_{11}. F is forward; R is reverse.

CONNECTEDNESS

G is a *connected network* if there exists a path between every pair of points in it. A network that is not connected is just two or more independent connected networks put side by side. The connected networks in it are called its *connected components.*

SUBNETWORKS AND PARTIAL NETWORKS

A *subnetwork* of G is a network F = $(\mathscr{N}, \bar{\mathscr{A}})$ with the same set of points, but with $\bar{\mathscr{A}} \subset \mathscr{A}$. A *partial network* of G is a network $(\hat{\mathscr{N}}, \hat{\mathscr{A}})$ in which the set of points is $\hat{\mathscr{N}} \subset \mathscr{N}$, and the set of lines is $\hat{\mathscr{A}}$, which is the set of all lines in G that have both their incidents points in $\hat{\mathscr{N}}$. A *partial subnetwork* of G is a partial network of a subnetwork of G.

FORESTS AND TREES

A *forest* in G is a partial subnetwork of G that contains no cycles. In Figure 12.1, the wavy partial subnetwork is a forest. When a forest is connected it is called a *tree.* A *spanning tree* in a network is a subnetwork that is a tree. In Figure 12.1 the subnetwork consisting of the thick lines is a spanning tree.

Problems

12.1 Let **T** be a tree. A point in **T** is called a *terminal node* or *end node* or *pendant node* if its degree in **T** is 1 (for example, in the tree consisting of the thick arcs in Figure 12.1 the point 1 is a terminal node while 2 is not). Prove that every tree has at least one terminal node.

12.2 Let **T** be a tree containing two or more lines. Prove that if a line, $\mathbf{e}_1$, is deleted from **T**, what is left is a forest. Also, if $\mathbf{e}_1$ is a line containing a terminal node of **T**, then what is left after deleting $\mathbf{e}_1$ from **T** is another tree.

12.3 Let G be connected and let $\mathscr{N}$ consist of n points. Prove that every spanning tree in G contains $n - 1$ lines. (*Hint.* Use induction on n and the results in previous problems.)

12.4 Let $\mathcal{N}$ consist of n points. Suppose **T** is a subnetwork of G containing $n - 1$ lines and **T** has no cycles in it. Prove that **T** must be a spanning tree. (You have to prove that **T** is connected. Prove that any subnetwork like **T** must have a point j whose degree in **T** is one. Use induction.)

12.5 If $(\mathbf{e}_1, \ldots, \mathbf{e}_r)$ is a set of r lines containing no cycles in a connected network, prove that there is at least one spanning tree containing all the lines $(\mathbf{e}_1, \ldots, \mathbf{e}_r)$.

12.6 Prove that the number of nodes of degree one is an even number in a network in which no point has degree greater than 2.

12.7 Prove that the number of lines in a network is one half the sum of the degrees of the points in the network.

12.8 Prove that the number of odd degree nodes in a network must be an even number.

CAPACITIES AND LOWER BOUNDS

The *capacity* of an arc (i, j), denoted by k_{ij}, is the maximum amount of material that can be transported from i to j along it. The *lower bound* on an arc (i, j), denoted by ℓ_{ij}, is the minimum amount of material that should be shipped from i to j along it. $\ell_{ij} \leqq k_{ij}$ always.

Capacity and lower bound constraints on edges have to be clearly specified. The edge $(i; j)$ can be treated as a pair of arcs (i, j) and (j, i) as in Figure 12.5. In single commodity flow problems, a flow of 10 units in the direction i to j and 6 units in the direction j to i, as in Figure 12.6, is equivalent to a net flow of 4 units from i to j, as in Figure 12.7. However this argument is not valid if there are two or more distinct commodities involved, and the flow from i to j is of one commodity and the flow from j to i is of another commodity.

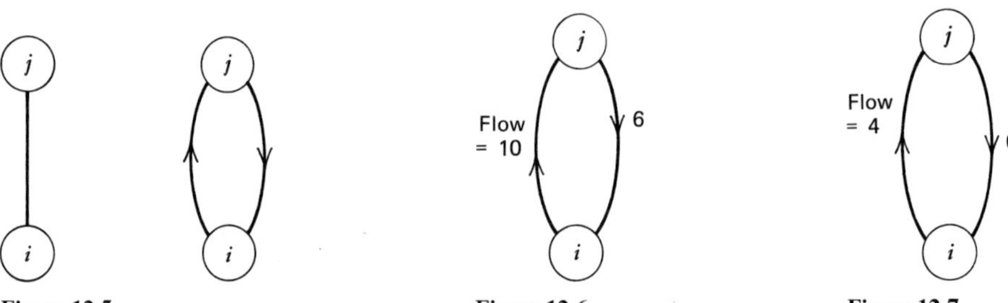

Figure 12.5　　Figure 12.6　　Figure 12.7

We make the following assumption. If $(i; j)$ is an edge in a single commodity flow problem, it can be used either as the arc (i, j) or as the arc (j, i), but it cannot be used in both directions simultaneously. Capacity restrictions for flow on this edge are assumed to apply in the direction in which it is used.

In most practical problems, lower bounds for the flow amount on arcs are zero. If there is a positive lower bound restriction on the flow along a line in the network, we assume that the line is an arc. Positive lower bounds for flows along an edge are hard to interpret and hard to handle, and in all further discussions we assume that the lower bounds for flows along all edges in the network are zero. Under these assumptions each edge in a network problem can be replaced by a pair of arcs (as in Figure 12.5) with the same data holding for each arc in the pair.

SOURCE, SINK, AND INTERMEDIATE POINTS

In most flow problems the network can be formulated so that it contains a *source node* from where all the material originates and a *sink node* into which all the material flows eventually. All the other points in the network are called *intermediate points.*

FEASIBLE FLOW VECTORS

A *flow vector* $f = (f_{ij})$ specifies the amount of flow on each arc of the network. For a single commodity flow problem on the directed network G with lower bounds $\ell = (\ell_{ij})$ and capacities $k = (k_{ij})$, a *feasible flow vector* satisfies

$$\ell_{ij} \leqq f_{ij} \leqq k_{ij} \qquad \text{for all } (i, j) \in \mathscr{A} \tag{12.1}$$

and

$$\begin{aligned} \sum_{j \in \mathbf{A}_i} f_{ij} - \sum_{j \in \mathbf{B}_i} f_{ji} &= v && \text{if } i \text{ is source} \\ &= -v && \text{if } i \text{ is sink} \\ &= 0 && \text{if } i \text{ is an intermediate point} \end{aligned} \tag{12.2}$$

where $\mathbf{A}_i = \{j\colon j \text{ such that } (i, j) \in \mathscr{A}\}$, $\mathbf{B}_i = \{j\colon j \text{ such that } (j, i) \in \mathscr{A}\}$, and v, the net amount of material leaving the source (and arriving at the sink) is known as the *value of the flow vector f*. The constraints (12.2) are known as *flow conservation equations.* They require that the total amount of material reaching, should be equal to total amount of material leaving, at every intermediate point.

Example 1

In the network in Figure 12.8, here are the constraints on the flow vector f for feasibility.

Conservation equation for node	f_{12}	f_{13}	f_{23}	f_{24}	f_{52}	f_{34}	f_{35}	f_{54}	f_{46}	f_{56}	$-v =$	
1	1	1									1	0
2	−1		1	1	−1							0
3		−1	−1			1	1					0
4				−1		−1		−1	1			0
5					1		−1	1		1		0
6									−1	−1	−1	0

$$f \geqq 0 \quad \text{and} \quad (f_{12}, f_{13}, f_{23}, f_{24}, f_{52}, f_{34}, f_{35}, f_{54}, f_{46}, f_{56}) \leqq k = (8, 2, 6, 5, 3, 5, 10, 6, 19, 7)$$

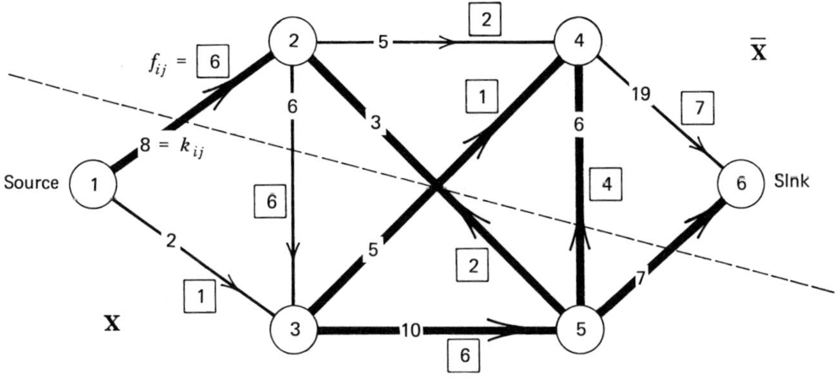

Figure 12.8 Capacities are entered on arcs. Lower bounds are zero.

The amount of flow on each arc in a flow vector is marked inside a small square "□" by the side of the arc, if that amount is nonzero. Verify that f is feasible [e.g., total amount reaching 5 is 6 units along (3, 5). Total amount leaving 5 is 6 units, 2 units along (5, 2) and 4 units along (5, 4). So conservation holds at 5, etc.] This flow vector has value 7 units.

NODE-ARC INCIDENCE MATRICES

Let G be a directed network with n points and m acrs. Let $\boldsymbol{E}$ be the $n \times m$ matrix that has a row associated with each point in G and a column associated with each arc in G, where the column associated with $(i, j) \in \mathscr{A}$ has only two nonzero entries, a "1" entry in the row associated with point i and a "-1" entry in the row associated with point j. $\boldsymbol{E}$ is known as the *node-arc incidence matrix* of G. The node-arc incidence matrix is only defined for directed networks. It can be verified that the coefficient matrix of the flow amounts f_{ij} in (12.2) is the node-arc incidence matrix.

Example 2

Verify that the coefficient matrix of the f_{ij} in Example 1 is the node-arc incidence matrix of the directed network in Figure 12.8.

Let $\boldsymbol{E}$ be the node-arc incidence matrix. Let $q = (q_i) \in \mathbf{R}^n$ be the column vector where $q_i = 1$ if i is the source, -1 if i is the sink, and 0 otherwise. Then the constraints (12.2) can be rewritten as

$$\boldsymbol{E}f - qv = 0 \tag{12.3}$$

Problem

12.9 Prove that the set of row vectors of a node-arc incidence matrix is linearly dependent (check their sum).

Consider the cycle 2, (2, 3), 3, (3, 4), and 4(2, 4), 2 in Figure 12.8. Let $\boldsymbol{E}_{.1}, \boldsymbol{E}_{.2}, \boldsymbol{E}_{.3}$ be, respectively, the columns in the node-arc incidence matrix of this network, associated with the arcs (2, 3) (3, 4), (2, 4), respectively. Verify that $\boldsymbol{E}_{.1} + \boldsymbol{E}_{.2} - \boldsymbol{E}_{.3} = 0$. Obtain a similar linear relationship among the columns of $\boldsymbol{E}$ associated with the arcs in the cycle 1, (1, 2), 2, (5, 2), 5, (3, 5) 3, (1, 3), 1. The general property is stated in the following problem.

Problems

12.10 Let G be a directed network with node-arc incidence matrix $\boldsymbol{E}$. Let x_1, $\mathbf{e}_1$, x_2, $\mathbf{e}_2, \ldots, \mathbf{e}_{r-1}, \mathrm{x}_r, \mathbf{e}_r, \mathrm{x}_1$ be a cycle in G. Treating $r + 1$ as 1, for $p = 1$ to r, let λ_p be 1, if $\mathbf{e}_p$ is the arc $(\mathrm{x}_p, \mathrm{x}_{p+1})$, or -1, if $\mathbf{e}_p$ is the arc $(\mathrm{x}_{p+1}, \mathrm{x}_p)$. Prove that by multiplying the column vector in $\boldsymbol{E}$ associated with arc $\mathbf{e}_p$ by λ_p and then adding over $p = 1$ to r, we get the zero vector.

12.11 Let G be a directed network with $\boldsymbol{E}$ as the node-arc incidence matrix. Suppose $(\mathbf{e}_1, \ldots, \mathbf{e}_r)$ is a subset of arcs containing no cycles. Prove that the set of column vectors in $\boldsymbol{E}$ associated with the arcs $\mathbf{e}_1, \ldots, \mathbf{e}_r$ is a linearly independent set. (Use the fact they form a forest. Thus, there must be at least one terminal node; use this argument repeatedly.)

12.12 Prove that the set of column vectors in $\boldsymbol{E}$ associated with the arcs in a spanning tree of G is linearly independent. (Use the fact that a tree has at least one terminal node repeatedly.)

12.13 If $\boldsymbol{E}$ is the node-arc incidence matrix of a connected directed network with n points, prove that the rank of $\boldsymbol{E}$ is $n - 1$.

12.14 Prove that the determinant of every square submatrix of $\boldsymbol{E}$ is either 0 or -1 or 1. This property is the *unimodularity property* discussed in Chapter 11.

OUT-OF-TREE AND IN-TREE ARCS

Let G be a connected directed network with the node-arc incidence matrix $\boldsymbol{E}$. Since the sum of the constraints in (12.3) leads to the equation "0 = 0," any one of the constraints in (12.3) can be eliminated and the remaining will be a linearly independent system. Suppose the row associated with some point, say, n, is erased from (12.3). Let the resulting equivalent system be

$$\bar{\boldsymbol{E}}f - \bar{q}v = 0 \tag{12.4}$$

The columns in any basis for $\bar{\boldsymbol{E}}$ are associated with the arcs in a spanning tree of G. Hence, in dealing with linear programming-type problems on directed networks, every basis corresponds to a spanning tree in the network and vice versa.

When considering a spanning tree **T**, an arc (i, j) in G is said to be an *in-tree arc* if it lies in **T**. Otherwise the arc (i, j) is called an *out-of-tree arc*.

When an out-of-tree arc (i, j) is included in **T**, a unique cycle is created. This is the cycle associated with the out-of-tree arc (i, j) when considering the tree **T**.

Example 3

Let **T** be the spanning tree consisting of the thick arcs in the network in Figure 12.1. The arc (6, 5) is an out-of-tree arc. The cycle associated with it is 6, (6, 5), 5, (4, 5), 4, (4, 2), 2, (2, 7), 7, (6, 7), 6. All the arcs on this cycle are in-tree arcs excepting (6, 5).

By replacing an in-tree arc in the cycle associated with the out-of-tree arc (i, j) by (i, j), a new tree is obtained. This operation of obtaining a new tree by adding an out-of-tree arc and dropping an in-tree arc in its cycle is the same as that of changing a basis for $\bar{\boldsymbol{E}}$ in (12.4) by bringing a nonbasic column into the basis to replace a basic column.

Example 4

As before, let **T** be the spanning tree consisting of the thick arcs in Figure 12.1. By adding the out-of-tree arc (6, 5) to the tree and dropping an in-tree arc in its cycle, say (2, 7), a new spanning tree is obtained.

Problem

12.15 Discuss how the spanning tree **T** can be stored using the predecessor indexing method discussed in Chapter 11. Explain how to find the cycle associated with a given out-of-tree arc using the predecessor indices. Also explain how the indices change when a new tree is obtained by bringing an out-of-tree arc and dropping an in-tree arc in its cycle.

CUTSETS SEPARATING THE SOURCE AND THE SINK

Let G be a directed network. Let **X** be a subset of points of G containing the source and not containing the sink. Let $\bar{\mathbf{X}}$ be the set of all the points of G that are not in **X**. The partition **X**, $\bar{\mathbf{X}}$ generates *a cutset separating the source and the sink.* The set of *forward arcs* of this cutset is $\{(i, j) : (i, j) \in \mathscr{A}, i \in \mathbf{X}, j \in \bar{\mathbf{X}}\}$. The set of *reverse arcs* of this cutset is $\{(i, j) : (i, j) \in \mathscr{A}, i \in \bar{\mathbf{X}}, j \in \mathbf{X}\}$. The cutset itself is denoted by $(\mathbf{X}, \bar{\mathbf{X}})$.

The capacity of the cutset $(\mathbf{X}, \bar{\mathbf{X}})$ separating the source and the sink is defined as

$$\sum_{\substack{(i,j)\\ \text{Forward}\\ \text{arc}}} k_{ij} - \sum_{\substack{(i,j)\\ \text{Reverse}\\ \text{arc}}} \ell_{ij}$$

Example 5

In Figure 12.8, let $\mathbf{X} = \{1, 3, 5\}$, $\bar{\mathbf{X}} = \{2, 4, 6\}$. The cutset is indicated by drawing a dashed line separating **X** and $\bar{\mathbf{X}}$. The set of forward arcs is $\{(1, 2), (3, 4), (5, 2), (5, 4), (5, 6)\}$ and the set of reverse arcs is $\{(2, 3)\}$ and the capacity of the cutset is 29.

Problems

12.16 Prove that every chain from the source to the sink must contain at least one forward arc of any cutset separating the source and the sink.

12.17 Prove that in the remaining network after all the forward arcs of a cutset separating the source and the sink are deleted, there exists no chain from the source to the sink.

FLOW AUGMENTING PATHS

Let G be a directed network with capacity vector k and lower bound vector ℓ. Let f be a feasible flow vector on G. A path, $\mathscr{P}$, from the source to the sink is said to be a *flow augmenting path* (FAP) *with respect to* f if it satisfies

$$f_{ij} < k_{ij} \qquad \text{whenever } (i, j) \text{ is a forward arc on the path } \mathscr{P}$$

$$f_{ij} > \ell_{ij} \qquad \text{whenever } (i, j) \text{ is a reverse arc on } \mathscr{P}$$

The reason for this name can be easily explained. Let v be the *value* of the flow vector f. Let

$$\begin{aligned} \varepsilon_1 &= \text{minimum } \{(k_{ij} - f_{ij}) : (i, j) \text{ a forward arc on } \mathscr{P}\} \\ &= +\infty, \text{ if all the arcs on } \mathscr{P} \text{ are reverse arcs} \\ \varepsilon_2 &= \text{minimum } \{(f_{ij} - \ell_{ij}) : (i, j) \text{ a reverse arc on } \mathscr{P}\} \\ &= +\infty, \text{ if all the arcs on } \mathscr{P} \text{ are forward arcs} \\ \varepsilon &= \text{minimum } \{\varepsilon_1, \varepsilon_2\} \end{aligned}$$

ε is obviously positive. Define a new flow vector $\hat{f}$ by

$$\begin{aligned} \hat{f}_{ij} &= f_{ij} && \text{if } (i, j) \text{ is not on the path } \mathscr{P} \\ &= f_{ij} + \varepsilon && \text{if } (i, j) \text{ is a forward arc on } \mathscr{P} \\ &= f_{ij} - \varepsilon && \text{if } (i, j) \text{ is a reverse arc on } \mathscr{P} \end{aligned}$$

Then, $\hat{f}$ is a feasible flow vector whose value is $\hat{v} = v + \varepsilon$. Hence, given a feasible flow vector and an FAP with respect to it, the FAP can be used to construct another feasible flow vector with a higher value.

Example 6

Consider the feasible flow vector in Figure 12.8. The thick path from source to sink is 1, (1, 2), 2, (5, 2), 5, (5, 4), 4, (3, 4), 3, (3, 5), 5, (5, 6), 6. Forward arcs are (1, 2), (5, 4), (3, 5), (5, 6) and reverse arcs are (5, 2), (3, 4). It is an FAP with respect to the present flow vector. $\varepsilon_1 =$ minimum $\{8 - 6, 6 - 4, 10 - 6, 7 - 0\} = 2$. $\varepsilon_2 =$ minimum $\{2 - 0, 1 - 0\} = 1$. $\varepsilon =$ minimum $\{\varepsilon_1, \varepsilon_2\} = 1$. The new flow vector obtained by increasing the flow amounts on forward arcs by ε and decreasing the flow amounts on reverse arcs by ε is

$$f = (f_{12}, f_{13}, f_{23}, f_{24}, f_{52}, f_{34}, f_{35}, f_{54}, f_{46}, f_{56}) = (7, 1, 6, 2, 1, 0, 7, 5, 7, 1),$$

and has a value 8.

Note An FAP is always defined with respect to some feasible flow vector. If the flow vector is changed, the path may no longer be an FAP.

12.3 SINGLE COMMODITY MAXIMUM FLOW PROBLEMS

Let G $= (\mathscr{N}, \mathscr{A})$ be a directed network with a source and a sink. Let k be the capacity vector for arc flows. We assume here that the lower bounds for all the arc flows are 0. The case in which lower bounds are positive is considered later. The problem is to find a feasible flow vector that has the maximum value.

$$\begin{aligned} &\text{Maximize} && v \\ &\text{Subject to} && Ef - qv = 0 && (12.5) \\ & && 0 \leqq f \leqq k && (12.6) \end{aligned}$$

It is an LP of special structure. It can be solved by the simplex method using the upper bounding routine (Chapter 10) in which each working basis corresponds to a spanning tree in the network. Here we will discuss an efficient network algorithm called the *labeling algorithm* for solving this maximum flow problem.

Notes (*i*) Suppose there are r sources in the problem, $1, 2, \ldots, r$, with a_i units of the commodity available at i, $i = 1$ to r. Introduce a new node, $\mathscr{s}$, called the *supersource*, and new arcs, $(\mathscr{s}, i)$ for $i = 1$ to r. See Figures 12.9 and 12.10. Leave the rest of the network unchanged. The problem stated on the new network, treating $\mathscr{s}$ as the single source, is equivalent to the original problem.

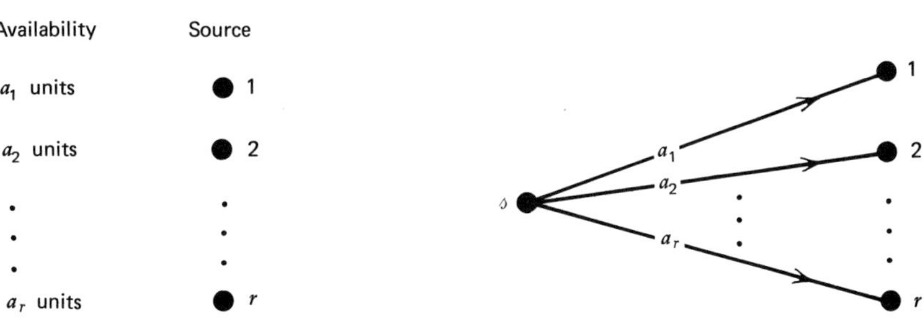

Figure 12.9

Figure 12.10 All lower bounds are zero. Capacities are marked on arcs.

(*ii*) Similarly suppose $n + 1, n + 2, \ldots, n + p$ are all the sinks into which the commodity has to flow. Introduce a new node, $\mathscr{t}$, called the *supersink* and new arcs $(n + 1, \mathscr{t})$, $(n + 2, \mathscr{t}), \ldots, (n + p, \mathscr{t})$, with lower bounds of zero, and capacity of $(n + j, \mathscr{t})$ equal to the minimal requirement at point $n + j$. The original problem is equivalent to the problem stated on the new network in which $\mathscr{t}$ is the single sink.

(*iii*) Let $(j_1, i), (j_2, i), \ldots, (j_b, i)$ be all the arcs incident into i, and let $(j_{b+1}, i), \ldots, (j_{b+d}, i)$ be all the arcs incident out of i and suppose there is a constraint that no more than c_i units of the commodity can pass through i. The portion of this network containing i is depicted in Figure 12.11. Change this portion as in Figure 12.12, leaving the rest of the network unchanged. In the new network i_1 represents

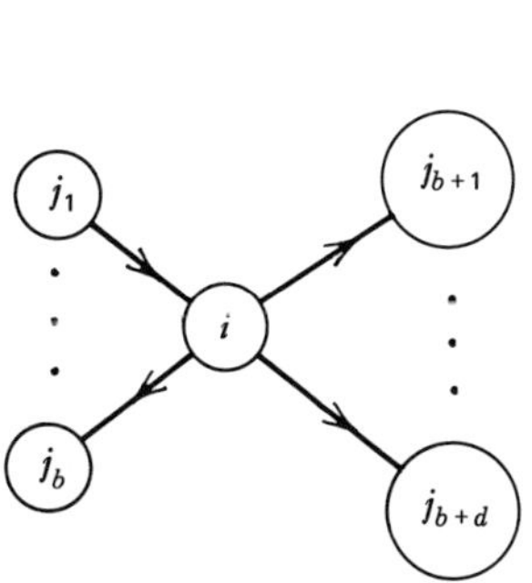

Figure 12.11 Point i with capacity c_i units of flow through it.

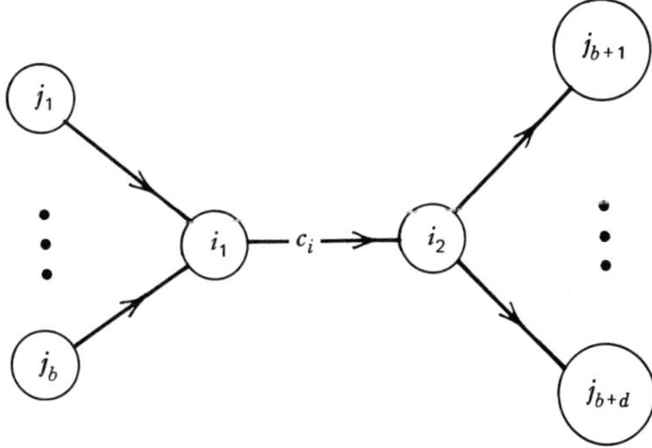

Figure 12.12

the *receiving end of point i* in the original network, and i_2 represents its *departing end.* In this manner the constraint that no more than c_i units can flow through point i is transformed into the capacity constraint on the new arc (i_1, i_2) in the new network.

12.3.1 Results

Theorem 1 Let f be a feasible flow vector of value v. Let $(\mathbf{X}, \bar{\mathbf{X}})$ be a cutset separating the source and the sink. The *net flow across the cutset* $(\mathbf{X}, \bar{\mathbf{X}})$ in the flow vector f, denoted by $f(\mathbf{X}, \bar{\mathbf{X}})$, is defined to be

$$\sum_{\substack{(i,\,j)\in\mathscr{A}\\ i\in\mathbf{X},\ j\in\bar{\mathbf{X}}}} f_{ij} - \sum_{\substack{(i,\,j)\in\mathscr{A}\\ i\in\bar{\mathbf{X}},\ j\in\mathbf{X}}} f_{ij}$$

This is equal to v.

Proof Since f is a feasible flow vector, it must satisfy the flow conservation Equations (12.5). Add all the conservation equations corresponding to $i \in \mathbf{X}$. $\mathbf{X}$ contains the source and not the sink. So on the right-hand side of this sum we get v. It can be verified that on the left-hand side of the sum we get $f(\mathbf{X}, \bar{\mathbf{X}})$. Hence $f(\mathbf{X}, \bar{\mathbf{X}}) = v$.

Theorem 2 The maximum value of v is less than or equal to the capacity of any cutset separating the source and the sink.

Proof Let $(\mathbf{X}, \bar{\mathbf{X}})$ be a cutset separating the source and the sink. Let f be a maximum value feasible flow vector, whose value is v. By Theorem 1, $v = f(\mathbf{X}, \bar{\mathbf{X}})$. But from the feasibility conditions (12.6), it is verified that $f(\mathbf{X}, \bar{\mathbf{X}})$ is less than or equal to the capacity of the cutset $(\mathbf{X}, \bar{\mathbf{X}})$.

Theorem 3 Let f be a feasible flow vector. If a cutset $(\mathbf{X}, \bar{\mathbf{X}})$ separating the source and the sink exists such that the value of f is equal to the capacity of the cutset $(\mathbf{X}, \bar{\mathbf{X}})$, then f is a maximum value feasible flow vector, and $(\mathbf{X}, \bar{\mathbf{X}})$ is a minimum capacity cutset separating the source and the sink.

Proof This is a direct consequence of Theorem 2.

Theorem 4 A feasible flow vector is a maximum value feasible flow vector iff there exists no FAP with respect to it.

Proof If a FAP exists with respect to a feasible flow vector f, using that FAP another flow vector of higher value can be constructed as in section 12.2. Hence, if f is a maximum value feasible flow vector, there cannot be an FAP with respect to it.

To prove the other part of the theorem, let f be a feasible flow vector with respect to which no FAP exists. Obtain a set of points $\mathbf{X}$ by the following rules.

(a) Source node is in $\mathbf{X}$.

(b) If $i \in \mathbf{X}$; j not yet in $\mathbf{X}$, and $(i, j) \in \mathscr{A}$ and $f_{ij} < k_{ij}$, include j also in $\mathbf{X}$.

(c) If $\mathrm{x} \in \mathbf{X}$, y not yet in $\mathbf{X}$, and $(\mathrm{y}, \mathrm{x}) \in \mathscr{A}$ and $f_{\mathrm{yx}} > 0$, include y also in $\mathbf{X}$.

Repeat (b) and (c) until no more points can be included in $\mathbf{X}$. It is clear that a FAP with respect to f does not exist iff the above procedure terminates with a set $\mathbf{X}$ that does not include the sink. Let $\bar{\mathbf{X}}$ be the set of all the points in the network not in **X**. Then $(\mathbf{X}, \bar{\mathbf{X}})$ is a cutset separating the source and the sink. From the manner in which **X** is obtained, we know that

$$f_{ij} = k_{ij} \qquad \text{for all } (i, j) \in \mathscr{A}, i \in \mathbf{X}, j \in \bar{\mathbf{X}}$$

$$f_{ij} = 0 \qquad \text{for all } (i, j) \in \mathscr{A}, i \in \bar{\mathbf{X}}, j \in \mathbf{X}$$

If these conditions are not satisfied, by rules (b) and (c) additional points could be included in the set **X**, contradicting the assumption that the procedure discussed above has terminated and no more points could be included in **X**.

These conditions imply that $f(\mathbf{X}, \bar{\mathbf{X}})$ is equal to the capacity of the cutset $(\mathbf{X}, \bar{\mathbf{X}})$ separating the source and the sink. By Theorem 3, f is a maximum value feasible flow vector.

Theorem 5 **Maximum Flow Minimum Cut Theorem** In the single commodity flow problem (12.5) and (12.6), the maximum value of feasible flow vectors is equal to the minimum capacity of cutsets separating the source and the sink.

Proof The problem is feasible since $f = 0$ is a feasible flow vector. Clearly the maximum value of v is infinite iff there exists a chain from the source to the sink such that all the arcs on the chain have infinite capacities. If such a chain exists, by the results in section 12.2, every cutset separating the source and the sink must contain an arc of such a chain as a forward arc and, hence, the capacities, of all those cutsets are also infinite and the theorem holds.

If the maximum value of v is finite, the result follows from the discussion in Theorems 4 and 3.

Note Theorem 5 has been stated and proved for single commodity flow problems. In a multicommodity flow problem, each commodity may originate at some specified source and end up at some specified sink. On such problems, the corresponding version of the maximum flow minimum cut theorem may not be true.

12.3.2 Labeling Algorithm

Assume that the lower bound vector is ℓ. The algorithm begins with an initial feasible flow vector. If $\ell = 0$, some known feasible flow vector or $f = 0$ can be used as the initial flow vector. An algorithm for generating an initial feasible flow vector when $\ell \neq 0$ is discussed in section 12.3.5.

Let $\tilde{f}$ be the present feasible flow vector. The labeling routine is a systematic method for checking whether a FAP with respect to $\tilde{f}$ exists. First, label the source as $(s, +)$. Continue labeling by the following labeling rules.

(i) If i is a labeled node and j is an unlabeled node at this stage, and $(i, j) \in \mathscr{A}$, and $\tilde{f}_{ij} < k_{ij}$, label j with $(i, +)$.

(ii) If x is a labeled node, and y is an unlabeled node at this stage and $(\text{y}, \text{x}) \in \mathscr{A}$, $\tilde{f}_{\text{yx}} > \ell_{\text{yx}}$, label y with (x, −).

Discussion At some stage of the algorithm there may be several different ways in which an unlabeled point can be labeled. Pick any one of these arbitrarily, label the point, and continue the labeling further. During a labeling cycle, once a point is labeled, do not change the label on it. At each stage, check whether the sink can be labeled and if so label it and terminate the labeling routine (even though there may be other unlabeled points that could be labeled further). If the sink gets labeled, the labeling routine is said to have terminated in a *breakthrough.* If the sink cannot be labeled, label some other point that can be labeled and continue. If the sink has not yet been labeled, and no other point can be labeled further, the labeling routine terminates in a *nonbreakthrough.*

Meaning of the Labels The labels are trying to trace an FAP from the source to the sink. The path can be obtained by reading the labels backwards. If the label on a point j is $(i, +)$, there is an FAP from the source to the point j, and the last arc on this path is (i, j), the "+" sign indicating that it is a forward arc. On the other hand, if the label on j were $(r, -)$, it indicates that the arc (j, r) is a reverse arc on the path, the "−" sign indicating that the arc is a reverse arc. Go back to the previous node whose number appears in the label on j; and trace the path from it backwards in the same manner until the source is reached.

Termination in a breakthrough implies that an FAP from the source to the sink has been found. Using the FAP obtain a new feasible flow vector of higher value as in section 12.2. Erase all labels and go back to another labeling cycle.

If termination occurs in a nonbreakthrough, the present feasible flow vector is a maximum flow vector. Let $\mathbf{X}, \bar{\mathbf{X}}$ be the sets of labeled and unlabeled nodes at this stage. $(\mathbf{X}, \bar{\mathbf{X}})$ is a minimum capacity cutset separating the source and the sink.

Numerical Example

In the network in Figure 12.13, all lower bounds on arc flows are zero. The capacity of each arc is entered on that arc. Flow amounts in an initial feasible flow vector are entered in small squares "□" by the side of the arcs. Zero flows amounts are not recorded. This flow vector has a value of 7. Label the source $(\emptyset, +)$; 2 can be labeled $(1, +)$; 5 can be labeled with $(2, -)$. Then the sink can be labeled with $(5, +)$. There is a breakthrough. The FAP is:

Forward arcs: (5, 6), (1, 2); reverse arc: (5, 2).

As in section 12.2, compute, $\varepsilon_1 = \text{minimum } \{7 - 0, 8 - 6\} = 2,$

$$\varepsilon_2 = \text{minimum } \{2\}, \qquad \varepsilon = \text{minimum } \{\varepsilon_1, \varepsilon_2\} = 2.$$

The new flow vector is obtained by subtracting 2 from the flow amounts on all reverse arcs and adding 2 to all the flow amounts on forward arcs. The new flow vector is in Figure 12.14.

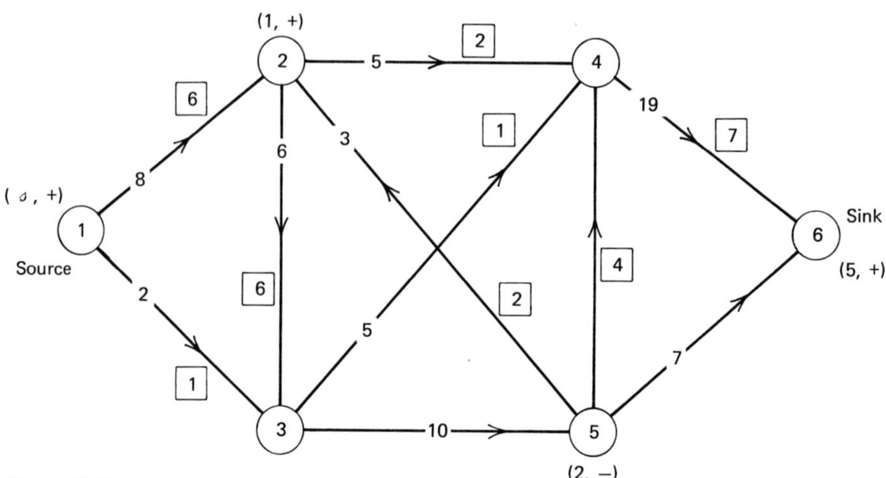

Figure 12.13

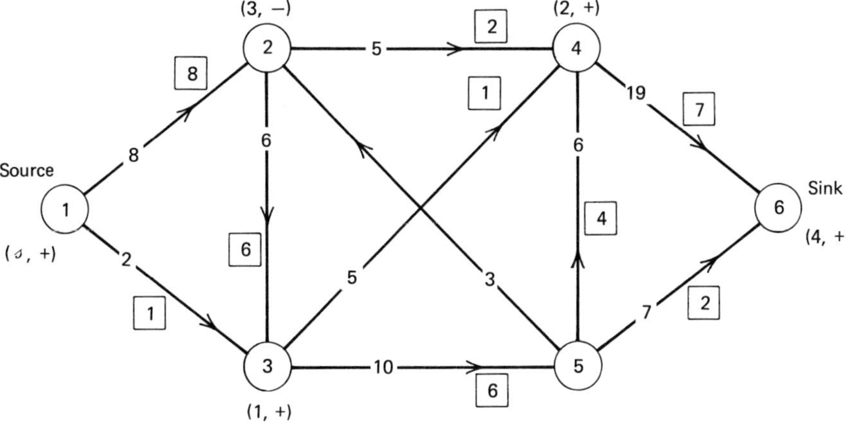

Figure 12.14

Erase the old labels and label again. There is a breakthrough again. The new FAP is

Forward arcs: (4, 6), (2, 4), (1, 3); reverse arc: (2, 3)

Changing the flow vector using this FAP leads to the flow vector in Figure 12.15. Erase all old labels and label again. Termination in nonbreakthrough occurs. This implies that the present flow vector is a maximum flow vector. The maximum value is 10. The set of labeled nodes is $\mathbf{X} = \{1\}$. The minimum capacity cutset separating the source and the sink, $(\mathbf{X}, \bar{\mathbf{X}})$, is marked in Figure 12.15 with a dashed line.

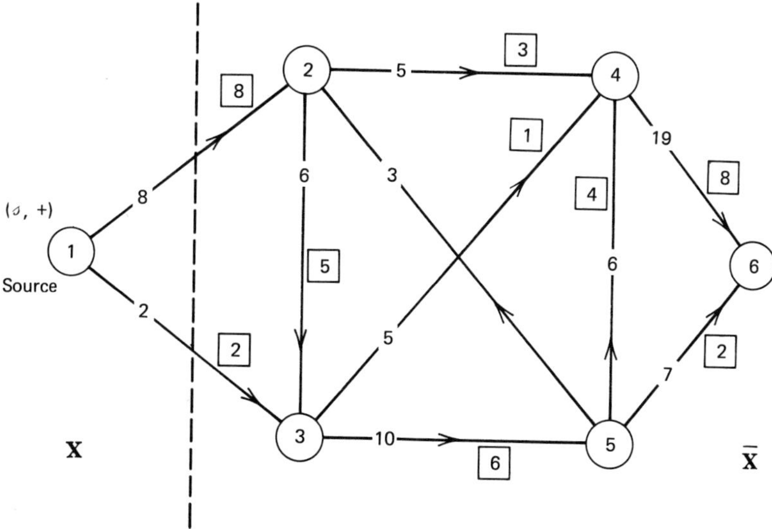

Figure 12.15

Problems

12.18 $G = (\mathcal{N}, \mathcal{A})$ is a directed network with source $\mathscr{s}$ and sink t. Prove that there exist r arc disjoint chains (i.e., chains containing no common arcs) from $\mathscr{s}$ to t iff there is no cutset separating $\mathscr{s}$ and t in which the number of forward arcs is strictly less than r. (Set up data on network appropriately and use the maximum flow minimum cut theorem.)

12.19 Write the dual of (12.5) and (12.6). In the dual problem, there is a dual variable associated with each point in the network. Using a minimum capacity cutset separating the source and the sink, construct an optimum solution for the dual problem. Using this solution derive the maximum flow minimum cut theorem as a special case of the duality theorem of linear programming.

12.20 Consider a club of m men and w women. In this club, *compatibility* is a mutual relationship in a man, woman pair. The subset of women compatible with the ith man, $i = 1$ to m are given. Find the maximum number of compatible couples that can be formed. Formulate as a maximum flow problem.

12.21 Find the maximum value flow vector from source to sink on the following networks. All lower bounds are zero and capacities are entered on the arcs.
(a)

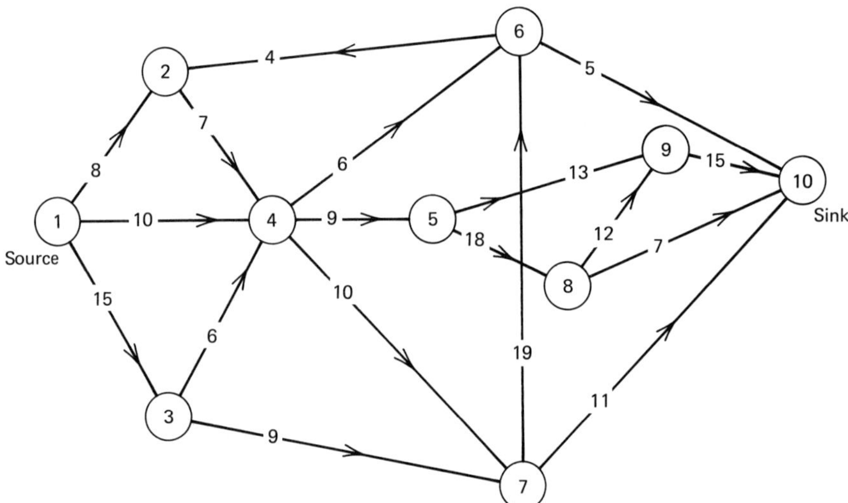

Figure 12.16

(b) This is an undirected network.

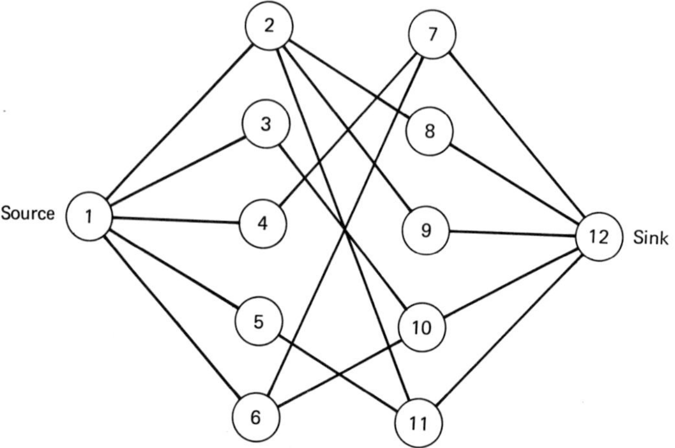

Figure 12.17 All arcs directed from left to right and all capacities are 1.

(c)

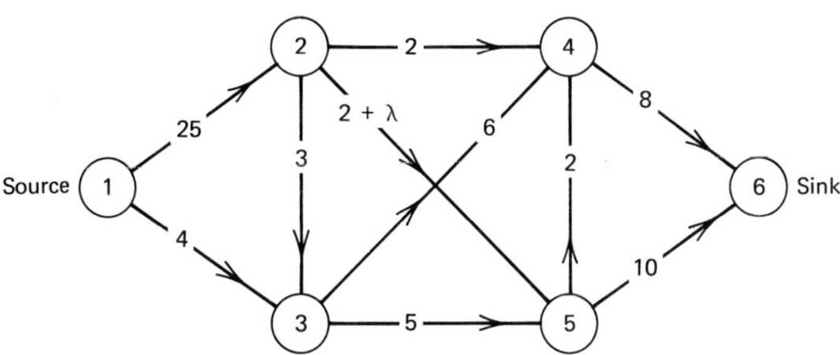

Figure 12.18 Find the maximum value flow vector from source to sink as a function of λ for all $\lambda \geqq 0$.

(d) This is an undirected network. If $(i; j)$ is an edge in a maximum flow problem with lower bound for flow amount of 0 and capacity for flow amount of k_{ij}, one way of handling it is to replace it by the pair of arcs (i, j) and (j, i), both with lower bounds of 0 and capacity k_{ij}. This increases the number of arcs in the network. Another way of handling edges in a single commodity maximum flow problem is the following. Select an arbitrary orientation, say from i to j, on the edge $(i; j)$, and treat it as an arc with the selected orientation, throughout the algorithm, that is, as arc (i, j). Let the capacity on this arc be k_{ij}, and let the lower bound for flow amount on this arc be $-k_{ij}$. Adopt the convention that if the flow amount on this arc turns out to be a negative number, $-\alpha$, this implies that the flow on this edge is $+\alpha$ in the direction opposite to the selected orientation for this edge. Use this approach to solve this problem.

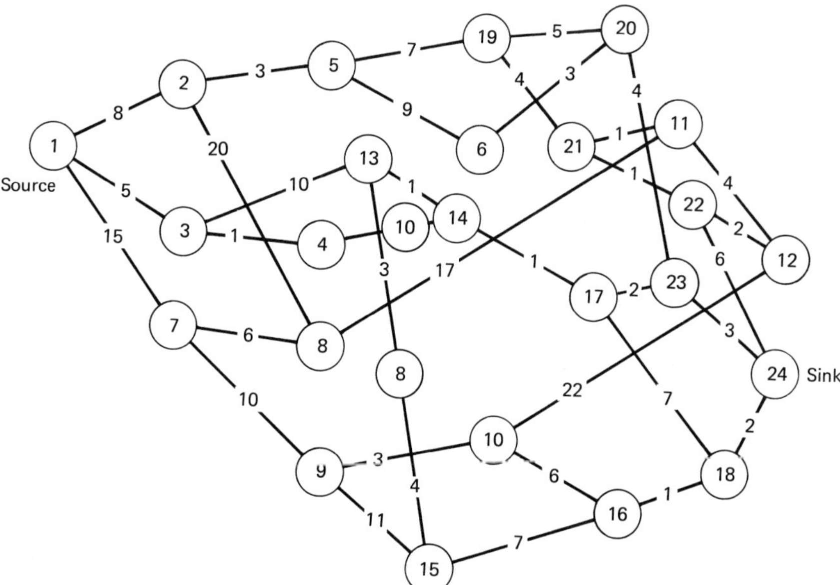

Figure 12.19

12.22 Discuss an algorithm for finding a feasible flow vector whose value is a specified number v^*.

12.3.3 Integer Property

Start the labeling algorithm with the initial flow vector $f = 0$. If k is an integer vector, all subsequent flow vectors obtained during the algorithm are integer vectors. Hence the algorithm terminates with a maximum flow vector that is an integer vector. Thus when k is an integer vector, (12.5); (12.6) has an optimum solution that is an integer vector.

12.3.4 Sensitivity Analysis

Consider a particular arc $(i, j) \in \mathscr{A}$ in the maximum flow problem (12.5) and (12.6). Suppose $k_{ij} = \xi$. Let $\bar{v}(\xi)$ be the maximum value of flow when the capacity of arc (i, j) is set at ξ. The capacities of all the other arcs in the network are assumed to remain unchanged. We wish to find $\bar{v}(\xi)$ in the region $\xi \geqq 0$.

Initially let ξ be 0. Let $\mathbf{U}_1$ be the set of all cutsets separating the source and the sink that contain (i, j) as a forward arc, and $\mathbf{U}_2$, the set of all cutsets separating the source and the sink that do not contain (i, j) as a forward arc. The capacity of every cutset in $\mathbf{U}_2$ is unaffected by changes in the capacity of the arc (i, j). Suppose the minimum capacity among cutsets in $\mathbf{U}_2$ is c_2.

Let c_1 be the minimum capacity of cutsets in $\mathbf{U}_1$ when the capacity of the arc (i, j) is 0. Since all cutsets in $\mathbf{U}_1$ contain (i, j) as a forward arc, the minimum capacity of cutsets in $\mathbf{U}_1$ is obviously $c_1 + \xi$ when the capacity of the arc (i, j) is ξ.

Hence the minimum capacity of cutsets separating the source and the sink is equal to minimum $\{c_1 + \xi, c_2\}$ as a function of ξ. If $c_1 \geqq c_2$, this minimum is always equal to c_2, for every $\xi \geqq 0$. In this case, there is a minimum capacity cutset separating the source and the sink, not containing the arc (i, j) as a forward arc, whatever the capacity of arc (i, j) may be.

If $c_1 < c_2$, for all values of $\xi \leqq c_2 - c_1$, there is a minimum capacity cutset separating the source and the sink, which contains (i, j) as a forward arc, and its capacity is $c_1 + \xi$. For all values of $\xi > c_2 - c_1$, the minimum capacity of cutsets separating the source and the sink is c_2, independent of the value of ξ. Using the maximum flow minimum cut theorem we arrive at the following conclusions.

1. If $c_1 \geqq c_2, \bar{v}(\xi) = c_2$, a constant for all $\xi \geqq 0$.
2. If $c_1 < c_2, \bar{v}(\xi) = c_1 + \xi$ for all ξ in the range $0 \leqq \xi \leqq c_2 - c_1$ and $\bar{v}(\xi) = c_2$ for $\xi > c_2 - c_1$. In this case the maximum flow value increases as the capacity of the arc (i, j) is increased from 0 up to $c_2 - c_1$, and beyond $c_2 - c_1$, it does not change. This number $c_2 - c_1$ is the therefore called the *critical capacity* of the arc (i, j). We will denote it by k^*_{ij}.
3. For any value of $\xi \geqq 0$

$$\bar{v}(\xi) = \text{minimum } \{\bar{v}(\xi = 0) + \xi, \bar{v}(\xi = \infty)\}$$

   Also the critical capacity of arc (i, j), k^*_{ij}, is $\bar{v}(\xi = \infty) - \bar{v}(\xi = 0)$. Hence to compute the critical capacity of an arc we have to solve two maximum flow problems, one with the capacity of that arc set equal to 0 and the other with the capacity of that arc set equal to infinity.

4. Whenever the capacity of the arc (i, j) is $k_{ij} < k^*_{ij}$, (i, j) is a forward arc in every minimum capacity cutset separating the source and the sink. If $k_{ij} > k^*_{ij}$,

(i, j) is not a forward arc in any minimum capacity cutset separating the source and the sink. If $k_{ij} = k_{ij}^*$, there exists a minimum capacity cutset separating the source and the sink that contains (i, j) as a forward arc, and another that does not contain (i, j) as a forward arc.

5. Destroying an arc in a network is equivalent to reducing its capacity to zero. The amount by which the maximum flow value would decrease if arc (i, j) is destroyed is equal to minimum $\{k_{ij}, k_{ij}^*\}$

Problems

12.23 The numbers on the arcs in the following network are the capacities. All lower bounds are zero. Find the maximum flow value from the source to the sink and identify a minimum capacity cutset separating the source and the sink. Evaluate the critical capacities of arcs (3, 5) and (2, 6).

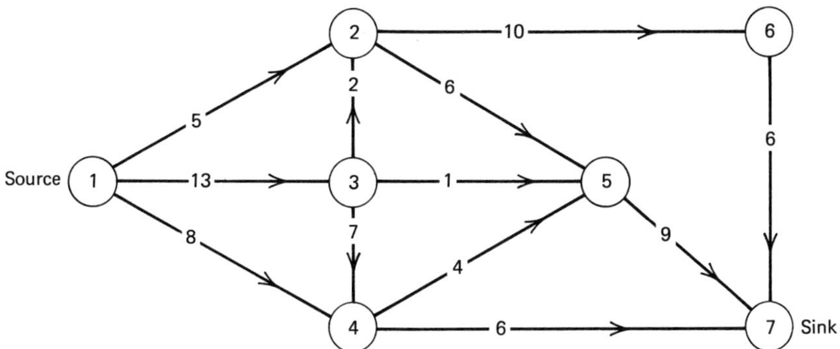

Figure 12.20

12.24 Discuss an efficient scheme for finding an arc, the destruction of which reduces the maximum flow value from the source to sink the most.

12.25 Let k_{ij}, k_{ij}^* be the present capacity and critical capacities, respectively, of arc (i, j). Maximum $\{k_{ij}^* - k_{ij}, 0\}$ is the *scope* for increasing the maximum flow value by developing the arc (i, j). Discuss an efficient scheme for finding an arc with *maximum scope* from the arcs in some specified subset of arcs.

12.3.5 Single Commodity Flow Problems with $\ell \neq 0$

Let $G = (\mathcal{N}, \mathcal{A})$ be a directed network with a source s and sink t. Let ℓ be the lower bound vector and k the capacity vector. Here we discuss an algorithm for finding a feasible flow vector on this network. Add an artificial arc (t, s) to the network, with lower bound of 0 and capacity infinity. Let $\mathcal{A}^1 = \mathcal{A} \cup \{(t, s)\}$ and $G^1 = (\mathcal{N}, \mathcal{A}^1)$. If f is a feasible flow vector on G with value $v(f)$, define

$$f_{ts}^1 = v(f) \qquad \text{and} \qquad f^1 = (f, f_{ts}^1)$$

Then f^1 is a feasible flow vector on G^1. If f^1 is implemented, the net amount flowing out of s is 0, since the artificial arc routes back $v(f)$ units from t to s. Thus the net amount of flow out of every point is zero if the vector f^1 is implemented. Hence, f^1 is known as a *feasible circulation* on G^1.

Given any feasible circulation on G^1, by deleting the amount of flow on the artificial arc (t, s) from it, it becomes a feasible flow vector on the original network

G. Hence the problem of finding a feasible flow vector on G is equivalent to the problem of finding a feasible circulation of G^1. We will transform the problem of finding a feasible circulation on G^1 into a maximum flow problem on an augmented network G^*. Add an artificial source s^* and an artificial sink t^* to G^1. Let $\mathcal{N}^* = \mathcal{N} \cup \{s^*, t^*\}$. Add artificial arcs (s^*, i) and (i, t^*) for each $i \in \mathcal{N}$. Let $\mathcal{A}^* = \mathcal{A}^1 \cup \{(s^*, i), (i, t^*)$: for all $i \in \mathcal{N}\}$. Let $G^* = (\mathcal{N}^*, \mathcal{A}^*)$. In G^*, make the lower bounds on all the arcs zero. Let $\mathbf{A}_i = \{r: (i, r) \in \mathcal{A}\}$, $\mathbf{B}_i = \{r: (r, i) \in \mathcal{A}\}$. Define capacities by the following (Figure 12.21):

Capacity of $(i, j) \in \mathcal{A}^1$	is	$k_{ij} - \ell_{ij}$
Capacity of (s^*, i)	is	$\sum_{r \in \mathbf{B}_i} \ell_{ri}$
Capacity of (i, t^*)	is	$\sum_{r \in \mathbf{A}_i} \ell_{ir}$

Find a maximum flow vector, f^* from s^* to t^*, starting with the zero flow vector, on G^*. The following conclusions are clear:

1. If the value of f^* on G^* (i.e., the net flow amount from s^* to t^* if f^* is implemented on G^*) is strictly less than $\sum_{(i,j) \in \mathcal{A}} \ell_{ij}$, there exists no feasible flow vector on the original network G.
2. If the value of f^* on G^* is equal to $\sum_{(i,j) \in \mathcal{A}} \ell_{ij}$, then let:

$$\hat{f}_{ij} = f^*_{ij} + \ell_{ij} \qquad \text{for all } (i, j) \in \mathcal{A}$$

The vector $\hat{f} = (\hat{f}_{ij})$ is a feasible flow vector on the original network G.

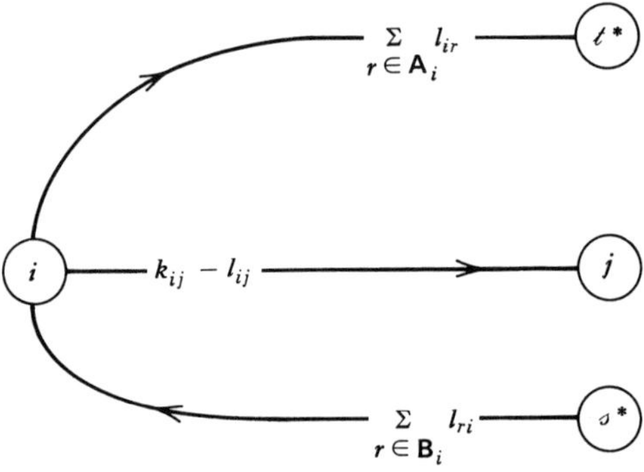

Figure 12.21 Demonstration of the capacities on arcs in G^* incident at node i. All Lower bounds are zero.

Numerical Example

Consider the network G in Figure 12.22. Data on the arcs is (i) — ℓ_{ij}, k_{ij} →— (j). The problem of finding a feasible flow vector on G is equivalent to the problem of finding a feasible circulation on G^1 (Figure 12.23). The augmented network G^*, the maximum flow vector from s^* to t^* in G^*, is in Figure 12.24. Since all the arcs incident out of s^* are *saturated* (an arc is said to be *saturated* if the flow amount on this arc is equal to its capacity in the present flow), it is clear that the maximum flow value in G^* is equal to $\sum_{(i,j) \in \mathcal{A}} \ell_{ij} = 11$. Hence the original problem is feasible, and a feasible flow vector on the original network is $(f_{12}, f_{13}, f_{23}, f_{32}, f_{24}, f_{34}) = (3, 4, 1, 2, 4, 3)$.

Problem

12.26 Find the maximum flow from 1 to 5 on the network in Figure 12.25. Data in Figure 12.25 are (i) — ℓ_{ij}, k_{ij} →— (j).

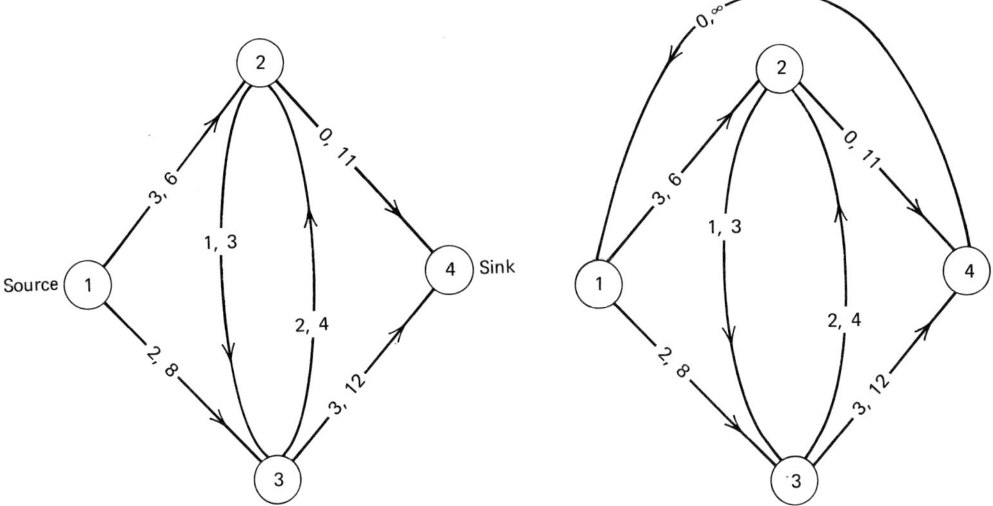

Figure 12.22 Network G.

Figure 12.23 Network G^1.

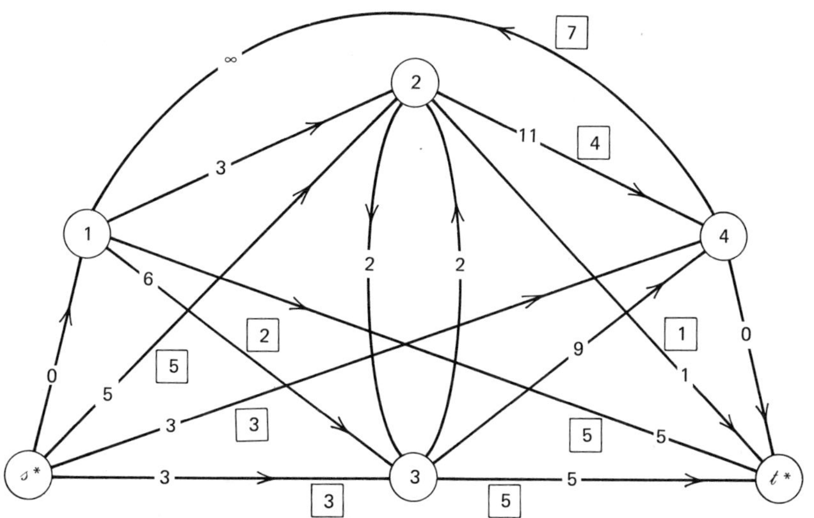

Figure 12.24 Network G*. Capacities entered on arcs. f_{ij}^* entered in "□" near arcs.

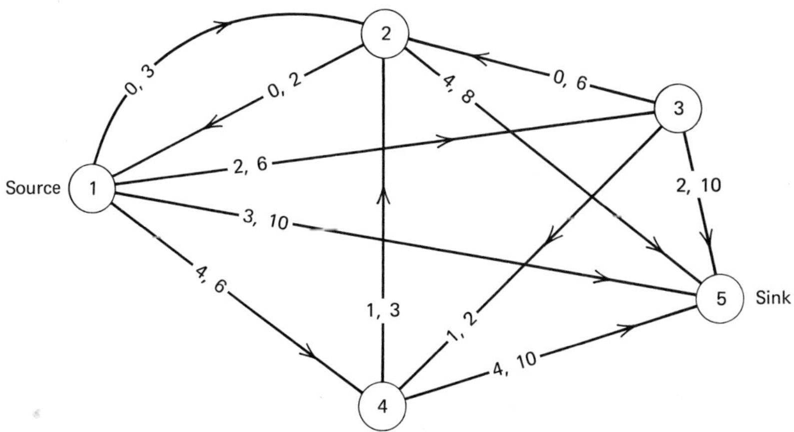

Figure 12.25

12.4 THE PRIMAL-DUAL APPROACHES FOR THE ASSIGNMENT AND TRANSPORTATION PROBLEMS

12.4.1 The Hungarian Method for the Assignment Problem

The problem where $C = (c_{ij})$ is a given cost matrix, known as an *assignment problem* is

$$\begin{aligned} \text{Minimize} \quad & z(x) = \sum_{i=1}^{n} \sum_{j=1}^{n} c_{ij} x_{ij} \\ \text{Subject to} \quad & \sum_{j=1}^{n} x_{ij} = 1 \quad \text{for } i = 1 \text{ to } n \\ & \sum_{i=1}^{n} x_{ij} = 1 \quad \text{for } j = 1 \text{ to } n \\ & x_{ij} \geqq 0 \quad \text{for all } i, j \end{aligned} \tag{12.7}$$

If $x = (x_{ij})$ is a BFS of (12.7), then $x_{ij} = 0$ or 1 for all i, j (Chapter 11). Any such x is known as an *assignment or* a *permutation matrix.* In an assignment $x = (x_{ij})$, the cell (i, j) is said to have an allocation if $x_{ij} = 1$. In this case row i is said to be allocated to (or matched with) column j. If $x_{ij} = 0$, the cell (i, j) has no allocation. There is only one allocation in each row and column. We will denote the assignment $x = (x_{ij})$ by the set $\{(i, j): x_{ij} = 1\}$, that is, the set of cells with allocations.

Represent each row and each column by a point. Join each row to each column by an edge. This results in a *bipartite network.* Treat c_{ij} as the cost of the edge joining row i with column j.

In any network a *matching* is a partial subnetwork in which each point has a degree less than or equal to 1. A *perfect matching* or a *complete matching* in a network is a subnetwork in which each point has degree 1. In Figure 12.26, the wavy subnetwork is a complete matching. Take a complete matching in this bipartite network. Define $x_{ij} = 1$ if the edge joining row i, column j is in the complete matching, $x_{ij} = 0$ otherwise. Then $x = (x_{ij})$ is an assignment. Likewise there is a complete matching corresponding to each assignment. Hence (12.7) is equivalent to the problem of finding a minimum cost complete matching in the bipartite network discussed above. That's why the assignment problem is also known as the *bipartite matching problem.*

Direct all the lines in the bipartite network from the row to the column. In this

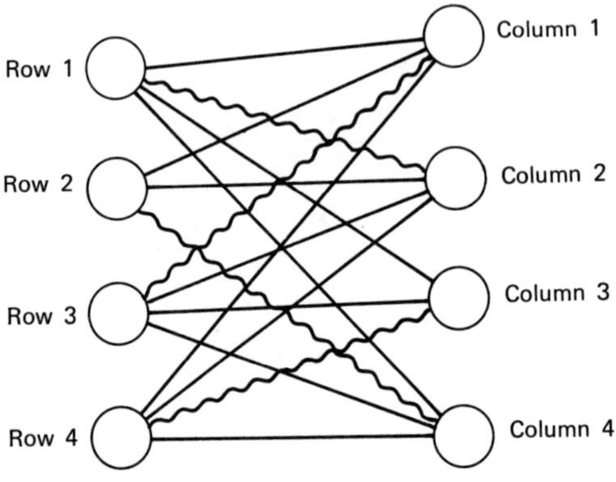

Figure 12.26 Bipartite network.

network treat all the rows as sources, each with a supply of one unit. Treat all the columns as sinks, each with a demand of one unit. Treat c_{ij} as a cost of transporting one unit on the arc (i, j). The assignment problem is the problem of finding a minimum cost solution of this transportation problem. The variable x_{ij} becomes the amount of flow on the arc (i, j) in the solution to this problem. The Hungarian method for finding an optimum assignment makes use of this idea.

THE HUNGARIAN METHOD

Denote the objective function as $z_C(x)$. It is the cost of the assignment x with respect to C as the cost matrix. Let C^1 be the matrix obtained by subtracting a real number α from every element in a row or a column of C. Then clearly

$$z_C(x) = \alpha + z_{C^1}(x) \qquad \text{for every assignment } x \tag{12.8}$$

This idea is used repeatedly. First, find the minimum element in each row of C and subtract it from every element in that row. In the resulting matrix, find the minimum element in each column and subtract it from every element in that column. The final matrix obtained is a nonnegative matrix called the *reduced matrix* at this stage. The total amount subtracted from the rows and the columns is known as the *reduction* at this stage. Let the reduced matrix be C^0 and let the reduction be r_0. The arguments used in obtaining (12.8) lead to

$$z_C(x) = r_0 + z_{C^0}(x) \tag{12.9}$$

C^0 is clearly a nonnegative matrix with at least one zero entry in each row and column. Hence $z_{C^0}(x) \geqq 0$. If we find an assignment x such that x has allocations only in those cells in which the entries in C^0 are zero, then $z_{C^0}(x) = 0$ and from (12.9) x is an optimal assignment. We try to find such an assignment by solving a maximum flow problem.

Define the cell (i, j) as an *admissible cell* only if the entry in that cell in the current reduced cost matrix is zero. Obtain a network by joining row i, column j by an arc only if (i, j) is an admissible cell. Introduce a supersource, s, and join it by an arc to all the rows. Also introduce a supersink, t, and join each column to t by an arc. The lower bounds on all the arcs are 0 and the capacities on all the arcs are 1. See Figure 12.27.

The value of the maximum flow on this network from s to t is the maximum number of allocations that can be made among the admissible cells at this stage. If this maximum flow saturates all the arcs from each column to t, these allocations

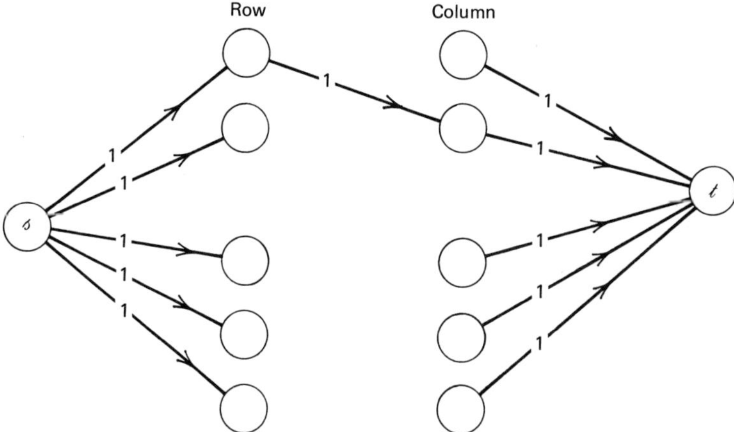

Figure 12.27 Admissible network.

define an optimum assignment. If the maximum flow does not saturate all the arcs from each column to t, it is impossible to find an assignment that has allocations only among admissible cells. In this case we go to a further reduction routine to create additional admissible cells.

Set up an array like the transportation array in Chapter 11. Identify all the admissible cells in this array. All the operations performed in finding the maximum flow on the admissible network can be performed directly on the array itself, by associating the cell (i, j) with the arc (i, j). If a flow of one unit occurs along this arc, it implies that an allocation is made in the associated cell in the array.

TO FIND THE INITIAL ALLOCATIONS

Allocations can be made among admissible cells only, and each row and each column can have at most one allocation. Look for a row or column in the array with a single admissible cell. Make an allocation in the admissible cell in that row or column and strike off all other admissible cells in the column or row of the allocated cell. Repeat the same procedure using admissible cells that are not yet struck off and rows and columns that have no allocation yet. At some stage, if each row and column without allocation so far has two or more admissible cells (which have not been struck off so far), make an allocation arbitrarily in one of those admissible cells, strike off all the other admissible cells in the row and the column of the allocated cell, and continue the procedure. The procedure terminates when all the admissible cells are either allocated or struck off. When the procedure terminates, remove all the strikings made on all the unallocated admissible cells. If each column has an allocation at the end of this procedure, these allocations define an assignment that is optimal and we terminate the algorithm. On the other hand, if some columns have no allocations, go to the labeling routine.

THE LABELING ROUTINE

1. Label every row that has no allocations at this stage with $(s, +)$.
2. If row i is labeled and column j is unlabeled so far, and if cell (i, j) is an admissible cell at this stage, label column j with (row i, $+$).
3. If column j is labeled and row i is unlabeled so far and if cell (i, j) has an allocation at this stage, label row i with (column j, $-$).

Go through labeling rules 2 and 3 as often as possible. In this process if a column without an allocation is labeled, we have a *breakthrough* which implies that it is possible to get one more allocation into the array. Go to the *allocation change routine.* If no column without an allocation can be labeled and the labeling cannot be continued any further, the labeling routine terminates in a *nonbreakthrough.* In this case, go to the *further reduction routine.*

Note Whenever a nonbreakthrough occurs in this algorithm, draw a line through each unlabeled row and through each labeled column of the array. Then every admissible cell in the array at this stage must have at least one line through it. In other words, the lines drawn above cover all the admissible cells in the array at this stage. Also the number of allocations made at this stage will be equal to the number of lines drawn. See problems 12.47 and 12.48.

ALLOCATION CHANGE ROUTINE

Suppose column j, which does not have an allocation, has been labeled. Let the label on it be (row i, $+$). The "$+$" sign inside the label indicates that a new allocation should be made in cell (i, j). Next, look at the label on row i. Suppose it is (column j_1, $-$), This implies that the present allocation in the cell (i, j_1) should be deleted. Then look at the label on column j_1. Continue adding a new allocation and deleting a present allocation in this manner until a row with the label $(s, +)$ is reached. At this stage the allocation change routine stops. If all the columns have allocations now, these allocations define an optimum assignment and we terminate. Otherwise erase the labels on all the rows and columns and go back to the labeling routine afresh.

FURTHER REDUCTION ROUTINE

Since labeling has terminated, the entries in the present reduced matrix in all the cells among labeled rows and unlabeled columns must be positive. Let δ be the minimum entry among these cells.

Subtract δ from each entry in the present reduced matrix in every labeled row. Subtract $-\delta$ from every entry in the resulting matrix in each labeled column. Whenever a quantity is subtracted from all the entries in a row or a column, add that quantity to the amount of reduction. Another method of carrying out these operations is the following:

1. Subtract δ from the entries in the cells in labeled rows and unlabeled columns of the reduced matrix.
2. Add δ to the entries in the cells in unlabeled rows and labeled columns of the reduced matrix.
3. Leave all the other entries unchanged. Let α = the number of labeled rows minus the number of labeled columns at this stage. Add $\delta\alpha$ to the amount of reduction.

Define an admissible cell as any cell that has a zero entry in the newly obtained reduced matrix. Some of the old admissible cells become inadmissible, and some new admissible cells are created. However, all cells that have allocations at present remain admissible. Now go back to the labeling routine by resuming the labeling from where it was left off at the occurrence of a nonbreakthrough.

TERMINATION

The algorithm terminates when each column obtains an allocation. The assignment defined by these allocations is an optimum assignment. Its objective value will be equal to the total amount of reduction at termination.

Numerical Example

Original cost matrix								Row minimum	Cost matrix after subtracting Row minimum from rows							
9	12	9	6	9	6	9	10	6	3	6	3	0	3	0	3	4
7	8	5	4	3	8	15	5	3	4	5	2	1	0	5	12	2
6	3	16	18	3	19	6	8	3	3	0	13	15	0	16	3	5
6	1	17	6	5	11	7	9	1	5	0	16	5	4	10	6	8
5	0	13	4	5	6	1	2	0	5	0	13	4	5	6	1	2
12	13	4	12	3	1	12	14	1	11	12	3	11	2	0	11	13
3	12	3	7	13	6	8	3	3	0	9	0	4	10	3	5	0
13	4	1	5	5	5	4	9	1	12	3	0	4	4	4	3	8
								Column minimum	0	0	0	0	0	0	1	0

Subtracting the column minimum from the columns on the right-hand side matrix leads to the first reduced matrix. Total reduction so far is 19.

First Reduced Matrix with Initial Allocations

Column $j =$	1	2	3	4	5	6	7	8	Row label
Row $i = 1$	3	6	3	□ 0	3	0	2	4	
2	4	5	2	1	□ 0	5	11	2	(Column 5, −)
3	3	□ 0	13	15	0	16	2	5	(Column 2, −)
4	5	0	16	5	4	10	5	8	(Δ, +)
5	5	0	13	4	5	6	□ 0	2	
6	11	12	3	11	2	□ 0	10	13	
7	□ 0	9	0	4	10	3	4	0	
8	12	3	□ 0	4	4	4	2	8	
Column label		(Row 4, +)			(Row 3, +)				

Allocations are indicated by putting a "□" in the allocated cells. Only seven allocations are made. Column 8 has no allocation. So enter the labeling routine. Row 4, which has no allocation, is labeled with the label (Δ, +). Then column 2 can be labeled with (row 4, +). Then row 3 can be labeled with (column 2, −). Continuing this way the labels entered on the array are obtained. Since column 8 is not labeled, we have a nonbreakthrough. Enter the further reduction routine. The minimum entry in the present reduced matrix among labeled rows and unlabeled columns is $\delta = 1$. The next reduced matrix is the second reduced matrix, with the previous allocations and labels copied. The new admissible cells have 0 entries in the new reduced matrix. The previous labeling is continued from where it was left off. The new labels are entered in the array under the heading "labels when continued." Labeling routine ends in nonbreakthrough again. The minimum entry in the present reduced matrix among the cells in labeled rows and unlabeled columns is $\delta = 1$. The next reduced matrix is the third reduced matrix. The new admissible cells have zero entries in this new reduced matrix. If we continue the previous labeling, a breakthrough occurs. Go to the allocation change routine. Since the label on column 8 is (row 2, +), put an allocation in the cell (2, 8). The label on row 2 is (column 5, −). So delete the allocation in the cell (2, 5). The label on column 5 is (row 3, +). So add an allocation in the cell (3, 5). The label on row 3 is (column 2, −). So delete the allocation in the cell (3, 2). The label in column 2 is (row 4, +). So add an allocation in the cell (4, 2). The label on row 4 is (Δ, +). So the allocation change routine terminates, leaving all the other allocations unchanged.

Now all the columns have allocations. The new allocation defines the assignment, {(1, 4), (2, 8), (3, 5), (4, 2), (5, 7), (6, 6), (7, 1), (8, 3)}, which is an optimum assignment. The net reduction so far is 21, which can be verified to be the objective value of this assignment.

Second Reduced Matrix

Column j =	1 •	2	3	4	5	6	7	8	Previous row labels	Labels when continued
Row i = 1	3	7	3	□ 0	4	0	2	4		(Column 4, −)
2	3	5	1	0	□ 0	4	10	1	(Column 5, −)	(Column 5, −)
3	2	□ 0	12	14	0	15	1	4	(Column 2, −)	(Column 2, −)
4	4	0	15	4	4	9	4	7	(s, +)	(s, +)
5	5	1	13	4	6	6	□ 0	2		
6	11	13	3	11	3	□ 0	10	13		(Column 6, −)
7	□ 0	10	0	4	11	3	4	0		
8	12	4	□ 0	4	5	4	2	8		
Previous column labels		(Row 4, +)			(Row 3, +)					
Labels when continued		(Row 4, +)		(Row 2, +)	(Row 3, +)	(Row 1, +)				

Third Reduced Matrix with Previous Allocations

Column j =	1	2	3	4	5	6	7	8	Previous row labels	Labels when continued
Row i = 1	2	7	2	□ 0	4	0	1	3	(Column 4, −)	(Column 4, −)
2	2	5	0	0	□ 0	4	9	0	(Column 5, −)	(Column 5, −)
3	1	□ 0	11	14	0	15	0	3	(Column 2, −)	(Column 2, −)
4	3	0	14	4	4	9	3	6	(s, +)	(s, +)
5	5	2	13	5	7	7	□ 0	2		(Column 7, −)
6	10	13	2	11	3	□ 0	9	12	(Column 6, −)	(Column 6, −)
7	□ 0	11	0	5	12	4	4	0		
8	12	5	□ 0	5	6	5	2	8		(Column 3, −)
Previous column labels		(Row 4, +)		(Row 2, +)	(Row 3, +)	(Row 1, +)				
Labels when continued		(Row 4, +)	(Row 2, +)	(Row 2, +)	(Row 3, +)	(Row 1, +)	(Row 3, +)	(Row 2, +)		

Problem

12.27 Solve the assignment problem with the following cost matrix.

$$C = \begin{bmatrix} 10 & 11 & 10 & 5 & 6 & 4 & 3 & 6 \\ 5 & 26 & 14 & 18 & 15 & 10 & 10 & 16 \\ 6 & 22 & 18 & 17 & 15 & 8 & 8 & 12 \\ 2 & 14 & 16 & 16 & 24 & 25 & 12 & 7 \\ 4 & 15 & 19 & 10 & 8 & 14 & 11 & 8 \\ 10 & 22 & 22 & 15 & 28 & 24 & 12 & 30 \\ 8 & 18 & 21 & 18 & 18 & 18 & 14 & 25 \\ 5 & 14 & 21 & 17 & 26 & 9 & 10 & 31 \end{bmatrix}$$

12.4.2 The Primal-Dual Algorithm For The Balanced Transportation Problem

This algorithm is a generalization of the Hungarian method to solve transportation problems. It starts with a dual feasible solution and tries to find a primal solution closest to primal feasibility while satisfying the complementary slackness conditions with the present dual feasible solution. If the primal solution obtained is primal feasible, it must bc optimal and the algorithm terminates. Otherwise the dual feasible solution is altered so as to allow the primal solution to move closer to primal feasibility. The new primal solution is found, and the algorithm continued in the same manner. The moment a primal feasible solution is obtained the algorithm terminates.

This approach is known as the *primal-dual approach.* It can be adopted to solve the general linear programming problem and this leads to the *primal-dual algorithm* for solving LPs. However, for general LPs this algorithm seems to offer no particular advantages over the primal simplex algorithm. But for the class of transportation problems, the primal-dual approach leads to an efficient algorithm. There are several reasons for this. It is easy to obtain a dual feasible solution for the transportation problem, and it is also easy to make changes in the dual feasible solution. Given a dual feasible solution, the problem of finding the primal solution closest to primal feasibility while satisfying the complementary slackness conditions with the present dual feasible solution turns out to be a maximum flow problem on a simple bipartite network, which can be solved by the labeling algorithm very efficiently. We discuss the primal-dual algorithm for solving a balanced transportation problem here.

The balanced transportation problem discussed in Chapter 11 is:

$$\text{Minimize} \quad z(x) = \sum_{i=1}^{m} \sum_{j=1}^{n} c_{ij} x_{ij}$$

$$\text{Subject to} \quad \sum_{j=1}^{n} x_{ij} = a_i \qquad i = 1 \text{ to m}$$

$$\sum_{i=1}^{m} x_{ij} = b_j \qquad j = 1 \text{ to n}$$

$$x_{ij} \geqq 0 \qquad \text{for all } i, j$$

x_{ij} is the amount of the commodity shipped from warehouse i to market j. a_i is the amount of the commodity available at warehouse i, and b_j is the amount of the commodity required in market j. $\sum a_i = \sum b_j$.

This can be viewed as a minimum cost flow problem on a bipartite network connecting the warehouses with the markets. It is the problem of finding a minimum cost flow vector saturating all the arcs leading to the sink in the following network. Data on the arcs is, lower bound, capacity, unit cost coefficient, in that

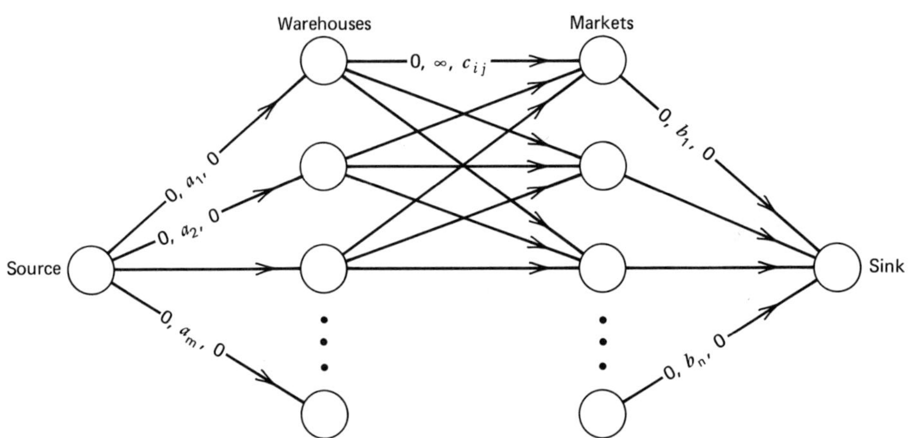

order. x_{ij} represents the amount of flow on the arc joining warehouse i with market j. By identifying the cell (i, j) in the transportation array with the arc joining warehouse i with market j in the above network, all the computations performed on the network can actually be carried out using the transportation arrays.

	Market j			Total flow out of warehouse	Total available
Warehouse i		x_{ij}		$\sum_j x_{ij}$	a_i
Total flow into market		$\sum_i x_{ij}$			
Total required in market		b_j			

Since it is much more convenient to use the transportation arrays, we will only discuss how to carry out the algorithm using them. The dual problem is

$$\begin{aligned} &\text{Maximize} \quad \textstyle\sum a_i u_i + \sum b_j v_j \\ &\text{Subject to} \quad u_i + v_j \leqq c_{ij} \qquad \text{for all } i, j \end{aligned}$$

Let $\bar{c}_{ij} = c_{ij} - u_i - v_j$, for $i = 1$ to m, $j = 1$ to n. The complementary slackness conditions for optimality are

$$x_{ij}\bar{c}_{ij} = 0 \qquad \text{for all } i, j$$

TO OBTAIN AN INITIAL-DUAL FEASIBLE SOLUTION

This can be done by selecting the dual solution to be

$$u_i = \text{minimum } \{c_{ij}: j = 1 \text{ to n}\}$$
$$v_j = \text{minimum } \{c_{ij} - u_i: i = 1 \text{ to m}\}$$

THE FLOW PROBLEM

If (u, v) is the present dual feasible solution, compute $\bar{c}_{ij} = c_{ij} - u_i - v_j$ for all i, j. By dual feasibility $\bar{c}_{ij} \geqq 0$ for all i, j. The cell (i, j) is said to be *admissible* in this stage if $\bar{c}_{ij} = 0$; otherwise it is *inadmissible*. Complementary slackness conditions require that the flow amounts should be zero in all the inadmissible cells. The flow problem in this stage is therefore to find a flow that maximizes the total amount reaching all the columns subject to the availabilities and the requirements, using only the admissible cells.

This flow problem can be solved by initiating the algorithm with the flow $x = 0$ or some other known feasible flow that uses only the admissible cells in this stage, and then using the labeling algorithm.

LABELING ROUTINE

Let $\tilde{x}$ be the present flow. Label the rows and columns of the array by the following rules:

(i) If i is such that $\sum_j \tilde{x}_{ij} < a_i$, some more material can be drawn out of row i. Label row i with the label $(\emptyset, +)$. Label all such rows similarly.

(ii) If row i is labeled and column j is not labeled yet, and (i, j) is an admissible cell at this stage, label column j with the label (row i, $+$).

(iii) If column j is labeled and row i is not yet labeled, and $\tilde{x}_{ij} > 0$, label row i with the label (column j, $-$).

Go through (ii) and (iii) as often as possible. If a column with unfulfilled requirements (i.e., a column j such that $\sum_i \tilde{x}_{ij} < b_j$) is labeled, there is a *breakthrough* and the labeling routine terminates. Go to the flow change routine. If none of the columns with unfulfilled requirements have been labeled and it is not possible to continue the labeling any further, the labeling routine terminates in a *nonbreakthrough.* Then go to the dual solution change routine.

FLOW CHANGE ROUTINE

A breakthrough implies that the present flow is not a maximum flow on the admissible network at this stage. To get the new flow, let column j_1 be the column, the labeling of which led to a breakthrough. Suppose the label on column j_1 is (row i_1, $+$). Add a "$+\theta$" to the present amount of flow in cell (i_1, j_1). Now look at the label on row i_1. Suppose it is (column j_2, $-$). Then add a "$-\theta$" to the present amount of flow in cell (i_1, j_2). Now look at the label on column j_2 and so on. Repeat adding a "$+\theta$" and a "$-\theta$" alternately to the flow amounts in the cells indicated by the labels until a row, say row i_k, with a label of $(\emptyset, +)$ is reached. Let

θ = Minimum $\{a_{i_k} - \sum_j \tilde{x}_{i_k, j}; \ b_{j_1} - \sum_i \tilde{x}_{i, j_1}; \ \tilde{x}_{ij}$ over all cells (i, j) with a $-\theta$ alteration$\}$. Substitute this value for θ and get the new flow, $\hat{x}$. If $\hat{x}$ fulfills the requirements in all the columns, it is an optimum feasible solution of the transportation problem and the algorithm is terminated. Otherwise erase the old labels on all the rows and the columns and return to the labeling routine once more.

DUAL SOLUTION CHANGE ROUTINE

If the labeling routine terminates in a nonbreakthrough, the present flow, $\tilde{x}$, is a maximum flow on the admissible network. To increase the flow value further, it is necessary to create additional admissible cells, particularly among the cells in labeled rows and unlabeled columns, because this will make it possible to label some more columns. Let $(\tilde{u}, \tilde{v})$ be the present dual solution. Let:

$$\delta = \text{Minimum } \{c_{ij} - \tilde{u}_i - \tilde{v}_j \text{: row } i \text{ is a labeled row, column } j \text{ is an unlabeled column}\}$$

δ must be positive, since otherwise additional columns could be labeled, contradicting the occurrence of a nonbreakthrough. Let the new dual feasible solution

Array I Summary of Position at Occurrence of Nonbreakthrough

	Block of labeled columns	Block of unlabeled columns	Supply position	Dual change
Block of labeled rows	$\hat{\bar{c}}_{ij} = \tilde{\bar{c}}_{ij}$ in this block, that is, the admissibility pattern remains unchanged in this block.	No admissible cells at present in this block; otherwise a column from here could be labeled contradicting occurrence of nonbreakthrough. $\delta = \text{Min } \{\tilde{\bar{c}}_{ij}\text{: } (i, j) \text{ in this block}\} > 0$. $\hat{\bar{c}}_{ij} = \tilde{\bar{c}}_{ij} - \delta$ in this block. Dual change creates new admissible cells here. Every admissible cell created here makes it possible to continue labeling by labeling the column in which it lies.	Material for shipment available from some rows in this block.	$\hat{u}_i = \tilde{u}_i + \delta$ in this block.
Block of unlabeled rows	Flows in all cells in this block are 0 at present; otherwise a row from here could be labeled, contradicting occurrence of nonbreakthrough. $\hat{\bar{c}}_{ij} = \tilde{\bar{c}}_{ij} + \delta$ in this block. All cells in this block become inadmissible in next stage.	$\hat{\bar{c}}_{ij} = \tilde{\bar{c}}_{ij}$ in this block, that is, the admissibility pattern remains unchanged in this block.	All available material used up from rows in this block in the present flow, that is, $\sum_j \tilde{x}_{ij} = a_i$ here. Otherwise a row from here could have been labeled.	$\hat{u}_i = \tilde{u}_i$ in this block.
Requirement position	Requirements in these columns are fulfilled, that is, $\sum_i \tilde{x}_{ij} = b_j$ for all j in this block. Otherwise there would have been a breakthrough.	Columns with unfulfilled requirements lie in this block.		
Dual change	$\hat{v}_j = \tilde{v}_j - \delta$ in this block.	$\hat{v}_j = \tilde{v}_j$ in this block.		

be $(\hat{u}, \hat{v})$, where:

$$\begin{aligned} \hat{u}_i &= \tilde{u}_i + \delta && \text{if row } i \text{ is labeled} \\ &= \tilde{u}_i && \text{if row } i \text{ is unlabeled} \\ \hat{v}_j &= \tilde{v}_j - \delta && \text{if column } j \text{ is labeled} \\ &= \tilde{v}_j && \text{if column } j \text{ is unlabeled} \end{aligned}$$

Verify that $(\hat{u}, \hat{v})$ is again dual feasible. Make cell (i, j) an admissible cell if $c_{ij} - \hat{u}_i - \hat{v}_j = 0$. Some of the old admissible cells loose their admissibility, and new admissible cells are created among labeled rows and unlabeled columns. Verify that any cell that carries a positive flow amount in the present flow $\tilde{x}$ remains admissible.

Let $\tilde{u}, \tilde{v}, \tilde{\bar{c}}$ refer to the dual solution and the dual slack vector at the occurrence of a nonbreakthrough before the dual variable change routine is applied. Let $\hat{u}, \hat{v}, \hat{\bar{c}}$ be the dual solution and the dual slack vector just after the dual variable change. Let $\tilde{x} = (\tilde{x}_{ij})$ be the present flow vector. The situation in the array at the occurrence of a nonbreakthrough is summarized in Array I.

Using the new admissible cells, continue the labeling routine from where it was left off at the occurrence of the nonbreakthrough. If breakthrough occurs, go to the flow change routine. If nonbreakthrough occurs again, go back to the dual solution change routine.

The algorithm terminates when a flow that fulfills the requirements in all the columns is obtained. That flow is an optimum feasible solution of the transportation problem.

Numerical Example

$j =$	1	2	3	4	5	6	a_i	Initial Dual Solution u_i
$i = 1$	$c_{11} = 5$	3	7	3	8	5	4	3
2	5	6	12	5	7	11	3	5
3	2	8	3	4	8	2	3	2
4	9	6	10	5	10	9	7	5
b_j	3	3	6	2	1	2		
Initial Dual Solution v_j	0	0	1	0	2	0		

$\bar{c}_{ij} = c_{ij} - u_i - v_j$ is entered in the upper left corner of the following array. All cells in which $\bar{c}_{ij} = 0$ are admissible cells and they are marked with a little square "□" in the middle.

An initial flow among the admissible cells is obtained by inspection. The flow amounts are entered in the little squares "□" in the first array.

First Array

$j =$	1	2	3	4	5	6	a_i	u_i	Row label
$i = 1$	$2 = \bar{c}_{11}$	0 [3]	3	0 [1]	3	2	4	3	(Column 4, −)
2	0 □ $+ \theta$	1	6	0 □	0 [1]	6	3	5	(s, +)
3	0 [3] $- \theta$	6	0 □ $+ \theta$	2	4	0 □	3	2	(Column 1, −)
4	4	1	4	0 [1]	3	4	7	5	(s, +)
b_j	3	3	6	2	1	2			
v_j	0	0	1	0	2	0			
Column label	(Row 2, +)		(Row 3, +)	(Row 2, +)	(Row 2, +)				

Rows 2, 4 have additional material to be shipped. So they are labeled with the label (s, +). Columns 3, 6 have unfulfilled requirements. The labeling routine is applied and it ends in a breakthrough, with column 3 labeled. So go to the flow change routine. Entries of $+\theta$ and $-\theta$ are made in admissible cells, as indicated by the labels. The value of θ should be minimum $\{6 =$ unfulfilled requirements in column 3, 2 = additional material available in row 2, 3 = flow amount in cell (3, 1) with a $-\theta$ entry$\} = 2$. The new flow vector is recorded in the second array.

Second Array

$j =$	1	2	3	4	5	6	a_i	u_i	Row label
$i = 1$	2	0 [3]	3	0 [1]	3	2	4	3	(Column 4, −)
2	0 [2]	1	6	0 □	0 [1]	6	3	5	
3	0 [1]	6	0 [2]	2	4	0 □	3	2	
4	4	1	4	0 [1]	3	4	7	5	(s, +)
b_j	3	3	6	2	1	2			
v_j	0	0	1	0	2	0			
Column label		(Row 1, +)		(Row 4, +)					

After erasing old labels, the labeling routine is applied afresh. It ends in a nonbreakthrough. So go to the dual solution change routine. δ = Minimum $\{\bar{c}_{ij}$: row i is a labeled row and column j is an unlabeled column$\}$ = 2. So the new dual feasible solution in $(\hat{u}, \hat{v})$ where $\hat{u}$ = (5, 5, 2, 7) and $\hat{v}$ = (0, −2, 1, −2, 2, 0). $c_{ij} - \hat{u}_i - \hat{v}_j$ is computed and entered in the top left-hand side of each cell in the third array. All cells in which $c_{ij} - \hat{u}_i - \hat{v}_j = 0$ are the new admissible cells.

Third Array

$j =$	1	2	3	4	5	6	a_i	$\hat{u}_i$
$i = 1$	0	0 [3]	1	0 [1]	1	0 [ ]	4	5
2	0 [2]	3	6	2	0 [1]	6	3	5
3	0 [1]	8	0 [2]	4	4	0 [ ]	3	2
4	2	1	2	0 [1]	1	2	7	7
b_j	3	3	6	2	1	2		
$\hat{v}_j$	0	−2	1	−2	2	0		

Notice that all the cells with positive flow amounts in them remain admissible. Using the new set of admissible cells, go to the labeling routine again. Continue in the same manner until an optimum feasible solution of the original transportation problem is obtained.

GENERAL COMMENTS

The primal dual algorithm for the transportation problem turns out to be an efficient algorithm for solving problems by hand. Recent experience indicates that in solving large problems using a digital computer, the primal algorithm discussed in Chapter 11 using the predecessor indices, etc. is superior to the primal-dual algorithm discussed here. However the primal-dual algorithm for the transportation problem is useful in doing sensitivity analysis when the a_i's, and b_j's change.

12.5 SINGLE COMMODITY MINIMUM COST FLOW PROBLEMS

Let G = $(\mathcal{N}, \mathcal{A})$ be a directed network with a source and sink and lower bound vector, ℓ, and capacity vector, k. The cost of transporting the commodity along $(i, j) \in \mathcal{A}$ from i to j is c_{ij} dollars/unit, and c is the vector of these cost coefficients (c_{ij}). The problem is to find a minimum cost feasible flow vector of specified value V.

$$\text{Minimize} \quad \sum_{(i,j)\in\mathcal{A}} c_{ij} f_{ij}$$

$$\text{Subject to} \quad \sum_{\substack{j \text{ such} \\ \text{that } (i,j)\in\mathcal{A}}} f_{ij} - \sum_{\substack{j \text{ such} \\ \text{that } (j,i)\in\mathcal{A}}} f_{ji} = \begin{cases} \text{V, if } i \text{ is source} \\ 0, \text{ if } i \text{ is an intermediate point} \\ -\text{V, if } i \text{ is sink.} \end{cases} \quad (12.10)$$

$$\ell_{ij} \leqq f_{ij} \leqq k_{ij} \qquad \text{for all } (i, j) \in \mathcal{A}$$

In the dual problem there is a dual variable associated with the flow conservation equation corresponding to each point in G. Let π_i be the dual variable associated with point $i \in \mathcal{N}$. The π_i's are called *node variables* or *node prices* or *node potentials.* The complementary slackness conditions for the optimality of a feasible flow vector f are the existence of a dual vector π satisfying the following (see section 4.5.12): For every $(i, j) \in \mathcal{A}$, if

$$
\begin{array}{ll}
\pi_j - \pi_i > c_{ij} & \text{then } f_{ij} = k_{ij} \\
\pi_j - \pi_i < c_{ij} & \text{then } f_{ij} = \ell_{ij} \\
\pi_j - \pi_i = c_{ij} & \text{then } \ell_{ij} \leqq f_{ij} \leqq k_{ij}
\end{array}
$$

These conditions can be illustrated on a diagram known as the *complementary slackness diagram for arc* (i, j). In Figure 12.28 the value of the flow amount on this arc, f_{ij}, is plotted on the horizontal axis. The value of $\pi_j - \pi_i$ from the current dual vector is plotted on the vertical axis.

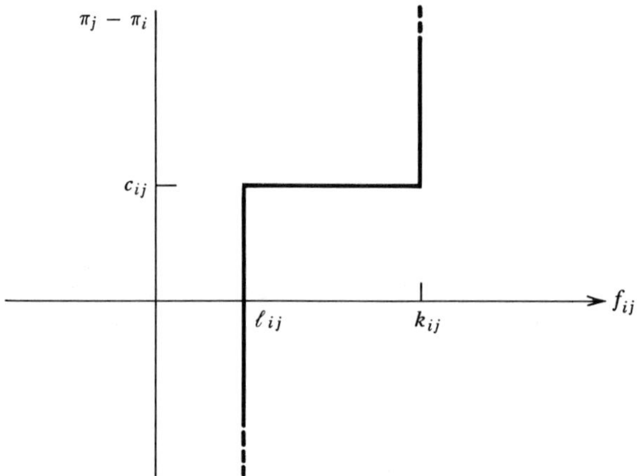

Figure 12.28 Complementary slackness diagram for arc (i,j).

The complementary slackness conditions are satisfied if the point $(f_{ij}, \pi_j - \pi_i)$ plotted on the above coordinate plane lies on the thick curve.

The out-of-kilter algorithm deals with pairs of primal feasible and dual feasible solutions and tries to change them one at a time until the complementary slackness conditions are satisfied.

12.5.1 The Out-of-Kilter Algorithm

Obtain a feasible flow vector of specified value V. See section 12.3. All subsequent flow vectors obtained during the algorithm will have value V.

Let f be the present feasible flow vector. Select a vector of node prices π, so that as many arcs as possible satisfy the complementary slackness conditions with f. If this is not possible, $\pi = 0$ is a satisfactory initial node price vector.

KILTER-STATUS OF ARCS

The status of the arc $(i, j) \in \mathscr{A}$ is defined as below.

Conditions			Status
$\pi_j - \pi_i < c_{ij}$	and	$f_{ij} = \ell_{ij}$	α-arc
$\pi_j - \pi_i = c_{ij}$	and	$\ell_{ij} \leqq f_{ij} \leqq k_{ij}$	β-arc
$\pi_j - \pi_i > c_{ij}$	and	$f_{ij} = k_{ij}$	γ-arc
$\pi_j - \pi_i < c_{ij}$	and	$f_{ij} > \ell_{ij}$	a-arc
$\pi_j - \pi_i > c_{ij}$	and	$f_{ij} < k_{ij}$	b-arc

The α-, β-, and γ-arcs are said to be *in kilter*. The a- and b-arcs are said to be *out of kilter*. If all the arcs in the network are in kilter, the present flow vector must be an optimum feasible flow vector and the algorithm terminates. In each step the algorithm either changes the flow vector while keeping the node price vector unchanged or changes the node price vector while keeping the flow vector unchanged, and tries to bring the out-of-kilter arcs into kilter without losing the in kilter status of any arc. The kind of changes required to bring an out-of-kilter arc into kilter are summarized below. See also Figure 12.29.

Status of arcs	Type of change	Required change
a-arc	Flow change	Decrease the flow amount, to the lower bound if possible.
a-arc	Node price change	Increase the value of $\pi_j - \pi_i$, up to c_{ij} if possible.
b-arc	Flow change	Increase the flow amount up to the capacity if possible.
b-arc	Node price change	Decrease the value of $\pi_j - \pi_i$ to c_{ij} if possible.

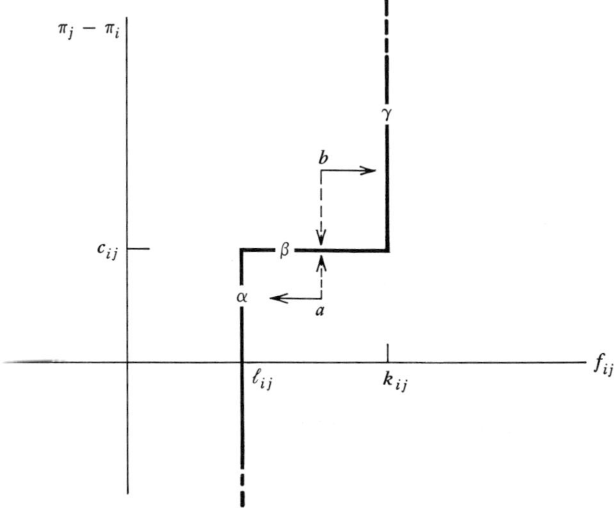

Figure 12.29 Kilter status and types of desired changes.

Once an arc comes into kilter, it remains in killer throughout the remaining portion of the algorithm. *In each stage, the algorithm selects an out-of-kilter arc and tries to bring it closer to the "in-kilter" status.*

FLOW CHANGES

Consider an out-of-kilter arc (i, j). If it is an a-arc, the flow amount on it has to be decreased. To decrease the flow amount on arc (i, j), the same amount of material must be sent from i to j by a different route. Hence this flow change operation is called a *flow rerouting*. The point i is the *rerouting source* and point j is the *rerouting sink* for this flow rerouting.

If (i, j) is a b-arc, the flow amount on this arc has to be increased. Any increase in the flow amount on this arc has to be rerouted back from j to i to keep feasibility. Hence in this case j is the rerouting source and i is the rerouting sink. (See Figure 12.30). The flow changes on arcs should be restricted to the following to make sure that the kilter status of every arc either improves or stays the same.

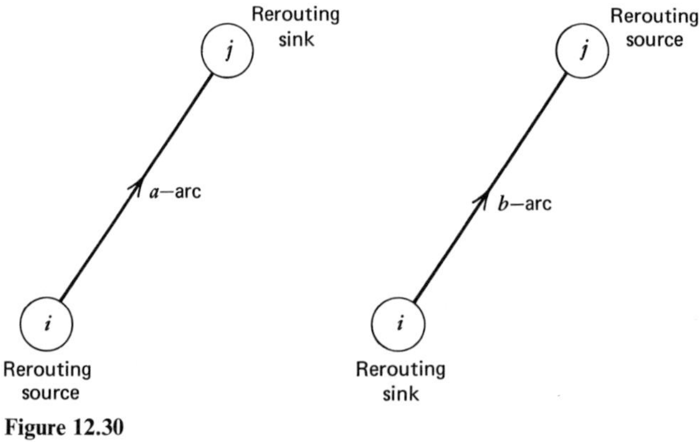

Figure 12.30

Arc status	Type of flow change
β-arc	Admissible for flow increase or decrease subject to the lower bound, capacity restrictions.
a-arc	Admissible for flow decrease only.
b-arc	Admissible for flow increase only.
α- and γ-arcs	Not admissible for flow change.

Rerouting is done by finding an admissible FAP from the rerouting source to the rerouting sink.

LABELING ROUTINE

Let f be the present feasible flow vector and let π be the present node price vector.

1. Label the rerouting source with the label $(\mathscr{s}, +)$.
2. If x is a labeled node, and y is an unlabeled node at this stage, $(x, y) \in \mathscr{A}$, $f_{xy} < k_{xy}$, and (x, y) is either a b-arc or β-arc, label y with the label $(x, +)$.

3. If x is a labeled node and y is an unlabeled node at this stage, $(\mathrm{y}, \mathrm{x}) \in \mathscr{A}$, $f_{\mathrm{yx}} > \ell_{\mathrm{yx}}$ and (y, x) is an a-arc or β-arc, label y with the label (x, $-$).

Repeat 2. and 3. as often as possible. If the rerouting sink gets labeled, go to the *flow rerouting routine.* If rerouting sink is not labeled and it is impossible to continue the labeling any further, go to the *node price change routine.*

FLOW REROUTING ROUTINE

Identify all the forward and reverse arcs on the FAP from the rerouting source to the rerouting sink using the labels. Let

$$\begin{aligned} \varepsilon_1 &= \text{minimum } \{k_{\mathrm{xy}} - f_{\mathrm{xy}}\colon (\mathrm{x}, \mathrm{y}) \text{ a forward arc on the FAP}\} \\ &= \infty \text{ if there are no forward arcs on the FAP} \end{aligned}$$

$$\begin{aligned} \varepsilon_2 &= \text{minimum } \{f_{\mathrm{xy}} - \ell_{\mathrm{xy}}\colon (\mathrm{x}, \mathrm{y}) \text{ a reverse arc on the FAP}\} \\ &= \infty \text{ if there are no reverse arcs on the FAP} \end{aligned}$$

If i_1 is the rerouting source, j_1 is the rerouting sink and (i_1, j_1) is the out-of-kilter arc selected for change of status, it must be an a-arc. Let:

$$\varepsilon = \text{minimum } \{\varepsilon_1, \varepsilon_2, f_{i_1, j_1} - \ell_{i_1, j_1}\}$$

Increase flow amounts on all the forward arcs of the FAP by ε. Decrease the flow amounts on all the reverse arcs of the FAP and on the arc (i_1, j_1) by ε. Leave flow amounts on all the other arcs unchanged.

On the other hand, if i_1 is the rerouting source, j_1 is the rerouting sink and (j_1, i_1) is the out-of-kilter arc selected for change of status, it must be a b-arc. Let:

$$\varepsilon = \text{minimum } \{\varepsilon_1, \varepsilon_2, k_{j_1, i_1} - f_{j_1, i_1}\}$$

Increase the flow amounts on all the forward arcs of the FAP and on the arc (j_1, i_1) by ε. Decrease the flow amounts on all the reverse arcs of the FAP by ε. Leave all the other flow amounts unchanged.

Let $\hat{f}$ be the new feasible flow vector. Erase all the labels on all the points. Determine the new kilter status of each arc using $\hat{f}$, π. If all the arcs are now in kilter, $\hat{f}$ is an optimum feasible flow vector and the algorithm terminates. Otherwise, select an out-of-kilter arc (the arc in the previous step can again be selected if it is still out of kilter) and go back to the labeling routine again.

NODE PRICE CHANGE ROUTINE

Let i_1, j_1 be the rerouting source and sink, respectively. Let $\mathbf{X}$ be the set of labeled nodes at this stage and $\bar{\mathbf{X}}$ the set of unlabeled nodes. Clearly, $i_1 \in \mathbf{X}$, $j_1 \in \bar{\mathbf{X}}$. Let:

$$\mathbf{A}^1 = \{(\mathrm{x}, \mathrm{y})\colon (\mathrm{x}, \mathrm{y}) \in \mathscr{A}, \mathrm{x} \in \mathbf{X}, \mathrm{y} \in \bar{\mathbf{X}} \text{ and } (\mathrm{x}, \mathrm{y}) \text{ is } a\text{-arc or } \alpha\text{-arc}\}$$

$$\mathbf{A}^2 = \{(\mathrm{x}, \mathrm{y})\colon (\mathrm{x}, \mathrm{y}) \in \mathscr{A}, \mathrm{x} \in \bar{\mathbf{X}}, \mathrm{y} \in \mathbf{X} \text{ and } (\mathrm{x}, \mathrm{y}) \text{ is } b\text{-arc or } \gamma\text{-arc}\}$$

$$\delta = \text{minimum } \{|\pi_{\mathrm{y}} - \pi_{\mathrm{x}} - c_{\mathrm{xy}}|\colon (\mathrm{x}, \mathrm{y}) \in \mathbf{A}^1 \cup \mathbf{A}^2\}$$

Define the new vector of node prices to be $\hat{\pi}$, where

$$\begin{aligned} \hat{\pi}_{\mathrm{x}} &= \pi_{\mathrm{x}} && \text{if} \quad \mathrm{x} \in \mathbf{X} \\ &= \pi_{\mathrm{x}} + \delta && \text{if} \quad \mathrm{x} \in \bar{\mathbf{X}} \end{aligned}$$

Verify that with respect to f, $\hat{\pi}$, the kilter status of every arc in the network is either the same as it was before, or better. All the arcs that were in kilter before stay in kilter.

Determine the new kilter status of each arc in the network with respect to f, $\hat{\pi}$. If all the arcs in the network are now in kilter, f is an optimum feasible flow vector, and the algorithm terminates. Otherwise erase all the old labels, select an out-of-kilter arc and go back to the labeling routine again.

The algorithm terminates when all the arcs in the network become in kilter. The final flow vector obtained is optimal.

Numerical Example

On the network in Figure 12.31, find a minimum cost feasible flow vector of value 12 units from point 1 to point 6. A feasible flow vector has been found and entered on the network. The serial number of each point is entered inside the point. An initial vector of node prices has been chosen and these node prices are entered by the side of the points. Data on the arcs is recorded in the order indicated. Select the a-arc (1, 2) for bringing into kilter. The rerouting source is 1 and the rerouting sink is 2. Labeling routine is applied, and the labels are entered by the side of the points. The rerouting sink, 2, is labeled. The FAP for rerouting is

Forward arcs: (1, 3), (3, 5)

Reverse arcs: (2, 5)

$$\varepsilon_1 = \text{minimum } \{(10 - 7, 4 - 3)\} = 1, \varepsilon_2 = \text{minimum } \{5\} = 5,$$

$$\varepsilon = \text{minimum } \{1, 5, 5\} = 1$$

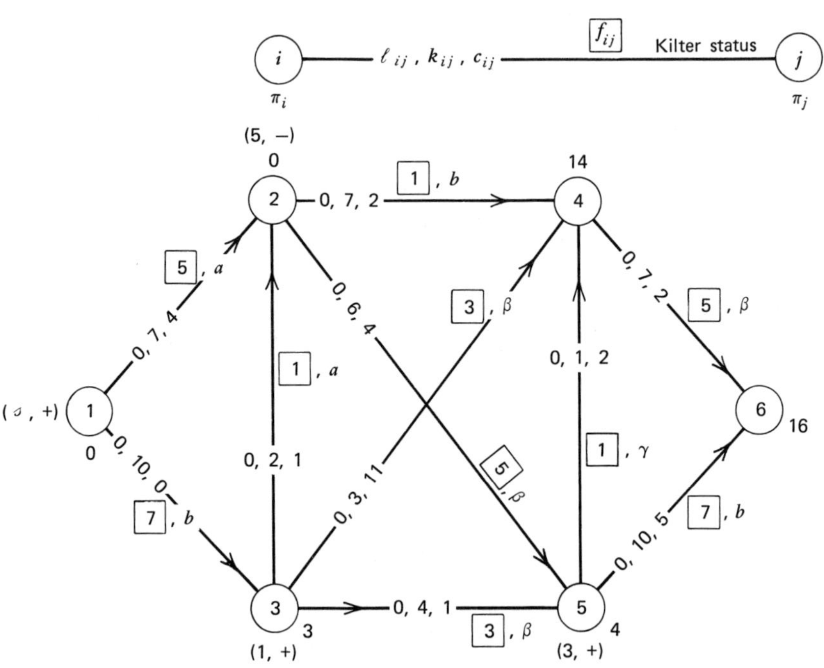

Figure 12.31

The new flow vector is entered in Figure 12.32. The new kilter status of each arc is also entered. Arc (1, 2) is still an a-arc. Select it again for bringing into kilter. Again the rerouting source is 1, and the rerouting sink is 2. All old labels are erased and the labeling routine is entered

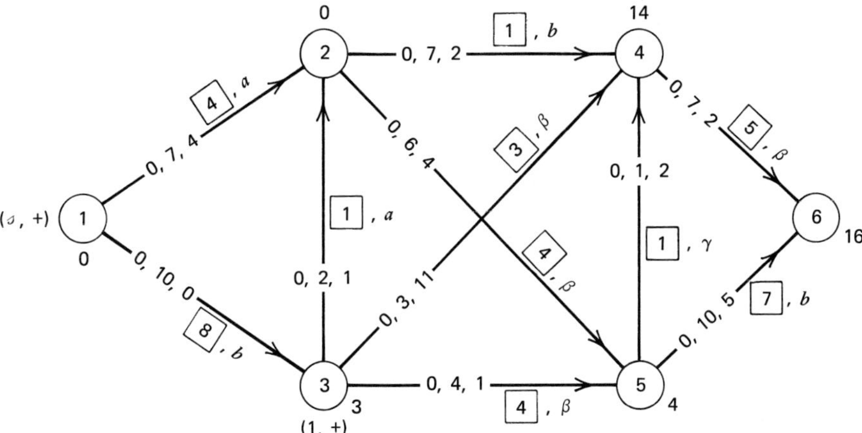

Figure 12.32

again. After labeling points 1, 3, labeling cannot be continued any further. The set of labeled nodes is $\mathbf{X} = \{1, 3\}$. $\bar{\mathbf{X}} = \{2, 4, 5, 6\}$. $\mathbf{A}^1 = \{(1, 2), (3, 2)\}$, and $\mathbf{A}^2 = \emptyset$. So $\delta =$ minimum $\{4, 4\} = 4$. The node prices are changed. The new node prices and the new kilter status of each arc are entered in Figure 12.33. Arc (1, 2) has now become a β-arc, in kilter. But arcs (1, 3),

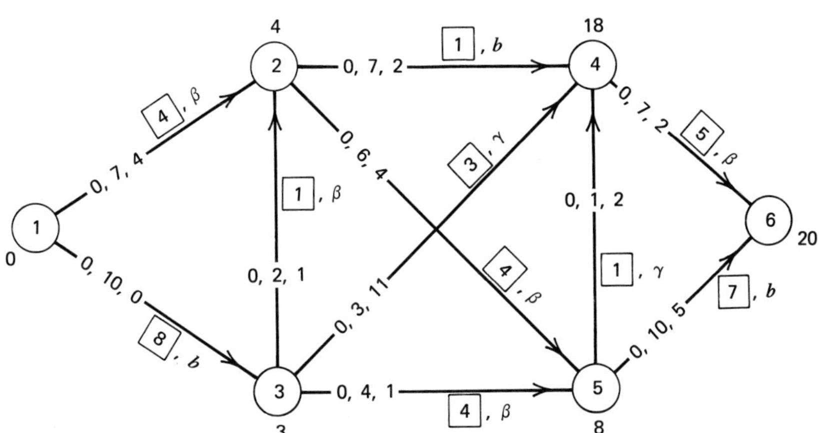

Figure 12.33

(2, 4), and (5, 6) are still out of kilter. One of these arcs can be selected for bringing into kilter and the algorithm continued until every arc becomes in kilter.

An optimum feasible flow vector for this problem is

$$\begin{aligned} f &= (f_{12}, f_{13}, f_{32}, f_{34}, f_{35}, f_{25}, f_{24}, f_{54}, f_{46}, f_{56}) \\ &= (6, 6, 2, 0, 4, 1, 7, 0, 7, 5) \end{aligned}$$

and an optimum node price vector is

$$\pi = (\pi_1, \pi_2, \pi_3, \pi_4, \pi_5, \pi_6) = (11, 15, 11, 21, 19, 24)$$

12.5.2 Parametric Analysis

In a minimum cost flow problem let 1 be the source node and n, the sink node. Let $\tilde{f}, \tilde{\pi}$ be the optimal feasible flow vector and the node price vector respectively for $v = \mathrm{V}$, the specified value. Suppose you have to find the minimum cost flow vectors for other possible values of v. Let $g(v)$ be the minimum objective value as a function of v.

Start with the value $v = \mathrm{V}$. $\tilde{f}$, $\tilde{\pi}$ together satisfy the complementary slackness conditions for all the arcs in the network. Find the kilter status of each arc. Every arc must be either an α, or β, or γ-arc. An arc is defined to be admissible for flow change only if it is a β-arc.

On this admissible network use the labeling routine and try to increase the flow value with 1 as *flow changing source* and n as *flow changing sink*. The labeling routine would involve labeling nodes by the following rules. Let f be the present flow vector (it is $\tilde{f}$ to start with).

1. Label the flow changing source by the label $(\mathscr{s}, +)$.
2. If i is a labeled node and j an unlabeled node at this stage, $(i, j) \in \mathscr{A}$, (i, j) is a β-arc, and $f_{ij} < k_{ij}$, label j with $(i, +)$.
3. If i is a labeled node and j an unlabeled node at this stage, $(j, i) \in \mathscr{A}$, (j, i) is a β-arc and $f_{ij} > \ell_{ij}$, label j with $(i, -)$.

Continue applying (2) and (3) as often as possible. If the flow changing sink gets labeled find out the admissible FAP from 1 to n. Let

$$\begin{aligned} \varepsilon_1 &= \text{minimum } \{k_{ij} - f_{ij}\colon (i, j) \text{ a forward arc on the FAP}\} \\ \varepsilon_2 &= \text{minimum } \{f_{ij} - \ell_{ij}\colon (i, j) \text{ a reverse arc on the FAP}\} \\ \varepsilon &= \text{minimum } \{\varepsilon_1, \varepsilon_2\}. \end{aligned}$$

For all $0 \leqq \theta \leqq \varepsilon$, define the flow vector $f(\theta)$ by

$$\begin{aligned} f_{ij}(\theta) &= f_{ij} + \theta && \text{if } (i, j) \text{ is a forward arc on the FAP} \\ &= f_{ij} - \theta && \text{if } (i, j) \text{ is a reverse arc on the FAP} \\ &= f_{ij} && \text{if } (i, j) \text{ is not on the FAP} \end{aligned}$$

Then $f(\theta)$ is an optimum feasible flow vector for $v = \mathrm{V} + \theta$ and for $0 \leqq \theta \leqq \varepsilon$. $f(\theta)$, and the present node price vector π, satisfy the complementary slackness conditions on all the arcs. In this range $\mathrm{V} \leqq v \leqq \mathrm{V} + \varepsilon$, and $g(v)$ is a linear function with slope $\pi_n - \pi_1$. To find optimum feasible flow vectors for values $v > \mathrm{V} + \varepsilon$, make $f(\varepsilon)$ the present flow vector, leave π as the present node price vector, find out the present kilter status of each arc in the network, redefine admissible arcs as the new β-arcs, erase all old labels on the nodes, and go back to the labeling routine. Repeat in the same manner.

If the labeling routine ends in a nonbreakthrough, let $\mathbf{X}$ be the set of labeled nodes and $\bar{\mathbf{X}}$ the set of unlabeled nodes. Let

$$\mathbf{A}^1 = \{(\mathrm{x}, \mathrm{y})\colon (\mathrm{x}, \mathrm{y}) \in \mathscr{A}, \mathrm{x} \in \mathbf{X}, \mathrm{y} \in \bar{\mathbf{X}} \text{ and } (\mathrm{x}, \mathrm{y}) \text{ an } \alpha\text{-arc}\}$$

$$\mathbf{A}^2 = \{(\mathrm{x}, \mathrm{y})\colon (\mathrm{x}, \mathrm{y}) \in \mathscr{A}, \mathrm{x} \in \bar{\mathbf{X}}, \mathrm{y} \in \mathbf{X} \text{ and } (\mathrm{x}, \mathrm{y}) \text{ a } \gamma\text{-arc}\}$$

If $\mathbf{A}^1 \cup \mathbf{A}^2 = \varnothing$, the present flow value is the maximum flow value in this network, and it is impossible to make the flow value any higher. If $\mathbf{A}^1 \cup \mathbf{A}^2 \neq \varnothing$, let

$$\delta = \text{minimum } \{|\pi_{\mathrm{y}} - \pi_{\mathrm{x}} - c_{\mathrm{xy}}|\colon (\mathrm{x}, \mathrm{y}) \in \mathbf{A}^1 \cup \mathbf{A}^2\}$$

Define the new node price vector to be

$$\hat{\pi}_x = \pi_x \qquad \text{if } x \in \mathbf{X}$$
$$= \pi_x + \delta \qquad \text{if } x \in \bar{\mathbf{X}}$$

Leaving the present flow vector unchanged, find the new kilter status of each arc in the network. Redefine new admissible arcs to be the new β-arcs. Erase all the old labels and go back to the labeling routine.

Continue in this manner until the maximum flow value is reached.

To find the optimum flow vectors for $v < V$, let $\tilde{f}$, $\tilde{\pi}$ be again the optimum flow vector and node price vector for $v = V$, and let n be the *flow changing source* and 1 be the *flow changing sink*. Go to the labeling routine and repeat the same process. Here the net flow value from the original source to the original sink keeps decreasing. This routine again terminates when the minimum possible value of v in the network is reached.

As v varies, the optimum objective values $g(v)$ varies in a piecewise linear manner. For all values of v between two node price changes in the above routine, $g(v)$ is linear with slope $\pi_n - \pi_1$.

12.5.3 Other Sensitivity Analysis

If changes take place in the value of the cost coefficients c_{ij}, redefine the kilter status of each arc with respect to the present flow vector and the present node price vector, using the new values of c_{ij}. If some arcs become out of kilter, apply the algorithm until they come into kilter.

If some of the capacities k_{ij} decrease and the present flow vector violates the new capacities, decrease the flow amounts on the arcs on which the capacity constraints are violated by using the flow rerouting routine. Changes in the lower bounds can be handled in a similar manner.

GENERAL COMMENTS

The out-of-kilter algorithm is a very efficient algorithm for solving minimum cost flow problems. It is also very convenient for doing sensitivity analysis on the problem.

The minimum cost flow problem (12.10) can also be solved by the primal simplex algorithm using the upper bounding routine (Chapter 10). A working basis for (12.10) corresponds to a spanning tree in the network. Changing the status of a nonbasic variable becomes an analysis on an out-of-tree arc. The operation of a nonbasic variable replacing a basic variable from the working basis corresponds to an out-of-tree arc replacing an in-tree arc in its cycle.

An initial spanning tree can be selected, and using some artificial arcs, if necessary, an artificially feasible flow vector can be constructed. The artificial flows enter the objective function with very large positive cost coefficients. Starting with such a solution, the primal algorithm can be applied to minimize the combined objective function. Recent computational experience using tree labeling methods seems to indicate that the primal approach is slightly superior to the out-of-kilter approach on large problems.

Problems

12.28 Find a minimum cost flow vector of value $V = 5$ in the network in Figure 12.34. Draw the curve depicting the optimum objective value as a function of v.

12.29 Find a maximum flow vector of minimum cost on the network in Figure 12.35. Data is ⓘ —— $\ell_{ij}, k_{ij}, c_{ij}$ ⟶ ⓙ.

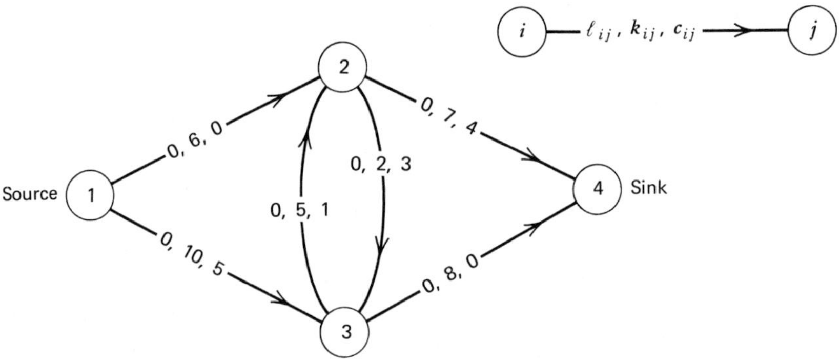

Figure 12.34

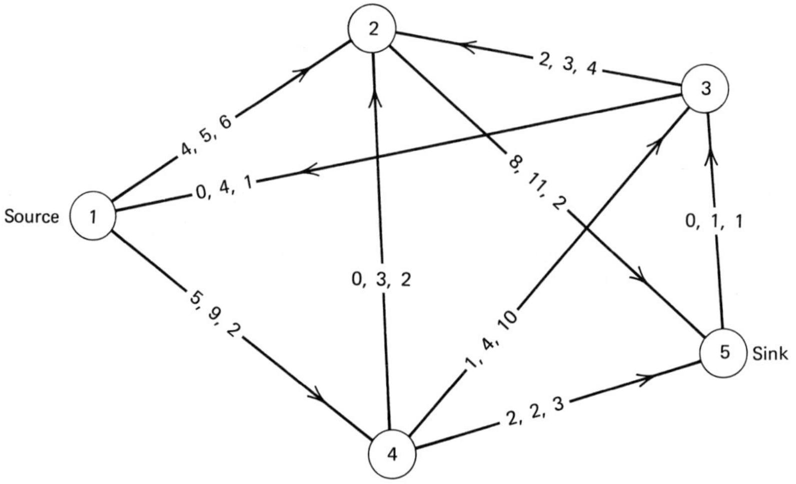

Figure 12.35

12.6 SHORTEST ROUTE PROBLEM

12.6.1 Shortest Chains from an Origin to all the Other Points

Let G = $(\mathcal{N}, \mathcal{A})$ be a directed network. Let node 1 be *the source* or *the origin*. c_{ij} is the *length* of the arc (i, j). For the moment assume that $c_{ij} \geqq 0$ for all $(i, j) \in \mathcal{A}$. Suppose you need to find the shortest chains from the origin to all the other points in the network.

Definitions A spanning tree, **T**, in G, is said to be an *out-tree with* 1 *as the root* if all the arcs in **T** are directed away from 1, that is, if the unique path in **T** from 1 to any node $j \neq 1$, is a chain from 1 to j.

For an example, see Figure 12.36. Hence **T** contains exactly one arc incident into j for each node $j \neq 1$. If (L_j, j) is the unique arc in **T** incident into node j, define the label on j to be L_j. Given the out tree **T**, the labels L_j on nodes are uniquely defined. Also given the labels on all the nodes, the set of arcs in the out tree is $\{(L_j, j) : j \in \mathcal{N}, j \neq 1, L_j \text{ is the label on } j\}$. Hence the out tree itself can be stored by storing the labels on the nodes obtained in this manner.

For each $j \neq 1$, suppose we are given a chain from 1 to j. When all these chains are put together, we obviously get an out tree with 1 as the root. The chains can

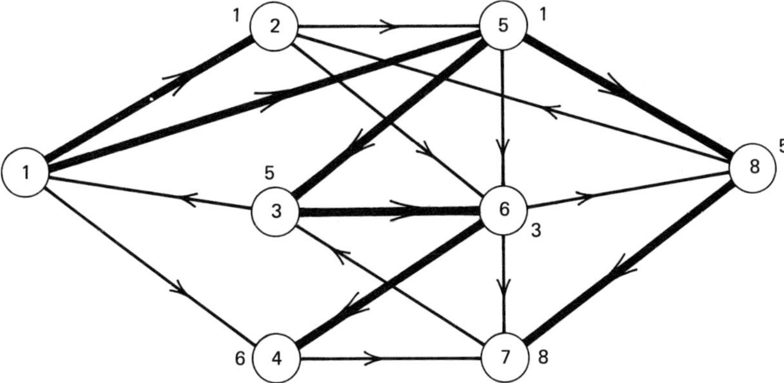

Figure 12.36 Out tree containing chains out of 1. Labels are recorded by the sides of the nodes.

be stored by storing the labels on the nodes generated by this out tree. If the labels on all the points (other than 1) are given, tracing the chain from the origin to any point j is done by the following:

> If the label on j is p, then (p, j) is the last arc on the current chain from the origin to j. Now look at the label on p to get the other arcs in this chain. Continue until the origin is reached.

Thus, out trees and the labels describing them provide a convenient method for storing the chains generated by the shortest chain algorithm from the origin to all the other nodes.

Let **T** be an out tree with 1 as the root. Let L_j be the label on j for $j \neq 1$, corresponding to **T**. The *set of immediate successors* of a node j is $\{i : j \text{ is the label on } i\}$. The *set of descendants* of a node j is the empty set if the set of immediate successors of j is empty. Otherwise the set of descendants of j is defined recursively to be the union of the set of immediate successors of node j and the sets of descendants of each immediate successor of j.

Example

Let **T** be the thick out tree in Figure 12.36. The set of immediate successors of 5 is $\{3, 8\}$. The set of descendants of 5 is $\{3, 8, 6, 4, 7\}$.

Clearly, the methods discussed in section 11.4 can be used to generate the set of descendants of any given point efficiently.

We discuss two algorithms for finding the shortest chains from 1 to all other nodes. The *tree changing algorithm* starts with an initial out tree with 1 as the root, *changes it by one arc in each step*, until an out tree containing all the shortest chains is obtained. The *tree building algorithm* builds the out tree containing all shortest chains by obtaining one new arc in it per step.

12.6.2 The Tree Changing Algorithm

If there are several arcs joining a pair of points i and j in G, obviously only the shortest among these will be used on the shortest routes. Call this the arc (i, j) and eliminate all other arcs from i to j from further consideration.

To start the algorithm, if there is no arc from 1 to i, introduce an artificial arc $(1, i)$ and make its length equal to ∞, or a very large positive number. Let the augmented network be called $G^1 = (\mathcal{N}, \mathcal{A}^1)$.

The out tree will be stored by labeling the points as in section 12.6.1. On each point, in addition to the label, also enter the total length of the current chain from the origin to the point, which is known as the *current distance* to the point. The ordered pair (the label, the distance) is the data on the point.

Let the data on the origin be (1, 0). Initially, the data on i are $(1, c_{1i})$ for all $i \neq 1$.

General Step At the end of the rth step, suppose the data on point i are (x_i, π_i^r), $i \in \mathcal{N}$. Examine the arcs in the network and find an arc $(i, j) \in \mathcal{A}$ satisfying

$$\pi_j^r > \pi_i^r + c_{ij}$$

If such an arc is found, change the data on j from the present (x_j, π_j^r) to $(i, \pi_i^r + c_{ij})$. Find the set of all descendants of j. If t is a descendant of j, leave its label unchanged, but change its current distance, from π_t^r to $\pi_t^r + \pi_i^r + c_{ij} - \pi_j^r$. Leave the rest of the data unchanged and go to the next step. If no such arc exists, that is, if $\pi_j^r \leqq \pi_i^r + c_{ij}$ for all $(i, j) \in \mathcal{A}$, then terminate the algorithm.

Note The change made in this step is exactly the same as deleting the arc (x_j, j) from the present out tree and introducing the arc (i, j) in its place. As a result of this change, the present chain from 1 to j of length π_j^r is replaced by the shorter chain of length $\pi_i^r + c_{ij}$. If p is any descendant of j, the current chain from 1 to p passes through j, and the portion of this chain from 1 to j is altered by this change. The new chain from 1 to p is shorter by $\pi_j^r - (\pi_i^r + c_{ij})$.

RESULTS AT TERMINATION

When termination occurs, the distance on each point is the length of the shortest chain from the origin to the point, and the shortest chain itself can be found by tracing the labels backwards from the point to the origin. At termination, if an artificial arc $(1, i)$ is contained on the chain from 1 to a point, then in the original network G there is no chain from 1 to that point.

Example

Let G be the network in Figure 12.37 with the c_{ij} entered on the arcs.

Artificial arcs (1, 3), (1, 6) of infinite length are introduced as wavy arcs as in Figure 12.38. The algorithm works only with the labels on the points, but for the purpose of illustration the

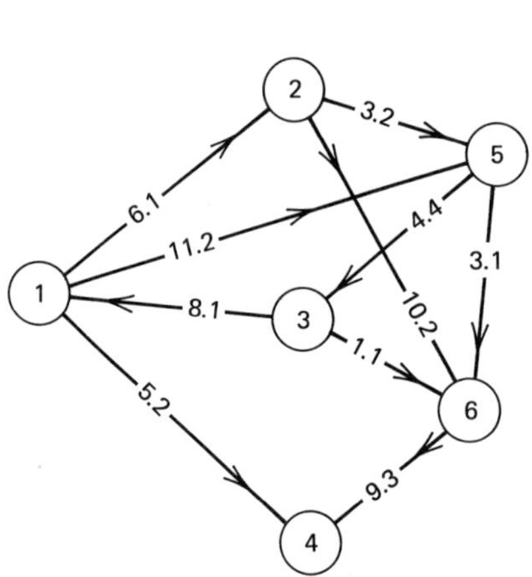

Figure 12.37 G

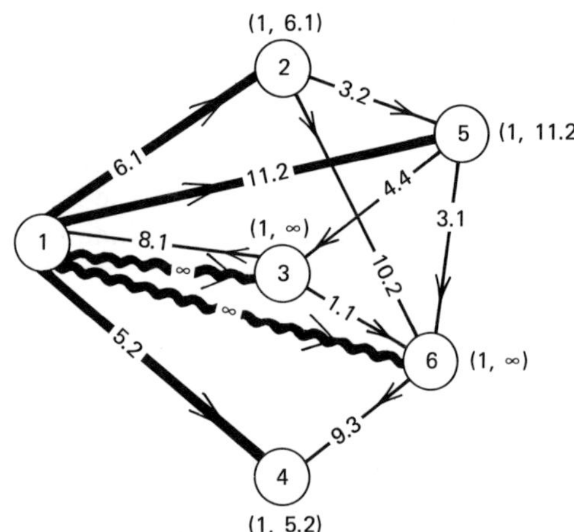

Figure 12.38 G^1 with initial labels.

current out tree is also marked with thick lines. Artificial arcs are deleted once they leave the tree. The data on the node is (label, current distance). Hence from Figure 12.41, the length of the shortest chain from 1 to 6 is 12.4. Using the labels we find that the arcs in this chain are (5, 6), (2, 5), (1, 2). Hence the chain itself is 1, (1, 2), 2, (2, 5), 5, (5, 6), 6. Similarly the shortest chains to all the other nodes can be read out using the labels in Figure 12.41. During the algorithm, the data changes as in the sequence of Figures 12.38 to 12.41.

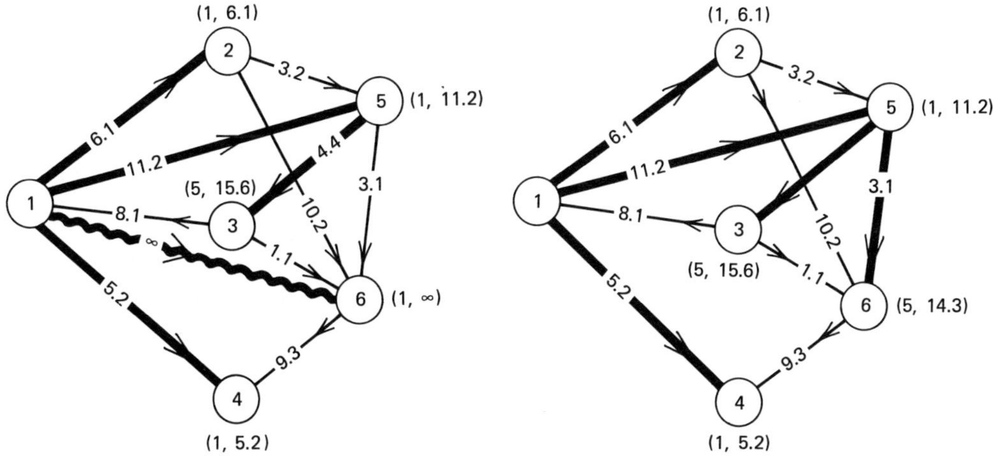

Figure 12.39

Figure 12.40

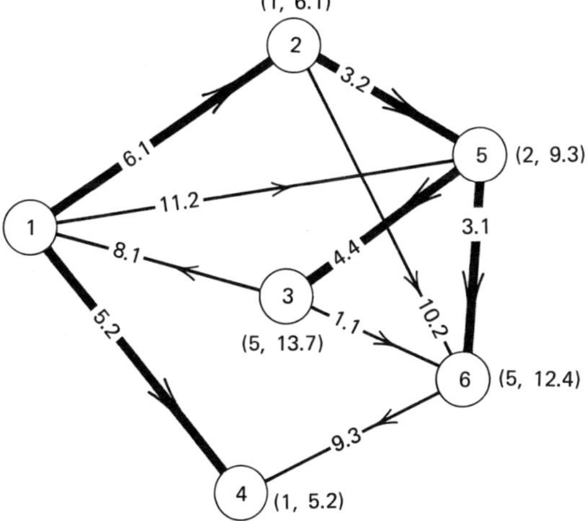

Figure 12.41 This out tree contains shortest chains from 1.

12.6.3 The Tree Building Algorithm

As before, let G = $(\mathcal{N}, \mathcal{A})$ be a directed network with arc lengths $c = (c_{ij})$ and 1 as the origin. *c should be a nonnegative vector*. As in section 12.6.2, assume that if there are several arcs joining a pair of points *i*, *j*, only the shortest among them is kept and all the others are deleted.

The tree building algorithm starts by building the out tree with 1 as the root, containing the shortest chains, one arc per step. Once an arc is included in the out tree, it will never be taken out again. In a sense, the arc that will be added to the out tree in a step can be interpreted as the best arc among all the out-of-tree arcs in that step. The arc selection in each step is known as a *greedy selection*, and the algorithm itself is known as a *greedy algorithm.*

The out tree is stored by labeling the points as before. All the labeled nodes in a step are the nodes that are in the tree at that step. As in section 12.6.2, the data on a node is recorded as the order pair "(label, distance)."

Initially, label the origin, 1, with the data (1, 0). The only labeled node is 1 in this step.

General Step Let $\mathbf{X}$ be the set of all labeled nodes, $\bar{\mathbf{X}}$ the set of unlabeled nodes at this stage. If $\bar{\mathbf{X}} = \emptyset$, all the nodes are labeled and the algorithm terminates. In this case the distances on the points are the lengths of the shortest chains from the origin to the points. The shortest chains themselves can be traced using the labels.

Suppose $\bar{\mathbf{X}} \neq \emptyset$. For each $i \in \mathbf{X}$, let the data on i be (x_i, π_i). If there are no arcs of the form $(x, y) \in \mathscr{A}$ with $x \in \mathbf{X}$, $y \in \bar{\mathbf{X}}$, there are no chains in G from 1 to any node in $\bar{\mathbf{X}}$. The labels on the points in $\mathbf{X}$ provide the shortest chains from 1 to all the points in $\mathbf{X}$, and there is no chain from 1 to y for all $y \in \bar{\mathbf{X}}$. The algorithm is then terminated.

If there are some arcs $(x, y) \in \mathscr{A}$, with $x \in \mathbf{X}$, $y \in \bar{\mathbf{X}}$ let (i, j) be an arc satisfying $i \in \mathbf{X}, j \in \bar{\mathbf{X}}$ and

$$\pi_i + c_{ij} = \text{minimum } \{\pi_x + c_{xy} : (x, y) \in \mathscr{A}, x \in \mathbf{X}, y \in \bar{\mathbf{X}}\}.$$

Then label point j with the data $(i, \pi_i + c_{ij})$ and go to the next step.*

Once a point is labeled in this algorithm, the shortest chain from the origin to that point can be traced using the labels at this stage. Thus, if the shortest chains from an origin to only a specified subset of points are desired, the algorithm can be terminated as soon as all the points in the specified subset are labeled.

Example

Apply this algorithm to find the shortest chains from 1 to all the other nodes in the network in Figure 12.37. The first arc to be added to the out tree is (1, 4). Verify that the arcs (1, 2), (2, 5), (5, 6), (5, 3) are then added to the out tree in that order. The out tree in Figure 12.41 is obtained.

Problems

12.30 Show that the tree building algorithm does not work unless $c_{ij} \geqq 0$ for all (i, j). Try this network.

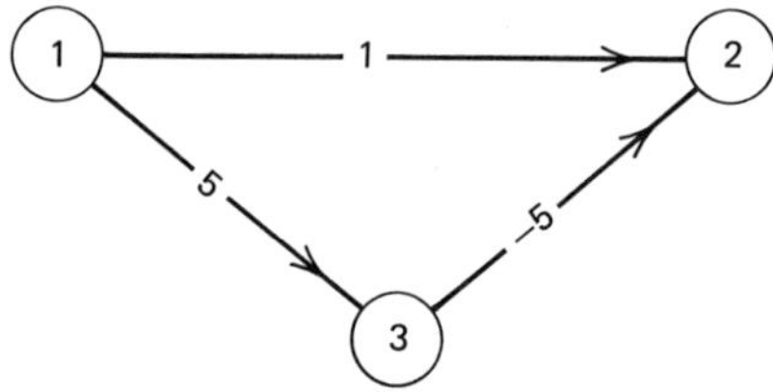

12.31 As long as the labels on the points of G define an out tree, prove that no point can lie in the set of its descendants.

* The efficiency of the algorithm can be improved by storing the set of arcs $\{(i, j): i \in \mathbf{X}, j \in \bar{\mathbf{X}}\}$ and updating this set after each step.

12.32 Let $G = (\mathcal{N}, \mathcal{A})$ be a directed network with c as the vector of arc lengths. Suppose c is not necessarily nonnegative, but is an arbitrary vector. Let 1 be the origin and n the destination. Define all arc lower bounds to be 0 and all arc capacities to be infinity. Show that the problem of finding the shortest chain from 1 to n is equivalent to the problem of finding a minimum cost feasible flow vector of value 1 unit, with 1 as the source, n as the sink, and c as the vector of cost coefficients. Write the dual of this flow problem. Prove that the dual problem is infeasible if there is a circuit in G whose total length is strictly negative. Interpret what this implies about the shortest chain problem when c is not necessarily nonnegative.

12.33 Let $G = (\mathcal{N}, \mathcal{A})$ be a directed network with 1 as the origin and c (an arbitrary vector, not necessarily nonnegative) as the vector of arc lengths. Suppose the shortest chains from 1 are being found using the tree changing algorithm. If there is a circuit containing 1, whose total length is negative, show that at some stage of the algorithm a point j with current data (L_j, π_j) will be found such that $(j, 1) \in \mathcal{A}$ and $\pi_j + c_{j1} < 0$.

On the other hand, if there is a negative length circuit not containing 1, show that at some stage a point will appear among the set of its descendants.

Also, the occurrence of either of these events during the algorithm signals the existence of a negative length circuit in G. Illustrate with a numerical example.

12.6.4 Comments

The tree building algorithm works if $c_{ij} \geqq 0$ for all (i, j). If the problem is such that some of the c_{ij} are negative, the tree changing algorithm works as long as there is no circuit of negative length in the network.

When shortest chain problems have to be solved on undirected or mixed networks, these algorithms can only be used if the lengths of all edges are nonnegative. In that case replace each edge by a pair of arcs as in section 12.2 (Figure 12.5) both of which are the same length as the edge, and transform the problem into a shortest chain problem on a directed network.

Let $G = (\mathcal{N}, \mathcal{A})$ be a directed network with c as the arc length vector. Suppose some c_{ij} are negative and there are some circuits in G of negative length. In this case suppose we want to find the shortest simple chain from an origin to a destination in G. This seems to be a very hard problem. The traveling salesman problem discussed in Chapters 13 and 14 can be posed as a problem of this type. At the moment such problems can only be solved by enumerative or branch and bound approaches.

12.6.5 Sensitivity Analysis

Let $G = (\mathcal{N}, \mathcal{A})$ be a directed network with c as the arc length vector and 1 as the origin and n as the destination. Let (x, y) be a particular arc in G. Let $c_{x,y} = \xi$. Let $\pi(\xi)$ denote the length of the shortest chain from 1 to n when $c_{x,y} = \xi$, and the lengths of all the other arcs remain unchanged.

Problem

12.34 Prove that for any $0 \leqq \xi < \infty$, $\pi(\xi) = \text{minimum } \{\pi(0) + \xi, \pi(\infty)\}$.

If (x, y) is contained on a shortest chain from 1 to n when $c_{xy} = \xi = 0$, then this shortest chain remains the shortest chain as ξ increases from 0 to $\pi(\infty) - \pi(0)$. If $\xi > \pi(\infty) - \pi(0)$, the shortest chain from 1 to n changes. The new shortest chain does not contain the arc (x, y) and it remains the shortest chain for all $\xi > \pi(\infty) - \pi(0)$.

This number $\pi(\infty) - \pi(0) = c^*_{xy}$ is known as the *critical length* of the arc (x, y).

Destroying an arc in a network is equivalent to making its length infinity. If the arc (x, y) is destroyed, the length of the shortest chain from 1 to n goes up by

$$\text{Maximum } \{0, c^*_{x,y} - c_{x,y}\}$$

Problems

12.35 Develop an efficient scheme for finding the arc in the network, the destruction of which increases the length of the shortest chain from 1 to n by as much as possible.

12.36 Suppose the length of an arc (i, j) is the driving time for traveling along that arc from i to j. By making improvements on the condition of the arc, this length can be reduced. Discuss how you will pick an arc for making improvements if the aim is to reduce the length of the shortest chain from 1 to n by as much as possible by making improvements on only one arc.

12.6.6 Shortest Chains between All Pairs of Points in a Network

Suppose you need to find the shortest chains between every possible pair of points in $G = (\mathcal{N}, \mathcal{A})$, a directed network with c as the vector of arc lengths. This problem can be solved by applying the algorithms of sections 12.6.2 or 12.6.3 with each point as the origin separately. But it is convenient to find all these shortest chains simultaneously by applying the *inductive algorithm* discussed below.

If there are several arcs joining a pair of points, all but the shortest among them is eliminated as before. The shortest chains themselves are stored using a *label* matrix. The rows and the columns of the label matrix are associated with points in the network. If the label in the row corresponding to point i and the column corresponding to point j is p, it implies that the last arc in the present chain from i to j is (p, j). The remaining arcs on this chain can be obtained by looking at the entry in the row associated with point i and the column associated with point p of the label matrix, and continuing in this manner until the point i is reached.

Simultaneously another matrix called the *distance matrix* is generated. The rows and the columns in the distance matrix are also associated with the points in the network. The entry in the distance matrix, in the row associated with point i and the column associated with point j, is the total length of the present chain from i to j.

If there is no arc between a pair of points i, j in G, add an artificial arc (i, j) and make its length equal to infinity or a very large positive number. Let G^1 be the augmented network. The algorithm works on G^1.

The algorithm proceeds inductively on the number of points. At the rth stage all the shortest chains between every pair of nodes in the partial network of G^1 consisting of the subset of nodes $1, \ldots, r$, are obtained. In the $r + 1$th stage the algorithm brings in node $r + 1$ and obtains the shortest chains between every pair of nodes in the partial network of G^1 in the subset of nodes $\{1, \ldots, r, r + 1\}$. The algorithm terminates when all the nodes of G^1 are included.

The algorithm works as long as the network does not contain a directed circuit of negative length. If the network does contain a directed circuit of negative length, the algorithm finds one such circuit and then terminates.

Initially the algorithm begins with the partial network of node 1 alone. The initial matrices are

Label Matrix	To 1
From 1	1

Distance Matrix	To 1
From 1	0

At the end of rth step suppose d_{ij} is the length of the shortest chain from i to j in the partial network of $\{1, \ldots, r\}$ for $i, j = 1$ to r. Let $\hat{d}_{ij}$ indicate the length of the shortest chain from i to j in the partial network of $\{1, \ldots, r, r + 1\}$, for $i, j = 1$ to $r + 1$. Then for $i = 1$ to r:

$$\hat{d}_{i,r+1} = \text{minimum } \{c_{i,r+1}; d_{ij} + c_{j,r+1} \quad \text{for } j = 1 \text{ to } r, j \neq i\} \tag{12.11}$$

$$\hat{d}_{r+1,i} = \text{minimum } \{c_{r+1,i}; c_{r+1,j} + d_{ji} \quad \text{for } j = 1 \text{ to } r, j \neq i\} \tag{12.12}$$

$$\hat{d}_{r+1,r+1} = \text{minimum } \{0; \hat{d}_{r+1,j} + \hat{d}_{j,r+1} \quad \text{for } j = 1 \text{ to } r\}$$

$\hat{d}_{r+1,r+1}$ is the length of the shortest circuit containing $r + 1$ in the partial network consisting of points $\{1, 2, \ldots, r, r + 1\}$. If $\hat{d}_{r+1,r+1}$ is strictly negative, the network contains a circuit of negative length and the algorithm is terminated. As discussed in section 12.6.4 the problem cannot be solved in this case.

If $\hat{d}_{r+1,r+1} \geqq 0$, the algorithm is continued. For $i = 1$ to r and $j = 1$ to r,

$$\hat{d}_{ij} = \text{minimum } \{d_{ij}; \hat{d}_{i,r+1} + \hat{d}_{r+1,j}\}$$

$(\hat{d}_{ij})$ with i, j varying from 1 to $r + 1$ is the new distance matrix. For $i = 1$ to r and $j = 1$ to r, the entry in the label matrix in the row associated with i and the column associated with j is the same as the previous entry if $\hat{d}_{ij} = d_{ij}$ or is changed to $r + 1$ otherwise. The entry in the label matrix in the row associated with i and the column associated with $r + 1$ is i if $\hat{d}_{i,r+1} = c_{i,r+1}$ or the j that attains the minimum in equation (12.11) otherwise. The entry in the label matrix in the row associated with $r + 1$ and the column associated with i is $r + 1$ if $\hat{d}_{r+1,i} = c_{r+1,i}$, or the j that attains the minimum in equation (12.12) otherwise.

When all the points in the network are included, the algorithm terminates. The final label matrix provides the shortest chains between all possible pairs of points. If one of these shortest chains contains an artificial arc of G^1, it implies that there is no chain between that pair of points in G. The final distance matrix provides the lengths of the shortest chains.

Problem

12.37 Find the shortest chains between all the pairs of points in the mixed network in Figure 12.42. The lengths are entered on the lines.

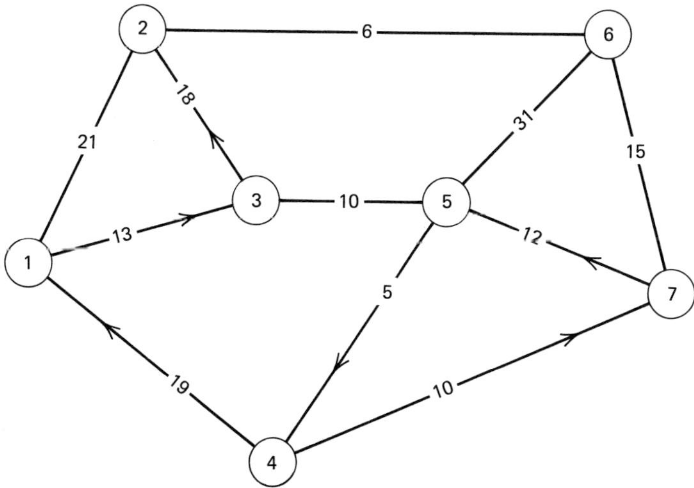

Figure 12.42

12.7 MINIMUM SPANNING TREE PROBLEMS

Given a set of points on the two-dimensional plane, the problem of finding the shortest connecting network, which connects all the points, arises in many practical applications. If the length between every pair of points is positive, the shortest connecting network will obviously be a spanning tree.

In general, we consider the following problem. Let $G = (\mathcal{N}, \mathcal{A})$ be a connected network with c as the vector of arc lengths. The length of a spanning tree in G is defined to be the sum of the lengths of the lines in it. You need to find the minimum length spanning tree in G.

Note Define the length of a subnetwork of G to be the sum of the lengths of the lines in it. Here G may be a directed, undirected, or mixed network and we are interested in finding the shortest subnetwork of G that contains a path (not necessarily a chain) between every pair of points in G. This gives the minimum length spanning tree in G. In some problems G is a directed or mixed network, an origin in G is specified, and we are required to find the shortest subnetwork of G that contains a chain from the origin to all the other points in G. This will be the shortest out tree with the origin as the root (also called an *optimum branching*, or *optimum spanning arborescence* in G). The algorithm discussed here can be used only to find minimum length spanning trees. The problem of finding an optimum branching in a directed or mixed network is a much harder problem, and we will not discuss it here. See references [2, 10, 19, and 22].

Since the orientation of the arcs in G is unimportant in this problem, we will disregard all the orientations and treat every line as an edge.

Problem

12.38 If **T** is a minimum length spanning tree and $(i; j)$ is an out-of-tree line, prove that $(i; j)$ is the maximum length line in the cycle associated with it.

This problem can be solved very efficiently by the following algorithms. c can be an arbitrary real vector. It need not be nonnegative. Let n be the number of points in G.

Algorithm 1

(i) Find out the line $(i_1; j_1)$ of minimum length in G. Include $(i_1; j_1)$ in the set **T** and go to the next step.

(ii) In the rth stage, let $(i_r; j_r)$ be a line of minimum length among the lines $\{(i; j): (i; j) \in \mathcal{A}, (i; j) \notin \mathbf{T}$ so far, and $(i; j)$ does not complete a cycle with lines that are already in $\mathbf{T}\}$. Include $(i_r; j_r)$ in **T** and go to the next stage.

The algorithm is terminated after $n - 1$ stages. Let $\mathbf{T} = \{(i_r; j_r) = r = 1$ to $n - 1\}$. Then **T** is the set of lines of a minimum length spanning tree in G.

Algorithm 2 Initiate the algorithm by selecting an arbitrary point, say, i_1, in G. Let j_1 be such that

$$c_{i_1; j_1} = \text{minimum } \{c_{i_1; j}: (i_1; j) \in \mathcal{A}\}$$

Include $(i_1; j_1)$ in a set **T**, make i_1, j_1 scanned nodes and go to be next stage.

In a general stage, let **X** be the present set of scanned nodes and $\bar{\mathbf{X}}$, the set of unscanned nodes. If $\bar{\mathbf{X}} = \emptyset$, the algorithm is terminated. In this case the present set **T** is the set of lines of a minimum length spanning tree in G. If $\bar{\mathbf{X}} \neq \emptyset$, let

$i_r \in \mathbf{X}$, $j_r \in \bar{\mathbf{X}}$ be such that

$$c_{i_r; j_r} = \text{minimum } \{c_{ij}: i \in \mathbf{X}\, j \in \bar{\mathbf{X}}, (i; j) \in \mathscr{A}\}$$

Include $(i_r; j_r)$ in the set $\mathbf{T}$, make j_r a scanned node and go to the next stage.

Problem

12.39 Let $G = (\mathscr{N}, \mathscr{A})$ be a connected network with n points and c the length vector. Suppose $\mathbf{T} = \{(i_r; j_r): r = 1 \text{ to } n - 1\}$ is the set of lines in a minimum length spanning tree in G Given $\mathbf{T}$, develop an algorithm for finding a minimum length spanning tree in G not containing the line $(i_1; j_1)$.

12.8 MULTICOMMODITY FLOW PROBLEMS

Let $G = (\mathscr{N}, \mathscr{A})$ be a directed network. Suppose there are p distinct commodities to be transported on this network. In general each commodity may originate at some specified source and may be required at some specified sink. Let s_r, t_r be the source and sink, respectively, for the rth commodity $r = 1$ to p.

Suppose the lower bounds on all the arcs are 0. Let k_{ij} be the capacity of arc (i, j) and suppose this capacity applies to the sum of the flow amounts of the different commodities on arc (i, j) from i to j.

Let $\boldsymbol{E}$ be the node-arc incidence matrix of G. Let q_r be the node-arc incidence vector of (s_r, t_r) for $r = 1$ to p. Let f^r be the vector of the flow amounts of the rth commodity on the arcs of the network. The combined vector $(f^1, \ldots, f^p)$ is feasible to the problem if flow conservation holds with respect to each commodity separately, and the lower bound and capacity constraints are satisfied on all the arcs. Indicating the value of the flow of the rth commodity by v_r, $(f^1, \ldots, f^p)$ is feasible if:

$$\boldsymbol{E}f^r - q_r v_r = 0 \qquad \text{for each } r = 1 \text{ to } p$$

and

$$\sum_{r=1}^{p} f^r_{ij} \leqq k_{ij} \qquad \text{for all } (i, j) \in \mathscr{A}$$

and

$$f^r_{ij} \geqq 0 \qquad \text{for all } r \text{ and } (i, j) \in \mathscr{A}$$

In tableau form, here are the constraints for feasibility where I is an identity matrix of the appropriate order.

f^1	v_1	f^2	v_2	$\ldots$	f^p	v_p	
$\boldsymbol{E}$	$-q_1$						$= 0$
		$\boldsymbol{E}$	$-q_2$				$= 0$
				$\ddots$			$\vdots$
					$\boldsymbol{E}$	$-q_p$	$= 0$
I		I			I		$\leqq k$

$f^r \geqq 0$ for all r

If it is desired to maximize the sum of the values of the flows of the various commodities, then we have a multicommodity maximum flow problem, in which $\sum_{r=1}^{p} v_r$ is to be maximized subject to the above constraints.

Suppose c^r_{ij} is the cost of shipping one unit of the rth commodity across arc $(i, j) \in \mathscr{A}$. Let c^r be the cost coefficient vector for the rth commodity. If a flow of V_r units of the rth commodity is required, the multicommodity minimum cost flow problem is the problem of minimizing $\sum_{r=1}^{p} c^r f^r$, subject to the above constraints, and the requirements $v_r = V_r$ for $r = 1$ to p.

When the number of commodities is large, multicommodity flow problems become LPs of a very large size. The structure of the problem can be exploited and the problem solved by structured linear programming techniques. See Appendix 1.

In traffic analysis each passenger or vehicle may originate at some specified source and want to go to some specified sink on the traffic network. It is necessary to treat the passenger corresponding to each source, sink pair as a distinct commodity. Thus many problems in traffic analysis are multicommodity flow problems.

12.9 OTHER NETWORK ALGORITHMS

Suppose $G = (\mathscr{N}, \mathscr{A})$ is an undirected network with c as the length vector. The *blossom algorithm* finds a minimum cost perfect matching in G. The problem of finding a minimum cost perfect matching can be posed as a symmetric assignment problem and solved by the branch and bound approach discussed in Chapter 15. Computational experience indicates that in most practical matching problems, the branch and bound approach is superior to the blossom algorithm. The blossom algorithm is not discussed here.

There are many network algorithms for solving a variety of other problems. See the references.

Concluding Remarks

The basic reference on network algorithms is by Ford and Fulkerson [15]; references [1, 3, 5, 13, 16, 17, 18, 22 and 24] are other books in the field. The original reference for the Hungarian method is the paper by Kuhn [21]. See references [8 and 9] for the blossom algorithm for finding a minimum cost perfect matching in a general undirected network. (This algorithm is not discussed here.) The inductive algorithm for finding all shortest chains in a network is by Dantzig [4]. References [7, 14, 23, and 27] are other papers on shortest chains; references [20 and 25] deal with minimum cost spanning trees.

PROBLEMS

12.40 There are n students in a course. There are m available projects. Each student has to work on precisely one project. At most b_i students can work on the ith project for $i = 1$ to m. If there are no students to work on a project, that project can be dropped without any harm. A total of r supervisors are available. Corresponding to each project a subset of one or more supervisors is specified, who can supervise the project. Each student working on a project has to be supervised by a supervisor who is in the subset of those eligible to supervise that project. k_p is the maximum number of students that supervisor p can handle, for $p = 1$ to r.

Each student picks a subset of projects on which he would like to work, and he ranks the projects in this subset as 1, 2 etc., in descending order of preference.

The objective is to assign students to projects and supervisors, so that each student gets to work on a project that has his most preferred ranking.

Taking as an objective function the sum of the rankings of the projects the students work on, formulate the problem of doing these assignments as a minimum cost flow problem.

12.41 CATERER PROBLEM A caterer is planning his linen napkin supply over a planning horizon consisting of days 1 to N. On the jth day he requires exactly r_j linen napkins, $r_j \geqq 0$ for $j = 1$ to N. Once a napkin is used, it is sent for laundering, which offers two types of service. The *quick service* takes used linen napkins on the ith day, washes them, and delivers them on the $i + m$th day for use again on that day or later on for $i = 1$ to $\mathrm{N} - m$. The *slow service* takes linen napkins used on the ith day, washes them, and delivers them on the $i + n$th day, for use again on that day or later on for $i = 1$ to $\mathrm{N} - n$. It costs q cents per napkin to send it for quick service and c cents per napkin to send it for slow service. $q > c$ and $n > m$. New linen napkins can be purchased on any day at a cost of p cents per napkin. $p > q$.

Starting with zero napkins, how does he operate so as to minimize his total cost, which is the cost of buying new napkins plus the costs of laundering used napkins over the planning horizon? Formulate it as a transportation problem. Solve the problem for:

$$\mathrm{N} = 10, m = 1, n = 2, q = 4, c = 2, p = 10$$

$$r = (4, 7, 8, 10, 15, 12, 10, 12, 15, 15).$$

—*W. Jacobs*

12.42 LIBRARY PROBLEM Books come in various heights and thicknesses. $h_1 < h_2 < \cdots < h_n$ are the possible values of book heights. L_i is the total thickness of all the books of height h_i that are to be shelved, $i = 1$ to n. A book of height h can be stored in a shelf of any height greater than or equal to h.

Let p_i denote the total length of shelf space of height h_i that is constructed, $i = 1$ to n, for storing these books. The associated cost is 0 if $p_i = 0$; $k_i + c_i p_i$ if $p_i > 0$, where k_i is a known fixed charge for construction and c_i is the cost per unit length after the fixed change is incurred. The problem is to determine the optimal $(p_1, \ldots, p_n)$ to shelve all the books at minimum total cost.

Let $\mathcal{N} = \{0, 1, \ldots, n\}$. Let $\mathcal{A} = \{(i, j): i \in \mathcal{N}, j \in \mathcal{N}, j > i\}$. Let the length of the arc $(i, j) \in \mathcal{A}$ be $\ell_{ij} = k_j + c_j \sum_{r=i+1}^{j} \mathrm{L}_r$. Here ℓ_{ij} is the cost of shelving all the books of heights $h_{i+1}, \ldots, h_j$ in shelf space of height h_j.

Show that the problem of constructing adequate shelving space for all the books is equivalent to the problem of finding the shortest chain from 0 to n on the network $(\mathcal{N}, \mathcal{A})$ with ℓ as the arc length vector. Given a shortest chain in this network, show how to derive the optimal $(p_1, \ldots, p_n)$ vector corresponding to it. Solve the problem for the following data:

$$(h_1, h_2, h_3, h_4) = (5, 7, 9, 12), (\mathrm{L}_1, \mathrm{L}_2, \mathrm{L}_3, \mathrm{L}_4) = (3, 4, 18, 12)$$

$$(k_1, k_2, k_3, k_4) = (7, 7, 9, 9), (c_1, c_2, c_3, c_4) = (6, 8, 9, 12)$$

—*A. Ravindran*

12.43 $G = (\mathcal{N}, \mathcal{A})$ is a directed network with a source and sink, and 0 as the lower bound vector and $k + \lambda\hat{k}$ as the capacity vector for the flow amounts on the arcs, where λ is a real-valued parameter and k, $\hat{k}$ are given vectors. Let $f(\lambda)$ denote a maximum value feasible flow vector as a function of the parameter and let $v(\lambda)$ be its value. Discuss how $f(\lambda)$ and $v(\lambda)$ can be determined for all values of λ in some given interval.

12.44 Consider the caterer problem discussed in Problem 12.41. Let $c = 0$, $p = 1$. Treating q as a parameter, discuss how an optimum solution of this problem can be obtained for all values of q in the interval $0 \leqq q \leqq 1$.

—*C. S. Ramakrishnan*

12.45 There is a club of 10 men and 10 women. If a man and woman are compatible with each other, it is indicated with a letter c in the following table.

	Women	1	2	3	4	5	6	7	8	9	10
Man	1		c	c							
	2			c				c			c
	3		c				c				
	4		c								
	5	c		c	c	c				c	
	6						c		c	c	
	7				c		c			c	
	8										c
	9				c			c			c
	10				c		c		c		

Find out the maximum number of compatiable couples that can be formed out of this club.

12.46 Consider the bounded variable transportation problem (11.20) discussed in section 11.7. Write down the dual of this problem. Devise a method for finding a dual feasible solution for this problem directly. Using it, develop a primal-dual approach for solving the transportation problem (11.20).

12.47 KÖNIG-EGERVÁRY THEOREM Consider an m × n transportation array. In this array a specified subset, **J**, of admissible cells is given. A subset $\mathbf{H} \subset \mathbf{J}$ is said to be an *independent subset of admissible cells* if no two cells in **H** lie in the same row or column. A subset of rows or columns of this array is said to *cover* **J**, if every admissible cell is contained in either a row and/or column from this subset. Prove that the cardinality of a maximum cardinality independent set of admissible cells is equal to the cardinality of a minimum cardinality set of lines covering all the admissible cells (use max flow min-cut theorem on an appropriately constructed network).

12.48 At every nonbreakthrough in the Hungarian method, prove that the total number of allocations at that stage is equal to the total number of unlabeled rows and labeled columns at that stage. Also prove that every admissible cell at that stage must lie in an unlabeled row or a labeled column. (Prove that the set of unlabeled rows and labeled columns at this stage is a minimum cardinality set of lines covering all the admissible cells at this stage and that the set of cells with allocations at this stage is a maximum cardinality independent subset of admissible cells.)

12.49 Discuss the usefulness (in practical applications) of the optimum node price vector in the minimum cost network flow model (12.10) and its economic interpretation.

12.50 MINIMUM CONVEX PIECEWISE LINEAR COST FLOWS Consider a single commodity directed network $G = (\mathcal{N}, \mathcal{A})$ with source $\mathscr{s}$, sink $\mathscr{t}$, and lower bound vector for arc flows ℓ, and capacity vector for arc flows k. Suppose the cost of flow on each arc (i, j) is a piecewise linear convex function, $c_{ij}(f_{ij})$. (That is, the interval ℓ_{ij} to k_{ij} is divided into a finite number, say r, of intervals, of the form

Interval	Slope of $c_{ij}(f_{ij})$ in the interval
ℓ_{ij} to x_{ij}^1	c_{ij}^1
$\vdots$	$\vdots$
x_{ij}^{r-1} to k_{ij}	c_{ij}^r

and in each interval $c_{ij}(f_{ij})$ is linear with the slope given as above. Also the slopes are strict monotone increasing, that is, $c_{ij}^1 < c_{ij}^2 < c_{ij}^3 \cdots < c_{ij}^r$. The various slopes and the break points at which $c_{ij}(f_{ij})$ changes slopes are all given for all $(i, j) \in \mathscr{A}$. Find a flow vector of specified value V, which minimizes the sum of the costs for flows on all the arcs. Using the ideas discussed in section 1.1.5, transform this into a minimum cost flow problem with a linear objective function on an enlarged network.

REFERENCES

1. C. Berge and A. Ghouila-Houri, *Programming, Games, and Transportation Networks*, Wiley, New York, 1965.
2. F. C. Bock, "An Algorithm to Construct a Minimum Directed Spanning Tree in a Directed Network," in *Developments in Operations Research*, B. Avi-Itzak (ed.), Gordon and Breach, 1971.
3. R. G. Busacker and T. L. Saaty, *Finite Graphs and Applications*, McGraw-Hill, New York, 1965.
4. G. B. Dantzig, *All Shortest Routes in a Graph.* O. R. House, Stanford University, 1966; Also in *Theorie des Graphes*, pp. 91–92, *Proceedings of the International Symposium*, Rome July 1966, published by Dunod, Paris.
5. J. B. Dennis, *Mathematical Programming and Electrical Networks*, Wiley, New York, 1959.
6. E. W. Dijkstra, "A Note on Two Problems in Connection with Graphs," *Numerische Mathematik*, *1*, 269–271, 1959.
7. S. E. Dreyfus, "An Appraisal of Some Shortest Path Algorithms," *Operations Research*, *17*, *3*, pp. 395–412, May–June 1969.
8. J. Edmonds, "Paths, Trees and Flowers," *Canadian J. Math.*, *17*, *3*, 449–467, May 1965.
9. J. Edmonds, "Matriods and the Greedy Algorithm," *Mathematical Programming*, *1*, *2*, pp. 127–136, 1971.
10. J. Edmonds, "Optimum Branchings," *Journal of Research of the National Bureau of Standards*, 71B, pp. 233–240, 1967. Article also appeared in *Mathematics and the Decision Sciences*, Part 1, G. B. Dantzig and A. F. Veinott, Jr. (eds.), American Mathematical Society Lectures in Applied Mathematics, *11*, 1968.
11. J. Edmonds and R. M. Karp, "Theoretical Improvements in Algorithmic Efficiency for Network flow problems," *J. ACM*, *19*, pp. 248–264, April 1972.
12. S. E. Elmaghraby, "The Theory of Networks and Management Science, Part I," *Management Science*, *17*, pp. 1–34, 1970.
13. S. E. Elmaghraby, "*Some Network Models in Management Science*," Lecture Notes in Operations Research and Mathematical Systems 29, Springer-Verlag, Berlin, 1970.
14. R. W. Floyd, "Algorithm 97: Shortest Path," *Communications of ACM*, *5*, *6*, 345, 1962.
15. L. R. Ford and D. R. Fulkerson, *Flows in Networks*, Princeton University Press, 1962.
16. H. Frank and I. T. Frisch, "*Communication, Transmission and Transportation Networks*," Addison-Wesley, Reading, Mass., 1971.
17. A. M. Geoffrin (Editor), *Perspectives on Optimization; A Collection of Expository Articles*, Addison-Wesley, Reading, Mass., 1972.
18. T. C. Hu., *Integer Programming and Network Flows*, Addison-Wesley, Reading, Mass., 1970.
19. R. M. Karp, "A Simple Derivation of Edmonds' Algorithm for Optimum Branchings," *Networks*, *1*, pp. 265–272, 1971.
20. J. B. Kruskal, Jr., "On the Shortest Spanning Subtree of a Graph and the Traveling Salesman Problems," *Proc. Am. Math. Soc.*, *7*, 48–50, 1956.
21. H. W. Kuhn, "The Hungarian Method for the Assignment Problem," *Naval Research Logistics Quarterly*, *2*, pp. 83–97, 1955.

22. E. L. Lawler, "*Combinatorial Optimization: Networks and Matroids*," to be published by Holt, Rinehart and Winston.

23. J. D. Murchland, "The Once-Through Method for Finding All Shortest Distances in a Graph from a Single Origin," Transportation Network Theorey Unit, London Graduate School of Business Studies Report LBS–TNT–56, August 1967.

24. R. B. Potts and R. M. Oliver, *Flows in Transportation Networks*, Academic Press, New York and London, 1972.

25. R. C. Prim, "Shortest Connection Networks and Some Generalizations," *Bell System Technical Journal*, *36*, 1389–1402, 1957.

26. J. W. Suurballe, "Disjoint Paths in a Networks," *Networks*, *4*, pp. 125–145, 1974.

27. S. Warshall, "A Theorem on Boolean Matrices," *J. ACM*, *9*, 11–12, 1962.

13 formulation of integer and combinatorial programming problems

13.1 INTRODUCTION

Thus far we have considered optimization problems in which the decision variables are all *continuous variables*, that is, they can assume all possible values within their range of variation. Here we consider problems, called *discrete optimization problems*, in which some or all the decision variables are restricted to assume values within specified discrete sets. This class of problems also includes various combinatorial optimization problems in which an optimum combination out of a possible set of combinations has to be determined. All these problems can be formulated as LPs with the additional restrictions that some or all of the decision variables can assume only integer values. Such problems are called *integer linear programming* (or *integer programming*) problems. They can be put into two classes: (1) *Pure (or all) integer programs*, in which all the decision variables in the problem are restricted to assume only integer values, and (2) *mixed integer programs*, in which there are some continuous decision variables and some integer decision variables. In each of these classes there are two subclasses: (a) integer programs in which all the integer decision variables are restricted to be either 0 or 1 and (b) general nonnegative integer decision variable problems.

The versatility of the integer programming model in applications stems from the fact that in many practical problems, activities and resources, like machines, ships, and operators are indivisible. Many problems require the determination of yes-no decisions, which can be considered as the 0-1 values of an integer variable so constrained. Also, most optimization problems of a combinatorial nature can be formulated as integer programs. Here we present various examples of problems that can be formulated as integer programs.

13.2 FORMULATION EXAMPLES

13.2.1 Discrete Valued Variables

Consider a model that is a linear programming model with the additional requirement that some variables can only lie in specified discrete sets. For example, in the problem of designing a water distribution system, one of the variables is the diameter of the pipe used, and this variable must be in the set {6 in., 8 in., 10 in., 12 in., 16 in., 20 in., 24 in., 30 in., 36 in.} because pipe is available only in these diameters.

In general, if x_1 is a decision variable in the model that is restricted to assume values in the specified discrete set $\{\alpha_1, \ldots, \alpha_k\}$, the constraint that $x_1 \in \{\alpha_1, \ldots, \alpha_k\}$ is equivalent to

$$x_1 - (\alpha_1 y_1 + \cdots + \alpha_k y_k) = 0$$

$$y_1 + \cdots + y_k = 1$$

$$y_j = 0 \quad \text{or} \quad 1 \quad \text{for each } j$$

Such constraints can be augmented to the other linear equality and inequality constraints in the model for each discrete value restricted variable. This transforms the problem into an integer program.

13.2.2 Batch Size Problems

In addition to the usual linear equality-inequality constraints in a linear programming model, suppose there are constraints of the form: variable x_j in the model can be either zero or if it is positive it must be greater than or equal to some specified lower bound ℓ_j. Constraints of this type arise when the model includes variables that represent the amounts of some raw materials used, and the suppliers for these raw materials will only supply in amounts greater than or equal to specified lower bounds.

Let α_j be a very large positive number or some practical upper bound for the value of x_j in an optimum solution of the problem. The constraint that x_j is either 0 or greater than or equal to ℓ_j is equivalent to

$$x_j - \ell_j y_j \geqq 0$$

$$x_j - \alpha_j y_j \leqq 0$$

$$y_j = 0 \quad \text{or} \quad 1$$

Constraints like this can be introduced into the model for each such batch size restricted variable in the model. This transforms the model into an integer program.

13.2.3 Fixed Charge Problems

In linear programming models the decision variables normally represent the levels at which the various activities are carried out. It is assumed that the objective function to be minimized is a linear function of the variables. In many practical problems, however, the cost of performing an activity as a function of the level at which it is performed may be in the form seen in Figure 13.1, that is, the cost of performing the activity, at level x_j is 0 if $x_j = 0$ and $f_j + c_j x_j$ if $x_j > 0$. f_j is a *set up cost* or a *fixed charge*, which is incurred only when the jth activity is performed at a positive level. For example, suppose the jth activity corresponds to trans-

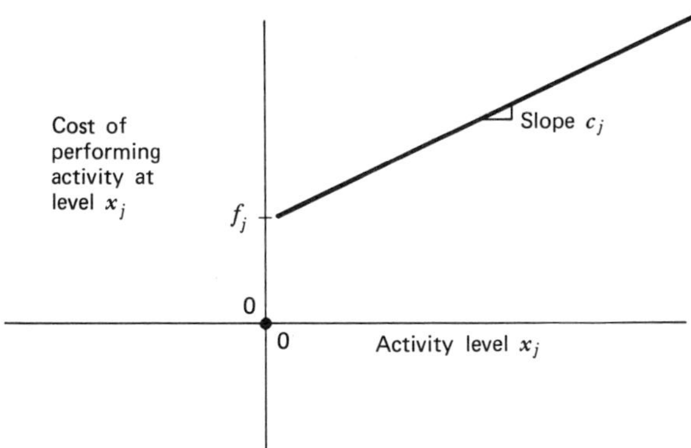

Figure 13.1

porting crude oil from an Alaskan well to a refinery in the south. The level x_j represents the total number of barrels transported from the well to the refinery over the lifetime of the well and f_j represents the cost of building a pipeline from the well to the refinery, if a decision is made to transport a positive quantity of crude from the well to the refinery. c_j represents the variable cost of shipping one barrel of crude from the well to the refinery after the pipeline is built. We assume that f_j are nonnegative for all j. Let the constraints on the decision variables be

$$Ax = b$$
$$x \geqq 0$$

The problem of finding a solution that minimizes the overall cost (the variable cost plus the fixed charges incurred for positive variable in the solution) is known as a *fixed charge linear programming problem.* The f_j are known as *fixed charges.* The c_j are the *variable cost coefficients.* The objective function is

$$z(x) = \sum_{j=1}^{n} c_j x_j + \sum_{j \in \mathbf{J}(x)} f_j$$

where for the feasible solution x, $\mathbf{J}(x)$ is the set $\{j: j \text{ such that } x_j > 0\}$. Clearly $z(x)$ is discontinuous at the origin. It is a step function. Using the fact that f_j is nonnegative for all j, it can be shown that $z(x)$ is a concave function defined on the nonnegative orthant of $\mathbf{R}^n$, that is, for any $x^1 \geqq 0$, $x^2 \geqq 0$ and $0 \leqq \alpha \leqq 1$, $z(\alpha x^1 + (1 - \alpha)x^2) \geqq \alpha z(x^1) + (1 - \alpha)z(x^2)$. Hence the fixed charge problem is the problem of minimizing a special concave objective function subject to linear constraints. Set up costs or fixed charges appear in many practical problems. However, because of the complicated nature of the objective function, these problems cannot be solved by the simplex method.

Fixed charge problems can be formulated as integer programs. Let α be a very large positive number or a practical upper bound for the values of the variables in an optimum solution of the problem. Consider the system of constraints:

$$x_j - \alpha y_j \leqq 0$$
$$x_j \geqq 0, \; y_j = 0 \quad \text{or} \quad 1$$

Clearly when y_j is equal to 0, x_j must be equal to 0 under these constraints. Alternatively y_j is forced to 1 when $x_j > 0$. This observation and the fact that f_j are all nonnegative implies that the fixed charge problem is equivalent to the following mixed integer LP:

$$\begin{aligned}
\text{Minimize} \quad & \textstyle\sum c_j x_j + \sum f_j y_j \\
\text{Subject to} \quad & Ax = b \\
& x_j - \alpha y_j \leqq 0 \qquad \text{for all } j \\
& x_j \geqq 0 \\
& y_j = 0 \quad \text{or} \quad 1 \qquad \text{for all } j
\end{aligned}$$

13.2.4 Plant Location Problems

Plant location (or *facility location* or *warehouse location*) problems form an important class of practical problems that can be formulated as integer programs. The simplest problem in this class has the following structure: there are n sites in a region that require a product. The demand for the product over the planning horizon in the area containing site i is d_i units, $i = 1$ to n. The demand has to be met by manufacturing the product within the region. You need to set up m or less plants for manufacturing the product to satisfy the demand, where m is specified. The set-up cost for building a manufacturing plant at site i is f_i dollars. If a plant is built at site i, k_i units is its production capacity over the planning horizon. The fixed charge for setting up a route for transporting the product from site i to site j is f_{ij} dollars. If a route is set up to transport the product from site i to j, k_{ij} units is its capacity over the planning horizon, and c_{ij} dollars is the cost for transporting one unit of the product from i to j. In practice m will be much smaller than n, and the product will be shipped from the sites where it is manufactured to all other sites in the region. The problem is to determine an optimal subset of sites for locating the plants and a shipping schedule over the entrie horizon so as to minimize the total cost of building the plants, setting up the routes and actually transporting the product. The problem of determining the subset of sites for locating the plants and that of setting up the necessary transportation routes is a combinatorial optimization problem. Once the optimum solution of this combinatorial problem is known, the problem of determining the optimal amounts to be transported along the various routes set up is a simple transportation problem.

This problem can be formulated as an integer program using decision variables with the following interpretations:

$y_i = 1$ if a plant is located at site i
$\quad = 0$ otherwise

$y_{ij} = 1$ if a transportation route is set up from site i to site j
$\quad = 0$ otherwise

$x_{ij} =$ amount (in units) of product transported from site i to site j over the planning horizon

Then the problem formulation is

$$\begin{aligned}
\text{Minimize} \quad & \sum_i \sum_j c_{ij} x_{ij} + \sum_i f_i y_i + \sum_i \sum_j f_{ij} y_{ij} \\
\text{Subject to} \quad & \sum_j x_{ij} - k_i y_i \leqq 0 \qquad \text{for all } i \\
& x_{ij} - k_{ij} y_{ij} \leqq 0 \qquad \text{for all } i, j \\
& \sum_i x_{ij} \geqq d_j \qquad \text{for all } j
\end{aligned}$$

$$\sum_i y_i \leqq m$$

$$x_{ij} \geqq 0 \qquad \text{for all } i, j;\ y_i, y_{ij} \text{ are all 0 or 1}$$

Other plant location problems may have more complicated constraints in them. They can be formulated as integer programs using similar ideas.

13.2.5 Procurement Problems

This class of problems has the following structure: There are m types of equipment available. There is already a stock of a_i units of type i, $i = 1$ to m. In addition a budget of α dollars is available to buy extra quantities of equipment. One unit of type i costs p_i dollars. Equipment can only be bought and allotted in integer number of units. There are n stations that use these equipments. The jth station can use a maximum of b_j units of all types of equipments put together. The profit accruing as a result of assigning one equipment unit of type i to the jth station is c_{ij} dollars. The problem is to determine how many pieces of equipment units of each type are to be purchased, and how to distribute this equipment among the stations. Define variables with the following interpretations:

y_i = number of units of equipment type i to be purchased

x_{ij} = number of units of equipment type i allotted to station j

Then the problem is

$$\text{Maximize} \quad \sum_i \sum_j c_{ij} x_{ij}$$

$$\text{Subject to} \quad \sum_j x_{ij} - y_i \leqq a_i \qquad \text{for } i = 1 \text{ to } m$$

$$\sum_i x_{ij} \leqq b_j \qquad \text{for } j = 1 \text{ to } n$$

$$\sum_i p_i y_i \leqq \alpha$$

$$x_{ij} \geqq 0,\ y_i \geqq 0, \text{ all variables integers}$$

13.2.6 "Either, Or," Constraints

In section 13.2.2, a restriction in which either the constraint "$x_j = 0$" or the constraint "$x_j \geqq \ell_j$" has to hold was discussed. These restrictions are known as *either, or constraints.*

A generalization of this restriction requires that k constraints out of a given set of r constraint have to hold, where k is a given number. Suppose the constraints are $g_1(x) \geqq 0, \ldots, g_r(x) \geqq 0$, where $g_1, \ldots, g_r$ are given functions of the decision variables x. Let L_j be a very large positive number such that $-L_j$ is a lower bound for $g_j(x)$ on the set of feasible solutions. The requirement that k of these constraints have to be satisfied is equivalent to

$$g_1(x) + L_1 y_1 \geqq 0$$

$$\vdots$$

$$g_r(x) + L_r y_r \geqq 0$$

$$y_1 + \cdots + y_r = r - k$$

$$y_j = 0 \quad \text{or} \quad 1 \text{ for each } j$$

Hence restrictions of this type can be formulated using 0-1 variables.

For example, consider the shaded region in Figure 13.2. This region is the set of feasible solutions of

$$x_1 \geqq 0 \qquad x_2 \geqq 0 \qquad x_1 \leqq 10 \qquad x_2 \leqq 10$$

and either $5 - x_1 \geqq 0$ or $5 - x_2 \geqq 0$. Taking $L_1 = L_2 = 15$, this system of restrictions is equivalent to

$$\begin{aligned} 5 - x_1 \qquad\quad + 15y_1 \qquad\quad &\geqq 0 \\ 5 \qquad - x_2 \qquad\quad + 15y_2 &\geqq 0 \\ y_1 + \quad y_2 &= 1 \\ x_1, x_2 \geqq 0, \qquad y_1, y_2 = 0 \text{ or } 1, \qquad & x_1, x_2 \leqq 10 \end{aligned}$$

Using similar arguments, we find that sets that are not necessarily convex but can be represented as the union of a finite number of convex polyhedra can be represented as the set of feasible solutions of a system of linear constraints in which some variables are restricted to be equal to 0 or 1.

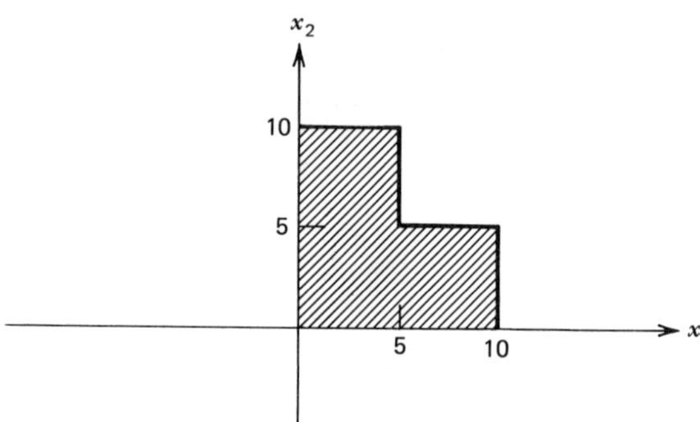

Figure 13.2

13.2.7 Project Selection Problems

This model deals with situations in which there are various projects awaiting approval and the problem is to determine an optimum subset of them to be actually approved. The planning horizon is n periods. If the ith project is approved it requires a budget of c_{it} dollars in the tth period, $t = 1$ to n. The total available funds in the tth period are c_t dollars. The total payoff (or profit) if the ith project is approved is p_i dollars. Define integer variables with the following interpretation:

$$\begin{aligned} x_i &= 1 \qquad \text{if project } i \text{ is approved} \\ &= 0 \qquad \text{otherwise} \end{aligned}$$

Then the problem is

$$\begin{aligned} \text{Maximize} \quad & \sum_i p_i x_i \\ \text{Subject to} \quad & \sum_i c_{it} x_i \leqq c_t \qquad t = 1 \text{ to } n \\ & x_i = 0 \text{ or } 1 \qquad \text{for all } i \end{aligned}$$

Additional constraints on the use of some other resources during each period can also be included in the model.

13.2.8 Multiple Choice Problems

Consider the usual linear programming model in which there are various possible activities. Let x_j be the level at which the jth activity is performed. The variables x_j have to be nonnegative and satisfy some linear constraints. In addition, suppose the activities are grouped into various groups and exactly one activity from each group has to be performed.

Let α be a large positive number that is a practical upper bound on the level of any activity. Suppose $\mathbf{J} = \{1, \ldots, k\}$ form a group of activities. Define

$$
\begin{aligned}
y_j &= 1 \quad \text{if activity } j \text{ is performed from group } \mathbf{J} \\
&= 0 \quad \text{otherwise}
\end{aligned}
$$

Then the constraint that exactly one activity in group **J** should be performed is equivalent to

$$
\begin{aligned}
x_j - \alpha y_j &\leqq 0 \qquad \text{for all } j \in \mathbf{J} \\
\sum_{j \in \mathbf{J}} y_j &= 1 \\
x_j \geqq 0,\ y_j &= 0 \text{ or } 1 \qquad \text{for all } j \in \mathbf{J}
\end{aligned}
$$

Similar constraints can be included corresponding to each group of activities from which only one activity has to be performed. The final model becomes a mixed integer program.

13.2.9 Sequencing Problems

These problems arise in determining an optimum sequence for processing jobs on a machine. There are n jobs. Let p_i be the processing time required by the ith job on the machine. Assume that the machine can only process one job at a time. Also when a job is put on the machine, its processing must be completed before the machine can take up another job. Let t_i represent the clocktime at which the ith job is started on the machine, $i = 1$ to n. The problem is to determine the constraints on the time variables t_i's so that they represent an implementable sequence on the machine.

If job i is started on the machine before job j, we must have $t_j \geqq t_i + p_i$. On the other hand, if job j is started on the machine before job i, we must have $t_i \geqq t_j + p_j$. Thus for each of the $n(n-1)/2$ possible pairs i, j, $i \neq j$, exactly one of the pairs of constraints

$$
\begin{aligned}
\text{Either} \quad & t_i \geqq t_j + p_j \\
\text{Or} \quad & t_j \geqq t_i + p_i
\end{aligned}
$$

must hold. Let α be a very large positive number. Define

$$
\begin{aligned}
y_{ij} &= 1 \quad \text{if job } i \text{ is started on the machine before job } j \\
&= 0 \quad \text{otherwise}
\end{aligned}
$$

Consider the system of constraints

$$
\begin{aligned}
\alpha y_{ij} + t_i - t_j &\geqq p_j \\
\alpha(1 - y_{ij}) + t_j - t_i &\geqq p_i \\
y_{ij} &= 0 \text{ or } 1 \qquad \text{for all } i, j
\end{aligned}
$$

If (t, y) is feasible to this system, the time vector can be implemented on the machine and conversely for every feasible time vector, t, there exists a y such that (t, y) is feasible to the above system. If there are precedence requirements in the processing of jobs they could be included. If there are any other restrictions they could also be included to complete the model.

Many models in which an optimum permutation of a set of jobs or some other objects has to be determined can be modeled as integer programs using similar ideas.

13.2.10 Knapsack Problems

This model refers to the following class of problems: Articles of n different types are available. Each article of type i has a weight of w_i kg and a value of v_i dollars. A knapsack that can hold a weight of at most w kg has to be loaded with these articles so as to maximize the total value of the articles included, subject to the knapsack's weight capacity. Articles cannot be broken; only a nonnegative integer number of articles of each type can be loaded.

Let x_j be the number of articles of type j included in the knapsack. The problem is

$$\begin{aligned} \text{Maximize} \quad & \sum_{j=1}^{n} v_j x_j \\ \text{Subject to} \quad & \sum_{j=1}^{n} w_j x_j \leqq w \\ & x_j \geqq 0 \quad \text{and integer for all } j \end{aligned}$$

This is known as the general *nonnegative integer knapsack problem.* In addition, if each x_j is required to be 0 or 1 we get the *0-1 knapsack problem.*

Knapsack problems are single constraint pure integer programs. They form a very important class of integer programs. Recent work in *aggregating the constraints* has shown that every pure integer program can be transformed into a knapsack problem. However, when this is done the resulting knapsack problem tends to have very large coefficients (i.e., the w_i and w) in its constraint, and it is probably no easier to solve than the original pure integer program. At any rate, this result indicates that the knapsack problem is a very important problem among integer programs. See Problem 13.25 and references [4, 9].

13.2.11 Set Representation, Set Covering, Set Partitioning, and Set Packing Problems

Let $\mathbf{\Gamma} = \{1, 2, \ldots, n\}$. Let $\mathbf{F} = \{\mathbf{F}_1, \ldots, \mathbf{F}_m\}$ be a class of nonempty subsets of $\mathbf{\Gamma}$. A subset $\mathbf{E} \subset \mathbf{\Gamma}$ is said to be a *represent* for $\mathbf{F}$ if $\mathbf{E}$ contains at least one common element with each subset $\mathbf{F}_i$, $i = 1$ to m. A represent that has the least cardinality among all represents is called a *minimum cardinality represent* (MCR) for $\mathbf{F}$. The problem of finding a MCR for $\mathbf{F}$ is known as a *minimum cardinality set representation problem.* Let c_j be the cost of including j in a represent. The problem of finding a represent that minimizes the total cost is known as a *weighted set representation problem.* Define integer variables:

$$\begin{aligned} x_j &= 1 \quad \text{if } j \text{ is included in the represent} \\ &= 0 \quad \text{otherwise} \end{aligned}$$

Then the problem is

$$\begin{array}{lll} \text{Minimize} & \sum_{j=1}^{n} c_j x_j & \\ \text{Subject to} & \sum_{j \in \mathbf{F}_i} x_j \geqq 1 & \text{for all } i \\ & x_j = 0 \text{ or } 1 & \text{for all } j \end{array}$$

Let

$$\begin{aligned} a_{ij} &= 1 \qquad \text{if } j \in \mathbf{F}_i \\ &= 0 \qquad \text{otherwise} \end{aligned}$$

$A = (a_{ij})$ is known as the *set-element incidence matrix* of the class $\mathbf{F}$. The row vector A_i is the *incidence vector* of the subset $\mathbf{F}_i$. Let e be column vector in $\mathbf{R}^m$ all of whose entries are one. The set representation problem is

$$\begin{array}{lll} \text{Minimize} & cx & \\ \text{Subject to} & Ax \geqq e & \\ & x_j = 0 \text{ or } 1 & \text{for all } j \end{array}$$

The transpose of every optimum solution, x, of this integer program is the incidence vector of an optimum represent. The set representation problem has many applications. Here are some of them.

1. *Delivery and Routing Problems* A warehouse has to deliver material to m clients in a region by a carrier. The carrier can combine a number of clients together to form a route. A list of several feasible routes is made. A client may receive delivery via a number of different routes. Suppose the list has n routes. Suppose c_j is the cost (this could be the length of the route) of route j, $j = 1$ to n. Let $\mathbf{F}_i = \{j\text{: route } j$ contains the ith client$\}$ $i = 1$ to m. Then the minimum cost represent for $\mathbf{F} = \{\mathbf{F}_1, \ldots, \mathbf{F}_m\}$ provides an optimum set of routes to service all the clients. Many routing problems can be formulated as set representation problems using similar ideas.

2. *Location of Fire Hydrants* Given a network of streets (edges in the networks) and traffic centers (nodes of the network), find a subset of nodes for locating fire hydrants so that each street contains at least one fire hydrant (Figure 13.3). Let the nodes be numbered $1, \ldots, n$. Suppose there are m edges in the network. Let c_j be

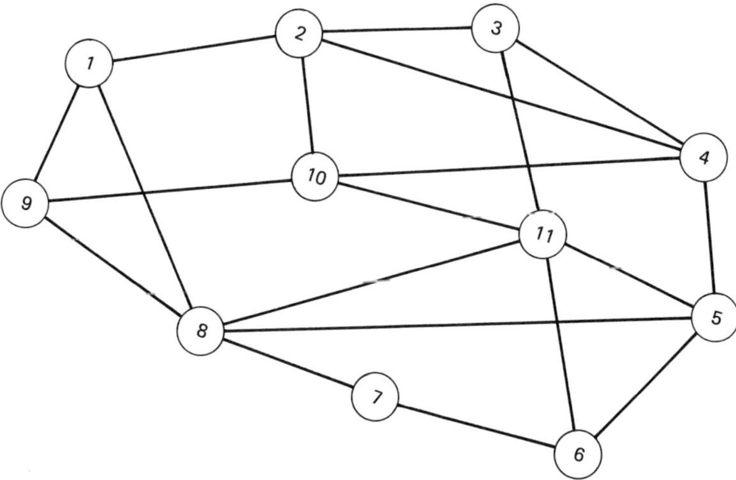

Figure 13.3

the cost of locating a fire hydrant at node j. Then the problem is to determine a minimum cost subset of nodes such that every edge contains at least one node from the subset. This is the problem of representing the edges of the network by a minimum cost subset of nodes. Let the set of nodes on the ith edge be $\mathbf{F}_i$. Then the minimum cost represent of $\mathbf{F} = \{\mathbf{F}_1, \ldots, \mathbf{F}_m\}$ provides an optimum solution of the problem.

3. *Facility Location Problem* There are m demand centers (or neighborhoods) in a region that requires the use of some facility (schools, fire stations, hospitals, snow removal equipment, banks, etc.). There are n possible locations for locating these facilities (some of these locations might be demand centers too). The distance between demand center i and location j is d_{ij} units. The cost of locating a facility in location j is c_j dollars. There is a restriction that every demand node should be within a distance of d units from its nearest facility. Let $\mathbf{F}_i = \{\text{location } j\colon d_{ij} \leqq d\}$ for $i = 1$ to m. Then a minimum cost represent for $\mathbf{F} = \{\mathbf{F}_1, \ldots, \mathbf{F}_m\}$ provides an optimum set of locations for the facilities.

4. *Airline Crew Scheduling* A *flight segment* is either a flight or a portion of a flight between two stations. There are m flight segments to be operated. A *trip* is a combination of one or more flight segments into duty periods for the crew, creating as many good days of work as possible, which meets the feasibility requirements. One of the feasibility requirements is that the first flight segment of a trip must depart from a station that is the same as the return station of the last flight segment. Another is that the flight segments must correspond to a good time sequence so that they can be combined into a trip. Also, if there is more than a day's work for a crew member in a trip, that trip must include enough layover periods.

The flight segments can be combined into various possible trips. Suppose a list of n trips is prepared. Each flight segment may appear on several trips in the list. Let c_j dollars be the cost of the jth trip (this may be the total wages for the crew including overtime and layover charges, if any). Let $\mathbf{F}_i = \{j\colon j\text{th trip contains the } i\text{th flight segment}\}$, $i = 1$ to m. Then a minimum cost represent for $\mathbf{F} = \{\mathbf{F}_1, \ldots, \mathbf{F}_m\}$ provides an optimum set of trips.

The set representation problem finds a vast number of such application in airline operations.

There are many other applications of the set representation problem in the design of switching circuits, in assembly line balancing and in various other areas.

SET COVERING PROBLEM

Let $\mathbf{N}$ be some nonempty finite set. Let $\mathbf{H} = \{\mathbf{H}_j\}$ be a given class of subsets of $\mathbf{N}$. A *cover* for $\mathbf{H}$ is a subclass $\mathscr{H} \subset \mathbf{H}$ satisfying

$$\bigcup_{\mathbf{H}_j \in \mathscr{H}} \mathbf{H}_j = \bigcup_{\mathbf{H}_j \in \mathbf{H}} \mathbf{H}_j$$

Let c_j be the cost of including the subset $\mathbf{H}_j$ in the cover. The problem of finding a minimum cost cover for $\mathbf{H}$ is known as a *set covering problem.*

Problem

13.1 Using integer variables with the interpretation

$$\begin{aligned} y_j &= 1 \quad \text{if } \mathbf{H}_j \text{ is included in the cover} \\ &= 0 \quad \text{otherwise} \end{aligned}$$

formulate the set covering problem as an integer program. Show that this integer program has the same form as the integer program corresponding to the set representation problem. Given any set representation problem, show that there is a set covering problem equivalent to it and vice versa. What is the set covering problem equivalent to the problem of finding a minimum cost represent for $\mathbf{F} = \{\{1,2,3,4,5,6\}, \{1,2,7,8,9\}, \{2,3,7,10\}, \{3,4,8,11,12\}\}$?

SET PARTITIONING PROBLEM

Let $\mathbf{N} = \{1, \ldots, m\}$. Let $\mathbf{H} = \{\mathbf{H}_1, \ldots, \mathbf{H}_n\}$ be a given class of subsets of $\mathbf{N}$. A *partition in* $\mathbf{H}$ is a subclass $\mathscr{H} \subset \mathbf{H}$ such that the subsets in $\mathscr{H}$ are mutually disjoint and their union is $\mathbf{N}$. Let c_j dollars be the cost of including $\mathbf{H}_j$ in the partition. Let

$$a_{ij} = 1 \quad \text{if } i \in \mathbf{H}_j$$
$$= 0 \quad \text{otherwise}$$

A is the element-subset incidence matrix of the class $\mathbf{H}$. $A_{.j}$ is the incidence vector of $\mathbf{H}_j$ written as a column vector. The problem of finding a minimum cost partition in $\mathbf{H}$ is equivalent to the integer program:

$$\begin{aligned} \text{Minimize} \quad & cx \\ \text{Subject to} \quad & Ax = e \\ & x_j = 0 \text{ or } 1 \qquad \text{for all } j \end{aligned}$$

where $e \in \mathbf{R}^m$ is a vector in which all the entires are one.

Notice the difference between the set representation problem (in which the constraints are of the form $Ax \geqq e$) and the set partitioning problem (in which the constraints are of the form $Ax = e$).

The set partitioning problem finds many applications. One of them is the following: Consider a region consisting of many sales areas. These areas have to be arranged into groups called sales districts. It should be possible for each sales district to be handled by a sales representative. Let $\mathbf{N} = \{1, \ldots, m\}$ be the sales areas in the region. Identify various subsets of $\mathbf{N}$, each of which is a feasible district (i.e., provides enough work for a sales representative and satisfies any other constraints that may be required). Let $\{\mathbf{H}_1, \ldots, \mathbf{H}_n\}$ be the various subsets generated. Let c_j be the cost of making $\mathbf{H}_j$ into a sales district. A minimum cost partition in $\{\mathbf{H}_1, \ldots, \mathbf{H}_n\}$ provides a set of optimal districts.

In a similar manner set partitioning problems can be applied in political districting, and in various other problems where a region has to be partitioned in an optimal manner.

SET PACKING PROBLEM

Let $\mathbf{N} = \{1, \ldots, m\}$ and $\mathbf{F} = \{\mathbf{F}_1, \ldots, \mathbf{F}_n\}$ be a family of subsets of $\mathbf{N}$. A *packing* for $\mathbf{F}$ is a subclass $\mathbf{D} \subset \mathbf{F}$ such that the subsets in $\mathbf{D}$ are mutually disjoint. Let d_j be the profit obtained by including $\mathbf{F}_j$ in a packing, $j = 1$ to n. Let A be the element-subset incidence matrix of $\mathbf{F}$. Then the set packing problem is

$$\begin{aligned} \text{Maximize} \quad & dx \\ \text{Subject to} \quad & Ax \leqq e \\ & x_j = 0 \text{ or } 1 \qquad \text{for all } j \end{aligned}$$

where e is again a vector, all the entries in which are one.

13.2.12 The Chinese Postman Problem

The street network of a town is given. It is a connected undirected network. It consists of edges (which can be traveled in both directions) and nodes (street intersections). The edges are $\mathbf{e}_1, \ldots, \mathbf{e}_m$ and c_i is the length of $\mathbf{e}_i$, $i = 1$ to m. All c_i are nonnegative. An *edge covering route* (ECR) is a route that begins at a node, travels along all the edges in the network in some sequence, and returns to the starting node in the end after passing through each edge of the network at least once. If an ECR, τ, passes ℓ_r times through edge $\mathbf{e}_r$, its length is $\sum_r \ell_r c_r$. The problem of finding a minimum length ECR is known as the *Chinese postman problem.*

An *Euler route* is an ECR that passes through each edge of the network exactly once. If an Euler route exists, it must be a minimal length ECR (since all c_r are nonnegative, and every ECR has to pass through each edge at least once, an ECR that passes through each edge exactly once must be optimal). In his famous theorem of 1736, Euler has proved that an Euler route exists on a connected undirected network iff every node has even degree.

Note The problem that aroused Euler's interest in this field is known as the *Königsberg bridge problem.* It is concerned with a river with two islands in the middle and seven bridges connecting the various land areas across the river as in Figure 13.4. The various land areas are denoted by 1, 2, 3, 4 and the bridges are denoted by $\mathbf{e}_1$ to $\mathbf{e}_7$. The problem that intrigued Euler is whether it is possible to start in one of the land areas and walk along each bridge exactly once, returning to the starting land area at the end. Represent each land area by a node of a network and each bridge by an edge connecting the nodes corresponding to the land areas that the bridge joins. The network is in Figure 13.5. Any walk that goes through each bridge exactly once, returning to the starting land area at the end, corresponds to an Euler route on this network. Work on this problem led Euler to his famous theorem. Since this network has odd degree nodes (all nodes are of odd degree) this network has no Euler route.

If all the nodes in a connected undirected network are of even degree, the Euler route in the network can be found by *Fleury's algorithm.* This algorithm is discussed in the problem below.

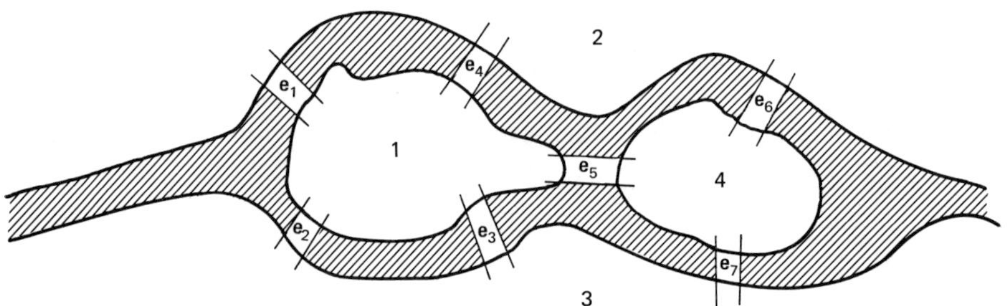

Figure 13.4

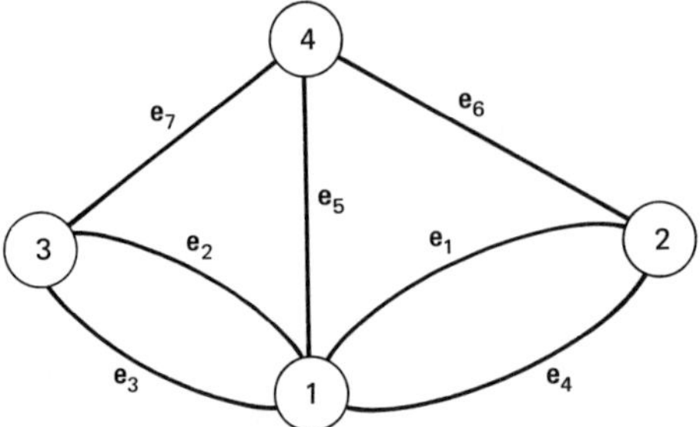

Figure 13.5

Problems

13.2 Let G be a connected undirected network, every node in which has even degree. Prove that an Euler route in G can be generated by the following method:

1. Start at any node, say node 1 and move along edges. Whenever an edge is traveled, erase that edge. When all the edges through a node are erased, erase that node too.
2. In each stage of the algorithm pick the edge for traveling as an edge that is not an *isthmus* of the remaining network (an *isthmus* at this stage is an edge, the removal of which leaves the remaining network disconnected, with two connected components, each of which contains at least one edge), containing the node at which you are at this stage and move to the other node on the edge picked.

Continue until all edges are traveled.

13.3 If G is a connected undirected network, prove that the total number of odd degree nodes in G is an even number. (Use the fact that every edge contains exactly two nodes.)

Let G be the original network. The edges in it are $\mathbf{e}_1, \ldots, \mathbf{e}_m$. If there are some odd degree nodes in G, let them be nodes $1, 2, \ldots, 2n$.

Suppose τ is an ECR that passes ℓ_r times through edges $\mathbf{e}_r$, $r = 1$ to m. Obtain the network G_τ by copying the edge $\mathbf{e}_r$ exactly ℓ_r times, $r = 1$ to m. Then τ must be an Euler route in G_τ and, hence, every node in G_τ must be an even degree node.

These facts and the fact that all c_r are nonnegative imply that an optimum ECR has the property that it passes through each edge of G at most twice. Also, $\mathscr{A}$, the set of edges through which it passes twice, satisfies the following property: The odd degree nodes $1, 2, \ldots, 2n$ can be partitioned into n pairs, say $(i_{11}, i_{12}), \ldots, (i_{n1}, i_{n2})$ such that there is a path from i_{r1} to i_{r2} among the set of edges $\mathscr{A}$, and $\mathscr{A}$ is the set of all edges on these paths.

Hence an optimum ECR in G can be obtained by the following method.

1. Find the shortest routes in G between all pairs of odd degree nodes in G. Let $\mathscr{P}(i, j)$ be a shortest route from i to j and let d_{ij} be its length for $i \neq j = 1$ to $2n$. Let $D = (d_{ij})$ be the $2n \times 2n$ shortest route distance matrix.
2. Find a way of partitioning the odd degree nodes $1, 2, \ldots, 2n$ into n pairs of the form $(i_{11}, i_{12}), \ldots, (i_{n1}, i_{n2})$, such that

$$\sum_{r=1}^{n} d_{i_{r1}, i_{r2}}$$

is minimized among all such pairings. Define integer variables with the following interpretation:

$$x_{ij} = 1 \quad \text{if odd degree nodes } i, j \text{ are paired}$$
$$= 0 \quad \text{otherwise}$$

Then the problem of partitioning the odd degree nodes into pairs optimaly is the integer program

$$\begin{aligned}
\text{Minimize} \quad & \textstyle\sum\sum d_{ij} x_{ij} \\
\text{Subject to} \quad & \sum_{i=1}^{2n} x_{ij} = 1 && \text{for all } j = 1 \text{ to } 2n \\
& \sum_{j=1}^{2n} x_{ij} = 1 && \text{for all } i = 1 \text{ to } 2n \\
& x_{ij} = x_{ji} && \text{for all } i, j \\
& x_{ij} = 0 \text{ or } 1 && \text{for all } i, j
\end{aligned}$$

where d_{ii} is defined to be a very large positive number to make sure that no odd degree node will be paired with itself in the above problem. This problem is the usual assignment problem with additional symmetry requirements (i.e., $x_{ij} = x_{ji}$ for all i, j). Hence this problem is known as a *symmetric assignment problem.* A feasible solution of this problem is known as a *symmetric assignment.*

Every extreme point of the usual assignment problem is known to be an integer vector. However, this integer property may not hold when symmetry requirements are augmented to the usual assignment problem. Hence the symmetric assignment problem is a nontrivial integer program.

3. If $(i_{11}, i_{12}), \ldots, (i_{n1}, i_{n2})$ is an optimal partition of the odd degree nodes in G into pairs, obtain a new network $\bar{G}$ by starting with G and duplicating all the edges along the shortest routes $\mathscr{P}(i_{r1}, i_{r2})$, $r = 1$ to n. The Euler route in $\bar{G}$ is an optimal ECR on G.

The Chinese postman problem finds many applications in determining optimum school bus routes, garbage collection vehicle routes, etc.

13.2.13 Traveling Salesman Problem

A salesman has to travel to cities $1, 2, \ldots, n$. The cost of traveling from city i to city j is c_{ij} dollars, for $i \neq j = 1$ to n. He wants to start in some city, visit each of the other cities exactly once in some order, and in the end return to the starting city. The problem is to determine an optimal order for traveling the cities so that the total cost is minimized.

Suppose he starts in city 1. If he travels to cities in the order i to $i + 1$, $i = 1$ to $n - 1$, and then from city n to city 1, this route can be represented by the order "$1, 2, \ldots, n; 1$." Such an order is known as a *tour*. So a tour is a *circuit* that leaves each city exactly once. Hence the starting city is immaterial and without any loss of generality we can fix it to be any city, say city 1. From city 1 he can go to any of the other $n - 1$ cities. So there are $n - 1$ different ways in which he can pick the city that he travels to from city 1. From that city he can travel to any of the remaining $n - 2$ cities etc. Thus the total number of possible tours in a n city traveling salesman problem is $(n - 1)(n - 2) \ldots 1 = (n - 1)!$

Let τ be any given tour. Define integer variables by

$$\begin{aligned} x_{ij} &= 1 \quad \text{if the salesman goes from city } i \text{ to city } j \text{ in tour } \tau \\ &= 0 \quad \text{otherwise} \end{aligned}$$

Then $x = (x_{ij})$ is obviously an assignment and it is an assignment corresponding to tour τ. Hence every tour is an assignment. But there are assignments that are not tours; for example, using the notation of section 12.4, we see that the assignment $\{(1, 2), (2, 1), (3, 4), (4, 3)\}$ is not a tour because the unit cells in this assignment do not form a single circuit spanning all cities but, instead, form two subtours (one spanning cities 1 and 2 and the other spanning cities 3 and 4). So the traveling salesman problem is:

$$\begin{aligned} \text{Minimize} \quad & \sum\sum c_{ij}x_{ij} \\ \text{Subject to} \quad & \sum_i x_{ij} = 1 \qquad j = 1 \text{ to } n \\ & \sum_j x_{ij} = 1 \qquad i = 1 \text{ to } n \end{aligned}$$

$$x_{ij} = 0 \text{ or } 1 \qquad \text{for all } i, j$$
$$x = (x_{ij}) \text{ is a tour assignment}$$

Here c_{ii} is defined to be a large positive number to make sure that all the x_{ii} will be zero. The last constraint that $x = (x_{ij})$ must be a tour assignment makes this a difficult problem to solve.

Various formulations of the traveling salesman problem as an integer program have been suggested. Here is one of them:

$$\begin{array}{lrl} \text{Minimize} & \sum\sum c_{ij}x_{ij} & \\ \text{Subject to} & \sum_i x_{ij} = 1 & j = 1 \text{ to } n \\ & \sum_j x_{ij} = 1 & i = 1 \text{ to } n \\ & u_i - u_j + nx_{ij} \leqq n - 1 & \text{for } 2 \leqq i \neq j \leqq n \\ & x_{ij} = 0 \text{ or } 1 & \text{for all } i, j \\ & u_i \text{ are arbitrary real numbers} & \end{array}$$

Problem

13.4 If (x, u) is a feasible solution of the above problem, prove that x must be a tour assignment. Conversely, if x is a tour assignment, prove that there exists a u such that (x, u) is feasible to the above mixed integer program.

13.2.14 Other Applications

Integer programming finds applications in *trim problems*. These are problems in which sheets of various sizes have to be cut from a big master sheet of given dimensions to minimize total wastage. Applications are also found in *container problems*, which deal with the problems of determining the optimal sizes and the minimal number of containers required for packing several objects of varying sizes. Also, since most logical constraints can be transformed into linear constraints involving 0-1 variables, combinatorial, graph theoretic, and network design problems can be posed as integer programs.

PROBLEMS

13.5 A chemical process consists of immersing components in four different tanks in serial order from tank 1 to tank 4. An overhead crane moves the components from one tank to the next. Each tank can hold only one component at a time. a_i, b_i (in seconds), respectively, are the minimum and maximum durations of time that a component can be held in tank i, $i = 1$ to 4. c_{ij} (in seconds) is the travel time for the empty crane from tank i to tank j. $c_{ii} = 0$ for all i and $c_{ij} = c_{ji}$. $c'_{ij} = c_{ij} + 20$ (in seconds) is the travel time for the crane carrying a component to move from tank i to tank j (the 20 seconds is the time required to lift the component at tank i and the time to immerse it in tank j). A cycle begins with the crane just removing a component out of tank 1 at time point 0. At time point 0 some of the other tanks may have components in them at varying stages of processing (i.e., with different elapsed times since those components were immersed in those tanks). The cycle ends (and the next cycle begins) when the crane is just removing a component out of tank 1 again next time with all the tanks in the same stages of processing as they were at the beginning of the cycle. Assume that during each cycle exactly one "component immersion" and "component taking out" take place in each tank (if a tank has a component in it at the beginning of the cycle, in that tank the "component taking out" operation takes place before "component immersion" operation occurs).

Note At the time a component is being taken out of tank i, tank $i + 1$ must be empty for $i = 1, 2, 3$. Also, if there are components in tanks i, $i + 1$, the component in tank $i + 1$ must be removed before the component in tank i can be.

Counting time in seconds beginning with $t = 0$ at the beginning of the cycle, let

n = number of tanks in the process. Here it is 4.

t_i = time at which a component is removed from tank i. Notice that $t_1 = 0$.

$$t_{max} = \text{maximum } \{t_1, \ldots, t_n\}$$

$$z_i = \begin{cases} 1 & \text{if } t_i = t_{max} \\ 0 & \text{otherwise} \end{cases}$$

$$y_{ij} = \left.\begin{cases} 1 & \text{if } t_i > t_j \\ 0 & \text{otherwise} \end{cases}\right\} \text{ for all } i > j$$

$c'_{n,j}$ = travel time for the crane to remove the component from the last tank, deposit it in inventory, and then come to tank j

(a) Prove that the cycle time is

$$t_{max} + \sum_{i=1}^{n} z_i(c'_{i,i+1} + c_{i+1,1})$$

Here $c_{n,n+1}$ is defined to be $c_{n,1}$.

(b) Using these variables and defining other variables that may be necessary, formulate the problem of minimizing the cycle time as an integer program.

(c) What is the integer programming formulation if the data is:

Tank (i)	a_i	b_i
1	120	∞
2	150	200
3	90	120
4	120	180

$(c_{ij}) =$		$j = 1$	2	3	4
	$i = 1$	0	12	16	18
	2	12	0	3	7
	3	16	3	0	3
	4	18	7	3	0

(d) Once an optimum (t_i) is obtained explain how it can be implemented as a work schedule for the crane.

—*L. Phillips and P. Unger*

13.6 OPERATOR SCHEDULING. The requirements of a process in terms of the number of operators during each hour of the day are given. d_i is the number of operators required from clock hour i to $i + 1$ for $i = 0$ to 24 (both $i = 0$ and $i = 24$ correspond to 12 midnight). Each operator has to follow a *shift*. A shift always begins on any hour of the day, yields a one hour break after either 3, 4, or 5 hours of work, and then resumes and continues until the total period worked becomes eight hours. With the break each shift spans a duration of nine hours. c_i dollars is the cost of an operator beginning his shift at clock hour i of the day. Obtain a minimum cost schedule specifying how many operators should work in what shifts to meet the given operator demands during the various hours of the day.

—*M. Segal*

13.7 MULTIPLE CHOICE PROBLEMS. There are k projects to be carried out. There are n_i alternate ways of carrying out project i, $i = 1$ to k. There are t resources required (these may be the dollars required in the various years of the planning horizon, other resources like raw materials, etc.) and b_p is the limit on the availability of the pth resource $p = 1$ to t. α_{ij} is the net profit if project i is carried out by alternative j. If project i is performed by alternative j, a_{ijp} units of resource p are required. Determine which alternative should be adopted for each project so that the total profit is maximized subject to resource availability. Formulate this as an integer program.

13.8 There are n centers in a region. Offices must be set up in k of these centers where k is a positive integer less than n. If an office is set up in center i, it generates a profit of a_{ii} dollars, $i = 1$ to n. If offices are located in centers i and j, a cost of a_{ij} dollars is incurred for communications between the offices. $A = (a_{ij})$ is a nonnegative matrix. Locate the offices optimally so as to maximize the net profit. Formulate it as an integer program.

13.9 The *quadratic assignment problem* is

$$\text{Minimize} \quad \sum_i \sum_j \sum_p \sum_q x_{ij} c_{ijpq} x_{pq} + \sum_i \sum_j d_{ij} x_{ij}$$

$$\text{Subject to} \quad \sum_{i=1}^{n} x_{ij} = 1 \qquad j = 1 \text{ to } n$$

$$\sum_{j=1}^{n} x_{ij} = 1 \qquad i = 1 \text{ to } n$$

$$x_{ij} = 0 \text{ or } 1 \qquad \text{for all } i, j$$

Define $y_{ijpq} = x_{ij} x_{pq}$. Clearly, y_{ijpq} are all 0-1 variables. Transform the quadratic assignment problem into an integer linear programming problem in terms of x_{ij}, y_{ijpq}.

13.10 Consider the optimization problem

$$\text{Minimize} \quad cx$$

$$\text{Subject to} \quad Ax = b$$

$$\text{Absolute value of } (d_{i1}x_1 + \cdots + d_{in}x_n - p_i) \geqq q_i, \qquad \text{for } i = 1 \text{ to } r$$

$$x_j \geqq 0 \qquad \text{for all } j$$

where q_i is a positive number for all $i = 1$ to r. Formulate this problem as an integer linear program.

13.11 There are n congressional committees. The jth committee has v_j vacancies, $j = 1$ to n. There are m congressmen who are eligible to fill these vacancies, where $m \leqq \sum v_j$. p_{ij} is the *preference* of congressman i to a vacancy in committee, j, $i = 1$ to m, $j = 1$ to n. Each vacancy has to be filled by a congressman. Let

$$x_{ij} = 1 \quad \text{if the } i\text{th congressman fills a vacancy in Committee } j$$

$$= 0 \quad \text{otherwise}$$

The preferences (p_{ij}) are all given, and they are properly scaled so that if $x = (x_{ij})$ is a *vacancy filling allocation*, then $\sum\sum x_{ij} p_{ij}$ is an appropriate measure of its performance. The following constraints have to be satisfied: (a) Each congressman can be allocated to fill the vacancies in at most two committees; (b) The committees are grouped into three groups called, exclusionary, semiexclusionary, and nonexclusionary committees. If a congressman is allocated to a vacancy in an exclusionary committee, he cannot fill any other vacancy. Also, if a congressman is allotted to fill a vacancy in a semi-exclusionary committee and another vacancy in some other committee, then the other committee must be a nonexclusionary committee.

Formulate the problem of finding an optimum vacancy filling allocation that maximizes the measure of performance subject to all the constraints, as an integer program. What is this formulation for the following numerical example?

The p_{ij} Matrix

	Committee j Agriculture	Bank	Education	Interstate	Judicial	Science	Interior	Post
Congressman								
$i =$ 1		1		$\frac{1}{2}$				$\frac{1}{2}$
2	$\frac{1}{3}$			1				1
3								
4				1	$\frac{1}{2}$			
5				1			$\frac{1}{2}$	
6				$\frac{1}{2}$	1	$\frac{1}{3}$		
7								
8				1	$\frac{1}{2}$			$\frac{1}{3}$
9				1				
10		1						
11				1				
12	$\frac{1}{2}$							
13	$\frac{1}{3}$	$\frac{1}{2}$		1	1			
Number of vacancies	5	3	1	1	1	2	1	1
Type	Semiex	Semiex	Semiex	Ex	Semiex	Semiex	Nonex	Nonex

All blank entries in the table are zero.

—*M. Cohen and S. Pollock*

13.12 A hospital prepares the work schedules for its nurses in two-week horizons. There must be at least three nurses on duty in each shift (day, evening, and night) of the 14-day scheduling horizon. Their nursing staff consists of 9 nurses, each working 5 shifts per week. A *schedule* for a nurse is a scheme that specifies the 10 shifts of the 42 shifts in a planning horizon that she should work.

Corresponding to each nurse a *list* of 30 possible schedules for her is made. The list of schedules for different nurses may be different.

Before the beginning of the planning horizon, each nurse provides a *desirability rating* for working according to each of the possible schedules in her list. Let c_{ij} be the desirability rating of the ith nurse ($i = 1$ to 9) for working her jth possible schedule ($j = 1$ to 30).

Given the list of schedules corresponding to each nurse and the 9×30 matrix of desirability ratings, select a schedule for each nurse so that the sum of the desirability ratings of the selected schedules is maximized. Formulate this as an integer program.

—*D. M. Warner*

13.13 Consider the problem:

$$\text{Minimize} \quad cx$$

$$\text{Subject to} \quad \sum_{j=1}^{n} a_{ij}x_j \quad \text{is either} \quad \leqq b_i'$$

$$\text{or} \quad \geqq b_i''$$

$$\text{for } i = 1 \text{ to } m$$

where all the data is given and $b_i' < b_i''$ for all i. Formulate this as an integer program. Pick a numerical example with $m = 3$, $n = 2$ and plot the set of all feasible solutions for this problem on the two-dimensional Cartesian plane.

13.14 Consider the balanced transportation model with m sources and n markets. The ith source can supply material to at most n_i markets, $i = 1$ to m.

The jth market can receive material from at most m_j sources, $j = 1$ to n. The unit transportation costs, the total availability at the sources, and the requirements at the markets are all given. Formulate the problem of determining a minimum cost shipping schedule subject to all the constraints as an integer program.

13.15 M is a given square matrix of order n. Consider the linear complementarity problem:

$$w - Mz = q$$

$$w \geqq 0, \; z \geqq 0$$

$$w^{\mathrm{T}}z = 0$$

Formulate this as an integer program.

13.16 (*Special basis problem*). The system of constraints

$$Ax = b$$

$$x \geqq 0$$

is given and it is known that $\{A_{.1}, \ldots, A_{.r}\}$ is a linearly independent set. Find a *feasible basis* for this system containing the columns $A_{.1}, \ldots, A_{.r}$, if such a basis exists. Formulate this problem as an integer program.

13.17 There are n jobs to be performed. The ith job requires the processing for p_i hours on a machine. It is available for processing at clock time a_i, and it must be finished by clock time b_i ($b_i \geqq p_i + a_i$), $i = 1$ to n. Formulate the problem of determining the minimum number of machines required to process the n jobs as an integer program.

13.18 In a standard balanced transportation problem with specified availabilities at the various sources and specified requirements at the various markets, we want to find a feasible solution $x = (x_{ij})$ in which the number of positive x_{ij} is as small as possible. Formulate this as an integer program.

13.19 The hospital director is faced with the problem of deciding the mix of rooms in a new facility to be built shortly. There are three types of rooms, single-bed, two-bed and three-bed rooms. The total number of rooms in the facility should be at most 70. There should be at least 100 beds in all. The percentage of single bed rooms in the facility should be between 15 and 30. The floor area required for a single-bed, two-bed, or three-bed room is 10, 14, and 17 square meters, respectively. Each patient in a two-bed or three-bed room requires the use of 80% of the man-hours of nonmedical personnel that are needed by a patient in a single bedroom. The profit (to the hospital) from a bed is inversely proportional to the number of beds in the room in which it is situated. Formulate the problem if the objective is (a) minimize the requirement of nonmedical personnal man-hours, (b) maximize total profit, (c) maximize the average profit per bed, and (d) minimize the total floor space needed at the facility.

13.20 A latin square of order n is a square matrix of order n, all entries in which are $1, 2, \ldots, n$ with the property that each number appear exactly once in each row and column. Two latin squares of order n are *orthogonal* if for each pair (r, s), $1 \leqq r \leqq n$ and $1 \leqq s \leqq n$, there exists exactly one position in the matrix, such that the entry in that position in one matrix is r and in the other matrix is s. Formulate the problem of finding a pair of orthogonal latin squares of order n as the problem of finding a feasible solution of an integer problem.

13.21 BALANCING THE MOVING WHEEL ON STEAM TURBINE ROTOR. There are n positions on the rotor and n blades, each one to be assembled in a position. r is the radius of the rim of the wheel. c_j is the distance from the filling end of blade j, to its center of gravity. w_j is the weight of blade j. $m_j = w_j(r + c_j)$ is called the moment of blade j about the axis of the wheel. Let

$$\alpha_i = \cos\left(\frac{2\pi i}{n}\right) \quad \text{and} \quad \beta_i = \sin\left(\frac{2\pi i}{n}\right)$$

Assign blades to positions in such a way that the sum of the horizontal (and vertical) components of the moment vectors are both as close to zero as possible. If θ_i is the moment of the blade in position i, then the horizontal component of the moment vector is $\sum_{i=1}^{n} \theta_i \alpha_i$, and the vertical component of the moment vector is $\sum_{i=1}^{n} \theta_i \beta_i$. Let

$$\begin{aligned} x_{ij} &= 1 \quad \text{if blade } j \text{ is assigned to position } i \\ &= 0 \quad \text{otherwise} \end{aligned}$$

Using these variables, formulate the problem of finding an optimum assignment of blades to positions as an integer program.

13.22 SWIMMING RELAY PROBLEM. The East Club has 12 swimmers. There are four types of swimming strokes: back, breast, butterfly, and freestyle, denoted by stroke type $i = 1$ to 4, respectively. t_{ij} is the time in seconds that the ith swimmer takes to cover 50 meters using stroke type j, and all the t_{ij} are given.

Group the swimmers into three teams of four swimmers each, to compete in a 200 meter relay against another club. The relay consists of four legs (each of 50 meters), each leg to be executed by a different stroke type and by a different member in the team. The *total time* of a team in the relay is the sum of the times taken by its members in the various legs.

(a) Formulate the problem of determining the best team (i.e., the one that uses the least possible total time) from the swimmers in the club, as an integer program.

(b) The East Club has to compete in the relay with three teams from another club, the West Club. The total times of the three teams from the West Club are t_1, t_2, and t_3.

After the relay race the six teams of both the clubs are ranked in increasing order of the total time. The scores given to the teams in positions 1 to 6 in this ranking are 8, 6, 4, 3, 2, 1, respectively. The scores of the club are the sum of the scores attained by its three teams. Given the total times of the teams from the West Club (i.e., t_1, t_2, t_3) how would you team the swimmers of the East Club so as to maximize its score?

—*S. Pollock*

13.23 SEPARABLE PIECEWISE LINEAR PROGRAMMING. Consider the problem

$$\begin{aligned} \text{Minimize} \quad & \sum_{j=1}^{n} z_j(x_j) \\ \text{Subject to} \quad & Ax = b \\ & x \geqq 0 \end{aligned}$$

where each $z_j(x_j)$ is piecewise linear, not necessarily convex. Using the ideas in section 1.1.5, formulate this as an integer program.

13.24 SEPARABLE PROGRAMMING. Consider the problem

$$\begin{aligned} \text{Minimize} \quad & \sum_{j=1}^{n} z_j(x_j) \\ \text{Subject to} \quad & Ax = b \\ & x \geqq 0 \end{aligned}$$

where each $z_j(x_j)$ is a nonlinear function. Explain how this problem can be approximated by a separable piecewise linear program, and solved as a mixed integer program. How can the precision of the approximation be improved?

13.25 Consider the following system of constraints:

$$\begin{aligned} \sum_{j=1}^{n} a_{1j}x_j &= b_1 \\ \sum_{j=1}^{2} a_{2j}x_j &= b_2 \\ x_j &\geqq 0 \quad \text{and integer, for all } j \end{aligned} \tag{1}$$

where a_{1j}, a_{2j}, b_1, b_2 are all positive integers. Prove that if (1) has a nonnegative integer solution, x, then there must exist a j such that $b_2(a_{1j}/a_{2j}) \geqq b_1$. Also, let t be a positive integer satisfying $t > b_2$ (maximum $\{a_{1j}/a_{2j}: j = 1 \text{ to } n\}) \geqq b_1$. Consider the system of constraints

$$\begin{aligned} \sum_{j=1}^{n} (a_{1j} + ta_{2j})x_j &= b_1 + tb_2 \\ x_j &\geqq 0 \quad \text{and integer, for all } j \end{aligned} \tag{2}$$

Prove that the sets of feasible integer solutions of systems (1), (2) are the same.

REFERENCES

1. J. P. Arabeyre, J. Fearnley, F. C. Steiger, and W. Teather, "The Airline Crew Scheduling Problem, A Survey," *Transportation Science*, *3*, pp. 140–163, May 1969.
2. K. R. Baker, "*Introduction to Sequencing & Scheduling*," Wiley New York, 1974.
3. M. L. Balinski and R. Quandt, "On an Integer Program for a Delivery Problem," *Operations Research 12*, pp. 300–304, 1964.
4. G. Bradley, "Transformation of Integer Programs to Knapsack Problems," *Discrete Mathematics*, *1*, pp. 29–45, 1971.
5. G. B. Dantzig, "On the Significance of Solving Linear Programming Problems with Some Integer Variables." *Econometrica*, *28*, 1, pp. 30–44, January 1960.
6. R. H. Day, "On Optimal Extracting from a Multiple File Data Storage System: An Application of Integer Programming," *Operations Research*, *13*, pp. 482–494, 1965.
7. J. Edmonds, "The Chinese Postman Problem," *Operations Research*, *13* Suppl. 1 (1965) 373.
8. J. Edmonds and E. L. Johnson, "Matching, Euler Tours and the Chinese Postman" *Mathematical Programming*, *5*, pp. 88–124, 1973.
9. S. Elmaghraby and M. Wig, "On the Treatment of Stock Cutting Problems as Diophantine Programs," O. R. Center Report Number 61, North Carolina State University at Raleigh, May 11, 1970.
10. S. L. Hakimi, "Optimum Distribution of Switching Centers in a Communication Network and Some Related Graph Theoretical Problems," *Operations Research*, *13*, pp. 462–475, 1965.

11. H. M. Markovitz and A. S. Manne, "On the Solution of Discrete Programming Problems," *Econometrika*, *25*, *1*, January 1957.

12. K. G. Murty, "The Symmetric Assignment Problem," Operations Research Center University of California, Berkeley, *ORC*-67-12, 1967.

13. R. J. Wilson, *Introduction to Graph Theory*, Academic Press, New York, 1972.

See also the references listed in Chapters 14 and 15.

14 cutting plane methods for integer programming

14.1 INTRODUCTION

14.1.1 Rounding Off the Continuous Optimum Solution to a Neighboring Integer Point

Consider the general pure integer program:

$$\begin{aligned} \text{Minimize} \quad & z(x) = cx \\ \text{Subject to} \quad & Ax = b \\ & x \geqq 0 \end{aligned} \tag{14.1}$$

$$x \text{ an integer vector} \tag{14.2}$$

where A is a matrix of order $m \times n$. The problem obtained by relaxing the integer requirements (14.2) is known as the *associated LP*.

Solve the associated LP by the simplex method. If its optimum solution, $\bar{x}$, satisfies the integer requirements, (14.2), it is an optimum solution of the integer program and we are done. On the other hand, if one or more of the $\bar{x}_j$ are noninteger, $\bar{x}$ is not feasible to the integer program.

One may be tempted to conclude that the optimum solution of the integer program can be obtained by *rounding off $\bar{x}$ to a neighboring integer solution.* This *rounding off method* may be a practical method to use, if all the variables in the model are integer variables representing the number of various indivisible items used or produced, and are expected to have fairly large values in an optimum solution. Also the constraints in the model should be such that it is easy to decide whether feasibility is preserved by rounding off a noninteger value of a variable to the next larger integer or the nearest integer lower than it. For example, if

x_1 represents the number of automobiles of a specific model assembled in a production month and if there is a lower bound restriction on x_1, a value $\bar{x}_1 = 1000.1$ can probably be rounded off safely to 1001 without losing feasibility. If all the variables in the model are of this type and if an easy method of rounding off a noninteger solution to a neighboring feasible integer solution can be developed, then a practical approach is to take the feasible integer solution obtained by rounding off $\bar{x}$ as a reasonably good solution of the integer program.

In many integer programming models there are usually several 0-1 variables. In some models these 0-1 variables may be the *combinatorial choice variables* to determine which of the possible alternatives are to be implemented. If $\bar{x}$ is an optimum solution of the associated LP in such a model and $\bar{x}$ is noninteger, rounding it to a neighboring integer vector does not make sense. Besides, $\bar{x} \in \mathbf{R}^n$ can have 2^n neighboring integer points, and it is not clear which of them should be chosen as the rounded off integer point corresponding to $\bar{x}$. Many of these 2^n neighboring integer points may be infeasible, and the problem of determining a feasible neighboring integer point of $\bar{x}$ is itself a very hard problem when n is large.

A two constraint problem is illustrated in Figure 14.1 Feasible points lie on the side of the arrow on each half-space and should be integer points. The grid of integer points is also plotted in the figure. The optimum solution of the associated LP, $\bar{x}$, is noninteger. None of the four neighboring integer points of $\bar{x}$ is feasible to the problem.

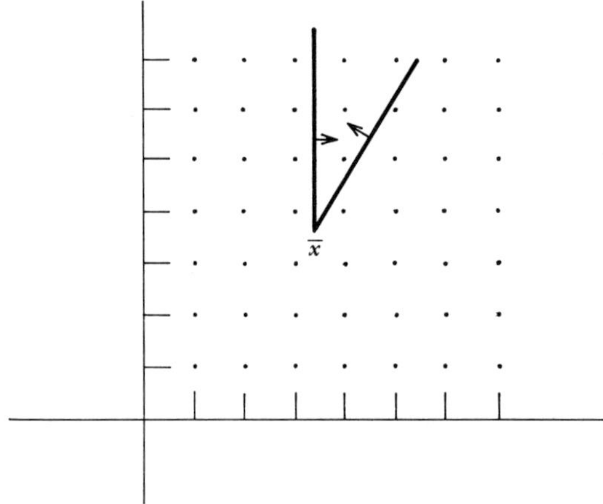

Figure 14.1 The dots (grid points) are the integer points.

Even if a feasible neighboring integer point of $\bar{x}$ is found, it may be quite far away from the optimum integer solution. In Figure 14.2, $\bar{x}$ is the optimum solution of the associated LP. x^1 is the only feasible neighboring integer point of $\bar{x}$. However, x^2 is the optimum integer solution.

Thus even though rounding off is a procedure that can be used in simple integer programming models, it cannot be used in models that contain combinatorial choice variables, or any integer variables whose values in an optimum solution of the problem are likely to be small.

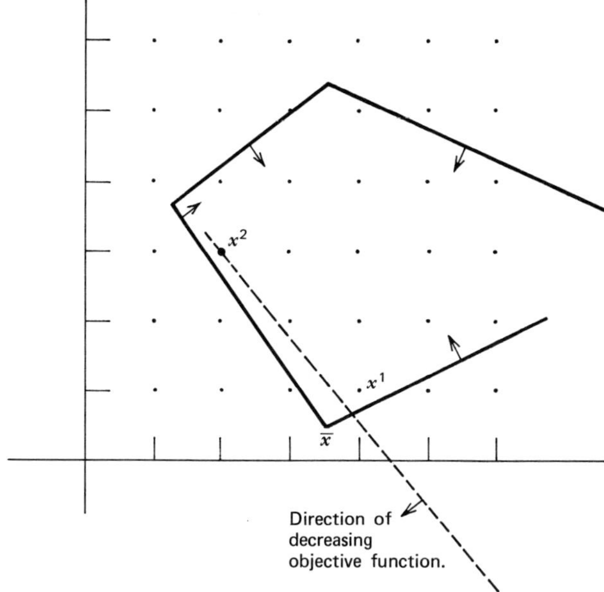

Figure 14.2

14.1.2 Cutting Planes

Let $\bar{x}$ be an optimum solution of the associated LP and suppose $\bar{x}$ is noninteger. A *cutting plane* is defined to be a hyperplane that *strictly separates* $\bar{x}$ from the set of integer feasible solutions of this problem, that is, it is a hyperplane with $\bar{x}$ on one side of it ($\bar{x}$ cannot lie on the hyperplane) and all the integer feasible solutions of the problem either on it or on the other side of it. Figure 14.3 contains

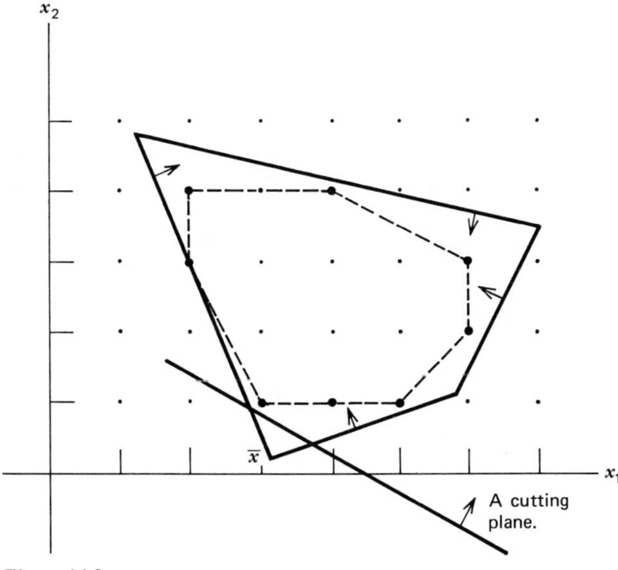

Figure 14.3

an illustration of a cutting plane. The associated LP has four constraints and $\bar{x}$ is an optimum solution for it. The convex hull of all the integer feasible solutions is the polytope outlined with dashed lines. $\bar{x}$ is on one side of the cutting plane and all the integer feasible solutions are on its other side.

In general, if $\sum_j a_{m+1,j}x_j = b_{m+1}$ is a cutting plane and if

$$\sum_j a_{m+1,j}\bar{x}_j < b_{m+1}$$

and

$$\sum_j a_{m+1,j}x_j \geqq b_{m+1} \qquad \text{for all integer feasible solutions,}$$

then the constraint $\sum_j a_{m+1,j}x_j \geqq b_{m+1}$ is known as a *cut*. A cut is a linear inequality constraint satisfying the following properties:

1. It eliminates the current noninteger optimum solution of the associated LP.
2. It is satisfied by all the integer feasible solutions of the original problem.

Example 1

Consider the following problem:

x_1	x_2	x_3	x_4	x_5	x_6	$-z$	b
1	0	0	1	−1	1	0	$\frac{3}{2}$
0	1	0	1	2	−2	0	$\frac{5}{2}$
0	0	1	−1	1	1	0	4
0	0	0	3	4	5	1	−20

$x_j \geqq 0$, and integer, for all j, z to be minimized

The optimum solution of the associated LP is $\bar{x} = (3/2, 5/2, 4, 0, 0, 0)^{\mathrm{T}}$. Consider the constraint

$$x_4 + x_5 + x_6 \geqq 1 \tag{14.3}$$

(x_1, x_2, x_3) is the optimum basic vector for the associated LP. x_4, x_5, x_6 are the nonbasic variables, and (14.3) requires that the sum of these nonbasic variables should be greater than or equal to one. All these nonbasic variables have zero values in $\bar{x}$. Hence (14.3) eliminates $\bar{x}$ from further consideration. Also, in every integer feasible solution of this problem at least one of these nonbasic variables has to be strictly positive, because $\bar{x}$ is the only solution of the associated LP in which all these nonbasic variables are zero, and $\bar{x}$ is noninteger. Hence (14.3) is satisfied by every integer feasible solution of this problem. The constraint (14.3) satisfies both properties 1. and 2. and is therefore a cut.

The addition of a cut to the constraints of the original problem slices off a portion of the set of feasible solutions of the associated LP. The sliced off portion includes the current optimum solution of the associated LP, but does not include any integer feasible solution.

14.1.3 A Cutting Plane Method

In solving an integer program, we find that the cutting plane method proceeds as follows:

1. Let the current problem be the current associated LP. Solve the current problem. In Step 1 of this method, the current problem is the associated LP of the original problem. It can be solved by the simplex method. In all subsequent steps an optimum solution of the current problem may be obtained efficiently by applying the dual simplex method or other techniques of sensitivity analysis, starting with an optimum solution of the previous problem.

   If the current problem is infeasible, the original integer program is also infeasible. Terminate.

   If the optimum solution of the current problem satisfies the integer requirements, it is an optimum solution of the original integer program. Terminate.

   If the optimum solution of the current problem violates some integer restrictions, go to 2.

2. Use the cut generation method and get a cutting plane. Add the cut to the constraints of the current problem. Go to 1.

   The method is repeated until termination occurs. We will only consider cutting plane methods satisfying the following property:

   (i) The method must terminate wither with the conclusion that the problem is infeasible or with an optimum integer solution in a finite number of steps.

The cutting plane method based on the cut generation scheme discussed in Example 1, section 14.1.2, was proposed by Dantzig [4]. However in [11] Gomory and Hoffman have constructed very simple examples in which this cutting plane method violates property (i). Cutting plane methods satisfying property (i) were first developed by Gomory [12 to 14] and these methods are discussed in section 14.2.

DEPTH OF CUTS

Let **K** be the set of feasible solutions of (14.1). Let $\mathbf{\Gamma}$ be the convex hull of the integer feasible solutions of (14.1). See Figure 14.3 for an illustration. Clearly every facet (see section 3.10 for definition) of $\mathbf{\Gamma}$ leads to a cutting plane.

Suppose

$$\sum_j a_{m+1,j}x_j \geqq b_{m+1} \qquad \text{and} \qquad \sum_j a_{m+1,j}x_j \geqq b'_{m+1}$$

are both cuts, where the coefficients, $a_{m+1,j}$, are the same in both the cuts, but $b_{m+1} > b'_{m+1}$. Clearly the first cut removes more of a region of **K** that is infeasible to the integer program than the latter cut. Hence the first cut is said to be a *deeper cut* than the latter cut.

In general, given two arbitrary cuts it is hard to compare them for *depth*. The deepest cuts are those corresponding to facets of $\mathbf{\Gamma}$. A cutting plane method with a cut generation scheme that generates deepest cuts can be expected to lead to a more efficient algorithm.

LEXICO DUAL SIMPLEX METHOD

In cutting plane literature, the simplex tableau is recorded in a special manner. We will discuss this notation first. Consider a general stage in solving the associated LP. For notational convenience, assume that the basic vector in this stage is $(x_1, \ldots, x_m)$. Let the canonical tableau with respect to this basic vector be as in Tableau 14.1.

Tableau 14.1

$x_1 \dots x_m$	$x_{m+1} \dots x_j \dots x_n$	$-z$	b
$1 \dots 0$	$\bar{a}_{1,m+1} \dots \bar{a}_{1,j} \dots \bar{a}_{1n}$	0	$\bar{b}_1$
$0 \quad 0$	$\bar{a}_{2,m+1} \dots \bar{a}_{2,j} \dots \bar{a}_{2n}$	0	$\bar{b}_2$
$\vdots \quad \vdots$	$\vdots \quad \vdots \quad \vdots$	$\vdots$	$\vdots$
$0 \dots 1$	$\bar{a}_{m,m+1} \dots \bar{a}_{m,j} \dots \bar{a}_{m,n}$	0	$\bar{b}_m$
$0 \dots 0$	$\bar{c}_{m+1} \dots \bar{c}_j \dots \bar{c}_n$	1	$-\bar{z}$

Tableau 14.1 represents the following system of constraints:

$$x_i = \bar{b}_i - \sum_{j=m+1}^{n} \bar{a}_{ij} x_j \qquad i = 1 \text{ to } m$$

$$-z = -\bar{z} - \sum_{j=m+1}^{n} \bar{c}_j x_j$$

In cutting plane literature, this system of constraints is recorded in tableau form as in Tableau 14.2.

Tableau 14.2

Variables	Constant terms	$-x_{m+1} \dots -x_{m+j} \dots -x_n$
$-z$	$-\bar{z}$	$\bar{c}_{m+1} \dots \bar{c}_{m+j} \dots \bar{c}_n$
x_1	$\bar{b}_1$	$\bar{a}_{1,m+1} \dots \bar{a}_{1,m+j} \dots \bar{a}_{1,n}$
x_2	$\bar{b}_2$	$\bar{a}_{2,m+1} \dots \bar{a}_{2,m+j} \dots \bar{a}_{2,n}$
$\vdots$	$\vdots$	$\vdots \quad \vdots \quad \vdots$
x_i	$\bar{b}_i$	$\bar{a}_{i,m+1} \dots \bar{a}_{i,m+j} \dots \bar{a}_{i,n}$
$\vdots$	$\vdots$	$\vdots \quad \vdots \quad \vdots$
x_m	$\bar{b}_m$	$\bar{a}_{m,m+1} \dots \bar{a}_{m,m+j} \dots \bar{a}_{m,n}$
x_{m+1}	0	$-1 \dots 0 \dots 0$
$\vdots$	$\vdots$	$\vdots \quad \vdots \quad \vdots$
x_{m+j}	0	$0 \dots -1 \dots 0$
$\vdots$	$\vdots$	$\vdots \quad \vdots \quad \vdots$
x_n	0	$0 \dots 0 \dots -1$

All the variables are recorded on the left-hand side of the tableau, and in all tableaus during the algorithm they are always recorded in the same order. Let the columns of Tableau 14.2 be numbered as column 0, 1, 2, . . . , $n - m$, and let the rows be numbered as rows 0, 1, 2, . . . , n. Column 0 contains the values of all the variables in the current solution. Each column from 1 to $n - m$ is associated with the negative of a nonbasic variable. These associated variables are recorded on the top of the tableau. In each iteration, one of these is changed.

Row 0 is the objective row. Each row from 1 to n is associated with a decision variable in the problem. If x_i is a basic variable, the row associated with x_i in the current tableau provides an expression for x_i as an affine function of the current nonbasic variables. If x_{m+j} is a nonbasic variable, the row associated with x_{m+j} in the current tableau leads to the equation "$x_{m+j} = (-1)(-x_{m+j})$." In the tableau, every column from columns 1 to $n - m$ contains a "-1" entry.

The tableau is primal feasible if all the entries in column 0 (excepting the entry in the objective row) are nonnegative. The tableau is dual feasible if all the entries in row 0 (excepting the entry in column 0) are nonnegative. The tableau is an optimum tableau if it is both primal and dual feasible.

As defined in Chapter 9, a column in the tableau is said to be *lexico positive* if the topmost nonzero entry in it is positive. The tableau is said to be *lexico dual feasible* if columns 1 to $n - m$ are all lexico positive.

Let x denote the column vector of all the variables in the problem including "$-z$", arranged in the order in which they appear on the left-hand side of the tableau. Let D_j be the jth column in the present tableau, $j = 0$ to $n - m$. Then the tableau can be written down as:

Variables	Constant terms	$-x_{m+1} \dots -x_{m+j} \dots -x_n$
x	D_0	$D_1 \quad \dots \quad D_j \quad \dots D_{n-m}$

A *pivot operation* on this tableau is the operation of replacing a variable in the present basic vector by a nonbasic variable. Suppose the *entering variable* for the pivot operation is x_{m+j}, let the *dropping variable* be x_i. Then D_j is the pivot column, row i is the *pivot row*, and the element in the pivot row and the pivot column is the *pivot element* for this pivot operation. The pivot operation transforms this tableau into:

Variables	Constant terms	$-x_{m+1} \dots -x_{m+j-1}$	$-x_i$	$-x_{m+j+1} \dots -x_n$
x	D_0^*	$D_1^* \quad \dots \quad D_{j-1}^*$	D_j^*	$D_{j+1}^* \quad \dots D_{n-m}^*$

where

$$D_j^* = \frac{-1}{\bar{a}_{i,\,m+j}} D_j$$

$$D_r^* = D_r - \frac{\bar{a}_{i,\,m+r}}{\bar{a}_{i,\,m+j}} D_j \qquad \text{for } r \neq j$$

This can be easily verified. Notice that the entries in D_j in the present tableau are $\bar{a}_{1,\,m+j}, \bar{a}_{2,\,m+j}, \dots, \bar{a}_{m,\,m+j}$ etc. Hence the pivot element is $\bar{a}_{i,\,m+j}$. The entry in row i and D_r is $\bar{a}_{i,\,m+r}$ in general. In this notation, $\bar{a}_{i,\,m}$ is used to denote the entry in row i and column 0 (i.e., the constant terms column). Also $\bar{a}_{0,\,m+r}$ denotes the entry in row 0 (i.e., the objective row) and column r, $r = 0$ to $n - m$.

In using cutting plane methods, it is preferable to use lexicographic minimum ratio rules for the choice of the pivot column. After adding several cutting planes, there is a possibility of cycling if the lexicographic minimum ratio rules are not used. We will briefly review the version of the lexico dual simplex algorithm that will be used in section 14.2.

Suppose the current tableau is lexico dual feasible, but not primal feasible. Identify a basic variable whose value in the present solution is negative. Suppose it is x_i. Then the row associated with x_i is the pivot row. If all the entries in columns

1 to $n - m$ in the pivot row are nonnegative, that is, $\bar{a}_{i,m+j} \geqq 0$ for $j = 1$ to $n - m$, the problem is infeasible. Terminate. Otherwise continue.

The pivot column is the column that is the lexico-minimum

$$\{D_j(1/\bar{a}_{i,m+j}): 1 \leqq j \leqq n - m \text{ and } j \text{ such that } \bar{a}_{i,m+j} < 0\}$$

Identifying this requires the following steps:

Step 0 Find minimum $\{\bar{c}_j/\bar{a}_{i,m+j}: 1 \leqq j \leqq n - m \text{ and } j \text{ such that } \bar{a}_{i,m+j} < 0\}$. If there are no ties for this minimum and it is attained by $j = r$, then the rth column (i.e., D_r) is the pivot column. If there are ties, go to the next step.

General Step *k* Find minimum $\{\bar{a}_{k,m+j}/\bar{a}_{i,m+j}: 1 \leqq j \leqq n - m, j \text{ is among those tied for the minimum in the previous step}\}$. If there are no ties for the minimum here, and it is attained by $j = r$, the rth column is the pivot column. If there are ties, go to the next step. Repeat until the pivot column is identified.

Perform the pivot and check whether the transformed tableau is primal feasible. If not continue in the same way. Clearly all these computations can be performed with the inverse tableaus using the results in Chapters 5 and 6. We leave it to the reader to figure out these details.

14.2 FRACTIONAL CUTTING PLANE METHOD FOR PURE INTEGER PROGRAMS

THE FRACTIONAL CUT

Consider the pure integer program (14.1) and (14.2). Solve the associated LP. For notational convenience, assume that the optimum basic vector $(x_1, \ldots, x_m)$ is obtained. Suppose the optimum tableau with respect to this basic vector is Tableau 14.2. $x_{m+1}, \ldots, x_n$ are the nonbasic variables in this tableau. If all the basic variables have integer values in the current solution, it is optimal to the integer program and we are done. Otherwise let x_i be a basic variable whose value in the present solution is noninteger. The row associated with x_i in the current tableau leads to the constraint

$$x_i = \bar{b}_i + \sum_{j=m+1}^{n} \bar{a}_{ij}(-x_j) \tag{14.4}$$

The value of x_i in the present solution is $\bar{b}_i$ and since this is noninteger, $\bar{b}_i > 0$. For any real number α, let $[\alpha]$ denote the largest integer less than or equal to α. Let $f_0 = \bar{b}_i - [\bar{b}_i]$ and $f_{ij} = \bar{a}_{ij} - [\bar{a}_{ij}]$, $j = m + 1$ to n. Since $\bar{b}_i$ is noninteger, $f_0 > 0$, f_0 is known as the *fractional part* of $\bar{b}_i$. From (14.4)

$$x_i = [\bar{b}_i] + f_0 + \sum_{j=m+1}^{n} [\bar{a}_{ij}](-x_j) + \sum_{j=m+1}^{n} f_{ij}(-x_j)$$

that is,

$$-f_0 - \sum_{j=m+1}^{n} f_{ij}(-x_j) = -x_i + [\bar{b}_i] + \sum_{j=m+1}^{n} [\bar{a}_{ij}](-x_j) \tag{14.5}$$

The right-hand side of (14.5) is an integer at every feasible solution of the integer program. So the left-hand side of (14.5) must be an integer too. However, when $x \geqq 0$, $-f_0 + \sum_{j=m+1}^{n} f_{ij}x_j \geqq -f_0$ because all f_{ij} are nonnegative. But $0 < f_0 < 1$. The conditions that $-f_0 + \sum_{j=m+1}^{n} f_{ij}x_j$ must be an integer and must be greater than or equal to $-f_0$ imply that it must be a nonnegative integer. So every integer feasible solution of the original problem satisfies

$$-f_0 + \sum_{j=m+1}^{n} f_{ij}x_j \geqq 0 \tag{14.6}$$

Since $x_{m+1}, \ldots, x_n$ are all nonbasic variables in the present tableau, their values in the present solution are zero. Hence the present solution of the associated LP does not satisfy (14.6). So (14.6) satisfies both properties 1. and 2. of section 14.1.2. Hence it is a cut. It is known as the *fractional cut* and the row associated with x_i is known as the *source row* from which this cut is derived. Let

$$s^1 = -f_0 + \sum_{j=m+1}^{n} (-f_{ij})(-x_j) \tag{14.7}$$

From previous arguments, at every feasible integer solution of the original problem, s^1 must be a nonnegative integer. So the augmented problem is also a pure integer program. Add (14.7) conveniently at the bottom of the present tableau. Treat s^1 as an additional basic variable. Here is the augmented tableau.

Variables	Constant terms	$-x_{m+1} \; \cdots \; -x_n$
$-z$	$-\bar{z}$	$\bar{c}_{m+1} \; \ldots \bar{c}_n$
x_1	$\bar{b}_1$	$\bar{a}_{1,m+1} \; \ldots \bar{a}_{1n}$
$\vdots$	$\vdots$	$\vdots \quad \vdots$
x_m	$\bar{b}_m$	$\bar{a}_{m,m+1} \; \ldots \bar{a}_{mn}$
x_{m+1}	0	$-1 \; \ldots \; 0$
$\vdots$	$\vdots$	$\vdots \quad \vdots$
x_n	0	$0 \; \ldots -1$
s^1	$-f_0$	$-f_{1,m+1} \cdots -f_{m,m+1}$

This tableau is dual feasible, but primal infeasible, because the basic variable s^1 has a negative value, $-f_0$, in the present solution. Therefore select the cut row just introduced as the pivot row and make a lexico dual simplex pivot. Just after this pivot s^1 becomes a nonbasic variable and becomes equal to zero in the next solution. Continue the application of the lexico dual simplex algorithm until either termination occurs due to infeasibility, or primal feasibility is achieved again. The solutions obtained during these computations are on the hyperplane corresponding to the cut (14.6) as long as s^1 remains a nonbasic variable and is equal to zero in the solution. If primal feasibility is attained after some dual simplex pivots, and the new optimum solution is again noninteger, obtain a similar cut from the tableau at that stage, add that to the tableau, and repeat the process.

Numerical Example

Consider the following problem

$$\begin{aligned} \text{Minimize} \quad & z(x) = 4x_1 + 5x_2 \\ \text{Subject to} \quad & 3x_1 + x_2 \geqq 2 \\ & x_1 + 4x_2 \geqq 5 \\ & 3x_1 + 2x_2 \geqq 7 \\ & x_1 \geqq 0, \quad x_2 \geqq 0 \quad x_1, x_2 \text{ integers} \end{aligned}$$

Let $x_3 = 3x_1 + x_2 - 2$, $x_4 = x_1 + 4x_2 - 5$, $x_5 = 3x_1 + 2x_2 - 7$ be the slack variables corresponding to the inequality constraints. Obviously these slack variables are also required to be nonnegative integers. The problem was transformed into standard form and the associated LP was solved. It led to the following tableau:

Variables	Constants terms	$-x_5$	$-x_4$
$-z$	$-\frac{112}{10}$	$\frac{11}{10}$	$\frac{7}{10}$
x_3	$\frac{42}{10}$	$-\frac{11}{10}$	$\frac{3}{10}$
x_4	0	0	-1
x_5	0	-1	0
x_1	$\frac{18}{10}$	$-\frac{4}{10}$	$\frac{2}{10}$
x_2	$\frac{8}{10}$	$\frac{1}{10}$	$-\frac{3}{10}$

The optimum solution of the associated LP is therefore: $x^1 = (x_1{}^1, x_2{}^1) = (18/10, 8/10)$. At this solution the slack variables are $(x_3{}^1, x_4{}^1, x_5{}^1) = (42/10, 0, 0)$. This solution is not an integer solution. Select the row associated with x_1 as the source row. It leads to the cut:

$$s^1 = -\frac{8}{10} - \frac{6}{10}(-x_5) - \frac{2}{10}(-x_4) \geqq 0 \tag{14.8}$$

Since $x_4 = x_1 + 4x_2 - 5$, $x_5 = 3x_1 + 2x_2 - 7$, the cut (14.8) in terms of the original variables, x_1 and x_2, is

$$s^1 = -0.8 + 0.6(3x_1 + 2x_2 - 7) + 0.2(x_1 + 4x_2 - 5) \geqq 0$$

that is,

$$2x_1 + 2x_2 - 6 \geqq 0$$

See Figure 14.4 where this is plotted as cut 1. The augmented tableau is as follows:

Variables	Constant terms	$-x_5$	$-x_4$	
$-z$	$-\frac{112}{10}$	$\frac{11}{10}$	$\frac{7}{10}$	
x_3	$\frac{42}{10}$	$-\frac{11}{10}$	$\frac{3}{10}$	
x_4	0	0	-1	
x_5	0	-1	0	
x_1	$\frac{18}{10}$	$-\frac{4}{10}$	$\frac{2}{10}$	
x_2	$\frac{8}{10}$	$\frac{1}{10}$	$-\frac{3}{10}$	
s^1	$-\frac{8}{10}$	$\left(-\frac{6}{10}\right)$	$-\frac{2}{10}$	Pivot row
		Pivot Column		

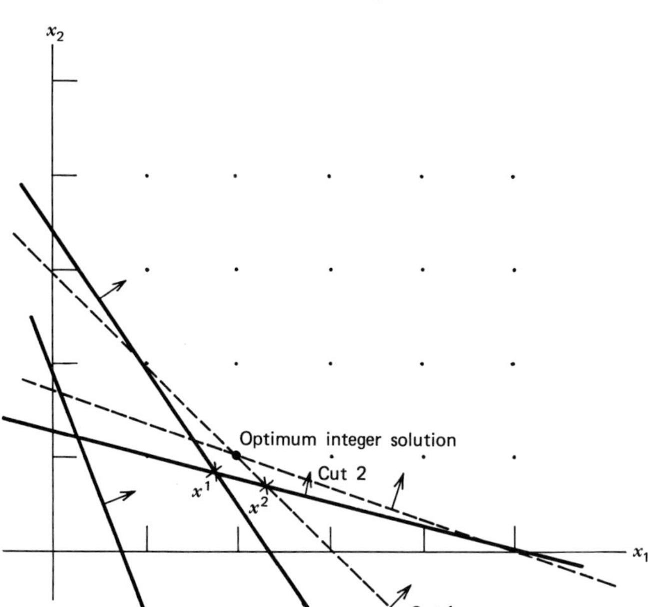

Figure 14.4

Performing the dual simplex pivot leads to:

Variables	Constant terms	$-s^1$	$-x_4$
$-z$	$-\frac{76}{6}$	$\frac{11}{6}$	$\frac{2}{6}$
x_3	$\frac{34}{6}$	$-\frac{11}{6}$	$\frac{4}{6}$
x_4	0	0	-1
x_5	$\frac{8}{6}$	$-\frac{10}{6}$	$\frac{2}{6}$
x_1	$\frac{14}{6}$	$-\frac{4}{6}$	$\frac{2}{6}$
x_2	$\frac{4}{6}$	$\frac{1}{6}$	$-\frac{2}{6}$
s^1	0	-1	0

This tableau is primal feasible and the optimum solution in this tableau, $x^2 = (x_1{}^2, x_2{}^2) = (14/6, 4/6)$, is again noninteger. Select the row associated with x_2 in this tableau as the source row. The cut is

$$s^2 = -\frac{4}{6} - \frac{1}{6}(-s^1) - \frac{4}{6}(-x_4) \geqq 0 \tag{14.9}$$

Since $s^1 = 2x_1 + 2x_2 - 6$ and $x_4 = x_1 + 4x_2 - 5$, this cut is $s^2 = x_1 + 3x_2 - 5 \geqq 0$ in terms of the original variables. This is plotted as cut 2 in Figure 14.4. Augmenting (14.9) to the present tableau leads to:

Variables	Constant terms	$-s^1$	$-x_4$
$-z$	$-\frac{76}{6}$	$\frac{11}{6}$	$\frac{2}{6}$
x_3	$\frac{34}{6}$	$-\frac{11}{6}$	$\frac{4}{6}$
x_4	0	0	-1
x_5	$\frac{8}{6}$	$-\frac{10}{6}$	$\frac{2}{6}$
x_1	$\frac{14}{6}$	$-\frac{4}{6}$	$\frac{2}{6}$
x_2	$\frac{4}{6}$	$\frac{1}{6}$	$-\frac{2}{6}$
s^1	0	-1	0
s^2	$-\frac{4}{6}$	$-\frac{1}{6}$	$\left(-\frac{4}{6}\right)$ Pivot row
			Pivot column

Performing the dual simplex pivot in the last row leads to:

Variables	Constant terms	$-s^1$	$-s^2$
$-z$	-13	$\frac{7}{4}$	$\frac{2}{4}$
x_3	5	-2	1
x_4	1	$\frac{1}{4}$	$-\frac{6}{4}$
x_5	1	$-\frac{7}{4}$	$\frac{2}{4}$
x_1	2	$-\frac{3}{4}$	$\frac{2}{4}$
x_2	1	$\frac{1}{4}$	$-\frac{2}{4}$
s^1	0	-1	0
s^2	0	0	-1

The tableau is again primal feasible. The present solution is $(x_1, x_2) = (2, 1)$. Since this satisfies the integer requirements, it is an optimum solution of the original integer program and we terminate.

FINITE TERMINATION PROPERTY

Gomory [13] has proved that this cutting plane method terminates in a finite number of steps either with the conclusion that the integer program is infeasible or with an optimum integer feasible solution. As it is not important for understanding the algorithm, the proof is not discussed here.

HOW LONG SHOULD A CUT BE KEPT IN THE TABLEAU?

Each cut adds a new variable (the slack variable corresponding to it) and a new row to the tableau. If all the cuts are retained in the tableau till termination, the tableau size can become unwieldy very quickly. Hence as newer cuts are added, old cuts should be discarded as soon as they become inactive.

Let s^k be the slack variable corresponding to a cut. In the first pivot after this cut is introduced in the tableau, s^k becomes a nonbasic variable. This implies that the value of s^k is zero in the solution obtained after that pivot. Hence that solution lies on the cutting plane corresponding to this cut. (The cut is $s^k \geqq 0$ and, hence, the corresponding cutting plane is $s^k = 0$) As long as s^k remains a nonbasic variable in the tableau, the solution in those steps is on the cutting plane corresponding to this cut. In all these steps this cut is therefore active. The variable s^k appears in the rest of the tableau with nonzero coefficients and the cut corresponding to it cannot be discarded.

At some stage later on, the algorithm may reach a pivot step in which s^k is selected as the entering variable. Just after this pivot step s^k becomes a basic variable in the tableau, and its value in the solution becomes positive. Hence this solution is not on the cutting plane ($s_k = 0$) corresponding to this cut and, therefore, this cut has become inactive. After the pivot that has made s^k a basic variable, s^k appears only in the row associated with s^k in the tableau and nowhere else. Thus after this pivot step, the cut can be discarded. This is achieved by erasing the row associated with the basic variable s^k and continuing the algorithm as usual with the rest of the tableau.

Hence a cut is discarded from the tableau the moment its associated slack becomes a basic variable. The tableau contains exactly $n - m$ nonbasic variables always. Hence, if this rule is followed, there will never be more than $n - m$ cuts at any stage of the algorithm.

Problems

14.1 Let h be an integer. Consider the source row

$$x_i = \bar{b}_i + \sum_{j=m+1}^{n} \bar{a}_{ij}(-x_j)$$

Multiplying both sides by h we get

$$hx_i = h\bar{b}_i + \sum_{j=m+1}^{n} ha_{ij}(-x_j) \tag{14.10}$$

Derive a cut using (14.10) as the source row by arguments similar to those used above.

Show that any integer linear combination of the rows of the tableau can be used as a source row for generating a cut in a similar manner.

14.2 Consider the pure 0-1 integer programming problem:

$$\begin{aligned} \text{Minimize} \quad & z(x) = cx \\ \text{Subject to} \quad & Ax = b \\ & x_j = 0 \text{ or } 1 \quad \text{for all } j \end{aligned} \tag{14.11}$$

The associated LP is

$$\begin{aligned} \text{Minimize} \quad & z(x) = cx \\ \text{Subject to} \quad & Ax = b \\ & x \geqq 0 \end{aligned} \tag{14.12}$$

$$x_j \leqq 1 \qquad \text{for all} \qquad j \tag{14.13}$$

The associated LP (14.12) and (14.13) becomes very cumbersome to solve if the upper bound constraints (14.13) have to be included explicitly in the tableau. This problem is solved much more efficiently using working bases and the upper bounding techniques discussed in Chapter 10. Develop the cutting plane method for solving (14.11) using these upper bounding techniques for solving the associated LPs at each stage. Show that this method can also be used for solving any bounded variable pure integer program.

14.3 CUTTING PLANE METHODS FOR MIXED INTEGER PROGRAMS

Consider the problem:

$$\begin{aligned} \text{Minimize} \quad & z(x) = cx \\ \text{Subject to} \quad & Ax = b \\ & x \geqq 0 \end{aligned} \tag{14.14}$$

$$x_j \text{ integer for all } j \in \mathbf{J} \tag{14.15}$$

where A is a matrix of order $m \times n$ and $\mathbf{J}$ is a given subset of $\{1, 2, \ldots, n\}$. $\mathbf{J}$ is the set of subscripts of all the variables that are restricted to integer values. All variables whose subscripts are not in $\mathbf{J}$ are continuous variables, and they can assume any nonnegative value within their range of variation.

Solve the associated LP and arrange the tableau as in sections 14.1 and 14.2. If all the variables x_j, $j \in \mathbf{J}$ are integer in the optimum solution of the associated LP, it is optimal to the integer program. Terminate. Otherwise add a cut, solve the augmented problem by using the dual simplex algorithm, and continue. The derivation of the cuts is discussed below.

For notational convenience, assume that $(x_1, \ldots, x_m)$ is the optimum basic vector obtained for the associated LP. So $\{m + 1, \ldots, n\}$ is the set of subscripts of nonbasic variables in this basis. Suppose the optimum tableau of the associated LP is Tableau 14.2. Select a basic variable x_i, $i \in \mathbf{J}$, whose value in the present solution is noninteger. The row associated with x_i in the current tableau is

$$x_i = \bar{b}_i + \sum_{j=m+1}^{n} \bar{a}_{ij}(-x_j) \tag{14.16}$$

As before let $f_0 = \bar{b}_i - [\bar{b}_i]$ and $f_{ij} = \bar{a}_{ij} - [\bar{a}_{ij}]$ for $j = m + 1$ to n, $j \in \mathbf{J}$. Let

$$\mathbf{H} = \{m + 1, \ldots, n\} \cap \mathbf{J} \qquad \mathbf{F} = \{m + 1, \ldots, n\} \backslash \mathbf{J}$$
$$\mathbf{P} = \{j : j \in \mathbf{F} \quad \text{and} \quad \bar{a}_{ij} \geqq 0\}$$
$$\mathbf{Q} = \{j : j \in \mathbf{F} \quad \text{and} \quad \bar{a}_{ij} < 0\}$$

Then (14.16) can be rewritten as:

$$\sum_{j \in \mathbf{H}} f_{ij} x_j + \sum_{j \in \mathbf{F}} \bar{a}_{ij} x_j = u + f_0 \tag{14.17}$$

where $u = [\bar{b}_i] + \sum_{j \in \mathbf{H}} [\bar{a}_{ij}](-x_j) - x_i$. u is clearly an integer at every feasible solution of the original mixed integer program (14.14) and (14.15). So u must be either nonnegative or less than or equal to -1. If $u \geqq 0$, from (14.17) we have

$$\sum_{j \in \mathbf{H}} f_{ij} x_j + \sum_{j \in \mathbf{P}} \bar{a}_{ij} x_j \geqq f_0 \tag{14.18}$$

If $u \leqq -1$, from (14.17) we have

$$\sum_{j \in \mathbf{H}} f_{ij} x_j + \sum_{j \in \mathbf{F}} \bar{a}_{ij} x_j \leqq -1 + f_0$$

This implies

$$\sum_{j \in \mathbf{F}} \bar{a}_{ij} x_j \leqq -1 + f_0$$

and hence

$$\sum_{j \in \mathbf{Q}} \frac{x_j f_0(-\bar{a}_{ij})}{1 - f_0} \geqq f_0 \tag{14.19}$$

Therefore (14.18) holds if $u \geqq 0$ and (14.19) holds if $u \leqq -1$. One of the inequalities (14.18) or (14.19) must always hold. Also the expressions on the left-hand side of both (14.18) and (14.19) must be nonnegative at every feasible solution of the problem. Hence

$$s^1 = -f_0 + \sum_{j \in \mathbf{H}} (-f_{ij})(-x_j) + \sum_{j \in \mathbf{P}} (-\bar{a}_{ij})(-x_j) + \sum_{j \in \mathbf{Q}} \frac{(-x_j) f_0 \bar{a}_{ij}}{1 - f_0} \geqq 0 \tag{14.20}$$

at every feasible solution of (14.14) and (14.15). The value of s^1 in the current solution is $-f_0$, which is negative. Hence the current solution violates (14.20). The inequality (14.20) satisfies both properties 1. and 2. of section 14.1.2 and, hence, it is a cut.

Add the cut (14.20) at the bottom of the tableau. Treating s^1 as an additional basic variable, use the dual simplex algorithm to find an optimum solution of the augmented LP. Continue in the same manner until termination occurs either with the conclusion that (14.14) and (14.15) is infeasible or with an optimum feasible solution for it.

Gomory [10] has proved that this cutting plane method terminates in a finite number of steps. The comments made in section 14.2 also apply here on the question of how long a cut should be retained in the tableau.

14.4 OTHER CUTTING PLANE METHODS

Cutting planes were first used by Dantzig, Fulkerson, and Johnson [5] for solving the traveling salesman problem. There is now a vast amount of literature on cutting plane methods in integer programming. Several new techniques for generating cuts have been proposed. See Gomory and Johnson [16] and Jeroslow [18, 19 and 20]. Rubin and Graves [23] discuss *strengthened Dantzig cuts. Convexity cuts* and other cuts are discussed by Glover [8] Balas [1] has proposed a new method of generating cuts based on convex analysis, and these cuts are called *intersection cuts.* In [7] and [25] cutting plane algorithms for solving an integer program by a *primal approach* (an approach that starts with an integer feasible solution and moves among integer feasible solutions until an optimum integer feasible solution is reach) are discussed.

Using arguments similar to those in Problem 14.1, it has been shown that a whole *group* of cuts can be constructed from the optimum tableau of the associated

LP, in which the optimum solution violates the integer requirements. This group is a finite additive Abelian group. Study of these groups has led to *group theoretic approaches* for solving integer programming problems. See Gomory [15]; also see Gorry and Shapiro [17] and [24].

14.5 HOW EFFICIENT ARE CUTTING PLANE METHODS

Even though it has been proved mathematically that the cutting plane algorithms discussed here solve integer programs in a finite number of steps, this finite number can be very large. In [21] simple problems are presented in which these algorithms require a very large number of steps before termination. The cutting plane methods are reported to work well on set representation problems [22], but in most other practical problems, these methods by themselves are not suitable. Extensive experience in solving integer programming problems that arise in practical applications indicates that branch and bound methods (Chapter 15) give much better results than cutting plane methods.

The greatest usefulness of cutting plane methods is probably within a branch and bound approach. In solving an integer program by the branch and bound approach using lower bounds based on relaxed linear programs, if a candidate problem is not fathomed (see Chapter 15 for definitions of these terms), several steps of a cutting plane method can be applied to see if it can be fathomed quickly using these methods. If the candidate problem is still not fathomed after a prescribed number of cuts are added, the cutting plane approach can be interrupted and the candidate problem branched. Such hybrid approaches using cutting plane methods in a branch and bound approach seem to work well.

PROBLEMS

14.3 Solve the following pure integer programs.

(a)

$$\begin{array}{ll} \text{Minimize} & 3x_1 - x_2 \\ \text{Subject to} & -10x_1 + 6x_2 \leqq 15 \\ & 14x_1 + 18x_2 \geqq 63 \\ & x_1, x_2 \geqq 0 \text{ and integer.} \end{array}$$

(b)

$$\begin{array}{lll} \text{Minimize} & \sum_{r=1}^{9} x_r & \\ \text{Subject to} & x_i + x_j \geqq 1 & \text{for all } (i, j) \in \mathscr{A} \\ & x_j = 0 \text{ or } 1 & \text{for all } j \end{array}$$

where $\mathscr{A} = \{(1, 2), (2, 3), (1, 3), (4, 5), (5, 6), (4, 6), (7, 8), (8, 9), (7, 9)\}$

REFERENCES

1. E. Balas, "Intersection Cut—A New Type of Cutting Planes for Integer Programming," *Operations Research*, *19*, pp. 19–39, 1971.
2. M. L. Balinski, "Integer Programming: Methods, Uses, Computation," *Management Science*, *12*, pp. 253–313, 1965.
3. M. L. Balinski and K. Spielberg, "Methods for Integer Programming: Algebraic, Combinatorial and Enumerative" in J. S. Aronofsky (editor), *Progress in Operations Research* Vol. III, Wiley, New York, 1969.

4. G. B. Dantzig, "Note on Solving Linear Programs in Integers," *Naval Research Logistics Quarterly*, *6*, pp. 75–76, 1959.
5. G. B. Dantzig, D. R. Fulkerson, and S. M. Johnson, "Solution of a Large Scale Traveling Salesman Problem," *Operations Research*, *2*, pp. 393–410, 1954.
6. F. Glover, "Stronger Cuts in Integer Programming," *Operations Research*, *15*, pp. 1174–1176, 1967.
7. F. Glover, "A New Foundation for a simplified Primal Integer Programming Algorithm," *Operations Research*, *16*, pp. 727–740, 1968.
8. F. Glover, "Convexity Cuts and Cut Search," *Operations Research*, *21*, pp. 123–134, 1973.
9. F. Glover, "On Polyhedral Annexation in Mixed Integer Programming," to appear in *Mathematical Programming*.
10. R. E. Gomory, "An Algorithm for the Mixed Integer Problem," *Rand Report*, P. 1885, February 1960.
11. R. E. Gomory and A. J. Hoffman, "On the Convergence of an Integer Programming Process," *Naval Research Logistics Quarterly*, *10*, pp. 121–123, 1963.
12. R. E. Gomory, "Outline of an Algorithm for Integer Solutions to Linear Programs," *Bulletin of Americal Mathematical Society*, *64*, pp. 275–278, 1958.
13. R. E. Goromy, "An Algorithm for Integer Solutions to Linear Programs," pp. 269–302, in *Recent Advances in Mathematical Programming*, R. L. Graves and P. Wolfe (eds.), MaGraw-Hill, New York, 1963.
14. R. E. Gomory, "All Integer Programming Algorithm," pp. 193–206 in J. F. Muth and G. L. Thompson (editors) *Industrial Scheduling*, Prentice-Hall, 1963.
15. R. E. Gomory, "Some Problems Related to Combinatorial Problems," *Linear Algebra and its Applications*, *2*, pp. 451–458, 1965.
16. R. E. Gomory and E. L. Johnson, "Some Continuous Functions Related to Corner Polyhedra," *Mathematical Programming*, *3*, pp. 23–85 and pp. 359–389, 1972.
17. G. A. Gorry and J. F. Shapiro, "An Adaptive Group Theoretic Algorithm for Integer Programming Problems," *Management Science*, *17*, pp. 285–306, 1971.
18. R. G. Jeroslow, "Cutting Planes for Relaxations of Integer Programs," *Management Science Research Report No. 347*, Graduate School of Industrial administration Cornegie-Mellon University, Pittsburgh, Pa. July 1974.
19. R. G. Jeroslow, "A Generalization of a Theorem of Chvâtal and Gomory," in *Nonlinear Programming II*, O. L. Mangasarian, R. R. Meyer, and S. M. Robinson (eds.), Academic Press, 1975.
20. R. G. Jeroslow, "The Principles of Cutting Plane Theory, II: Algebraic Methods, Disjunctive Methods," Management Science Research Report No. 370, Graduate School of Industrial Administration, Carnegie-Mellon University, Pittsburgh, Pennsylvania, September 1975.
21. R. G. Jeroslow and K. O. Kortanek, "On an Algorithm of Gomory," *SIAM*, *J. Applied Mathematics*, *21*, pp. 55–60 1971.
22. G. T. Martin, "An Accelerated Euclidean Algorithm for Integer Linear Programming" in R. L. Graves and P. Wolfe (eds.) *Recent Advances in Mathematical Programming*, McGraw-Hill, New York, 1963.
23. D. S. Rubin and R. L. Graves, "Strengthened Dantzig Cuts for Integer Programming," *Operations Research*, *20*, pp. 173–177, 1972.
24. H. A. Taha, *Integer Programming Theory, Application and Computations*, Academic Press, New York, 1975.
25. R. D. Young, "A Simplified Primal (All-Integer) Integer Programming Algorithm," *Operations Research*, *16*, pp. 750–782, 1968.

15 the branch and bound approach

15.1 INTRODUCTION

Branch and bound is a very useful approach for solving discrete optimization, combinatorial optimization, and integer programming problems in general. If the total number of feasible solutions for the problem being solved is small, one can evaluate each feasible solution individually and select the optimum by comparing them with each other. This is the *total or exhaustive enumeration method.* In most real life problems the total enumeration method is impractical, as the number of feasible solutions often turns out to be very large.

Branch and bound provides a methodology to search for an optimum feasible solution by doing only a *partial enumeration.* In the course of applying the branch and bound approach, the overall set of feasible solutions is partitioned into many simpler subsets. (This is what one would do in practice, if one is looking, say, for a needle in a haystack. The haystack is big and it is impossible to search all of it simultaneously. So one divides it visually into approximately its right and left halves and selects one of the halves to search for the needle first, while keeping the other half aside, to be pursued later, if necessary.) In each stage, one promising subset is chosen and an effort made to find the best feasible solution from it. If it is found, that subset is said to be *fathomed.* If the best feasible solution in it is not found, that subset is again partitioned into two or more simpler subsets (this operation is called *branching*) and the same process is repeated again.

The execution of the branch and bound approach is aided by the computation of either the optimum objective value (if it can be computed without too much effort) or a bound for it in each subset of the partition generated. The bounds are used in selecting promising subsets in the search for the optimum, and to discard some subsets that cannot possibly contain the optimum feasible solution. Sometimes the bounding technique applied on a subset produces fortuitously the best feasible solution in that subset, thus fathoming it.

Enumerative methods of optimization are in contrast with *algebraic methods* (i.e., those based on theoretically proven necessary and sufficient conditions for optimality; for example, the simplex method for solving linear programs, which tries to find a basis that is both primal and dual feasible) in that the former examine all feasible solutions enumeratively to search for the optimum, while the later do not. Enumerative methods are thus generally applied to discrete rather than continuous problems.

There are several ingredients in a branch and bound approach. We will discuss them with reference to a problem in which an objective function, z, has to be minimized. If an objective function z^1 is required to be maximized, we solve the equivalent problem of minimizing $z = -z^1$ subject to the same constraints. The ingredients are

1. A lower bounding strategy.
2. A branching strategy.
3. A search strategy.

15.2 THE LOWER BOUNDING STRATEGY

15.2.1 Purpose of the Lower Bounding Strategy

The minimum value of z is obtained precisely if the original problem is solved, but it may be very hard to solve. The lower bounding strategy computes a lower bound for the minimum value of z. It should be relatively easy to implement and computationally very efficient. Among several lower bounding strategies, the one which gives a bound closest to the minimum value of z, without undue increase in the computational effort is likely to lead to the most efficient algorithm. Some of the commonly used lower bounding strategies are discussed below. For examples illustrating the development of lower bounding strategies, see sections 15.6 to 15.10.

15.2.2 Relaxing Difficult Constraints

In this method all the difficult constraints are relaxed and z is minimized subject only to the remaining (easy to handle) constraints. The minimum value of z in the *relaxed problem* is a lower bound for the minimum value of z in the original problem.

15.2.3 Modified Objective Function

In this method a modified objective function, f, is constructed satisfying the properties:

1. $f \leqq z$ at all feasible solutions of the original problem.
2. Computationally, it is easy to minimize f subject to the constraints on the original problem.

If a function f satisfying these properties can be constructed, the minimum value of f is a lower bound for the minimum value of z.

15.2.4 Lagrangian Relaxation

Suppose the original problem is

$$\begin{aligned} &\text{Minimize} && z(x) && && \\ &\text{Subject to} && g_i(x) \geqq 0 && i = 1 \text{ to } r && (15.1) \\ &\text{and} && x \in \mathbf{X} && && (15.2) \end{aligned}$$

where $g_i(x) \geqq 0$, $i = 1$ to r, are the constraints that are difficult to handle, and $\mathbf{X}$ is the set of feasible solutions of the remaining constraints. When the objective function is in the minimization form, if we want to obtain a lower bound for z by relaxing some constraints using Lagrangian relaxation, all the inequality constraints to be relaxed should be written in the form "$g_i(x) \geqq 0$." A new objective function, known as the *relaxed Lagrangian* is then constructed by multiplying the left-hand side expression in each relaxed inequality by a nonnegative multiplier, and subtracting their sum from the original objective function. The multipliers used are called *Lagrange multipliers.* If $u = (u_1, \ldots, u_r) \geqq 0$ is the Lagrange multiplier vector used for relaxing (15.1), the relaxed Lagrangian for the above problem for relaxing (15.1) is $\mathrm{L}(x, u) = z(x) - \sum_{i=1}^{r} u_i g_i(x)$. The *relaxed problem* is

$$\begin{aligned} &\text{Minimize} && \mathrm{L}(x, u) \\ &\text{Subject to} && x \in \mathbf{X} \end{aligned}$$

The minimum objective value in the relaxed problem is clearly a function of the Lagrange multiplier vector u. Let it be $\mathrm{L}(u)$. It can be proved that for any $u \geqq 0$, $\mathrm{L}(u)$ is a lower bound for the minimum value of $z(x)$ in the original problem. The relaxed constraints are a subset of the set of constraints in the original problem. To determine which constraints to relax, the following guidelines should be used.

1. If $\mathbf{X}$ is the set of feasible solutions of the remaining unrelaxed constraints, then for any $u \geqq 0$, the computation of $\mathrm{L}(u)$ [i.e., the minimum value of $\mathrm{L}(x, u)$ over $x \in \mathbf{X}$)], should take very little effort.
2. The subset of relaxed constraints should be the smallest subset satisfying 1.

Once the subset of relaxed constraints is determined, if an algorithm can be developed for finding the maximum value of $\mathrm{L}(u)$ over $u \geqq 0$, or at least to find a value of $\mathrm{L}(u)$ close to its maximum value over $u \geqq 0$ with a reasonable amount of computational effort, that algorithm can be used to obtain a good Lagrange multiplier vector, u, and a good lower bound for the minimum value of z in the original problem.

The constraints to be relaxed and the algorithm to be used for finding a good Lagrange multiplier vector depend on the structure of the original problem. If good techniques for handling these aspects can be developed, this method can produce very satisfactory lower bounds.

Note Suppose the original problem is

$$\begin{aligned} &\text{Minimize} && z(x) && && \\ &\text{Subject to} && g_i(x) \geqq 0 && i = 1 \text{ to } r && (15.3) \\ &\text{and} && h_t(x) = 0 && t = 1 \text{ to } s && (15.4) \\ &\text{and} && x \in \mathbf{X} && && (15.5) \end{aligned}$$

and it is required to relax both the inequality constraints (15.3) and the equality constraints (15.4). Associate a Lagrange multiplier u_i with the relaxed constraint, $g_i(x) \geqq 0$, $i = 1$ to r, and a Lagrange multiplier v_t with the relaxed constraint, $h_t(x) = 0$, $t = 1$ to s. The relaxed Lagrangian is $L(x, u, v) = z(x) - \sum_{i=1}^{r} u_i g_i(x) - \sum_{t=1}^{s} v_t h_t(x)$. The Lagrange multiplier vector u has to be nonnegative, but the multiplier vector v is unrestricted. For any $u \geqq 0$, the relaxed problem is

$$\begin{array}{ll} \text{Minimize} & L(x, u, v) \\ \text{Subject to} & x \in \mathbf{X} \end{array}$$

If the optimum objective value in this problem is $L(u, v)$, it is a lower bound on the minimum objective value in the original problem. The best lower bound can be obtained by solving

$$\begin{array}{ll} \text{Maximize} & L(u, v) \\ \text{Subject to} & u \geqq 0 \\ & v \text{ unrestricted} \end{array}$$

15.3 THE BRANCHING STRATEGY

Let $\mathbf{S}$ be a set. A partition of $\mathbf{S}$ is a division of it into subsets $(\mathbf{S}_1, \ldots, \mathbf{S}_r)$ such that

(i) $r \geqq 2$, and

(ii) the subsets of $\mathbf{S}_1, \ldots, \mathbf{S}_r$ are mutually disjoint and their union is $\mathbf{S}$.

For example, $\{\{1\}, \{2, 3, 4\}\}$ is a partition of $\{1, 2, 3, 4\}$. The branching strategy partitions the set of feasible solutions of the original problem into two or more subsets. Each subset in the partition is the set of feasible solutions of a problem obtained by imposing additional constraints on the original problem. These problems are called *candidate problems.* The candidate problems must be generated in such a way that the lower bounding strategy can be applied on each of them.

In the process of applying the lower bounding strategy on a candidate problem, it may happen that an optimum solution for that candidate problem is obtained. A solution is an optimum solution of a candidate problem if it satisfies all the constraints in the candidate problem and has an objective value equal to the lower bound corresponding to this candidate problem. If this happens, that candidate problem is said to be *fathomed.*

If a candidate problem is not fathomed, we only have a lower bound for the minimum objective value in it. It should be possible to apply the branching strategy on any candidate problem and generate two or more candidate subproblems such that the following properties hold:

1. Each candidate subproblem is obtained by imposing additional constraints on the candidate problem.
2. The sets of feasible solutions of the candidate subproblem form a partition of the set of feasible solutions of the candidate problem.
3. The lower bounds on the minimum objective values in the candidate subproblems are as high as possible.

The operation of partitioning the set of feasible solutions of a candidate problem by generating the candidate subproblems is called *branching.* If Q is a candidate subproblem generated when candidate problem P is branched, P is known as the *parent problem* of Q. It turns out to be convenient if the branching strategy generates two candidate subproblems whenever it is applied. They are called *binary branching strategies* and they seem to lead to efficient algorithms. In our examples we will only discuss binary branching strategies. The overall efficiency of a branch and bound algorithm is very much dependent on the performance of the branching strategy employed regarding property 3 above.

15.3.1 To Branch Using A 0-1 Variable

In a candidate problem suppose there is a variable, x_1, which is restricted to be either 0 or 1. By adding one more constraint, "$x_1 = 0$" or "$x_1 = 1$," respectively, to the constraints on the candidate problem, two candidate subproblems are generated. In this branching strategy, the variable x_1 is known as the *branching variable.* Among the eligible set of variables, the branching variable should be selected so as to satisfy property 3 above as much as possible. The branching strategy must provide good selection criteria for branching variables.

15.3.2 To Branch Using an Integer Variable

Consider a candidate problem in which a linear function of the variables, $a_1x_1 + \cdots + a_nx_n$, has to be an integer. Let α be an integer. By adding one more constraint, "$a_1x_1 + \cdots + a_nx_n \leqq \alpha$" or "$a_1x_1 + \cdots + a_nx_n \geqq \alpha + 1$," respectively, to the constraints of the candidate problem, a branching into two candidate subproblems occurs.

15.4 THE SEARCH STRATEGY

15.4.1 Initial Stages

Initially, the original problem is the only candidate problem. Begin by applying the lower bounding strategy on it. Represent this by drawing a node corresponding to this problem and enter the lower bound by its side.

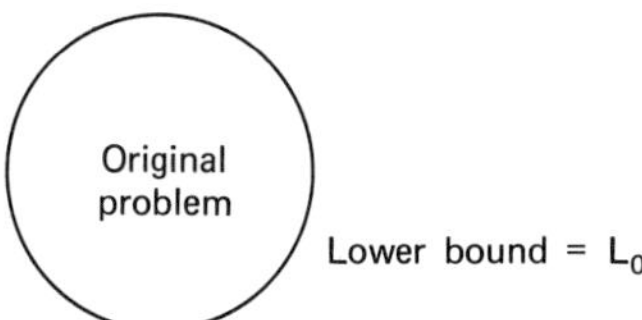

Figure 15.1

If an optimum solution of the problem is obtained, in the process of applying the lower bounding strategy, the algorithm terminates. Otherwise the branching strategy is applied on the original problem. Suppose this generates candidate problems 1 and 2. Apply the lower bounding strategy on these candidate problems. Figure 15.2 represents the situation at this stage. The node corresponding to the original problem had already been branched. The nodes corresponding to candidate problems 1, 2 have not yet been branched and these are known as the *terminal nodes at this stage* of the algorithm. The *list* is the collection of all the unfathomed candidate problems corresponding to the terminal nodes at this stage.

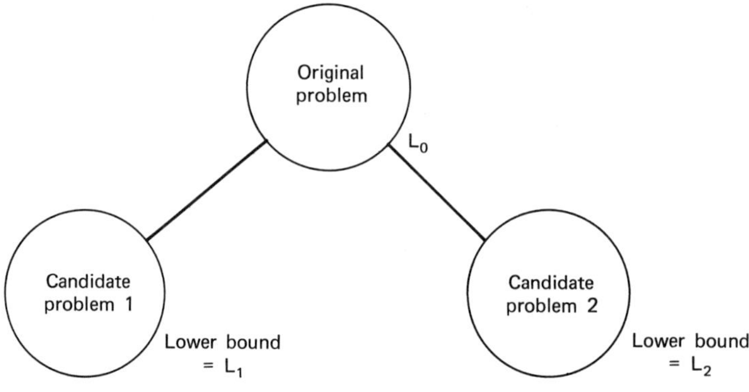

Figure 15.2

Suppose $L_1 < L_2$. The sets of feasible solutions of candidate problems 1, 2 form a partition of the set of feasible solutions of the original problem. Since $L_1 < L_2$, there is a possibility that the optimum solution of candidate problem 1 corresponds to an objective value less than L_2; if this is true, it is optimal to the original problem. So temporarily set aside candidate problem 2 and try to identify an optimum solution of candidate problem 1. If it has already been obtained during the lower bound computation, L_1 must be the objective value corresponding to it, hence it is optimal to the original problem and the algorithm terminates. Otherwise apply the branching strategy on candidate problem 1. Suppose this generates two new candidate subproblems *called candidate problems 11 and 12*. Remove candidate problem 1 from the list, and add the new candidate problems 11 and 12 to the list. Compute L_{11}, L_{12}, the lower bounds corresponding to them. The diagram at this stage is seen in Figure 15.3. This diagram is known as the *search tree* at this stage. The list at this stage consists of candidate problems 2, 11, and 12.

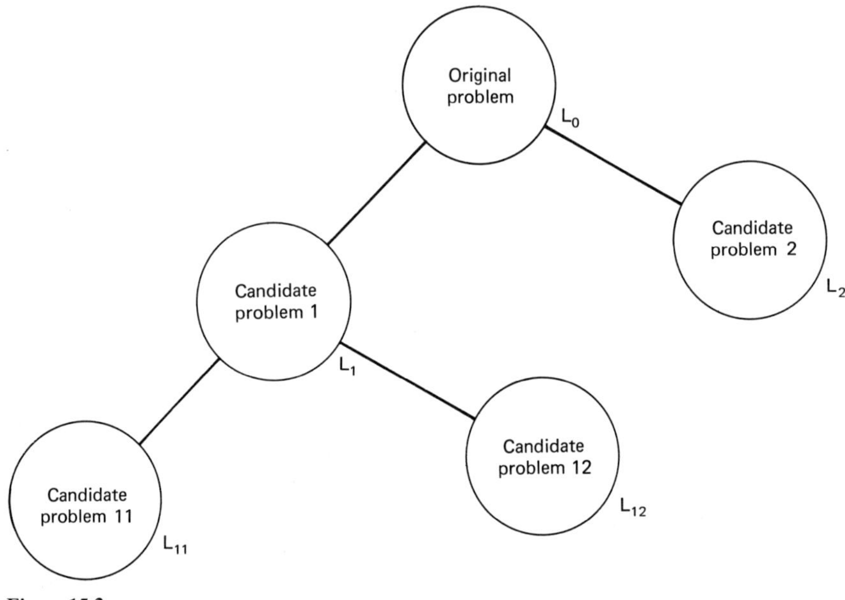

Figure 15.3

15.4.2 Incumbent and Pruning

At some stage of the algorithm if one or more of the candidate problems are fathomed, the feasible solution that has the least objective value among all the feasible solutions obtained until this stage is known as the *incumbent*. In the beginning we may not have an incumbent, but the moment a candidate problem is fathomed we have an incumbent.

Let $\bar{z}$ be the objective value corresponding to the incumbent. At this stage if there is a candidate problem in the list associated with a lower bound L where $L \geqq \bar{z}$, every feasible solution of that candidate problem makes the objective function greater than or equal to L (thus greater than or equal to $\bar{z}$); hence it is not better than the current incumbent. So delete that candidate problem from the list. This operation of deleting a candidate problem from the list is called *pruning*.

In the process of applying the lower bounding strategy on a candidate problem, we may find out that it is infeasible. Such infeasible candidate problems are also pruned from the list.

When a new candidate problem is generated, suppose the lower bounding strategy fathoms it. Suppose the optimum solution in this candidate problem is $\hat{x}$, which has an objective value of $\hat{z}$. Let $\bar{z}$ be the objective value of the current incumbent. If $\hat{z} < \bar{z}$, $\hat{x}$ is a better solution of the original problem than the current incumbent. Hence discard the current incumbent and make $\hat{x}$ the new incumbent. This is the operation of "*updating the incumbent*." On the other hand, if $\hat{z} \geqq \bar{z}$, that candidate problem is pruned. So whenever a newly generated candidate problem is fathomed, it is never added to the list. The optimum solution in it is used to update the incumbent.

Therefore at any stage of the algorithm, the list consists of all the unpruned, unfathomed, and unbranched candidate problems at that stage. The following properties are satisfied:

(i) The set of feasible solutions of the candidate problems in the list are mutually disjoint.

(ii) If there is an incumbent at this stage, any feasible solution of the original problem that is strictly better than the current incumbent is a feasible solution of some candidate problem in the list.

(iii) If there is no incumbent at this stage, the union of the sets of feasible solutions of the candidate problems in the list is the set of feasible solutions of the original problem.

15.4.3 The Search Continued

At the end of section 15.4.1, there were three terminal nodes, those corresponding to candidate problems 2, 11, and 12. If any one of these is fathomed, we would have an incumbent and we may be able to prune.

Identify a candidate problem that is associated with the least lower bound among all candidate problems in the list at this stage. Let this be candidate problem P. Delete P from the list and apply the branching strategy on it. Apply the lower bounding strategy on the candidate subproblems generated. If any of them turn out to be infeasible, prune them. If any of them is fathomed, update the incumbent. If there is a change in the incumbent, look through the list and prune. Among the newly generated candidate subproblems, add the unpruned and unfathomed ones to the list. Then go to the next stage.

Suppose candidate problem 2 is branched, producing candidate problems 21 and 22, associated with lower bounds L_{21} and L_{22}, respectively. In Figure 15.4 is the current search tree. If L_{12} is the least among L_{11}, L_{12}, L_{21}, and L_{22}, candidate problem 12 is the one to be branched in the next stage. After this branching, the search tree will appear as in Figure 15.5. The search tree is drawn in this discussion only to illustrate how the search is progressing. In practice the algorithm can be operated with the list of candidate problems and the incumbent (whenever it is obtained), and updating these after each stage.

The algorithm terminates when the list of candidate problems becomes empty. At termination, if there is an incumbent, it is an optimum feasible solution of the

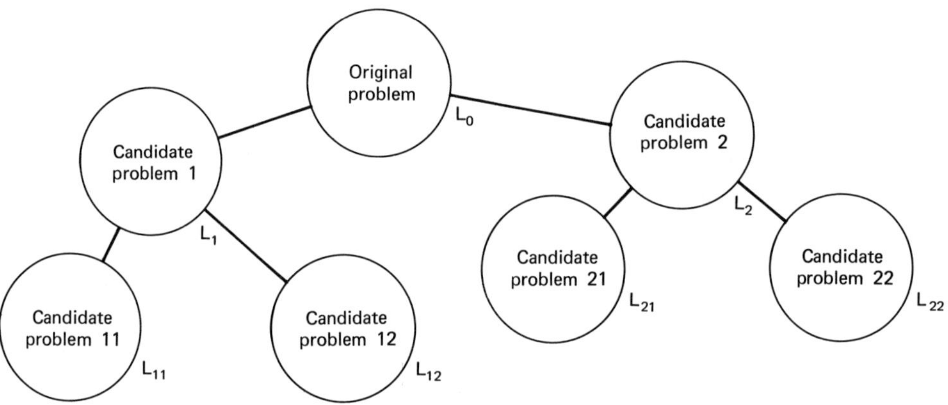

Figure 15.4

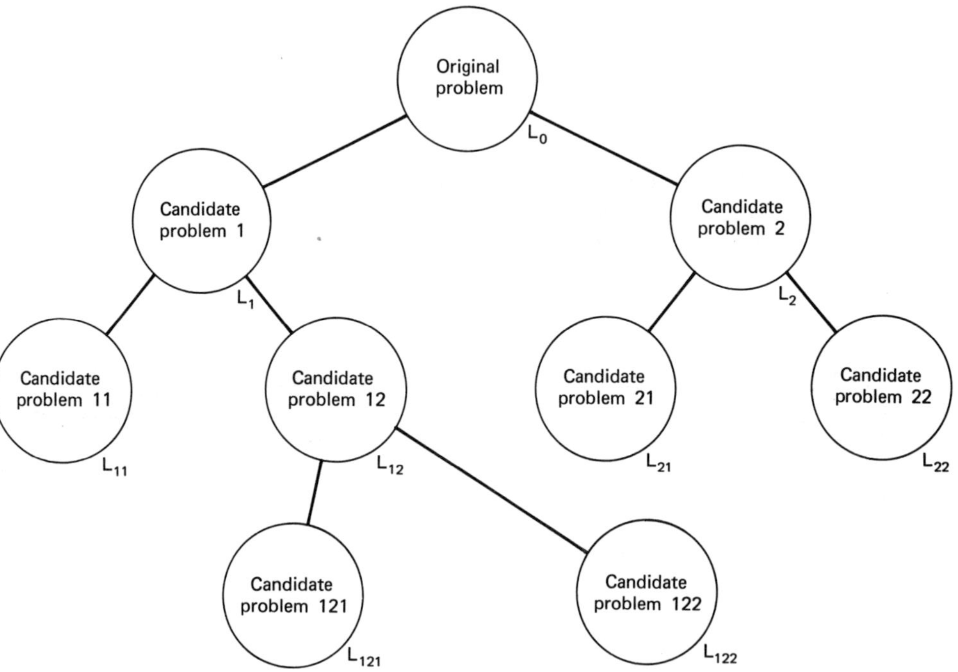

Figure 15.5

original problem. If there is no incumbent at termination, the original problem must be infeasible.

In general the search strategy dictates the sequence in which the generated nodes will be examined. The search strategy discussed above is a *jump-track strategy*, because it always jumps to examine the node with the least lower bound among those in the list at this stage. It is also known as a *priority strategy* with the least lower bound as the *priority criterion.*

15.4.4 Different Search Strategies

The search strategy based on the *least lower bound criterion* is a good search strategy, but it may involve too much work. Another search strategy, known as a *back-track strategy*, keeps one of the candidate problems from the list for the purpose of the search and calls it the *current candidate problem.* The other candidate problems in the list constitute the *stack.*

If the current candidate problem is fathomed, the incumbent is updated and the current candidate problem is discarded. If a change in the incumbent occurs, the necessary pruning is carried out in the stack. Then a candidate problem is selected from the stack and made the new current candidate problem. The algorithm then continues in a similar manner.

If the current candidate problem is not fathomed, the branching strategy is applied on it, and the lower bounding strategy applied on the candidate subproblems generated. If both the newly generated candidate subproblems are fathomed, the incumbent is updated, these candidate subproblems are dropped, the stack is pruned, and a new current candidate problem selected from the stack. If only one of the candidate subproblems generated is fathomed, the incumbent is updated, pruning is carried out and, if the other candidate subproblems is unpruned, it is made the new current candidate problem and the algorithm continued. If neither of the candidate subproblems generated is fathomed, the more promising one among them (may be the one associated with the least lower bound among them) is made the new current candidate problem, the other candidate subproblem is added to the stack and the algorithm continued.

When the algorithm has to select a candidate problem from the stack, the best *selection criterion* seems to be that of choosing the most recent problem added to the stack. This selection criterion is called LIFO (Last in First Out).

If this search strategy is employed, the algorithm terminates when it becomes necessary to select a candidate problem from the stack and the stack is found empty at that time. If there is an incumbent at this stage, it is an optimum solution of the original problem. If there is no incumbent, the original problem must be infeasible.

15.5 REQUIREMENTS FOR EFFICIENCY

For the branch and bound approach to work well, the bounding strategy must provide a lower bound fairly close to the minimum objective value in the problem but with very little computational effort. The branching strategy must generate candidate subproblems that have lower bounds as high as possible. Also the lower bounds in the candidate subproblems generated should be computable fairly easily.

A well-designed branch and bound method makes it possible to do extensive (and effective) pruning throughout, and this enables location of the optimum by examining only a very small fraction of the overall set of feasible solutions. That is why branch and bound methods are known as *partial enumeration methods*.

Practical experiences indicates that a backtrack search allows for extensive pruning and, hence, leads to more efficient algorithms.

In most well-designed branch and bound algorithms, it often happens that an optimum feasible solution of the original problem turns up as an incumbent at an early stage. A good heuristic in such situations is to terminate the algorithm at an intermediate stage (when the limit on the available computer time is reached) with the current incumbent as a near optimum solution.

The application of the branch and bound approach will now be illustrated with several examples.

15.6 THE 0-1 KNAPSACK PROBLEM

Consider a knapsack problem in which the capacity of the knapsack (by weight) is 35 units. There are 7 objects available, data on which are tabulated below. It is required to determine which objects are to be loaded into the knapsack to maximize the value-loaded subject to the weight capacity constraint.

Object number	Weight (units)	Value (dollars)	Value per unit weight (dollars/unit)
1	3	12	4
2	4	12	3
3	3	9	3
4	3	15	5
5	15	90	6
6	13	26	2
7	16	112	7

As in section 13.2.10, define the variables x_j with the interpretation:

$$x_j = 1 \quad \text{if object } j \text{ is loaded into the knapsack}$$
$$= 0 \quad \text{otherwise}$$

The problem is:

$$\begin{aligned}
\text{Minimize} \quad z(x) &= -12x_1 - 12x_2 - 9x_3 - 15x_4 - 90x_5 - 26x_6 - 112x_7 \\
\text{Subject to} \quad & 3x_1 + 4x_2 + 3x_3 + 3x_4 + 15x_5 + 13x_6 + 16x_7 \leqq 35 \\
& x_j \geqq 0 \quad \text{for all } j \qquad (15.6) \\
& x_j \leqq 1 \quad \text{for all } j \\
& x_j = 0 \text{ or } 1 \quad \text{for all } j \qquad (15.7)
\end{aligned}$$

Since we always work with minimization, the problem is stated in minimization form above. The associated LP (15.6) is a one-constraint bounded variable LP.

For the lower bounding strategy, relax (15.7) and solve the remaining problem. From the results in Chapter 10 it is clear that the optimum solution of (15.6) is obtained by making $x_j = 1$ for j in decreasing order of the value per unit weight until the weight capacity of the knapsack is reached, and at that stage by making the last variable equal to a fraction until the weight capacity is exactly used up. All the remaining variables are made equal to zero. The optimum solution of the relaxed problem, (15.6), obtained by this method is

$$x_7 = x_5 = x_4 = 1 \qquad x_1 = \frac{1}{3} \qquad x_2 = x_3 = x_6 = 0$$

$$\text{Objective value} = -221$$

If this solution satisfies (15.7), it would be an optimum solution of the original knapsack problem. However, since it does not (x_1 is not an integer), the original problem is not fathomed. A lower bound is -221 for the minimum objective value in the original problem.

Now the original problem has to be branched. For branching, use the strategy discussed in section 15.3.1. Clearly the branching variable can be chosen from the variables whose value in the optimum solution of the relaxed problem is noninteger. Here there is exactly one such variable. Hence this identifies the branching variable unambiguously.

The lower bounds in the candidate problems generated are also determined by solving the corresponding relaxed problem in which the integer requirements on the variables are relaxed. Select a candidate problem in the list associated with the least lower bound at this stage for branching.

The entire search tree for the algorithm is in Figure 15.6. Beside each node, the optimum solution of the relaxed problem is recorded by giving the values of the variables that are nonzero in the solution. If that solution is integer, it is an optimum solution of the candidate problem. All the additional constraints in a candidate problem (over those of the original problem) are entered inside the node representing that candidate problem. The candidate problems are all designated so as to make it easy to trace where they come from; for example, candidate problems 221 and 222 are generated when candidate problem 22 is branched. The following abbreviations are used.

L.B. = Lower Bound

B.V. = Branching Variable

C.P. = Candidate Problem

The various stages in the algorithm are discussed below.

Stage 1 The original problem is branched into candidate problems 1, 2 using x_1 as the branching variable. In candidate problem 1, the additional restriction is $x_1 = 1$. Since object 1 weighs 3 units, the question in this problem is to determine what subset of the remaining objects 2 to 7 should be loaded in the remaining capacity of 32 units in the knapsack. This is a knapsack problem with knapsack weight capacity of 32 units and choice limited to objects 2 to 7 only.

In candidate problem 2, the additional constraint is $x_1 = 0$. This is a knapsack problem with knapsack weight capacity of 35 units, but choice limited to objects 2 to 7 only.

Stage 2 Of the candidate problems 1, 2 in the list at this stage, candidate problem 2 is associated with the least lower bound. So select it for branching. The variable x_3, whose value in the relaxed optimum solution is a fraction, is the branching variable. Candidate problems 21, 22 are generated.

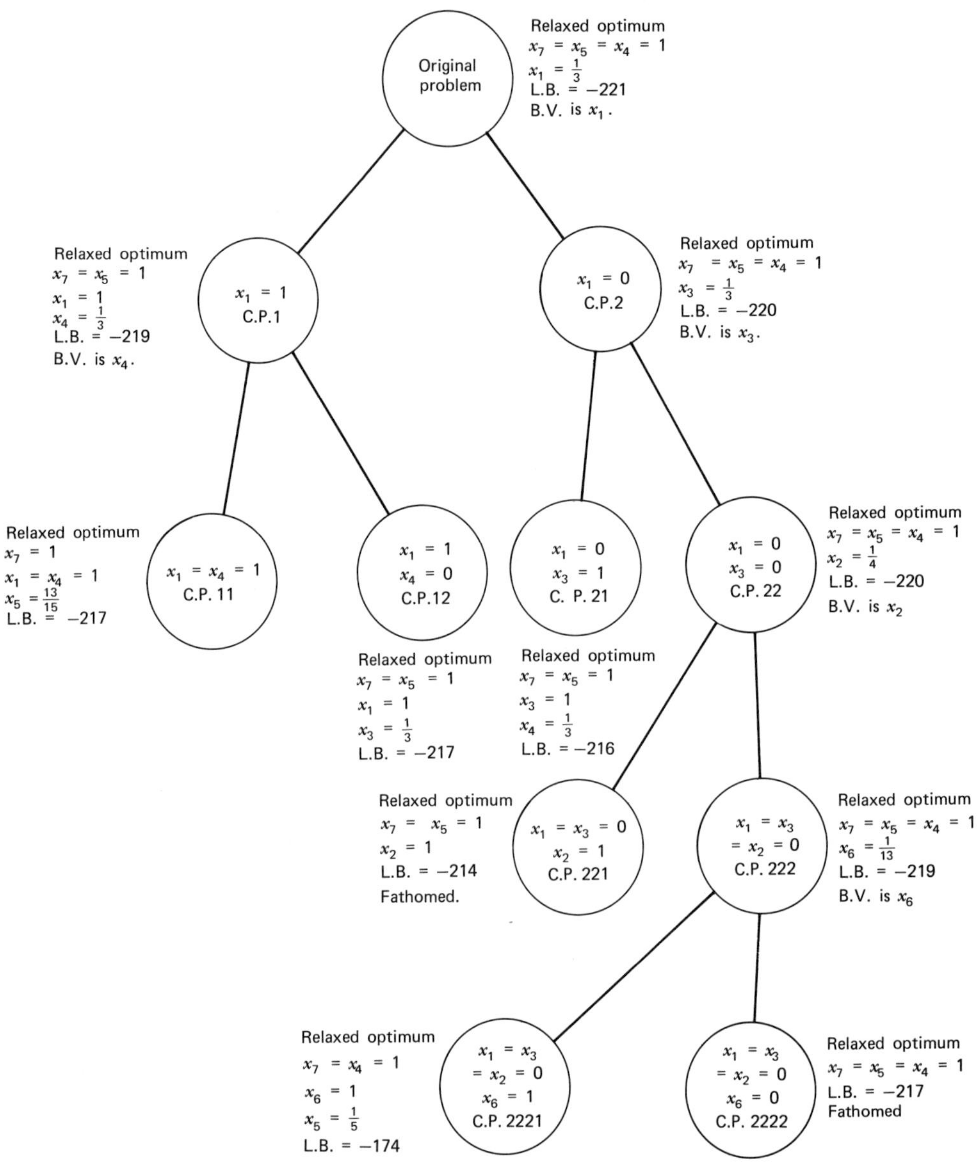

Figure 15.6

Stage 3 The list now consists of candidate problems 1, 21, and 22. Candidate problem 22 is associated with the least lower bound and so we select it for branching. Candidate problems 221 and 222 are generated. The relaxed problem corresponding to candidate problem 221 turns out to have an integer optimum solution. This solution is optimal to candidate problem 221, and it is the incumbent. Its objective value is −214. No pruning is possible.

Stage 4 The list now consists of candidate problems 1, 21, and 222. Candidate problem 222 is the one to be branched. Candidate problems 2221 and 2222 are generated. Since the relaxed optimum in candidate problem 2222 is an integer vector, that problem is fathomed. The objective value of the optimum solution is −217, less than the objective value of the current incumbent. So it becomes the new incumbent. Candidate problems 2221 and 21 are pruned.

Stage 5 The list now contains only candidate problem 1. Candidate problems 11 and 12 are generated. The lower bounds in both these problems turn out to be -217, the objective value of the current incumbent. Hence both these problems are pruned.

The list is now empty and hence the incumbent is an optimum solution of the original problem. It is

$$x_7 = x_5 = x_4 = 1, \qquad x_1 = x_2 = x_3 = x_6 = 0$$
$$\text{Objective value} = -217$$

This implies that an optimum choice is to load objects 7, 5, and 4 into the knapsack.

Problems

15.1 Check whether a different branching variable selection strategy would lead to a more efficient algorithm.

15.2 Solve the 0-1 knapsack problems with the following data:

(a)

Object	Weight (units)	Value (dollars)
1	1	4
2	4	9
3	17	14
4	2	3
5	3	9
6	4	10
7	13	14
8	3	2
Knapsack capacity	22	Maximize value

(b)

Object	Weight (units)	Value (dollars)
1	8	80
2	6	48
3	2	14
4	3	18
5	2	10
Knapsack capacity	11	Maximize value

15.7 THE TRAVELING SALESMAN PROBLEM

Consider the traveling salesman problem with the following cost matrix

Cost of traveling from city i to city j

$j =$	1	2	3	4	5	6
$i = 1$	×	27	43	16	30	26
2	7	×	16	1	30	25
3	20	13	×	35	5	0
4	21	16	25	×	18	18
5	12	46	27	48	×	5
6	23	5	5	9	5	×

Here $n = 6$. The problem is (section 13.2.13)

$$\begin{aligned} &\text{Minimize} \quad z_C(x) = \sum_i \sum_j c_{ij} x_{ij} \\ &\text{Subject to} \quad \sum_j x_{ij} = 1 \quad \text{for all } i \\ &\qquad \sum_i x_{ij} = 1 \quad \text{for all } j \\ &\qquad x_{ij} = 0 \text{ or } 1 \quad \text{for all } i, j \end{aligned} \tag{15.8}$$

$$x = (x_{ij}) \text{ is a tour assignment} \tag{15.9}$$

It makes no sense to say that the salesman goes from city i to i itself. Hence all x_{ii} must be equal to zero in any tour assignment. This is achieved by making the cost coefficients of all x_{ii} (i.e., the c_{ii}) very large positive numbers that are denoted by "$\times$" in the cost matrix.

If the constraints (15.9) are relaxed, the remaining problem is the standard assignment problem, which can be solved efficiently by the Hungarian method (see section 12.4). The minimum objective value in the assignment problem (15.8) is a lower bound for the minimum cost of a tour. We will use this as a lower bounding strategy.

We will indicate an assignment x by the set $\{(i, j): i, j \text{ such that } x_{ij} = 1\}$. For our problem the optimum assignment turns out to be: $x^0 = \{(1, 4), (2, 1), (3, 5), (4, 2), (5, 6), (6, 3)\}$ with an objective value of 54. There are two subtours in x^0 and, hence, it is not a tour assignment. The problem is not fathomed and 54 is a lower bound on the minimum tour cost. The final reduced matrix after the assignment problem is solved is as follows

	$j =$	1	2	3	4	5	6
	$i = 1$	×	7	23	0	10	11
	2	0	×	11	0	25	25
$C^0 =$	3	13	8	×	34	0	0
	4	3	0	9	×	2	7
	5	0	36	17	42	×	0
	6	16	0	0	8	0	×

If x is any tour assignment, we have

$$z_C(x) = 54 + z_{C^0}(x) \tag{15.10}$$

Since the original problem is not fathomed it has to be branched. Use the same type of branching strategy as in section 15.3.1. If x_{st} is selected as the branching variable, candidate problems 1 and 2 are generated by adding one additional constraint, "$x_{st} = 0$" or "$x_{st} = 1$," respectively, to the original problem. In candidate problem 1, x_{st} is constrained to be zero. This is the same as the original problem, with the additional constraint that the salesman cannot go from city s to t. Let C_1 be the matrix obtained from C^0 by changing $c^0(s, t)$ to ∞. Then, if x is any tour assignment feasible to this problem,

$$z_C(x) = 54 + z_{C_1}(x) \tag{15.11}$$

To get a lower bound in candidate problem 1, solve the assignment problem with C_1 as the cost matrix. Let r_1 be the objective value of the optimum assignment here and let C^1 be the final reduced matrix obtained after the assignment problem is solved. For any tour assignment, x, feasible to this problem

$$z_C(x) = 54 + r_1 + z_{C^1}(x)$$

Since $C^1 \geqq 0$, $54 + r_1$ is a lower bound in candidate problem 1. So r_1 is the amount by which the lower bound in candidate problem 1 is higher than the lower bound in the original problem.

We would like to select our branching variable in such a way that this quantity r_1 is as large as possible. If this is done we know that the lower bound of at least one of the candidate problems generated after the branching is as high as possible. However, determining the quantity r_1 for each eligible branching variable requires the solving of an assignment problem, which is too much computational effort. It is necessary to develop a criterion for selecting the branching variable, which does not involve too much computation, but which makes a reasonably good selection.

Define the *evaluation* of the branching variable x_{st} with respect to the reduced cost matrix C^0 to be the sum of the minimum entries in row s and column t of C^0 excluding $c^0(s, t)$. The quantity r_1, the amount by which the lower bound in candidate problem 1 is higher than the lower bound in the original problem, is greater than or equal to the evaluation of the branching variable. Hence a reasonably good selection rule for the branching variable is to select the variable that has the highest evaluation. We will use this selection rule.

Since C^0 has at least one zero entry in each row and column, if $c^0(i, j) > 0$, the evaluation of x_{ij} will be zero. So it is only necessary to compute the evaluations of variables corresponding to zero entries in C^0. They are given in the following table.

Branching variable	Evaluation
x_{14}	7
$x_{21}, x_{24}, x_{35}, x_{36}, x_{51}, x_{56}, x_{62}, x_{65}$	0
x_{42}	2
x_{63}	9

So select x_{63} as the branching variable. Candidate problem 1 has the additional restriction "$x_{63} = 0$." C_1 is the matrix obtained by changing $c^0(6, 3)$ to ∞. Solve the assignment problem with C_1 as the cost matrix. The optimum assignment is found to be, $\bar{x}$, which is $\{(1, 4), (4, 3), (3, 5), (5, 6), (6, 2), (2, 1)\}$. $z_{C_1}(\bar{x}) = 9$. Also, $\bar{x}$ is a tour assignment. Hence it is an optimum tour assignment in this candidate problem and this problem is fathomed. $\bar{x}$ is the incumbent. $z_C(\bar{x}) = 63$ is the objective value of the incumbent.

Candidate problem 2 has the additional restriction "$x_{63} = 1$." When the salesman goes from 6 to 3 he cannot go from 3 to 6, as this completes a subtour. Hence this additional constraint automatically implies "$x_{36} = 0$." So the additional constraints in this candidate problem are "$x_{63} = 1$, $x_{36} = 0$." Let C_2 be the matrix obtained by striking off row 6 and column 3 from C^0 and changing $c^0(3, 6)$ to ∞. Let x be any tour assignment feasible to candidate problem 2 and let x^2 be the matrix obtained by striking off row 6 and column 3 from x. Then

$$z_C(x) = 54 + z_{C_2}(x^2)$$

Obtain the optimum assignment with C_2 as the cost matrix. Let r_2 be the optimal objective value in this assignment problem and let C^2 be the final reduced matrix obtained. Then

$$z_C(x) = 54 + r_2 + z_{C^2}(x^2)$$

for any tour assignment x feasible to this candidate problem. Thus $54 + r_2$ is a lower bound in candidate problem 2.

	$j =$	1	2	4	5	6
	$i = 1$	×	7	0	10	11
	2	0	×	0	25	25
$C_2 =$	3	13	8	34	0	×
	4	3	0	×	2	7
	5	0	36	42	×	0

The optimum assignment with C_2 as the cost matrix (and with its rows and columns numbered as in C_2) is $\{(1, 4), (4, 2), (2, 1), (3, 5), (5, 6)\}$ with an objective value of zero. Thus after relaxing the tour restrictions in candidate problem 2, the optimum assignment is $\{(1, 4), (4, 2), (2, 1), (3, 5), (5, 6), (6, 3)\}$, which is not a tour. The lower bound in this problem is 54 and it is not fathomed. The final reduced cost matrix corresponding to this problem is $C^2 = C_2$. The search tree at this stage is seen in Figure 15.7. Candidate problem 2 is the only problem in the list, so branch it. To select the branching variable, compute the evaluation in C^2 of all the variables corresponding to positions in which C^2 has a zero entry. These evaluations are:

Branching variable	Evaluation
x_{35}	10
x_{21}, x_{24}, or x_{51}	0
x_{42}	9
x_{56}, x_{14}	7

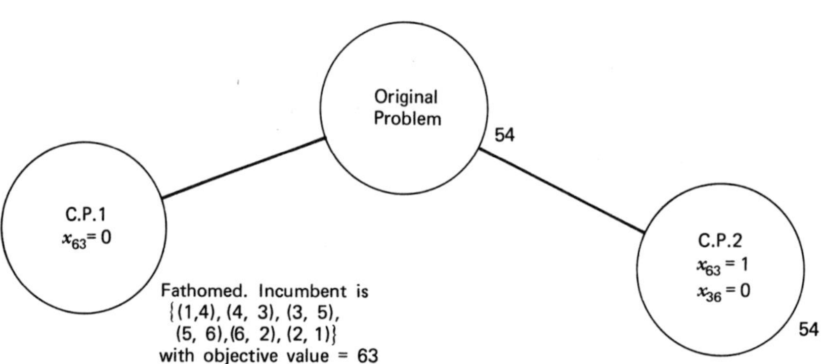

Figure 15.7

So select x_{35} as the branching variable. Two candidate problems are generated. Candidate problem 21 has the additional restriction "$x_{35} = 0$" over those restrictions in candidate problem 2. The lower bound in this candidate problem can be computed by applying the lower bounding strategy as before, but clearly this lower bound will be greater than or equal to 54 + evaluation of $x_{35} = 54 + 10 = 64$. Since the objective value of the incumbent is 63, this candidate problem is pruned.

Candidate problem 22 has the additional restriction "$x_{35} = 1$" over those in candidate problem 2. In this problem both x_{63} and x_{35} are restricted to be 1. These restrictions automatically imply that $x_{56} = 0$. Let C_{22} be the matrix obtained by striking off the row corresponding to $i = 3$ and the column corresponding to $j = 5$ in C^2 and changing $c^2(5, 6)$ to ∞. It is as follows:

	$j =$	1	2	4	6
	$i = 1$	×	7	0	11
$C_{22} =$	2	0	×	0	25
	4	3	0	×	7
	5	0	36	42	×

The lower bounding strategy consists of solving the relaxed problem, which is an assignment problem with C_{22} as the cost matrix. The objective value of an optimum assignment with respect to C_{22} as the cost matrix turns out to be 11. So the lower bound in this candidate problem is $54 + 11 = 65$. This is greater than 63, the objective value of the incumbent. So this candidate problem gets pruned too. The search tree at this stage is seen in Figure 15.8. The list is now empty. Hence the incumbent, $\{(1, 4), (4, 3), (3, 5), (5, 6), (6, 2), (2, 1)\}$, which is the tour 143562; 1, is an optimum tour with an objective value of 63.

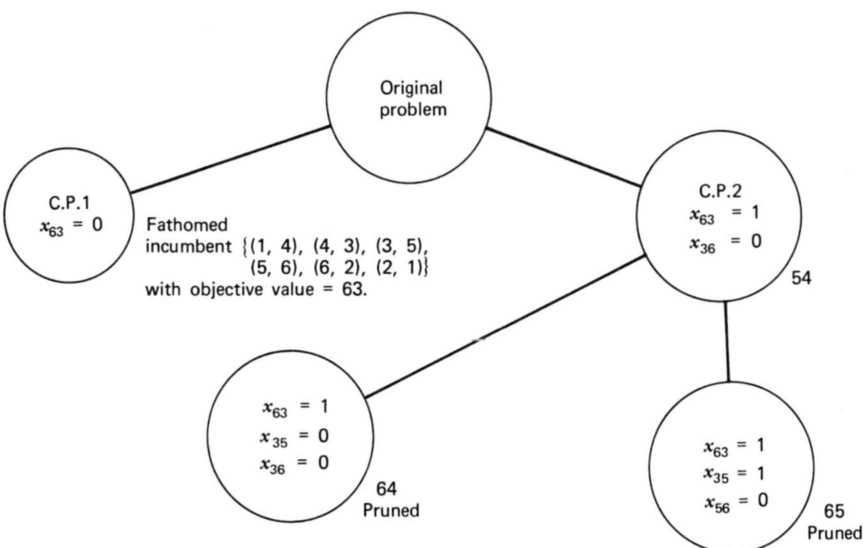

Figure 15.8

Problems

15.3 Solve the traveling salesman problems with the following cost matrices.

(a)

$j =$	1	2	3	4	5	6	7
$i = 1$	×	13	9	19	22	3	7
2	6	×	7	8	39	4	8
3	18	3	×	5	18	4	3
4	4	16	6	×	5	13	16
5	14	22	99	17	×	8	10
6	3	21	7	3	10	×	9
7	11	19	8	19	8	3	×

(b)

$j =$	1	2	3	4	5	6	7	8	9	10
$i = 1$	×	51	55	90	41	63	77	69	0	23
2	50	×	0	64	8	53	0	46	73	72
3	30	77	×	21	25	51	47	16	0	60
4	65	0	6	×	2	9	17	5	26	42
5	0	94	0	5	×	0	41	31	59	48
6	79	65	0	0	15	×	17	47	32	43
7	76	96	48	27	34	0	×	0	25	0
8	0	17	9	27	46	15	84	×	0	24
9	56	7	45	39	0	93	67	79	×	38
10	30	0	42	56	49	77	72	49	23	×

(c) Suppose the salesman starts in city 1 and returns to city 1 at the end of the tour. The cost matrix is the same as in (b). However, he wants a minimum cost tour that goes through cities 2, 3, 4 before going to 9. Find this by a branch and bound approach.

15.4 THE SYMMETRIC ASSIGNMENT PROBLEM. There are 10 objects that must be grouped into 5 pairs. Every object has to be in exactly 1 pair and each pair must consist of 2 distinct objects. The cost of pairing objects i, j is d_{ij}, which is tabulated below for $j > i$.

$j =$	1	2	3	4	5	6	7	8	9	10
$i = 1$	×	4	6	110	116	126	118	120	116	114
2		×	8	118	114	124	106	102	118	106
3			×	112	116	112	110	122	116	124
4				×	2	4	218	216	226	230
5					×	6	212	212	234	232
6						×	338	306	316	308
7							×	4	2	16
8								×	6	8
9									×	10
10										×

(a) Let $c_{ii} = \infty$ for $i = 1$ to 10 and let c_{ij} be real numbers satisfying $c_{ij} + c_{ji} = d_{ij}$ for all $j > i$. Prove that the problem of finding a minimum cost pairing is equivalent to the symmetric assignment problem:

$$\text{Minimize} \quad \sum\sum c_{ij}x_{ij}$$

$$\text{Subject to} \quad \sum_j x_{ij} = 1 \qquad i = 1 \text{ to } 10$$

$$\sum_i x_{ij} = 1 \qquad j = 1 \text{ to } 10$$

$$x_{ij} = x_{ji}, \qquad \text{for all } i, j$$

$$x_{ij} = 0 \text{ or } 1 \qquad \text{for all } i, j$$

(b) Obtain a lower bound for the minimum objective value in the symmetric assignment problem using a relaxed assignment problem. Verify that the best lower bound is obtained by taking $c_{ij} = c_{ji} = d_{ij}/2$ for all $j > i$.

(c) Develop a branch and bound algorithm for solving the symmetric assignment problem. Develop *evaluations* (as in the traveling salesman problem) for selecting good branching variables in this algorithm. Solve the above pairing problem using this algorithm.

15.5 There are 6 students in a project course. It is required to combine them into groups of at most 2 students each. Here are the data.

Cost of forming students i, j into a group for $j \leq i$

$j =$	1	2	3	4	5	6
$i = 1$	16	10	8	58	198	70
2		10	6	72	50	32
3			15	26	198	24
4				15	14	18
5					13	6
6						10

Find a minimum cost grouping.

15.6 Let n be an even positive integer. Prove that the total number of distinct ways of combining n objects into $n/2$ distinct pairs is $n!/[(n/2)!2^{(n/2)}]$.

15.7 Let n be any positive integer. You need to combine n objects into groups of at most 2 each. Prove that the total number of distinct ways of doing this is:

$$\sum_{\substack{r=1 \\ r \text{ odd}}}^{n} \binom{n}{r} (n-r)!/(((n-r)/2)!2^{(n-r)/2}) \qquad \text{if } n \text{ is odd}$$

$$\sum_{\substack{r=1 \\ r \text{ even}}}^{n} \binom{n}{r} (n-r)!/(((n-r)/2)!2^{(n-r)/2}) \qquad \text{if } n \text{ is even}$$

15.8 Consider the undirected network in Figure 15.9. There are 11 vertices and the serial numbers of the vertices are entered inside them. A *clique* is a subset of vertices in this network such that every pair of vertices in that subset is connected by an edge of the network. For example, $\{1, 10, 9, 2\}$ is not a clique because 1 and 9 are not connected by an edge of the network. $\{1, 2, 3, 5\}$ is a clique.

The cost of including each vertex in a clique is recorded by the side of that vertex. Notice that all the cost coefficients are negative. The total cost of a clique is the sum of the cost coefficients of the nodes in the clique. Develop a branch and bound algorithm for finding a minimum cost clique. Find the minimum cost clique in the network in Figure 15.9 using your algorithm.

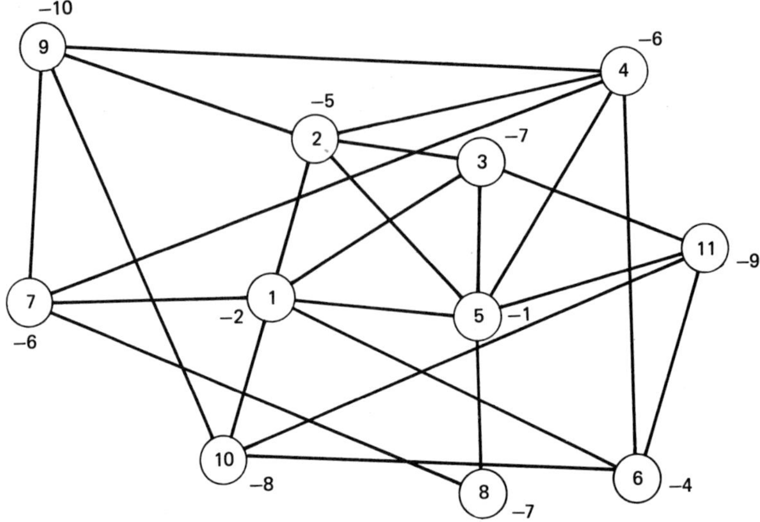

Figure 15.9

15.8 THE GENERAL MIXED INTEGER LINEAR PROGRAM

Consider the general mixed integer program:

$$\begin{aligned} \text{Minimize} \quad & z(x, y) = cx + dy \\ \text{Subject to} \quad & Ax + Dy = b \\ & x \geqq 0, y \geqq 0 \end{aligned} \tag{15.12}$$

$$y \text{ integer vector} \tag{15.13}$$

If there are no continuous variables in the problem, it is a pure integer program. A lower bounding strategy for this problem is to relax the integer requirements (15.13) and solve the relaxed problem (15.12). Let the optimum solution of the relaxed problem be (x^0, y^0) with an objective value of z^0. If y^0 satisfies (15.13), (x^0, y^0) is an optimum solution of the original problem and we terminate. Otherwise z^0 is a lower bound for the minimum objective value in the original problem.

The branching strategy consists of selecting an integer variable y_j such that y_j^0 is noninteger.

1. If y_j is a 0-1 variable, generate two candidate problems by imposing one additional constraint "$y_j = 0$" or "$y_j = 1$."
2. If y_j is a general integer variable, let $[y_j^0]$ be the maximum integer less than or equal to y_j^0. Generate two candidate problems by imposing one additional constraint "$y_j \leqq [y_j^0]$" or "$y_j \geqq [y_j^0] + 1$," respectively.

If there are several j's such that y_j^0 is noninteger, the branching variable is selected from them, so that the lower bounds of the candidate subproblems generated after branching are as high as possible. The data in the optimum simplex tableau can be used to get estimates (called *penalties*) of the amount by which the lower bounds in the candidate subproblems are greater than the lower bound of the candidate problem. For lack of space, these things are not discussed here. See references [10, 6, and 24].

The lower bounds in the newly generated candidate problems are computed by solving the LPs obtained by relaxing the integer requirements on the y's in them.

The relaxed problem corresponding to a candidate problem contains just one additional constraint over its relaxed parent problem. Since the optimum solution of the relaxed parent problem is already known, an optimum solution of the relaxed candidate problem can be obtained from it by using sensitivity analysis techniques.

A candidate problem is fathomed when the optimum solution of the relaxed problem corresponding to it satisfies the integer requirements on the y's. When an incumbent is obtained, any candidate problem in the list associated with a lower bound greater than or equal to the objective value of the current incumbent is pruned.

A candidate problem for branching is selected from the list using the least lower bound criterion. The algorithm terminates when the list becomes empty. The incumbent at that time is an optimum solution of the original problem. If there is no incumbent at that time, the original problem must be infeasible.

Numerical Example

Consider the following mixed integer program:

y_1	y_2	x_1	x_2	x_3	x_4	$-z$	b
1	0	0	1	-2	1	0	$\frac{3}{2}$
0	1	0	2	1	-1	0	$\frac{5}{2}$
0	0	1	-1	1	1	0	4
0	0	0	3	4	5	1	-20

$y_1 \geqq 0$, $y_2 \geqq 0$, and integer; x_1 to $x_4 \geqq 0$, z to be minimized

The optimum solution of the relaxed problem does not satisfy the integer requirements on y_1, y_2. So the original problem is not fathomed. It has to be branched.

The search tree for the branch and bound algorithm is given in Figure 15.10. The constraints inside a node corresponding to a candidate problem are the additional constraints

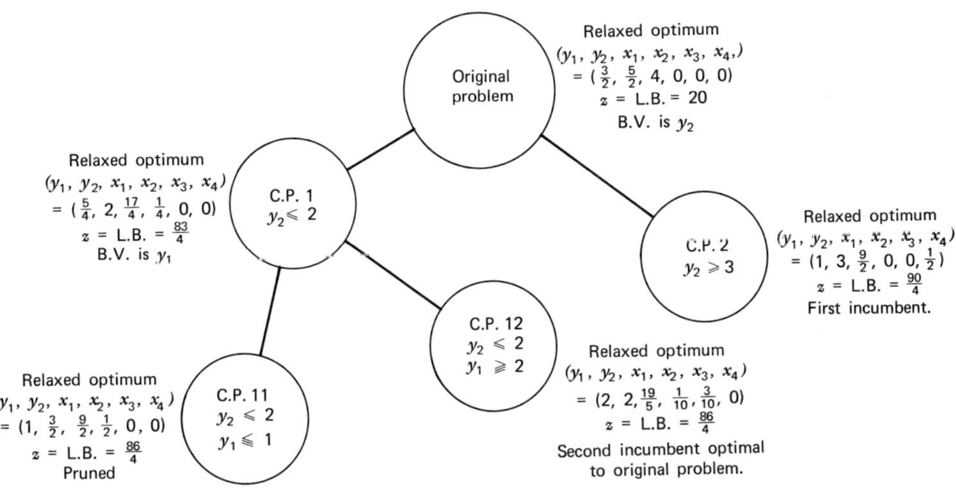

Figure 15.10

in it over those of the original problem. An optimum solution of the relaxed problem is entered beside the node. The optimal solution of the original problem is $(y_1, y_2, x_1, x_2, x_3, x_4) = (2, 2, 19/5, 1/10, 3/10, 0)$, with an objective value of 86/4.

15.9 THE SET REPRESENTATION PROBLEM

15.9.1 The Problem

In this problem (section 13.2.11) a set $\mathbf{\Gamma} = \{1, \ldots, n\}$, and a class $\mathbf{F} = (\mathbf{F}_1, \ldots, \mathbf{F}_m)$ of nonempty subsets of $\mathbf{\Gamma}$ are given. A *represent* for $\mathbf{F}$ is a subset of $\mathbf{\Gamma}$ that contains at least one element in common with each $\mathbf{F}_i$. The cost of including the element j in a represent is c_j, $j = 1$ to n. Find a minimum cost represent. If all the c_j are 1, the problem is to find a minimum cardinality represent. Let $A_{i.}$ be the incidence vector of $\mathbf{F}_i$, $i = 1$ to m. The incidence vector of a minimum cost represent for $\mathbf{F}$ is an optimum solution of

$$\begin{aligned} \text{Minimize} \quad & cx \\ \text{Subject to} \quad & Ax \geqq e \\ & x_j = 0 \text{ or } 1 \qquad \text{for all } j \end{aligned} \tag{15.14}$$

where e is a column vector of all 1's in $\mathbf{R}^m$ and vice versa.

The subset $\mathbf{F}_i \in \mathbf{F}$ leads to the ith constraint in (15.14). Hence we will refer to the *subset* $\mathbf{F}_i$ in the class $\mathbf{F}$ as the *ith constraint* in the problem. An *element j is said to represent* $\mathbf{F}_i$ if $j \in \mathbf{F}_i$.

15.9.2 Reduction

Clearly every minimum cost represent must include all the j such that $c_j < 0$. Also, if there are some j's such that $c_j = 0$, there exists a minimum cost represent that contains all of them. So let $\mathbf{E}_1 = \{j : c_j \leqq 0\}$. Eliminate all elements in $\mathbf{E}_1$ from $\mathbf{\Gamma}$ and eliminate all constraints represented by elements in $\mathbf{E}_1$. If $\mathbf{E}_2$ is a minimum cost represent in the remaining smaller size problem, called the *reduced problem*, $\mathbf{E}_1 \cup \mathbf{E}_2$ is a minimum cost represent for the original problem. In subsequent discussion we will assume that all the cost coefficients are positive.

15.9.3 Unordered Cartesian Products

In the algorithm we will use *unordered Cartesian products*, abbreviated as UCP. If $\mathbf{D}_1, \ldots, \mathbf{D}_r$ are nonempty, mutually disjoint sets, the UCP of $\mathbf{D}_1, \ldots, \mathbf{D}_r$ denoted by $\mathbf{D}_1 \,\&\, \mathbf{D}_2 \,\&\, \ldots \,\&\, \mathbf{D}_r$ is $\{\{i_1, \ldots, i_r\} : i_t \in \mathbf{D}_t,\ t = 1 \text{ to } r\}$. The sets $\mathbf{D}_1, \ldots, \mathbf{D}_r$ are known as the *factors of this* UCP. *In any* UCP, the *factors are always nonempty and mutually disjoint. In the remaining part of this chapter, the symbol* "&" *is exclusively used to denote this* UCP *operator.*

The branch and bound algorithm discussed here is an *inductive algorithm*. At each stage of the algorithm only a subset of the original constraints are considered explicitly. These are called the *included constraints* at that stage. The remaining constraints are called the *unincluded constraints* at that stage. The unincluded constraints are relaxed using a Lagrangian relaxation approach. In each stage one unincluded constraint gets transferred into the set of included constraints (this is the inductive feature in the algorithm). Each candidate problem in the algorithm is associated with:

1. A set of unincluded constraints.
2. A UCP.
3. A set of banned elements.

Every subset in the UCP represents all the included constraints. If j is contained in the set of banned elements, this implies that "$x_j = 0$" is an additional restriction in the candidate problem.

15.9.4 Lower Bounding Strategy

Before discussing how the candidate problems are generated, we will discuss how the lower bounding strategy is applied on any candidate problem. In a typical candidate problem, suppose the UCP is $\mathbf{D}_1 \,\&\, \ldots \,\&\, \mathbf{D}_r$; the unincluded constraints are all $\mathbf{F}_i$ for $i \in \mathbf{I}$, where $\mathbf{I}$ is given; and suppose $\mathbf{H}$ is the set of all banned elements. This candidate problem is:

$$\begin{aligned} &\text{Minimize} \quad z(x) = \sum_j c_j x_j \\ &\text{Subject to} \quad \sum_{j \in \mathbf{D}_t} x_j \geqq 1 && t = 1 \text{ to } r \\ &\qquad x_j = 0 && \text{for all } j \in \mathbf{H} \qquad (15.15) \\ &\qquad x_j = 0 \text{ or } 1 && \text{for all } j \end{aligned}$$

$$\sum_{j \in \mathbf{F}_i} x_j \geqq 1 \qquad \text{for all } i \in \mathbf{I} \qquad (15.16)$$

By the manner in which the candidate problem is generated, the following remarks will hold.

1. $\mathbf{D}_1, \ldots, \mathbf{D}_r$ are all nonempty and mutually disjoint. Also, every subset in $\mathbf{D}_1 \,\&\, \ldots \,\&\, \mathbf{D}_r$ represents all the included constraints, that is, all $\mathbf{F}_i$ for $i \notin \mathbf{I}$.
2. $\mathbf{D}_1 \cup \cdots \cup \mathbf{D}_r$ and $\mathbf{H}$ are disjoint.

Notice that a solution of this candidate problem can have x_j equal to 1, for more than one j in any $\mathbf{D}_t$(The constraint in system (15.15) is $\sum_{j \in \mathbf{D}_t} x_j \geqq 1$ and not $\sum_{j \in \mathbf{D}_t} x_j = 1$). Constraints (15.16) make the candidate problem difficult to solve. We will relax the constraints in (15.16) using Lagrangian relaxation. Associate a nonnegative Lagrange multiplier u_i for the ith constraint in (15.16), $i \in \mathbf{I}$. Let $u(\mathbf{I})$ be the row vector of these Lagrange multipliers. Let $\mathbf{G}_j = \{i: i \in \mathbf{I}, \mathbf{F}_i \text{ contains } j\}$. The relaxed Lagrangian is

$$\begin{aligned} \mathrm{L}(u(\mathbf{I}), x) &= \sum_j c_j x_j - \sum_{i \in \mathbf{I}} u_i \left(\sum_{j \in \mathbf{F}_i} x_j - 1 \right) \\ &= \sum_{i \in \mathbf{I}} u_i + \sum_j \left(c_j - \sum_{i \in \mathbf{G}_j} u_i \right) x_j \end{aligned}$$

As before let $a_{ij} = 1$ if $j \in \mathbf{F}_i$, and $a_{ij} = 0$ otherwise. Using this fact the relaxed Lagrangian is $\mathrm{L}(u(\mathbf{I}), x) = \sum_{i \in \mathbf{I}} u_i + \sum_j (c_j - \sum_{i \in \mathbf{I}} a_{ij} u_i) x_j$. The *relaxed problem* is

$$\begin{aligned} &\mathrm{L}(u(\mathbf{I})) = \underset{x}{\text{minimum}}\ \mathrm{L}(u(\mathbf{I}), x) \\ &\text{Subject to} \quad \sum_{j \in \mathbf{D}_t} x_j \geqq 1 && \text{for all } t = 1 \text{ to } r \\ &\qquad x_j = 0 && \text{for all } j \in \mathbf{H} \qquad (15.17) \\ &\qquad x_j = 0 \text{ or } 1 && \text{for all } j \end{aligned}$$

Result For any $u(\mathbf{I}) \geqq 0$, the minimum objective value in the relaxed problem, $L(u(\mathbf{I}))$, is a lower bound on the minimum objective value in the candidate problem (15.15) and (15.16).

Proof Let x^* be an optimum solution of the candidate problem (15.15) and (15.16). Then $z(x^*) = \sum c_j x_j^*$ is the minimum objective value in the candidate problem. x^* is a feasible solution for (15.17), but may not be optimal to it. Therefore,

$$\begin{aligned} L(u(\mathbf{I})) &\leqq L(u(\mathbf{I}), x^*) \\ &= z(x^*) - \sum_{i \in \mathbf{I}} u_i \left(\sum_{j \in \mathbf{F}_i} x_j^* - 1 \right) \\ &\leqq z(x^*), \end{aligned}$$

because $u_i \geqq 0$, and $\sum_{j \in \mathbf{F}_i} x_j^* - 1 \geqq 0$ by the feasibility of x^* to the candidate problem.

15.9.5 Solving the Relaxed Problem

The fact that $\mathbf{D}_1, \ldots, \mathbf{D}_r$ are mutually disjoint makes the relaxed problem very easy to solve. Given $u(\mathbf{I})$, an optimum solution, $\bar{x}(u(\mathbf{I}))$ for the relaxed problem is obtained by the following simple rules:

$$\text{Compute } \bar{c}_j = c_j - \sum_{i \in \mathbf{I}} a_{ij} u_i \qquad \text{for all } j \notin \mathbf{H}. \text{ Then,}$$

(i) $\bar{x}_j(u(\mathbf{I})) = 0$ for all $j \in \mathbf{H}$

(ii) $\bar{x}_j(u(\mathbf{I})) = 1$ for all $j \notin \mathbf{H}$ such that $\bar{c}_j \leqq 0$

(iii) For $t = 1$ to r, if none of the $\bar{x}_j(u(\mathbf{I}))$ for $j \in \mathbf{D}_t$ have been set equal to 1 by rule (ii), then find a $p \in \mathbf{D}_t$ such that $\bar{c}_p = \text{minimum } \{\bar{c}_j : j \in \mathbf{D}_t\}$. Set $\bar{x}_p(u(\mathbf{I})) = 1$ and $\bar{x}_j(u(\mathbf{I})) = 0$ for all other $j \in \mathbf{D}_t$.

(iv) If j is such that $\bar{x}_j(u(\mathbf{I}))$ has not been given a value by (i), (ii), (iii), make $\bar{x}_j(u(\mathbf{I})) = 0$.

The optimum objective value in the relaxed problem is

$$L(u(\mathbf{I})) = L(u(\mathbf{I}), \bar{x}(u(\mathbf{I})) = \sum_{i \in \mathbf{I}} u_i + \sum_j \bar{c}_j \bar{x}_j(u(\mathbf{I})).$$

Example

Suppose $r = 3$, $\mathbf{D}_1 = \{1, 2, 3\}$, $\mathbf{D}_2 = \{4, 5\}$, $\mathbf{D}_3 = \{6, 7, 8\}$, and $\mathbf{H} = \{13, 14\}$. Suppose the Lagrange multiplier vector is $u(\mathbf{I}) = (5, 1, 6.9, 4.1)$. Suppose

$$\bar{c} = (1, 2, 3, 0, -4, 0, 0, 1, -3, -4, 4, 0).$$

Since $\mathbf{H} = \{13, 14\}$, x_{13} and x_{14} are zero in this candidate problem. In the above vector $\bar{c}$, the values of $\bar{c}_j$ for $j = 1$ to 12 only are given. Then the relaxed problem is

$$L(u) = \sum u_i + \text{minimum } (x_1 + 2x_2 + 3x_3 - 4x_5 + x_8 - 3x_9 - 4x_{10} + 4x_{11})$$

$$\begin{aligned} \text{Subject to } \quad x_1 + x_2 + x_3 &\geqq 1 \\ x_4 + x_5 &\geqq 1 \\ x_6 + x_7 + x_8 &\geqq 1 \\ x_{13} = x_{14} &= 0 \\ x_j &= 0 \text{ or } 1 \qquad \text{for all } j \end{aligned}$$

Obviously an optimum solution for this problem is the vector

$$\bar{x} = (1, 0, 0, 1, 1, 1, 1, 0, 1, 1, 0, 1, 0, 0)^T,$$

which is the solution obtained by the rules (i) to (iv) above. In this example $L(u)$ turns out to be 6.1.

15.9.6 Subgradient Optimization Algorithm for Obtaining a Good Lagrange Multiplier Vector

We will now discuss how to obtain a good Lagrange multiplier vector for the candidate problem discussed in sections 15.9.4 and 15.9.5. Given any $u(\mathbf{I}) \geqq 0$, let $\bar{x}(u(\mathbf{I}))$ be the optimum solution of the relaxed problem obtained as in section 15.9.5. For $i \in \mathbf{I}$, let

$$v_i(\mathbf{I}) = 1 - \sum_{j \in \mathbf{F}_i} \bar{x}_j(u(\mathbf{I})) = 1 - \sum_j a_{ij}\bar{x}_j(u(\mathbf{I})) \tag{15.18}$$

Let $v(\mathbf{I})$ be the row vector of $v_i(\mathbf{I})$, where the entries $v_i(\mathbf{I})$ are recorded in the same order in which the $u_i(\mathbf{I})$ are recorded in the vector $u(\mathbf{I})$. In the rest of this section we will denote $u(\mathbf{I})$ by u, $\bar{x}(u(\mathbf{I}))$ by $\bar{x}$ and $v(\mathbf{I})$ by v for notational simplicity. Both x and v are functions of u, and v, u are now vectors of the same dimension. (This dimension is equal to the number of unincluded constraints in this candidate problem.) The ideal Lagrange multipler vector is an optimum solution of the problem

$$\begin{aligned} &\text{Maximize} \quad L(u) \\ &\text{Subject to} \quad u \geqq 0 \end{aligned} \tag{15.19}$$

Starting with an initial Lagrange multiplier vector, u^0, the *subgradient optimization algorithm* obtains a sequence of vectors $u^0, u^1, u^2, \ldots$. This sequence converges asymptotically to an optimum solution of (15.19). However, there is no guarantee that $L(u^p)$ is monotone increasing with p. It may oscillate.

Since $L(u)$ is only used as a lower bound for the candidate problem in the branch and bound approach, (15.19) need not be solved exactly. Our aim is to find a Lagrange multiplier vector u, which gives as high a value of $L(u)$ as possible, with little computational effort. So we will apply the subgradient optimization algorithm until it yields s vectors in the sequence, that is, $u^0, u^1, \ldots, u^s$, and pick the best Lagrange multiplier vector for this candidate problem as that vector u^p that satisfies $L(u^p) = \text{maximum } \{L(u^0), \ldots, L(u^s)\}$. The number s should be a fairly small number to keep the computational effort small. The choice of s will be discussed later on. The algorithm is based on the following results.

1. $L(u)$ is a concave function in the region $u \geqq 0$, that is, for any $\hat{u} \geqq 0$, $\tilde{u} \geqq 0$, and $0 \leqq \alpha \leqq 1$, $L(\alpha\hat{u} + (1 - \alpha)\tilde{u}) \geqq \alpha L(\hat{u}) + (1 - \alpha)L(\tilde{u})$.
2. Given a $\hat{u} \geqq 0$, a vector w is known as a *subgradient of the concave function* $L(u)$ *at* $\hat{u}$ if $L(u) \leqq L(\hat{u}) + w(u - \hat{u})$ for all $u \geqq 0$. The vector $\hat{v}$ corresponding to $\hat{u}$ [see equation (15.18)], is a subgradient of $L(u)$ at $\hat{u}$.

Since the subgradient optimization algorithm is only a tool used here, and not something central to the discussion of the branch and bound approach, we will not prove these results. For a detailed discussion of the subgradient optimization algorithm see reference [17].

We need a sequence of nonincreasing positive numbers, $\rho_0, \rho_1, \ldots$. The requirement is that $2 \geqq \rho_0 \geqq \rho_1 \geqq \rho_2 \geqq \ldots$ etc. and that ρ_p converges to 0 as p tends to infinity. A good choice for this sequence is discussed later on.

Let $\bar{L}$ be a good estimate of the maximum objective value of (15.19). Suitable choice for $\bar{L}$ is discussed later on. The subgradient optimization algorithm consists of the following steps.

1. Select an initial Lagrange multiplier vector $u^0 \geqq 0$. When this method is applied for getting a lower bound for the original problem, the vector u^0 is taken to be 0.
   Let **I** be the index set of unincluded constraints in a candidate problem and let $u^0(\mathbf{I})$ be the best Lagrange multiplier vector for this problem. When this candidate problem is branched, suppose a candidate subproblem is obtained in which the index set of unincluded constraint is $\mathbf{I}_1 = \mathbf{I}\backslash\{k\}$. Let $u^0(\mathbf{I}_1)$ be the vector obtained by suppressing $u_k{}^0$ from $u^0(\mathbf{I})$. Then use $u^0(\mathbf{I}_1)$ as the initial Lagrange multiplier vector when applying the subgradient optimization algorithm on the candidate subproblem.
2. Suppose $u^0, u^1, \ldots, u^p$, in the sequence have already been obtained. To obtain the next vector in the sequence, u^{p+1}, solve the relaxed problem with u^p as the Lagrange multiplier vector. Let $\bar{x}^p$ be the optimum solution of the relaxed problem obtained as in section 15.9.5. Let v^p be the subgradient vector for $L(u)$ at u^p obtained as in (15.18). Let:

$$\lambda_p = \rho_p(\bar{L} - L(u^p))/\|v^p\|^2$$

and

$$u_i^{p+1} = \text{maximum}\ \{0, u_i{}^p + \lambda_p v_i{}^p\} \qquad \text{for } i \in \mathbf{I} \tag{15.20}$$

   Then u^{p+1} is the vector of u_i^{p+1}, with the entries u_i^{p+1} recorded in the same order in which the corresponding entries are recorded in u^p.

A good choice for the numbers ρ_p is to start with $\rho_0 = 2$. If no improvement in the value of $L(u^p)$ occurs in the last d steps, then the value of ρ is halved in the next step. A good choice for d is a number from 2 to 6. When this algorithm is used to obtain a lower bound for the original problem, d can be taken to be 5 or 6. On candidate problems generated after one or more branchings, d can be taken to be 2 or 3.

The number, s, the number of steps through which the subgradient optimization algorithm is carried out, can be between 60 to 100 when this algorithm is applied to get a lower bound for the original problem. In subsequent candidate problems, s need not be greater than 10 to 20.

At the beginning of the branch and bound algorithm, some good known represent is chosen as the incumbent. If no other good represent is known, $\mathbf{\Gamma}$ can be chosen as the initial incumbent (the incumbent will actually be the incidence vector of the represent). Whenever a relaxed problem is solved, the optimum solution $\bar{x}$ for it is tested for feasibility to the original problem. If it is feasible to the original problem, and if its objective value, $z(\bar{x})$, is less than the objective value of the current incumbent, this $\bar{x}$ is made the new incumbent. In the subgradient optimization algorithm, $\bar{L}$ can be conveniently taken to be the objective value of the current incumbent.

Example

Consider a problem in which $\mathbf{\Gamma} = \{j: j = 1 \text{ to } 13\}$. In this problem consider a candidate problem for which $\mathbf{D}_1 = \{7, 9, 10, 13\}$, $\mathbf{D}_2 = \{3, 6, 8\}$, $\mathbf{D}_3 = \{2, 4, 5, 12\}$, $\mathbf{H}$ = set of banned elements = $\{11\}$, and $\mathbf{I}$ = set of indices of unincluded constraints = $\{2, 3, 4, 6, 8, 9, 10\}$ where $\mathbf{F}_i$ for $i = 1$ to 10 are given in section 15.9.10. $c_j = 1$ for all j. So this candidate problem is

$$\begin{aligned}
\text{Minimize} \quad & \sum_{j=1}^{13} x_j \\
\text{Subject to} \quad & x_7 + x_9 + x_{10} + x_{13} \geqq 1 \\
& x_3 + x_6 + x_8 \geqq 1 \\
& x_2 + x_4 + x_5 + x_{12} \geqq 1 \qquad (15.21) \\
& x_{11} = 0 \\
& x_j = 0 \text{ or } 1 \quad \text{for all } j
\end{aligned}$$

$$\sum_{j \in F_i} x_j \geqq 1 \qquad \text{for all } i \in \mathbf{I} \qquad (15.22)$$

Relax (15.22). The relaxed Lagrangian is $L(u, x) = \sum_{i \in \mathbf{I}} u_i + \sum_{j \neq 11} \bar{c}_j x_j$, where $(\bar{c}_j: j = 1$ to $10, 12, 13) = (1 - u_9 - u_{10}, 1 - u_2 - u_9, 1 - u_3 - u_6, 1 - u_4 - u_8 - u_9, 1 - u_4 - u_8 - u_{10},$ $1 - u_6 - u_8, 1 - u_6 - u_{10}, 1 - u_2 - u_4, 1 - u_2 - u_3 - u_4, 1 - u_3 - u_6, 1 - u_3 - u_{10},$ $1 - u_2 - u_8)$. The relaxed problem is

$$\begin{aligned}
L(u) = & \ \underset{x}{\text{minimum}} \ L(u, x) \\
\text{Subject to} \quad & x_7 + x_9 + x_{10} + x_{13} \geqq 1 \\
& x_3 + x_6 + x_8 \geqq 1 \\
& x_2 + x_4 + x_5 + x_{12} \geqq 1 \\
& x_{11} = 0 \\
& x_j = 0 \text{ or } 1 \quad \text{for all } j
\end{aligned}$$

If we pick $u^0 = (u_2{}^0, u_3{}^0, u_4{}^0, u_6{}^0, u_8{}^0, u_9{}^0, u_{10}{}^0) = (0.23, 0.09, 0.22, 0.09, 0.31, 0.40, 0.42)$ as the initial Lagrange multiplier vector, the corresponding $\bar{c}$ is (0.18, 0.37, 0.82, 0.07, 0.05, 0.60, 0.49, 0.55, 0.46, 0.82, 0.49, 0.46). By section 15.9.5, an optimum solution for the relaxed problem is

$$\bar{x}^0 = (0, 0, 0, 0, 1, 0, 0, 1, 1, 0, 0, 0, 0)^T$$

So $L(u^0) = 2.82$. The subgradient vector

$$v^0 = (v_2{}^0, v_3{}^0, v_4{}^0, v_6{}^0, v_8{}^0, v_9{}^0, v_{10}{}^0) = (-1, 0, -2, 1, 0, 1, 0).$$

If an estimate for $\bar{L}$ is 4 and if the value of $\rho_0 = 0.25$, we see that $\lambda_0 = 0.25(4 - 2.82)/\|v^0\|^2 = 0.04$. Hence the next Lagrange multiplier vector in the sequence is given by $u_i{}^1 = \max\{0, u_i{}^0 + (0.04)v_i{}^0\}$ for $i \in \mathbf{I}$. So $u^1 = (0.19, 0.09, 0.14, 0.13, 0.31, 0.44, 0.42)$. The algorithm can be repeated in the same manner for as many steps as necessary.

15.9.7 Fathoming

Consider the candidate problem (15.15) and (15.16) again. Suppose u is the best Lagrange multiplier vector for this problem obtained under the subgradient optimization routine. Using u as the Lagrange multiplier vector, suppose $\bar{x}$ is the optimum solution and $L(u)$ the minimum objective value in the relaxed problem (15.17).

$L(u)$ is a lower bound for the minimum objective value in the candidate problem (15.15) and (15.16). If all the c_j are positive integers, the minimum objective value in this candidate problem must be a positive integer. Hence the lower bound for the minimum objective value in the candidate problem can be taken to be $<L(u)>$, which is the smallest integer greater than or equal to $L(u)$.

Even if $\bar{x}$ is feasible to this candidate problem [i.e., if $\bar{x}$ satisfies the relaxed constraints (15.16)], there is no guarantee that $\bar{x}$ is an optimum solution of the candidate problem. We can only conclude that $\bar{x}$ is an optimum solution for this candidate problem if $z(\bar{x})$ is equal to the lower bound for the objective valve in this candidate problem.

15.9.8 Branching Strategy

Suppose the candidate problem (15.15) and (15.16) has to be branched. We will use the following branching strategy. Select one of the unincluded constraints in this candidate problem, say the kth, where $k \in \mathbf{I}$. The algorithm will be more efficient if k is chosen so that the cardinality of $(\mathbf{D}_1 \cup \mathbf{D}_2 \cup \cdots \cup \mathbf{D}_r) \cap \mathbf{F}_k$ is the least among the cardinalities of $(\mathbf{D}_1 \cup \cdots \cup \mathbf{D}_r) \cap \mathbf{F}_i$ for $i \in \mathbf{I}$. There are several cases to consider.

Case 1 Suppose $(\mathbf{D}_1 \cup \cdots \cup \mathbf{D}_r) \cap \mathbf{F}_k = \emptyset$. In this case if $\mathbf{F}_k \backslash \mathbf{H} = \emptyset$, that is, the constraint $\mathbf{F}_k$ contains no elements other than the banned elements, this candidate problem is infeasible. It is pruned.

If $\mathbf{F}_k \backslash \mathbf{H} \neq \emptyset$, change this candidate problem into another for which the UCP is $\mathbf{D}_1$ & ... & $\mathbf{D}_r$ & $(\mathbf{F}_k \backslash \mathbf{H})$; the banned set is the same, $\mathbf{H}$; and the set of unincluded constraints is $\mathbf{I} \backslash \{k\}$. No branching is done in this step. The new candidate problem is obtained by deleting the set $\mathbf{F}_k$ from the set of unincluded constraints and generating an additional factor in the UCP from it.

If $\mathbf{I} \backslash \{k\} = \emptyset$, let j_t be such that $c_{j_t} = \text{minimum } \{c_j : j \in \mathbf{D}_t\}$, $t = 1$ to r. Let j_{r+1} be such that $c_{j_{r+1}} = \text{minimum } \{c_j : j \in \mathbf{F}_k \backslash \mathbf{H}\}$. Then $\{j_1, \ldots, j_r, j_{r+1}\}$ is an optimum represent for this candidate problem and its incidence vector is an optimum solution. So this candidate problem is fathomed.

If $\mathbf{I} \backslash \{k\} \neq \emptyset$, apply the lower bounding strategy on the new candidate problem and continue.

Case 2 Suppose $(\mathbf{D}_1 \cup \cdots \cup \mathbf{D}_r) \cap \mathbf{F}_k \neq \emptyset$. Let p be the smallest value of i for which $\mathbf{D}_i \cap \mathbf{F}_k \neq \emptyset$. Suppose $\mathbf{D}_p \backslash \mathbf{F}_k = \emptyset$. This can only happen if $\mathbf{D}_p \subset \mathbf{F}_k$. So every feasible solution of (15.15) automatically satisfies the unincluded constraint $\mathbf{F}_k$. If $\mathbf{I} \backslash \{k\} = \emptyset$, let $j_t = \text{minimum } \{c_j : j \in \mathbf{D}_t\}$ for $t = 1$ to r. Then $\{j_1, \ldots, j_r\}$ is an optimum represent and its incidence vector is an optimum solution for this candidate problem. The candidate problem is therefore fathomed.

If $\mathbf{I} \backslash \{k\} \neq \emptyset$, change this candidate problem into a new one that corresponds to the same UCP, same banned element set, and has the set of unincluded constraints as $\mathbf{I} \backslash \{k\}$. No branching is done in this step. Apply the lower bounding strategy on the new candidate problem and continue.

Case 3 Suppose $(\mathbf{D}_1 \cup \cdots \cup \mathbf{D}_r) \cap \mathbf{F}_k \neq \emptyset$, and $\mathbf{D}_p$ is defined as in Case 2. Suppose $\mathbf{D}_p \backslash \mathbf{F}_k \neq \emptyset$. Consider two candidate subproblems. Candidate subproblem 1 corresponds to the UCP, $\mathbf{D}_1$ & ... & $\mathbf{D}_{p-1}$ & $(\mathbf{D}_p \cap \mathbf{F}_k)$ & $\mathbf{D}_{p+1}$ & ... & $\mathbf{D}_r$, the same banned set $\mathbf{H}$, and the set of unincluded constraints $\mathbf{I} \backslash \{k\}$. Candidate subproblem 2, corresponds to the UCP, $\mathbf{D}_1$ & ... & $\mathbf{D}_{p-1}$ & $(\mathbf{D}_p \backslash \mathbf{F}_k)$ & $\mathbf{D}_{p+1}$ & ... & $\mathbf{D}_r$, the banned set $\mathbf{H} \cup (\mathbf{D}_p \cap \mathbf{F}_k)$ and the same set of unincluded constraints $\mathbf{I}$. Candidate problem (15.15) and (15.16) is branched into these candidate subproblems 1 and 2.

We will use the type of search strategy discussed in section 15.4.4. Suppose the candidate problem just branched is the current candidate problem. If $\mathbf{I} \backslash \{k\} = \emptyset$, candidate subproblem 1 can be fathomed as in Case 2. The incumbent is then updated, and any pruning is carried out. Then candidate subproblem 2 is made the new current candidate problem, the lower bounding strategy is applied on it and the algorithm is continued.

If $\mathbf{I} \backslash \{k\} \neq \varnothing$, the lower bounding strategy is applied on both candidate subproblems 1 and 2, and candidate subproblem 1 is made the new current candidate problem. The incumbent is updated, pruning is carried out, and if candidate subproblem 2 is unpruned, it is added to the stack and the algorithm continued with the new current candidate problem.

15.9.9 How to Start the Algorithm

The algorithm begins by treating the original problem (15.14) in the form of a candidate problem with $\mathbf{F}_1$ as a single factor UCP, the set of banned elements as the empty set, and the set of indices of unincluded constraints as $\mathbf{I} = \{2, \ldots, m\}$. Thus in the beginning, the original problem, posed in this form, is the current candidate problem and the stack (using the terminology of the search strategy discussed in section 15.4.4) is empty. As discussed in section 15.4.4, a LIFO strategy is adopted whenever a candidate problem has to be removed from the stack.

The algorithm terminates when the stack becomes empty, and the current candidate problem itself is either pruned or fathomed. The final incumbent is an optimum solution of the original problem.

15.9.10 Example

We consider the set representation problem in which $\mathbf{\Gamma} = \{j: j = 1 \text{ to } 13\}$. The class $\mathbf{F}$ to be represented is $\{\mathbf{F}_i: i = 1 \text{ to } 10\} = \{\{7, 9, 10, 13\}, \{2, 8, 9, 13\}, \{3, 9, 10, 12\}, \{4, 5, 8, 9\}, \{3, 6, 8, 11\}, \{3, 6, 7, 10\}, \{2, 4, 5, 12\}, \{4, 5, 6, 13\}, \{1, 2, 4, 11\}, \{1, 5, 7, 12\}\}$. The cost coefficients c_j are 1 for all j. The subgradient optimization algorithm was applied for only seven steps for getting a good Lagrange multiplier vector for the lower bounding of the original problem. In all subsequent candidate problems, the subgradient optimization algorithm was applied for only two steps.

The search tree for the algorithm is given in Figure 15.11. The UCP, set of banned elements, $\mathbf{H}$, and the index set, $\mathbf{I}$, of unincluded rows corresponding to each candidate problem are entered in the node of the search tree representing that candidate problem. The branching constraint (abbreviated as B.C.) used for branching a candidate problem is also entered inside the node. Since all c_j are equal to 1 here, the actual lower bound can be taken to be the smallest integer greater than or equal to the L.B. value entered inside the node.

Each candidate problem is given a serial number. Also the stage number when the candidate problem is the current candidate problem in the algorithm is entered. When a new incumbent is obtained during the application of the lower bounding strategy on a candidate problem, it is entered under the node representing the candidate problem.

Whenever the branching strategy on a candidate problem led to the occurrence of Cases 1 or 2 of section 15.9.8, no branching occurs, but the candidate problem is modified and the analysis repeated with the modified candidate problem. This information is also indicated inside the nodes.

Some candidate problems generated during the algorithm are stored in the stack. They may be pruned before they have a chance to become the current candidate problem. Such problems carry only a serial number, but no stage number.

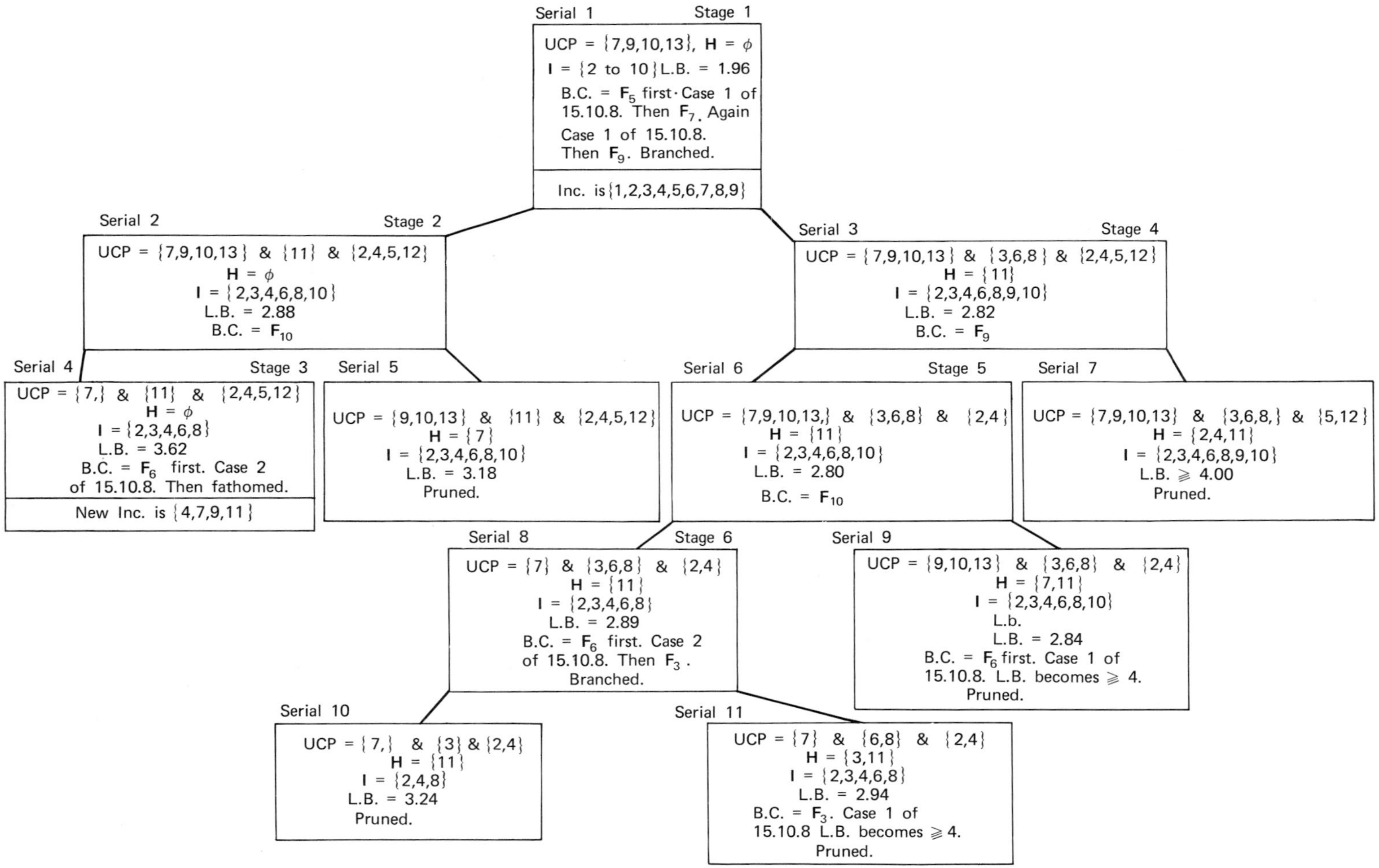

Serial 1 — Stage 1

UCP = {7,9,10,13}, **H** = ϕ
I = {2 to 10} L.B. = 1.96
B.C. = $\mathbf{F}_5$ first · Case 1 of 15.10.8. Then $\mathbf{F}_7$. Again Case 1 of 15.10.8. Then $\mathbf{F}_9$. Branched.
Inc. is {1,2,3,4,5,6,7,8,9}

Serial 2 — Stage 2

UCP = {7,9,10,13} & {11} & {2,4,5,12}
H = ϕ
I = {2,3,4,6,8,10}
L.B. = 2.88
B.C. = $\mathbf{F}_{10}$

Serial 3 — Stage 4

UCP = {7,9,10,13} & {3,6,8} & {2,4,5,12}
H = {11}
I = {2,3,4,6,8,9,10}
L.B. = 2.82
B.C. = $\mathbf{F}_9$

Serial 4 — Stage 3

UCP = {7,} & {11} & {2,4,5,12}
H = ϕ
I = {2,3,4,6,8}
L.B. = 3.62
B.C. = $\mathbf{F}_6$ first. Case 2 of 15.10.8. Then fathomed.
New Inc. is {4,7,9,11}

Serial 5

UCP = {9,10,13} & {11} & {2,4,5,12}
H = {7}
I = {2,3,4,6,8,10}
L.B. = 3.18
Pruned.

Serial 6 — Stage 5

UCP = {7,9,10,13,} & {3,6,8} & {2,4}
H = {11}
I = {2,3,4,6,8,10}
L.B. = 2.80
B.C. = $\mathbf{F}_{10}$

Serial 7

UCP = {7,9,10,13} & {3,6,8,} & {5,12}
H = {2,4,11}
I = {2,3,4,6,8,9,10}
L.B. ⩾ 4.00
Pruned.

Serial 8 — Stage 6

UCP = {7} & {3,6,8} & {2,4}
H = {11}
I = {2,3,4,6,8}
L.B. = 2.89
B.C. = $\mathbf{F}_6$ first. Case 2 of 15.10.8. Then $\mathbf{F}_3$.
Branched.

Serial 9

UCP = {9,10,13} & {3,6,8} & {2,4}
H = {7,11}
I = {2,3,4,6,8,10}
L.b.
L.B. = 2.84
B.C. = $\mathbf{F}_6$ first. Case 1 of 15.10.8. L.B. becomes ⩾ 4.
Pruned.

Serial 10

UCP = {7,} & {3} & {2,4}
H = {11}
I = {2,4,8}
L.B. = 3.24
Pruned.

Serial 11

UCP = {7} & {6,8} & {2,4}
H = {3,11}
I = {2,3,4,6,8}
L.B. = 2.94
B.C. = $\mathbf{F}_3$. Case 1 of 15.10.8 L.B. becomes ⩾ 4.
Pruned.

Figure 15.11

15.10 0-1 PROBLEMS

Consider the following problem:

$$\begin{aligned} \text{Minimize} \quad & z(x) = cx \\ \text{Subject to} \quad & Ax \leqq b \\ & x_j = 0 \text{ or } 1 \qquad \text{for all } j = 1 \text{ to } n \end{aligned} \tag{15.23}$$

If c_j is negative, eliminate x_j from the problem by substituting $x_j = 1 - y_j$. In the rest of the section we will assume that $c \geqq 0$.

In the algorithm to be discussed below, candidate problems are obtained by selecting a subset of the variables x_j and fixing them at value 0 or 1. Any variable that is fixed at 0 is called a *0-variable* and any variable fixed at 1 is called a *1-variable*. The 0-variables at value 0 and the 1-variables at value 1, constitute a *partial solution.* Each candidate problem generated in the algorithm corresponds to a partial solution. Variables that are not included in the partial solution are called *free variables* in the candidate problem. Given a partial solution, a *completion of it* is obtained by giving values of 0 or 1 to each of the free variables.

We will first discuss the analysis to be performed on a typical candidate problem. Consider the candidate problem in which $\mathbf{U}_0, \mathbf{U}_1, \mathbf{U}_f$ are the sets of subscripts of the 0 and 1-variables and the free variables, respectively. $\mathbf{U}_f = \{1, \ldots, n\} \backslash (\mathbf{U}_0 \cup \mathbf{U}_1)$. Compute the vector $y = b - \sum_{j \in \mathbf{U}_1} A_{.j}$. The *fathoming criterion* is $y \geqq 0$. If $y \geqq 0$, the completion obtained by giving the value 0 to all the free variables is obviously optimal to this candidate problem, and the optimum objective value in it is $\sum_{j \in \mathbf{U}_1} c_j$.

If $y \ngeqq 0$, several *tests* are available to check whether the candidate problem is infeasible or whether it has a feasible solution better than the current incumbent. Let $\bar{z}$ be the objective value of the incumbent, or ∞, if there is no incumbent at this stage. For applying these tests on the candidate problem, the system of constraints is

$$\begin{aligned} \sum_{j \in \mathbf{U}_f} a_{ij} x_{ij} \leqq b'_i = b_i - \sum_{j \in \mathbf{U}_1} a_{ij}, \qquad i = 1 \text{ to } m \\ \sum_{j \in \mathbf{U}_f} a_{m+1,j} x_j \leqq b'_{m+1} = \bar{z} - \sum_{j \in \mathbf{U}_1} a_{m+1,j} \end{aligned} \tag{15.24}$$

$$x_j = 0 \text{ or } 1 \qquad \text{for all } j \in \mathbf{U}_f \tag{15.25}$$

where, for notational convenience we denote c_j by $a_{m+1,j}$. The tests usually examine one of the constraints in (15.24) and check whether it can be satisfied in 0-1 variables. For example, one of the tests is the following. In the ith constraint in (15.24), if $\sum_{j \in \mathbf{U}_f} (\text{minimum } \{a_{ij}, 0\}) > b'_i$, obviously it cannot be satisfied; *and* hence (15.24) and (15.25) are infeasible, and the candidate problem can be pruned.

The tests may also determine that some of the free variables must have a specific value in $\{0, 1\}$ for (15.24), (15.25) to be feasible. Here is an example of such a test. Let $k \in \mathbf{U}_f$. Suppose in the ith constraint in (15.24) we have $\sum_{j \in \mathbf{U}_f} (\text{minimum } \{a_{ij}, 0\}) + |a_{ik}| > b'_i$. Then obviously x_k must be 0 if $a_{ik} > 0$ and 1 if $a_{ik} < 0$ for (15.24), (15.25) to be feasible. If such variables are identified by the tests, they are included in the sets of 0- or 1-variables accordingly.

Let $\pi = (\pi_1, \ldots, \pi_{m+1})$ be a nonnegative vector. Any solution satisfying (15.24) must obviously satisfy

$$\sum_{i=1}^{m+1} \pi_i \left(\sum_{j \in \mathbf{U}_f} a_{ij} x_j \right) \leqq \sum_{i=1}^{m+1} \pi_i b'_i.$$

A constraint like this, obtained by taking a nonnegative linear combination of (15.24), is known as a *surrogate constraint*. For example, from the system

$$x_1 - 3x_2 \leqq -2 \qquad x_2 \leqq 0$$

we get the surrogate constraint $x_1 - x_2 \leqq -2$ by taking the multiplier vector to be (1, 2). From this surrogate constraint we clearly see that the system has no 0-1 solution, even though we cannot make this conclusion by considering any one of the two original constraints individually. Quite often, a surrogate constraint enables us to make some conclusions about the system (15.24), (15.25), which are not apparent from anyone of the constraints considered individually. For a detailed discussion of useful tests, and methods for generating useful surrogate constraints etc., see references [3, 13, 14, and 30].

If these tests determine that (15.24), (15.25) is infeasible, the candidate problem is pruned. Otherwise let $\mathbf{U}'_0$, $\mathbf{U}'_1$ be the set of subscripts of the 0- and 1-variables, respectively, in the candidate problem, after augmenting the 0- and 1-variables, determined by the tests to $\mathbf{U}_0$ and $\mathbf{U}_1$, respectively. The set of free variables is $\mathbf{U}'_f = \{1, \ldots, n\} \backslash (\mathbf{U}'_0 \cup \mathbf{U}'_1)$. A lower bound for the minimum objective value in the candidate problem is $\sum_{j \in \mathbf{U}'_1} c_j$.

THE ALGORITHM

The algorithm uses the backtrack search strategy discussed in section 15.4.4, with a LIFO selection criterion. Initially the current candidate problem is the original problem, in which both the subscript sets of 0- and 1-variables are empty. The stack is empty initially.

In a general stage of the algorithm suppose the current candidate problem has the subscript sets $\mathbf{U}_0$, $\mathbf{U}_1$ for 0- and 1-variables, respectively.

1. If the current candidate problem is fathomed, update the incumbent (or if there is no incumbent so far declare the optimum solution of the current candidate problem as the incumbent) prune the stack and go to (2).

2. If the stack is found empty at this stage, the incumbent is an optimum solution of the original problem and we terminate. If the stack is found empty and there is no incumbent at this stage, the original problem is infeasible and we terminate. Otherwise, retrieve a candidate problem from the stack according to the LIFO selection criterion, and make it the updated current candidate problem, and continue.

3. If the current candidate problem is not fathomed, apply the tests on it. If the current candidate problem is pruned by the tests, go to (2). If it is not pruned by the tests, let $\mathbf{U}'_0$, $\mathbf{U}'_1$ be the subscript sets of 0- and 1-variables in the problem after augmenting the 0- and 1-variables, determined by the tests to $\mathbf{U}_0$, $\mathbf{U}_1$, respectively.

   The subscript set of free variables is $\mathbf{U}'_f = \{1, \ldots, n\} \backslash (\mathbf{U}'_0 \cup \mathbf{U}'_1)$. If $\mathbf{U}'_f = \emptyset$, the current candidate problem is fathomed; go to (1). If $\mathbf{U}'_f \neq \emptyset$, select a $j \in \mathbf{U}'_f$ and make x_j the branching variable. Two candidate subproblems are generated. Candidate problem 1 has $\mathbf{U}'_0$, $\mathbf{U}'_1 \cup \{j\}$ as the subscript sets for 0- and 1-variables, respectively. Candidate subproblem 2 has $\mathbf{U}'_0 \cup \{j\}$, $\mathbf{U}'_1$ as the subscript sets for 0- and 1-variables, respectively. Add candidate subproblem 2 to the stack. Make candidate subproblem 1 the updated current candidate problem and continue.

For efficient branching variable selection criteria in this algorithm see references [3, 13, and 30].

15.11 ADVANTAGES AND LIMITATIONS

In this chapter we have discussed the general nature of the branch and bound approach and its application to solve a variety of problems. The various methods for developing bounding, branching, and the search strategies have also been illustrated in the applications. The examples provide an insight into how branch and bound algorithms can be developed for solving combinatorial optimization problems.

It is not expected that the branch and bound approach will solve all combinatorial optimization problems very efficiently. Simple integer programs have been constructed on which the branch and bound approach performs very poorly (Problems 15.11 and 15.12). However, in many practical applications of integer programming and combinatorial optimization, branch and bound approach has given very satisfactory performance. If the problem being solved is very large, it is probably too much to expect that branch and bound (or any other) approach will produce a mathematically proven optimum solution with very little expenditure of computer time. But on such problems the branch and bound approach will produce very good incumbents very early in the algorithm; even though these early incumbents cannot be proved to be optimal to the problem, in most problems they turn out to be very close to the optimum. This is what makes the branch and bound approach very useful in practical applications.

In solving integer programs by the branch and bound approach using the lower bounding strategy based on relaxed LPs, the approach can be combined with the cutting plane methods (Chapter 14). If a candidate problem is not fathomed during the branch and bound approach, a small number of iterations of a cutting plane method can be applied on it to try to fathom it. If it is still not fathomed, it can be branched as in the branch and bound approach. Such hybrid approaches seem to yield very satisfactory results.

Problems

15.9 Consider the set partitioning problem discussed in section 13.2.11. The problem is

$$\begin{aligned} \text{Minimize} \quad & z(x) = cx \\ \text{Subject to} \quad & Ax = e \\ & x_j = 0 \text{ or } 1 \qquad \text{for all } j \end{aligned}$$

where A is a 0-1 matrix of order $m \times n$, and e is the column vector of all 1's in $\mathbf{R}^m$. Let $\boldsymbol{\Gamma} = \{1, \ldots, n\}$ and $\mathbf{F}_i = \{j: a_{ij} = 1\}$ for $i = 1$ to m. A *partition* is a subset $\mathbf{E} \subset \boldsymbol{\Gamma}$ such that $\mathbf{E}$ has exactly one element in common with each $\mathbf{F}_i$, $i = 1$ to m. An optimum solution of this problem is the incidence vector of a minimum cost partition.

1. Let $\mathbf{G}(j) = \{i: j \in \mathbf{F}_i\}$. If $\mathbf{G}(p) \cap \mathbf{G}(t) \neq \emptyset$, prove that x_p and x_t cannot both be equal to 1 in any feasible solution of the set partitioning problem.
2. Consider a candidate problem of the form:

$$\begin{aligned} \text{Minimize} \quad & z(x) \\ \text{Subject to} \quad & \sum_{j \in \mathbf{D}_t} x_j = 1 && t = 1 \text{ to } r \\ & x_j = 0 && \text{for all } j \in \mathbf{H} && (15.26) \\ & x_j = 0 \text{ or } 1 && \text{for all } j \\ & \sum_{j \in \mathbf{F}_i} x_j = 1 && \text{for all } i \in \mathbf{I} && (15.27) \end{aligned}$$

where $\mathbf{D}_1, \ldots, \mathbf{D}_r$ are all nonempty and mutually disjoint, $\mathbf{H}$ is the set of banned elements that is disjoint from $\mathbf{D}_1 \cup \cdots \cup \mathbf{D}_r$, and $\mathbf{I}$ is the index set of unincluded constraints. Relaxing the constraints (15.27), the relaxed Lagrangian for this candidate problem is

$$L(x, \pi) = z(x) - \sum_{i \in \mathbf{I}} \pi_i \left(\sum_{j \in \mathbf{F}_i} x_j - 1 \right)$$

Since (15.27) are equality constraints, the Lagrange multipliers, π_i's are all unrestricted in sign. For a given Lagrange multiplier vector the relaxed problem is

$$\begin{aligned} L(\pi) = \ & \underset{x}{\text{minimum}}\ L(x, \pi) \\ \text{Subject to} \ & \sum_{j \in \mathbf{D}_t} x_j = 1 \qquad t = 1 \text{ to } r \\ & x_j = 0 \qquad \text{for } j \in \mathbf{H} \\ & x_j = 0 \text{ or } 1 \qquad \text{for all } j \end{aligned} \tag{15.28}$$

Develop a simple solution method for solving the relaxed problem (15.28) and for computing $L(\pi)$. Obtain the subgradient of $L(\pi)$ at the given vector π, using ideas similar to those in sections 15.9.5 and 15.9.6. Prove that $L(\pi)$ provides a lower bound for the minimum objective value in the candidate problem (15.26), (15.27),

3. The subgradient optimization algorithm can be used for finding the best Lagrange multiplier vector, π, for this relaxed problem using ideas similar to those in section 15.9.6. The only difference is that here the Lagrange multipliers are unrestricted in sign. So if v^p is the subgradient vector at the pth step of the subgradient optimization routine, the next Lagrange multiplier vector in the sequence is obtained by $\pi_i^{p+1} = \pi_i^p + \lambda_p v_i^p$ for $i \in \mathbf{I}$. Using this information, develop a branch and bound algorithm for the set paritioning problem, similar to the one for the set representation problem.

—J. Etcheberry

15.10 Consider the set packing problem discussed in section 13.2.11. It is a problem of the form

$$\begin{aligned} \text{Maximize} \quad & z^1(x) = dx \\ \text{Subject to} \quad & Ax \leqq e \\ & x_j = 0 \text{ or } 1 \qquad \text{for all } j \end{aligned} \tag{15.29}$$

where A is a 0-1 matrix of order $m \times n$. Let $e, \Gamma, \mathbf{F}_i$ have the same meaning as in the previous problem. Using the ideas in section 15.9 and the previous problem, develop a branch and bound algorithm for this problem. Make sure that the candidate problems obtained during the algorithm are of the form:

$$\begin{aligned} \text{Minimize} \quad & z(x) = -dx \\ \text{Subject to} \quad & 1 - \sum_{j \in \mathbf{D}_t} x_j \geqq 0 \qquad t = 1 \text{ to } r \\ & x_j = 0 \qquad \text{for } j \in \mathbf{H} \\ & x_j = 0 \text{ or } 1 \qquad \text{for all } j \end{aligned} \tag{15.30}$$

$$1 - \sum_{j \in \mathbf{F}_i} x_j \geqq 0 \qquad \text{for all } i \in \mathbf{I} \tag{15.31}$$

where $\mathbf{D}_1, \ldots, \mathbf{D}_r$ are all nonempty and mutually disjoint, $\mathbf{H}$ is a set of banned elements, and $\mathbf{I}$ is the index set of unincluded constraints in the candidate problem. The lower bounding strategy for this candidate problem should relax the constraints (15.31) using Lagrangian relaxation.

—J. Etchberry

15.11 Let n be an odd positive integer. $\langle n/2 \rangle$ is the smallest integer greater than or equal to $n/2$. Consider the problem

$$\text{Minimize} \quad z(x) = -x_1$$

$$\text{Subject to} \quad 2x_1 + 2x_2 + \cdots + 2x_n = n$$
$$x_j = 0 \text{ or } 1 \qquad \text{for all } j$$

Prove that this problem is infeasible. Also prove that the branch and bound algorithm of section 15.8 will examine at least $2^{\langle n/2 \rangle}$ candidate problems before it terminates, when it is applied on this problem.

—*R. G. Jeroslow*

15.12 Consider the integer program

$$\text{Minimize} \quad z(x) = x_{n+1}$$
$$\text{Subject to} \quad 2x_1 + 2x_2 + \cdots + 2x_n + x_{n+1} = n$$
$$x_j = 0 \text{ or } 1 \qquad \text{for all } j$$

where n is an odd positive integer. Prove that this problem is feasible. Find the optimum solution by inspection. Prove that the branch and bound algorithm of section 15.8 will have to examine at least $2^{\langle n/2 \rangle - 1}$ candidate problems before it can solve this problem.

—*R. G. Jeroslow*

15.13 $G = (\mathcal{N}, \mathcal{A})$ is a given undirected graph. A subset $\mathbf{P} \subset \mathcal{N}$ is said to be a *vertex packing* in G, if no pair of vertices of $\mathbf{P}$ are connected by an edge in $\mathcal{A}$. w_i is the profit by including the vertex i in a packing, and all w_i, $i \in \mathcal{N}$ are given. Develop a branch and bound algorithm for finding a maximum profit vertex packing in G.

15.14 STEINER PROBLEM. $G = (\mathcal{N}, \mathcal{A})$ is an undirected connected network. $\mathbf{Y} \subset \mathcal{N}$ is a specified subset of vertices of G. c_{ij} is the length of the edge $(i; j) \in \mathcal{A}$. All c_{ij} are given and are positive. Given a subset $\mathbf{S} \subset \mathcal{A}$, the length of $\mathbf{S}$, $c(\mathbf{S})$, is defined to be the sum of the lengths of the edges in $\mathbf{S}$. A subset $\mathbf{S}$ is a *Steiner subset* if every pair of vertices in $\mathbf{Y}$ is connected by a path composed of edges from $\mathbf{S}$ only. Develop an algorithm for finding a Steiner subset of minimal length.

15.12 RANKING METHODS

In these methods the set of feasible solutions of a combinatorial problem are arranged in a sequence, such that the value of a specified objective function is monotone increasing (actually it is nondecreasing) along the sequence. If necessary the sequence can be continued until all the feasible solutions are in it, or it should be possible to terminate the method after enough solutions in the sequence are obtained. The method normally obtains one solution in the ranked sequence in each step. In most practical applications, each solution obtained in the sequence is examined to see whether it satisfies certain desirability properties (these may be constraints that were not included in the original model) and the method is terminated the moment a desirable solution is obtained in the sequence.

Ranking methods can be developed when the set of feasible solutions to be ranked is a finite set. Example of ranking methods on some problems and their applications are discussed here briefly.

15.12.1 Ranking the Assignments in Increasing Order of Cost

Consider an assignment problem of order n with $C = (c_{ij})$ as the cost matrix. Suppose you have to rank all the assignments in increasing order of cost. An assignment is a square matrix $x = (x_{ij})$ of order n with the property that it has a single nonzero entry of 1 in each row and column. The assignment, x, will also be denoted by the set $\{(i, j) : (i, j) \text{ such that } x_{ij} = 1\}$. Correspondingly we shall write $(i, j) \in x$ or $(i, j) \notin x$ to indicate that $x_{ij} = 1$ or 0, respectively.

Let $\mathscr{a}(1)$ denote a minimum cost assignment. If $\mathscr{a}(1), \mathscr{a}(2), \ldots, \mathscr{a}(r), \ldots$, is the ranked sequence of assignments, then for $r \geqq 2$

$$\begin{aligned}\mathscr{a}(r) = {} & \text{minimum cost assignment excluding} \\ & \mathscr{a}(1), \ldots, \mathscr{a}(r-1)\end{aligned}$$

In the algorithm we will use subjsets of assignments called *nodes.* A *node* is a nonempty subset of assignments $\mathscr{a}$ of the form

$$\mathbf{N} = \{\mathscr{a}\colon (i_1, j_1) \in \mathscr{a}, \ldots, (i_r, j_r) \in \mathscr{a}; (m_1, p_1) \notin \mathscr{a}, \ldots, (m_u, p_u) \notin \mathscr{a}\} \quad (15.32)$$

The cells $(i_1,j_1), \ldots, (i_r,j_r)$ are *specified to be contained in*, and the cells $(m_1, p_1), \ldots,$ (m_u, p_u) are *specified to be excluded from* each assignment in node $\mathbf{N}$. We will use the notation

$$\mathbf{N} = \{(i_1, j_1), \ldots, (i_r, j_r); (\overline{m_1, p_1}), \ldots, (\overline{m_u, p_u})\}$$

to denote the node in (15.32). The matrix obtained by striking off rows $i_1, \ldots, i_r$ and columns $j_1, \ldots, j_r$ from C and replacing the entries in positions $(m_1, p_1), \ldots,$ (m_u, p_u) by infinity (or a very large positive number) is known as the *remaining cost matrix* corresponding to node $\mathbf{N}$ and is denoted by $C_{\mathbf{N}}$.

Every assignment in $\mathbf{N}$ of (15.32) contains all the cells $(i_1, j_1), \ldots, (i_r, j_r)$ and does not contain any of the cells $(m_1, p_1), \ldots, (m_u, p_u)$. Hence it is clear that a minimum cost assignment in $\mathbf{N}$ can be found by solving an assignment problem of order $n - r$ with $C_{\mathbf{N}}$ as the cost matrix. Let $x_{\mathbf{N}}$ denote a minimum cost assignment in $\mathbf{N}$ and let $z_{\mathbf{N}}$ be its objective value.

The algorithm will be so organized that in all the nodes generated during the computations, all the "specified to be excluded" cells belong to the same row of the matrix.

One of the operations performed on the nodes during the algorithm is that of *partitioning it using a minimal cost assignment* in it. Let $\mathbf{N}$ be the node in (15.32) and let $x_{\mathbf{N}} = \{(i_1, j_1), \ldots, (i_r, j_r), (s_1, t_1), \ldots, (s_{n-r}, t_{n-r})\}$ be an optimum assignment in it. From the definition of $\mathbf{N}$, each of $(s_1, t_1), \ldots, (s_{n-r}, t_{n-r})$ should be distinct from $(m_1, p_1), \ldots, (m_u, p_u)$. Let

$$\begin{aligned}
\mathbf{N}_1 &= \{(i_1, j_1), \ldots, (i_r, j_r); (\overline{m_1, p_1}), \ldots, (\overline{m_u, p_u}), (\overline{s_1, t_1})\}\\
\mathbf{N}_2 &= \{(i_1, j_1), \ldots, (i_r, j_r); (s_1, t_1); (\overline{m_1, p_1}), \ldots, (\overline{m_u, p_u})(\overline{s_2, t_2})\}\\
\mathbf{N}_3 &= \{(i_1, j_1), \ldots, (i_r, j_r), (s_1, t_1)(s_2, t_2); (\overline{m_1, p_1}), \ldots, (\overline{m_u, p_u})(\overline{s_3, t_3})\}\\
&\;\;\vdots\\
\mathbf{N}_{n-r-1} &= \{(i_1, j_1), \ldots, (i_r, j_r), (s_1, t_1, \ldots, (s_{n-r-2}, t_{n-r-2});\\
&\qquad (\overline{m_1, p_1}), \ldots, (\overline{m_u, p_u}), (\overline{s_{n-r-1}, t_{n-r-1}})\}
\end{aligned}$$

The partitioning of $\mathbf{N}$ using $x_{\mathbf{N}}$ generates the subnodes $\mathbf{N}_1, \ldots, \mathbf{N}_{n-r-1}$, and the partition itself is:

$$\mathbf{N} = \{x_N\} \cup \bigcup_{v=1}^{n-r-1} \mathbf{N}_v \quad (15.33)$$

It can be verified easily that the subnodes $\mathbf{N}_1, \ldots, \mathbf{N}_{n-r-1}$ are all nonempty and mutually disjoint and that (15.33) is correct.

At each stage, the algorithm maintains a *list*, which is a set of nodes. Consider the stage in which $\mathscr{a}(1), \ldots, \mathscr{a}(r)$ in the ranked sequence have already been obtained. Suppose the list of nodes in this stage is $\mathbf{M}_1, \mathbf{M}_2, \ldots, \mathbf{M}_\ell$. From the manner in which these are generated the following property will be satisfied.

(i) $\mathbf{M}_1, \ldots, \mathbf{M}_\ell$ are mutually disjoint and their union is the set of all assignments excluding $\mathscr{a}(1), \ldots, \mathscr{a}(r)$.

Let $x_{\mathbf{M}_d}$ be an optimum assignment in the node $\mathbf{M}_d$ and let $z_{\mathbf{M}_d}$ be its objective value. From property (i), it is clear that the next assignment in the ranked sequence can be obtained from

$$\mathscr{a}(r+1) = x_{\mathbf{M}_d} \text{ where } d \text{ satisfies } z_{\mathbf{M}_d} = \text{minimum } \{z_{\mathbf{M}_1}, \ldots, z_{\mathbf{M}_\ell}\}$$

If $\mathscr{a}(r+1)$ is the last assignment in the ranked sequence that is required, the algorithm terminates here. If another assignment in the ranked sequence is required, the node $\mathbf{M}_d$ is partitioned using $x_{\mathbf{M}_d}$. Let $\mathbf{M}_{d,1}, \ldots, \mathbf{M}_{d,f}$ be the subnodes generated.

Delete $\mathbf{M}_d$ from the list, add $\mathbf{M}_{d,1}, \ldots, \mathbf{M}_{d,f}$ to the list and go to the next stage.

INITIAL STAGES

The algorithm is initiated by finding a minimum cost assignment, $\mathscr{a}(1)$, using the Hungarian method (Chapter 12). If $\mathscr{a}(1) = \{(1, j_1), \ldots, (n, j_n)\}$, let the list at this stage be

$$\{\{(\overline{1, j_1})\}, \{(1, j_1); (\overline{2, j_2})\}, \ldots, \{(1, j_1), \ldots, (n-2, j_{n-2}); (\overline{n-1, j_{n-1}})\}\}$$

This completes the initial stage and the algorithm moves to the next stage. The algorithm can be continued until as many assignments in the sequence of ranked assignments are obtained as desired. If the algorithm is not terminating in a stage, the minimal cost assignments in the nodes generated in that stage are computed by solving the corresponding assignment problems. Each node in the list is stored together with the minimum cost assignment in it and its objective value. If a predetermined number, k, of assignments in the ranked sequence are required, only the k nodes that are associated with the least objective values are stored in the list and the rest are pruned. If it is desired to obtain all the assignments whose cost is less than or equal to some predetermined number, α, only those nodes in which the minimum objective value is less than or equal to α are stored in the list, and the rest are pruned.

Example

Consider the assignment problem with the following data.

	$j =$	1	2	3	4	5	6	7	8	9	10
	$i =$ 1	7	51	52	87	38	60	74	66	0	20
	2	50	12	0	64	8	53	0	46	76	42
	3	27	77	0	18	22	48	44	13	0	57
	4	62	0	3	8	5	6	14	0	26	39
	5	0	97	0	5	13	0	41	31	62	48
$C =$	6	79	68	0	0	15	12	17	47	35	43
	7	76	99	48	27	34	0	0	0	28	0
	8	0	20	9	27	46	15	84	19	3	24
	9	56	10	45	39	0	93	67	79	19	38
	10	27	0	39	53	46	24	69	46	23	1

The minimum cost assignment in this problem is

$$\mathscr{a}(1) = \{(1,9), (2,7), (3,3), (4,8), (5,6), (6,4), (7,10), (8,1), (9,5), (10,2)\}$$

with an objective value of 0. The list at the end of the initial stage consists of the following nodes. The minimum objective value in each node is also recorded.

$$\mathbf{M}_1 = \{\overline{(1, 9)}\}, \; z_{\mathbf{M}_1} = 10$$
$$\mathbf{M}_2 = \{(1, 9), \overline{(2, 7)}\}, \; z_{\mathbf{M}_2} = 14$$
$$\mathbf{M}_3 = \{(1, 9), (2, 7), \overline{(3, 3)}\}, \; z_{\mathbf{M}_3} = 14$$
$$\mathbf{M}_4 = \{(1, 9), (2, 7), (3, 3), \overline{(4, 8)}\}, \; z_{\mathbf{M}_4} = 1$$
$$\mathbf{M}_5 = \{(1, 9), (2, 7), (3, 3), (4, 8), \overline{(5, 6)}\}, \; z_{\mathbf{M}_5} = 15$$
$$\mathbf{M}_6 = \{(1, 9), (2, 7), (3, 3), (4, 8), (5, 6), \overline{(6, 4)}\}, \; z_{\mathbf{M}_6} = 53$$
$$\mathbf{M}_7 = \{(1, 9), (2, 7), (3, 3), (4, 8), (5, 6), (6, 4), \overline{(7, 10)}\}, \; z_{\mathbf{M}_7} = 45$$
$$\mathbf{M}_8 = \{(1, 9), (2, 7), (3, 3), (4, 8), (5, 6), (6, 4), (7, 10), \overline{(8, 1)}\}, \; z_{\mathbf{M}_8} = 47$$
$$\mathbf{M}_9 = \{(1, 9), (2, 7), (3, 3), (4, 8), (5, 6), (6, 4), (7, 10), (8, 1), \overline{(9, 5)}\}, \; z_{\mathbf{M}_9} = 56$$

Comparing the values of $z_{\mathbf{M}_1}$ to $z_{\mathbf{M}_9}$, we find that $a(2)$ is the minimal cost assignment in $\mathbf{M}_4$. It is

$$a(2) = \{(1, 9), (2, 7), (3, 3), (4, 2), (5, 6), (6, 4), (7, 8), (8, 1), (9, 5), (10, 10)\}$$

with an objective value of 1. If it is required to find $a(3)$, then $\mathbf{M}_4$ should be partitioned using $a(2)$, and the algorithm continued as before.

Note This algorithm was developed in 1962 and it appeared in reference [27]. It can be used to solve problems in which a minimal cost assignment satisfying several additional constraints is required. The additional constraints are stored separately and the assignments ranked in increasing order of cost. The first assignment in the ranked sequence that satisfies all the additional constraints is an optimum assignment for the problem.

Problems

15.15 Find $a(3)$ to $a(6)$ in this numerical example.

15.16 Let $G = (\mathcal{N}, \mathcal{A})$ be a connected directed network with $c = (c_{ij})$ as the vector of arc lengths. s is the origin and t is the destination. Develop an algorithm for ranking the chains from s to t in G in increasing order of length.

15.17 Let $G = (\mathcal{N}, \mathcal{A})$ be a connected undirected network with $c = (c_{ij})$ as the vector of edge lengths. Let $\mathbf{T}$ be a minimum length spanning tree in G. Let (i_1, j_1) be an edge in $\mathbf{T}$. The edge (i_1, j_1) generates a cutset $(\mathbf{X}, \bar{\mathbf{X}})$ with the property that the unique path in $\mathbf{T}$ from any point in $\mathbf{X}$ to any point in $\bar{\mathbf{X}}$ contains the edge (i_1, j_1). Prove that (i_1, j_1) must be a minimal length edge in the cutset $(\mathbf{X}, \bar{\mathbf{X}})$.

If (i_1, j_1) is the only edge in the cutset $(\mathbf{X}, \bar{\mathbf{X}})$, prove that every spanning tree in G must contain the edge (i_1, j_1).

If there are some other edges in the cutset $(\mathbf{X}, \bar{\mathbf{X}})$ besides (i_1, j_1) let (r, s) be a minimal length edge in it excluding (i_1, j_1). Show that the spanning tree obtained by replacing the edge (i_1, j_1) by the edge (r, s) in $\mathbf{T}$ is a minimal length spanning tree in G not containing the edge (i_1, j_1).

Using this, develop an algorithm for ranking the spanning trees in G in increasing order of length.

15.12.2 Methods Using Extreme Point Ranking

Consider the problem

$$\begin{aligned} \text{Minimize} \quad & \theta(x) \\ \text{Subject to} \quad & Ax = b \\ & x \geqq 0 \end{aligned} \tag{15.34}$$

where $\theta(x)$ may be a nonlinear function. Let **K** be the set of feasible solutions of (15.34). If $\theta(x)$ is a concave function on **K** (i.e., if $\theta[\alpha x^1 + (1 - \alpha)x^2] \geqq \alpha\theta(x^1) + (1 - \alpha)\theta(x^2)$ for all feasible x^1, x^2, and $0 \leqq \alpha \leqq 1$), then an optimum solution of this problem occurs at an extreme point of **K**. Such problems can be solved by a search among the extreme points of **K**.

There are many practical problems that are of the form (15.34), but with the additional restriction that the best extreme point among all the extreme points of **K** is required. The 0-1 integer programming problems and quadratic assignment problems are examples of such problems. These problems can also be solved by a search among the extreme points of **K**.

One method of searching among the extreme points of **K** for solving (15.34) proceeds as follows. Construct a linear function $z(x) = cx$ satisfying the property

$$z(x) \leqq \theta(x) \qquad \text{for all } x \in \mathbf{K} \tag{15.35}$$

Methods for constructing such *lower bounding linear functions* are discussed later on.

Rank the extreme points of **K** in increasing order of $z(x)$. An algorithm for doing this is discussed in Problem 3.49. Let $x^1, x^2, \ldots,$ be the ranked sequence of extreme points. The ranking algorithm obtains one additional extreme point in the sequence in each step.

Step 1 In this step, x^1, the extreme point of **K** that minimizes $z(x)$ is obtained. Since $z(x)$ is linear, this can be done using the simplex method. x^1 is the *incumbent* in this step. Let $U_1 = \theta(x^1)$. Clearly, the minimum objective value in (15.34) has to be less than or equal to U_1. Let $L_1 = \text{minimum } \{z(x^1), U_1\}$. Obviously $L_1 \leqq U_1$, and from (15.35), L_1 is a lower bound for the minimum objective value in (15.34).

If $L_1 = U_1$, the incumbent must be an optimum solution of (15.34) and the algorithm terminates. If $U_1 > L_1$, the algorithm moves to the next step. In each step the lower bound and the upper bound for the minimum objective value in (15.34) and the incumbent are updated.

General Step Suppose $x^1, x^2, \ldots, x^r$ in the ranked sequence of extreme points [in increasing order of $z(x)$] have already been obtained and suppose the algorithm did not terminate yet. The incumbent at this stage is x^i, satisfying $\theta(x^i) = \text{minimum } \{\theta(x^1), \ldots, \theta(x^r)\}$. The current upperbound, U_r, is the value of $\theta(x)$ at the incumbent. The current lower bound, L_r, is minimum $\{z(x^r), U_r\}$. Since the algorithm did not terminate yet, L_r must be strictly less then U_r. Obtain the next extreme point in the ranked sequence, x^{r+1}.

If $\theta(x^{r+1}) < U_r$, x^{r+1} is made the new incumbent and the new upper bound is $U_{r+1} = \theta(x^{r+1})$. If $\theta(x^{r+1}) \geqq U_r$ the current incumbent is left unchanged and $U_{r+1} = U_r$. The new lower bound is $L_{r+1} = \text{minimum } \{z(x^{r+1}), U_{r+1}\}$. If $L_{r+1} = U_{r+1}$, the current incumbent is an optimum solution of (15.34) and the algorithm terminates. If $L_{r+1} < U_{r+1}$, the algorithm moves to the next step.

The lower bound keeps steadily increasing and the upper bound keeps steadily decreasing during the algorithm. If a near optimum solution is satisfactory, the algorithm can be terminated whenever the difference between the current lower and upper bounds becomes small enough. In this case, the incumbent in the terminal step is a near optimum solution of the problem.

HOW TO OBTAIN A LOWER BOUNDING LINEAR FUNCTION

Suppose $\theta(x)$ is separable, that is, $\theta(x) = \sum_j \theta_j(x_j)$. Also suppose that each $\theta_j(x_j)$ is a concave function. Let α_j be a practical upper bound for the value of the variable x_j in the problem. Obtain the slope c_j, as in Figure 15.12. Then $c_j x_j \leqq \theta_j(x_j)$ for all x_j in the range 0 to α_j. Hence $z(x) = \sum c_j x_j$ is a lower bounding linear function for $\theta(x)$ in this case. See reference [4]. As another example, consider the case where $\theta(x) = \sum\sum x_i d_{ij} x_j + \sum c_j x_j$, a quadratic function. Let α_j = minimum $\{\sum_i x_i d_{ij}: x \in \mathbf{K}\}$, which can be found by solving an LP. Then $z(x) = \sum(\alpha_i + c_j)x_j$ is a lower bounding linear function in this case. See reference [37]. Similar methods can be used to construct lower bounding linear functions in other cases. The search among the extreme points of **K** will be more efficient if the lower bounding linear function is a close approximation of $\theta(x)$. In most practical applications, by using good lower bounding linear functions, it turns out that the optimum solution of the problem is obtained as an incumbent after only a few steps of the ranking method, even though the algorithm may go through many more steps before terminating with the conclusion that the incumbent is in fact an optimum solution.

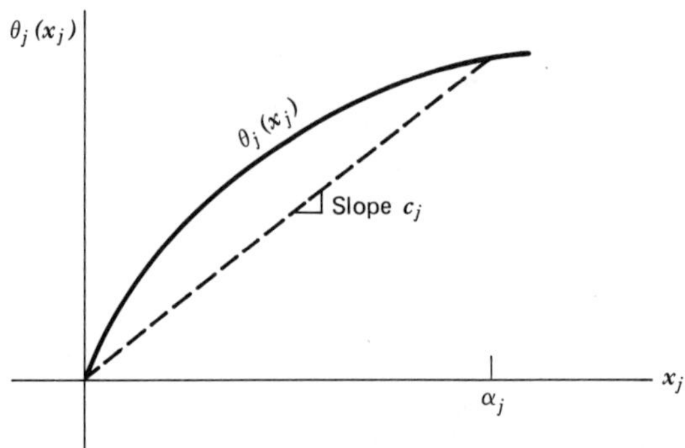

Figure 15.12

Problems

15.18 Develop an algorithm for solving the fixed charge problem discussed in section 13.2.3, using extreme point ranking methods.

15.19 Let x be an assignment of order n. Let $C^1, C^2, \ldots, C^k$ be k given cost matrices. Let $z^i(x)$ be the cost of the assignment x with C^i as the cost matrix, $i = 1$ to k. Let

$$\theta(x) = \text{maximum } \{z^1(x), \ldots, z^k(x)\}$$

Find an assignment that minimizes $\theta(x)$. Develop an algorithm for solving this problem using an assignment ranking approach.

15.20 Let D_1, D_2 be two columns consisting of n_1, n_2 numbers, respectively. Suppose $n_2 \leqq n_1$. A *policy* is an ordered pair (i_1, i_2) such that $1 \leqq i_1 \leqq n_1$ and $1 \leqq i_2 \leqq n_2$. The cost of the policy (i_1, i_2) is the sum of the i_1th entry in D_1 and the i_2th entry in D_2. Develop an algorithm for ranking the policies in increasing order of cost, such that in each step the computational effort required to obtain an additional policy in the ranked sequence is no more than n_2 additions and n_2 comparisons.

15.21 Let $D_1, \ldots, D_k$ be k columns of numbers with D_r containing n_r numbers, $r = 1$ to k. In this problem, a *policy* is an ordered vector $(i_1, \ldots, i_k)$ with the property that $1 \leqq i_r \leqq n_r$ for $r = 1$ to k. The cost of the policy $(i_1, \ldots, i_k)$ is the sum of the i_rth entry in D_r over $r = 1$ to k. Develop an efficient algorithm for ranking the policies in increasing order of cost using the algorithm developed in Problem 15.20 $k - 1$ times. Using the algorithm, obtain for the following data the first 25 policies in increasing order of cost.

D_1	D_2	D_3	D_4
5	10	10	5
10	35	20	15
20	40	30	25
25		40	30

—*K. G. Murty, R. Saigal, and P. Unger*

MULTIPLE CHOICE PROBLEMS

Consider a multiple choice problem in which k projects are to be carried out. Each project is independent of the others and the rth project can be performed by any one of n_r alternate ways. c_{ij} is the cost of performing the ith project by its jth alternative, $j = 1$ to n_i, $i = 1$ to k. There are t resources required in carrying these projects and the requirements of each resource, if the rth project is performed by its jth alternative is given. Let b_p be the total amount of resource p available, $p = 1$ to t. In this problem, a *policy* is an ordered vector that specifies the alternative used for performing the projects. The policy $(i_1, \ldots, i_k)$ implies that the rth project is performed by its i_rth alternative. The problem is to determine an optimum policy that minimizes the total cost subject to the resource availability constraints.

This problem can be formulated as an integer program and the optimum policy can be found by solving the integer program. However, the minimum objective value in this integer program (which is the minimum cost of performing all the projects) is not necessarily a continuous function of the resources availabilities. *Small changes in b_p may result in substantial changes in the minimum objective value.* In practice, resource availabilities can be augmented by purchasing additional quantities of resources, and a small amount of money invested in this might lead to substantial savings in the minimum objective value in the integer program. But it is hard to perform sensitivity analysis of this type using the integer programming model.

The following approach is probably a better approach for solving practical problems of this type. Keep all the resource constraints separately. Rank all the policies in increasing order of cost as in Problem 15.21. As each policy in the ranked sequence is obtained, compute the resource consumptions if that policy is implemented and compare them with the resource availabilities. The total cost of augmenting the supplies of the resources whose consumptions under that policy exceed the availabilities can also be assessed.

The cost of the policies in the ranked sequence is monotone increasing. The increasing cost can be compared to the cost of augmenting the supplies of the resources under these policies. The algorithm can be terminated whenever a policy that seems to be satisfactory in all respects is reached in the ranked sequence.

15.13 NOTES ON THE BRANCH AND BOUND APPROACH

The branch and bound algorithm for the traveling salesman problem first appeared in the unpublished article [29]. The branch and bound approach for solving integer programs was discussed in [19]. The branching strategy in section 15.3.2 and used in section 15.8 was proposed in [7]. An algorithm for the set representation problem based on unordered Cartesian products appeared in [28]. The algorithm for the set representation problem discussed in section 15.9 is a branch and bound adoption of the UCP algorithm and this is developed by J. Etcheberry in his Ph.D dissertation [8]. Subgradient optimization algorithm and its uses in a branch and bound approach are discussed in [16], [17], and, [9]. Some of the terminology used in this chapter is from [12]. *Implicit enumeration* is the general term used for branch and bound approaches to solve 0-1 problems. The implicit enumeration algorithm discussed in section 15.10 is by Balas [2], and he called it the *additive algorithm* because it requires only additions and comparisons. Methods for generating good surrogate constraints and their usefulness in branch and bound algorithms are discussed in [14] and [13]. For a discussion on the efficiency of branch and bound approaches, see [18], [33], [9], [10], and [30]. The literature on integer programming and branch and bound is very extensive, and it is impossible to discuss all the contributions in this limited space. For good bibliographies see [28], [17], [39], [11], and [5]. Extreme point ranking methods are discussed in [26] and [37].

Problems

15.22 Consider the following problem

$$\begin{aligned} \text{Minimize} \quad & \textstyle\sum \theta_j(x_j) \\ \text{Subject to} \quad & Ax = b \\ & x \geqq 0 \end{aligned}$$

where $\theta_j(x_j)$ is a piecewise linear function as in Figure 15.13 for each j. The point x_j^1 where $\theta_j(x_j)$ changes slope and the slopes c_j^1 and c_j^2 are given for all j. $c_j^1 > c_j^2$ for all j. Develop an algorithm for solving this problem.

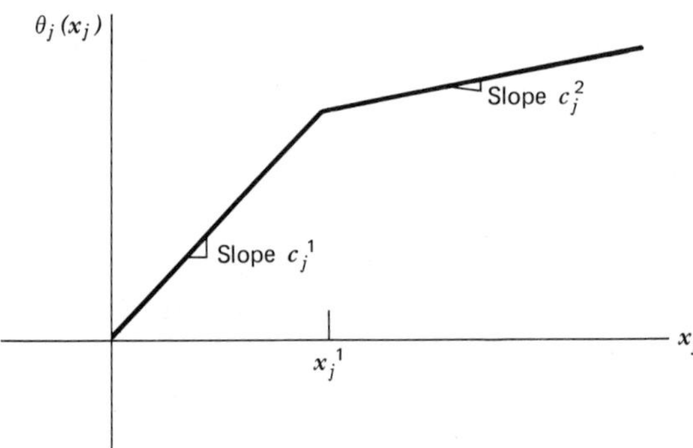

Figure 15.13

15.23 G = $(\mathcal{N}, \mathcal{A})$ is a connected, undirected network, with $c = (c_{ij})$ as the vector of edge lengths. Let **T** be a spanning tree in G and let $d_{ij}(\mathbf{T})$ be the length of the path between i and j in **T**.

(i) $(i_1, i_1), \ldots, (i_r, j_r)$ are given pairs of points in G and $V_{i_1 j_1}, \ldots, V_{i_r j_r}$ are the specified amounts of flow between these respective pairs of points. Find a spanning tree **T** that minimizes $\sum_{t=1}^{r} V_{i_t j_t} d_{i_t j_t}(\mathbf{T})$

(ii) 1 is the specified origin in $\mathcal{N}$. Find a spanning tree **T** that satisfies $d_{1j}(\mathbf{T}) \leqq d$, for all $j \in \mathcal{N}$, where d is a specified number.

(iii) 1 is the specified origin; our problem is to ship a specified amount of a single commodity, V_i, along the path from 1 to i in the spanning tree **T**, and V_i are given for all $i \in \mathcal{N}$. The total flow on any edge in $\mathcal{A}$ cannot exceed a specified amount b. Find a spanning tree **T** that minimizes $\sum d_{1i} V_i$ subject to these constraints.

Develop algorithms for solving these problems.

—*M. Segal*

REFERENCES

1. J. Abadie (ed.), *Integer and Nonlinear Programming*, North-Holland Publishing Co., Amsterdam, 1970.
2. E. Balas, "An Additive Algorithm for Solving Linear Programs with Zero-One Variables," *Operations Research*, *13*, pp. 517–546, 1965.
3. E. Balas, "Discrete Programming by the Filter Method," *Operations Research*, *15*, 5, pp. 915–957, 1967.
4. M. L. Balinski, "Fixed Cost Transportation Problems," *Naval Research Logistics Quarterly*, *8*, 1, 1961.
5. M. L. Balinski, "On Recent Developments in Integer Programming," in H. W. Khun (ed.), *Proceedings of the Princeton Symposium on Mathematical Programming*, Princeton University Press, 1970.
6. E. M. L. Beale and R. E. Small, "Mixed Integer Programming by Branch and Bound Techniques," in W. A. Kalenich (ed.), *Proceedings of IFIP Congress*, *2*, Spartan Press, Washington, D.C. pp. 450–451, 1965.
7. R. J. Dakin, "A Tree Search Algorithm for Mixed Integer Programming Problem," *Computer Journal*, *8*, 3, pp. 250–255, 1965.
8. J. Etcheberry, "The Set Representation Problem," Ph.D Dissertation, Department of Industrial and Operations Engineering, University of Michigan, August 1974.
9. M. L. Fisher and P. F. Shapiro, "Constructive Duality in Integer Programming," *SIAM Journal of Applied Mathematics*, *27*, pp. 31–52, 1974.
10. J. J. H. Forrest, J. P. H. Hirst, and J. A. Tomlin, "Practical Solution of Large Mixed Integer Programming Problems with UMPIRE," *Management Science*, *20*, 5, pp. 736–773, January 1974.
11. R. S. Garfinkel and G. L. Nemhauser, *Integer Programming*, Wiley, 1972.
12. A. M. Geoffrion, "Lagrangian Relaxation for Integer Programming" in *Mathematical Programming Study 2: Approaches to Integer Programming*, M. L. Balinski (editor), North Holland Publishing Co., Amsterdam, 1974.
13. A. M. Geoffrion, "An Improved Implicit Enumeration Approach for Integer Programming," *Operations Research*, *17*, 3, 437–454, 1969.
14. F. Glover, "Surrogate Constraints," *Operations Research*, *16*, 4, 741–749, 1968.
15. H. Greenberg, *Interger Programming*, Academic Press, New York, 1971.
16. M. Held and R. M. Karp, "The Traveling Salesman Problem and Minimum Spanning Trees. Part II. *Mathematical Programming*, *1*, 1, pp. 6–25, October 1971.
17. M. Held, P. Wolfe, and H. Crowder, "Validation of Subgradient Optimization," *Mathematical Programming*, *6*, 1, pp. 62–88, February 1974.

18. R. G. Jeroslow, "Trivial Integer Programs Unsolvable by Branch and Bound," *Mathematical Programming*, *6*, 1, pp. 105–109, February 1974.
19. A. H. Land and A. G. Doig, "An Automatic Method for Solving Discrete Programming Problems," *Econometrika 28*, pp. 497–520, 1960.
20. E. L. Lawler, "Procedure for Computing the k-best Solutions to Discrete Optimization Problems and Its Applications to Shortest Path Problem," *Management Science*, *18*, 7, pp. 401–405, March 1972.
21. E. L. Lawler and D. E. Wood, "Branch and Bound Methods; A Survey," *Operations Research*, *14*, pp. 699–719, 1966.
22. C. E. Lemke and K. Spielberg, "Direct Search Algorithms for Zero-One Mixed Integer Programming," *Operations Research*, *15*, 5, pp. 892–914, September-October, 1967.
23. R. E. Marsten, "An Algorithm for Large Set Partitioning Problems," *Management Science*, *20*, 5, pp. 774–787, January 1974.
24. G. Mitra, "Investigation of Some Branch and Bound Strategies for the Solution of Mixed Integer Linear Programs," *Mathematical Programming*, *4*, 2, pp. 155–170, April 1973.
25. L. G. Mitten, "Branch and Bound Methods; General Formulation and Properties," *Operations Research*, *18*, pp. 24–34, 1970.
26. K. G. Murty, "Solving the Fixed Charge Problem by Ranking the Extreme Points," *Operations Research*, *16*, 2, pp. 268–279, March-April, 1968.
27. K. G. Murty, "An Algorithm for Ranking All the Assignments in Order of Increasing Cost," *Operations Research*, *16*, 3, pp. 682–687, May-June 1968.
28. K. G. Murty, "On the Set of Representation and Set Covering Problems," in S. E. Elmaghraby (editor), *Proceedings of the Symposium on the Theory of Scheduling and Its Applications*, Springer Verlag, Heidelberg, 1973, pp. 143–163.
29. K. G. Murty, C. Karel, and J-D. C. Little, "The Traveling Salesman Problem; Solution by a Method of Ranking Assignments," Case Institute of Technology, 1962 (unpublished).
30. C. Petersen, "Computational Experience with Variants of the Balas Algorithm Applied to the Selection of R & D Projects," *Management Science*, *13*, 9, 736–750, 1967.
31. D. R. Plane and C. McMillan "*Discrete Optimization: Inetger Programming and Network Analysis to Management Decisions*," Prentice-Hall, Englewood Cliffs, N.J., 1971.
32. M. Raghavachari, "On Connections between Zero-One Integer Programming and Concave Programming under Linear Constraints," *Operations Research*, *17*, pp. 680–684, 1969.
33. B. Roy, R. Benayoun, and J. Tergny, "From S. E. P. Procedure to the Mixed Ophelie Program" in J. Abadie (ed.), *Integer and Nonlinear Programming* Americal Elsevier Publishing Co., 1970, pp. 419–436.
34. H. Salkin, *Integer Programming*, Addison-Wesley, Reading, Mass., 1975.
35. A. J. Scott, "*Combinatorial Programming, Spatial Analysis and Planning*," Methuen & Co. Ltd., London, 1971.
36. J. A. Tomlin "An Improved Branch and Bound Method for Integer Programming," *Operations Research*, *15*, pp. 1070–1074, 1971.
37. A. Victor Cabot and R. L. Francis, "Solving Certain Nonconvex Quadratic Minimization Problems by Ranking the Extreme Points," *Operations Research*, *18*, pp. 82–86, January-February, 1970.
38. J. Y. Yen, "Finding the *k*-shortest loopless Paths in a Network," *Management Science*, *17*, 11, July 1971.
39. S. Zionts, "*Linear and Integer Programming*," Prentice-Hall, Englewood Cliffs, N.J., 1974.

16 complementarity problems

16.1 INTRODUCTION

In section 4.6.5, we discussed the formulation of an LP as a *Linear Complementarity Problem* (abbreviated as LCP). Recently an algorithm known as the *complementary pivot algorithm* has been developed for solving LCPs. Computational experience on this new algorithm is limited, but indicates that it may be superior to the simplex method for solving LPs. The new algorithm has the additional advantage that it can also solve quadratic programs and bimatrix game problems. Hence the theory of the LCP serves as a unified theory for studying linear and quadratic programs and bimatrix games. Here we discuss the main results in linear complementarity and their applications.

Let M be a given square matrix of order n and q a column vector in $\mathbf{R}^n$. In earlier chapters we used the symbols w, z to denote the Phase I and II objectives, respectively, in an LP. But throughout this chapter we will use the symbols $\mathrm{w}_1, \ldots, \mathrm{w}_n; \mathrm{z}_1, \ldots, \mathrm{z}_n$ to denote the variables in the problem. *There is no objective function in an* LCP. Consider the problem: find $\mathrm{w}_1, \ldots, \mathrm{w}_n; \mathrm{z}_1, \ldots, \mathrm{z}_n$ satisfying

$$\mathrm{w} - M\mathrm{z} = q$$
$$\mathrm{w} \geqq 0 \qquad \mathrm{z} \geqq 0 \qquad \text{and} \qquad \mathrm{w}_i\mathrm{z}_i = 0 \qquad \text{for all } i$$

As a specific example, let $n = 2$, $M = \begin{pmatrix} 2 & 1 \\ 1 & 2 \end{pmatrix}$, $q = \begin{pmatrix} -5 \\ -6 \end{pmatrix}$. For this case, the above problem is to solve:

$$\begin{aligned} \mathrm{w}_1 - 2\mathrm{z}_1 - \mathrm{z}_2 &= -5 \\ \mathrm{w}_2 - \mathrm{z}_1 - 2\mathrm{z}_2 &= -6 \\ \mathrm{w}_1, \mathrm{w}_2, \mathrm{z}_1, \mathrm{z}_2 \geqq 0 \qquad \text{and} \qquad \mathrm{w}_1\mathrm{z}_1 &= \mathrm{w}_2\mathrm{z}_2 = 0 \end{aligned} \tag{16.1}$$

Such problems are known as *linear complementarity problems.* Problem (16.1) can be expressed in the form of a vector equation as

$$\mathrm{w}_1\begin{pmatrix} 1 \\ 0 \end{pmatrix} + \mathrm{w}_2\begin{pmatrix} 0 \\ 1 \end{pmatrix} + \mathrm{z}_1\begin{pmatrix} -2 \\ -1 \end{pmatrix} + \mathrm{z}_2\begin{pmatrix} -1 \\ -2 \end{pmatrix} = \begin{pmatrix} -5 \\ -6 \end{pmatrix} \tag{16.2}$$

$$\mathrm{w}_1, \mathrm{w}_2, \mathrm{z}_1, \mathrm{z}_2 \geqq 0 \qquad \text{and} \qquad \mathrm{w}_1\mathrm{z}_1 = \mathrm{w}_2\mathrm{z}_2 = 0 \tag{16.3}$$

In any solution satisfying (16.3), at least one of the variables in each pair (w_1, z_1), (w_2, z_2) has to equal zero. One approach for solving this problem is to pick one variable from each of the pairs (w_1, z_1), (w_2, z_2) and to fix them at zero value in (16.2). The remaining variables in the system may be called *active variables*. After eliminating the zero variables from (16.2), if the remaining system has a solution in which the active variables are nonnegative, that would provide a solution to (16.2) and (16.3).

Pick w_1, w_2 as the zero-valued variables. After setting w_1, w_2 equal to 0 in (16.2), the remaining system is

$$z_1 \begin{pmatrix} -2 \\ -1 \end{pmatrix} + z_2 \begin{pmatrix} -1 \\ -2 \end{pmatrix} = \begin{pmatrix} -5 \\ -6 \end{pmatrix} = \begin{pmatrix} q_1 \\ q_2 \end{pmatrix} = q \tag{16.4}$$

$$z_1 \geqq 0, \qquad z_2 \geqq 0$$

Equation (16.4) has a solution iff the vector q can be expressed as a nonnegative linear combination of the vectors $(-2, -1)^{\mathrm{T}}$ and $(-1, -2)^{\mathrm{T}}$. The set of all nonnegative linear combinations of $(-2, -1)^{\mathrm{T}}$ and $(-1, -2)^{\mathrm{T}}$ is a cone in the q_1, q_2-space as in Figure 16.1. Only if the given vector $q = (-5, -6)^{\mathrm{T}}$ lies in this cone, does the LCP (16.1) have a solution in which the active variables are z_1, z_2. We verify that the point $(-5, -6)^{\mathrm{T}}$ does lie in the cone, that the solution of (16.4) is $(z_1, z_2) = (4/3, 7/3)$ and, hence, a solution for (16.1) is

$$(w_1, w_2, z_1, z_2) = (0, 0, 4/3, 7/3).$$

The cone in Figure 16.1 is known as a *complementary cone* associated with the LCP (16.1). Complementary cones are generalizations of the well-known class of quadrants or orthants (see the next section).

The LCP (16.1) is of order 2. In a LCP of order n, there will be $2n$ variables.

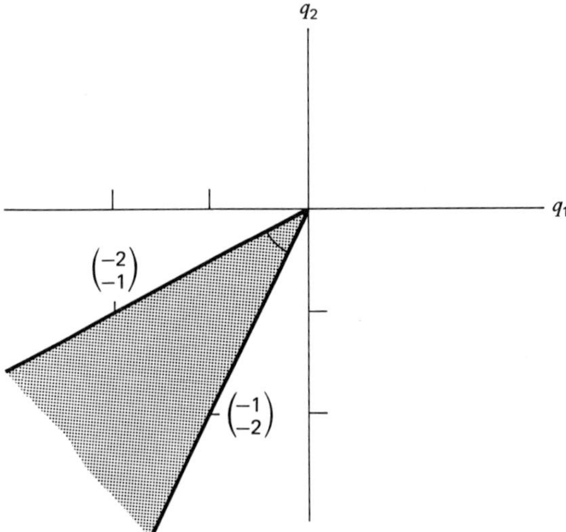

Figure 16.1

16.2 GEOMETRIC INTERPRETATION

COMPLEMENTARY CONES

Let M be a given square matrix of order n. For obtaining $\mathscr{C}(M)$, the class of complementary cones corresponding to M, the pair of column vectors $(I_{.j}, -M_{.j})$ is known as the jth *complementary pair of vectors*, $1 \leqq j \leqq n$. Pick a vector from the pair $(I_{.j}, -M_{.j})$ and denote it by $A_{.j}$. The ordered set of vectors $(A_{.1}, \ldots, A_{.n})$ is known as a *complementary set of vectors*. The cone pos $(A_{.1}, \ldots, A_{.n})$ is known as a *complementary cone* in the class $\mathscr{C}(M)$. Clearly there are 2^n complementary cones.

Example 1

Let $n = 2$ and $M = I$. In this case, the class $\mathscr{C}(I)$ is just the class of orthants in $\mathbf{R}^2$. In general for any n, $\mathscr{C}(I)$ is the class of orthants in $\mathbf{R}^n$. Thus the class of complementary cones is a generalization of the class of orthants. See Figure 16.2. Figures 16.3 and 16.4 provide some more examples of complementary cones. In the example in Figure 16.5 since $\{I_{.1}, -M_{.2}\}$ is a

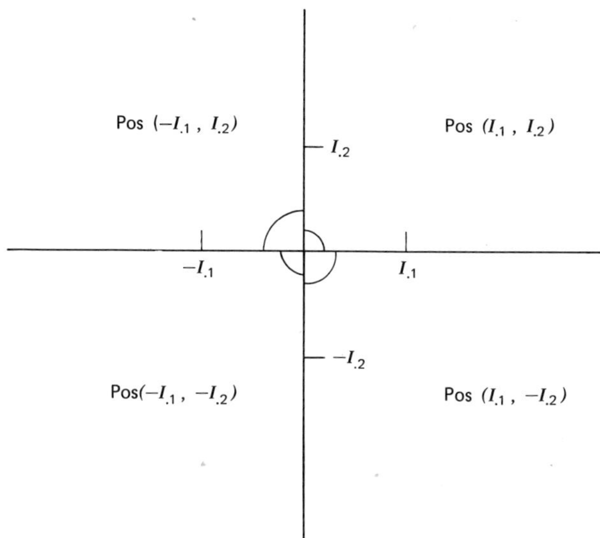

Figure 16.2

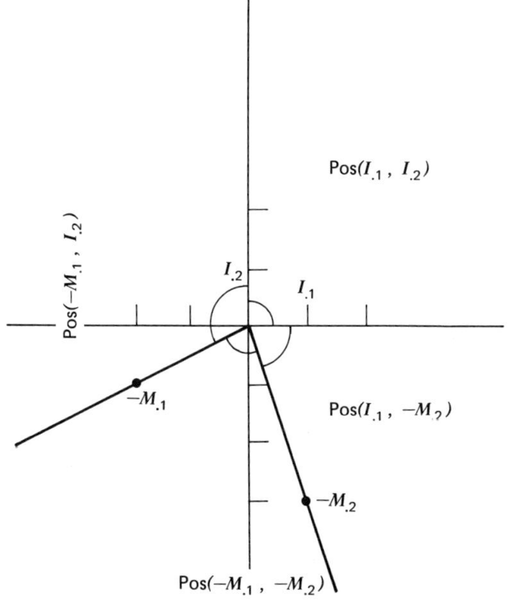

Figure 16.3 Complementary cones when $M = \begin{pmatrix} 2 & -1 \\ 1 & 3 \end{pmatrix}$.

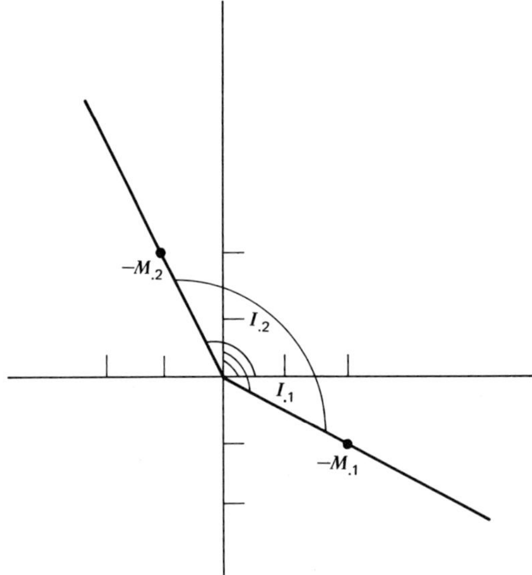

Figure 16.4 Complementary cones when $M = \begin{pmatrix} -2 & 1 \\ 1 & -2 \end{pmatrix}$.

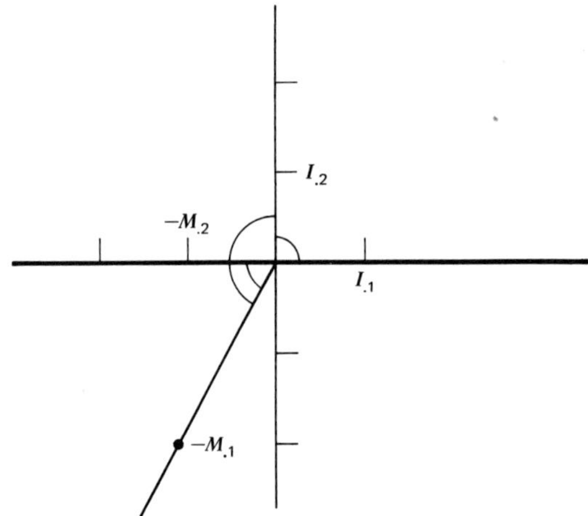

Figure 16.5 Complementary cones when $M = \begin{pmatrix} 1 & 1 \\ 2 & 0 \end{pmatrix}$.

linearly dependent set, the cone pos $(I_{.1}, -M_{.2})$ has an empty interior. It consists of all the points on the horizontal axis in Figure 16.5 (the thick axis). The remaining three complementary cones have nonempty interiors.

THE LINEAR COMPLEMENTARY PROBLEM

Given the square matrix M of order n and the column vector $q \in \mathbf{R}^n$, the *LCP* (q, M), is to find a complementary cone in $\mathscr{C}(M)$ that contains the point q, that is, to find a complementary set of column vectors $(A_{.1}, \ldots, A_{.n})$ such that

(i) $A_{.j} \in \{I_{.j}, -M_{.j}\}$ for $1 \leqq j \leqq n$

(ii) q can be expressed as a nonnegative linear combination of $(A_{.1}, \ldots, A_{.n})$

This is equivalent to finding $w \in \mathbf{R}^n$, $z \in \mathbf{R}^n$ satisfying

$$\sum_{j=1}^{n} I_{.j} w_j - \sum_{j=1}^{n} M_{.j} z_j = q$$

$$w_j \geqq 0 \qquad z_j \geqq 0 \qquad \text{for all } j$$

$$\text{either} \quad w_j = 0 \qquad \text{or } z_j = 0 \qquad \text{for all } j$$

In matrix notation this is

$$w - Mz = q \tag{16.5}$$

$$w \geqq 0 \qquad z \geqq 0 \tag{16.6}$$

$$w_j z_j = 0 \qquad \text{for all } j \tag{16.7}$$

Because of (16.6), the condition (16.7) is equivalent to $\sum_{j=1}^{n} w_j z_j = w^T z = 0$; this condition is known as the *complementarity constraint.* In any solution of the LCP (q, M), if one of the variables in the pair (w_j, z_j) is positive, the other should be zero. Hence, the pair (w_j, z_j) is known as a *complementary pair of variables* and each variable in this pair is in the *complement* of the other. A *complementary vector of variables* in this problem is a vector $(y_1, \ldots, y_n)$ with the property that y_j is either w_j or z_j for all $j = 1$ to n. It is a *complementary basic vector of variables,* if the set of column vectors corresponding to y_j, $j = 1$ to n in (16.5) is a linearly independent set.

16.3 APPLICATIONS IN LINEAR PROGRAMMING

Every LP can be expressed in the *symmetric form*

$$\begin{aligned} \text{Minimize} \quad & cx \\ \text{Subject to} \quad & Ax \geqq b \\ & x \geqq 0 \end{aligned} \tag{16.8}$$

HOW TO EXPRESS THE PROBLEM IN SYMMETRIC FORM

In Chapter 2 we discussed how to transform any given LP into standard form. Suppose when the problem is put in standard form it becomes

$$\begin{aligned} \text{Minimize} \quad & \theta(y) = gy \\ \text{Subject to} \quad & Dy = d \\ & y \geqq 0 \end{aligned} \tag{16.9}$$

Transform this into row echelon normal form as in section 3.3.3. Enter the objective row as $gy - \theta = 0$, and price out all the dependent variables in this row. Suppose this transforms the 0 on the right-hand side of this row into "$-\overline{\theta}$." Denote the vector of independent variables by x and the vector of dependent variables by s. In the transformed system suppose the right-hand side vector is "$-b$." (This "$-b$" may no longer be nonnegative.) Let "$-A$" be the coefficient matrix of the independent variables. Let c be the vector of coefficients of independent variables in the objective row. Then the problem can be rewritten as

$$\begin{aligned} \text{Minimize} \quad & \theta = \overline{\theta} + cx \\ \text{Subject to} \quad & -Ax + Is = -b \\ & x \geqq 0 \qquad s \geqq 0 \end{aligned}$$

Since $\overline{\theta}$ is a constant, minimizing θ is equivalent to minimizing $z(x) = cx$. Also the constraints $-Ax + s = -b, s \geqq 0$ are together equivalent to $Ax \geqq b$. So the LP (16.9) is equivalent to the problem in symmetric form (16.8).

THE CORRESPONDING LCP

In the LP (16.8) suppose A is a matrix of order $m \times \mathrm{N}$. If x is an optimum solution of (16.8) by the results in section 4.6.5, there exists a dual vector $y \in \mathbf{R}^m$ and slack vectors $v \in \mathbf{R}^m$, $u \in \mathbf{R}^{\mathrm{N}}$ such that u, v, x, and y together satisfy

$$\begin{pmatrix} u \\ \cdots \\ v \end{pmatrix} - \begin{pmatrix} 0 & \vdots & -A^{\mathrm{T}} \\ \cdots & \vdots & \cdots \\ A & \vdots & 0 \end{pmatrix} \begin{pmatrix} x \\ \cdots \\ y \end{pmatrix} = \begin{pmatrix} c^{\mathrm{T}} \\ \cdots \\ -b \end{pmatrix}$$

$$\begin{pmatrix} u \\ \cdots \\ v \end{pmatrix} \geqq 0 \quad \begin{pmatrix} x \\ \cdots \\ y \end{pmatrix} \geqq 0 \quad \text{and} \quad \begin{pmatrix} u \\ \cdots \\ v \end{pmatrix}^{\mathrm{T}} \begin{pmatrix} x \\ \cdots \\ y \end{pmatrix} = 0 \tag{16.10}$$

Conversely, if u, v, x, y together satisfy all the conditions in (16.10), x is an optimum solution of (16.8).

Note In (16.10) all the vectors and matrices are written in *partitioned form.* For example, $\begin{pmatrix} u \\ \cdots \\ v \end{pmatrix}$ is the vector $(u_1, \ldots, u_{\mathrm{N}}, v_1, \ldots, v_m)^{\mathrm{T}}$. If $n = m + \mathrm{N}$,

$$w = \begin{pmatrix} u \\ \cdots \\ v \end{pmatrix}, z = \begin{pmatrix} x \\ \cdots \\ y \end{pmatrix}, M = \begin{pmatrix} 0 & \vdots & -A^{\mathrm{T}} \\ \cdots & \vdots & \cdots \\ A & \vdots & 0 \end{pmatrix}, q = \begin{pmatrix} c^{\mathrm{T}} \\ \cdots \\ -b \end{pmatrix},$$

(16.10) is seen to be an LCP of order n of the type (16.5) to (16.7). Solving the LP (16.8) can be achieved by solving the LCP (16.10).

16.4 QUADRATIC PROGRAMMING

Using the methods discussed in Chapter 2 and in section 16.3, any problem in which a quardratic objective function has to be minimized subject to linear equality and inequality constraints can be transformed into a problem of the form

$$\begin{aligned} \text{Minimize} \quad & \mathrm{Q}(x) = cx + \frac{1}{2} x^{\mathrm{T}} D x \\ \text{Subject to} \quad & Ax \geqq b \\ & x \geqq 0 \end{aligned} \tag{16.11}$$

where A is a matrix of order $m \times \mathrm{N}$, and D *is a square symmetric matrix of order* N.

Note There is no loss of generality in assuming that D is a symmetric matrix, because if it is not symmetric, replacing D by $(D + D^{\mathrm{T}})/2$ (which is a symmetric matrix) leaves $\mathrm{Q}(x)$ unchanged.

Problem

16.1 Prove that $\mathrm{Q}(x)$ is a convex function over $\mathbf{R}^{\mathrm{N}}$ [i.e., $\mathrm{Q}(\alpha x^1 + (1-\alpha)x^2) \leqq \alpha \mathrm{Q}(x^1) + (1-\alpha)\mathrm{Q}(x^2)$ for every pair of points x^1, x^2 in $\mathbf{R}^{\mathrm{N}}$ and for every $0 \leqq \alpha \leqq 1$] iff D satisfies the property $y^{\mathrm{T}} D y \geqq 0$ for all $y \in \mathbf{R}^{\mathrm{N}}$.

16.4.1 Review on Positive Semidefinite Matrices

A square matrix $F = (f_{ij})$ of order n, whether it is symmetric or not, is said to be a *positive semidefinite matrix* if $y^{\mathrm{T}} F y \geqq 0$ for all $y \in \mathbf{R}^n$. It is said to be a *positive definite matrix* if $y^{\mathrm{T}} F y > 0$ for all $y \neq 0$. We will use the abbreviations PSD, PD for "positive semidefinite" and "positive definite," respectively.

Let F be a square matrix of order n. Let $\{i_1, \ldots, i_r\}$ be a nonempty subset of $\{1, 2, \ldots, n\}$. Erase all the entries in F in row i and column i for all $i \notin \{i_1, \ldots, i_r\}$. What remains is a square submatrix of F of order r. Such submatrices are called *principal submatrices of F*. The determinant of a principal submatrix of F is called a *principal subdeterminant* of F. Since the number of distinct nonempty subsets of F is $2^n - 1$, there are $2^n - 1$ principal submatrices of F.

Example
Let

$$F = \begin{pmatrix} 0 & -1 & 2 \\ 1 & 3 & 4 \\ 1 & 5 & -3 \end{pmatrix}.$$

The principal submatrix corresponding to the subset $\{1, 3\}$ is $\begin{pmatrix} 0 & 2 \\ 1 & -3 \end{pmatrix}$. The principal submatrix corresponding to the subset $\{2\}$ is 3, the second element in the principal diagonal of F.

Several results useful in studying P(S)D matrices will now be discussed.

Result 1 If F is a PD matrix all its principal submatrices must also be PD.

Proof Consider the principal submatrix, G, generated by the subset $\{1, 2\}$.

$$G = \begin{pmatrix} f_{11} & f_{12} \\ f_{21} & f_{22} \end{pmatrix}. \text{ Let } t = \begin{pmatrix} y_1 \\ y_2 \end{pmatrix}$$

Pick $y = (y_1, y_2, 0, 0, \ldots, 0)^T$. Then $y^T F y = t^T G t$. However, since F is PD, $y^T F y > 0$ for all $y \neq 0$. So $t^T G t > 0$ for all $t \neq 0$. Hence, G is PD too. A similar argument can be used to prove that every principal submatrix of F is also PD.

Result 2 If F is PD, $f_{ii} > 0$ for all i. This follows as a corollary of result 1.

Result 3 If F is a PSD matrix, all principal submatrices of F are also PSD. This is proved using arguments similar to those in result 1.

Result 4 If F is PSD matrix, $f_{ii} \geqq 0$ for all i. This follows from result 3.

Result 5 Suppose F is a PSD matrix. If $f_{ii} = 0$, then $f_{ij} + f_{ji} = 0$ for all j.

Proof To be specific let f_{11} be 0 and suppose that $f_{12} + f_{21} \neq 0$. By result 3 the principal submatrix

$$\begin{pmatrix} f_{11} & f_{12} \\ f_{21} & f_{22} \end{pmatrix} = \begin{pmatrix} 0 & f_{12} \\ f_{21} & f_{22} \end{pmatrix}$$

must be PSD. Hence, $f_{22}y_2^2 + (f_{12} + f_{21})y_1 y_2 \geqq 0$ for all y_1, y_2. Since $f_{12} + f_{21} \neq 0$, take $y_1 = (-f_{22} - 1)/(f_{12} + f_{21})$ and $y_2 = 1$. The above inequality is violated since the left-hand side becomes equal to -1, leading to a contradiction.

Result 6 If D is a symmetric PSD matrix and $d_{ii} = 0$, then $D_{.i} = D_{i.} = 0$. This follows from result 5.

Result 7 Let D be a square symmetric matrix of order $n \geqq 2$. Suppose D is PD. Subtract suitable multiples of row 1 from each of the other rows so that all the entries in column 1 excepting the first are transformed into zero. That is, transform

$$D = \begin{bmatrix} d_{11} & \ldots & d_{1n} \\ d_{21} & \ldots & d_{2n} \\ \vdots & & \vdots \\ d_{n1} & \ldots & d_{nn} \end{bmatrix} \quad \text{into} \quad D_1 = \begin{bmatrix} d_{11} & & \ldots & d_{1n} \\ 0 & \tilde{d}_{22} & \ldots & \tilde{d}_{2n} \\ \vdots & \vdots & & \vdots \\ 0 & \tilde{d}_{n2} & \ldots & \tilde{d}_{nn} \end{bmatrix}$$

where clearly $\tilde{d}_{ij} = d_{ij} - d_{1j}d_{i1}/d_{11}$ for all $i, j \geqq 2$. E_1, the matrix obtained by striking off the first row and the first column from D_1, is also symmetric and PD.

Also, if D is an arbitrary square matrix, it is PD iff $d_{11} > 0$ and the matrix E_1 obtained as above is PD.

Proof Since D is symmetric $d_{ij} = d_{ji}$ for all i, j. Therefore,

$$y^T D y = \sum_{i=1}^{n} \sum_{j=1}^{n} y_i y_j d_{ij} = d_{11} y_1^2 + 2y_1 \sum_{j=2}^{n} d_{1j} y_j + \sum_{i,\, j \geqq 2} y_i y_j d_{ij}$$

$$= d_{11} \left(y_1 + \left(\sum_{j=2}^{n} d_{1j} y_j \right) \Big/ d_{11} \right)^2 + \sum_{i,\, j \geqq 2} y_i \tilde{d}_{ij} y_j$$

Letting $y_1 = -(\sum_{j=2}^{n} d_{1j} y_j)/d_{11}$, we verify that if D is PD, then $\sum_{i,\, j \geqq 2} y_i \tilde{d}_{ij} y_j > 0$ for all $(y_2, \ldots, y_n) \neq 0$, which implies that E_1 is PD. The fact that E_1 is also symmetric is clear since $\tilde{d}_{ij} = d_{ij} - d_{1j}d_{i1}/d_{11} = \tilde{d}_{ji}$ by the symmetry of D. If D is an arbitrary symmetric matrix, the above equation clearly implies that D is PD iff $d_{11} > 0$ and E_1 is PD.

16.4.2 Algorithm for Testing Positive Definiteness

Let $F = (f_{ij})$ be a given square matrix of order n. Find $D = (F + F^T)$. F is PD iff D is.

(i) If any of the principal diagonal elements in D are nonpositive, D is not PD. Terminate.

(ii) Subtract suitable multiples of row 1 from all the other rows, so that all the entries in column 1 and rows 2 to n of D are transformed into zero. That is, transform D into D_1 as in result 7 of section 16.4.1. If any diagonal element in the transformed matrix, D_1, is nonpositive, D is not PD. Terminate.

(iii) In general, after r steps we will have a matrix D_r of the form:

$$\begin{bmatrix} d_{11} & d_{12} & & & \ldots & d_{1n} \\ 0 & \tilde{d}_{22} & & & \ldots & d_{2n} \\ & 0 & \ddots & & & \vdots \\ & & \bar{d}_{rr} & & \ldots & \bar{d}_{rn} \\ & & 0 & \hat{d}_{r+1,\, r+1} & \ldots & \hat{d}_{r+1,\, n} \\ \vdots & \vdots & \vdots & \vdots & & \vdots \\ 0 & 0 & 0 & \hat{d}_{n,\, r+1} & \ldots & \hat{d}_{nn} \end{bmatrix}$$

Subtract suitable multiples of row $r + 1$ in D_r from rows i for $i > r + 1$, so that all the entries in column $r + 1$ and rows i for $i > r + 1$ are transformed into 0. This transforms D_r into D_{r+1}. If any element in the principle diagonal of D_{r+1} is nonpositive, D is not PD. Terminate. Otherwise continue the algorithm in the same manner for $n - 1$ steps, until D_{n-1} is obtained, which is of the form

$$\begin{bmatrix} d_{11} & d_{12} & \ldots & d_{1n} \\ 0 & \tilde{d}_{22} & \ldots & \tilde{d}_{2n} \\ & 0 & & \\ \vdots & \vdots & & \vdots \\ 0 & 0 & \ldots & \bar{d}_{nn} \end{bmatrix}$$

D_{n-1} is upper triangular. That's why this algorithm is called the *superdiagonalization algorithm*. If no termination has occurred earlier and all the diagonal elements of D_{n-1} are positive, D, and, hence, F is PD.

Example 1

Test whether $F = \begin{bmatrix} 3 & 1 & 2 & 2 \\ -1 & 2 & 0 & 2 \\ 0 & 4 & 4 & \frac{5}{4} \\ 0 & -2 & -\frac{13}{3} & 6 \end{bmatrix}$ is PD.

Symmetrizing, $D = F + F^T = \begin{bmatrix} 6 & 0 & 2 & 2 \\ 0 & 4 & 4 & 0 \\ 2 & 4 & 8 & -\frac{8}{3} \\ 2 & 0 & -\frac{8}{3} & 12 \end{bmatrix}$.

All the entries in the principal diagonal of D (i.e., the entries d_{ii} for all i) are strictly positive. So apply the first step in superdiagonalization.

$$D_1 = \begin{bmatrix} 6 & 0 & 2 & 2 \\ 0 & 4 & 4 & 0 \\ 0 & 4 & \frac{22}{3} & -\frac{10}{3} \\ 0 & 0 & -\frac{10}{3} & \frac{34}{3} \end{bmatrix}$$

Since all elements in the principal diagonal of D_1 are strictly positive, continue. The matrices obtained in the order are:

$$D_2 = \begin{bmatrix} 6 & 0 & 2 & 2 \\ 0 & 4 & 4 & 0 \\ 0 & 0 & \frac{10}{3} & -\frac{10}{3} \\ 0 & 0 & -\frac{10}{3} & \frac{34}{3} \end{bmatrix} \qquad D_3 = \begin{bmatrix} 6 & 0 & 2 & 2 \\ 0 & 4 & 4 & 0 \\ 0 & 0 & \frac{10}{3} & -\frac{10}{3} \\ 0 & 0 & 0 & 8 \end{bmatrix}$$

The algorithm terminates now. Since all diagonal entries in D_3 are strictly positive, conclude that D and, hence, F is PD.

Example 2

Test whether $D = \begin{pmatrix} 1 & 0 & 2 & 0 \\ 0 & 2 & 4 & 0 \\ 2 & 4 & 4 & 5 \\ 0 & 0 & 5 & 3 \end{pmatrix}$ is PD.

D is already symmetric, and all its diagonal elements are positive. The first step of the algorithm requires performing the operation: (row 3) − 2(row 1) on D. This leads to

$$D_1 = \begin{pmatrix} 1 & 0 & 2 & 0 \\ 0 & 2 & 4 & 0 \\ 0 & 4 & 0 & 5 \\ 0 & 0 & 5 & 3 \end{pmatrix}$$

Since the third diagonal element in D_1 is not strictly positive, D is not PD.

16.4.3 Algorithm for Testing Positive Semidefiniteness

Let $F = (f_{ij})$ be the given square matrix. Obtain $D = F + F^T$. If any diagonal element of D is 0, all the entries in the row and column of the zero diagonal entry must be zero. Otherwise D (and hence F) is not PSD and we terminate. Also, if any diagonal entries in D are negative, D cannot be PSD and we terminate. If termination has not occurred, reduce the matrix D by striking off the rows and columns of zero diagonal entries.

Start off by performing the row operations as in (ii) of section 16.4.2, that is, transform D into D_1. If any diagonal element in D_1 is negative, D is not PSD. Let E_1 be the submatrix of D_1 obtained by striking off the first row and column of D_1. Also, if a diagonal element in E_1 is zero, all entries in its row and column in E_1 must be zero. Otherwise D is not PSD. Terminate. Continue if termination does not occur.

In general, after r steps we will have a matrix D_r as in (iii) of section 16.4.2. Let E_r be the square submatrix of D_r obtained by striking off the first r rows and columns of D_r. If any diagonal element in E_r is negative, D cannot be PSD. If any diagonal element of E_r is zero, all the entries in its row and column in E_r must be zero; otherwise D is not PSD. Terminate. If termination does not occur, continue.

Let d_{ss} be the first nonzero (and, hence, positive) diagonal element in E_r. Subtract suitable multiples of row s in D_r from rows i, $i > s$, so that all the entries in column s and rows i, $i > s$ in D_r, are transformed into 0. This transforms D_r into D_s and we repeat the same operations with D_s. If termination does not occur until D_{n-1} is obtained and, if the diagonal entries in D_{n-1} are nonnegative, D and hence F are PSD.

Note In the process of obtaining D_{n-1}, if all the diagonal elements in all the matrices obtained during the algorithm are strictly positive, D and hence F is not only PSD but actually PD.

Example 1

Is the matrix $F = \begin{bmatrix} 0 & -2 & -3 & -4 & 5 \\ 2 & 3 & 3 & 0 & 0 \\ 3 & 3 & 3 & 0 & 0 \\ 4 & 0 & 0 & 8 & 4 \\ -5 & 0 & 0 & 4 & 2 \end{bmatrix}$ PSD?

If so, is it PD?

If we symmetrize, $D = F + F^T = \begin{bmatrix} 0 & 0 & 0 & 0 & 0 \\ 0 & 6 & 6 & 0 & 0 \\ 0 & 6 & 6 & 0 & 0 \\ 0 & 0 & 0 & 16 & 8 \\ 0 & 0 & 0 & 8 & 4 \end{bmatrix}$

$D_{.1}$ and $D_{1.}$ are both zero vectors. So we eliminate them, but we will call the remaining matrix by the same name D. All the diagonal entries in D are nonnegative. Thus we apply the first step in superdiagonalization. This leads to

$$D_1 = \begin{pmatrix} 6 & 6 & 0 & 0 \\ 0 & 0 & 0 & 0 \\ 0 & 0 & 16 & 8 \\ 0 & 0 & 8 & 4 \end{pmatrix}$$

Here,

$$E_1 = \begin{pmatrix} 0 & 0 & 0 \\ 0 & 16 & 8 \\ 0 & 8 & 4 \end{pmatrix}$$

The first diagonal entry in E_1 is 0, but the first column and row of E_1 are both zero vectors. Also all the remaining diagonal entries in D_1 are strictly positive. So continue with superdiagonalization. Since the second diagonal element in D_1 is zero, move to the third diagonal element of D_1. This step leads to

$$D_3 = \begin{pmatrix} 6 & 6 & 0 & 0 \\ 0 & 0 & 0 & 0 \\ 0 & 0 & 16 & 8 \\ 0 & 0 & 0 & 0 \end{pmatrix}$$

All the diagonal entries in D_3 are nonnegative. D and hence F is PSD but not PD.

Example 2

Is the matrix D in Example 2, section 16.4.2 PSD? Referring to Example 2 of 16.4.2, after the first step in superdiagonalization, we have

$$E_1 = \begin{pmatrix} 2 & 4 & 0 \\ 4 & 0 & 5 \\ 0 & 5 & 3 \end{pmatrix}$$

The second diagonal entry in E_1 is 0, but the second row and column of E_1 are not zero vectors. So D is not PSD.

16.4.4 Necessary Conditions for Optimality

We will now resume our discussion of the quadratic program (16.11).

Theorem If $\bar{x}$ is an optimum solution of (16.11), $\bar{x}$ is also an optimum solution of the LP:

$$\begin{aligned} \text{Minimize} \quad & (c + \bar{x}^{\mathrm{T}} D)x \\ \text{Subject to} \quad & Ax \geqq b \\ & x \geqq 0 \end{aligned} \tag{16.12}$$

Proof The cost coefficients in the LP (16.12) depend on $\bar{x}$. The constraints in both (16.11) and (16.12) are the same. The set of feasible solutions is a convex polyhedron. Let $\hat{x}$ be any feasible solution. By convexity $x_\lambda = \lambda\hat{x} + (1 - \lambda)\bar{x} = \bar{x} + \lambda(\hat{x} - \bar{x})$ is also a feasible solution for any $0 < \lambda < 1$. Since $\bar{x}$ is an optimum feasible solution of (16.11), $Q(x_\lambda) - Q(\bar{x}) \geqq 0$, that is, $\lambda(c + \bar{x}^{\mathrm{T}}D)(\hat{x} - \bar{x}) + (1/2)\lambda^2(\hat{x} - \bar{x})^{\mathrm{T}}D(\hat{x} - \bar{x}) \geqq 0$ for all $0 < \lambda < 1$. Dividing both sides by λ leads to $(c + \bar{x}^{\mathrm{T}}D)(\hat{x} - \bar{x}) \geqq (-\lambda/2)(\hat{x} - \bar{x})^{\mathrm{T}}D(\hat{x} - \bar{x})$ for all $0 < \lambda < 1$. This obviously implies

$$(c + \bar{x}^{\mathrm{T}}D)(\hat{x} - \bar{x}) \geqq 0$$

that is,

$$(c + \bar{x}^{\mathrm{T}}D)\hat{x} \geqq (c + \bar{x}^{\mathrm{T}}D)\bar{x}$$

Since this must hold for an arbitrary feasible solution $\hat{x}$, $\bar{x}$ must be an optimum feasible solution of (16.12).

Corollary If $\bar{x}$ is an optimum feasible solution of (16.11), there exist vectors $\bar{y} \in \mathbf{R}^m$ and slack vectors $\bar{u} \in \mathbf{R}^N$, $\bar{v} \in \mathbf{R}^m$ such that $\bar{x}, \bar{y}, \bar{u}, \bar{v}$ together satisfy

$$\begin{pmatrix} \bar{u} \\ \cdots \\ \bar{v} \end{pmatrix} - \begin{pmatrix} D & \vdots & -A^{\mathrm{T}} \\ \cdots & \cdots & \cdots \\ A & \vdots & 0 \end{pmatrix} \begin{pmatrix} \bar{x} \\ \cdots \\ \bar{y} \end{pmatrix} = \begin{pmatrix} c^{\mathrm{T}} \\ \cdots \\ -b \end{pmatrix}$$

$$\begin{pmatrix} \bar{u} \\ \cdots \\ \bar{v} \end{pmatrix} \geqq 0 \qquad \begin{pmatrix} \bar{x} \\ \cdots \\ \bar{y} \end{pmatrix} \geqq 0 \qquad \text{and} \qquad \begin{pmatrix} \bar{u} \\ \cdots \\ \bar{v} \end{pmatrix}^{\mathrm{T}} \begin{pmatrix} \bar{x} \\ \cdots \\ \bar{y} \end{pmatrix} = 0 \tag{16.13}$$

Proof From the above theorem $\bar{x}$ must be an optimum solution of the LP (16.12). The corollary follows by using the results of section 16.3 on this fact.

Note The conditions in the above corollary are known as the *Kuhn-Tucker necessary optimality conditions* (or the *Karush-Kuhn-Tucker necessary optimality conditions*) for the optimality of $\bar{x}$ to (16.11). Any feasible solution $\bar{x}$ that satisfies the conditions in the above corollary is known as a *Kuhn-Tucker point* (or the *Karush-Kuhn-Tucker point*) for the quadratic program (16.11). See references [9, 10 and 11] at the end of this chapter.

Theorem If D is PSD and $\bar{x}$ is a Kuhn-Tucker point of (16.11), $\bar{x}$ is an optimum feasible solution of (16.11).

Proof From the definition of a Kuhn-Tucker point and the results in section 16.3, if $\bar{x}$ is a Kuhn-Tucker point for (16.11), it must be an optimum feasible solution of the LP (16.12). Let x be any feasible solution of (16.11).

$$Q(x) - Q(\bar{x}) = (c + \bar{x}D)(x - \bar{x}) + \frac{1}{2}(x - \bar{x})^T D(x - \bar{x})$$

The first term on the right-hand side expression is nonnegative since $\bar{x}$ is an optimal feasible solution of (16.12). The second term in that expression is also nonnegative since D is PSD. Hence, $Q(x) - Q(\bar{x}) \geqq 0$ for all feasible solutions, x, of (16.11). This implies that $\bar{x}$ is an optimum feasible solution of (16.11).

Clearly (16.13) is a LCP. An optimum solution of (16.11) must be a Kuhn-Tucker point for it. Solving (16.13) provides a Kuhn-Tucker point for (16.11) and if D is PSD, this Kuhn-Tucker point is an optimum solution of (16.11). [If D is not PSD and if a Kuhn-Tucker point is obtained when (16.13) is solved, it may not be an optimum solution of (16.11).]

Example Minimum Distance Problem

Let **K** denote the shaded convex polyhedral region in Figure 16.6. Let P_0 be the point $(-2, -1)$. Find the point in **K** that is closest to P_0 (in terms of the usual Euclidean distance). Such problems appear very often in engineering and in operations research applications.

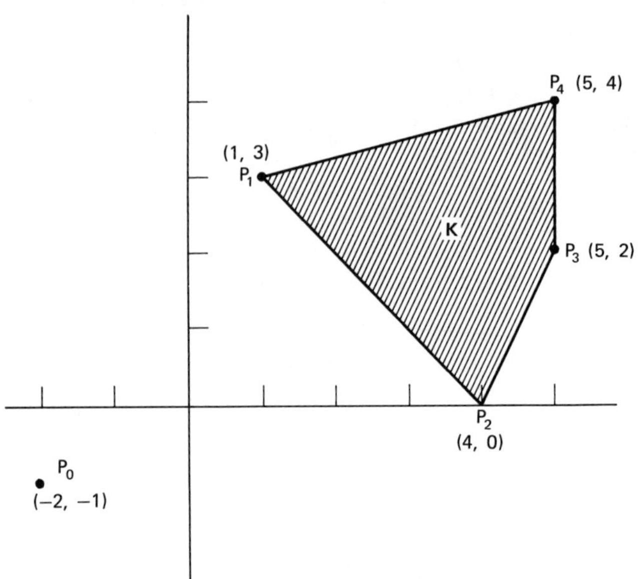

Figure 16.6

Every point in **K** can be expressed as a convex combination of its *extreme points* (or *corner points*) P_1, P_2, P_3, P_4. That is, the coordinates of a general point in **K** are: $(\lambda_1 + 4\lambda_2 + 5\lambda_3 + 5\lambda_4, 3\lambda_1 + 0\lambda_2 + 2\lambda_3 + 4\lambda_4)$ where the λ_i satisfy $\lambda_1 + \lambda_2 + \lambda_3 + \lambda_4 = 1$ and $\lambda_i \geqq 0$ for all i. Hence, the problem of finding the point in **K** closest to P_0 is equivalent to solving:

$$\begin{array}{ll} \text{Minimize} & (\lambda_1 + 4\lambda_2 + 5\lambda_3 + 5\lambda_4 - (-2))^2 + (3\lambda_1 + 2\lambda_3 + 4\lambda_4 - (-1))^2 \\ \text{Subject to} & \lambda_1 + \lambda_2 + \lambda_3 + \lambda_4 = 1 \\ & \lambda_i \geqq 0 \quad \text{for all } i \end{array}$$

λ_4 can be eliminated from this problem by substituting the expression $\lambda_4 = 1 - \lambda_1 - \lambda_2 - \lambda_3$ for it. Doing this and simplifying, leads to the quadratic program

$$\begin{array}{ll} \text{Minimize} & (-66, -54, -20)\lambda + \left(\frac{1}{2}\right)\lambda^{\mathrm{T}} \begin{pmatrix} 34 & 16 & 4 \\ 16 & 34 & 16 \\ 4 & 16 & 8 \end{pmatrix} \lambda \\ \text{Subject to} & -\lambda_1 - \lambda_2 - \lambda_3 \geqq -1 \\ & \lambda \geqq 0 \end{array}$$

where $\lambda = (\lambda_1, \lambda_2, \lambda_3)^{\mathrm{T}}$. Solving this quadratic program is equivalent to solving the LCP

$$\begin{pmatrix} u_1 \\ u_2 \\ u_3 \\ v_1 \end{pmatrix} - \begin{pmatrix} 34 & 16 & 4 & 1 \\ 16 & 34 & 16 & 1 \\ 4 & 16 & 8 & 1 \\ -1 & -1 & -1 & 0 \end{pmatrix} \begin{pmatrix} \lambda_1 \\ \lambda_2 \\ \lambda_3 \\ y_1 \end{pmatrix} = \begin{pmatrix} -66 \\ -54 \\ -20 \\ 1 \end{pmatrix}$$

All variables $u_1, u_2, u_3, v_1, \lambda_1, \lambda_2, \lambda_3, y_1 \geqq 0$

and $u_1\lambda_1 = u_2\lambda_2 = u_3\lambda_3 = v_1 y_1 = 0$.

Let $(\tilde{u}_1, \tilde{u}_2, \tilde{u}_3, \tilde{v}_1, \tilde{\lambda}_1, \tilde{\lambda}_2, \tilde{\lambda}_3, \tilde{y}_1)$ be a solution to this LCP. Let $\tilde{\lambda}_4 = 1 - \tilde{\lambda}_1 - \tilde{\lambda}_2 - \tilde{\lambda}_3$. Then $\tilde{x} = (\tilde{\lambda}_1 + 4\tilde{\lambda}_2 + 5\tilde{\lambda}_3 + 5\tilde{\lambda}_4, 3\tilde{\lambda}_1 + 2\tilde{\lambda}_3 + 4\tilde{\lambda}_4)$ is the point in **K** that is closest to P_0.

16.5 TWO PERSON GAMES

In each play of the game, player I picks one out of a possible set of his m choices and independently player II picks one out of a possible set of his N choices. In a play, if player I has picked his choice, i, and player II has picked his choice j, then player I loses an amount a'_{ij} dollars and player II loses an amount b'_{ij} dollars, where $A' = (a'_{ij})$ and $B' = (b'_{ij})$ are given *loss matrices*.

If $a'_{ij} + b'_{ij} = 0$ for all i and j, the game is known as a *zero sum game*; in this case it is possible to develop the concept of an *optimum strategy* for playing the game using *Von Neumann's Minimax theorem.* Games that are not zero sum games are called *nonzero sum games* or *bimatrix games.* In bimatrix games it is difficult to define an optimum strategy. However, in this case, an *equilibrium pair of strategies* can be defined (see next paragraph) and the problem of computing an equilibrium pair of strategies can be transformed into a LCP.

Suppose player I picks his choice i with a probability of x_i. The column vector $x = (x_i) \in \mathbf{R}^m$ completely defines player I's strategy. Similarly let the probability vector $y = (y_j) \in \mathbf{R}^{\mathrm{N}}$ be player II's strategy. If player I adopts strategy x and player II adopts strategy y, the expected loss of player I is obviously $x^{\mathrm{T}}A'y$ and that of player II is $x^{\mathrm{T}}B'y$.

The strategy pair $(\bar{x}, \bar{y})$ is said to be an *equilibrium pair* if no player benefits by unilaterally changing his own strategy while the other player keeps his strategy in the pair $(\bar{x}, \bar{y})$ unchanged, that is, if

$$\bar{x}^{\mathrm{T}}A'\bar{y} \leqq x^{\mathrm{T}}A'\bar{y} \qquad \text{for all probability vectors } x \in \mathbf{R}^m$$

and

$$\bar{x}^{\mathrm{T}}B'\bar{y} \leqq \bar{x}^{\mathrm{T}}B'y \qquad \text{for all probability vectors } y \in \mathbf{R}^{\mathrm{N}}$$

Let α, β be arbitrary positive numbers such that $a_{ij} = a'_{ij} + \alpha > 0$ and $b_{ij} = b'_{ij} + \beta > 0$ for all i, j. Let $A = (a_{ij})$, $B = (b_{ij})$. Since $x^{\mathrm{T}}A'y = x^{\mathrm{T}}Ay - \alpha$ and $x^{\mathrm{T}}B'y = x^{\mathrm{T}}By - \beta$ for all probability vectors $x \in \mathbf{R}^m$ and $y \in \mathbf{R}^{\mathrm{N}}$, if $(\bar{x}, \bar{y})$ is an equilibrium pair of strategies for the game with loss matrices A', B', then $(\bar{x}, \bar{y})$ is an equilibrium pair of strategies for the game with loss matrices A, B, and vice versa. So without any loss of generality, consider the game in which the loss matrices are A, B.

Since x is a probability vector, the condition $\bar{x}^{\mathrm{T}}A\bar{y} \leqq xA\bar{y}$ for all probability vectors $x \in \mathbf{R}^m$ is equivalent to the system of constraints

$$\bar{x}^{\mathrm{T}}A\bar{y} \leqq A_{i.}\bar{y} \qquad \text{for all } i = 1 \text{ to } m$$

Let e_r denote the column vector in $\mathbf{R}^r$ in which all the elements are equal to 1. In matrix notation the above system of constraints can be written as $(\bar{x}^{\mathrm{T}}A\bar{y})e_m \leqq A\bar{y}$. In a similar way the condition $\bar{x}^{\mathrm{T}}B\bar{y} \leqq \bar{x}^{\mathrm{T}}By$ for all probability vectors $y \in \mathbf{R}^{\mathrm{N}}$ is equivalent to $(\bar{x}^{\mathrm{T}}B\bar{y})e_{\mathrm{N}} \leqq B^{\mathrm{T}}\bar{x}$. Hence the strategy pair $(\bar{x}, \bar{y})$ is an equilibrium pair of strategies for the game with loss matrices A, B iff

$$\begin{aligned} A\bar{y} &\geqq (\bar{x}^{\mathrm{T}}A\bar{y})e_m \\ B^{\mathrm{T}}\bar{x} &\geqq (\bar{x}^{\mathrm{T}}B\bar{y})e_{\mathrm{N}} \end{aligned} \tag{16.14}$$

Since A, B are strictly positive matrices, $\bar{x}^{\mathrm{T}}A\bar{y}$ and $\bar{x}^{\mathrm{T}}B\bar{y}$ are strictly positive numbers. Let $\bar{\xi} = \bar{x}/\bar{x}^{\mathrm{T}}B\bar{y}$ and $\bar{\eta} = \bar{y}/\bar{x}^{\mathrm{T}}A\bar{y}$. Introducing slack variables corresponding to the inequality constraints, (16.14) is equivalent to

$$\begin{gathered} \begin{pmatrix} \bar{u} \\ \cdots \\ \bar{v} \end{pmatrix} - \begin{pmatrix} 0 & \vdots & A \\ \cdots & & \cdots \\ B^{\mathrm{T}} & \vdots & 0 \end{pmatrix} \begin{pmatrix} \bar{\xi} \\ \cdots \\ \bar{\eta} \end{pmatrix} = \begin{pmatrix} -e_m \\ \cdots \\ -e_{\mathrm{N}} \end{pmatrix} \\ \begin{pmatrix} \bar{u} \\ \cdots \\ \bar{v} \end{pmatrix} \geqq 0 \quad \begin{pmatrix} \bar{\xi} \\ \cdots \\ \bar{\eta} \end{pmatrix} \geqq 0 \quad \begin{pmatrix} \bar{u} \\ \cdots \\ \bar{v} \end{pmatrix}^{\mathrm{T}} \begin{pmatrix} \bar{\xi} \\ \cdots \\ \bar{\eta} \end{pmatrix} = 0 \end{gathered} \tag{16.15}$$

Conversely, it can easily be shown that if $(\bar{u}, \bar{v}, \bar{\xi}, \bar{\eta})$ is a solution of the LCP (16.15), then an equilibrium pair of strategies for the original game is $(\bar{x}, \bar{y})$ where $\bar{x} = \bar{\xi}/(\sum \bar{\xi}_i)$ and $\bar{y} = \bar{\eta}/(\sum \bar{\eta}_j)$. Thus an equilibrium pair of strategies can be computed by solving the LCP (16.15).

Example

Consider the game in which the loss matrices are

$$A' = \begin{pmatrix} 1 & 1 & 0 \\ 0 & 1 & 1 \end{pmatrix} \qquad B' = \begin{pmatrix} -1 & 1 & 0 \\ 0 & -1 & 1 \end{pmatrix}$$

Player I's strategy is a probability vector $x = (x_1, x_2)^{\mathrm{T}}$ and player II's strategy is a probability vector $y = (y_1, y_2, y_3)^{\mathrm{T}}$. Add 1 to all the elements in A' and 2 to all the elements in B', to make all the elements in the loss matrices strictly positive. This leads to

$$A = \begin{pmatrix} 2 & 2 & 1 \\ 1 & 2 & 2 \end{pmatrix} \qquad B = \begin{pmatrix} 1 & 3 & 2 \\ 2 & 1 & 3 \end{pmatrix}$$

The LCP corresponding to this game problem is

$$\begin{bmatrix} u_1 \\ u_2 \\ v_1 \\ v_2 \\ v_3 \end{bmatrix} - \begin{bmatrix} 0 & 0 & 2 & 2 & 1 \\ 0 & 0 & 1 & 2 & 2 \\ 1 & 2 & 0 & 0 & 0 \\ 3 & 1 & 0 & 0 & 0 \\ 2 & 3 & 0 & 0 & 0 \end{bmatrix} \begin{bmatrix} \xi_1 \\ \xi_2 \\ \eta_1 \\ \eta_2 \\ \eta_3 \end{bmatrix} = \begin{bmatrix} -1 \\ -1 \\ -1 \\ -1 \\ -1 \end{bmatrix} \tag{16.16}$$

$$u, v, \xi, \eta \geqq 0 \text{ and } u_1\xi_1 = u_2\xi_2 = v_1\eta_1 = v_2\eta_2 = v_3\eta_3 = 0.$$

OTHER APPLICATIONS

Besides these applications, LCP has been used to study problems in the plastic analysis of structures, problems in the inelastic flexural behavior of reinforced concrete beams, in the free boundary problems for journal bearings, in the study of finance models, and in several other areas.

16.6 COMPLEMENTARY PIVOT ALGORITHM

LCPs of order 2 can be solved by drawing all the complementary cones in the q_1-, q_2-plane as discussed in section 16.1.

Example 1

Let $q = \begin{pmatrix} 4 \\ -1 \end{pmatrix}$, $M = \begin{pmatrix} -2 & 1 \\ 1 & -2 \end{pmatrix}$ and consider the LCP (q, M). The class of complementary cones corresponding to this problem is in Figure 16.4.

w_1	w_2	z_1	z_2	q
1	0	2	−1	4
0	1	−1	2	−1

(16.17)

$w_1, w_2, z_1, z_2 \geqq 0$, $w_1 z_1 = w_2 z_2 = 0$

q lies in two complementary cones pos $(-M_{.1}, I_{.2})$ and pos $(-M_{.1}, -M_{.2})$. This implies that the sets of active variables (z_1, w_2) and (z_1, z_2) lead to solutions of the LCP.

Putting $w_1 = z_2 = 0$ in (16.17) and solving the remaining system for the values of the active variables (z_1, w_2) lead to the solution $(z_1, w_2) = (2, 1)$. Hence, $(w_1, w_2, z_1, z_2) = (0, 1, 2, 0)$ is a solution of this LCP. Similarly putting $w_1 = w_2 = 0$ in (16.17) and solving it for the values of the active variables (z_1, z_2) leads to the second solution $(w_1, w_2, z_1, z_2) = (0, 0, 7/3, 2/3)$ of this LCP.

Example 2

Let $q = \begin{pmatrix} -1 \\ -1 \end{pmatrix}$ and $M = \begin{pmatrix} -2 & 1 \\ 1 & -2 \end{pmatrix}$ and consider the LCP (q, M). The class of complementary cones corresponding to this problem is seen in Figure 16.4. Verify that q is not contained in any complementary cone. Hence this LCP has no solution.

This graphic method can be conveniently used only for LCPs of order 2. In LCPs of higher order, in contrast to the graphic method where all the complementary cones were generated, we seek only one complementary cone in which q lies. We now discuss the *complementary pivot algorithm*, which can be used to find it.

In the LCP (16.5) to (16.7) if $q \geqq 0$, $(w; z) = (q; 0)$ is a solution and we are done. So we assume $q \ngeqq 0$.

BASES

An artificial variable, z_0 associated with the column vector $-e_n$ (e_n is the column vector in $\mathbf{R}^n$ in which all the entries are equal to 1) is introduced into the system (16.5) to get a feasible basis for starting the algorithm. In detached coefficient tableau form, (16.5) then becomes

w	z	z_0	
I	$-M$	$-e_n$	q
$w \geqq 0$	$z \geqq 0$	$z_0 \geqq 0$	

(16.18)

In each stage, the algorithm deals with a *basis*, which is a square nonsingular submatrix of order n of the coefficient matrix in (16.18).

The solution of (16.18) corresponding to a given basis is obtained by setting all the nonbasic variables equal to zero, and then solving the remaining system for the values of the basic variables. The basis is a *feasible basis* if the values of all the basic variables in the solution turn out to be nonnegative. The algorithm deals only with *feasible bases.*

16.6.1 Pivot Operations

The primary computational step used in the algorithm is the *pivot step*, which is also the main step in the *simplex algorithm.*

In each stage of the algorithm, the basis is changed by bringing into the basic vector exactly one nonbasic variable known as the *entering variable.* Its updated column vector is the *pivot column* for this basis change. The *dropping variable* has to be determined according to the *minimum ratio test* to guarantee that the new basis obtained after the pivot step will also be a feasible basis.

For example, assume that the present basic vector is $(y_1, \ldots, y_n)$ with y_r as the rth basic variable, and let the entering variable be x_s. (The variables in (16.18) are $w_1, \ldots, w_n; z_1, \ldots, z_n, z_0$. Exactly n of these variables are the present basic variables. For convenience in reference, we assume that these basic variables are called $y_1, \ldots, y_n$). After we rearrange the variables in (16.18), if necessary, the canonical form of (16.18), with respect to the present basis is of the form:

Basic variable	$y_1 \cdots y_n$	x_s	Other variables	Right-hand constant vector
y_1	$1 \ldots 0$	$\bar{a}_{1s}$	$\ldots$	$\bar{q}_1$
$\vdots$	$\vdots \quad \vdots$	$\vdots$		$\vdots$
y_n	$0 \ldots 1$	$\bar{a}_{ns}$	$\ldots$	$\bar{q}_n$

Keeping all the nonbasic variables other than x_s, equal to zero, and giving the value λ to the entering variable, x_s, leads to the new solution:

$$\begin{aligned} x_s &= \lambda \\ y_i &= \bar{q}_i - \lambda \bar{a}_{is} \qquad i = 1 \text{ to } n \\ &\text{All other variables} = 0 \end{aligned} \tag{16.19}$$

There are two possibilities here.

1. The pivot column may be nonpositive, that is, $\bar{a}_{is} \leqq 0$ for all $1 \leqq i \leqq n$. In this case, the solution in (16.19) remains nonnegative for all $\lambda \geqq 0$. As λ varies from 0 to ∞, this solution traces an *extreme half-line* (or an *unbounded edge*) of the set of feasible solutions of (16.18).
2. There is at least one positive entry in the pivot column. In this case, if the solution in (16.19) should remain nonnegative, the maximum value that λ can take is $\theta = \bar{q}_r/\bar{a}_{rs} = \text{minimum } \{\bar{q}_i/\bar{a}_{is}: i \text{ such that } \bar{a}_{is} > 0\}$. y_r drops from the basic vector and x_s becomes the rth basic variable in its place. The rth row is the *pivot row* for this pivot step. The pivot leads to the canonical tableau with respect to the new basis.

16.6.2 Initial Basic Vector

The artificial variable z_0 has been introduced into (16.18) for the sole purpose of obtaining a feasible basis to start the algorithm.

Identify row t such that $q_t = \text{minimum } \{q_i: 1 \leqq i \leqq n\}$. Since we assumed $q \ngeq 0$, $q_t < 0$. When a pivot is made in (16.18) with the column vector of z_0 as the pivot column and the tth row as the pivot row, the right-hand side constant vector becomes a nonnegative vector. The result is the canonical tableau with respect to the basic vector $(w_1, \ldots, w_{t-1}, z_0, w_{t+1}, \ldots, w_n)$. This is the initial basic vector for starting the algorithm.

16.6.3 Properties of Basic Vectors

The initial basic vector satisfies the following properties:

(i) There is at most one basic variable from each complementary pair of variables (w_j, z_j).

(ii) It contains exactly one basic variable from each of $(n - 1)$ complementary pairs of variables, and both the variables in the remaining complementary pair are nonbasic.

(iii) z_0 is a basic variable in it.

A feasible basic vector for (16.18) in which there is exactly one basic variable from each complementary pair (w_j, z_j) is known as a *complementary feasible basic vector*. A feasible basic vector for (16.18) satisfying properties (i), (ii), and (iii) above is known as an *almost complementary feasible basic vector*. All the basic vectors obtained in the algorithm with the possible exception of the final basic vector are almost complementary feasible vectors. If at some stage of the algorithm, a complementary feasible basic vector is obtained, it is a final basic vector and the algorithm terminates.

ADJACENT ALMOST COMPLEMENTARY FEASIBLE BASIC VECTORS

Let $(y_1, \ldots, y_{j-1}, z_0, y_{j+1}, \ldots, y_n)$ be an almost complementary feasible basic vector for (16.18), where $y_i \in \{w_i, z_i\}$ for each $i \neq j$. Both the variables in the complementary pair (w_j, z_j) are not in this basic vector. Adjacent almost complementary feasible basic vectors can only be obtained by picking as the entering variable either w_j or z_j. Thus from each almost complementary feasible basic vector there are exactly two possible ways of generating adjacent almost complementary feasible basic vectors.

In the initial almost complementary feasible basic vector, both w_t and z_t are nonbasic variables. In the canonical tableau with respect to the initial basis, the updated column vector of w_t can be verified to be $-e_n$, which is negative. Hence, if w_t is picked as the entering variable into the initial basic vector, an extreme half-line is generated. Hence, the initial almost complementary BFS is at the end of an almost complementary ray.

So there is a unique way of obtaining an adjacent almost complementary feasible basic vector from the initial basic vector, and that is to pick z_t as the entering variable.

16.6.4 Complementary Pivot Rule

In the subsequent stages of the algorithm there is a unique way to continue the algorithm, which is to pick as the entering variable, the complement of the variable that just dropped from the basic vector. This is known as the *complementary pivot rule.*

The main property of the path generated by the algorithm is the following. Each BFS obtained in the algorithm has two almost complementary edges containing it. We arrive at this solution along one of these edges. And we leave it by the other edge. So the algorithm continues in a unique manner. It is also clear that a basic vector that was obtained in some stage of the algorithm can never reappear.

16.6.5 Termination

There are exactly two possible ways in which the algorithm can terminate.

1. At some stage of the algorithm, z_0 may leave the basic vector and a complementary feasible basis is then obtained. The solution of (16.18) corresponding to this final basis is a solution of the LCP (16.5) to (16.7).
2. At some stage of the algorithm the pivot column may turn out to be nonpositive [discussed in (1) of section 16.6.1] and in this case the algorithm terminates with a ray. This is called *ray termination.* When this happens, the algorithm is unable to solve the LCP. It is possible that the LCP (16.5) to (16.7) may not have a solution, but if it does have a solution, the algorithm is unable to find it. If ray termination occurs the algorithm is also unable to determine whether a solution to the LCP exists in the general case. However, when M satisfies some conditions, it can be proved that ray termination in the algorithm will only occur, when the LCP has no solution. See section 16.8.

DEGENERACY

The argument that each almost complementary feasible basis has at most two adjacent almost complementary feasible bases is used in developing the algorithm. This guarantees that the path taken by the algorithm continues unambiguously in a unique manner till termination occurs in one of the two possibilities. This property that each almost complementary feasible basis has at most two adjacent almost complementary feasible bases holds when (16.18) is nondegenerate. If (16.18) is degenerate, the dropping variable during some pivots may not be uniquely determined. In such a pivot step, by picking different dropping variables, different adjacent almost complementary feasible bases may be generated. If this happens, the almost complementary feasible basis in this step may have more than two adjacent almost complementary feasible bases. The algorithm can still be continued unambiguously according to the complementary pivot rule, but the path taken by the algorithm may depend on the dropping variables selected during the pivots in which these variables are not uniquely identified by the minimum ratio test. All the arguments mentioned in earlier sections are still valid, but in this case termination may not occur in a finite number of steps if the algorithm keeps cycling along a finite sequence of degenerate pivot steps. This can be avoided by using the concept of *lexico feasibility* of the solution as in Chapter 9. This requires the use of the *lexico minimum ratio test* of Chapter 9 instead of the usual minimum ratio test. In this case the algorithm deals with *almost complementary lexico feasible bases* throughout. In each pivot step the lexico minimum ratio test determines the dropping variable unambiguously and, hence, each almost complementary lexico feasible basis can have at most two adjacent almost complementary lexico feasible bases. With this, the path taken by the algorithm is again unique and unambiguous, no cycling can occur and termination occurs after a finite number of pivot steps.

INTERPRETATION

B. C. Eaves has given a simple haunted house interpretation of the path taken by the complementary pivot algorithm. A man who is afraid of ghosts has entered a haunted house from the outside through a door in one of its rooms. The house has the following properties.

(i) It has a finite number of rooms.

(ii) Each door is on a boundary wall between two rooms or on a boundary wall of a room on the outside.

(iii) Each room may have a ghost in it or may not. However, every room which has a ghost has exactly two doors.

All the doors in the house are open initially. The man's walk proceeds according to the following property.

(iv) When the man walks through a door, it is instantly sealed permanently and he can never walk back through it.

The man finds a ghost in the room he has entered initially, by properties (iii) and (iv) this room has exactly one open door when the man is inside it. In great fear he runs out of the room through that door. If the next room that he has entered has a ghost again, it also satisfies the property that it has exactly one open door when the man is inside it, and he runs out through that as fast as he can. In his walk, every room with a ghost satisfies the same property. He enters that room through one of its doors and leaves through the other. A *sanctuary* is defined to be either a room that has no ghost, or the outside of the house. The man keeps running until he finds a sanctuary. Property (i) guarantees that the man finds a sanctuary after running through at most a finite number of rooms. The sanctuary that he finds may be either a room without a ghost or the outside of the house.

We leave it to the reader to construct parallels between the ghost story and the complementary pivot algorithm and to find the walk of the man through the haunted house in Figure 16.7 where each door is marked by an "x." A "G" inside a room signifies a ghost. The man walks into the house initially from the outside through the door numbered with "1."

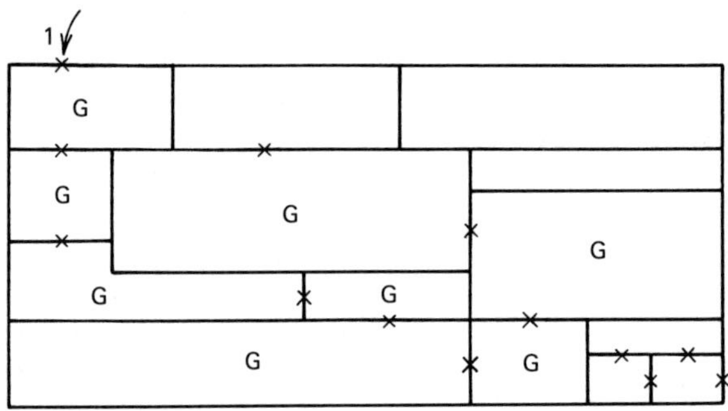

Figure 16.7 Haunted house.

Example 1

Consider the following LCP. (This is not an LCP corresponding to an LP.)

w_1	w_2	w_3	w_4	z_1	z_2	z_3	z_4	q
1	0	0	0	−1	1	1	1	3
0	1	0	0	1	−1	1	1	5
0	0	1	0	−1	−1	−2	0	−9
0	0	0	1	−1	−1	0	−2	−5

$$w_i \geqq 0, z_i \geqq 0, w_i z_i = 0 \quad \text{for all } i$$

When we introduce the artificial variable z_0 the tableau becomes:

w_1	w_2	w_3	w_4	z_1	z_2	z_3	z_4	z_0	q
1	0	0	0	−1	1	1	1	−1	3
0	1	0	0	1	−1	1	1	−1	5
0	0	1	0	−1	−1	−2	0	(−1)	−9
0	0	0	1	−1	−1	0	−2	−1	−5

The most negative q_i is q_3. Therefore pivot in the column vector of z_0 with the third row as the pivot row.

Basic variables	w_1	w_2	w_3	w_4	z_1	z_2	z_3	z_4	z_0	q	Ratios
w_1	1	0	−1	0	0	2	3	1	0	12	$\frac{12}{3}$
w_2	0	1	−1	0	2	0	3	1	0	14	$\frac{14}{3}$
z_0	0	0	−1	0	1	1	2	0	1	9	$\frac{9}{2}$
w_4	0	0	−1	1	0	0	(2)	−2	0	4	$\frac{4}{2}$

As in section 16.6.3, pick z_3 as the entering variable. The column vector of z_3 is the pivot column. w_4 drops from the basic vector.

Basic variables	w_1	w_2	w_3	w_4	z_1	z_2	z_3	z_4	z_0	q	Ratios
w_1	1	0	$\frac{1}{2}$	$-\frac{3}{2}$	0	2	0	$\textcircled{4}$	0	6	$\frac{6}{4}$
w_2	0	1	$\frac{1}{2}$	$-\frac{3}{2}$	2	0	0	4	0	8	$\frac{8}{4}$
z_0	0	0	0	-1	1	1	0	2	1	5	$\frac{5}{2}$
z_3	0	0	$-\frac{1}{2}$	$\frac{1}{2}$	0	0	1	-1	0	2	

Since w_4 has dropped from the basic vector, its complement, z_4 is the entering variable for the next step. w_1 drops from the basic vector.

Basic variables	w_1	w_2	w_3	w_4	z_1	z_2	z_3	z_4	z_0	q	Ratios
z_4	$\frac{1}{4}$	0	$\frac{1}{8}$	$-\frac{3}{8}$	0	$\frac{1}{2}$	0	1	0	$\frac{6}{4}$	
w_2	-1	1	0	0	$\textcircled{2}$	-2	0	0	0	2	$\frac{2}{2}$
z_0	$-\frac{1}{2}$	0	$-\frac{1}{4}$	$-\frac{1}{4}$	1	0	0	0	1	2	$\frac{2}{1}$
z_3	$\frac{1}{4}$	0	$-\frac{3}{8}$	$\frac{1}{8}$	0	$\frac{1}{2}$	1	0	0	$\frac{14}{4}$	

Since w_1 has dropped from the basic vector, its complement, z_1, is the new entering variable. Now w_2 drops from the basic vector.

Basic variables	w_1	w_2	w_3	w_4	z_1	z_2	z_3	z_4	z_0	q	Ratios
z_4	$\frac{1}{4}$	0	$\frac{1}{8}$	$-\frac{3}{8}$	0	$\frac{1}{2}$	0	1	0	$\frac{6}{4}$	3
z_1	$-\frac{1}{2}$	$\frac{1}{2}$	0	0	1	-1	0	0	0	1	
z_0	0	$-\frac{1}{2}$	$-\frac{1}{4}$	$-\frac{1}{4}$	0	$\textcircled{1}$	0	0	1	1	1
z_3	$\frac{1}{4}$	0	$-\frac{3}{8}$	$\frac{1}{8}$	0	$\frac{1}{2}$	1	0	0	$\frac{14}{4}$	7

Since w_2 has dropped from the basic vector, its complement, z_2, is the entering variable. Now z_0 drops from the basic vector.

Basic variables	w_1	w_2	w_3	w_4	z_1	z_2	z_3	z_4	z_0	q
z_4	$\frac{1}{4}$	$\frac{1}{4}$	$\frac{1}{4}$	$\frac{1}{4}$	0	0	0	1	$-\frac{1}{2}$	1
z_1	$-\frac{1}{2}$	0	$-\frac{1}{4}$	$-\frac{1}{4}$	1	0	0	0	1	2
z_2	0	$-\frac{1}{2}$	$-\frac{1}{4}$	$-\frac{1}{4}$	0	1	0	0	1	1
z_3	$\frac{1}{4}$	$\frac{1}{4}$	$-\frac{1}{4}$	$\frac{1}{4}$	0	0	1	0	$-\frac{1}{2}$	3

Since the present basis is a complementary feasible basis, the algorithm terminates. The corresponding solution of the LCP is $w = 0$, $(z_1, z_2, z_3, z_4) = (2, 1, 3, 1)$.

Example 2

w_1	w_2	w_3	z_1	z_2	z_3	q
1	0	0	1	0	3	-3
0	1	0	-1	2	5	-2
0	0	1	2	1	2	-1

$$w_i \geqq 0, z_i \geqq 0, w_i z_i = 0 \quad \text{for all } i$$

The tableau with the artificial variable z_0 is:

w_1	w_2	w_3	z_1	z_2	z_3	z_0	q
1	0	0	1	0	3	(-1)	-3
0	1	0	-1	2	5	-1	-2
0	0	1	2	1	2	-1	-1

The initial canonical tableau is:

Basic variables	w_1	w_2	w_3	z_1	z_2	z_3	z_0	q	Ratios
z_0	-1	0	0	-1	0	-3	1	3	
w_2	-1	1	0	-2	2	2	0	1	
w_3	-1	0	1	(1)	1	-1	0	2	$\frac{2}{1}$

The next tableau is:

Basic variables	w_1	w_2	w_3	z_1	z_2	z_3	z_0	q
z_0	−2	0	1	0	1	−4	1	5
w_2	−3	1	2	0	4	0	0	5
z_1	−1	0	1	1	1	−1	0	2

The entering variable here is z_3. The pivot column is nonpositive. Hence, the algorithm stops here with ray termination. The algorithm has been unable to solve this LCP.

16.7 CONDITIONS UNDER WHICH THE ALGORITHM WORKS

We define several classes of matrices that are useful in the study of the LCP. Let M be a square matrix of order n. It is said to be a:

Copositive matrix if $y^T My \geqq 0$ for all $y \geqq 0$.

Strict copositive if $y^T My > 0$ for all $y \geq 0$.

Copositive pus matrix if it is a copositive matrix and whenever $y \geqq 0$, and satisfies $y^T My = 0$, we have $y^T(M + M^T) = 0$.

P-matrix if all its principal subdeterminants are positive.

Q-matrix if the LCP (q, M) has a solution for every $q \in \mathbf{R}^n$.

Nondegenerate matrix if all its principal subdeterminants are nonzero.

Degenerate matrix if it is not a nondegenerate matrix.

Problem

16.2 Prove that every PSD matrix is also a copositive plus matrix.

Theorem If M is a copositive plus matrix and the system of constraints (16.5) and (16.6) has a feasible solution, then the LCP (16.5) to (16.7) has a solution and the complementary pivot algorithm of section 16.6 will terminate with a complementary feasible basis. Conversely, when M is a copositive plus matrix, if the complementary pivot algorithm applied on (16.5) to (16.7) terminates in ray termination, the system of constraints (16.5), (16.6) must be infeasible.

Proof Assume that either (16.5) is nondegenerate or that the lexico feasibility requirements discussed in section 16.6 are being used throughout the algorithm. This implies that each almost complementary feasible (or lexico feasible) basis obtained during the algorithm has exactly two adjacent almost complementary feasible (or lexico feasible) bases, excepting the initial and terminal bases, which have exactly one such adjacent basis only. The complementary pivot algorithm operates on the system (16.18).

The initial basic vector is $(w_1, \ldots, w_{t-1}, z_0, w_{t+1}, \ldots, w_n)$ (as in section 16.6.2). The corresponding BFS is $z = 0$, $w_t = 0$, $z_0 = -q_t$, and $w_i = q_i - q_t$ for all $i \neq t$. If w_t is taken as the entering variable into this basic vector, it generates the half-line (called the *initial extreme half-line*)

$$\begin{aligned} w_i &= q_i - q_t + \lambda \qquad \text{for all } i \neq t \\ w_t &= \lambda \\ z &= 0 \\ z_0 &= -q_t + \lambda \end{aligned}$$

where $\lambda \geqq 0$. (This can be seen by obtaining the canonical tableau corresponding to the initial basic vector.) This initial extreme half-line contains the initial BFS of (16.18) as its end point. Among the basic vectors obtained during the algorithm, the only one that can be adjacent to the initial basic vector is the one obtained by introducing z_t into it. Once the algorithm moves to this adjacent basic vector, the initial basic vector will never again appear during the algorithm. Hence, if the algorithm terminates with ray termination, the extreme half-line obtained at termination cannot be the initial extreme half-line.

At every point on the initial extreme half-line all the variables w, z_0 are strictly positive. From the conditions for a feasible solution to be on an edge (discussed in Chapter 3), it is clear that the only edge of (16.18) that contains a point in which all the variables w, z_0 are strictly positive is the initial extreme half-line.

Suppose the algorithm terminates in ray termination without producing a solution of the LCP. Let B_k be the terminal basis. When the complementary pivot algorithm is continued from this basis B_k, the updated column vector of the entering variable must be nonpositive resulting in the generation of an extreme half-line. Let the terminal extreme half-line be

$$\{(w, z, z_0) = (w^k + \lambda w^h, z^k + \lambda z^h, z_0{}^k + \lambda z_0{}^h) : \lambda \geqq 0\}$$

where $(w^k, z^k, z_0{}^k)$ is the BFS of (16.18) with respect to the terminal basis B_k, and $(w^h, z^h, z_0{}^h)$ is a homogeneous solution corresponding to (16.18), that is,

$$\begin{aligned} w^h - Mz^h - e_n z_0{}^h &= 0 \\ w^h \geqq 0, z^h \geqq 0, z_0{}^h &\geqq 0 \end{aligned} \tag{16.20}$$

$(w^h, z^h, z_0{}^h) \neq 0$. If $z^h = 0$, (16.20) and the fact that $(w^h, z^h, z_0{}^h) \neq 0$ together imply that $w^h \neq 0$ and hence $z_0{}^h > 0$, and consequently $w^h > 0$. Hence, if $z^h = 0$, points on this terminal extreme half-line have all the variables w, z_0 strictly positive, which by earlier arguments would imply that the terminal extreme half-line is the initial extreme half-line, a contradiction. So $z^h \neq 0$.

Since every solution obtained under the algorithm satisfies the complementarity constraint, $w^T z = 0$, we must have $(w^k + \lambda w^h)^T (z^k + \lambda z^h) = 0$ for all $\lambda \geqq 0$. This implies that $(w^k)^T z^k = (w^k)^T z^h = (w^h)^T z^k = (w^h)^T z^h = 0$. From (16.20) $(w^h)^T = (Mz^h + e_n z_0{}^h)^T$. Hence from $(w^h)^T z^h = 0$, we can conclude that $(z^h)^T M^T z^h = (z^h)^T M z^h = -e_n{}^T z^h z_0{}^h \leqq 0$. Since $z^h \geq 0$ and M is copositive plus by the hypothesis, $(z^h)^T M z^h = 0$, and this implies that $(z^h)^T (M + M^T) = 0$. So $(z^h)^T M = -(z^h)^T M^T$. Also since $-e_n{}^T z^h z_0{}^h = (z^h)^T M z^h = 0$, $z_0{}^h$ must be zero (since $z^h \geq 0$). Since $(w^k, z^k, z_0{}^k)$ is the BFS of (16.18) with respect to the feasible basis B_k, $w^k = Mz^k + q + e_n z_0{}^k$. Now

$$\begin{aligned} 0 = (w^k)^T z^h &= (Mz^k + q + e_n z_0{}^k)^T z^h \\ &= (z^k)^T M^T z^h + q^T z^h + z_0{}^k e_n{}^T z^h \\ &= (z^h)^T M z^k + (z^h)^T q + z_0{}^k e_n{}^T z^h \\ &= -(z^h)^T M^T z^k + (z^h)^T q + z_0{}^k e_n{}^T z^h \\ &= -(z^k)^T M z^h + (z^h)^T q + z_0{}^k e_n{}^T z^h \\ &= -(z^k)^T w^h + (z^h)^T q + z_0{}^k e_n{}^T z^h \\ &= (z^h)^T q + z_0{}^k e_n{}^T z^h \end{aligned}$$

So $(z^h)^T q = -z_0{}^k e_n{}^T z^h$. Since $z^h \geq 0$ and $z_0{}^k > 0$ [otherwise (w^k, z^k) would be a solution of the LCP], $z_0{}^k e_n{}^T z^h > 0$. Hence, $(z^h)^T q < 0$. Hence, if $\pi = (z^h)^T$ we have

$$\pi q < 0$$
$$\pi \geqq 0$$
$$\pi(-M) = -(z^h)^T M = (z^h)^T M^T = (w^h)^T \geqq 0$$

that is,

$$\pi q < 0$$
$$\pi(I \;\vdots\; -M) \geqq 0$$

By Farkas lemma discussed in section 4.6.7, this implies that the system:

$$(I \;\vdots\; -M)\begin{pmatrix} w \\ \cdots \\ z \end{pmatrix} = q$$
$$\begin{pmatrix} w \\ \cdots \\ z \end{pmatrix} \geqq 0$$

has no feasible solution. Hence, if the complementary pivot algorithm terminates in ray termination, the system (16.5) and (16.6) has no feasible solutions in this case and thus there cannot be any solution to the LCP.

This also implies that whenever (16.5) and (16.6) have a feasible solution, the LCP (16.5) to (16.7) has a solution in this case and the complementary pivot algorithm finds it.

The following results can be derived as corollaries.

Result 1 In the LCPs corresponding to LPs and convex quadratic programs, the matrix M is PSD and hence copositive plus. Hence, if the complementary pivot algorithm applied to the LCP corresponding to an LP or a convex quadratic program terminates in ray termination, that LP or convex quadratic program must either be infeasible, or if it is feasible, the objective function must be unbounded below on the set of feasible solutions of that problem.

Problem

16.3 Suppose the complementary pivot algorithm has terminated in ray termination when applied on the LCP corresponding to a LP (or a convex quadratic program). In this case develop an efficient method to check whether the problem is infeasible or whether it is feasible and the objective function is unbounded below in it.

Hence the complementary pivot algorithm works when used to solve LPs or convex quadratic programs.

Result 2 If M is strict copositive the complementary pivot algorithm applied on (16.5) to (16.7) terminates with a solution of the LCP.

Proof If the complementary pivot algorithm terminates in ray termination, as seen in the proof of the above theorem there exists a $z^h \geq 0$ such that $(z^h)^T M z^h = 0$, contradicting the hypothesis that M is strict copositive.

Thus all strict copositive matrices are Q-matrices. Also, if $M = (m_{ij}) \geqq 0$ and $M_{ij} > 0$ for all i, M is strict copositive and hence a Q-matrix.

Problem

16.4 Suppose $M \geqq 0$ and $m_{11} = 0$. Prove that if $q = (-1, 1, 1, \ldots, 1)^T$, the LCP (16.5) to (16.7) cannot have a solution.

Result 3 If M is a P-matrix, the complementary pivot algorithm terminates with a complementary feasible basis for the LCP (16.5) to (16.7).

Proof By the theorem of Gale and Nikaido [5], when M is a P matrix, the system $z_i^h(Mz^h)_i = z_i^h M_{i.} z^h \leqq 0$ for $i = 1$ to n, $z^h \geq 0$ cannot have a solution. This implies that when M is a P-matrix, ray termination cannot occur in the complementary pivot algorithm.

Problem

16.5 Prove that the LCP (16.5) to (16.7) has a unique solution corresponding to every $q \in \mathbf{R}^n$ iff M is a P-matrix.

Note When the complementary pivot alogirthm is applied on an LCP in which the matrix M is not a copositive plus matrix or a P-matrix, it is still possible that the algorithm terminates with a complementary feasible basis for the problem. However, in this general case it is also possible that the algorithm stops with ray termination even if a solution to the LCP exists.

THE BIMATRIX GAME PROBLEM

The LCP corresponding to the problem of finding an equilibrium pair of strategies for a bimatrix game is (16.15), where A, B^T are positive matrices. The complementary pairs of variables in this problem are (u_i, ξ_i) $i = 1$ to m and (v_j, η_j); $j = 1$ to N. Here a feasible basis can be obtained directly and there is no need to introduce an artificial variable z_0. For this problem we define an *almost complementary feasible basic vector* to be a feasible basic vector that contains exactly one basic variable from each complementary pair excepting two pairs. Both variables of one of these pairs are basic variables, and both variables in the other pair are nonbasic variables.

Initially make the variable ξ_1 a basic variable and the variables $\xi_2, \ldots, \xi_m$ nonbasic variables. Make ξ_1 equal to $\xi_1{}^0$, the smallest positive number such that $v^0 = -e_N + (B^T)_{.1}\xi_1{}^0 \geq 0$. At least one of the components in v^0, say, $v_r{}^0$ is zero. Make v_r a nonbasic variable too. The complement of v_r is η_r. Make the value of η_r to be the smallest positive value, $\eta_r{}^0$, such that $u^0 = A_{.r}\eta_r{}^0 - e_m \geq 0$. At least one of the components in u^0, say $u_s{}^0$ is 0. If $s = 1$, the basic vector $(u_2, \ldots, u_m, v_1, \ldots, v_{r-1}, v_{r+1}, \ldots, v_N, \xi_1, \eta_r)$ is a complementary feasible basic vector, and the feasible solution corresponding to it is a solution of the LCP (16.15). Terminate.

If $s \neq 1$, the basic vector

$$(u_1, \ldots, u_{s-1}, u_{s+1}, \ldots, u_m, v_1, \ldots, v_{r-1}, v_{r+1}, \ldots, v_N, \xi_1, \eta_r)$$

is a feasible basic vector. Both the variables in the complementary pair (u_1, ξ_1) are basic variables in it. And both variables in the complementary pair (u_s, ξ_s) are nonbasic variables. Hence the initial basic vector is an almost complementary feasible basic vector. All the basic vectors obtained during the algorithm (excepting the terminal complementary feasible basic vector) will be almost complementary feasible basic vectors containing both the variables in the pair (u_1, ξ_1) as basic variables.

When u_s is made as the entering variable into the initial basic vector, an almost complementary extreme half-line is generated. Hence the BFS of (16.15) with respect to the initial basic vector is an almost complementary BFS at the end of an almost complementary extreme half-line.

The algorithm begins by taking ξ_s as the entering variable into the initial basic vector. In all subsequent steps, the entering variable is picked by the complementary pivot rule. The algorithm terminates when one of the variables in the pair (u_1, ξ_1) drops from the basic vector. It can be proved that termination occurs after at most a finite number of pivots. The terminal basis is a complementary feasible basis. In this algorithm if degeneracy is encountered, it should be resolved using the lexico minimum ratio rule (see section 16.6).

Example

We will solve the LCP (16.16) corresponding to the 2 person game in the example in section 16.5. In tableau form it is

u_1	u_2	v_1	v_2	v_3	ξ_1	ξ_2	η_1	η_2	η_3	q
1	0	0	0	0	0	0	−2	−2	−1	−1
0	1	0	0	0	0	0	(−1)	−2	−2	−1
0	0	1	0	0	(−1)	−2	0	0	0	−1
0	0	0	1	0	−3	−1	0	0	0	−1
0	0	0	0	1	−2	−3	0	0	0	−1

$$u, v, \xi, \eta \geqq 0 \quad \text{and} \quad u_1\xi_1 = u_2\xi_2 = v_1\eta_1 = v_2\eta_2 = v_3\eta_3 = 0$$

Making $\xi_2 = 0$, the smallest value of ξ_1 that will yield nonnegative values to the v's is 1. When $\xi_2 = 0$, $\xi_1 = 1$ the value of v_1 is 0. Hence, v_1 will be made a nonbasic variable. The complement of v_1 is η_1. So make η_2 and η_3 nonbasic variables. The smallest value of η_1 that will make the u's nonnegative is $\eta_1 = 1$. When $\eta_1 = 1$ with $\eta_2 = \eta_3 = 0$, u_2 becomes equal to 0. So make u_2 a nonbasic variable. The canonical tableau with respect to the initial basic vector is therefore obtained as below by performing pivots in the columns of ξ_1 and η_1 with the circled elements as pivot elements.

Basic variables	u_1	u_2	v_1	v_2	v_3	ξ_1	ξ_2	η_1	η_2	η_3	q	Ratios
u_1	1	−2	0	0	0	0	0	0	2	3	1	
η_1	0	−1	0	0	0	0	0	1	2	2	1	
ξ_1	0	0	−1	0	0	1	2	0	0	0	1	$\frac{1}{2}$
v_2	0	0	−3	1	0	0	(5)	0	0	0	2	$\frac{2}{5}$, minimum
v_3	0	0	−2	0	1	0	1	0	0	0	1	$\frac{1}{1}$

The algorithm continues by selecting ξ_2, the complement of u_2, as the entering variable. v_2 drops from the basic vector.

Basic variables	u_1	u_2	v_1	v_2	v_3	ξ_1	ξ_2	η_1	η_2	η_3	q
u_1	1	-2	0	0	0	0	0	0	(2)	3	1
η_1	0	-1	0	0	0	0	0	1	2	2	1
ξ_1	0	0	$\frac{1}{5}$	$-\frac{2}{5}$	0	1	0	0	0	0	$\frac{1}{5}$
ξ_2	0	0	$-\frac{3}{5}$	$\frac{1}{5}$	0	0	1	0	0	0	$\frac{2}{5}$
v_3	0	0	$-\frac{7}{5}$	$-\frac{1}{5}$	1	0	0	0	0	0	$\frac{3}{5}$

Since v_2 has dropped from the basic vector, its complement η_2 is the next entering variable. There is a tie in the minimum ratio when η_2 is the entering variable, since it can replace either u_1 or η_1 from the basic vector. Such ties should be resolved by the lexico minimum ratio test, but in this case we will let u_1 drop from the basic vector, since that leads to a complementary feasible basis to the problem.

Basic variables	u_1	u_2	v_1	v_2	v_3	1	2	1	2	3	q
η_2	$\frac{1}{2}$	-1	0	0	0	0	0	0	1	$\frac{3}{2}$	$\frac{1}{2}$
η_1	-1	1	0	0	0	0	0	1	0	-1	0
ξ_1	0	0	$\frac{1}{5}$	$-\frac{2}{5}$	0	1	0	0	0	0	$\frac{1}{5}$
ξ_2	0	0	$-\frac{3}{5}$	$\frac{1}{5}$	0	0	1	0	0	0	$\frac{2}{5}$
v_3	0	0	$-\frac{7}{5}$	$-\frac{1}{5}$	1	0	0	0	0	0	$\frac{3}{5}$

The present basic vector is a complementary feasible basic vector. The solution

$$(u_1, u_2, v_1, v_2, v_3; \xi_1, \xi_2, \eta_1, \eta_2, \eta_3) = \left(0, 0, 0, 0, \frac{3}{5}; \frac{1}{5}, \frac{2}{5}, 0, \frac{1}{2}, 0\right)$$

is a solution of the LCP. In this solution $\xi_1 + \xi_2 = 3/5$ and $\eta_1 + \eta_2 + \eta_3 = 1/2$. Hence the probability vector $x = \xi/(\sum \xi_i) = (1/3, 2/3)^{\mathrm{T}}$ and $y = \eta/(\sum \eta_j) = (0, 1, 0)^{\mathrm{T}}$ constitute an equilibrium pair of strategies for this game.

16.8 REMARKS

An experimental study (see references [22 and 23]) has shown that the complementary pivot algorithm for solving LCPs is superior to the simplex method for solving LPs. It is possible that in the near future, computer programs for solving the LCP will be used routinely to solve LPs.

It has also been shown that the arguments used in the complementary pivot algorithm can be generalized, and these generalizations have led to algorithms that can compute approximate Brouwer and Kakutani fixed points! Until now, the greatest single contribution of the complementarity problem is probably the insight that it has provided for the development of fixed point computing algorithms (see next section). In mathematics, fixed point theory is very highly developed, but the absence of efficient algorithms for computing these fixed points has so far frustrated all attempts to apply this rich theory to real life problems. With the development of these new algorithms, fixed point theory is finding numerous applications in mathematical programming, in mathematical economics, and in various other areas.

16.8.1 Extensions to Fixed Point Computing Methods

Consider a map that associates a subset $\mathbf{F}(x) \subset \mathbf{R}^n$ for each point $x \in \mathbf{R}^n$. A map like this is known as a *point to set map.* A *Kakutani fixed point* of this map is a point $y \in \mathbf{R}^n$ that satisfies $y \in \mathbf{F}(y)$.

Algorithms for computing these fixed points efficiently have been developed recently. These algorithms use *triangulations* of $\mathbf{R}^n$. A triangulation is a partition of $\mathbf{R}^n$ into *simplices.* Each simplex in $\mathbf{R}^n$ is the convex hull of $n + 1$ points called its *vertices.* Each vertex is given a label from the set $\{1, \ldots, n + 1\}$. The labeling is done in such a way that if a simplex whose vertices have all distinct labels is found, any point in that simplex is an approximation of the fixed point. So the problem reduces to that of finding a simplex in the triangulation, all of whose vertices have distinct labels. Since each simplex has exactly $n + 1$ vertices, the set of labels on the vertices of a distinctly labeled simplex must be $\{1, \ldots, n + 1\}$. Hence, such simplices are called *completely labeled simplices.*

The algorithm begins with a specially constructed simplex that is *almost completely labeled,* that is, the vertices of this simplex have distinct labels with the exception of exactly one pair of vertices that have the same label. For example, the set of labels on the vertices of the starting simplex might be $\{1, 1, 2, \ldots, n\}$. The labels on the vertices of an almost completely labeled simplex satisfy the following properties.

(i) Exactly one label appears twice.

(ii) Exactly one label from the set $\{1, \ldots, n + 1\}$ is missing.

The properties of the triangulation used guarantee that each almost completely labeled simplex has at most two adjacent simplices that are either completely labeled or almost completely labeled. The starting simplex has exactly one adjacent almost completely labeled simplex and the algorithm moves to that. All the simplices obtained along the path will be almost completely labeled excepting the terminal one, which will be completely labeled. We arrive at a simplex on the path from an adjacent almost completely labeled simplex. This simplex has exactly one other adjacent simplex that is either almost completely labeled or completely labeled, and the algorithm moves to that. Thus, the path continues in a unique, unambiguous manner until it terminates with a completely labeled simplex.

The resemblance of the almost completely labeled simplices path obtained in the algorithm here to the almost complementary bases path obtained in the algorithm for the LCP is quite clear.

It has been shown that nonlinear programming problems (either constrained or unconstrained) can be transformed into problems of computing fixed points. The algorithms for computing fixed points hold a great promise for wide applicability in mathematical programming and its applications. For detailed description of these algorithms see references [28 to 34].

16.8.2 Unsolved Problem in Linear Complementarity

The only methods that are guaranteed to solve an LCP when M is a general matrix are enumerative algorithms. Even when M belongs to a special class of matrices for which the complementary pivot method works, it will only produce one solution of the LCP when solutions exist. If M is known to be a P-matrix, this solution is unique. Otherwise one would like to know whether this solution is unique or whether alternate solutions exist. If alternate solutions exist, one would like to compute them. The enumerative algorithms available at present to answer these questions and to solve the LCP when M is a general matrix are not very efficient when n is large.

An algorithm for a problem is said to be a *good algorithm* if the number of computations required by it is bounded above by a polynomial in n, the order (or size) of the problem.

The algorithm described in section 16.4 for testing whether a given square matrix is PD or PSD is a good algorithm. At present there are no good algorithms to test whether a general square matrix is copositive, copositive plus, P-matrix or a Q-matrix. Also for a general matrix there are no simple necessary and sufficient conditions known for it to be a Q-matrix.

Besides these there are a large number of unsolved theoretical problems on the spanning properties of complementary cones and on the matrix theoretic properties associated with the LCP. Some of the special algorithms, and properties of the LCP are discussed in the problems that follow this section.

CONCLUDING REMARKS

The complementary pivot algorithm for computing equilibrium strategies in bimatrix games is due to C. E. Lemke and J. T. Howson [14]. The complementary pivot algorithm discussed in section 16.6 is due to C. E. Lemke[12].Various applications of the LCP, and a discussion of principal pivoting methods and the theorem in section 16.7 are discussed by R. W. Cottle and G. B. Dantzig in [2]. For theoretical results on the LCP see [4, 6, 13, 16, 17, 19, 20, 21, 24, 25 and 26]. For results on the *nonlinear complementary problem* (not discussed here), see [8]. The pioneering paper on fixed point computing methods is [32].

A portion of this chapter has appeared in the article on "Complementarity Problems" in the *Encyclopedia of Computer Science and Technology*.

PROBLEMS

16.6 A square matrix M is said to be a Z matrix if $m_{ij} \leqq 0$ for all $i \neq j$. Suppose M is a Z matrix. Consider the LCP.

w	z	
I	$-M$	q

$$w \geqq 0, z \geqq 0, w^T z = 0$$

Consider the following scheme for solving this LCP proposed by R. Chandrasekharan and R. Saigal.

Step 0 Start with the initial tableau with the identity matrix corresponding to w as the initial basis. If this is a feasible basis (i.e., if $q \geqq 0$) it is a complementary feasible basis. Terminate. Otherwise go to Step 1.

Step 1 Pick a row, say row p, such that the present entry of the right-hand side constant vector in this row, q_p, is negative. If $-m_{p,p} \geqq 0$, there is no feasible solution to the problem and, hence, there exists no solution to the LCP. Terminate. Otherwise, if $-m_{p,p} < 0$, make a pivot with row p as the pivot row and the column vector of z_p as the pivot column.

We will now discuss what happens in a general step.

Step 2 If no termination has occurred so far and if the present right-hand side constant vector $\bar{q}$ is nonnegative, the present basis is a complementary feasible basis. Terminate. Otherwise select a row, say row t, such that $\bar{q}_t < 0$. Let $-\bar{m}_{t,t}$ be the entry in the column vector of z_t and in row t in the present tableau.

If $-\bar{m}_{t,t} \geqq 0$, the problem is infeasible and hence no solution exists for the LCP. Terminate. Otherwise make a pivot with the column vector of z_t as the pivot column, row t as the pivot row, and go to the next step.

Prove the following about this algorithm:

(a) Once a row is picked as a pivot row, the right-hand side constant in that row remains nonnegative in all subsequent steps.

(b) Once a variable z_i is made a basic variable in the algorithm, it stays as a basic variable in all subsequent steps.

(c) The first feasible basis found under the algorithm will be a complementary feasible basis.

(d) The algorithm has to terminate in at most n pivot steps.

(e) Using this algorithm solve the LCP corresponding to

$$M = \begin{pmatrix} 1 & -2 & 0 & -2 & -1 \\ -1 & 0 & -1 & -2 & 0 \\ -2 & -3 & 3 & 0 & 0 \\ 0 & -1 & -1 & -2 & -1 \\ -2 & 0 & -1 & -2 & 3 \end{pmatrix} \qquad q = \begin{pmatrix} -4 \\ -4 \\ -2 \\ -1 \\ -2 \end{pmatrix}$$

(g) Prove the following using this algorithm: Let M be a Z matrix. Keep M fixed. The set of all $q \in \mathbf{R}^n$ for which the LCP (q, M) has a solution is Pos $(I \ \vdots \ -M)$.

16.7 Let $-M$ be a Z matrix. A well-known theorem states that if there exists an $x \geqq 0$ such that $x^{\mathrm{T}}M < 0$ in this case, then M^{-1} exists and $-M^{-1} \geqq 0$. Using this theorem, prove the following:

(a) If M satisfies all the above properties, there exist $y_{ij} \geqq 0$ for all i, j such that

$$I_{.j} = \sum_{i=1}^{n} (-y_{ij})M_{.i}, \text{ for all } j.$$

(*Hint*. Use the fact that $M^{-1} \leqq 0$.)

(b) Under the same conditions on M, Pos $(I \ \vdots \ -M) =$ Pos $(-M)$.

(c) Under the same conditions on M the LCP (q, M) has a solution iff $-M^{-1}q \geqq 0$. Also, if $-M^{-1}q \geqq 0$, then a solution to the LCP is $(\mathrm{w}, \mathrm{z}) = (0, -M^{-1}q)$.

—*R. Saigal*

16.8 Let M be a square matrix of order n satisfying the property "if $Mx \leqq 0$, then x must be nonnegative." Prove the following.

(a) M^{-1} must exist.

(b) $-M^{-1} \geqq 0$ [*Hint*. Use the fact that $(M(M^{-1}))_{.j} = I_{.j} \geqq 0$.]

(c) In this case Pos $(-M) \supset$ Pos (I).

16.9 Let M be an arbitrary square matrix of order n. Consider the LCP (q, M). Prove that the following property "the LCP has a solution whenever q is such that the system $w - Mz = q, w \geqq 0, z \geqq 0$ has a feasible solution and for all such q the LCP has a solution in which $w = 0$" holds iff Pos $(-M) \supset$ Pos (I) [i.e., Pos $(I \;\vdots\; -M) =$ Pos $(-M)$].
Also prove that this property holds iff for all x such that $Mx \leqq 0$, x must be nonnegative.

—*A. K. Rao*

16.10 If D is PSD (not necessarily symmetric) and $x^{\mathrm{T}}Dx = 0$ then $(D^{\mathrm{T}} + D)x = 0$.

16.11 If D is PD, its determinant is positive. Also all its principal subdeterminants are positive.

16.12 If D is PSD, its determinant is nonnegative. Also all its principal subdeterminants are nonnegative.

16.13 Consider a two person game with loss matrices A', B'. Suppose $A' + B' = 0$. That is, the game is a zero sum game. In this case prove that every optimal pair of strategies (in the minimax sense discussed in section 4.6.6) is an equilibrium pair of strategies (as defined in section 16.5) and vice versa.

16.14 If the LCP (q, M) has an infinite number of distinct solutions, prove that there must exist a complementary set of column vectors that is linearly dependent. In this case by expanding the determinant of the matrix consisting of these column vectors, prove that there exists a principal subdeterminant of M whose value is 0.

16.15 Consider the quadratic program:

$$\begin{aligned} \text{Minimize} \quad & Q(x) = cx + \frac{1}{2}x^{\mathrm{T}}Dx \\ \text{Subject to} \quad & Ax \geqq b \\ & x \geqq 0 \end{aligned}$$

where D is a symmetric matrix. **K** is the set of feasible solutions for this problem. $\bar{x}$ is an interior point of **K** (i.e., $A\bar{x} > b$ and $\bar{x} > 0$).

(a) What are the necessary conditions for $\bar{x}$ to be an optimum solution of the problem?

(b) Using the above conditions, prove that if D is not PSD, $\bar{x}$ could not be an optimum solution of the problem.

16.16 For the following quadratic program write down the corresponding LCP.

$$\begin{aligned} \text{Minimize} \quad & -6x_1 - 4x_2 - 2x_3 + 3x_1^2 + 2x_2^2 + \frac{1}{3}x_3^2 \\ \text{Subject to} \quad & x_1 + 2x_2 + x_3 \leqq 4 \\ & x_j \geqq 0 \quad \text{for all } j \end{aligned}$$

If it is known that this LCP has a solution in which all the variables x_1, x_2, x_3 are positive, find it.

16.17 Consider the bimatrix game problem with given loss matrices A, B. Let $x = (x_1, \ldots, x_m)^{\mathrm{T}}$ and $y = (y_1, \ldots, y_n)^{\mathrm{T}}$ be the probability vectors of the two players. Let $X = (x_1, \ldots, x_m, x_{m+1})^{\mathrm{T}}$ and $Y = (y_1, \ldots, y_n, y_{n+1})^{\mathrm{T}}$. Let e_r be the column vector in $\mathbf{R}^r$ all of whose entries are 1. Let $\mathbf{S} = \{X: B^{\mathrm{T}}x - e_n{}^{\mathrm{T}}x_{m+1} \geqq 0, e_m{}^{\mathrm{T}}x = 1, x \geqq 0\}$ and

$\mathbf{T} = \{Y: Ay - e_m{}^{\mathrm{T}}y_{n+1} \geqq 0, e_n{}^{\mathrm{T}}y = 1, y \geqq 0\}$. Let $Q(X, Y) = x^{\mathrm{T}}(A + B)y - x_{m+1} - y_{n+1}$. If $(\bar{x}, \bar{y})$ is an equilibrium pair of strategies for the game and $\bar{x}_{m+1} = \bar{x}^{\mathrm{T}}B\bar{y}$, $\bar{y}_{n+1} = \bar{x}^{\mathrm{T}}A\bar{y}$, prove that $(\bar{X}, \bar{Y})$ minimizes $Q(X, Y)$ over $\mathbf{S} \times \mathbf{T} = \{(X, Y): X \in S, Y \in \mathbf{T}\}$.

—*O. L. Mangasarian*

16.18 The LCP (q, M) has a solution for all $q \in \mathbf{R}^n$ if M is a nonsingular matrix such that $M^{-1} > 0$.

16.19 Write down the LCP corresponding to

$$\text{Minimize} \quad cx + \frac{1}{2}x^{\mathrm{T}}Dx$$

$$\text{Subject to} \quad x \geqq 0$$

16.20 Let

$$M = \begin{pmatrix} -2 & 1 \\ 1 & -2 \end{pmatrix} \qquad q = \begin{pmatrix} 1 \\ 1 \end{pmatrix}$$

Show that the LCP (q, M) has four distinct solutions. For $n = 3$, construct a square matrix M of order 3 and a $q \in \mathbf{R}^3$ such that (q, M) has eight distinct solutions.

Hint. $$\text{Try } -M = \begin{pmatrix} 2 & -1 & -1 \\ -1 & 3 & -1 \\ -1 & -1 & 4 \end{pmatrix} \qquad q = \begin{pmatrix} 1 \\ 1 \\ 1 \end{pmatrix}$$

16.21 Let

$$M = \begin{pmatrix} 0 & 0 & 1 \\ 0 & 0 & 1 \\ 0 & 0 & 0 \end{pmatrix} \qquad q = \begin{pmatrix} 0 \\ -1 \\ 0 \end{pmatrix}$$

Find out a solution of the LCP (q, M) by inspection. However prove that there exists no complementary feasible basis for this problem.

—*L. Watson*

16.22 Test whether the following matrices are PD, PSD, or not PSD by using the algorithms described in section 16.4.

$$\begin{pmatrix} 0 & 1 & -1 \\ 0 & 0 & -2 \\ 1 & 2 & 1 \end{pmatrix}, \begin{pmatrix} 4 & 3 & -7 \\ 0 & 0 & -2 \\ 0 & 0 & 6 \end{pmatrix}, \begin{pmatrix} 4 & 100 & 2 \\ 0 & 2 & 10 \\ 0 & 0 & 1 \end{pmatrix}, \begin{pmatrix} 5 & -2 & -2 \\ 0 & 5 & -2 \\ 0 & 0 & 5 \end{pmatrix}$$

16.23 Let $Q(x) = (1/2)x^{\mathrm{T}}Dx - cx$. If D is PD, prove that $Q(x)$ is bounded below.

16.24 $\{A_{.1}, \ldots, A_{.m}\}$ is a linearly independent set of vectors and b is a point in $\mathbf{R}^m$. Let $\mathbf{K} = \text{Pos}\,(A_{.1}, \ldots, A_{.m})$. Find the point in $\mathbf{K}$ that is closest to b in terms of the Euclidean distance. Set up this problem as a quadratic program and discuss an efficient algorithm for solving it. Discuss how the algorithm specializes in this problem and describe its execution in the simplest possible terms. Determine the maximum computational effort (in terms of the number of additions, multiplications, divisions, and comparisons required) that may be required to solve the problem.

16.25 Consider Problem 16.24 again. Prove that it has a unique solution. Also if $x \in \mathbf{K}$, prove that there exists a unique $\lambda = (\lambda_1, \ldots, \lambda_m) \geqq 0$, such that $x = \lambda_1 A_{.1} + \cdots + \lambda_m A_{.m}$. Let $\bar{x}$ be the optimum solution for Problem 16.24 and let $\bar{\lambda}$ correspond to it. Prove that $\bar{\lambda}_i = 0$ for at least one i. Is it possible to develop an algorithm for checking whether $\bar{\lambda}_1 = 0$ or positive without the need for computing $\bar{x}$?

16.26 Consider Problem 16.24 again. Suppose $A_{.1}, \ldots, A_{.m-1}$ are given and fixed, but $A_{.m}$ can be any vector from the finite set $\mathbf{S} = \{Y_{.1}, \ldots, Y_{.r}\}$. Choose $A_{.m}$ from $\mathbf{S}$, so that the minimum distance from b to $\mathbf{K}$ is as small as possible. Develop an efficient algorithm for doing it.

16.27 Let M be a square matrix of order n and $q \in \mathbf{R}^n$. Let $z \in \mathbf{R}^n$ be a vector of variables. Define: $f_i(z)$ = minimum $\{z_i, M_{i.}z + q_i\}$, that is

$$\begin{aligned} f_i(z) &= I_{i.}z && \text{if} \quad (M_{i.} - I_{i.})z + q_i \geqq 0 \\ &= M_{i.}z + q_i && \text{if} \quad (M_{i.} - I_{i.})z + q_i \leqq 0 \end{aligned}$$

for each $i = 1$ to n.

(a) Show that $f_i(z)$ is a piecewise linear concave function defined on $\mathbf{R}^n$.

(b) Consider the system of equations

$$f_i(z) = 0 \qquad i = 1 \text{ to } n$$

Let $\bar{z}$ be a solution of this system. Let $\bar{w} = M\bar{z} + q$. Prove that $(\bar{w}, \bar{z})$ is a complementary feasible solution of the LCP (q, M).

(c) Using (b) show that every LCP is equivalent to solving a system of piecewise linear equations.

—*R. Saigal*

16.28 Consider the LCP, (q, M), where M is a given square matrix of order n. In solving it we deal with the following system of equations in nonnegative variables.

w	z	
I	$-M$	q

(16.21)

$w \geqq 0 \quad z \geqq 0$

Let $(y_1, \ldots, y_n)$ be a complementary basic vector of variables for (16.21). Let t_j be the complement of y_j (i.e., t_j is z_j if $y_j = w_j$, or $t_j = w_j$ if $y_j = z_j$), $j = 1$ to n. Obtain the canonical tableau of (16.21) with respect to the basic vector $(y_1, \ldots, y_n)$ and rearrange the columns in the tableau so that the variables appear in the order $y_1, \ldots, y_n, t_1, \ldots, t_n$ from left to right. Let $-D$ be the matrix consisting of the column vectors corresponding to $t_1, \ldots, t_n$, in that order, in this canonical tableau.

$y_1 \ldots y_n$	$t_1 \ldots t_n$	
I	$-D$	$\bar{q}$

Then the matrix D is known as a *principal pivot transform* of the matrix M. Prove the following.

1. M is nondegenerate iff all complementary vectors of variables are basic vectors for (16.21).
2. If M is a P-matrix, every principal pivot transform of it is also a P-matrix.
3. M is a P-matrix iff all complementary vectors of variables are basic vectors for (16.21), and all the diagonal entries in all the principal pivot transforms of M are positive.

16.29 STRICT SEPARATION PROPERTY Consider the LCP, (q, M), where M is a P-matrix of order n. Prove the following.

(i) All the complementary sets of vectors for this problem are linearly independent.

(ii) Let $(A_{.1}, \ldots, A_{.n})$ be a complementary set of vectors for this problem. Let $\mathbf{H}$ be the hyperplane that is the linear hull of $\{A_{.1}, \ldots, A_{.j-1}, A_{.j+1}, \ldots, A_{.n}\}$ in $\mathbf{R}^n$. Prove that the vectors $I_{.j}$ and $-M_{.j}$ are strictly on opposite sides of $\mathbf{H}$. (These facts imply that the complementary cones $\mathbf{K}_1 = \text{pos}\,(A_{.1}, \ldots, A_{.j-1}, I_{.j}, A_{.j+1}, \ldots, A_{.n})$ and $\mathbf{K}_2 = \text{pos}\,(A_{.1}, \ldots, A_{.j-1}, -M_{.j}, A_{.j+1}, \ldots, A_{.n})$ have both nonempty interiors, and that the intersection $\mathbf{K}_1 \cap \mathbf{K}_2$ is their common facet pos $(A_{.1}, \ldots, A_{.j-1}, A_{.j+1}, \ldots, A_{.n})$. If this holds for every complementary set of vectors and for all j, the LCP, (q, M) is said to satisfy the *strict separation property*.

(iii) Conversely if the LCP (q, M) satisfies the strict separation property prove that M must be a P-matrix.

(Reference [17].)

16.30 PARAMETRIC LCP Consider the LCP $(q(\lambda), M)$, where $q_i(\lambda) = b_i + \lambda b_i^*$ for $i = 1$ to n, b_i, b_i^* are given numbers, λ is a real-valued parameter, and M is a given P-matrix of order n. Suppose it is required to find its solution as a function of the parameter λ for all values of λ. This is known as a *parametric* LCP. The following algorithm has been developed for this problem in reference [20]. Here we deal with the following system of constraints.

w	z	b	b^*
I	$-M$		

$\mathrm{w} \geqq 0 \quad \mathrm{z} \geqq 0$ (16.22)

The ith complementary pair here is $\{\mathrm{w}_i, \mathrm{z}_i\}$, $i = 1$ to n.

(i) Solve the LCP, $(q(\lambda), M)$, for a fixed value of λ, say λ_0. You can use the complementary pivot algorithm of section 16.6, with $q = q(\lambda_0)$, for this. Let $(y_1, \ldots, y_n)$ be the complementary feasible basic vector obtained for λ_0, where y_j is either w_j or z_j for each $j = 1$ to n. Go to (ii) with $(y_1, \ldots, y_n)$ as the current complementary basic vector.

(ii) Find out the range of values of the parameter λ, for which the current basic vector remains feasible for (16.22). This is determined as in parametric right-hand side LPs. To do it, let B be the basis for (16.22) corresponding to the present basic vector. Let $\bar{b} = B^{-1}b$ and $\bar{b}^* = B^{-1}b^*$ be the updated right-hand side constant vectors with respect to the basis B. Let

$$\begin{aligned} \underline{\lambda} &= \text{Maximum}\,\{-\bar{b}_i/\bar{b}_i^*: i \text{ such that } \bar{b}_i^* > 0\} \\ &= -\infty \qquad \text{if } \bar{b}_i^* \leqq 0 \text{ for all } i \end{aligned} \tag{16.23}$$

$$\begin{aligned} \bar{\lambda} &= \text{Minimum}\,\{-\bar{b}_i/\bar{b}_i^*: i \text{ such that } \bar{b}_i^* < 0\} \\ &= +\infty \qquad \text{if } \bar{b}_i^* \geqq 0 \text{ for all } i \end{aligned} \tag{16.24}$$

The present complementary basic vector remains feasible to (16.22) for all λ in the closed interval $\underline{\lambda} \leqq \lambda \leqq \bar{\lambda}$. In this interval, the solution of the LCP, $(q(\lambda), M)$, is

$$\text{Present basic vector} = \bar{b} + \lambda\bar{b}^*$$

$$\text{All nonbasic variables} = 0$$

If $\bar{\lambda}$ is finite, and it is required to find solutions of the LCP, $(q(\lambda), M)$ for $\lambda > \bar{\lambda}$, go to (iii). If $\underline{\lambda}$ is finite, and it is required to find the solutions of the LCP, $(q(\lambda), m)$ for $\lambda < \underline{\lambda}$, go to (iv). If solutions of the LCP have been obtained for all the desired values of λ, terminate.

(iii) Find out an i that attains the minimum in (16.24). Suppose it is $i = r$. Replace the present basic variable from the rth complementary pair $\{w_r, z_r\}$ by its complement. Go back to (ii) with the new basic vector as the current complementary basic vector.

(iv) Find out an i that attains the maximum in (16.23). Suppose it is $i = s$. Replace the present basic variable from the sth complementary pair $\{w_s, z_s\}$, by its complement. Go back to (ii) with the new basic vector as the current complementary basic vector.

Do the following.

1. Using the fact that M is a P-matrix, prove that all the pivot elements for the basis changes in steps (iii) and (iv) will always be negative.

2. Prove that this algorithm solves the parametric LCP for all values of λ in a finite number of pivot steps.

3. Solve the parametric LCP in which

$$M = \begin{pmatrix} 2 & 1 & -1 \\ -1 & 3 & 0 \\ 0 & 1 & 4 \end{pmatrix} \qquad q(\lambda) = \begin{pmatrix} 1 - \lambda \\ 2 + \lambda \\ 3 - 2\lambda \end{pmatrix}$$

using this algorithm.

4. Discuss what happens when this algorithm is applied to solve a parametric LCP in which M is not a P-matrix.

16.31 Consider the LCP (q, M) where M is a P-matrix of order n, and the original tableau is (16.21) in problem 16.28. Consider the following algorithm for solving this LCP. The algorithm deals only with complementary basic vectors. The moment a complementary feasible basic vector is obtained, the algorithm terminates. If $q \geqq 0$, w is a complementary feasible basic vector and we terminate. Suppose $q \ngeqq 0$.

(i) Let w be the initial complementary basic vector. Using it as the current complementary basic vector, go to (ii).

(ii) Let B be the basis for (16.21) corresponding to the current complementary basic vector. Let $\bar{q} = B^{-1}q$ be the updated right-hand side constant vector of (16.21) with respect to the basis B. If $\bar{q} \geqq 0$, the current complementary basic vector is feasible and we terminate. Otherwise go to (iii).

(iii) Find r = minimum $\{i$: i such that $\bar{q}_i < 0\}$. Replace the present basic variable from the rth complementary pair $\{w_r, z_r\}$ by its complement. With the new basic vector as the current complementary basic vector, go back to (ii).

Prove that this algorithm cannot cycle and that it solves the LCP, (q, M), in a finite number of steps. Solve the LCP, (q, M), where

$$M = \begin{pmatrix} 2 & 1 & -1 \\ -1 & 3 & 0 \\ 0 & 1 & 4 \end{pmatrix} \qquad q = \begin{pmatrix} -1 \\ -2 \\ -3 \end{pmatrix}$$

using this algorithm.

(Reference [21])

16.32 Let M be a given square matrix of order n. Let $\mathbf{\Gamma} = \{1, \ldots, n\}$. Let f be a real-valued function defined on subsets of $\mathbf{\Gamma}$, by the following rules. If $\mathbf{S} \subset \mathbf{\Gamma}$

$$f(\mathbf{S}) = 1 \quad \text{if} \quad \mathbf{S} = \varnothing$$

= determinant of the principal submatrix of M corresponding to the subset $\mathbf{S}$

The matrix M is said to be a *weak separation matrix* if there exists no subset $\mathbf{S} \subset \mathbf{\Gamma}$, satisfying the property that for some $j \in \mathbf{S}$, $f(\mathbf{S})$ and $f(\mathbf{S}\backslash\{j\})$ are both nonzero and have strictly opposite signs.

The LCP, (q, M) is said to satisfy the *weak separation property* if for every complementary set of column vectors, $(A_{.1}, \ldots, A_{.n})$, and for each $j = 1$ to n, there exists a hyperplane, $\mathbf{H}$, of dimension $n - 1$ in $\mathbf{R}^n$ containing the linear hull of $\{A_{.1}, \ldots, A_{.j-1}, A_{.j+1}, \ldots, A_{.n}\}$, such that either

(i) At least one of the points in the complementary pair of column vectors $\{I_{.j}, -M_{.j}\}$ lies in $\mathbf{H}$, or

(ii) If neither of the points in the pair $\{I_{.j}, -M_{.j}\}$ lie on $\mathbf{H}$, then $I_{.j}$ and $-M_{.j}$ lie strictly on opposite sides of $\mathbf{H}$.

Prove the following.

1. The LCP, (q, M) satisfies the weak separation property iff M is a weak separation matrix.
2. Every nondegenerate weak separation matrix is a P-matrix.
3. If the LCP, (q, M) does not satisfy the weak separation property, there must exist a principal subdeterminant of M which is negative.

(Reference [19]).

16.33 Let M be a square nondegenerate matrix. Prove that the number of complementary feasible solutions for the LCP, (q, M), is either even for all q that are nondegenerate in (16.21), or odd for all q that are nondegenerate in (16.21).

(Reference [17]).

REFERENCES

1. R. Chandrasekaran, "A Special Case of the Complementary Pivot Problem," *Opsearch*, *7*, pp. 263–268, 1970.
2. R. W. Cottle and G. B. Dantzig, "Complementary Pivot Theory of Mathematical Programming," *Linear Algebra and its Applications*, *1*, pp. 103–125, 1968.
3. O. De Donato and G. Maier, "Mathematical Programming Methods for the Inelastic Analysis of Reinforced Concrete Frames Allowing for Limited Rotation Capacity," *International Journal for Numerical Methods in Engineering*, *4*, pp. 307–329, 1972.
4. B. C. Eaves, "The Linear Complementarity Problem," *Management Science*, *17*, pp. 612–634, 1971.
5. D. Gale and H. Nikaido, "The Jacobian Matrix and Global Univalence of Mappings," *Mathematische Annalen*, *159*, pp. 81–93, 1965.
6. C. B. Garcia, "Some Classes of Matrices in Linear Complementarity Problem," *Mathematical Programming*, *5*, pp. 299–310, 1973.
7. R. L. Graves, "A Principal Pivoting Simplex Algorithm for Linear and Quadratic Programming," *Operations Research*, *15*, *3*, pp. 482–494, May-June 1967.
8. S. Karamardian, "The Complementarity Problem," *Mathematical Programming*, *2*, pp. 107–129, 1972.
9. W. Karush, "Minima of Functions of Several Variables with Inequalities as Side Conditions," M. S. dissertation, Department of Mathematics, University of Chicago, December 1939.
10. H. W. Kuhn, "Nonlinear Programming: A Historical View" in *Nonlinear Programming: Proceedings of the Joint SIAM-AMS Symposium on Applied Mathematics*, R. W. Cottle and C. E. Lemke (eds.), held in New York City March 1975, *9*, American Mathematical Society, Providence, R. I., 1976.

11. H. W. Kuhn and A. W. Tucker, "Nonlinear Programming," *Proceedings of Second Berkeley Symposium on Mathematical Statistics and Probability*, J. Neyman (editor) University of California Press, Berkeley, pp. 481–492, 1951.

12. C. E. Lemke, "Bimatrix Equilibrium Points and Mathematical Programming," *Management Science*, *11*, pp. 681–689, 1965.

13. C. E. Lemke, "Recent Results on Complementarity Problems," in J. B. Rosen, O. L. Managasarian, and K. Ritter (eds.), *Nonlinear Programming*, pp. 349–384, Academic Press, New York, 1970.

14. C. E. Lemke and J. T. Howson, "Equilibrium Points of Bimatrix Games," *SIAM Journal of Applied Mathematics*, *12*, pp. 413–423, 1964.

15. O. L. Managsarian, *Nonlinear Programming*, McGraw-Hill, New York, 1969.

16. K. G. Murty, "On a Characterization of P-matrices," *SIAM Journal on Applied Mathematics*, *20*, 3, pp. 378–384, May 1971.

17. K. G. Murty, "On the Number of Solutions of the Complementarity Problem and Spanning Properties of Complementarity Cones," *Linear Algebra and Its Applications*, *5*, pp. 65–108, 1972.

18. K. G. Murty, "Algorithm for finding all the Feasible Complementary Bases for a Linear Complementarity Problem," *Tech. Report No. 72-2*, Dept. I.O.E., University of Michigan, February 1972.

19. K. G. Murty, "On Two Related Classes of Complementary Cones," Technical Report 70-7, Department of Industrial and Operations Engineering, The University of Michigan, March 1970.

20. K. G. Murty, "On the Parametric Complementarity Problem," *Engineering Summer Conference Notes*, The University of Michigan, 1971.

21. K. G. Murty, "Note on a Bard-Type Scheme for Solving the Complementarity Problem," *Opsearch*, *11*, 2–3, pp. 123–130, June-September 1974.

22. A. Ravindran, "A Comparison of the Primal Simplex and Complementary Pivot Methods for Linear Programming," *Naval Research Logistics Quarterly*, *20*, 1, pp. 95–100, March 1973.

23. A. Ravindran, "A Computer Routine for Quadratic and Linear Programming Problems (H)," *Communications of ACM*, *15*, 9, pp. 818–820, September 1972.

24. R. Saigal, "A Note on a Special Linear Complementarity Problem," *Opsearch*, *7*, 3, pp. 175–183, September 1970.

25. R. Saigal, "On the Class of Complementary Cones and Lemke's Algorithm," *SIAM Journal on Applied Mathematics*, *23*, 46–60, 1972.

26. H. Samelson, R. M. Thrall, and O. Wesler, "A Partition Theorem for Euclidean *n*-Space," *Proceedings of American Mathematical Society*, *9*, pp. 805–807, 1958.

27. G. Zountendizk, *Methods of Feasible Directions*, Elsevier Publishing Co., Amsterdam, 1960.

Reference on Fixed Point Computing Methods

28. B. C. Eaves and R. Saigal, "Homotopies for Computation of Fixed Points on Unbounded Regions," *Mathematical Programming*, *3*, 225–237, 1972.

29. H. W. Kuhn, "Approximate Search for Fixed Points," in *Computing Methods in Optimization Problems*, *2*, Academic Press, New York, 1969.

30. O. H. Merrill, "A Summary of Techniques for Computing Fixed Points of Continuous Mappings," in *Mathematical Topics in Economic Theory and Computation*, *SIAM*, Philadelphia, Pa., 1972.

31. O. H. Merrill, "Applications and Extensions of an Algorithm That Computes Fixed Points of Upper Semicontinuous Point to Set Mappings," Ph.D. dissertation, Dept. I.O.E., The University of Michigan, 1972. Available from University Microfilms, 300 N. Zeeb Road, Ann Arbor, Michigan 48106, under number 73–6876.

32. H. Scarf, "The Approximation of Fixed Points of a Continuous Map," *SIAM Journal on Applied Mathematics*, *15*, pp. 1328–1343, 1967.

33. H. Scraf, *The Computation of Economic Equalibria*, Yale University Press, New Haven, 1973.

34. M. J. Todd, "Union Jack Triangulations," to appear in *Fixed Points: Algorithms and Applications*, S. Karamardian (editor), Academic Press, New York.

17 numerically stable forms of the simplex method

17.1 REVIEW

Consider the LP in standard form:

$$\begin{aligned} \text{Minimize} \quad & z(x) = cx \\ \text{Subject to} \quad & Ax = b \\ & x \geqq 0 \end{aligned} \tag{17.1}$$

where A is a matrix of order $m \times n$. In the process of solving this problem by the simplex algorithm, consider a step in which the basic vector is x_B. Let B be the corresponding basis, c_B the associated basic cost vector, and π_B the corresponding dual solution. The computations to be performed in this step are summarized below.

1. Compute the values of the basic variables in the basic solution of (17.1) corresponding to the basic vector x_B, by solving

$$Bx_B = b \tag{17.2}$$

In the notation of Chapters 2 and 5, the solution x_B of this system is denoted by $\bar{b}$. Normally (17.2) is not solved afresh in each step. The $\bar{b}$ vector in the present step is normally obtained by updating the $\bar{b}$ vector in the previous step. Hence either (17.2) has to be solved or the $\bar{b}$ vector has to be updated in each step.

2. Compute π_B by solving

$$\pi_B B = c_B \tag{17.3}$$

3. Compute $\bar{c}_j = c_j - \pi_B A_{.j}$ for each j.
4. If $\bar{c}_j \geqq 0$ for all j, the algorithm terminates in this step. Otherwise find an s such that $\bar{c}_s < 0$ and select x_s as the entering variable.

5. Compute $\bar{A}_{.s}$, the updated column vector of x_s, by solving:

$$B\bar{A}_{.s} = A_{.s} \tag{17.4}$$

6. If $\bar{A}_{.s} \leqq 0$, the algorithm terminates in this step. Otherwise, find an r satisfying

$$\frac{\bar{b}_r}{\bar{a}_{rs}} = \text{minimum}\left\{\frac{\bar{b}_i}{\bar{a}_{is}}: i \text{ such that } \bar{a}_{is} > 0\right\}$$

Then drop the rth column vector in the basis B and include $A_{.s}$ in its place. This gives the new basis. Then go to the next step, in which all these computations are repeated with the new basis.

In the revised simplex algorithm all these computations are performed using B^{-1} either explicitly or as a product of a sequence of pivot matrices. The major drawback in this approach is that roundoff errors accumulate as the algorithm moves from step to step. After several steps, the actual basis inverse may be quite different from the explicit inverse in storage now or the product of the sequence of pivot matrices in storage now. To reduce the effect of these roundoff errors, it is often recommended that periodically, after a certain number of steps of the algorithm, the basis should be identified from the original tableau and its inverse actually computed afresh, or the pivot matrices whose product is the basis inverse should be recomputed by pivoting on the basis. The old basis inverse or the old sequence of pivot matrices are discarded and are replaced by the freshly computed ones. Even with such periodic reinversion, roundoff error accumulation poses a serious problem, undermining the ability of the simplex algorithm to produce accurate answers.

There are two different methods for carrying out the computations in the simplex algorithm that have been proposed to alleviate this drawback. These are (1) the *LU* Decomposition and (2) the Cholesky factorization. Here we discuss these methods briefly.

17.2 THE *LU* DECOMPOSITION:

Let B be the basis under consideration. This method generates an upper triangular matrix U, a sequence of lower triangular pivot matrices $\Gamma_1, \Gamma_2, \ldots, \Gamma_k$ and a sequence of permutation matrices $Q_1, \ldots, Q_k$ (some of these permutation matrices may be equal to the unit matrix) satisfying the equation

$$\Gamma_k Q_k \cdots \Gamma_2 Q_2 \Gamma_1 Q_1 B = U. \tag{17.5}$$

The name '*LU decomposition*' stems from the customary practice of denoting the product $\Gamma_k Q_k \cdots \Gamma_1 Q_1$ in the decomposition, by the symbol L^{-1}. If all the permutation matrices $Q_k, \ldots, Q_1$, are equal to the unit matrix, L^{-1} will be a lower triangular matrix, but in general, in the decomposition of a basis obtained in the simplex algorithm after one or more basic variable changes, L^{-1} may not be lower triangular, or even triangular. L^{-1} is never computed explicitly. It is always stored in product form by storing the pivot matrices and the permutation matrices in their proper order. The diagonal entries of all the pivot matrices in the decomposition will always be equal to one. Since each pivot matrix is also lower triangular, it can be stored in a very compact fashion. Clearly, it is only necessary to store the permutation matrices which differ from the unit matrix.

First we discuss how to perform all the computations in the simplex method using such a decomposition, before discussing methods for obtaining and updating the decomposition.

17.2.1 How to Perform the Computations Using the *LU* Decomposition

Suppose you have to solve a system of equations like

$$By = d \tag{17.6}$$

for y. Let the basis B have the decomposition as in (17.5). First compute

$$t = \Gamma_k Q_k \cdots \Gamma_1 Q_1 d. \tag{17.7}$$

Since each Q_i is a permutation matrix and each Γ_i is a lower triangular pivot matrix, this multiplication can be carried out very efficiently from the right by including one new matrix in the product on the left at a time, until t is obtained. After the vector t is computed, solve the system

$$Uy = t \tag{17.8}$$

for y, by backsubstitution.

Both systems (17.2) and (17.4) are of the form (17.6), hence they can be solved as above using the decomposition (17.5). To solve (17.3), first find the vector h from

$$hU = c_B \tag{17.9}$$

by backsubstitution. Then find the vector π_B using the equation

$$\pi_B = h\Gamma_k Q_k \cdots \Gamma_1 Q_1. \tag{17.10}$$

where the multiplication in (17.10) can be carried out very conveniently by starting from the left and including one matrix in the product on the right hand side at a time until π_B is obtained.

17.2.2 How to Obtain the *LU* Decomposition for a Given Basis

Let the basis be B. When the following steps 1 to $m - 1$ are carried out on B, it gets transformed into U.

In step 1, the present matrix is $B^{(0)} = B$. At the end of the rth step suppose B is transformed into $B^{(r)}$, $r = 1$ to $m - 1$.

Step r: At the beginning of this step, the present matrix is $B^{(r-1)} = (b_{ij}^{(r-1)})$. Find i_1 such that $|b_{i_1,r}^{(r-1)}| = \text{maximum } \{|b_{i,r}^{(r-1)}|: i = r \text{ to } m\}$.

Interchange the rth row in the present matrix with row i_1. This operation is equivalent to multiplying the present matrix, $B^{(r-1)}$, on the left by Q_r, which is the identity matrix with row r and row i_1 interchanged. After the interchange, add suitable multiples of row r to row i to transform the entry in row i and column r to zero for each $i \geqq r + 1$. This is known as *Gaussian Elimination* in linear algebra. This is equivalent to multiplying the present matrix on the left by an elementary matrix of the form given in Figure 17.1, where $|g_{i,r}| \leqq 1$ for each $i \geqq r + 1$ (after multiplying $B^{(r-1)}$ on the left by Q_r, let the rth column in the resulting matrix be $(\bar{a}_{1r}, \ldots, \bar{a}_{mr})$. Then $g_{ir} = -\bar{a}_{ir}/\bar{a}_{rr}$, for $i = r + 1$ to m). Since we are only reducing the rth column of $B^{(r-1)}$ to upper triangular form, the matrix Γ_r is not a pivot matrix in the sense of chapter 3 (a pivot matrix in the sense of chapter 3 would reduce this column to the rth column vector of the unit matrix). However Γ_r has the property that it differs from the unit matrix in just one column, like the pivot matrices of chapter 3, and we will refer to matrices like Γ_r also as pivot matrices.

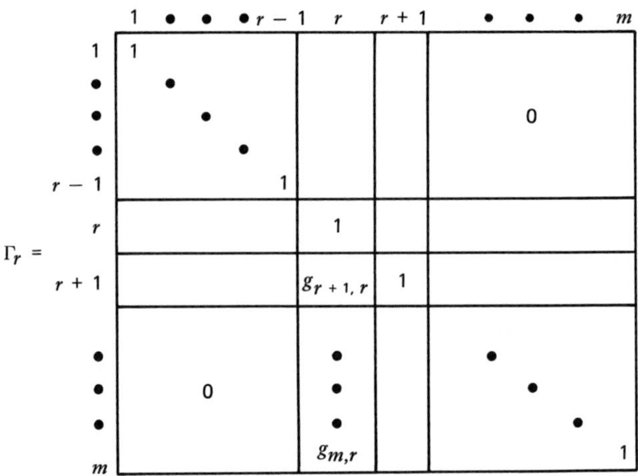

Figure 17.1

Then go to step $r + 1$ if $r < m - 1$.

At the end of step $m - 1$, the matrix is transformed into the upper triangular matrix U. Also, from the manner in which these matrices are generated, it is clear that

$$\Gamma_{m-1} Q_{m-1} \cdots \Gamma_1 Q_1 B = U$$

which is the decomposition required for the basis B. Since each of these matrices $\Gamma_{m-1}, \ldots, \Gamma_1, Q_{m-1}, \ldots, Q_1$, is a matrix of special structure, it can be stored in a compact manner. U and the lower triangular pivot matrices $\Gamma_1, \ldots, \Gamma_{m-1}$, can be stored in the space required for storing B. Additional locations may be required for storing the permutation matrices.

17.2.3 Updating the *LU* Decomposition When One Basic Column Vector Changes

Let the present basis be

$$B = (A_{.j_1}, \ldots, A_{.j_m})$$

Let the present decomposition be

$$\Gamma_t Q_t \cdots \Gamma_1 Q_1 B = U$$

In this pivot step suppose $A_{.s}$ is the entering column and suppose that $A_{.j_r}$ is the leaving basic column. In updating the decomposition, the entering column is always treated as the rightmost column in the new basis. All the columns in B to the right of the leaving column are recorded in the same order in which they appear in B, but they are shifted by one position to the left, that is, the new basis is recorded as

$$\hat{B} = (A_{.j_1}, \ldots, A_{.j_{r-1}}, A_{.j_{r+1}}, \ldots, A_{.j_m}, A_{.s})$$

Let $f = \Gamma_t Q_t \cdots \Gamma_1 Q_1 A_{.s}$. Notice that f is obtained as a by-product when (17.4) is solved by the method discussed in section 17.2.1. Hence the vector f is obtained without doing any additional computation. Then

$$\begin{aligned} \Gamma_t Q_t \cdots \Gamma_1 Q_1 \hat{B} &= (U_{.1}, \ldots, U_{.r-1}, U_{.r+1}, \ldots, U_{.m}, f) \\ &= \hat{H} \end{aligned}$$

where the $U_{.j}$ is the jth column vector of U. $\hat{H}$ is a matrix with zeros below the diagonal in the first $r - 1$ columns and zeros in rows that are two or more rows below the diagonal in columns r to m.

A matrix like this is known as an *upper Hessenberg matrix*. See Figure 17.2. $\hat{H}$ can be transformed into an upper triangular matrix by doing row operations as in section 17.2.2 to zero out all the elements just below the diagonal in columns r to m. This is achieved by performing the following operations for $k = r$ to $m - 1$ in that order.

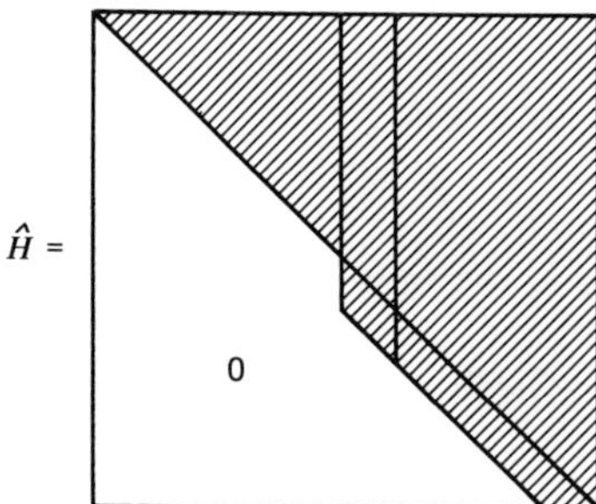

Figure 17.2

1. Let the diagonal entry in column k of the present matrix be h_{kk} and let the entry just below it be $h_{k+1,k}$.
2. Since the previous matrix U was nonsingular, $h_{k+1,k} \neq 0$. Find out which of $|h_{k,k}|$ and $|h_{k+1,k}|$ is larger. If $|h_{k+1,k}| > |h_{k,k}|$, interchange rows k and $k + 1$ of the present matrix. This is equivalent to multiplying the present matrix on the left by $Q_k^{(1)}$, where $Q_k^{(1)}$ is the identity matrix with its rows k and $k + 1$ interchanged. If $|h_{k,k}| \geqq |h_{k+1,k}|$, let $Q_k^{(1)} = I$. Now subtract a suitable multiple of the present kth row from the $k + 1$th row to transform the element in the present $k + 1$th row and column k into zero. This is equivalent to multiplying the present matrix on the left by an elementary matrix of the form given in Figure 17.3, where $g_k^{(1)} = -\bar{h}_{k+1,k}/\bar{h}_{k,k}$, if the kth and $k + 1$th

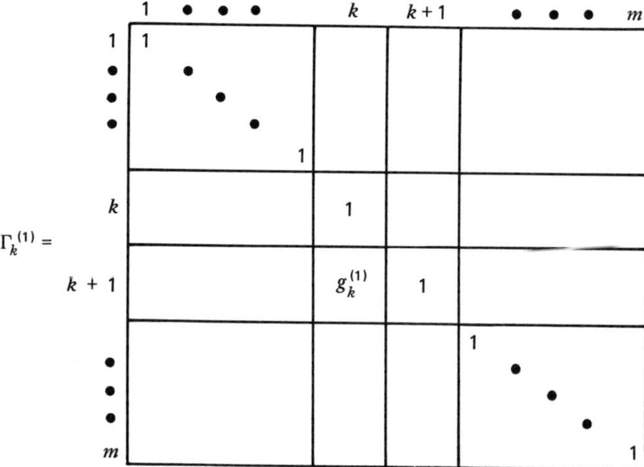

Figure 17.3

entries in column k of the present matrix, after the interchange if necessary, are $\bar{h}_{k,k}$ and $\bar{h}_{k+1,k}$ respectively.

After these operations are performed for $k = r + 1$ to $m - 1$, the matrix $\hat{H}$ is transformed into an upper triangular matrix, $\hat{U}$. The decomposition for the new basis $\hat{B}$ is

$$\Gamma_{m-1}^{(1)} Q_{m-1}^{(1)} \cdots \Gamma_r^{(1)} Q_r^{(1)} \Gamma_t Q_t \cdots \Gamma_1 Q_1 \hat{B} = \hat{U}$$

The simplex algorithm can now be continued using the decomposition of $\hat{B}$.

If $\tilde{B}$ is the basis under consideration at some stage of the simplex algorithm, the implementation computes the upper triangular matrix $\tilde{U}$, in the decomposition of $\tilde{B}$, explicitly. In each pivot step of the simplex algorithm, several lower triangular pivot matrices like $\Gamma_j^{(k)}$, and permutation matrices $Q_j^{(k)}$ are generated. All these matrices are stored in the order in which they are generated. All the computations needed in this stage of the simplex algorithm are carried out using $\tilde{U}$, and the lower triangular pivot matrices and the permutation matrices generated so far, as in section 17.2.1.

The Simplex method using the *LU* decomposition implemented in this manner is called the Simplex method using the *Elimination Form of the Inverse*.

Numerical Example

Here we will illustrate how to obtain the *LU* decomposition of a given basis, and how to update the decomposition when a column vector in this basis is changed. Let the initial basis be

$$B = \begin{pmatrix} 1 & -1 & 1 & 2 \\ 2 & 0 & 0 & 1 \\ 1 & 1 & 0 & 1 \\ -1 & 2 & 2 & 1 \end{pmatrix}$$

Going to step 1 of section 17.2.2, let $B^{(0)} = B$. Rows 1 and 2 have to be interchanged. This leads to

$$Q_1 = \begin{pmatrix} 0 & 1 & 0 & 0 \\ 1 & 0 & 0 & 0 \\ 0 & 0 & 1 & 0 \\ 0 & 0 & 0 & 1 \end{pmatrix}, \quad \Gamma_1 = \begin{bmatrix} 1 & 0 & 0 & 0 \\ -\frac{1}{2} & 1 & 0 & 0 \\ -\frac{1}{2} & 0 & 1 & 0 \\ \frac{1}{2} & 0 & 0 & 1 \end{bmatrix}.$$

$$B^{(1)} = \Gamma_1 Q_1 B^{(0)} = \begin{bmatrix} 2 & 0 & 0 & 1 \\ 0 & -1 & 1 & \frac{3}{2} \\ 0 & 1 & 0 & \frac{1}{2} \\ 0 & 2 & 2 & \frac{3}{2} \end{bmatrix}$$

Rows 2 and 4 of $B^{(1)}$ have to be interchanged. The matrices generated in step 2 are

$$Q_2 = \begin{pmatrix} 1 & 0 & 0 & 0 \\ 0 & 0 & 0 & 1 \\ 0 & 0 & 1 & 0 \\ 0 & 1 & 0 & 0 \end{pmatrix}, \quad \Gamma_2 = \begin{bmatrix} 1 & 0 & 0 & 0 \\ 0 & 1 & 0 & 0 \\ 0 & -\frac{1}{2} & 1 & 0 \\ 0 & \frac{1}{2} & 0 & 1 \end{bmatrix}$$

$$B^{(2)} = \Gamma_2 Q_2 B^{(1)} = \begin{bmatrix} 2 & 0 & 0 & 1 \\ 0 & 2 & 2 & \frac{3}{2} \\ 0 & 0 & -1 & -\frac{1}{4} \\ 0 & 0 & 2 & \frac{9}{4} \end{bmatrix}.$$

In step 3, we have to interchange rows 3 and 4 in $B^{(2)}$. The matrices generated are

$$Q_3 = \begin{pmatrix} 1 & 0 & 0 & 0 \\ 0 & 1 & 0 & 0 \\ 0 & 0 & 0 & 1 \\ 0 & 0 & 1 & 0 \end{pmatrix}, \qquad \Gamma_3 = \begin{bmatrix} 1 & 0 & 0 & 0 \\ 0 & 1 & 0 & 0 \\ 0 & 0 & 1 & 0 \\ 0 & 0 & \frac{1}{2} & 1 \end{bmatrix}$$

$$U = B^{(3)} = \Gamma_3 Q_3 B^{(2)} = \begin{bmatrix} 2 & 0 & 0 & 1 \\ 0 & 2 & 2 & \frac{3}{2} \\ 0 & 0 & 2 & \frac{9}{4} \\ 0 & 0 & 0 & \frac{7}{8} \end{bmatrix}.$$

The decomposition for B is

$$\Gamma_3 Q_3 \Gamma_2 Q_2 \Gamma_1 Q_1 B = U$$

where the various matrices are given above. Here, for illustration, all the matrices are written out in full. In the implementation, they will be stored in a compact fashion, as discussed earlier.

Now suppose a new basis $\hat{B}$ is obtained by introducing the column vector $A_{.5} = (0, -2, 0, 4)^T$ into B, and dropping the second column vector of B. We compute

$$f = \Gamma_3 Q_3 \Gamma_2 Q_2 \Gamma_1 Q_1 A_{.5} = \left(-2, 3, \frac{5}{2}, \frac{3}{4}\right)^T$$

So

$$\hat{H} = \begin{bmatrix} 2 & 0 & 1 & -2 \\ 0 & 2 & \frac{3}{2} & 3 \\ 0 & 2 & \frac{9}{4} & \frac{5}{2} \\ 0 & 0 & \frac{7}{8} & \frac{3}{4} \end{bmatrix}$$

We have to reduce columns 2 to 4 of $\hat{H}$ to upper triangular form. We begin by reducing column 2 first, as in section 17.2.3. No interchange of rows is necessary.

$$Q_2^{(1)} = I, \qquad \Gamma_2^{(1)} = \begin{pmatrix} 1 & 0 & 0 & 0 \\ 0 & 1 & 0 & 0 \\ 0 & -1 & 1 & 0 \\ 0 & 0 & 0 & 1 \end{pmatrix}$$

$$\Gamma_2^{(1)} Q_2^{(1)} \hat{H} = \begin{bmatrix} 2 & 0 & 1 & -2 \\ 0 & 2 & \frac{3}{2} & 3 \\ 0 & 0 & \frac{3}{4} & -\frac{1}{2} \\ 0 & 0 & \frac{7}{8} & \frac{3}{4} \end{bmatrix}$$

Now we have to reduce column 3 of $\Gamma_2^{(1)}Q_2^{(1)}\hat{H}$ to upper triangular form. Rows 3 and 4 of this matrix are interchanged.

$$Q_2^{(2)} = \begin{pmatrix} 1 & 0 & 0 & 0 \\ 0 & 1 & 0 & 0 \\ 0 & 0 & 0 & 1 \\ 0 & 0 & 1 & 0 \end{pmatrix} \qquad \Gamma_2^{(2)} = \begin{bmatrix} 1 & 0 & 0 & 0 \\ 0 & 1 & 0 & 0 \\ 0 & 0 & 1 & 0 \\ 0 & 0 & -\frac{6}{7} & 1 \end{bmatrix}$$

$$\hat{U} = \Gamma_2^{(2)}Q_2^{(2)}\Gamma_2^{(1)}Q_2^{(1)}\hat{H} = \begin{bmatrix} 2 & 0 & 1 & -2 \\ 0 & 2 & \frac{3}{2} & 3 \\ 0 & 0 & \frac{7}{8} & \frac{3}{4} \\ 0 & 0 & 0 & -\frac{8}{7} \end{bmatrix}$$

The decomposition for the new basis $\hat{B}$ is

$$\Gamma_2^{(2)}Q_2^{(2)}\Gamma_2^{(1)}Q_2^{(1)}\Gamma_3Q_3\Gamma_2Q_2\Gamma_1Q_1\hat{B} = \hat{U}$$

where the various matrices in the decomposition are obtained above.

17.2.4 Advantages

The LU implementation provides much better numerical accuracy in the results obtained in the simplex algorithm compared to the implementation using either the explicit inverse or the product form of the inverse. Also, it helps to preserve the *sparsity* in the basis, that is, the number of nonzero entries in U and the lower triangular pivot matrices generated, is not expected to be much larger than the number of nonzero entries in B. The actual inverse of B does not normally preserve sparsity. To keep the discussion brief, we will not go into the reasons for these facts; the interested reader should look up the references listed at the end of the chapter.

The LU decomposition can also be periodically re-evaluated using the method in section 17.2.2 for obtaining even better numerical accuracy.

If the simplex algorithm begins with the unit matrix as the basis, U is I initially, and there are no lower triangular pivot matrices or permutation matrices in the beginning step. If the algorithm begins with a known basis, B, different from I, its LU decomposition is obtained as in section 17.2.2 and the algorithm continued from there. If B is sparse, the rows and columns of B can be permuted to make it as close to either upper or lower triangular form as possible (before obtaining the LU decomposition) in order to preserve sparsity in the LU decomposition. See references [6, 7, 8 and 12] for generating these row and column permutations to preserve sparsity.

17.3 THE CHOLESKY FACTORIZATION

Let B be a basis. Then BB^{T} is a positive definite matrix. It is a well-known result in linear algebra that there exists a lower triangular matrix L such that

$$BB^{\mathrm{T}} = LL^{\mathrm{T}} \tag{17.11}$$

L is known as the *Cholesky factor of* BB^{T}. We will refer to L as the *Cholesky factor associated with the basis B.*

A square matrix Q is said to be an *orthogonal* matrix if $QQ^{\mathrm{T}} = I$. If B is a nonsingular matrix, it is well-known that there exists an orthogonal matrix Q such that

$$QB^{\mathrm{T}} = R$$

where R is a nonsingular upper triangular matrix. $R^{\mathrm{T}} = L$ is the Cholesky factor associated with B. The orthogonal matrix Q is only used in the discussion of the method; it is neither stored nor updated at any stage of the algorithm.

We will first discuss how to perform all the computations in the simplex algorithm using the Cholesky factors before discussing methods for obtaining and updating the Cholesky factors.

17.3.1 How to Perform the Computations in the Simplex Algorithm Using Cholesky Factors

Let B be the present basis and let L be the Cholesky factor associated with it. To solve a system of equations of the form (17.6), first find v from

$$Lv = d \tag{17.12}$$

by back substitution, and then solve for u in

$$L^{\mathrm{T}}u = v \tag{17.13}$$

again by back substitution. Then $y = B^{\mathrm{T}}u$ is the solution of (17.6).

To solve the system (17.3) first solve for w from

$$wL^{\mathrm{T}} = c_B B^{\mathrm{T}} \tag{17.14}$$

and then solve for π_B from

$$\pi_B L = w \tag{17.15}$$

Since L is lower triangular, (17.14) and (17.15) are both solvable by back substitution. Hence all the computations in a step of the simplex algorithm can be performed as here, by using the Cholesky factor associated with the basis.

17.3.2 To Obtain the Cholesky Factor Associated with a Given Basis

We discuss two commonly used methods for doing this.

Method 1: Using Givens Matrices

Let $a = (a_1, \ldots, a_m)^{\mathrm{T}}$ be a given column vector. Consider the element a_i. For some $j \neq i$, if there is an element $a_j \neq 0$, it can be reduced to zero by multiplying the column vector a on the left by an orthogonal matrix P_j^i as in Figure 17.4, where $\gamma = \sqrt{a_i^2 + a_j^2}$, $c = a_i/\gamma$ and $s = a_j/\gamma$. Then

$$P_j^i a = (a_1, \ldots, a_{i-1}, \gamma, a_{i+1}, \ldots, a_{j-1}, 0, a_{j+1}, \ldots, a_m)^{\mathrm{T}}$$

A matrix like P_j^i is known as a *Givens matrix*. It is a symmetric orthogonal matrix with a very special structure.

To find the Cholesky factor associated with a given basis, B, start with the matrix B^{T}. By multiplying the matrix on the left by a sequence of Givens matrices, of the form P_j^1 transform all the entries in column 1 below the diagonal into zero. Now by multiplying the resulting matrix on the left by a sequence of Givens matrices of the form P_j^2 transform all the entries in column 2 below the diagonal into zero. Then go to column 3. In a similar manner, repeat the process until the matrix is transformed into an upper triangular matrix. The transpose of the final upper triangular matrix is the Cholesky factor associated with the basis B.

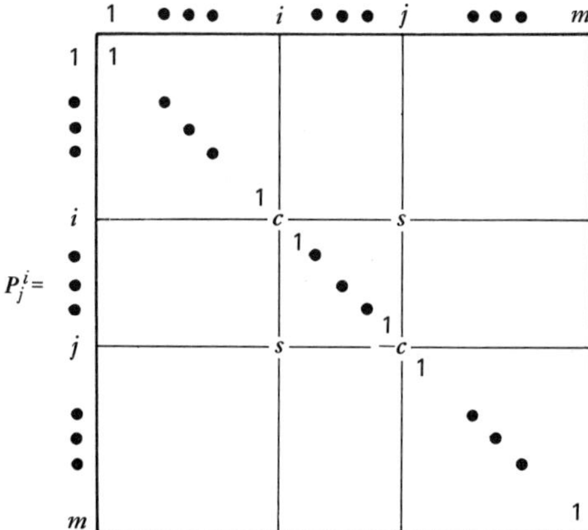

Figure 17.4

Method 2: Using Householder Matrices

Let $a = (a_1, a_2, \ldots, a_m)^T$ be a given column vector. Suppose $(a_{\ell+2}, \ldots, a_m) \neq 0$. Then the bottom $m - \ell - 1$ elements of a can be reduced to zero by multiplying it on the left by a matrix P_ℓ as in Figure 17.5, where,

$$
\begin{aligned}
u &= (u_1 u_2, \ldots, u_{m-\ell})^T \\
u_1 &= a_{\ell+1} + s \\
u_j &= a_{\ell+j}, \text{ for } j = 2, \ldots, m - \ell. \\
s &= f(a_{\ell+1})\sqrt{a_{\ell+1}^2 + \cdots + a_m{}^2} \\
\gamma &= 1/(u_1 s)
\end{aligned}
$$

$$
f(a_{\ell+1}) = \begin{cases} 1 & \text{if } a_{\ell+1} > 0 \\ -1 & \text{if } a_{\ell+1} \leqq 0. \end{cases}
$$

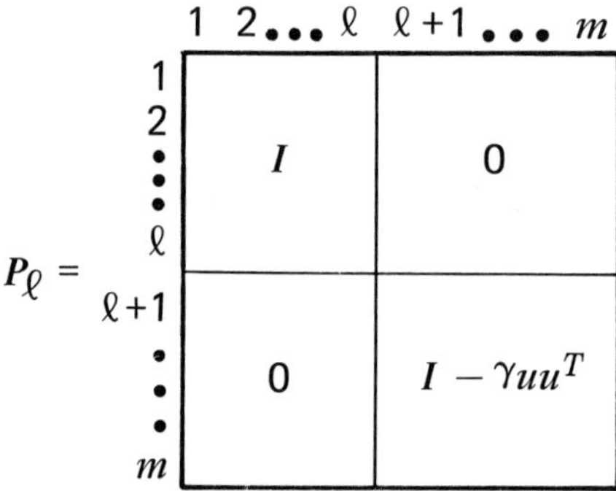

Figure 17.5

A matrix like P_ℓ is known as a *Householder matrix*. It is a symmetric orthogonal matrix with a special structure. Then

$$P_\ell a = (a_1, \ldots, a_\ell, -s, 0, \ldots, 0)^T$$

To find the Cholesky factor associated with the basis B, start with the matrix $B^{(0)} = B^T$. Go through steps $\ell = 0, 1, \ldots, m - 2$, as discussed below.

Step ℓ: If $\ell = 0$, the present matrix is $B^{(0)}$. If $\ell \geqq 1$, suppose this matrix was transformed into $B^{(\ell)}$ at the end of the previous step. Transform the bottom $m - \ell - 1$ entries in the $\ell + 1$th column of $B^{(\ell)}$ into zeros, by multiplying $B^{(\ell)}$ on the left by a Householder matrix of the form P_ℓ. Let $B^{(\ell+1)} = P_\ell B^{(\ell)}$. If $\ell = m - 2$, terminate. Otherwise go to the next step.

The matrix $B^{(m-1)}$ at the end of step $m - 2$, is uppertriangular. Its transpose is the Cholesky factor associated with the basis B.

Justification: It is well known that a product of orthogonal matrices is an orthogonal matrix. In both these methods, B^{T} was transformed into an upper triangular matrix, R, by multiplying B^{T} on the left by a sequence of orthogonal matrices. Let Q be the product (in the proper order) of these orthogonal matrices. So Q is an orthogonal matrix itself. So

$$
\begin{aligned}
QB^{\mathrm{T}} &= R. \\
\therefore BQ^{\mathrm{T}}QB^{\mathrm{T}} &= R^{\mathrm{T}}R \\
&= BB^{\mathrm{T}}
\end{aligned}
$$

Since R is uppertriangular, R^{T} is lower triangular, and this shows that it is the Cholesky factor associated with B.

17.3.3 Updating the Cholesky Factor When One Basic Column Vector Changes

Let B be the present basis and L be its Cholesky factor. Suppose $A_{.s}$ is the entering column $A_{.r}$ is the leaving column. Let $\hat{B}$ be the new basis. Then

$$\hat{B}\hat{B}^{\mathrm{T}} = BB^{\mathrm{T}} - A_{.r}A_{.r}^{\mathrm{T}} + A_{.s}A_{.s}^{\mathrm{T}}$$

Each of the matrices $A_{.r}A_{.r}^{\mathrm{T}}$ and $A_{.s}A_{.s}^{\mathrm{T}}$ are matrices of rank one. Hence $\hat{B}\hat{B}^{\mathrm{T}}$ is obtained from BB^{T} by making two changes of rank one each. The Cholesky factor of the new basis is obtained by updating the present Cholesky factor in two stages.

Stage 1 Here the Cholesky factor of the matrix obtained by deleting $A_{.r}$ from B is computed. Let $R = L^{\mathrm{T}}$ be the transpose of the Cholesky factor of B. Find p by solving

$$R^{\mathrm{T}}p = A_{.r} \tag{17.16}$$

by back substitution. Let F be the matrix as in Figure 17.6. F is a square matrix of order $m + 1$. By multiplying F on the left by a sequence of Givens matrices

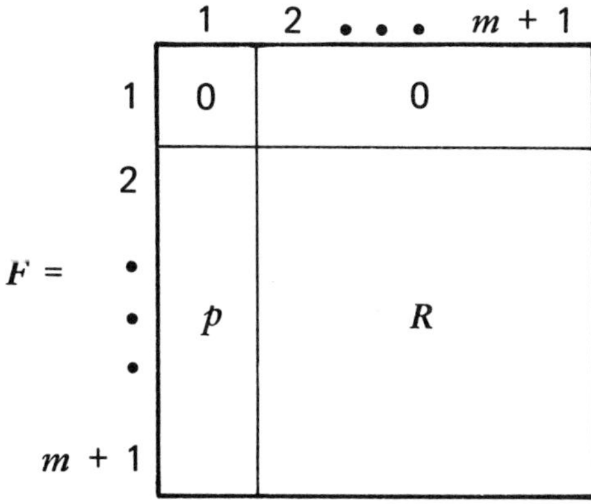

Figure 17.6

(discussed in section 17.3.2) of the form $P_j^{\,1}$, $j = m + 1, m, \ldots, 2$, in that specific order, transform all the entries in column 1, rows $m + 1$ to 2 of F into zeros. When the operations are carried out in this specific order, the first row of F gets filled up from right to left. R is modified row by row, but retains its upper triangular structure. Notice that the Givens matrices used here are all of order $m + 1$. At the end, suppose F gets transformed into

$$\begin{pmatrix} \delta & \vdots & v \\ \cdots & \cdots & \cdots \\ 0 & \vdots & \bar{R} \end{pmatrix}$$

The number δ here should be equal to 1. Furthermore, the vector v should be $A_{.r}{}^{\mathrm{T}}$. Then $\bar{R}^{\mathrm{T}}$ is the Cholesky factor of the matrix obtained after the column $A_{.r}$ has been deleted from B. Hence $\bar{R}$ is singular.

Stage 2 Here the Cholesky factor, $\bar{R}$, obtained at the end of stage 1 is modified into $\hat{R}$, the Cholesky factor of $\hat{B}$. Let E be the $(m + 1) \times m$ matrix

$$E = \begin{pmatrix} \bar{R} \\ \cdots \\ A_{.s}{}^{\mathrm{T}} \end{pmatrix}$$

By multiplying E on the left by a sequence of Givens matrices (each of order $m + 1$) of the form P_{m+1}^{i}, in the order $i = 1$ to m, transform all the entries in the last row of E into zero. At the end suppose E is transformed into

$$\begin{pmatrix} \hat{R} \\ \cdots \\ 0 \end{pmatrix}.$$

$\hat{R}$ is nonsingular and upper triangular. $\hat{R}$ is the transpose of the Cholesky factor associated with the new basis $\hat{B}$.

Justification for Stage 1 Since R^{T} is the Cholesky factor associated with the basis B, there exists an orthogonal matrix Q such that

$$QB^{\mathrm{T}} = R$$

Since Q is an orthogonal matrix, $Q^{\mathrm{T}} = Q^{-1}$. By the above equation, this implies that $R^{\mathrm{T}}Q = B$. Assume that $A_{.r}$, the dropping column, is the ℓth column of B. From equation (17.16) and the fact that $R^{\mathrm{T}}Q = B$, we conclude that $p = Q_{.\ell}$, the ℓth column of this orthogonal matrix Q. Since Q is orthogonal, $(Q_{.\ell})^{\mathrm{T}}Q_{.\ell} = 1$, and this implies that $p^{\mathrm{T}}p = 1$. In this stage, F is transformed by multiplying it on the left by an orthogonal matrix, say, $\bar{Q}$. So

$$\bar{Q}\begin{pmatrix} 0 & \vdots & 0 \\ \cdots & \cdots & \cdots \\ p & \vdots & R \end{pmatrix} = \begin{pmatrix} \delta & \vdots & v \\ \cdots & \cdots & \cdots \\ 0 & \vdots & \bar{R} \end{pmatrix}$$

Since $\bar{Q}^{\mathrm{T}}\bar{Q} = I$, we have

$$\begin{pmatrix} 0 & \vdots & p^{\mathrm{T}} \\ \cdots & \cdots & \cdots \\ 0 & \vdots & R^{\mathrm{T}} \end{pmatrix}\begin{pmatrix} 0 & \vdots & 0 \\ \cdots & \cdots & \cdots \\ p & \vdots & R \end{pmatrix} = \begin{pmatrix} \delta & \vdots & 0 \\ \cdots & \cdots & \cdots \\ v^{\mathrm{T}} & \vdots & \bar{R}^{\mathrm{T}} \end{pmatrix}\begin{pmatrix} \delta & \vdots & v \\ \cdots & \cdots & \cdots \\ 0 & \vdots & \bar{R} \end{pmatrix}$$

By writing down the products of these partitioned matrices on both sides, and equating corresponding terms, we conclude that

$$\delta^2 = p^{\mathrm{T}}p = 1.$$

From section 17.3.2, this implies that $\delta = 1$. From the above matrix product, we have

$$p^{\mathrm{T}}R = \delta v = v$$

and hence, $v^T = A_{.r}$, the dropping column. Also

$$R^TR = v^Tv + \bar{R}^T\bar{R}$$
$$\bar{R}^T\bar{R} = R^TR - v^Tv$$
$$= BB^T - A_{.r}A_{.r}^T$$

Hence $\bar{R}^T$ is the Cholesky factor of the matrix obtained after the rank one correction in stage 1.

The justification for stage 2 is based on a similar argument, and it is left to the reader to figure out.

This method of updating is discussed in Saunders [10] and it is based on the fact that each Givens matrix used in the updating process is an orthogonal matrix.

There are several other methods for updating the factorization. In some of these methods, the factorization itself is expressed as

$$BB^T = LDL^T$$

where D is a positive diagonal matrix and L is unit lower triangular, in order to minimize the number of square roots and divisions per iteration. The methods discussed here have been chosen because they are efficient and easy to understand, and because they illustrate the main features of the algorithm using these factorizations, in a simple manner. See references listed at the end of the chapter for other methods of implementing the factorization.

The simplex algorithm can begin with a unit basis, by taking $L = I$ initially. The algorithm can also begin with an arbitrary known basis by computing its Cholesky factor as in section 17.3.2. After each step the Cholesky factor of the new basis is obtained by updating in two stages as above.

Cholesky factorization is numerically stable, but it does not preserve sparsity as well as LU decomposition.

17.4 REINVERSIONS

Even under these implementations, the roundoff errors accumulate from step to step. At regular intervals (e.g., every 25 or 50 steps) the quantities $\|b - Bx_B\|$ and $\|c_B - \pi_B B\|$ should be computed. If both these numbers are smaller than a specified tolerance (e.g., 10^{-6} or 10^{-7}), the algorithm is continued with the existing LU decomposition or the Cholesky factor. Otherwise the basis corresponding to the present basic vector is read off from the original tableau and its LU decomposition or the Cholesky factor associated with it is computed afresh as in section 17.2.2 or section 17.3.2. The present LU decomposition or the Cholesky factor in storage now is replaced by the freshly computed one and the algorithm continued as before. This operation is called *reinversion.* Also in the method using Cholesky factors, reinversion is performed if the quantities $|\delta^2 - 1|$ and $\|A_{.r}^T - v\|$ (obtained during stage 1 of updating the Cholesky factor in any step) are not smaller than specified tolerances.

To keep the presentation brief, we have not discussed methods for permuting the rows and columns of the basis at each "reinversion" for preserving sparsity, for stability and for keeping the computational effort to a minimum. See references [3, 6, 7, 8, 12, and 13] for these methods, and for methods for preprocessing the input-output coefficient matrix A in the problem before applying the simplex method on it, to take advantage of the sparsity of A.

17.5 USE IN COMPUTER CODES:

The application of *LU* implementation and Cholesky factorization to linear programs discussed in this chapter has been developed recently. They may be used in modern Mathematical Programming Systems in the near future.

REFERENCES

1. R. H. Bartels, "A Stabilization of the Simplex Method," *Numerische Mathematik*, *16*, pp. 414–434, 1971.
2. R. H. Bartels, G. Golub, and M. A. Saunders, "Numerical Techniques in Mathematical Programming," pp. 123–176 in *Nonlinear Programming*, J. B. Rosen, O. L. Mangasarian and K. Ritter (Editors), Academic Press, New York, 1970.
3. J. J. H. Forrest and J. A. Tomlin, "Updating Triangular Factors of the Basis to Maintain Sparsity in the Product Form Simplex Method," *Mathematical Programming*, 2, pp. 263–278, 1972.
4. P. E. Gill, G. H. Golub, W. Murray, and M. A. Saunders, "Methods for Modifying Matrix Factorizations," *Mathematics of Computation*, 28, pp. 505–535, 1974.
5. P. E. Gill and W. Murray, "A Numerically Stable Form of the Simplex Algorithm," *Journal of Linear Algebra and Its Applications*, *6*, 1973.
6. P. E. Gill and W. Murray, (Editors), *Numerical Methods for Constrained Optimization*, Academic Press, London, 1974.
7. E. Hellerman and D. Rarick, "Reinversion with the Pre-assigned Pivot Procedure," *Mathematical Programming*, 1, pp. 195–216, 1971.
8. E. Hellerman and D. Rarick, "The Partitioned Pre-assigned Pivot Procedure (p4)," pp. 67–76 in *Sparse Matrices and Their Applications*, by D. J. Rose and R. A. Willoughby (Editors), Plenum Press, New York, 1972.
9. W. Orchard-Hays, *Advanced Linear Programming Computing Techniques*, McGraw-Hill, New York, 1968.
10. M. A. Saunders, "Large-Scale Linear Programming Using the Cholesky Factorization," Stan-cs-72-252, Computer Sciences Department, Stanford University, January 1972.
11. M. A. Saunders, "Product Form of the Cholesky Factorization for Large Scale Linear Programming," Stan-cs-72-301, Computer Sciences Department, Stanford University, August 1972.
12. M. A. Saunders, "A Fast, Stable Implementation of the Simplex Method Using Bartels-Golub Updating," Tech. Report SOL 75–25, Systems Optimization Laboratory, Department of Operations Research, Stanford University, Stanford, California, September, 1975.
13. J. A. Tomlin, "Modifying Triangular Factors of the Basis in the Simplex Method," pp. 77–85 in *Sparse Matrices and Their Applications*, D. J. Rose and R. A. Willoughby (Editors), Plenum Press, New York, 1972.
14. J. H. Wilkinson, *The Algebraic Eigenvalue Problem*, Clarendon Press, Oxford, 1965.

18 computational efficiency

18.1 HOW EFFICIENT IS THE SIMPLEX ALGORITHM?

Since the development of the simplex algorithm, one of the main questions studied in the theory of linear programming is the following. Given a general LP

$$\begin{aligned} \text{Minimize} \quad & z(x) = cx \\ \text{Subject to} \quad & Ax = b \\ & x \geqq 0 \end{aligned} \tag{18.1}$$

where A is an $m \times n$ matrix, how much computational effort is involved in solving it by the simplex method?

The main computational step in the simplex algorithm is the pivot step. Hence the effort involved in solving (18.1) by the simplex method can be measured by the number of pivots made before the algorithm terminates. If a feasible basis for (18.1) is not known, the simplex method applies the simplex algorithm twice, first on a phase I problem that decides whether (18.1) is feasible, and provides a feasible basis for it if it is feasible and then on (18.1) itself starting with the feasible basis provided by the phase I problem. So from a practical point of view the computational effort required for solving an LP for which a feasible basis is not known, is approximately twice the effort required for solving a comparable problem in which a feasible basis is available for starting the simplex algorithm. Hence in subsequent discussion we will assume that an initial feasible basis for (18.1) is given. The question that we study here is: Starting with a known feasible basis for (18.1), how many pivot steps are required before the simplex algorithm terminates?

The answer depends on m, n, the actual numbers in the matrices A, b, c and the initial feasible basis used in the algorithm. The number of pivot steps required before termination is expected to go up as m and n go up. Even when m, n are fixed, the number of pivot steps before termination depends on the actual data in A, b, c. A practitioner who uses the simplex algorithm in his daily work would probably pose the question in the following manner: As a function of m and n, what is the average number of pivot steps required before termination, when the simplex algorithm is applied on the class of LPs encountered in practical applications?

This is a statistical question and it is hard to answer precisely, because the kind of LPs that practical applications do generate is hard to specify. However, from the computational experience on a large number of LPs solved over several years, an empirical answer to this question has emerged. It indicates that the average number of pivot steps required before the termination of the simplex algorithm is a linear function of m that seems to be less than $3m$. See Dantzig's book [1], page 160. Experience gained after this book was written supports this observation.

The *size* of the LP (18.1) is determined by m and n. A more challenging question is: Given m, n what is a mathematical upperbound on the number of pivot steps required for solving any LP of type (18.1) by the simplex algorithm starting with a known feasible basis? Probably this question is not of much practical significance, but it is theoretically more interesting.

An algorithm for solving a problem is said to be a *polynomially bounded algorithm* or a *good algorithm* if the computational effort required for solving the problem using it, is bounded above by a polynomial in the size of the problem. To be specific, we will study the following problem: Is the simplex algorithm a *good algorithm* in this sense?

We will briefly review how the simplex algorithm solves (18.1). Let **K** be the set of feasible solutions of (18.1). Starting with a BFS of (18.1), the algorithm moves to an adjacent BFS in each step, such that the objective value decreases monotonically as the path progresses. A path like this on **K** is known as an *isotonic path* with respect to the objective function $z(x)$. It is a sequence of points in **K**; $x^0, x^1, \ldots, x^\ell$ satisfying:

1. Each point in the sequence is a BFS.
2. Consecutive points in the sequence are adjacent BFSs.
3. $z(x^0) > z(x^1) > z(x^2) > \cdots > z(x^\ell)$.

The length of this path is defined to be ℓ, the number of points in it excluding the initial point. Clearly, if an isotonic path is found on **K**, the simplex algorithm can be made to follow that path by making an appropriate entering variable selection in each pivot step. So an answer to our question can be obtained by studying the following question: "What is the maximum length of an isotonic path in a problem of type (18.1) as a function of m and n?".

Klee and Minty [5] have provided an answer to this question. First, we will discuss a sequence of examples constructed by them. Let $0 < \varepsilon < 1/2$.

Consider the following problem, the set of feasible solutions of which is the slightly perturbed unit square in $\mathbf{R}^2$.

$$\begin{array}{ll} \text{Minimize} & -x_2 \\ \text{Subject to} & x_1 \geqq 0 \\ & x_1 \leqq 1 \\ & x_2 \geqq \varepsilon x_1 \\ & x_2 \leqq 1 - \varepsilon x_1 \end{array} \tag{18.2}$$

The set of feasible solutions of this problem is plotted in Figure 18.1. The BFSs of this problem are:

$$x^0 = (0, 0) \qquad x^2 = (1, 1 - \varepsilon)$$
$$x^1 = (1, \varepsilon) \qquad x^3 = (0, 1)$$

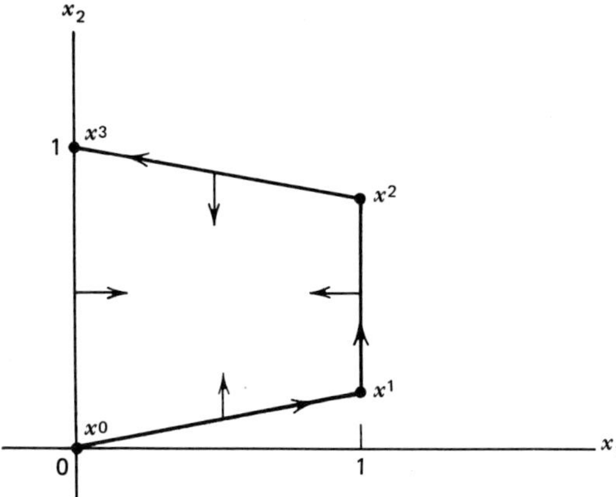

Figure 18.1

Clearly x^0, x^1, x^2, x^3 is an isotonic path for this problem. The length of this path is $3 = 2^2 - 1$. Consider the next problem:

$$\begin{array}{ll} \text{Minimize} & -x_3 \\ \text{Subject to} & x_1 \geqq 0 \\ & x_1 \leqq 1 \\ & x_2 \geqq \varepsilon x_1 \\ & x_2 \leqq 1 - \varepsilon x_1 \\ & x_3 \geqq \varepsilon x_2 \\ & x_3 \leqq 1 - \varepsilon x_2 \end{array} \tag{18.3}$$

The set of feasible solutions of this problem is the slightly perturbed cube in $\mathbf{R}^3$ (Figure 18.2). The BFSs of this problem are:

$$\begin{array}{ll} x^0 = (0, 0, 0) & x^4 = (0, 1, 1 - \varepsilon) \\ x^1 = (1, \varepsilon, \varepsilon^2) & x^5 = (1, 1 - \varepsilon, 1 - \varepsilon(1 - \varepsilon)) \\ x^2 = (1, 1 - \varepsilon, \varepsilon(1 - \varepsilon)) & x^6 = (1, \varepsilon, 1 - \varepsilon^2) \\ x^3 = (0, 1, \varepsilon) & x^7 = (0, 0, 1) \end{array}$$

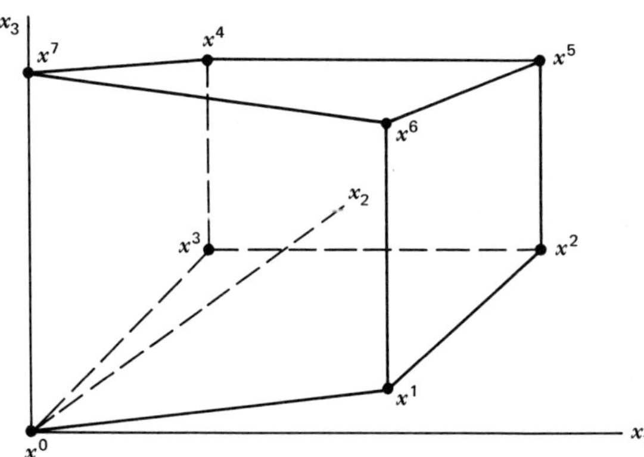

Figure 18.2

Clearly $x^0, x^1, x^2, x^3, x^4, x^5, x^6, x^7$ is an isotonic path for this problem. It includes all the BFSs of this problem. Its length is $7 = 2^3 - 1$.

In general consider the dth problem in the sequence

$$\begin{aligned} \text{Minimize} \quad & -x_d \\ \text{Subject to} \quad & x_1 \geqq 0 \\ & x_1 \leqq 1 \\ & x_2 \geqq \varepsilon x_1 \\ & x_2 \leqq 1 - \varepsilon x_1 \\ & x_3 \geqq \varepsilon x_2 \\ & x_3 \leqq 1 - \varepsilon x_2 \\ & \vdots \\ & x_d \geqq \varepsilon x_{d-1} \\ & x_d \leqq 1 - \varepsilon x_{d-1} \end{aligned} \tag{18.4}$$

The set of feasible solutions of (18.4) is the slightly perturbed unit cube in $\mathbf{R}^d$. Its BFSs can be arranged in an isotonic path of the form $x^0, x^1, \ldots, x^{2^d-1}$ where the first $d - 1$ coordinates in $x^0, x^1, \ldots, x^{2^{d-1}-1}$ are the same as those in the points in the isotonic path for the $(d - 1)$th problem in the sequence. The dth coordinate for the points $x^0, x^1, \ldots, x^{2^{d-1}-1}$ is ε times the $(d - 1)$th coordinate. The first $d - 1$ coordinates in the remaining points written in reverse order, that is, in $x^{2^d-1}, \ldots, x^{2^{d-1}+1}, x^{2^{d-1}}$ are again the same as those in the points in the isotonic path for the $(d - 1)$th problem in the sequence. The dth coordinate in each of these points is obtained by subtracting ε times its $(d - 1)$th coordinate from 1. This isotonic path contains all the BFSs of (18.4) and its length is $2^d - 1$.

When posed in standard form by introducing slack variables and eliminating unrestricted variables from the system, (18.4) becomes a problem of the form (18.1) with $m = d$, $n = 2d$. The isotonic path constructed above for this problem has length $2^d - 1$ and this number is not bounded above by any polynomial in m and n. Thus, when the simplex algorithm is applied on (18.4), and the entering variable selection rule is adopted so that the algorithm follows the isotonic path constructed above, the algorithm passes through all the BFSs and the number of pivot steps required before termination is not bounded above by any polynomial in m and n.

The examples of Klee and Minty clearly show that the simplex algorithm is not a polynomially bounded algorithm. Jeroslow has also shown that under any given entering variable selection rule, a sequence of problems can be constructed on which the simplex algorithm with that entering variable selection rule can be verified to be not polynomially bounded.

Thus on these specially constructed LPs, the simplex algorithm takes a very large number of pivot steps before termination. However on LPs encountered in practical applications it seems to perform very well. It is still the most efficient algorithm known for solving LPs.

OUTSTANDING THEORETICAL PROBLEMS

1. When it is known that the simplex algorithm is not in general a polynomially bounded algorithm, develop a theory to explain why it performs so well on almost every practical problem.

2. Develop an algorithm for solving LPs of the type (18.1) for which it can be proved that the computational effort required before settling the problem is bounded above by a polynomial in m, n. Otherwise conclusively prove that no such algorithm exists.

3. Lot A be of rank m in (18.1) and let **K** be the set of feasible solutions of (18.1). Assume that **K** is bounded. Let B_0 and B_* be two feasible bases for (18.1). In 1957, W. M. Hirsch has conjectured that by using Pivot steps of the type used in the simplex algorithm, one can get from the feasible basis B_0 to the feasible basis B_* in at most m pivot steps. See Dantzig's book [1], pages 160, 168. This conjecture is known as the *Hirsch Conjecture* or the *m-step conjecture.*
Let V_0 and V_* be any two extreme points of **K**. What does the Hirsch conjecture say about the lengths of edge paths connecting V_0 and V_*?
Let $\mathbf{\Gamma}$ be a convex polytype in $\mathbf{R}^d$ with n facets. Prove that the Hirsch conjecture is equivalent to stating that any two extreme points of $\mathbf{\Gamma}$ are connected by an edge path consisting of at most $n - d$ edges of $\mathbf{\Gamma}$.
Klee and Walkup [6] have constructed examples to show that the Hirsch conjecture is false if **K** is not bounded. They also prove the conjecture for all polytopes for which $n - m \leqq 5$.
Prove this conjecture or construct a counterexample for it.

4. Consider an LP of the form (18.1). Using pivot steps of the type used in the simplex algorithm is it possible to move from any feasible basis to any other feasible basis, such that the total number of pivot steps required is bounded above by a polynomial in m, n?

5. Consider the LP (18.1). Let $r \geqq 2$ and $\{A_{.j_1}, \ldots, A_{.j_r}\}$, be a specified linearly independent subset of column vector of A. Find whether there exists a feasible basis containing all the columns $A_{.j_1}, \ldots, A_{.j_r}$, and if so find such a basis. Develop a polynomially bounded algorithm for this problem.

6. Is the complementary pivot algorithm (discussed in section 16.6) a polynomially bounded algorithm?

7. Karp [3] has recently proved that a large number of combinatorial problems like the traveling salesman problem, the set representation problem, the general 0-1 integer programming problem, etc., all belong to a class of problems called the NP-*complete class.* If a polynomially bounded algorithm can be developed for solving one problem in this class, then every other problem in the class can be solved by a polynomially bounded algorithm. Develop a polynomially bounded algorithm for one of these problems (say, the traveling salesman problem or the set representation problem) or prove that no such algorithm exists.

8. Let e_r be the column vector in $\mathbf{R}^r$, all of whose entries are 1. Consider the problem:

$$\begin{aligned} \text{Minimize} \quad & e_n{}^{\mathrm{T}}x \\ \text{Subject to} \quad & Ax \geqq e_m \\ & x \geqq 0 \end{aligned}$$

where A is an $m \times n$, 0-1 matrix. Develop simple necessary and sufficient conditions on the matrix A under which this problem has an optimum solution that is an integer vector.

REFERENCES.

1. G. B. Dantzig, *Linear Programming and Extensions*, Princeton University Press, Princeton, N.J. 1963.
2. R. G. Jeroslow, "The Simplex Algorithm with the Pivot Rule of Maximizing Criterion Improvement," *Discrete Mathematics*, 4, pp. 367–378, 1973.
3. R. M. Karp, "Reducibility among combinatorial problems" in R. E. Miller and J. W. Thatcher (eds.) *Complexity of Computer Computations*, Plenum Press, New York, 1972.
4. V. L. Klee "Paths on polytopes: A survey," in B. Grunbaum *Convex Polytopes*, Wiley, New York, 1966.

5. V. Klee and G. J. Minty, "How good is the simplex algorithm" in O. Shisha (ed.) *Inequalities III*, Academic Press, New York, 1972, pp. 159–175.

6. V. Klee and D. W. Walkup "The d-step conjecture for polyhedra of dimension $d < 6$," *Acta Mathematica*, *117*, pp. 53–78, 1967.

7. K. G. Murty, "A Fundamental Problem in Linear Inequalities with Applications to the Traveling Salesman Problem," *Mathematical Programming*, *2*, 3 pp. 296–308, June 1972.

appendix 1

the decomposition principle of linear programming

This method can be used to solve LPs with large number of constraints, by solving a sequence of smaller size problems. It is based on the resolution theorem for convex polyhedra discussed in Chapter 3. Consider the LP

$$\begin{aligned} \text{Minimize} \quad z(x) &= cx \\ \text{Subject to} \quad Ax &= b \\ x &\geqq 0 \end{aligned} \tag{1}$$

where A is a matrix of order $m \times n$. To solve this problem by the simplex method, we have to deal with bases of order m. If m is very large the problem becomes unwieldy and it cannot be handled even by large computers. Decomposition principle can be used to break up such a large problem into smaller problems each of which contains fewer number of constraints.

While the decomposition principle can be applied on a general LP, its application can be expected to be fruitful only when the coefficient matrix is a matrix with a special structure. It has been observed that it gives most satisfactory results when the matrix A can be blocked as in the diagram below, such that all the entries in it outside of those blocks are zero. This structure appears often in modeling problems of a corporation with many plants that are mostly autonomous. For example, in modeling the product mix problem for such a corporation, the smaller blocks (in the bottom) represent the constraints of the plants that are independent of the activities of the other plants. The top block consists of the linking constraints at the corporate level that tie the plants together. We will discuss the decomposition principle only as it relates to problems of this structure.

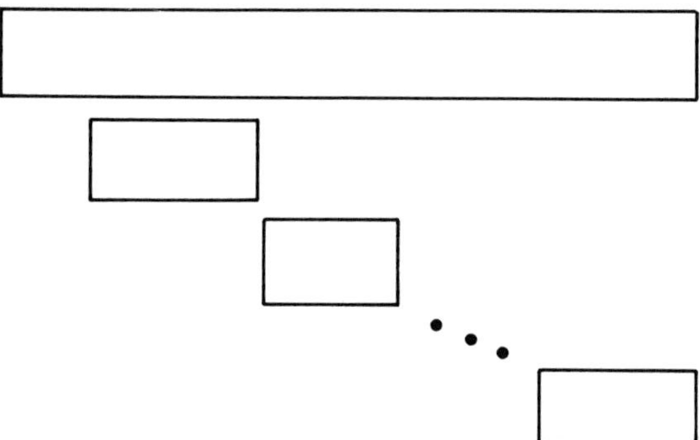

In particular we consider the problem

$$
\begin{array}{llll}
\text{Minimize} & c^0x^0 + c^1x^1 + c^2x^2 + \cdots + c^kx^k & & \\
\text{Subject to} & A^{00}x^0 + A^{01}x^1 + A^{02}x^2 + \cdots + A^{0k}x^k & = b^0 & \\
\hline
& A^1x^1 & = b^1 & \quad (2) \\
& \qquad A^2x^2 & = b^2 & \\
& \qquad\qquad \ddots & \vdots & \\
& \qquad\qquad\qquad A^kx^k & = b^k & \\
& x^t \geqq 0 \text{ for all } t & &
\end{array}
$$

where A^{0t} is of order $m_0 \times n_t$ and A^t is of order $m_t \times n_t$ for each t. The number of variables in the vector x^t, corresponding to the tth independent block is n_t (these can be interpreted as the levels of the activities at the tth plant of the corporation) and the number of constraints in this block is m_t, for $t = 1$ to k, where k is the number of independent blocks (plants). The number of linking constraints (which can be interpreted as the constraints at the corporate level that express the interconnections among the plants) is m_0.

For applying the decomposition principle the constraints in the overall model should be divided into two groups. One group constitutes the constraints of a problem called the *subprogram.* The other group *leads to* a problem called the *master program.* The division of the constraints into the two groups has to be done carefully, taking the structure of the matrix into account, so that the number of constraints in both the subprogram and the master program can be handled by the available computer. For (2), it is most advantageous to have the linking constraints on the top generate the master program and all the remaining constraints constitute the subprogram. An advantage of decomposing the problem this way is that the subprogram breaks up into k independent smaller size problems (one corresponding to each plant), called the *subproblems.* When the decomposition principle is applied in this manner, the problem is said to have been decomposed along the dashed line drawn in (2). Then the constraints in the subproblem are

$$
\begin{array}{ll}
A^1x^1 & = b^1 \\
\qquad A^2x^2 & = b^2 \\
\qquad\qquad \ddots & \vdots \\
\qquad\qquad\qquad A^kx^k & = b^k \\
x^t \geqq 0 \quad \text{for} \quad t = 1 \text{ to } k &
\end{array}
$$

This breaks up into k independent subproblems, the constraints on the tth subproblem being

$$\begin{aligned} A^t x^t &= b^t \\ x^t &\geqq 0 \end{aligned} \tag{3}$$

for $t = 1$ to k. Since the number of constraints in it is m_t [compared to the original problem (2), which had $\sum_{t=0}^{k} m_t$ constraints], it should be much easier to handle.

While the constraints and the variables in the subprogram are all taken from those in the original problem, both the constraints and the variables in the master program are derived from an analysis outlined below. When decomposed as above, the constraints of the original problem that generate the masterprogram are

$$\begin{aligned} A^{00}x^0 + A^{01}x^1 + \cdots + A^{0k}x^k &= b^0 \\ x^0 &\geqq 0 \end{aligned} \tag{4}$$

There are m_0 constraints in (4). The master program will be a problem with $m_0 + k$ constraints (k here is the number of independent subproblems into which the subprogram breaks up).

DERIVATION OF THE MASTERPROGRAM

If any of the subproblems, (3), is infeasible, the original problem is infeasible, and we terminate. So we assume that each subproblem is feasible. Let $x^{t,1}, \ldots, x^{t,p_t}$ be all the BFSs of (3) and $y^{t,1}, \ldots, y^{t,q_t}$ be all the extreme homogeneous solutions corresponding to (3). By resolution theorem II (see section 3.7), every feasible solution, x^t, of this subproblem can be expressed as

$$\begin{aligned} x^t = \alpha_{t,1}x^{t,1} + \cdots + \alpha_{t,p_t}x^{t,p_t} + \beta_{t,1}y^{t,1} + \cdots + \beta_{t,q_t}y^{t,q_t} \\ \text{where } \alpha_{t,1} + \cdots + \alpha_{t,p_t} = 1 \\ \text{and all } \alpha_{t,1}, \ldots, \alpha_{t,p_t}, \beta_{t,1}, \ldots, \beta_{t,q_t} \geqq 0. \end{aligned} \tag{5}$$

The variables in the masterprogram are those in x^0 (these are the original variables that do not appear in any subproblem) and the multipliers $\alpha_{t,j}$ and $\beta_{t,j}$ corresponding to the extreme points and the extreme homogeneous solutions of the subproblems.

Every feasible solution of the original problem must satisfy (5) and (4). Substituting (5) into (4) and the original objective function leads to

$$\begin{aligned}
\text{Minimize} \quad z = c^0x^0 &+ \sum_{j=1}^{p_1} (c^1x^{1,j})\alpha_{1,j} + \cdots + \sum_{j=1}^{p_k} (c^kx^{k,j})\alpha_{k,j} \\
&+ \sum_{j=1}^{q_1} (c^1y^{1,j})\beta_{1,j} + \cdots + \sum_{j=1}^{q_k} (c^ky^{k,j})\beta_{k,j} \\
\text{Subject to} \quad A^{00}x^0 &+ \sum_{j=1}^{p_1} (A^{01}x^{1,j})\alpha_{1,j} + \cdots + \sum_{j=1}^{p_k} (A^{0k}x^{k,j})\alpha_{k,j} \\
&+ \sum_{j=1}^{q_1} (A^{01}y^{1,j})\beta_{1,j} + \cdots + \sum_{j=1}^{q_k} (A^{0k}y^{k,j})\beta_{k,j} = b^0 \\
&\sum_{j=1}^{p_t} \alpha_{t,j} = 1 \qquad \text{for} \qquad t = 1 \text{ to } k \\
x^0 \geqq 0 \qquad &\text{and} \qquad \alpha_{t,j} \qquad \text{and} \qquad \beta_{t,j} \geqq 0 \qquad \text{for all } t, j
\end{aligned} \tag{6}$$

Problem (6) is the master program. Every feasible solution (x^0, α, β) of (6) corresponds to a unique feasible solution of the original problem (2) obtained by using (5). Also every feasible solution of (2) corresponds to at least one feasible solution of (6) (possibly many), because there may be many different ways in which a feasible solution, x^t, of (3) can be expressed as in (5). Hence the master program is equivalent to the original problem.

The decomposition principle solves the original problem by solving the master program. The number of constraints in (6) is $m_0 + k$. So (6) can be solved by the revised simplex method, using bases of order $m_0 + k$ each. Since $m_0 + k$ is much smaller than the number of constraints in the original problem (which is $\sum_{t=0}^{k} m_t$), it should be much easier to handle the master program than the original problem. However, the number of variables in the master program is $n_0 + \sum_{t=1}^{k} p_t + \sum_{t=1}^{k} q_t$, and this is likely to be a very large number. If we have to deal with all the variables in the master program explicitly, this approach would be useless. But, to solve (6) by the revised simplex method, we need only the $m_0 + k$ basic columns at each stage, and the other columns can be generated as they are needed. Hence this kind of approach is also known as a *column generation approach.*

THE ALGORITHM

If $x^{t,j}$ is a BFS of (3), we will denote $A^{0t}x^{t,j}$ by $S^{t,j}$, and $c^t x^{t,j}$ by $u^{t,j}$. If $y^{t,j}$ is an extreme homogeneous solution corresponding to (3), we will denote $A^{0t}y^{t,j}$ by $T^{t,j}$ and $c^t y^{t,j}$ by $v^{t,j}$.

Find a BFS for each subproblem (by a Phase I approach, if necessary). If any of the subproblems is infeasible, the original problem is infeasible. Terminate.

Otherwise let $x^{t,1}$, $t = 1$ to k be the BFSs obtained for the subproblems. As before $S^{t,1} = A^{0,t}x^{t,1}$ and $u^{t,1} = c^t x^{t,1}$. Let $\xi_1, \ldots, \xi_{m_0}$ be artificial variables. Let w be the Phase I objective. Here is the part of the master program corresponding to the initial basis for the Phase I problem

Basic Variables

$\alpha_{1,1}$	$\alpha_{2,1} \ldots \alpha_{k,1}$	$\xi_1 \ldots \xi_{m_0}$	$-z$	$-w$	
		$\pm 1 \ldots 0$	0	0	
$S^{1,1}$	$S^{2,1} \ldots S^{k,1}$	0			b^0
		$\vdots \quad \ddots \quad \vdots$	$\vdots$	$\vdots$	
		$0 \ldots \pm 1$	0	0	
1	$\ldots 0$	$0 \ldots 0$	0	0	1
$\vdots$	$\vdots$	$\vdots \quad \vdots$	$\vdots$	$\vdots$	$\vdots$
0	$\ldots 1$	$0 \ldots 0$	0	0	1
$u^{1,1}$	$\ldots u^{k,1}$	$0 \ldots 0$	1	0	0
0	$\ldots 0$	$1 \ldots 1$	0	1	0

Here the coefficient of each artificial variable in the tableau is chosen to be $+1$ or -1, after pivoting in the columns of $\alpha_{1,1}, \ldots, \alpha_{k,1}$, so that all the artificial variables are nonnegative in the basic solution. (This is why they are recorded as ± 1). Obtain the inverse tableau corresponding to this basis as in Chapter 5.

For each α or β variable in the basic vector, remember to store the associated BFS or the extreme homogeneous solution of the subproblem carefully. These are needed to get the solution of the original problem corresponding to the present solution of the master program (using equation (5)).

In a general stage during Phase I suppose the inverse tableau is

Basic variables	Inverse tableau				Basic values
					$\bar{b}_1$ $\vdots$ $\bar{b}_{m_0+k}$
$-z$	$-\pi_1$	$-\pi_{m_0+k}$	1	0	$-\bar{z}$
$-w$	$-\sigma_1$	$-\sigma_{m_0+k}$	0	1	$-\bar{w}$

1. The Phase I relative cost coefficient of $x_j{}^0$ with respect to the present basis in (6) is $\bar{d}_j{}^0 = (-\bar{\sigma}_1, \ldots, -\sigma_{m_0})A^{00}_{.j}$. If $\bar{d}_j{}^0 < 0$, $x_j{}^0$ can be chosen as the entering variable into the basic vector for the master program. The pivot column is its updated column and it is obtained by mutliplying the present inverse tableau on the right, by the column vector of $x_j{}^0$ in (6).
If $\bar{d}_j{}^0 \geqq 0$, $x_j{}^0$ is not a candidate to enter the basic vector. If $\bar{d}_j{}^0 \geqq 0$ for all $j = 1$ to n_0, go to 2.
2. Check whether any of the α or β variables corresponding to one of the subproblems can be the entering variable. The Phase I relative cost coefficient of $\beta_{t,j}$ is $(-\sigma_1, \ldots, -\sigma_{m_0})T^{t,j} = (-\sigma_1, \ldots, -\sigma_{m_0})A^{0t}y^{t,j}$. If this is negative, $\beta_{t,j}$ can be taken as the entering variable. In this case, by the results in section 3.7, the objective function in the LP (7) must be unbounded below.

$$\begin{aligned} &\text{Minimize} \quad (-\sigma_1, \ldots, -\sigma_{m_0})A^{0t}x^t \\ &\text{Subject to} \qquad\qquad A^t x^t = b^t \qquad\qquad (7) \\ &\qquad\qquad\qquad\qquad\quad x^t \geqq 0 \end{aligned}$$

Conversely, when the subproblem (7) is solved, if the objective function is unbounded below, find $y^{t,j}$, the extreme homogenous solution corresponding to the extreme half-line along which the objective function diverges to $-\infty$ in (7), associate $y^{t,j}$ with a multiplier $\beta_{t,j}$, and choose $\beta_{t,j}$ as the entering variable into the basic vector for the master program. The column vector corresponding to this $\beta_{t,j}$ in the Phase I problem is

$$\begin{bmatrix} A^{0t}y^{t,j} \\ \hdashline \mathbf{0} \\ \hdashline v^{t,j} \\ \hdashline 0 \end{bmatrix}$$

The "**0**" in this vector is the zero vector in $\mathbf{R}^k$ (since the β's do not appear in the constraints that the α-multipliers in each subproblem should sum to 1 in (6)). $v^{t,j}$ is the Phase II original cost coefficient of $\beta_{t,j}$ in (6). The bottom entry in this column is the Phase I original cost coefficient of $\beta_{t,j}$ in the Phase I problem, and since $\beta_{t,j}$ is a variable in (6), this is zero.

The updated column vector is obtained by multiplying this column vector on the left by the inverse tableau and this is the pivot column.

If (7) has a finite optimum solution, then none of the β variables corresponding to this subproblem are eligible to be the entering variable. Let the optimum solution of (7) be $x^{t,j}$. Associate the multiplier $\alpha_{t,j}$ with $x^{t,j}$. The Phase I relative cost coefficient of $\alpha_{t,j}$ is $-\sigma_{m_0+t} + (-\sigma_1, \ldots, -\sigma_{m_0})A^{0t}x^{t,j}$. If this is negative, choose $\alpha_{t,j}$ as the entering variable into the basic vector for the master program. The column vector of $\alpha_{t,j}$ in the Phase I problem is

$$\begin{bmatrix} A^{0t}x^{t,j} \\ \hline I_{.t} \\ \hline u^{t,j} \\ \hline 0 \end{bmatrix}$$

where $I_{.t}$ is the tth column vector of the unit matrix of order k (since $\alpha_{t,j}$ appears with a coefficient of one in the constraint that all the α-multipliers associated with the tth subproblem should sum to one in (6)). $u^{i,j}$ is the Phase II original cost coefficient of $\alpha_{t,j}$ in (6) and the bottom entry in this column is zero, the Phase I original cost coefficient of $\alpha_{t,j}$ in the Phase I problem.

The updated column vector is obtained by multiplying this column vector on the left by the inverse tableau and this is the pivot column.

If $-\sigma_{m_0+t} + (-\sigma_1, \ldots, -\sigma_{m_0})A^{0t}x^{t,j} \geqq 0$, then obviously none of the α or β variables corresponding to this subproblem are eligible to be the entering variables in this stage.

Begin solving the tth subproblem (7) from $t = 1$ to k. If an entering variable into the basic vector for the master program is identified, stop solving the subproblems. Get the new inverse tableau corresponding to the new basis for the master program and go to the next stage.

At any stage of the algorithm, the part of the master program remaining after dropping the columns of all the variables except those of the present basic variables and the entering variable is known as the *restricted master program.*

Notice that the objective functions for the subproblems depend on the Phase I dual vector for the master program during the stage and hence may change from stage to stage.

PHASE I: TERMINATION CRITERIA

Phase I for the master program terminates when the following conditions are satisfied.

1. $\bar{d}_j^{\,0} \geqq 0$ for all $j = 1$ to n_0.
2. All the subproblems (7) for $t = 1$ to k have finite optimum solutions.
3. For the tth subproblem (7), $-\sigma_{m_0+t}$ + the minimum value of $(-\sigma_1, \ldots, -\sigma_{m_0})A^{0t}x^t$ in (7) is nonnegative for all $t = 1$ to k.

When Phase I terminates, if the value of w is positive, the original problem is infeasible, and we terminate. If the value of w is zero, and if all the artificial variable have left the basic vector, the Phase I objective row and the w column are deleted from the inverse tableau and we go to Phase II.

If some of the artificial variables are still in the basic vector at Phase I termination, they are left in the basic vector, but care must be taken to see that all the entering variables during Phase II have zero phase I relative cost coefficients. Do the following before entering phase II for this.

1. All x_j^0 that have $\bar{d}_j^0 > 0$ in the final Phase I iteration should be set equal to zero and eliminated from consideration during Phase II.
2. Let $(-\bar{\sigma}_1, \ldots, -\bar{\sigma}_{m_0+k}, 0, 1)$ be the last row in the final Phase I inverse tableau. Add the constraint

$$(-\bar{\sigma}_1, \ldots, -\bar{\sigma}_{m_0})A^{0t}x^t + (-\bar{\sigma}_{m_0+t}) = 0$$

   to the system of constraints (3) in the tth subproblem for each t. This will make all the α and β variables that have positive Phase I relative cost coefficients ineligible from becoming entering variables during Phase II. However for notational convenience we will continue to refer to the system of constraints in the tth subproblem by (3) itself. Now delete the Phase I objective row and the w column from the final Phase I inverse tableau and go over to Phase II.

COMPUTATIONS DURING PHASE II

The computations during Phase II are similar to those in Phase I with the only change that Phase II relative cost coefficients should replace Phase I relative cost coefficients. Let the Phase II objective row in the inverse tableau during some stage in Phase II be $(-\pi_1, \ldots, -\pi_{m_0+k}, 1)$. In this stage.

1. The Phase II relative cost coefficient of x_j^0 is $\bar{c}_j^0 = c_j^0 + (-\pi_1, \ldots, -\pi_{m_0})A_j^0$
2. If x_j^0 have nonnegative Phase II relative cost coefficients for all $j = 1$ to n_0, then solve the subproblem:

$$\begin{aligned} \text{Minimize} \quad & (c^t + (-\pi_1, \ldots, -\pi_{m_0})A^{0t})x^t \\ \text{Subject to} \quad & A^t x^t = b^t \\ & x^t \geqq 0 \end{aligned} \tag{8}$$

If the objective function is unbounded below in (8), the entering column is the column in (6) corresponding to the extreme homogeneous solution associated with the extreme half-line along which the objective function in (8) diverges to $-\infty$. If (8) has a finite optimum solution, let $x^{t,j}$ be an optimal BFS of it and let $\alpha_{t,j}$ be the multiplier associated with it. The Phase II relative cost coefficient of $\alpha_{t,j}$ is

$$(-\pi_{m_0+t}) + (c^t + (-\pi_1, \ldots, -\pi_{m_0})A^{0t})x^{t,j}$$

If this is negative, $\alpha_{t,j}$ is chosen as the entering variable. If this is nonnegative, all the α and β variables corresponding to this subproblem have nonnegative Phase II relative cost coefficients, and are not eligible to be entering variables.

Solve the subproblems from $t = 1$ to k until an entering variable with a negative Phase II relative cost coefficient is identified.

PHASE II: TERMINATION CRITERIA

If at some stage of Phase II, the updated column vector of the entering variable is nonpositive, the objective function is unbounded below in the original problem. If unboundedness did not occur so far, Phase II terminates in this stage if:

1. $\bar{c}_j^{\,0} \geqq 0$ for all $j = 1$ to n_0.
2. All the subproblems (8) for $t = 1$ to k have finite optimum solutions.
3. $c^t x^{t,j} + (-\pi_{m_0+t}) + (-\pi_1, \ldots, -\pi_{m_0})A^{0t}x^{t,j} \geqq 0$ in subproblem (8), for all $t = 1$ to k. (Here $x^{t,j}$ is the optimum solution of subproblem (8) in this stage).

Notice that the objective function in the subproblem (8) depends on the Phase II dual solution in that stage. Hence as we move from stage to stage, the objective functions in the subproblems change, but the subproblem constraints remain unaltered. Hence after the first stage, the subproblems can be solved efficiently by parametric cost methods. Each subproblem may have to be solved several times, each time with a different objective function.

From the optimal Phase II inverse tableau an optimum solution of the original problem is obtained using (5). This is why the BFS or the extreme homogeneous solution associated with each α or β variable in the basic vector should be stored in each stage.

Example

Consider the following problem.

x_1^1	x_2^1	x_3^1	x_4^1	x_1^2	x_2^2	x_3^2	x_4^2	x_5^2	
3	−1	−3	2	1	2	0	−1	1	1
1	0	−1	1						3
0	1	−1	−1						4
				1	1	1	1	0	1
				2	1	−2	0	1	2
1	−1	−3	3	20	30	7	1	1	minimize

All variables nonnegative. Blank entries in tableau are zero.

We decompose along the dashed line. The subprogram breaks up into two independent subproblems. We use the following BFSs of the subproblems initially.

Associated multiplier	Solution
$\alpha_{1,1}$	$x^{1,1} = (3, 4, 0, 0)^T$
$\alpha_{1,2}$	$x^{1,2} = (0, 7, 0, 3)^T$
$\alpha_{2,1}$	$x^{2,1} = (0, 0, 0, 1, 2)^T$

We can verify that $(\alpha_{1,1}, \alpha_{1,2}, \alpha_{2,1})$ constitutes a feasible basic vector for the master program; therefore there is no need to do a Phase I solution. The portion of the master program corresponding to this basic vector is

$\alpha_{1,1}$	$\alpha_{1,2}$	$\alpha_{2,1}$	$-z$	
5	−1	1	0	1
1	1	0	0	1
0	0	1	0	1
−1	2	3	1	0

The inverse tableau corresponding to this basic vector is

First Inverse Tableau for the Master Program

Basic variables	Inverse Tableau				Basic values	Pivot column $\beta_{1,1}$
$\alpha_{1,1}$	$\frac{1}{6}$	$\frac{1}{6}$	$-\frac{1}{6}$	0	$\frac{1}{6}$	$-\frac{1}{18}$
$\alpha_{1,2}$	$-\frac{1}{6}$	$\frac{5}{6}$	$\frac{1}{6}$	0	$\frac{5}{6}$	$\frac{1}{18}$ (circled)
$\alpha_{2,1}$	0	0	1	0	1	0
$-z$	$\frac{1}{2}$	$-\frac{3}{2}$	$-\frac{7}{2}$	1	$-\frac{9}{2}$	$-\frac{21}{18}$

The objective function in subproblem 1 in this stage is: $(1/2)(3x_1^{\ 1} - x_2^{\ 1} - 3x_3^{\ 1} + 2x_4^{\ 1}) + (x_1^{\ 1} - x_2^{\ 1} - 3x_3^{\ 1} + 3x_4^{\ 1}) = (5/2)x_1^{\ 1} - (3/2)x_2^{\ 1} - (9/2)x_3^{\ 1} + 4x_4^{\ 1}$. With this objective function, the canonical tableau for this subproblem with $(x_1^{\ 1}, x_2^{\ 1})$ as the basic vector is

$x_1^{\ 1}$	$x_2^{\ 1}$	$x_3^{\ 1}$	$x_4^{\ 1}$		
1	0	−1	1	3	
0	1	−1	−1	4	
0	0	$-\frac{7}{2}$	0	$-\frac{3}{2}$	Objective row

From the column vector of $x_3^{\ 1}$ we see that this problem is unbounded below. The corresponding extreme homogeneous solution is

Associated multiplier	Extreme homogeneous solution
$\beta_{1,1}$	$y^{1,1} = \left(\frac{1}{3}, \frac{1}{3}, \frac{1}{3}, 0\right)$

$\beta_{1,1}$ is the entering variable. Its column vector in the master tableau is $(-1/3, 0, 0, -1)^{\mathrm{T}}$. The updated column vector is the pivot column and this is entered on the first inverse tableau.

Second Inverse Tableau for the Master Program

Basic variables	Inverse Tableau				Basic values	Pivot column $\alpha_{2,2}$
$\alpha_{1,1}$	0	1	0	0	1	0
$\beta_{1,1}$	−3	15	3	0	15	−9
$\alpha_{2,1}$	0	0	1	0	1	①
$-z$	−3	16	0	1	13	−1

The objective function in subproblem 1 in this stage is $-3(3x_1^1 - x_2^1 - 3x_3^1 + 2x_4^1) + (x_1^1 - x_2^1 - 3x_3^1 + 3x_4^1) = -8x_1^1 + 2x_2^1 + 6x_3^1 - 3x_4^1$. It can be verified that the optimum solution for subproblem 1 with this objective function is $x^{1,1}$ with an objective value of −16. The relative cost coefficient in its column is $-16 + 16 = 0$. This implies that none of the α or β variables from this subproblem are eligible to enter the basic vector at this stage. Therefore move to subproblem 2.

The objective function in subproblem 2 in this stage is $-3(x_1^2 + 2x_2^2 - x_4^2 + x_5^2) + (20x_1^2 + 30x_2^2 + 7x_3^2 + x_4^2 + x_5^2) = 17x_1^2 + 24x_2^2 + 7x_3^2 + 4x_4^2 - 2x_5^2$. Hence subproblem 2 at this stage is

x_1^2	x_2^2	x_3^2	x_4^2	x_5^2	
1	1	1	1	0	1
2	1	−2	0	1	2
17	24	7	4	−2	Minimize

Here is the optimum solution of this subproblem with an objective value of −1.

Associated multiplier	Optimum solution
$\alpha_{2,2}$	$x^{2,2} = (0, 0, 1, 0, 4)^T$

So the relative cost coefficient of $\alpha_{2,2}$ is $-1 + 0 = -1$. Hence $\alpha_{2,2}$ is the entering variable. Its column vector in the master program is $(4, 0, 1, 11)^T$ and the updated column vector is the pivot column entered on the second inverse tableau.

Third Inverse Tableau for the Master Program

Basic variables	Inverse Tableau				Basic values
$\alpha_{1,1}$	0	1	0	0	1
$\beta_{1,1}$	−3	15	12	0	24
$\alpha_{2,2}$	0	0	1	0	1
$-z$	−3	16	1	1	14

It can be verified that the optimality criterion is now satisfied. Hence the solution in the third inverse tableau is the optimum solution for the master program. The optimum solution of the original problem is

$$x^1 = (3, 4, 0, 0)^{\mathrm{T}} + 24\left(\frac{1}{3}, \frac{1}{3}, \frac{1}{3}, 0\right)^{\mathrm{T}} = (11, 12, 8, 0)^{\mathrm{T}}$$

$$x^2 = (0, 0, 1, 0, 4)^{\mathrm{T}}$$

$$z = -14$$

INTERPRETATION IN SUPPORT OF DECENTRALIZED PLANNING

Consider a corporation with k plants that are mostly autonomous and let problem (2) be the model that optimizes the operations of this corporation. *Centralized planning* requires solving this overall model and then handing over the optimum solution to the plants to be implemented.

However the manner in which this model is solved by applying the decomposition principle can be interpreted as suggesting a decentralized planning system that also leads to an optimum solution for the overall model. In this set up the corporate headquarters, (HQ) would handle the master program. Each plant would handle the subproblem corresponding to it. In each stage of the algorithm the HQ generates an objective function for each subproblem (corresponding to that stage of the master program) and sends a directive to each plant that it should come up with an optimum solution of its subproblem which minimizes that objective function. The HQ does not need any knowledge as to how the plant solves the subproblem (it does not even need any knowledge of the constraints in the subproblem, that is, the constraints under which the plant is operating). When the optimum solutions from all the plants are received, the HQ combines this information, check for optimality to the master program, and moves to the next stage in solving the master program, if optimality is not reached yet. In this set up the HQ is acting purely as a coordinating agency giving guidance. The plants do the optimization of their subproblems all by themselves according to the guidance obtained from the HQ. (This guidance comes in the form of the objective function in that stage, which the plant is asked to minimize. No quotas or any such additional constraints are ever imposed.) They then send their optimum plans to the HQ. When this process is repeated, after a finite number of iterations the system moves to a solution that is optimal for the whole corporation.

Problem 1

Consider the multicommodity minimum cost flow problem discussed in section 12.8.

f^1	f^2	$\ldots$	f^p	
I	I	$\ldots$	I	$\leqq k$
E				$= q_1 V_1$
	E			$= q_2 V_2$
		$\ddots$		$\vdots$
			E	$= q_p V_p$
c^1	c^2	$\ldots$	c^p	$=$ minimize

$$f^r \geqq 0 \text{ for all } r$$

where $V_1, \ldots, V_p$ are the specified values of flow for the various commodities. Explain how to solve this problem by decomposing it along the dashed line. Show that the subprogram breaks up into simple network flow problems.

Comments The decomposition principle has been developed by G. B. Dantzig and P. Wolfe, inspired by the work of L. R. Ford, D. R. Fulkerson and W. S. Jewell on the multicommodity flow problem.

REFERENCES

See Chapter 18 in Dantzig's book [1] at the end of Chapter 18. Also see

[1]. M. R. Rao and S. Zionts, "Allocation of Transportation Units to Alternative Trips—A Column Generation Scheme with Out-of-Kilter Subproblems," *Operations Research*, *16*, pp. 52–63, 1968.

appendix 2

benders' decomposition for mixed integer programs

This method can be used to decompose a mixed integer program into its integer and continuous parts, each of which can be solved by appropriate specialized algorithms. This method is also based on the resolution theorem for convex polyhedra discussed in Chapter 3. Consider the problem

$$\begin{aligned} \text{Minimize} \quad & z(x, y) = cx + dy \\ \text{Subject to} \quad & Ax + Dy \leqq b \\ & x \geqq 0, y \geqq 0 \\ & y \text{ an integer vector} \end{aligned} \tag{9}$$

Let $\mathbf{S} = \{y: y \geqq 0 \text{ and } y \text{ is an integer vector}\}$. For a fixed $y \in \mathbf{S}$, the problem (9) is equivalent to

$$\begin{aligned} dy + \text{minimum} \quad & cx \\ \text{Subject to} \quad & Ax \leqq b - Dy \\ & x \geqq 0 \end{aligned} \tag{10}$$

$y \in \mathbf{S}$ is said to be *admissible* if there exists an x such that (x, y) satisfies (9), that is, if (10) is feasible in x. If there is no admissible y, (9) is infeasible.

The dual of (10) is

$$\begin{aligned} v(y) = \underset{\pi}{\text{maximum}} \quad & v(\pi, y) = dy + \pi(Dy - b) \\ \text{Subject to} \quad & -\pi A \leqq c \\ & \pi \geqq 0 \end{aligned} \tag{11}$$

The constraints in (11) are independent of y. If (11) is infeasible, by the duality theorem, the objective function in (10) is unbounded below for every admissible $y \in \mathbf{S}$, that is, either (9) is infeasible or $z(x, y)$ is unbounded below in it. Hence (9) does not have a finite optimum solution in this case. [If it is still required to determine whether (9) is feasible in this case, change $z(x, y)$ to $0x + 0y$ and proceed as below.]

So we assume that (11) is feasible. Since for any fixed $y \in \mathbf{S}$, (9) is equivalent to (10), and (11) is the dual of (10), we conclude from the duality theorem that (9) is equivalent to

$$\begin{array}{ll} \text{Minimize} & v(y) \\ \text{Over admissible} & y \in \mathbf{S} \end{array} \tag{12}$$

Since (11) is feasible, the objective function in (11) is unbounded above, iff (10) is infeasible (by the duality theorem), that is, iff y is inadmissible. By the results in sections 3.7, the objective function in (11) is unbounded above in this case iff there exists an extreme homogeneous solution, μ, corresponding to (11), which satisfies $\mu(Dy - b) > 0$. Let $\mu^1, \ldots, \mu^p$ be all the extreme homogeneous solutions corresponding to (11). These facts together imply that $y \in \mathbf{S}$ is admissible iff

$$\begin{array}{ll} \mu^t(Dy - b) \leqq 0 & \text{for all } t = 1 \text{ to } p \\ y \in \mathbf{S} & \end{array} \tag{13}$$

Hence from the previous results and (12), we conclude that (9) is equivalent to

$$\begin{array}{lll} \text{Minimize} & v(y) & \\ \text{Subject to} & \mu^t(Dy - b) \leqq 0 & \text{for all } t = 1 \text{ to } p \\ & y \in \mathbf{S} & \end{array} \tag{14}$$

Let $\pi^1, \ldots, \pi^\ell$ be all the BFSs of (11). From section 3.7, we conclude that if (13) is satisfied, then

$$v(y) = \text{maximum } \{dy + \pi^j(Dy - b) : j = 1 \text{ to } \ell\} \tag{15}$$

From (15) and (14) we conclude that the original problem (9) is equivalent to:

$$\begin{array}{llr} \text{Minimize} & z_0 & \\ \text{Subject to} & z_0 - dy - \pi^j(Dy - b) \geqq 0 & \text{for all } j = 1 \text{ to } \ell \\ & \mu^t(Dy - b) \leqq 0 & \text{for all } t = 1 \text{ to } p \\ & y \geqq 0 & \text{and integer} \end{array} \tag{16}$$

We make the following observations.

1. The problem (16) is a pure integer program that is equivalent to the original problem. If $(\bar{z}_0, \bar{y})$ are optimal to (16), $\bar{z}_0$ is the minimum objective value in (9). Let $\bar{x}$ be the optimum solution of (10) after inserting $\bar{y}$ for y. Then $(\bar{x}, \bar{y})$ is an optimum solution of (9).
2. The number of constraints in (16) is $\ell + p$ where ℓ, p are the number of BFSs of (11) and the number of extreme homogeneous solutions corresponding to (11), respectively. These are likely to be very large numbers. Hence, if all the constraints in (16) have to be used explicitly, this transformation is useless. However, we will discuss below an algorithm for solving (16) using only a small subset of the constraints in (16) in any stage and generating the other constraints as they are needed.
3. The constraints in (16) are infeasible only if the constraints (13) are inconsistent.

THE ALGORITHM

In each stage the algorithm deals with a *restricted problem* that is obtained by considering only a subset of the constraints in (16) and neglecting all the others.

To start with select a $\bar{y} \in \mathbf{S}$, say $\bar{y} = 0$, and solve (11) for that $\bar{y}$. If (11) has an optimum solution in this case, let π^1 be an optimum BFS for it. On the other hand, if the objective function in (11) is unbounded above in this case, let μ^1 be the extreme homogeneous solution corresponding to the extreme half-line along which $v(\pi, \bar{y})$ diverges to $+\infty$. In the initial stage the restricted problem is (16) after eliminating all the constraints in it excepting those corresponding to some known BFSs π^j, and the one constraint corresponding to π^1 or μ^1 as the case may be, and $y \in \mathbf{S}$.

In a general stage of the algorithm, suppose the restricted problem is

$$\begin{array}{lll} \text{Minimize} & z_0 & \\ \text{Subject to} & z_0 - dy - \pi^j(Dy - b) \geqq 0 & \text{for } j = 1 \text{ to } r \\ & \mu^t(Dy - b) \leqq 0 & \text{for } t = 1 \text{ to } k \\ & y \geqq 0 \text{ and integer} & \end{array} \tag{17}$$

The computations involved in this stage are the following:

(i) Solve (17) by using some pure integer programming method. If (17) is infeasible, (16) and hence the original problem must be infeasible. Terminate. Otherwise go to (ii).

(ii) Let $(\hat{z}_0, \hat{y})$ be an optimum solution of (17). $\hat{z}_0$ is a lower bound for the minimum objective value in (16) or (9).

Solve (11) for $y = \hat{y}$. In this case if (11) has the objective function unbounded above, let μ^{k+1} be the extreme homogeneous solution corresponding to the extreme half-line along which $v(\pi, \hat{y})$ diverges to $+\infty$. Include the constraint "$\mu^{k+1}(Dy - b) \leqq 0$" in (17) and with this as the new restricted problem go to the next stage.

On the other hand if (11) has a finite optimum solution in this case, let π^{r+1} be an optimum BFS for it. The optimum objective value is $v(\pi^{r+1}, \hat{y}) = v(\hat{y})$. If $\hat{z}_0 \geqq v(\hat{y})$, then $\hat{z}_0$ is the optimum objective value in (16) or (9). Let $\hat{x}$ be an optimum solution of (10) with $y = \hat{y}$. Then $(\hat{x}, \hat{y})$ is an optimum solution of (9), Terminate.

If $\hat{z}_0 < v(\hat{y})$, the optimum objective value of (16) or (9) lies in the interval $[\hat{z}_0, v(\hat{y})]$. Include the constraint "$z_0 - dy - \pi^{r+1}(Dy - b) \geqq 0$" in (17) and with this as the new restricted problem go to the next stage. The new constraint included here prevents $\hat{y}$ from appearing as an optimum solution of the restricted problem in subsequent stages.

Comments Bender's decomposition has been used successfully in solving sequencing and scheduling problems and on other special classess of problems. See references below.

REFERENCES

1. J. F. Benders, "Partitioning Procedures for Solving Mixed-Variables Programming Problems," *Numerische Mathematik*, *4*, 238–252, 1962.

2. A. M. Geoffrion and G. W. Graves, "Multicommodity Distribution System Design by Benders Decomposition," *Management Science*, *20*, 5, 822–844, January 1974.

appendix 3

cramer's rule

This rule gives the solution of a square nonsingular system of simultaneous linear equations. Consider this system

$x_1 \ldots x_m$	
$a_{11} \ldots a_{1,m}$	$= b_1$
$\vdots \quad \vdots$	$\vdots$
$a_{m1} \ldots a_{mm}$	$= b_m$

Here $A = (a_{ij})$ is nonsingular. Let det $(A_{.1}, \ldots, A_{.j-1}, b, A_{.j+1}, \ldots, A_{.m})$ be the determinant of a matrix whose column vectors are listed inside the brackets. Let

$$\bar{x}_j = \frac{\det(A_{.1}, \ldots, A_{.j-1}b, A_{.j+1}, \ldots, A_{.m})}{\det A}$$

Then $\bar{x} = (\bar{x}_1, \ldots, \bar{x}_m)^T$ is a solution of the above system.

selected references in linear programming

BOOKS

1. M. L. Balinski (editor), *Mathematical Programming Study 1, Pivoting and Extensions: In Honor of A. W. Tucker*, North-Holland Publishing Company, Amsterdam, 1974.
2. M. L. Balinski and E. Hellerman (Editors), *Mathematical Programming Study 4, Computational Practice in Mathematical Programming*, North-Holland Publishing Company, Amsterdam, 1976.
3. E. M. L. Beale, *Mathematical Programming*, Isaac Pitman, London, 1968.
4. A. Charnes and W. W. Cooper, *Management Models and Industrial Applications of Linear Programming*, Wiley, New York, 1961.
5. A. Charnes, W. W. Cooper and A. Henderson, *An Introduction to Linear Programming*, Wiley, New York, 1953.
6. L. Cooper and D. Steinberg, *Introduction to Methods of Optimization*, Saunders, Philadelphia, 1970.
7. G. B. Dantzig, *Linear Programming and Extensions*, Princeton University Press, Princeton, N.J., 1963.
8. R. Dorfman P. A. Samuelson, and R. M. Solow, *Linear Programming and Economic Analysis*, McGraw-Hill, New York, 1958.
9. N. J. Driebeck, *Applied Linear Programming*, Addison-Wesley, 1969.
10. D. Gale, *Theory of Linear Economic Models*, McGraw-Hill, New York, 1960.
11. W. W. Garvin, *Introduction to Linear Programming*, McGraw-Hill, New York, 1960.
12. S. I. Gass, *Linear Programming: Methods and Applications*, 4th Edition, McGraw-Hill, New York, 1975.
13. D. P. Gaver and G. L. Thompson *Programming and Probability Models in Operations Research*, Brookes/Cole Publishing Co., Monterey, California, 1973.
14. G. Hadley, *Linear Programming*, Addision-Wesley, Reading, Mass., 1962.
15. F. S. Hillier and G. J. Lieberman *Introduction to Operations Research*, 2nd Edition Holden-Day, San Francisco, 1974.

16. T. C. Koopmans (Editor), *Activity analysis of Production and Allocation*, Wiley, New York, 1959.
17. H. W. Kuhn and A. W. Tucker (eds.) *Linear Inequalities and Related Systems*, Princeton University Press, Princeton, N.J. 1956.
18. L. S. Lasdon *Optimization Theory for Large Systems*, Macmillan, New York, 1970.
19. R. W. Llewellyn, *Linear Programming*, Holt, Rinehart and Winston, New York, 1964.
20. D. G. Luenberger, *Introduction to Linear and Non-Linear Programming*, Addison-Wesley, Reading, Mass., 1973.
21. B. Noble, *Applied Linear Algebra*, Prentice-Hall, New York, 1969.
22. D. P. Phillips, A. Ravindran and J. Solberg, *Operations Research: Principles and Practice*, Wiley, New York, 1976.
23. T. L. Saaty, *Mathematical Methods of Operations Research*, McGraw Hill, New York, 1959.
24. H. M. Salkin and J. Saha, *Studies in Linear Programming*, North-Holland Publishing Co., 1975.
25. M. Simonnard, *Linear Programming*, translated from French by W. S. Jewell, Prentice-Hall, Englewood Cliffs, N.J., 1966.
26. W. A. Spivey and R. M. Thrall, *Linear Optimization*, Holt, Rinehart and Winston, 1970.
27. J. E. Strum, *Introduction to Linear Programming*, Holden-Day, San Francisco, 1972.
28. S. Vajda *Mathematical Programming*, Addison-Wesley, Reading, Mass., 1961.
29. C. Van de Panne, *Methods for Linear and Quadratic Programming*, North-Holland, Amsterdam, 1975.
30. H. M. Wagner, *Principles of Operations Research with Applications to Managerial Decisions*, Prentice-Hall, Englewood Cliffs, N.J., 1969.
31. M. Zeleny, *Linear Multiobjective Programming*, Springer-Verlag, 1974.
32. S. I. Zukhovitskiy and L. I. Avdeyeva, *Linear and Convex Programming*, W. B. Saunders Company, 1966.

References to Track Down Available Computer Programs

1. W. W. White, "A Status Report on Computing Algorithms for Mathematical Programming," *Computing Surveys*, *5*, 3, pp. 135–166, September 1973.
2. *SIGMAP Newsletter*, Issues *10* (October 1971), *11* (May 1972), *12* (June 1972), *13* (January 1973), *14* (May 1973), Publication of the ACM.
3. J. L. Kuester and J. H. Mize, *Optimization Techniques with Fortran*, McGraw-Hill, 1973.
4. D. A. Pierre and M. J. Lowe, *Mathematical Programming Via Augmented Lagrangians: An Introduction with Computer Programs*, Addison-Wesley, 1975.
5. IBM Mathematical Programming System Extended/370 (MPSX/370) General Information Manual GH 19-1090-0, IBM, 1974.

index

Active variable, 482
Activities, 2, 5
  levels of, 2
Additive algorithm, 478
Additivity assumption, 2
Adjacency, 111–113
  path, 113
Adjacent, basis, 261
  basis path, 262
  extreme point, 113, 261
  extreme point path, 113
  vertex methods, 124
Admissible cells or arcs, 299, 361, 369
Affine, hull, 77–78
  space, 125
Aggregating, 404, 418
Airline crew scheduling, 406
Algebraic methods, 437
Algorithm, 36, 131
Allocation change routine, 362–363
Almost complementary, 497
Alternate optimum solutions, in LP, 56, 125–128, 167–168
  in LCP, 510
  in transportation problem, 321 (footnote)
Apex, 315–316
Arborescence, 390
Arc, 341
  forward, 342–343, 347
  reverse, 342–343, 347
Assignment, problem, 10, 155, 308, 360–367
  ranking, 471–474
Associated LP, in integer programming, 419
Augmented predecessor indexing method, 321

Back substitution, 291–292, 523, 529
Backtrack strategy, 445
Balanced transportation problem, 155, 289–325, 367–373
Basic, column vector, 42, 102
  feasible solution (BFS), 41, 97–107, 272–273
  part, 102
  set for transportation, 293–295
  solution for LP, 102, 103, 272
  solution for bounded variable LP, 272–273
  solution for transportation, 294–295
  solution for bounded variable transportation, 329
  variable in a row, 42, 272
  vector, 42, 102
Basis, 43, 102–103
  adjacent, 261
  degenerate, 259
  dual feasible, 131, 161–162, 201
  for $\mathbf{R}^n$, *see* notation in front
  full artificial, 44
  nondegenerate, 259
  optimal, 161, 201
  primal feasible, 102–103, 201
  relationship to tree in networks, 347
  terminal, 201
  triangular, 292
  working, 272, 274
Batch size problems, 398
Benders decomposition, 553–555
BFS, 41, 97–107
Big M method, 65–67, 165, 253
Bipartite matching problem, 360
Bipartite network, 360
Blossom algorithm, 392
Bottleneck transportation problem, 336–337
Bounded variable, LP, 271–287
  transportation problem, 329–334
Boundedness, of objective function in LP, 118–119
  of convex polyhedra, 116
Bounding (lower), in branch and bound, 438–440
  in ranking, 475
Branch and bound, 437–471
Branching, (a directed tree), 390
  binary, 441
  constraint, 465
  operation, 437
  strategy, 440–441
  variable, 441, 447
Breakthrough, 352, 362, 369
Brother indices, 312–317
Brouwer fixed point, 509

Candidate problem, 440, 447
Canonical form, 41
Canonical tableau, 40, 43, 130–131
  optimum, 52
Capacitated transportation problem, 329–334
Capacities, of arcs or edges, 344
  of cutsets, 347
Caterer problem, 393
Chain, 342
Characteristic interval, 222, 224–225, 227, 229
Characteristic values, lower, 222
  upper, 222
Chinese postman problem, 407–410

Cholesky, factorization, 528–533
  factor, 528
Circuit, 342
Circulation, 357
Clique, 455
Column generation methods, 544
Column rank, 93
Column vector, basic, 42, 102
  nonbasic, 42, 102
Combination, affine, 77
  convex, 79
  linear, 75
Combinatorial choice variable, 397, 420
Combinatorial optimization, 397
Commutative, 85
Complement, 485
Complemementary constraint, 485
Complementary, basic vector, 485
  cone, 482–484
  feasible basic vector, 497
  pair, 173, 482
  pivot algorithm, 169, 481, 495–506
  pivot rule, 498
  set, 483
  slackness diagram, 374–375
  slackness property, 165–169
  slackness theorem, 165–169
    its economic interpretation, 168
  vector, 485
Completion, 467
Computational efficiency, 535
Concave function, 11–12, 227–228
Cone, 82–84
  simplicial, 132–135
Connected, graph, 309
  network, 343
Consistent system of equations, 94
Constraints, 36, 150
Container problems, 411
Convex function, 11, 227–228
Convex, cone, 82–84
  polyhedral cone, 82–84
  polyhedral set, 82
  polyhedron, 82, 116, 263
  polytope, 82, 116, 263
  set, 81–82
Convexity cuts, 433
Copositive, 503
  strict, 503, 505
Copositive plus, 503
Corner points, 41
Cost coefficients, original, 40
  phase, I,II, 46
  relative, 43, 47, 53, 162–163
Cost ranging, basic, in LP, 250–251
    in transportation, 323–324
  nonbasic, in LP, 249–250
    in transportation, 323
Cost row, 40
Cover, 406
Cramer's rule, 307, 557
Critical capacity, 356
Critical length, 387
Current candidate problem, 445
Curve fitting, 18–19, 24, 33
Cut, 422
Cutset, 347
Cutting plane, 421–434
Cycle, 309
Cycling, in simplex algorithm, 62, 108, 264, 267
  resolution of, 108, 263–270

Dantzig cuts, 423
  strengthened, 433
Dantzig property, 299
Decentralized planning, 551
Decomposition principle, 118, 541–552
Deeper cuts, 423
Degeneracy, 62, 259–270, 498–499
Degree, 342
Delivery and routing problem, 405
Degenerate, BFS, 103
  basis, 103
  matrix, 503
  pivot step, 62–63, 108, 112
  problem, 260
Depth of a cut, 423
Descendent, 312–314, 383
Diagonal entries, 85
Diet Problem, 9, 23, 31–32, 147–149, 168
Dimension, 124–125
Discrete, optimization, 397–398
  valued variables, 398
Distance matrix, in shortest chains, 388
Donor cells, 298 (footnote)
Dropping variable, 51, 425, 496
Dual feasibility, 161–162
  criterion, 162
  of a basis, 131, 161–162
Dual infeasibility, 163–164
Duality, theorem (fundamental), 159–161
  theory, 155–176
Dual problem, 147–155
Dual simplex algorithm, 162, 169, 201–208
Dual simplex method, 208–216, 423–426
Dual simplex pivot step, 203
Dual variable, 149–155

Economic interpretation, of complementary
    slackness property, 168
  of dual problem, 147–149, 155, 170–171
Edge, in networks, 309, 341
  bounded (in LP), 111, 261
  unbounded (in LP), 119–124
Edge covering route (ECR), 408
Edge path, 109, 113
Either, or, constraints, 401–402
Elder brother, 313
  index, 313
Eldest son, 313
Elementary matrix, 129
Elementary row operations, 41

Elimination form of the inverse, 526
Endnode, 310, 343
Entering variable, 50, 276, 425, 496
Enumeration, implicit, 478
  methods, 437
  partial, 437, 446
  total or exhaustive, 437
Equilibrium, price theorem, 165–168
  price vector, 165
  strategy, in 2 person games, 493
Equivalent system, 40
Eta vector, 197
Euclidean vector space, 75–84
Euler route, 408
Evaluation, 451
Even node, 408–409
Excess, in modelling, 16
Explicit form of the inverse, 183
Extreme half line, 64, 119–124, 262, 497, 503
Extreme homogeneous solution, 116–124
Extreme point, 41, 262
Extreme point ranking, 141–142, 474–476

Face, 127–128
Facet, 127–128
Facility location, 400–401, 406
Factor of UCP, 458
Farkas lemma, 175–176
Fathom, 437, 440
Feasible solution, 19, 40, 97
Finite termination property, in integer programming, 431
  in LP, 107–108
Fire hydrant location, 405–406
Fixed charge problem, 398–402, 476
Fixed point computing methods, 509–510
Fleury's algorithm, 408–409
Flow augmenting path (FAP), 348
Flow conservation, 345
Flow vector, 345
  feasible, 345
  value of, 345
Flow rerouting, 376–377
Forest, 343
Formulation, of integer programs, 397–417
  of LP's by direct approach, 2–5
  of LP's by input-output approach, 5–8
Forward substitution, 292
Fractional cut, 426–427
Fractional part, 426
Fractional programming, 142
Free variable, 467
Further reduction routine, 362–363

Games, zero sum, 173–174
  bimatrix, 493–494, 506–508
  solution of, 173–174, 506–508
Gauss-Jordan elimination, 95
Gaussian elimination, 523
Geometrical argument, in LP, 124
Ghost story, 499
Givens matrix, 529
Good algorithm, 510, 536

Graph, 309
Greedy, algorithm, 386
  selection, 386
Group, 433–434
  theoretic approach, 434

Half line, 80
Half space, 80–81
Haunted house, 499–500
Hessenberg matrix, 525
Hirsch conjecture, 539
Homogeneous solution, 114–124
Householder matrix, 530
Housewife's problem, 147–149, 170–171
Hull, affine, 77–78
  convex, 79
  linear, 76–77
Hungarian method, 361–366
Hyperplane, 80
  separating, 174–176
  supporting, 127

Identity matrix, 85
Immediate successor, 312–313, 383
Implicit enumeration, 478
Inadmissible cells, 299
Incidence matrix, set element, 405
  element subset, 407
  node arc, 346
Incidence vector, 405
Incident, 309, 341
  into, 341
  out of, 341
Inconsistent system of equations, 94
Incumbent, 443, 475
Index, elder brother, 313
  predecessor, 311
  successor, 312
  younger brother, 313
Inductive algorithm, for shortest chains, 388–389
  for set representation problem, 458
Inequalities, right type, 150–151
Inequalities transportation problem, 325–328
Infeasibility criterion, 53, 188, 203, 226
  in bounded variable LP, 281
  in branch and bound, 445
Infeasibility form, 44
In-kilter arc, 375
Input-output approach, 5–8
Input output coefficients, 40
  changes in, 252–253
Integer programming, pure, 397, 426–431
  mixed, 397, 432–433, 456–458, 553–555
  0,1, 397, 467–468
Integer property, in transportation problem, 293
  in network flows, 356
Intermediate point, in networks, 344–345
In-tree arc, 347
Intersection cuts, 433
Inverse, of a matrix, 89–91
Inverse matrix, in LP, 131
Inverse tableau, for an LP, 183–184

Isotonic path, 536–538
Isthmus, 409
Items, 5

Jumptrack strategy, 445

Kakutani fixed point, 509
Karush-Kuhn-Tucker necessary conditions, 169, 492
Kilter status, of an arc, 375
Klee and Minty examples, 536–538
Knapsack problem, 404, 446–449
Kőnig-Egerváry theorem, 394
Konigsberg bridge problem, 408
Kuhn Tucker, Lagrange multipliers, 169–170
  Lagrangian, 169
  necessary conditions, 169–170, 492
  point, 492
  saddle point, 170
  theory, 35

LCP, 481
LIFO, 445, 465
Label matrix, in shortest chains, 388
Labelling algorithm for max flow, 349–359
Labelling methods in transportation, 298, 309–322
Lagrange multipliers, 169, 439–440
Lagrangian relaxation, 439–440
Latin squares, 416
  orthogonal, 416
Least lower bound criterion, 445
Leaving variable, 51, 185
Lexico (or Lexiographic), decrease, 108
  dual feasible basis, 425
  dual simplex method, 423–426
  feasible basis, 267
  minimum, 266
  minimum ratio rule, 108, 267–268
  negative, 265
  positive, 265, 425
Library problem, 393
Line, extreme half, 119
  half, 80
  in networks, 309, 341
  segment, 79
  straight, 79
Linear complementarity problem (LCP), 172–173, 415, 481–518
Linear dependence, 86–89, 296
Linear independence, 86–89, 293, 297
Linearly independent system of equations, 94
Linearity assumptions, proportionality and additivity, 2–3
List, 441, 472
Location problems, 400–401
Loss matrices, 493
Lower bounding linear function, 475
Lower bounding strategy, 439–440
Lower bounds, for arc or edge flows, 344
  in branch and bound, 437
Lower characteristic value, 222
LU-decomposition, 522–528, 533

m step conjecture, 539
Marginal analysis, 170–171
Marginal rates, 170–171
Marginal values, 150
Marriage problem, 308
Master program, 542–543
Matching, 360
  complete or perfect, 360
Material balance equation or inequality, 3
Matrices, 84–86
Matrix multiplication, 85–86
Max flow min cut theorem, 351
Max flow problem, single commodity, 348–359
  multi commodity, 391–392, 551–552
Maximal linearly independent subset, 91–94, 124–125
Min cost flow problem, single commodity, 373–382
  multi commodity, 391–392, 551–552
  piecewise linear convex, 394–395
Minimal linearly dependent set, 297
Minmax theorem, 176, 493
Min max strategy, 176, 493
Minimum cardinality represent (MCR), 404
Minimum cost spanning tree, 390–391
Minimum distance problem, 492–493
Minimum ratio, in dual simplex method, 203
  in primal simplex method, 51, 112, 185
  test, 51, 185, 496
Mixed stragies, 173
Modelling, or Formulation, 1–19, 22–23, 397–417
Multiple choice problems, 403, 412–413, 477

Negative part, 15
Net flow across cutset, 350
Network, directed, 25, 341
  mixed, 341
  undirected, 341
New activity, to introduce, 244–245
Node, in branch and bound and ranking methods, 441, 472
  or point, in networks, 309, 341
Node arc incidence matrix, 346
Node price change routine, 377–378
Node prices, or variables, or potentials, 374
Nonbasic, column vector, 42, 102
  part, 102
Nonbreakthrough, 352, 362, 369
Nondegeneracy, 103, 259
Nondegenerate, basis, 103, 259
  BFS, 103
  LP, 107
  matrix, 503
Nonlinear complementarity problem, 510
Nonsingular linear transformations, 40
Nonsingular square matrix, 89
NP-complete class, 539
Numerically stable, 198, 521–522, 528, 533

Objective function, phase I, 44, 46
  piecewise linear, 11–14, 394, 416
  value, change in, 51–52, 266

Objective row, 40
Phase I, 44
Phase II, 44
Odd node, 408–409
Off-diagonal entries, 85
One-variable, 467
Operator scheduling, 412
Optimality criterion, in bounded variable LP, 273
in dual simplex algorithm, 202
in primal simplex algorithm, 47, 161, 185
in transportation problem, 304
Optimality interval, 222, 225, 323
Optimum branching, 390
Optimum feasible solution, 19, 47
Optimum spanning arborescence, 390
Optimum strategy, 174, 493
Orthant, 482
Orthogonal latin squares, 416
Orthogonal matrix, 529
Out of kilter algorithm, 169, 374–381
Out of tree arc, 347
Out tree, 382–383

P-matrix, 503, 506, 514–516
Parallelogram law of addition, 76
Parametric, analysis in network flows, 379–381
cost problem, 219–225
cost transportation problem, 324
LCP, 515–516
problem, combined cost and right hand side, 239
right hand side constant problem, 219–220, 225–227
representation of a straight line, 79–80
Parent problem, 441
Partial enumeration, 437, 446
Partial network, 343
Partial solution, 467
Partial subnetwork, 343
Partition, 407, 440, 469
Partitioning a node, 472
Path, 309, 342–343
Payoff matrix, 173
Penalties, 456
Pendant node, 310, 343
Permutation matrix, 89, 360
Perturbed problem, 264
Phase I problem, 44–46, 136–138, 329–332
Phase I, II, 43–46, 53, 276
Piecewise linear, 11, 394, 416, 478
Pill maker problem, 147–149
Pivot, column, 50, 52, 87, 90, 128, 185, 186, 188, 425, 496
column, its choice rule, 50, 185, 188
element, 51–52, 87, 90, 128, 203
matrix, 128–130, 186
operation, 52, 128, 425
row, 51–52, 87, 90, 128, 186, 188, 203, 265–266, 425
row choice in dual simplex, 203, 425–426
step, degenerate, 62–63, 108, 112
in dual simplex, 203
nondegenerate, 62–63, 108, 112
Plant location, 400–401
Point, in a graph or network, 309, 341
Point to set map, 509
Pointwise supremum function, 14
Policy, 476–477
Political districting, 407
Polynomially bounded algorithm, 536
Pos cone, 82–84, 132–135
Positive definite (PD), 486–489
Positive part, 15
Positive semi definite (PSD), 486–491
Post optimality analysis, 243
Predecessor index, 311, 316
Prices, or dual variables, 149–150
Pricing out, 42, 52
Primal, dual pair, 149
feasibility, 45–46
infeasibility, 203, 226
problem, 149
simplex algorithm, 46–52, 201
Primal dual approach, 360–373
Principal, diagonal, 85
pivot transform, 514
subdeterminant, 487
submatrix, 487
Priority, criterion, 445
strategy, 445
Procurement problems, 401
Product-form of the inverse, 192–198
Product mix model, 1
Project selection problems, 402
Proportionality assumption, 2
Pruning, 443

Q-Matrix, 503, 505
Quadrant, 482
Quadratic assignment problem, 413
Quadratic programming, 486–493

Ranging, of a cost coefficient, basic, 250–251, 323–324
nonbasic, 249–250, 323
of a right hand side constant, 251–252, 325
input output coefficient, 252
Rank, of a matrix, 91–94
of a set of vectors, 94
Ranking methods, 141–142, 471–477
Ray, 80
Ray termination, 498, 502–506
Recipient cell, 298 (footnote)
Reduced matrix, 361, 450
Reduced problem, 458
Redundant equation, 54–55, 95
Reduction, 361
Reinversion, 533
Relative cost coefficients, 43, 47, 162–163, 303, 318–319
Relaxed Lagrangian, 439
Relaxed problem, 438–440
Relaxing constraints, 438

Represent, 404, 458
Representation, 91
Rerouting, sink, 376
  source, 376
Resolution of cycling, 108, 263–270
Restrictions, nonnegativity, 3, 36, 150
Resolution theorems, 114–121
Restricted master program, 546
Revised dual simplex method, 202
Revised simplex algorithm, primal, 183–186
Revised simplex method, primal, 187–189
Right hand side constants, 40
  ranging of, 251–252
Right type of inequality, 150–151
Rooted tree, 311
Round off errors, 522
Rounding off method, 419–420
Row echelon normal form, 94–96
Row operations, 41
Row rank, 93

Saddle point, 170
Sanctuary, 499
Saturated arc, 358
Scope, 357
Search strategy, 438, 441–445
Search tree, 442
Self-dual LP's, 155, 180
Self-loop, 341
Semipositive, 75
Sensitivity analysis, in LP, 243–258
  in maximum flow, 356–357
  in minimum cost flows, 379–381
  in shortest chains, 387–388
  in transportation, 321–325
Separable objective function, 2–3, 11, 33, 416–417
Separating hyperplane, 174–175
  theorem, 174–175
Separation of objective values, 160
Sequencing problems, 403–404
Set covering problem, 406
Set element incidence matrix, 405
Set packing problem, 407, 470
Set partitioning problem, 407, 469–470
Set representation problem, minimum cardinality, 404
  weighted, 404–406, 458–466
Set up cost, 398
Shadow costs, 162
Shadow prices, 150
Shortage, in modelling, 16
Shortest chains, 382–387
Shortest route problem, 382–387
Simple convex, polyhedron, 263
  polytope, 263
Simple, chain, 342
  circuit, 342
  cycle, 309, 343
  path, 309, 314, 343
Simplex, 132
  algorithm, 36, 43, 46–52, 107–113, 132–135
  almost completely labelled, 509
  completely labelled, 509
  method, 43, 53–56
  multipliers, 150
Simplicial cone, 132–135
Simultaneous linear equations, consistent, 94
  incomsistent, 94
  solvability of, 94
  triangular, 291–292
Singular matrix, 89
Sink, 344
Size, 510, 536
Solution, for LP's in standard form, 97
Source, 344, 382
Source row, 427
Sparse, matrix, 197
  vector, 197
Sparsity, 197, 528, 533
Special basis problem, 415
Square matrix, 85
Stack, 445
Standard form, 36–40
  for bounded variable LP, 271–272
Steiner, problem, 471
  subset, 471
Strict separation property, 514–515
Subgradient, 461
  optimization algorithm, 461–463
Submatrix, 91
  principal, 91, 487
Subnetwork, 343
Subproblem, 542
Subprogram, 542
Subspace, 76, 124–125
Successor, 312
  immediate, 312–314, 383
  index, 312–313
Sufficient optimality criterion, in LP, 158–159
Superdiagonalization algorithm, 489
Supersink, 349
Supersource, 349
Supporting hyperplane, 127
Surrogate constraint, 468, 478
Symmetric assignment, 410
  problem, 410, 454–455
Symmetric form, 485

Terminal node, 310, 343, 441
Termination criteria, in branch and bound, 444–445
  in bounded variable LP, 276, 281, 282
  in complementary pivot algorithm, 498
  in dual simplex algorithm, 202
  in primal simplex algorithm, 47, 108, 185, 188
  in transportation problem, 304, 331–332
Tests, 467
Theorems of alternatives, 174–176
Theta loop (θ-loop), 295–298, 310, 315
Three dimensional transportation problem, 24
Tight constraint, 180
Tour, 410–411
  assignment, 410–411
Translate, 75

Transportation array, 293
Transportation paradox, 338–339
Transportation problem, balanced, 9, 155, 289–325, 367–373
  dual of, 154–155, 303–304
  unbalanced, 326–328
  primal algorithm for, 299–321
Transpose, of a matrix, 85
Traveling salesman problem, 410–411, 449–454
Tree, 309, 343
  rooted, 311–312
  spanning, 309, 343, 390–391
Tree building algorithm, 383, 385–387
Tree changing algorithm, 383–385
Triangular matrix, 291–292
  lower, 291–292
  upper, 291–292
Triangulations, 509
Trim problems, 411
Trip, 406

Unbounded, objective in LP, 49, 122–124, 161
  edge, 119, 262, 497
Unboundedness criterion, 48–49, 122–124, 161
Unimodularity, 307–309, 346
Unit matrix, 85
Unordered cartesian product (UCP), 458
Updated column vector, 131
Updated entries, 47
Updating, brother indices, 317
  cholesky factorization, 531–533
  column vector, 185
  cost, 42
  dual vector in transportation, 318–319
  incumbent, 443
  inverse tableau, 186, 189
  LU decomposition, 524–528
  objective row, 52
  predecessor indices, 316
  rooted tree, 316
  successor indices, 317
Upper characteristic value, 222
Upper Hessenberg matrix, 525
Uses, Used, 97, 103, 114

Variable, as level of an activity, 2
  artificial, 43, 55
  basic, 42, 95
  continuous, 3
  cost coefficient, in fixed charge problem, 399
  discrete, 398
  dependent, 42, 95
  dual slack, 162
  independent, 42, 96
  nonbasic, 42, 96
  slack, 2, 37
  surplus, 37
  unrestricted, 39
Vector, eta, 197
  nonnegative, 75
  positive, 75
  sparse, 197
  semipositive, 75
Vertex packing, 471
Vertices, in graphs or networks, 309, 341

Warehouse location, 400–401
Weak duality theorem, 156–159
Weak separation matrix, 517
Weak separation property, 517
Working basis, 272, 274

Younger brother, 313
  index, 313

Zero variable, 467
Z-matrix, 510–511